AF616178

REMOTE SENSING IN THE 21ST CENTURY:
ECONOMIC AND ENVIRONMENTAL APPLICATIONS

PROCEEDINGS OF THE 19TH EARSeL SYMPOSIUM ON REMOTE SENSING IN THE 21ST CENTURY / VALLADOLID / SPAIN / 31 MAY – 2 JUNE 1999

Remote Sensing in the 21st Century

Economic and Environmental Applications

Edited by
José Luis Casanova
University of Valladolid, Spain

A.A. BALKEMA / ROTTERDAM / BROOKFIELD / 2000

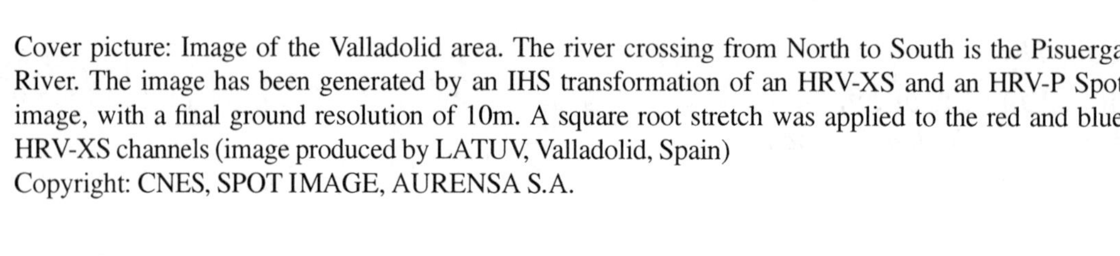

Cover picture: Image of the Valladolid area. The river crossing from North to South is the Pisuerga River. The image has been generated by an IHS transformation of an HRV-XS and an HRV-P Spot image, with a final ground resolution of 10m. A square root stretch was applied to the red and blue HRV-XS channels (image produced by LATUV, Valladolid, Spain)

The texts of the various papers in this volume were set individually by typists under the supervision of each of the authors concerned.

Published by
A.A. Balkema, P.O. Box 1675, 3000 BR Rotterdam, Netherlands
Fax: +31.10.413.5947; E-mail: balkema@balkema.nl; Internet site: www.balkema.nl
A.A. Balkema Publishers, Old Post Road, Brookfield, VT 05036-9704, USA
Fax: 802.276.3837; E-mail: info@ashgate.com

ISBN 90 5809 096 5

Printed in the Netherlands

Remote Sensing in the 21st Century: Economic and Environmental Applications, Casanova (ed.)
© 2000 Balkema, Rotterdam, ISBN 90 5809 096 5

Table of contents

2 *Meteorology*

3 *Hydrology*

4 *Methodology*

5 *Forest fires*

9 *Geology*

10 *Radar and hyperspectral techniques*

Miscellaneous

Remote Sensing in the 21st Century: Economic and Environmental Applications, Casanova (ed.)
© 2000 Balkema, Rotterdam, ISBN 90 5809 096 5

Preface

Exactly twenty years ago, in 1979, EARSeL held a General Assembly for the first time in Spain, to be exact, in Barcelona. More recently, in Enschede, the Bureau decided, and the General Assembly endorsed, that the millennium should close with a Symposium in Spain; to take place in Valladolid in May-June 1999, along with a Joint ISPRS/EARSeL Workshop. The theme of the Symposium was 'Remote Sensing in the 21st Century: Economic and Environmental Applications' and it dealt with what will be the principal concerns at the start of the next century; that are growth of the economy for the general well-being, and the environment.

The end of the millennium is a good point at which to take stock. Since its creation, EARSeL has developed considerably, having passed from its six original laboratories to nearly 300 research institutions it has today. At this moment, 39 countries are represented in EARSeL through universities, commercial and industrial laboratories or consultancies. Its founder members should feel deeply satisfied that the seed sown almost a quarter of a century ago has been transformed into the fine organisation it is today.

This volume is the result of a great effort which demonstrates the Association's high level of competence in the field of remote sensing. As the reader will learn from the following pages, the scientific level of the work is very high, due to the fact that EARSeL's efforts are carried out at the limits of human knowledge with a view to extending the frontiers of Science.

There were 180 participants at the Symposium, including those at the Workshop, from 29 countries, with more than one hundred papers being presented. In addition, there were addresses given by representatives from the Council of Europe, the European Space Agency and the European Commission. The Satellite Centre of the Western European Union and the Syrian Space Project were also presented.

The different sections of this volume are organised according to the structure of the symposium, covering almost all the operational applications and developments of remote sensing: Agriculture and Forestry and Meteorology, each of which were dealt with in two technical sessions, Hydrology, Desertification, Methodology, which also occupied two technical sessions, and Forest Fires, Yield Estimation, Hazards, Oceanography, Geology and Radar and Hyperspectral Techniques.

The enormous steps forward in remote sensing demonstrated during the symposium is only the prelude to the splendour on the way next century. Launching pads are ever-increasing and the observations from the satellites that are launched are more and more frequent; the geographic location of images is becoming more and more accurate; sensors are increasing the number of their measuring bands and spatial resolutions, and the advent of high-resolution hyperspectral sensors is heralding a new era. All of this will increase a hundredfold information obtainable from space. Only our imagination will allow us to get to where remote sensing can take us. One thing, however, is certain; in that both monumental and marvellous task which awaits us next century, EARSeL will continue growing and its members will continue to show proof of their scientific expertise and intellectual capacity.

It only remains for me to express gratitude to EARSeL on behalf of LATUV and myself for their entrusting to us the organisation of the 19th Symposium. I should also like to thank the Instituto Nacional de Investigaciones Agrarias (INIA) and the Comisión Interministerial de Ciencia y Tecnologia (CICYT) for their valuable support. My sincerest thanks also go to Mme Madeleine Godefroy, who was always willing to guide and stimulate us with her advice, to the members of the Scientific Committee for their assistance, and to Dr Robin Vaughan for the assurance we felt knowing that, as Symposium Chairman, he was constantly supervising our efforts.

Valladolid, June 1999
José Luis Casanova
President of the Organising Committee

Keynote lectures

Remote Sensing in the 21st Century: Economic and Environmental Applications, Casanova (ed.)
© 2000 Balkema, Rotterdam, ISBN 90 5809 096 5

The future ESA Earth-Observation Strategy and ESA's 'Living Planet' programme

D.J.Southwood
Directorate of Applications Programmes, European Space Agency, Paris, France

ABSTRACT: The European Space Agency's future strategy for Earth Observation marks a big departure with the past. Following the series, ERS-1, ERS-2, Envisat, which have established no doubt Europe's technical capability, the new programme will focus on smaller but more frequent spacecraft. Individual missions will be much more focused for both scientific purposes and applications. Information from space Earth Observation is an essential element in Europe's strategic armoury to meet the environmental and economic challenges of the next century. For the space community, the challenge will be to show that the small focused approach really can meet the requirements for a much-reduced budget.

1 INTRODUCTION

ESA's new 'Living Planet' programme [as described for example by Arend et al., 1998] is designed to start off a new era in space Earth Observation work in Europe. The first steps in the new programme have been taken. Four missions are vying for selection in Autumn 1999 for implementation as 'Core' missions in the new Earth Explorer science-based programme. Two missions have already been selected as low-cost 'Opportunity' missions in the same programme. New applications missions will be done in the frame of the Earth Watch programme, which are intended to be carried out in partnerships with entities (public or private) which will be the ultimate operator. Industry has been challenged to indicate what they see as the target market for a European Earth Observation Applications system and how they might invest in its development. For all programmes, budgets are going to be tight and there are a great many challenges to be met.

History

Europe began to establish its position in Earth Observation (EO) in the 1970s, both through ESRO/ESA initiatives (initially with Meteosat) and via national programmes (most notably, through the French SPOT programme). It has developed strong EO capabilities since, and through programmes like ERS-1 and ERS-2, Envisat and the Meteosat series, as well as other national programmes, most notably the French SPOT programme. However, as Envisat comes close to launch, a decision had to be made on what would follow. The Living Planet programme is the response. The challenge that it sets up is to move from a programme based on the launch of large multi-purpose spacecraft every 5 years or so, to a regime where focused spacecraft missions are the order of the day and the launch rate is closer to once per year.

The Envelope Programme Concept

The new ESA programme is an 'envelope' programme. This is a new development in ESA. In the traditional 'optional' programme approach, each nation subscribes to an individual mission according to their aspirations and/or perceived (industrial, technical or scientific) benefit. The weakness of optionality is that new ESA EO missions have required in the past separate subscription from each Member State at outset and retain a large element of managerial separation from other missions during implementation. Only the Science Programme of the Agency did not follow the optional approach; in that programme all member nations (compulsorily) subscribed every three years to a rolling programme, in proportion to their gross domestic product (GDP).

The new programme takes a new form. The Envelope (EOEP) programme is a hybrid of the previous types of programme. It is a multiple activity multi-mission programme like the science programme. Member States (and closely collaborating nations like Canada) subscribe to a programme, whose content includes not only multiple missions and allied activities but also the expectation that the programme will roll forward on a similar basis in the future. The envelope introduces some of the flexibility of the Agency's science programme without constraining nations to subscribe according to their GDP. The financial approach brings a new flexibility, and one clear change is that much smaller ESA missions become thinkable.

The gains in efficiency of the envelope approach are not just in economies of scale. Fluctuations in one element of the programme can be offset against those in other parts. The envelope means a long-term plan can now be made and one can plan against it. With that plan, technology can be pre-targeted, an instrument pre-development programme instituted and risk consequently can be much reduced during individual mission implementation. Moreover, instituting international co-operative programmes, where there are common interests with say Japan or the USA, is now much more straightforward as one can plan against an expectation of a budget.

The Development of the New Programme

The new programme can accommodate smaller missions and thus can be built up of more focused missions than in the past. This makes the programme much more able to be 'user-driven'. The principle of a user-driven programme based on two types of (smaller) Earth Observation missions had already been adopted at the ESA Council of Ministers in Toulouse:

> *Earth Explorer:* research/demonstration missions to advance the understanding of the different Earth system processes, including the demonstration of new observation techniques
>
> *Earth Watch:* prototype operational missions serving the operational applications-oriented needs of the market.

The Agency could only produce a 'user-driven' programme through consultation. In the past two years, the scientific, industrial and user Earth Observation communities as well as Member State Delegations, and Europe-wide entities like the European Commission (EC), Eumetsat, the Western European Union, and the European Environmental Agency have contributed to the debate. The overall consensus was that the days of technical demonstration are over; Europe has already proved its capabilities. New approaches are now needed, new ways of working, new user-orientated technical solutions, and new procedures for rapid implementation. The new ESA 'Living Planet' programme is the result. It fleshes out the Earth Explorer and Earth Watch ideas and makes it clear how both will be realised in a European context.

2 THE CONTENT OF THE NEW PROGRAMME

Earth Explorers

The Earth Explorers will be science-driven and will comprise two types of missions, namely:

> Core missions (major missions led by ESA to cover primary research objectives)
>
> Opportunity missions (smaller missions providing a quick reaction capability).

The programme itself will run in many respects like the ESA science programme. As in the Science programme, Core missions will undergo competitive Phase A study. At present the first four are nearing completion. These are:

> *Atmospheric Dynamics*: A mission based on the Doppler wind lidar technique for characterisation of winds in 3-D.
>
> *Gravity*: A mission to characterise the Earth's gravity field and thereby improve knowledge of ocean circulation.
>
> *Earth Radiation*: A mission to delimit the processes governing the absorption and reflection of solar radiation by the atmosphere and the heat emission by the Earth and its atmosphere.
>
> *Land Processes*: A mission using measurements in multiple spectral bands to understand processes and develop techniques to monitor vegetation and land cover.

The missions will be presented to the scientific community in Autumn 1999 on completion of Phase A studies. Two will be recommended for implementation to the ESA Earth Observation Programme Board.

The *Opportunity Missions* are an entirely new venture for ESA. The first are already selected. A call was issued in July 1998 and 27 responses were received in December. In late May 1999, the ESA PB-EO selected two missions, Cryosat (a mission to assess the polar ice) and SMOS (Soil moisture and salinity) for implementation. Strict financial constraints will be applied and a third mission (ACE, for atmospheric profiling) is being held in reserve as a 'hot spare'. A firm financial limit will be placed on the missions (cost to ESA less than 80 Mc). Opportunity missions will be a test bed for new ways of working, which are going to be essential if these missions are going to be implemented fast and effectively. The lessons learned by industry, the Agency and even the scientists involved will no doubt feed through in approaches elsewhere, in particular perhaps in realising the Earth Watch element.

Earth Watch

Earth Watch-type missions are expected to bridge the transition from science and the demonstration of potential applications to really operational applications system and service provision. Finding the way to achieve these transitions is a grand challenge.

Europe can argue that it has successfully done so in meteorological services from space. The Meteosat spacecraft and its successors in geostationary orbit MSG (Meteosat Second Generation) are now operated fully by the European Organisation for Exploitation of Meteorological Satellites (EUMETSAT). ESA remains involved in the development of new systems and improving technology. The first spacecraft for the new European polar meteorological system, Metop, is being developed jointly between ESA and EUMETSAT but operational responsibility belongs to EUMETSAT.

ESA is a research and development agency and the sharing and handing over of responsibility between ESA-EUMETSAT is a model to be sought in other fields where long-term service provision is the ultimate aim. This idea underpins the Earth Watch concept. The issue now is to identify the form of the operating entity and how it is financed. At present, industry has been given the chance to be an intermediary between ESA and the market for EO applications.

The Agency's call for Earth Watch Outline Mission Proposals received in 1998, showed industry's willingness to exploit past investments in radar, multi-spectral techniques and visible and infrared sensor technology, and to move to the provision of Earth Observation operational services. Major themes for service development identified by the industry's response to the call are natural hazard management (such as fires, floods, landslides), land use (agriculture, forestry, desertification) and mapping, coastal zone management).

The question now remains how to orchestrate developments. Ideally the market would drive but the diffuse nature of the EO market, the multiplicity of applications, the dependence in most sectors on publicly funded use, as well as the large areas of military common interest.

As a next step, early in 1999, the major players in Europe's space industry have been set the grand challenge of assessing its own investment strategy and of developing a plan for how Earth Watch will contribute to a European Earth Observation Applications system. The challenge will be to bring together all actors in an effective response to opportunities opening on a global market.

One should not preclude focused Earth Watch calls coming in the future. What is clear, is that the Agency will only enter Earth Watch development plans with a clear indication of a path to an operational system and a significant level of investment from the proposed operators. At the same time, as it does for EUMETSAT, the Agency is continuing a strong programme of support for development of technology for many potential applications uses

Earth Watch type missions as well as the Earth Explorers require cost-effective implementation approaches. Here analysis needs to be made of the best means in the ESA context.

3 THE DEVELOPMENT/EXPLOITATION COMPONENT

An important issue is early technology development and no technically qualified person is in doubt of its necessity for faster mission implementation. It has already been adopted implicitly in the Explorer Core Missions. Instrument pre-development removes risk. The biggest spend rate in space missions is when the largest number of people are working during phase C/D. Disaster hits when something blocks the critical path then. There should be no technical surprises from the payload in phases C/D, hence the use of instrument pre-development.

Technology development, particularly directed at instruments/sensors, is a major part of the final element of the programme, the Development and Exploitation Component. Overall the element encompasses:

Instrument Pre-development
(for identified/agreed candidates for Earth Explorer and Earth Watch type missions),

Mission Exploitation
(including already approved missions)/Market Development.

The second 'Exploitation' component is designed to motivate continual scrutiny of the programme for potential operational applications, in the belief that an operational use can lie at the end of most techniques.

4 CONCLUSIONS

The major aims of the new programme, the ESA Living Planet programme, are:

cheaper, more focused missions,

a user-driven approach through improved communication with scientists and service users,

coordination of a common European Earth Observation Strategy through partnership between ESA, the European Commission, Eumetsat and other European entities, and co-ordination with national programmes,

co-ordination and co-operation on a global scale through strategic cooperation with, for example, NASA and NASDA-, and active co-ordination, for example, in the frame of the Committee on Earth Observation Satellites (CEOS),

a partnership approach for applications missions through partnership mechanisms and co-funding schemes with other public bodies and industry,

new modes of dialogue with industry, to stimulate a user-driven, cost-effective and entrepreneurial European industry that is competitive in global markets,

higher effectiveness of the Agency and its delegate bodies in the decision-making process and in implementation,

stable financial planning,

flexibility, e.g. the possibility for extension of satellite operations without requesting a new supplementary budget,

inherent incentives for cost savings within the individual programme elements,

continuity, e.g. for long-term archiving of ESA mission data,

a long-term plan against which industry can plan strategically,

ease of setting up co-operative missions with other agencies.

Whether all aims are achieved is going to test the Agency, industry, the scientific and user community. One looks forward to the future with interest.

ACKNOWLEDGEMENT

This text has been adapted from a keynote address originally presented at the IAA symposium on small satellites for Earth Observation held in Berlin in April 1999.

REFERENCE

Arend, H., R. Bonnefoy and D.J. Southwood, The Future ESA Earth Observation Strategy and ‘Living Planet’ Programme, ESA Bulletin 94, May 1998, p.4-12.

Remote Sensing in the 21st Century: Economic and Environmental Applications, Casanova (ed.)
© 2000 Balkema, Rotterdam, ISBN 90 5809 096 5

The Council of Europe's EUR-OPA Major Hazards Agreement

J.P. Massué
EUR-OPA Major Hazards Agreement, Council of Europe, Strasbourg, France

C. Negru
STRIM Project, Council of Europe, Strasbourg, France

ABSTRACT: In March 1987, the Committee of Ministers of the Council of Europe adopted a Resolution establishing an intergovernmental Open Partial Agreement, aiming to reinforce co-operation between Member States in a multi-disciplinary context to ensure better prevention, protection and organisation of relief in the event of major natural or technological disasters. The most advanced present day resources and knowledge are called upon to ensure an efficient and interdependent management of major disasters, in particular the use of space techniques. The paper sets out the contents of this Agreement.

1 INTRODUCTION

Interactions between human society and the natural environment are showing increasing signs of risk vulnerability to hazardous events. Coping with risk situations can be complex and controversial. Governments and international organisations have devoted considerable resources to developing and applying techniques of risk analysis and risk assessment. Their main aim : to take better informed and more trustworthy decisions on hazards to human health, welfare and the environment.

2 THE AGREEMENT, A WIDE FRAMEWORK FOR CO-OPERATION IN RISK MANAGEMENT

2.1. *General presentation : structure, objectives and activities*

The trans-boundary effects of natural and technological disasters stress the international dimension of ensuring safety.. For more than 10 years, the EUR-OPA Major Hazards Agreement has progressed in becoming a wide framework for co-operation in the field of major natural and technological disasters, open to both member and non-member States of the Council of Europe.

The Agreement counts to date 23 member States from the Euro-Mediterranean area, given that the countries of the Mediterranean basin have always been privileged partners in economicc, political and cultural exchanges. Among its members : Albania, Algeria, Armenia, Azerbaijan, Belgium, Bulgaria, France, Georgia, Greece, Italy, Lebanon, Luxembourg, Malta, as well as the Republic of Moldavia, Monaco, Morocco, the former Yugoslav Republic of Macedonia, Portugal, Russia, San Marino, Spain, Turkey, Ukraine.

In addition, the European Commission, UNESCO, the World Health Organisation and the Office for co-ordination of Humanitarian Affairs (OCHA) of the United Nations, the International Federation of Red Cross and Red Crescent Societies participate in the Agreement or are associated in its work.

In March 1987, the Committee of Ministers of the Council of Europe adopted the Resolution establishing an intergovernmental Open Partial Agreement. In fact, this decision has given to the Agreement the concrete and challenging task of pursuing a policy that:

- on the one hand, will promote co-operation between member States to ensure better **prevention, protection and organisation** of relief in the event of major natural or technological disasters, and
- on the other hand, will offer a suitable platform for common action between Europe, in its geographical sense, and the South of the Mediterranean, by calling upon present day resources and knowledge.

The activities carried out within the Agreement are situated at three levels : political, scientific and technical or operational level, and last but not least the specific programmes.

- *(the political level)* The Agreement has been mainly called upon to consider how to improve political decisions about risks to public health, safety and environmental quality. The periodical meetings of the Ministers dealing with civil protection and environment and the meetings of the Committee of Permanent Correspondents aim to promote the exchange of information and know-how among the member States.

- *(the scientific and technical level)* One of the focal points of the Agreement is to have sought to ensure a direct participation of the member States by creating a Network of European Specialised Centres. Among the 21 Centres, allow me to mention one or two : the European Centre for Geodynamics and Seismology (Luxembourg), the European Centre for the use of new technologies for Risk Management (Russia), the European Centre on Prevention and Forecasting of Earthquakes (Greece), the European Centre for Seismic and Geomorphologic Hazards (France), the European Centre for Urban Risks (Portugal), the European Centre on Information for Emergency Situations (Spain), etc.

Within the same framework, the **"European Advisory Evaluation Committee for Earthquake Prediction"** was set up in reply to requests forwarded by the government. Seismologists, aware of the problems raised by the dissemination in the media of unreliable forecasting, had themselves adopted in 1991 a European Code of Ethics on information relating to earthquakes.

The **European Warning System** is another structure set up within the Agreement. Its objective is to support common efforts made by member States in the assistance provided to a State hit by a disaster, in particular earthquakes, but not exclusively. Its intervention is limited to the dissemination of technical data. Other international bodies then ensure humanitarian aid. The problem was of sufficiently broad interest that the programme received support from 33 networks of seismometers, situated in 24 countries world-wide.

- *(the specific programmes)* A series of initiatives has been developed in the framework of three *specific programmes* which are:
  - the European programme on training in risk sciences, known as the **FORM-OSE programme,**
  - the **EDRIM programme**, EDRIM for Electronic Discussion group for Risk Management consisting of a permanent telecommunications network for national authorities in charge with risk management, in particular in emergency situations, and
  - the **STRIM** Programme for Space Technology for Risk management.

2.2 *FORM-OSE programme*

In practice, risk management requires a combination of source-oriented and impact-oriented measures, involving operators of the hazardous activity, relevant public authorities and the public.

All of these depend on the existence of a strong **culture of risk prevention**. The development of training programmes in risk sciences, from school to university levels, including continuing education and sandwich training constitutes the core of the FORM-OSE programme.

Various analyses show that **space sciences and technologies** can significantly assist risk evaluation and management. There is a good reason to promote the study of space applications in higher education programmes dealing with risk, with a holistic, multidisciplinary and trans-boundary approach.

The findings pointed to an increase in risk-specialist posts in various sectors. This trend was correlated to the current development of the European environment law. Particularly the law on classified installations is both raising safety standards and giving people more right to be kept informed of and to be party to decision.

Bearing in mind these aspects, the Agreement is working jointly with a wide network of European universities on the definition of pre-doctoral and doctoral programmes in the field of intersection between space sciences and technologies and risk management.

In this context, special attention is given to key concepts such as: models of the environmental damages of accidents, risk quantification, source safety, emergency response planning, integrated risk assessment and management, as well as to GIS, Earth observation, new information and communication technologies (NICTs). As you know, together with EARSeL, the European Space Agency and in the framework of a joint programme with the DG XXII of the European Commission, the EURO-STRIM programme has been launched in October 1998.

2.3. *EDRIM programme*

Prevention begins, of course, with information. Furthermore, the Seveso Directive (to be implemented in the domestic law by 1999), obliges the member States to exchange information on major accidents in order to improve the basis for prevention. This legislation will be sooner or later integrated by the Eastern European countries. In this respect, the **EDRIM programme** is aimed at setting up a permanent space telecommunications network for national risk managers. Based on the use of the Internet system, it facilitates communication at several levels:

firstly, it facilitates traditional communications: audio and visual conferencing facilities making it possible to hold "virtual" meetings with: text presentations, images, animated images; simultaneous work on a shared document;

secondly, it gives access to pooled knowledge sources, on the principle of "distributed universal library" for basic data processing, in particular using GIS and MIS, modelling, etc. EDRIM is based on the concept of distributed resources accessible to any authorised person, opening the way to all kinds of distance learning. The authorisation system at all levels is supposed to ensure protection and security in the exchange of information in restricted networks, In addition, the use of space telecommunication systems enables the EDRIM system to operate even when traditional telecommunication systems are destroyed or saturated.

This programme aims to help national authorities in decision making and to strengthen co-operation in the case of emergency situations.

2.4 *STRIM programme*

A great effort has been made in the field of space technologies, particularly as far as Earth observation, telecommunications and global positioning are concerned. It is a question as to whether the space tool can contribute in an integrated way with ground systems to risk management, thus in relation to the overall spectrum of risk management.

In 1996, the prospective study into the use of space technologies for risk management conducted by the European Commission, the European Space Agency and the EUR-OPA Major Hazards Agreement, demonstrated that current Earth observation systems are able to meet users' needs to approximately 45 %, a figure which could be increased to almost 80 % with a radar satellite network.

These conclusions led to the implementation of a joint programme supported by the above-mentioned partners and the CNES (French Centre for Space Studies) to promote the use of space technologies to aid risk management, STRIM. This programme which is based on the pooling of Europe's space resources, slots perfectly into the permanent telecommunications network for national risk managers, known as EDRIM.

In fact, the STRIM Programme is conceived as a "basket" common to all the partners. The idea is the following: each of the partners places a certain number of programmes into the "basket" which they consider useful for management at international, bilateral or national levels. At the top there is a technical co-ordination unit, managed by the European Commission and situated in Ispra, and a Programme Committee made up of all the partners.

In practice, these programmes are implemented in four sectors of activity : training, information and public awareness raising of the space tool, which is co-ordinated by the EUR-OPA Major Hazards Agreement., evolution of the satellite systems, support and pre-operational activities, and finally, pilot projects.

The current satellite systems offer already a great number of potential applications for risk management, classified into three categories : prevention, alert and crisis management, post-crisis and rehabilitation management. Particular attention is given to the field of communications, meteorology, navigation and data capture with different space resolutions and spectral flames.

3 CONCLUSIONS

To conclude, I would like to point out that it is with great pleasure that we have learned of the positive conclusions of the Minister's session of the European Space Agency, held recently in Brussels, which focused on encouraging prospects for the Earth Observation programme.

Furthermore, on the occasion of the meeting organised a few days ago by the European Commission at Baveno in Italy, the participants underlined the major role that space technologies can and could play for environmental security.

The Council of Europe, through the EUR-OPA Major Hazards Agreement will make its contribution to these common efforts and we are sure that EARSeL will join us in the spirit of Baveno.

1 Agriculture and forestry

Remote Sensing in the 21st Century: Economic and Environmental Applications, Casanova (ed.)

Space-vegetation software: A remote sensing tool for environmental applications at regional level

César Carmona-Moreno, José-Andrés Moreno-Ruiz, Etienne Bartholome & Thierry Negre
Joint Research Centre, Space Applications Institute, Ispra, Italy

ABSTRACT: The VEGETATION Programme allows a daily monitoring of terrestrial vegetation cover through remote sensing, at regional and global levels. The instrument, funded by the EC, is part of the SPOT-4 payload. This instrument, launched the 24th Mars 1998, deliver measurements which were specifically tailored to monitor land surfaces' parameters with a frequency of about once a day on a global basis and a medium spatial resolution of 1 kilometre (2200 km swath width and 300 m of geolocation accuracy). Two types of ground processing services were foreseen: the Central/Global Processing facility (CTIV – Centre de Traitement d'Images VEGETATION in Mol –Belgium) and the Local Processing facilities for local/regional data processing.
The CTIV is now operational and the Joint Research Centre – Space Applications Institute, is developing the VEGETATION Local Processing system. The European Commission will deliver the processing software called "SPACE-VEGETATION software" free of charge to the VEGETATION programme, and this will be distributed freely to officially VEGETATION registered users for local/regional remote sensing applications. This paper deals with the system and data characteristics.

1 INTRODUCTION

The associated ground services for processing VEGETATION data and their distribution are operational since 1998. Two ground processing services were foreseen:

1) Central/Global Processing is performed by the CTIV (Centre de Traitement d'Images VEGETATION in Mol – Belgium). This Centralise Ground Segment processes instrument measurements to offer standard products to the general users' community. The VEGETATION data (VGT) are downlinked via VEGETATION X-Band receiving station in Kiruna (Sweden) and transmit to the CTIV for processing.
2) Local Processing facility is performed by local L-Band receiving stations that will provide by a specific pre-processing system. The level of quality of VGT L-Band products (with all characteristics of standard VGT P products but with a 1Km of geolocation accuracy) is equivalent to what is commonly available for similar satellite data (e.g. AVHRR HRPT). The software will be available early November 1999.

The Joint Research Centre – Space Application Institute is developing the local L-Band pre-processing system. The European Commission will deliver the software called "SPACE-VEGETATION software" free of charge to the VEGETATION programme, and this will be distributed freely to officially VEGETATION registered users. This software is inspired by SPACE-AVHRR software also developed by the JRC-SAI and distributed by the OSS. The applications addressed by this system are agriculture and environment monitoring.
In a general way, local L-Band receiving stations demodulate and store local VGT Level 0 image data files in Fi5 format from the L-Band transmission of the VEGETATION instrument. The software will process these local VGT Level 0 data files to produce a pre-processed product in the SPACE-VEGETATION data format. This product will be called VGT-SV product. The solar and zenith angles associated to each pixel will also be stored.
A general presentation of the system, the scientific objectives, data input/output description, image radiometric/geometric accuracy, processing algorithms implemented and distribution policy are presented in this paper.

2 SPACE-VEGETATION SYSTEM GENERAL DESCRIPTION

2.1 *Software Dissemination policy*

The European Commission will provide SPACE-VEGETATION software free of charge to the VEGETATION program. It will then be distributed to VEGETATION (VGT) registered L-Band stations (SRBL) by SPOT IMAGE, in the framework of a yearly agreement, including as well the reception of L-Band telemetry, and the basic setting up of this SPACE VGT software.

2.2 *SPACE-VEGETATION system general overview*

For one day of VEGETATION telemetry acquired by a local L-Band receiving station (SRBL), it is possible to get two or more image segments, depending of the latitude of the station. The circle of visibility of the receiving station has an approximate radius of 2500 km.

The development of the SPACE-VGT software assumes that the acquisition and pre-processing tasks are performed on separate hardware and software systems. Standard interfaces with the SPACE-VGT software will be soon available via Internet for the receiving station constructors in order to develop hardware/software for VGT telemetry acquisition producing VGT Level 0 data format compatible with the SPACE-VGT software. The current NOAA-AVHRR receiving stations may be easily upgraded for receiving/processing VEGETATION data.

The following sketch (Fig. 1) illustrates the general functional architecture of the SPACE-VEGETATION software.

The data flow within the pre-processing system is the following :

- input segments control and archiving
- pre-processing of input segments
- output mosaics control and archiving or edition

2.3 *USER Parameters*

The environment and processing parameters relative to the SPACE-VGT software shall be handled through one or two files that shall be easily edited and modified.

The software administrator shall manage the access rights through Windows NT services.

The related parameters can be managed in two sets:

- a first set for parameters that shall not be modified frequently or that shall require a scientific knowledge: such parameters will be modified under the Station Administrator authority.
- a second set that concerns directly the daily production, such as the user-defined window or the selection of the cartographic projection.

2.4 *VGT Level 0 Segment ingestion*

This function corresponds to the conversion of Level 0 input segments to Level 1A data. For an L-Band receiving station, the segments shall be uninterrupted during the acquisition time, except when we shall have a modification of the instrument gains, which shall behave as two contiguous segments.

The system shall have the capacity to ingest automatically Level 0 segments, from an input buffer.

This function shall generate a quick-look of input segments.

The image format of the quick-look shall be compliant with JPEG format.

The file resulting from the quick-look of input segment shall be stored in the system for a later use, at least until the generation of the daily product.

2.5 *Cataloguing and Archiving*

The SPACE-VGT software system will feature one catalogue comprising two parts, the first for input segments and their quick-looks, and the second for output mosaics and their quick-looks.

The catalogue database shall be based on the MS Access software. The archival sub-system is based on a 'commercial' device (hardware + Windows NT utility), that shall communicate with SPACE-VGT through a shared directory.

The SPACE-VGT software shall include a utility allowing to monitor the volume of the products to be recorded, in order to generate a warning whenever this volume will oversize the capacity of a CD-ROM.

2.5.1 *Input segments*

The input segments are those that have already been successfully ingested and converted to Level 1A.

The system shall enable the user to print a quick-look of input data, with sufficient annotations and it shall enable him to check the input segment thanks to the quick-look.

The system shall also enable the user to accept or reject the cataloguing of an input segment and activate the pre-processing of an input segment.

Once declared in the catalogue, the system shall enable the user to archive these segments together with their quick-look through the recording of a CD-ROM.

The system shall have the capacity to ingest a CD-ROM identifier, entered by the user, before the archiving process. CD-ROM identifiers shall be referenced in the database.

2.5.2 *Mosaic VGT-SV products*

A mosaic consists on a compositing of different segments included in a pre-defined ROI. The selection of pixels within the recovering areas of segments is based on the NDVI MVC (Maximum Value Composite) process.

The system shall produce a quick-look of the output mosaic. The file resulting from the quick-look of output data shall be stored for a later use.

The quick-look shall enable the user to check a coarse quality of the mosaic in order accept or destroy it.

The cataloguing and archiving tasks will be activated once a processing has been completed to the generation of a VGT-SV product (generated from one or more input segments).

The accepted mosaics and associated quick-look will be catalogued automatically excepted if the user wants only to generate/edit the product without declaring it in the catalogue.

The system shall have the capacity to ingest a CD-ROM identifier, entered by the user, before the archiving process. CD-ROM identifiers shall be referenced in the database.

2.5.3 *Catalogue consultation*

The user shall be able to query the catalogue according to specific parameters such like acquisition date, VEGETATION number and time window, CD-ROM number and ROI.

The catalogue could be consulted to select the type of data: Level 1A segment or output mosaic (VGT-SV product).

The user could display the quick-look of a selected image retrieved after a consultation and select one or many input segments within the list resulting from the query.

The user shall be able to transfer selected input segments in the list of input Level 1A segments to be pre-processed.

2.6 *Pre-processing Functions :*

2.6.1 *Manual processing*

As required previously, the VGT-SPACE software system shall handle automatically the ingestion of Level 0 segments, transferred from the acquisition system to the pre-processing system, and convert them into Level 1A.

The user could select the Level 1A segments to be used for generating the output mosaic and a multiple selection shall be possible. The system will verify that the input segments are catalogued and online and, in this moment, he will be enable to launch manually the processing.

2.6.2 *Automatic processing*

The user shall have the capacity to activate an automatic processing mode: every day, at a given time, and with selected parameters, the software will process the list of 1A segments accepted for processing. This mode will be active until a closing date that shall be entered by the user.

2.6.3 *Processing parameters*

The user shall define a ROI (Region Of Interest) by entering the geographical co-ordinates: latitude, longitude of top left and bottom right corners.

The system will verify the coherence of the geographic co-ordinates values in relation with the circle of the visibility of the antenna.

The user shall define a cartographic projection from the following list of possibilities: Plate Carrée, UTM, Albers Equal Area, Lambert, Mercator Oblique.

2.6.4 *Execution of the pre-processing*

For each VGT-Level 0 input segment, the following functions will be applied :

1. ***Reading of input data***
    - 1.1. Verification of the image telemetry coherence
    - 1.2. Line timing
    - 1.3. Functional classification of the data :
        - 1.3.1. Header + orbital data
        - 1.3.2. Image and timing
2. ***Computation of ephemeris data using MADRAS software;***
    - 2.1. Computation of the satellite data position P and velocity V every minute
    - 2.2. $(l,p) \to (x,y)_{h\ (min,max)}$ computation
    - 2.3. Computation of a (P, V) for each l
    - 2.4. View direction vector computation for each p
    - 2.5. Computation of the intersection between the earth model and the line defined by the view direction vector and the point p.
    - 2.6. $(x,y) \to (l,p)_{h\ (min,max)}$ computation
3. ***Computation of the useful area, from the ROI, for the segment to be processed***
4. ***Equalization computation***
    - 4.1. Radiometric calibration parameters from the Parameters Characteristic of the Instrument (PCI) are applied to the different image bands
5. ***Geometric modellisation***
    - 5.1. Definition of block size (optimization of the computation)
    - 5.2. In each block, for each node of a super grid (16 km x 16 km):
        - 5.2.1. $(l,p)_{h\ (min,max)}$ computation
        - 5.2.2. Sun and view direction angles computation for each super grid node
    - 5.3. In each block, for each pixel
        - 5.3.1. Bilinear interpolation of $(l,p)_{h\ (min,max)}$

5.3.2. Sun and view direction angles
5.3.3. Computation of the h elevation from DTM
5.3.4. Linear interpolation to get $(l,p)_h$ from $(l,p)_h$ and h
5.4. Computation of the smallest image area containing all $(l,p)_h$ couples
6. ***For each spectral band, extraction of the equalized image area***
7. ***Resampling and computation of the TOA values.***
7.1. ToA formulae and radiometric coefficients from the Parametric Characteristic of the Instrument are applied to the image.

Where :
(l, p) -> image co-ordinates.
l – line number and,
p – pixel number in the line
(x, y) -> projection co-ordinates

Following these steps, the software shall execute:

- Mosaicking: the selection of pixels within the recovering areas of segments shall be based on the maximum NDVI process.
- Generation of the quick-look of the mosaic
- Formatting the output product: BSQ format

2.6.5 *Product archiving or edition*

When a SV product is to be archived locally, it shall be declared in the catalogue and archived in the BSQ data format.

2.6.6 *Destruction of products*

After the automated ingestion of Level 0 segments, these input data would be automatically deleted.

The Level 1A segments and SV products will be deleted from the disk only on manual input.

The quick-looks of Level 1A segments shall be as far as possible kept on line.

2.7 *Display Function :*

This function shall allow the visualization of the quick-look of segments or mosaics and as well of the mosaic product in full resolution for image validation. It shall have at minimum basic features for image enhancements and display (zoom, pan, ...), and shall allow to display the cartographic co-ordinates related to the cursor position.

3 PERFORMANCE REQUIREMENTS

3.1 *Products specifications*

3.1.1 *Output mosaics*

- Output products shall be generated on the ROI window.
- The software will execute system-level corrections comparable to those applied to VGT-P, but an image quality restricted to this software (see below).
- The sun and observation angles shall be generated for each pixel.

3.1.2 *Quick-looks*

- Quick-looks shall be produced as color composite and in a JPEG format.

3.2 *Catalogue*

- The catalogue shall include as a minimum: acquisition date, processing date, VEGETATION number, geographic location, CD-ROM number, PCI reference, log files.

3.3 *Precision of SV products*

3.3.1 *Geometry*

- **Multitemporal registration** : The RMS error for registration is less than 1.5 km (the objective is 1 km).
- **Multispectral registration** : The RMS error for registration between spectral bands is less than 0.7 pixels RMS for all bands, and 0.3 pixels between B2 and B3.
- **Absolute location accuracy** : is less than 1 km

3.3.2 *Radiometry*

- **Radiometry** : Specifications are identical to the P products of the CTIV standard.

4 COMPUTING ENVIRONMENT

- The software will be developed for running in PC platforms with Windows NT environment system.
- The programming language will be C++ for scientific algorithm development and software integration.
- Microsoft ACCESS software will be used for building the Data Base Management System.
- The system shall be completed by a commercial CD-ROM recording sub-system.
- Input and output data can be transferred from or to another machine, via a LAN, without reducing the processing capacity of the system.
- The software shall be modular, to facilitate future evolutions, namely related to new geometric and radiometric corrections algorithms.
- The software will use the MADRAS library (developed by the CNES-French Space Agency) for the computation of the position and velocity of the satellite at instant t.

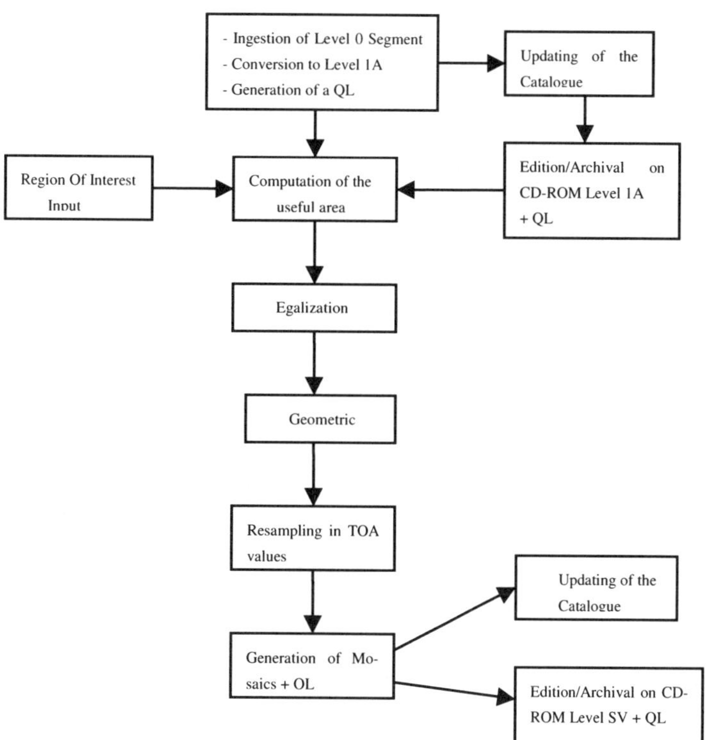

Figure 1. SPACE-VEGETATION Data flow.

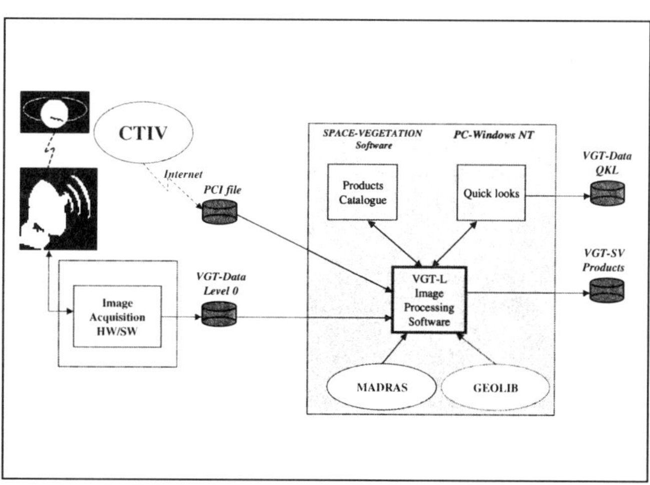

Figure 2. This figure shows the interfaces between the SPACE-VEGETATION software and the other external systems.

- The software will use the GEOLIB software library (developed by the CNES-French Space Agency), which will limit the development of functions related to the geometric component of the pre-processing

R:SWIR G: NIR B: Red

Figure 3. Image from the VEGETATION instrument. This image was taken on 28th July, 1998 over the area along the border between Angola and Zambia. At that time of the year dry season conditions prevail in the area. The color combination is: Red for Short Wave Infrared, Green for Near Infrared and Blue for Red.

The greenish area in the NW area is the so-called 'Miombo' woodland (1).

The yellowish corridor marked 2 is a savanna strip extending along a swampy valley bottom, while the one under 3 was left dry.

The same contrast appears between the alluvial plain along the Luanguinga (4), a tributary of the Zambezi River, that shows dark tones, and the Barotse floodplain (5), along the Zambezi itself. (Colour plate, see p. 595).

5 CONCLUSIONS

SPACE-VEGETATION software has been presented and its characteristics (software architecture, available functions, software policy distribution, ...) described. This software will be delivered to the VEGETATION program by the European Commission (JRC-SAI) early 2000 and distributed to VEGETATION (VGT) registered L-Band stations (SRBL) by SPOT IMAGE, in the framework of a yearly agreement, including as well the reception of L-Band telemetry, and the basic setting up of this SPACE VGT software.

Some examples of VEGETATION images and some thematic aspects have been presented and commented.

Remote Sensing in the 21st Century: Economic and Environmental Applications, Casanova (ed.)
© 2000 Balkema, Rotterdam, ISBN 90 5809 096 5

Unsupervised classification's optimization

C. Pinilla & F.J. Ariza
Departimento de Ingeniería Cartográfica, Geodésica y Fotogrametría, EPS, Universidad de Jaén, Spain

M. Sánchez de la Orden
Departimento de Ingeniería Gráfica e Ingeniería y Sistemas de Información Cartográfica ETSIAM, Universidad de Córdoba, Spain

ABSTRACT: This work shows first results of a research in order to give advice about how many iterations are needed in an unsupervised classification to get a predefined number of classes and a predefined coincidence index between this classification and the better one. The experimental work shows that the procedure has a linear behavior. Future work must be continued with a greater number of images and looking for an statistical parameter from which it could be derived the number of iterations and the percentage of agreement.

1 INTRODUCTION

In the context of the digital analysis of images, unsupervised classification is a kind of automatic statistical classification that consist to determine the internal structure of the information provided by some bands belonging to a multispectral image and elevate each group of coherent information to a class category. From the cartographic production point of view, classifications are generalizations procedures known as numerical categorization (MacMaster, 1992). This definition is very interesting for us due to it holds all the sense of the generalization process, which is understood as an art for many authors (Robinson, 1987). That means, as it occurs in generalization, that the classification process is complex, subjective and depends on a prestablished goal for the cartography.

Within the scope of the classifications always are referenced two main processes (Jensen, 1996; Schowengerdt, 1997): i) supervised, ii) unsupervised. Actually, both of them require operator's intervention, but only are considered within the first kind those in which the operator feed the system with statistical information belonging to the classes we wants to obtain (spectral signatures) (Pinilla, 1995). Unsupervised process are those in which it does not occur, but it is needed some information in order to guide the statistical process, for example, the number of classes, rules for classe's merging or spliting, etc.

Both approaches are, and have been, largely used in remote sensing applications to the environment: geology (Qari, 1992), land cover detection (Camara, 1988), inventories (Augenstein, 1991), etc.

For a Supervised classifications it is needed a field knowledge. Visits or other sources of greater accuracy are needed for the establishment of adequate training and comprobation fields (Congalton, 1998). For unsupervised classifications someone can think we need no knowledge, but it is no so. Actually it is not necessary to feed the system whit spectral signatures but, at least, process has to be oriented with the number of classes, because, otherways, system can detect so many classes as digital levels are in the band (in general, $256^{\text{Number of bands}}$), which is a nonsense. On the other hand, it can detect an insufficient number of classes. For these reasons, it is very important to have some knowledge about the field, and also to have clear idea of the propose of the cartography and about how the statistical process works.

In this context, it is also very important to know the use of unsupervised classifications, because of by this way are better understood the accuracy requirements that can be imposed. One of the more extended uses of the unsupervised classifications is like a preprocess to the supervised classification, in order to get better understanding about the statistical and spatial structure of the image data for the training fields establishment.

Given attention to practical aspects, there are no many references about how to guide setting of iterations number. This aspect bias computation time. Nowadays personal computers and work stations allow users to run and execute process in second or in minutes, but days in which for same process were needed days or weeks are no so far away. Such in many others practical aspect, there is lack and for this reason present work is devoted to show some concrete examples in order to help decision making of students and neophyte in relation to unsupervised classifications.

2 MATERIAL AND METHOD

One of the most robust unsupervised classifiers is the ISODATA (Duda y Hart, 1973), whose algorithm (Richards, 1994) calculates class means evenly distributed in the data space and then iteratively clusters the remaining pixels using minimum distance techniques. Each iteration recalculates means and reclassifies pixels with respect to the new means. Iterative class splitting, merging, and deleting is done based on input threshold parameters. All pixels are classified to the nearest class unless a standard deviation or distance threshold is specified. This process continues until the number of pixels in each class changes by less than the selected pixel change threshold or the maximum number of iterations is reached.

Software commonly used in remote sensing and digital image processing (g.e. ENVI, Er-Mapper Erdas, etc.) have the ISODATA classifier incorporated. It is also common to give the possibility to limitate the number of iteration. A problem arises: how many iterations are needed, and what implies more or less iterations.

In this work we have used the ENVI-IDL (Better Solutions Consulting, 1995, 1997) software tool. IDL allows to use resources of ENVI, for instance the functions that execute unsupervised classification. The options that can be customized by user are:

- Classes: interval of number of classes that are desired (minimum number and maximum number).
- Iterations: maximum number of iterations to run. It can be or not reached depending on the next parameter.
- Percentage of change: express the maximum threshold of change between iterations in the number of pixels that below to a class and that it is allow to change to other class. It is a way to limit the iteration process. For example, if we admit a 5% of change, if we were set iterations to 100, but between two successive iterations change are < 5%, the process is stopped.
- Pixels in class: minimum number of pixels per class.
- Maximum standard deviation: if a class past this value it is splited.
- Distance between classes: distance between means in order to merge classes.

In this work the number of classes have been 8. This value has a great importance for the process and for the cartographic sense, both aspect are tie. On one hand, the number of cartographic classes depends on the study area, but should be in agreement with work scale and objective, and with sensor characteristics.

The iterative process has been executed by means of a IDL program developed for this work. This program executes the classifications and comparations saving the results.

Because of the actual number of iterations depends not only on the setup value for the iterations but also on the percentage of change, the last was set to the value of 0.01%. This a restrictive value that force alarge number of iterations.

Steps of the process are as follow. The first is to determine the minimum number of iterations I_{min} that give a classification the adequate number of classes ($N = 8$). Later on, it has been determined the number of iterations I_{max} for which the percentage of change is less than 0.01%. The image corresponding to this number of iterations is considered as the comparison's pattern later because is the best possible result of the process.

The program allow us to derive automatically the values of the percentage of agreement (*Pa*) and Kappa (*K*) index (Veregin, 1994; Congalton, 1998), between the classification result for an iteration $I \in (I_{min}, I_{max})$, and the pattern by means of the confusion matrix.

For the development of the experiment have been used diverse subscenes bellowing to the province of Jaén, each of then with 400x400 pixels and 6 bands:

- Im1 = mountains (Fig.I.a). typical scene of mediterranean mountainous land cover with median range topography.
- Im2 = plain (Fig. I.b). This scene shows the proper characteristics of a river plain zone near to mountains. Irrigated and not irrigated crop fields from little to medium size, and mountainous vegetation dominate the scene.
- Im3 = pond (Fig. I.c). very similar to Im1, but in this case the topography is more complex and also there is a pond with a water layer and free cover boundary.
- Im4 = village (Fig. I.d). in some way similar to Im2. It shows a river plain more little than in Im2, but a more extensive dry crop field on an undulated topography and a urban settlement bellowing to a middle sized village.

3 RESULTS AND DISCUSSION

Study of results's behavior has been done analysing the evolution of *Pa* versus *I*. In Fig.II it is shown the result for the Im3. It is only shown the evolution of the percentage of agreement (*Pa*), because of the evolution of the *K* index is very similar.

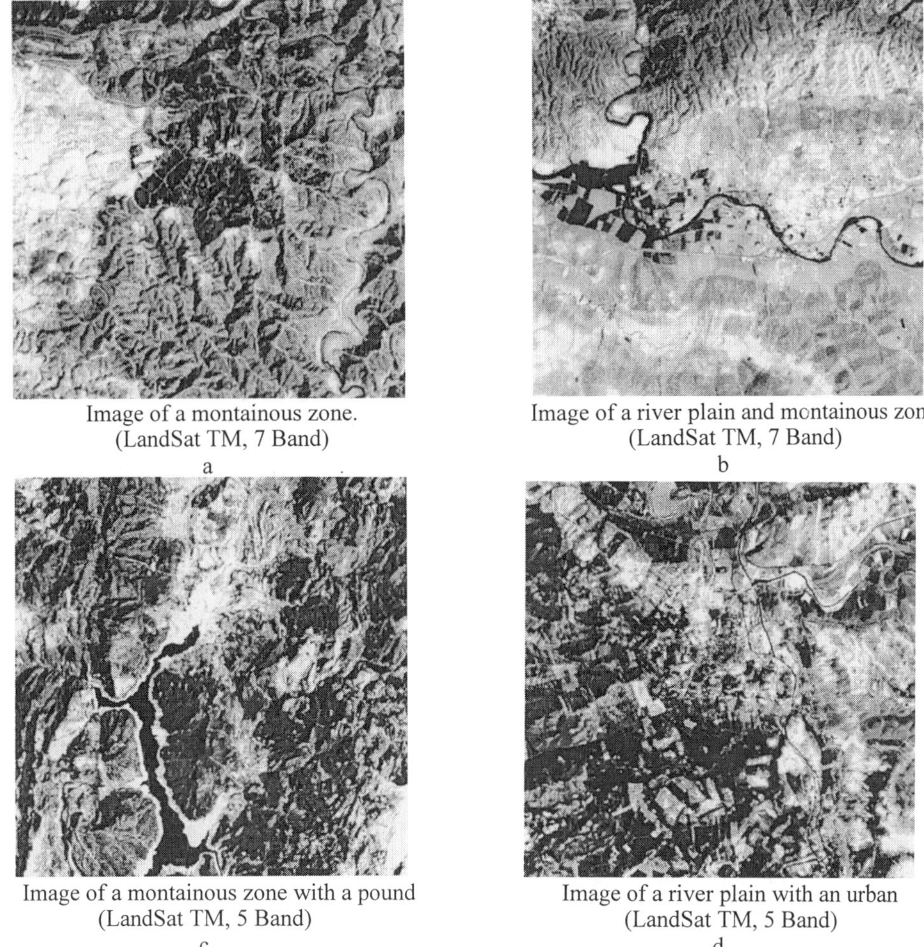

Image of a montainous zone.
(LandSat TM, 7 Band)
a

Image of a river plain and montainous zone
(LandSat TM, 7 Band)
b

Image of a montainous zone with a pound
(LandSat TM, 5 Band)
c

Image of a river plain with an urban
(LandSat TM, 5 Band)
d

Fig. I . Images used in the study

Values of *K* index are little smaller than *Pa*, reason because of *K* index evolution graphics are like a down translation of the *Pa/I* representations. It is important to notice that curves show only the evolution in the interval of $I \in (I_{min}, I_{max})$. For this interval we obtain always a classification with $N = 8$ classes but with different agreement with the pattern.

The behavior of classifications is very similar for all images, and can be very well represented by a common analytical function such as $Pa = A\ Ln\ (I) + B$. The values of *A* are very similar for all images but values of *B* can be very different depending on the initial behavior of the *Pa* for the I_{min}. A global adjustment (Fig. III) which equation *is* $Pa = 13.34\ Ln\ (I) + 33.56$, for $I \in (I_{min}=5,\ I_{max}=150)$, with a R^2 better than 95%. This adjustment has been successfully tested and used with these subscenes and a set of other images of similar characteristics. As it has been said, the *A* coefficient is very stable but it does not occur the same for the *B* constant. For this reason, we think that it would be also of great interest to represent the evolution of *Pa* versus an estimation of the complexity of the image, which is related to *Pa* at I_{min}, in order to link two parameters.

I_{max} is always ranging between *150* and *300* iterations, but another practical conclusion is that with a number of iterations similar to the number of desired classes we can assure the existence of these classes, nevertheless, the *Pa* can be very different for each case.

Results of other works show that when number of classes is greater, values of I_{min} and I_{max} increase. It is like a translation of the above explained curve.

The classification process could have been improved by means of histogram enhancement, or also by means of the use of other kind of transformations

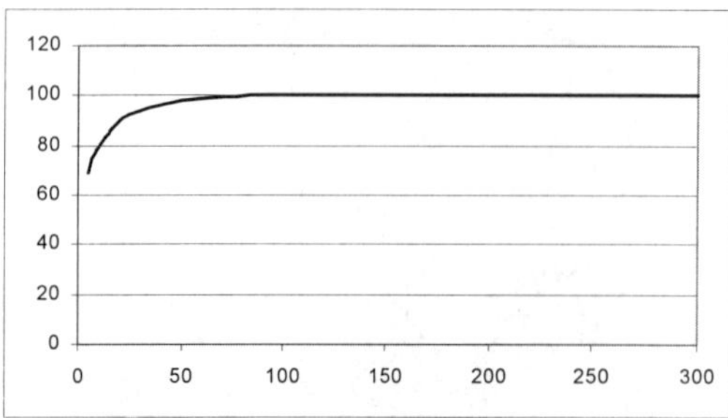

Fig. II . Evolution of the percentage of agreement versus number of iterations for the unsupervised procedure

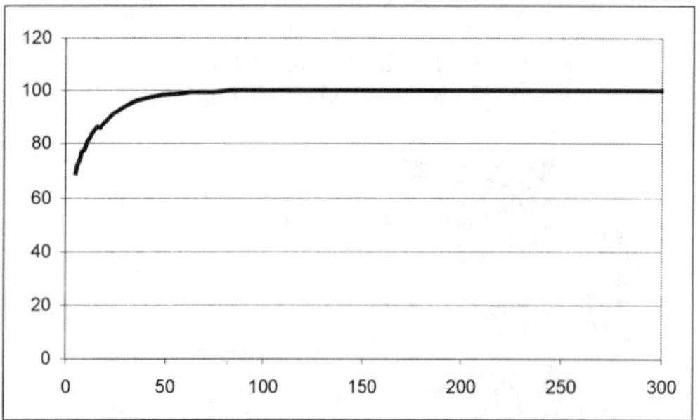

Fig.III. Empirical adjustment for the *Pa* index evolution versus number of iterations

such us PCA, the later also reduces the volume of data and processing time. But we have try to not supervised to much the unsupervised process and also to use original data in order to have a clean process without side effects. Because of the constrains of the work (little number of images, only one software, etc.) results have to be understood only as what they are. The work has been developed with the idea to show some practical aspects that do not appear in program's manual, and that also are not presented in remote sensing manuals because they are more devoted to fundamental aspect than to practical ones.

4 BIBLIOGRAPHY

Augenstein, E. (1991). Evaluation of SPOT HVR-XS data for kelp resource inventories. In PE&RS, vol 57, nº 5.

Better Solutions (1995). Interactive Data Language. IDL User's Guide. Researc Systems. Boulder, Colorado.

Better Solutions (1997). Interactive Data Language. IDL Reference Guide. Research Systems. Boulder, Colorado.

Camara, M.J (1988). Clasificación de recursos terrestres y geológicos por medio de observaciones de satélite. En Teledetección y planificación del territorio. MOPU, Madrid.

Congalton, R.G., Green, K. (1999). Asessing the Accuracy of Remotely Sensed Data: Principles and Practices. Lewis Publishers. London.

Jensen, J.R. (1996). Introductory Digital Image Processing. Prentice Hall.

McMaster, R., Shea, K. (1992). Generalization in Digital Cartography. AAG.

Pinilla, C. (1995). Elementos de Teledetección. RA-MA. Madrid.

Qari, M.Y. (1992). Application of Land Sat TM data to geologycal studies. In PE&RS, vol 58, nº 4.

Richards, J.A. (1994). Remote Sensing Digital Image Analysis. An Introduction. Springer Verlag. Berlín.

Robinson, R (1987). Elemento de Cartografía. Ed. Omega. Barcelona.

Schowengerdt, R.A. (1997). Remote sensing. Models and Methods for image processing.

Tou, J.Y. González, R.C. (1974). Pattern Recognition Principles. Addison Wesley Publishing Company. Reading, Massachusetts.

Veregin, H. Giordano, A. (1994). Il controllo di qualitá nei sistemi informativi territoriali. Il Cardo, Venezia.

Remote Sensing in the 21st Century: Economic and Environmental Applications, Casanova (ed.)
© 2000 Balkema, Rotterdam, ISBN 90 5809 096 5

A methodological approach for agricultural statistics by remote sensing

D.Yousfi
Centre National des Techniques Spatiales, Arzew, Algeria

ABSTRACT: The development of the agricultural activity requires a system of periodic evaluation, reliable and efficient, constituting a coherent instrument of observation on agriculture. The study should lead to the evaluation of the land and to the outputs of main cultivations. Facing this objective and regarding the failure of the classic methodologies of statistical collection, a new technology based on geographic studies, which is remote sensing, was used. The study was carried out on the region of Oum El Bouaghi (Algeria), the area of Ain Kercha has also been choseen, to carry out this research. Cereal cultivations have been studied. They represent the biggest part in the region. The development of a methodology of collection on the agricultural statistics let us work on different spatial scales locating, in the first time, the agricultural zones and in the second the cultivated zones. The investigation was done on high resolution SPOT data. A digital image processing methodology was developped in order to discriminate the different cultivated species. The stratification of the region into homogenous zones let us improve the results precision. On the zone of study, the setting up of a survey plan was done and the different parameters of its application were discussed. In order to improve slightly the estimations, the satellite data processed were compared to those issued from investigation by means of a regression estimator. A number of methods were tested, they are based on the combination of satellite data and terrain informations. The results obtained told us that there were over estimations (5,88 % for irrigated areas) and under estimations (0,86 % for cereal cultivation class with low efficiency). In fact, the approach developped gave us a result for which the error margin over all the sampled segments about 2,53 %.

1. PROBLEMATIC

Agriculture must be considered like having priority among the developing politics. Or, how can one formulate a plan of restoration, a politic of development without knowing the productions and needs of a region? it is necessary to establish a "diagnosis" before establishing a plan. More than everywhere elsewhere, the inventories and the statistics are therefore basis tools for our politicians.
The programs of agricultural development require some agricultural reliable statistics and updating regularly. However these measurements must be in adequacy with the used tool, which is Remote Sensing. Indeed, the repetitivity of image taking, the spatial resolution and the spectral discrimination of some images permit the acquirement of the spatial and temporal information on some cultivated surfaces in appropriate time. Currently some statistics are published, but the methods used for collection do not allow to value the reliability of these data.

The Algerian statistical system, especially in matter of agricultural and socioeconomic data, suffers from the following problems:

⇒ the bad quality of the data;

⇒ long delays which separate their collection and their final disposition;

⇒ their elevated costs with regard to their reduced utility.

2. OBJECTIVES OF THE STUDY

The objective of this survey is to set up a methodology of agricultural reliable inventory the results of which are provided in appropriate time.
Indeed, the development of a methodology, which is based on the combination of ground data and remote sensing, for better statistic and cartography of agricultural space, constitutes a long term objective of a capable system :

⇒ distinguish, identify and measure the more quickly possible and with the maximum of precision, the agricultural surfaces;

⇒ evaluate the achieved agricultural productions, this task needs the evaluation of the surfaces and measures of outputs;

⇒ achieve some classes of harvest forecasting, which constitutes the most delicate case, since it requires a precocious evaluation of some surfaces and a forecast of the outputs. In the setting of this survey, it is the first two points that are to be developped. The methodology developed in the setting of this survey is based on the following points:

1) the determination of a strategy of sampling of the soil using area investigations on the basis of geometric segments.

2) clarification of techniques of preprocessing and processing of images (IHS*, CC**, geometric corrections, classifications, vegetation indices, etc.). For some classifications and some vegetation indices, the combination of a certain number of processings is imposed. Indeed, it should first mask all which is «not agricultural» in order to avoid ambiguities between classes.

3) constitution of a data bank of the region of survey and setup of a data base on the segments tested.

This survey should lead to an evaluation of the surfaces and some outputs of the main cultures.

3. PRESENTATION OF THE ZONE OF STUDY:

We worked on a zone around the city of Ain Kercha, which is located in the Wilaya of Oum El Bouaghi. The zone of survey is a plain zone the mean altitude is of 900 meters surrounded of mountains. Indeed, the plain extends over 80% of the surface of the Wilaya. The Geo-ecological conditions of territory is favorable on the whole to agriculture.

The results of the climatic survey allowed us to classify this climate as an attenuated climate subdesertic, presenting a humid season in winter.

We will limit ourselves to the cereals statistics on the region of survey because they represent the majority theme to the level of the Wilaya, the intended earths for cereals represent 56,14% [Benhamouda and Al. 94] of the surface of the Wilaya.

The knowledge of the agricultural calendar of wheat, allows us to make an optimal choice on key dates of images acquisition. The analysis of different stages and some step mark of the life of wheat allowed us to identify the phase or the chlorophyll activity and the density of recovery are the most important. We notice that this phase it spreads from March 15 to April 15, it is for this reason that the images acquired on this period are supposed to give a maximum information concerning the cereals.

* Intensité Hue and Saturation.

** Color Composite.

4. METHODOLOGICAL APPROACH ADOPTED :

The approach developped in the course of this study is based on the information from satellite data (SPOT) as well as on the information collected on the ground. Indeed, a double approach was conducted on the region of survey from some satellite data.

In a first time the space meshing in big categories of land cover was achieved. In order to optimize this stratification and to make it adaptable to the objectives fixed while taking into account some limits of remote sensing, we decided upon the stratification «Corine Land Cover» which was modified slightly for the specificities of our country. The stratification realised allowed us to get to the following nomenclature:

Nomenclature kept		Stratification «corine land cover»
1) Cereals G.Y*	=	Arable land.
2) Cereals W.Y**	=	Arable land.
3) Fallow land	=	Arable land.
4) Market garde	=	Cultural systems.
5) Course on bare soil	=	Natural lawns and grazings
6) Bare soil	=	Scattered vegetation.
7) Forest	=	Forests of resinous.
8) Maquis	=	Vegetation (sclérophyll).
9) Agglomeration	=	Urban continuous tissu.

The methodological approach is based on the photo interpretation of images, for the stratification of the region of survey in big homogeneous units, and on some technics of automatic classification either by pixel (classification by maximum Likelihood or classification by minimum of distance) by parcel (Fig.1). This approach refers to some others data as agricultural statistics, climatic data, etc.

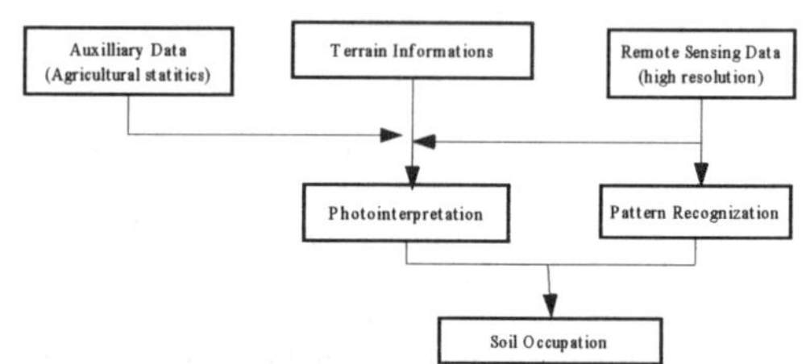

Fig.1: Methodological approach.

The classification being important, we defined several layers in which will be taken different samples to investigate. Several criterias as the type of met cultures, the parcels structure, the localization, allowed us to establish this stratification, which allowed us to define 7 layers in a first time:

* Cereals good yielding

** Cereals weak yielding

⇨ cereal zones;
⇨ markets zones;
⇨ Courses;
⇨ Bare soil;
⇨ Forests;
⇨ Reforestations;
⇨ agglomerations.

In order to get optimal classification, the image processing is done according to two approaches which were compared:

⇨ a so-called supervised approach (classification by maximum Likelihood or classification by minimum of distance);

⇨ an approach no supervised (dynamic cloud).

For the first approach, the selection of the test parcels was made on one hand, to leave some information collected on the ground and on the other hand, from the image of color composite. Two algorithms of supervised classification were tested, maximum Likelihood and minimum distance (Fig. 2). The radiometric characteristic of the different cultures was determined from the analysis of the whole test parcels defined previously. We choose the most homogeneous possible and rejected samples which present an important type gap, in order to avoid the risks of confusion.

4.1 The results obtained:

The supervised classification chosen is the one of maximum Likelihood of the XS channels with a mask which allowed us to eliminate all what is not vegetation in our region of survey. Indeed, this mask was achieved by using a vegetation index (NDVI). This indication allowed us to conceal all the non agricultural part where the chlorophyll vegetation is absent (image 1).

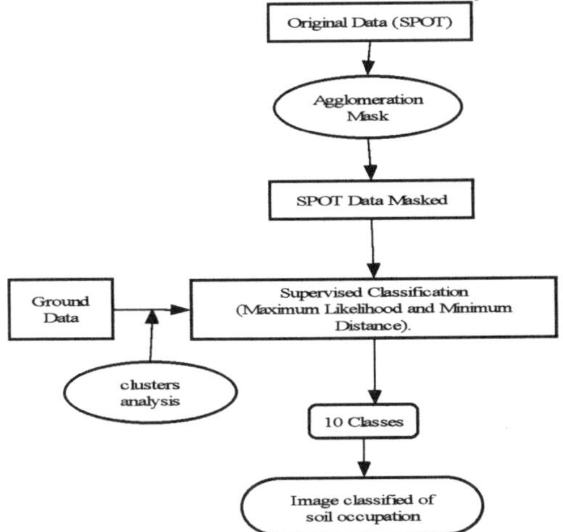

Fig. 2: Supervised classification (treatment by pixel).

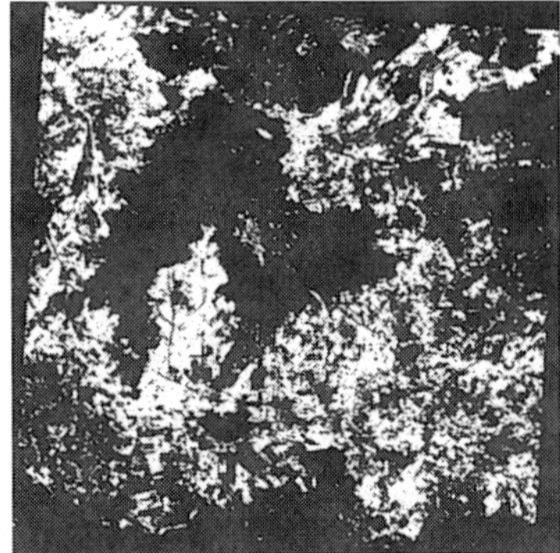
Image 1: NDVI range between 140-180.

We determined a threshold on this image thereafter in such a way to take into consideration the classes which interest us. This classification gave the following results:

Table 1: Result of the classification by maximum Likelihood of the XS channels (+ NDVI mask).

N°	Classes	Nbre. Pts	% image	Strate
0	Non classified	677258	64.59%	-
1	Cereals G.Y	52436	5.00%	14.12%
2	Cereals W.Y	303531	28.95%	81.74%
3	Courses	1	0.00%	0.00%
5	Bare soil	1639	0.16%	0.44%
6	Forest	776	0.07%	0.21%
7	Irriguated zones	12935	1.23%	3.48%
	Total	**1048576**	**100%**	**100%**

The masked region represents 677 258 pixels of the total image ie. : 64.59% (Table 1). On the 371318 classified pixels we notice that 95.86% is occupied by the two classes of cereals. These results reflect the cereal aspect of the zone.

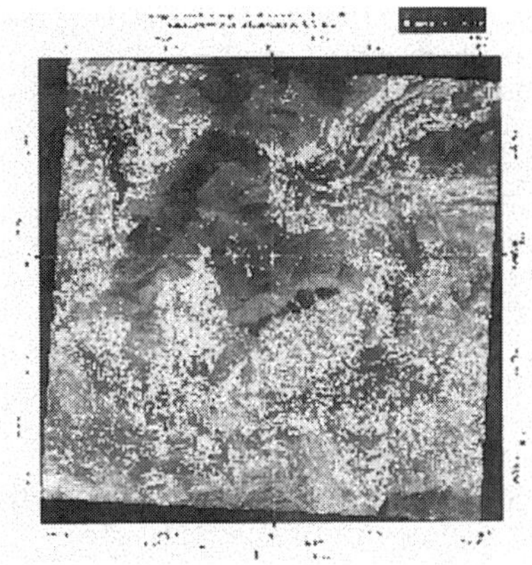
Image 2: Supervised classification (treatment by pixel).

The numbered results of this image, show us that the cereals in a general manner was well represented with a surface of 33.95% of the image, distributed in two classes; cereals good output (5,00%) and cereals weak output (28,95%). Note that this classification gave good results in spite of the error rate which is of 1.46% (Table 1).

4.2 Estimation of the classification:

We calculate with the aid of the matrix of confusion, the percentage of pixels well classified, as well as the statistics of classification. We notice that the percentage of pixels well classified for the class cereal good yield is of 73.74% (Table 3), a number of pixels of this class are matched with the classes cereal weak yield and reforestation (Table 2). The other classes appear well represented with some percentages of pixels well classified variable between 79.31% and 91.75%.

Table 2: Confusion matrix of classification by maximum Likelihood.

répartition classes	1	2	3	4	5	6	7	8	Répartition
Cereals G.Y	**73**	-	-	-	-	18	-	8	99
Cereals W.Y	15	**92**	-	-	-	-	-	9	116
Courses	-	-	**32**	12	8	-	-	-	52
Fallow land	-	6	5	**68**	-	-	-	-	79
Bare soil	-	-	8	-	**89**	-	-	-	97
Forest	12	-	-	-	-	**90**	8	-	110
Maquis	-	9	-	-	-	-	**83**	-	92
Irriguated zones	-	8	-	-	-	2	1	**91**	102
Résult. classif.	**100**	**115**	**45**	**80**	**97**	**110**	**92**	**108**	747

With the aid of this matrix we calculate some statistics of classification [FOURNIER et al.] (Table 3):

⇨ « statistics* »: inform us on the classes which were overestimated as well as the one that was under estimated by the classification.

⇨ «% of well classified** »: give us the percentage of pixels well classified by class.

⇨ «thematic significance*** »: allow us to have an idea on the part of pixels assigned to the class in question with regard to the result of the classification.

The global percentage of pixels well classified is in the order of 81.71% this represents a good rate. The assessment of this result allowed to note a certain number of errors and imprecisions (confusions, overestimates and under evaluations). Many of these problems are du to the fact that the given satellitales monodates data present a strong level of spectral confusion between the different types of covering. A good precision will be with a multitemporal approach. These imprecisions led us to test a method of classification by parcel (fig.3).

Table 3: Descended counts of the matrix of confusion.

Statistics Classes	Statistics	% well classified	Thematic significance
Cereals G.Y	101,01%	73,74%	73,00%
Cereals W.Y	99,14%	79,31%	80,00%
Courses	86,54%	61,54%	71,11%
Fallow land	101,27%	86,08%	85,00%
Bare soil	100,00%	91,75%	91,75%
Forest	100,00%	81,82%	81,82%
Maquis	100,00%	90,22%	90,22%
Irriguated zones	105,88%	89,22%	84,26%

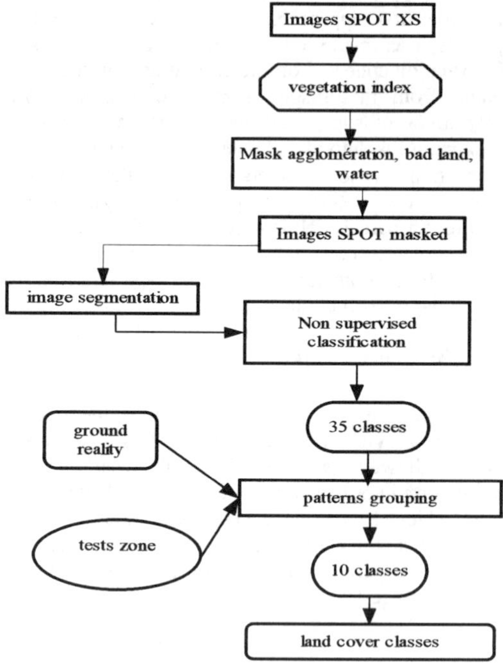

Fig.3: Automatic classification (parcels processing).

With regard to the previous approach, we masked the zones of uncovered soil and water surfaces so that to focus on the following processing on the only zones of interest (Fig.3). The following phase consists of segmenting image in order to get an image in which the homogeneous regions separated by continuous contours. An automatic classification by "dynamic clouds" is then achieved taking into account, the XS channels of SPOT and the segment image that we has just created.

This method proceeds by iterations in order to create 35 classes, this number was set has in prior to take sufficiently into account the radiométric variability of the zone.

* Statistics = Marge of classification / Marge of ground truth.
** Percent of well classified = Diagonal / Marge of ground truth.
*** Thematic significance = Diagonal / Marge of result classification.

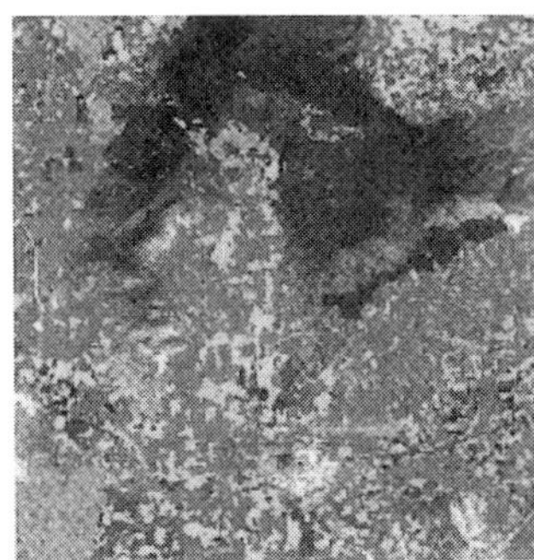

Image 3: Resulting image obtained by automatic classification (parcels).

5. POLL PLAN :

We then proceed to a thematic calibration of this classified image. To leave some zones of practice of which the work of soil is known and of the general knowledge of the region, we assigned the 35 initial classes a thematic content.

Once the classification achieved, we setup a polling plan. For this, we localize our segments on the satellital image as well as on the topographic map and proceed with ground data collection contained in these segments as well as of their surfaces. In our case, the area units is some segments (700* 700 m) in order to define them, a grid is superimposed on the image. Every element of the grid corresponds to a block of 100 segments, the sample is thus in pulling K numbers at random between 1 and 100 (Fig.4).

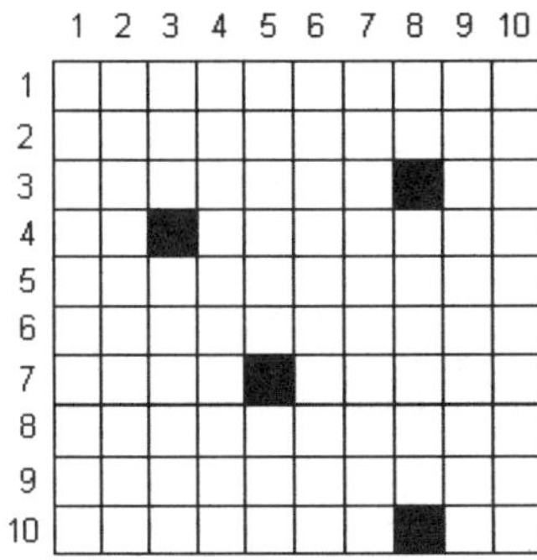

Fig. 4: Block representation (100 segments).

This method allows us to get a systematic sample aligned of K repetitions. Remote Sensing with some images of the year, represents an important asset for «frame» the results of ground obtained through the summary some segments. This means that the agricultural statistics are always finally achieved on the ground, and that remote sensing is only a support in order to give back these results of ground more «reliable» in the course of time.

The whole theoretical segments, which cover the totality of the territory serves a basis of poll for the pull of a sample, through which could take up statistics of surface.

In the case of a sample no stratified, or with a stratification found on some steady criterias in the time, the basis of poll does not degrade from a year to another.

6. GEOREFERENCING OF DATA TO OBTAIN AGRICULTURAL STATISTICS:

Because of the diversity of the basis information used, errors and imprecisions of all nature can take place at various levels of the georeferenced system.

⇨ The quality of content of the achieved stratification depends on, one hand, on the nomenclature and, on the other hand, on the identification of the land cover on some images straightened geometrically. The geographical precision of the stratification depends mainly on some constraints fixed by the sampling plan.

⇨ The radiometric quality of the images is essential in order to reach good performances in classification. The geographical precision of images results of course of the spatial resolution of sensor and of the precision of the geometric correction.

⇨ The structure of the investigated segments is bipolar, with a vectorial information of drawing some parcels included in the segment and a list of referenced attributes to the parcels.

6.1 Minimisation of the superimposing errors:

It concerns the adequacy between the position of the segments established from investigation documents and their coordinates in the straightened satellital data. The geographical superimposing of segments in the images allow to reduce errors, which will have a deterioration of the final quality of the regression.

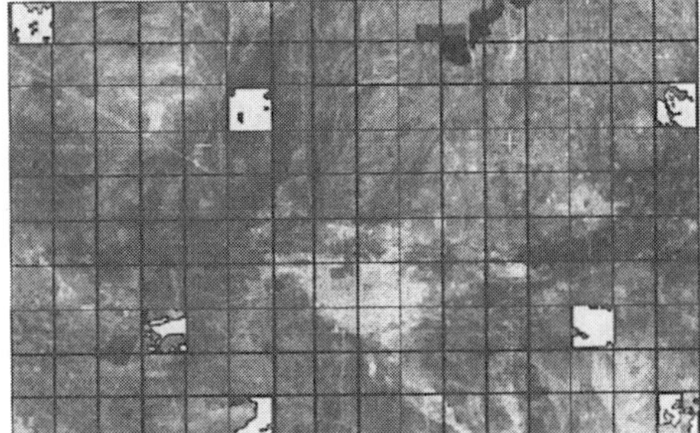

Image 4: Superimposing ground segments data on the image.

The approach adopted for the inventory of cultures by remote sensing passes by a certain number of stages of processing and control. Indeed, it includes four levels of information, which are the stratification, the satellital data, the segments as well as the ground statistics. The segments are superimposed on the image by using of the geometric correction in order to limit errors du to localization of drawing of reference marks. The crossing of the stratification with the satellital data allow us to get new layers, on which we choose some segments.

Some classifications are achieved, thereafter we achieve the matrix of confusion which is a stage of control. The results of the classification are thereafter compared to the ground statistics by a regression. Finally, an indicator is calculated in order to measure the efficiency of remote sensing.

6.2 Regression estimator:

In order to determine a given surface by remote sensing, we determine a law of a whole "classification by remote sensing" to "ground truth." This law could be determined by the least squares line. In remote sensing, we satisfy ourselves by estimating this law by a certain number of ground truth and their associate classification (following figure).

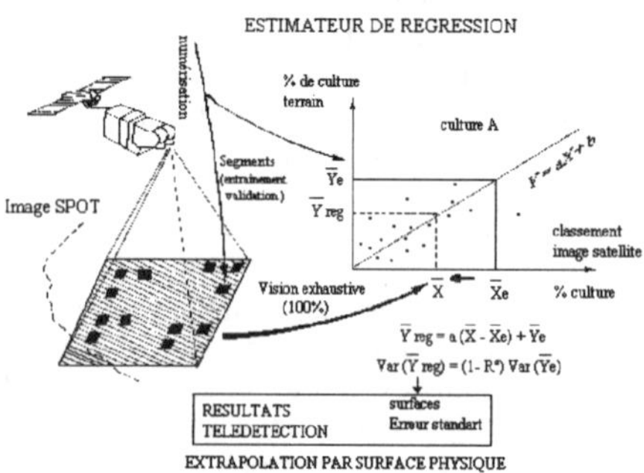

The classified data could reasonably be described by a linear line of type Y = aX + b + e, where the variable X is the surface obtained by classification. In these conditions, the line of regression of Y in X has for equation **Y = 0.9788 X + 1.5896**.

The sample segments are taken on some blocks of (10x10) segments of systematic uncertain way with a rate of poll of 4%. This rate is chosen as a function of degree of precision. The positions of pull of the segments were taken at random on the first block, then repeated on the following blocks.

The surface of the image is 44100 hectares, the one obtained by classification for the cereals class are of 14238.68 hectares. The evaluation which could be given from ground data is of the order of 15267.42 hectares. The difference between the result of the classification and the evaluation obtained to leave some ground data, could be explained by the confusion that we note between the cereals, which represent a rate of very weak cover, and the naked soils. Nevertheless, in a global manner, we can note that on the 36 drawn segments the difference between the classified data and the ground data is not meaningful (2,53%). The regression line calculated is presented like follows:

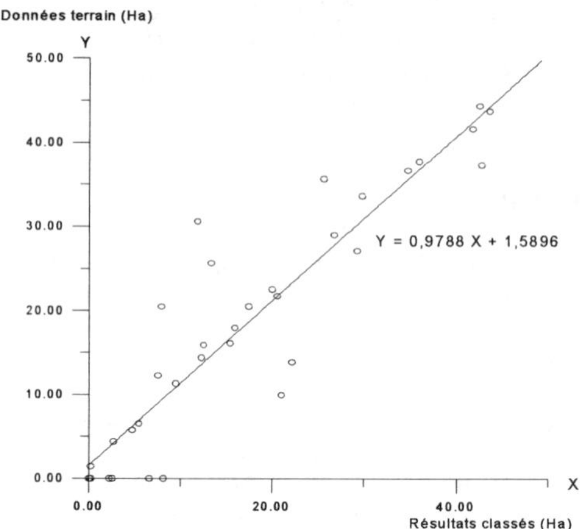

Fig. 5 : Line of regression obtained by crossing the ground data and remote sensing data.

Note that the interpretation of the regression line is linked to the importance of the existing correlation between ground data and the classified data. In our case is high since it is worth 0.92. The coefficient of determination r^2 is worth 0,848 then, this means that 84.8% of variations between the inventoried segments could be explained by the linear influence of classified data.

We can, because of to the variance obtained by the regression, calculate the efficiency of remote sensing [MEYER ROUX and al.- 87], which could be calculated by the following formula:

$$\frac{Var(\bar{y}_e)}{Var(\bar{y}_{reg})} = \frac{1}{1-r^2}$$

We notice that the efficiency of remote sensing increases r^2 with the growth of the coefficient of determination.

In our case, the coefficient of determination is of 0.848 this gives us an efficiency of 6,58%, allows us to affirm that the introduction of remote sensing improves the agricultural statistics. The efficiency of remote sensing begins to increase quickly from a coefficient of determination r^2 of 0,81.

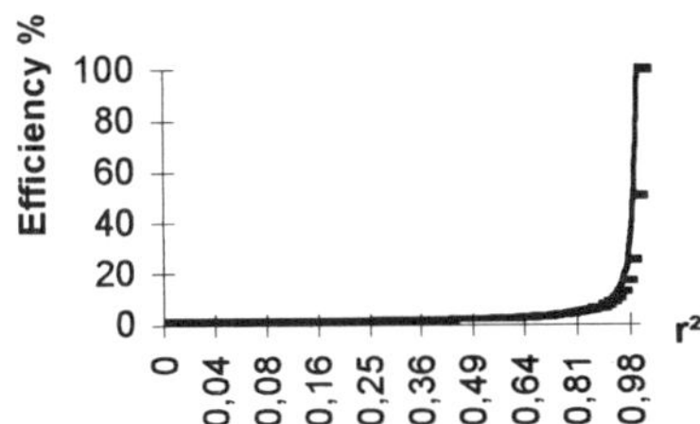

Fig. 6 : Line represent efficiency of remote sensing.

7. GENERALE CONCLUSION:

Agriculture must be based on some agricultural reliable statistics for an optimal set up. However, the present statistics do not answer the needs. It is therefore necessary to attempt to improve these data, unfortunately little useful in their present state. The contribution of remote sensing in the improvement of the agricultural statistics - in terms of surfaces- was studied on the region of Oum El Bouaghi. The accent was put on the collection of the surfaces cultivated by cereals. The results of this survey allowed us to note imprecisions. In order to improve the evaluations appreciably, the results of the image processing was confronted to the results of the ground reality with regression estimator. Indeed, this method allowed us to get a result for which the margin of errors on the whole of the segments samples are of 2,53%. The image processing combined to ground data permitted the development of a methodology of the following cultivated surfaces. Remote sensing is a tool appropriated for the cartography of agricultural domain to the regional scale. The evaluation of the agricultural surfaces is reliable, but then some outputs prove out to be delicate in the absence of measurement on the segments. Remote sensing permits the constitution of a basis of area poll also for some terrain investigations. The stratification of the region in homogeneous zones improve the precision of the evaluations. The perspectives of this study must first confirm the results obtained concerning the regression estimator and especially go toward the improvement of the operational processing. This will make in a worry of the generalization of this approach to some agricultural different contexts.

REFERENCES:

[BENHAMOUDA et al. 94]: F. Benhamouda, A. Hassani, T. Mostephaoui, A. Dif, L. Kebir, Z. Zebbar, A. Saad (1994):«évaluation des potentialités agricoles de la wilaya d'Oum El Bouaghi par télédétection» Rapp.N°1,CNTS/DSA d'Oum El Bouaghi de Mai 1994; 42 p.

[FOURNIER et al. 86]: Ph. Fournier et H. Geroyannis "Approche méthodologique pour la statistique agricole à partir de données satellitaires: application au département de l'Essonne".

[HIEDERER et al. - 93]: R. Hiederer; J.C. Favard; D. Guedes et M. Sharman, 1993. Estimating European crop surfaces from SPOT and Landsat TM data. Présenté au symposium International "Operationalization of Remote Sensing" du 19 au 23 avril 1993, ITC, HOLLAND.

[MEYER ROUX et al. - 87]: J. Meyer Roux; R. Sigman; M. Craig et M. Ozga, 1987. - Cartographie et statistique en télédétection. Extrait du cahier de statistique agricole n° 6, Nov. - Dec 1987, pp.(9-14).

[WILMET et al. - 89]: J. Wilmet et al., 1989. - Amélioration des statistiques agricoles par télédétection. Rapport réalisé au laboratoire de télédétection et d'analyse régionale, université catholique de Louvain. p. 4-18.

[D. YOUSFI - 97] : D. Yousfi, 1997. Approche méthodologique pour la statistique agricole à partir des données satellitales. Application à la région d'Oum El Bouaghi. Thèse de Magister, CNTS, Arzew, 106 pages.

Remote Sensing in the 21st Century: Economic and Environmental Applications, Casanova (ed.)
© 2000 Balkema, Rotterdam, ISBN 90 5809 096 5

The role of TM thermal band in describing the state/stress of vegetation

S.K.Alavi Panah
Iran Desert Research Centre, University of Tehran, Iran

R.Goossens
Geography Department, University of Gent, Belgium

ABSTRACT: The role of TM thermal band as a surrogate of surface temperature in describing the state/stress of the vegetation type has not been widely discussed. In this study, 7 TM bands of four playa margins in the Iranian Deserts, soil, soil salinity and other data and maps were used - the Playa margins located in Qom, Abarkooh, North of Siah Kooh and South of Siah Kooh in the arid regions. To study the role of TM 6 in describing the state /stress of the plants, the following attempts were made: 1) to carry out field work, 2) to remove the noisy pixels, 3) to analyse the two dimensional Feature Space of the training samples, 4) to compare the conventional and TM thermal dependent soil lines, 5) application of Principal Component Analysis, 6) to find the relationship between TM thermal and reflective bands on land cover types. These results shows that in this vegetated area, TM reflective bands are good indicators of surface temperature. The result obtained by the relationship between NDVI and TM thermal band can be confirmed by the FS between TM thermal band and NDVI. The FS between NDVI and TM thermal band shows that healthy vegetation has the highest NDVI and the lowest temperature and affected vegetation shows the reverse trend. It means that affected vegetation shows the lowest NDVI and highest temperature. We may also conclude that temperature as a fundamental thermodynamic quantity can be used to describe the state of vegetation and understanding some physical, chemical and biological systems in the study areas, therefore TM thermal band can be useful. In other words TM thermal band contains information complementary to the TM reflective bands and this combination of the TM thermal and the TM reflective bands provides a strong tool for land use and vegetation studies. Further research is needed to evaluate the possibility of modelling the relationship between TM thermal band and evaporation in those soil conditions of the study area.

1 INTRODUCTION

The understanding of the energy interaction with vegetation is crucial to the successful interpretation of remote sensing data as well as soil salinity investigation in areas under planted areas. The spectral reflectance characteristics of the leaves may be affected by various kinds of stresses as well as by the nutrient and salinity, due to change of leaf internal and external structure under these conditions. The leaf structure varies from one plant to another or even between plant species and consequently the spectral reflectance characteristics vary too. These characteristics are usually the clue to crop identification by means of remote sensing techniques. Spectral reflectance characteristics of non healthy vegetation is different from healthy vegetation. Knipling (1967) found that when leaves are senescent, their light reflectance usually increases markedly in the green visible light wavelength region peaking at 0.55 μm, because of the chlorophyll degradation. The decrease in near infrared light reflectance, however is not nearly as great as the increase in the reflectance of the visible (Elvidge, 1990). Thomas *et al.* (1966) found that increasing the chloride concentration of the nutrient solution has resulted in decreasing the reflectance in the visible region of the spectrum and increasing it in the near infrared wavelength from single cotton leaves. The salt precipitation on the leaves of some halophytic plants may increase plant reflectance. Many soils under the irrigation systems may acquire accumulation of the salts within the root zones, without evidence on the soil surface. Plants are sometimes good indicators of conditions that occur below the soil surface, therefore salinity and its severity may be apparent from some plant species. The collection of vegetation information using *in situ* techniques is often expensive, time consuming and difficult especially in the playa margin. The measurement of the vegetation amount, based on satellite imagery, can be an alternative. Nemanni

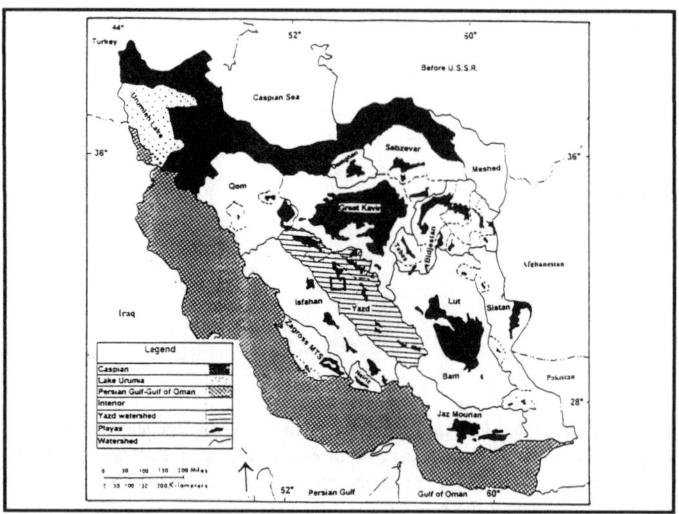

Figure 1 : Situation of the study area

and Running (1989) found surface temperatures ranging from 26-41°C from a 20*25 Km AVHRR scene of a Montana forested watershed on August 6, 1985. The variation in temperature was produced by thermal radiances from surfaces exposed by changes in Leaf Area Index (LAI) and canopy closure throughout the watershed. Salisbury and Aria (1992) showed that freshly fallen leaves, bare wood and dry grass will have the spectrum of cellulose. Canopy scattering must be taken into account in any discussion of the thermal behaviour of vegetation. The above result shows that there are different factors affecting the spectral emission of vegetation types.

2 STUDY AREA

Most investigations on the soils have taken place in arid and semiarid regions, because there are more suitable conditions for remote sensing applications with an absence or near absence of vegetation cover, usually clear skies, a shortage of rainfall and thus low soil moisture contents (Mulder 1987). The selection of the Landsat satellite images was based on the quality of the images, the preferred season and date of acquisition and availability of the images for the purposes of this study. The selected areas are located in the Central Iranian Deserts (Figure 1).

3 MATERIALS AND METHODS

Landsat TM images with 7 bands of three areas for extracting general information and TM images of the Ardakan area for extracting more detailed and specific information related to the dry season were used. Other materials used in this study are: soil, soil salinity and other data and maps, 3) computer softwares including ILWIS, IDRISI, Statgraphic, etc. To study the role of TM 6 in describing the state /stress of the plants, the following attempts were made: 1) to carry out field work, 2) to remove the noisy pixels, 3) to analyse the two dimensional Feature Space Analysis of the training samples, 4) to compare the conventional and TM thermal dependent soil lines, 5) application of Principal Component Analysis (PCA), 6) to study the relationship between TM thermal and reflective bands on different areas and also different land cover types.

3.1 . *Field work*

Field work as one of the most important steps was carried out. In order to choose representative training sites and to overcome some problems encountered in the variation within the field, the following attempts were made: 1) the Photo Morphic Unit (PMU) consideration, 2) use of several forms of field data, such as geologic and soil information, 3) confirmation of field data with the members of the team. In this study, the classes were primarily defined based on field observation, soil and soil salinity maps, geologic map, some soil physico-chemical properties, explanatory reports, interviewing with farmers and a general land use map of the study area.

3.2 *Two dimensional Feature Space Analysis of the training samples*

The FS can provide a fundamental understanding of the relationship between TM thermal band as a function of surface temperature and TM reflective bands. In this study the separability of the classes were evaluated by using two dimensional FS between the most informative TM reflective bands and the FS between TM thermal and TM reflective bands.

3.3. *Comparison between the conventional and TM thermal dependent soil lines*

The FS between red and near infrared produces a triangular shaped region. The base of the triangle corresponds to the soil line (Richardson and Weigand, 1977). The apex of the triangle is defined as the point at which additional layers of leaves produce no further increase in the near infrared reflectance, nor red absorbance. The position of the apex is essentially unaffected by soil background because the canopy is fully developed at this point (Jackson, 1983 b). The FS 3&6 gives a totally different shape in comparison to the FS 3&4. This is due to the effect of TM thermal band. In FS between the TM thermal and TM reflective bands, no soil line is formed, but the soil clusters have an ellipse shaped cluster or they are located along a tube shaped cluster (non-soil material classes). These two clusters are almost joined to each other.

In order to compare the conventional soil line and the TM thermal based soil line the following efforts were made: 1) using the NDVI threshold the vegetation cover was masked, 2) to obtain the TM thermal dependent soil line (soil ellipse), non soil material (mountain, vegetation and urban areas) was masked. Now the comparison between the conventional soil line and TM thermal dependent soil line shows: a) in the conventional soil line the TM thermal and TM infrared bands are highly dependent on each other, while in the TM thermal dependent soil line, the red and thermal bands are almost independent of each other, b) the TM thermal dependent soil line makes an ellipse shaped cluster which comprises the soil materials. It means that the urban and mountainous clusters are mostly isolated from the soil clusters. The position of training areas in a conventional soil line (bands 3&4) and TM thermal dependent soil line (bands 3 &6) shows that the vegetation classes are nearly isolated from soil classes in the conventional soil line. The salt affected vegetation class (no. 1) is very close to the base of the conventional soil line, but it is located nearly in the middle of the soil ellipse with a big distance to other vegetation classes. This is due to the temperature effect. The reason for a higher temperature in the salt affected vegetation class in comparison to other classes can be due to the effect of a) underlying soil, and b) moisture and salt stress in the salt affected plants. Some investigations show that salt affected vegetation has a higher temperature than the normal plants. Myers (1970) indicated that the range in leaf temperature associated with variations in soil salinity from 0.5 to 15 dS/m was 5.4 °C. He also stated that remote sensing of plant temperature is most successful when some physiological moisture stress (due to soil salinity) has developed especially under high evaporative demand conditions.

3.4. *Relationship between TM thermal and reflective bands on land cover types*

In this study, it was attempted to determine the relationship between the TM thermal and TM reflective bands on some broad land cover types in four study areas. This relationship can be useful to understand how some land cover types, such as vegetated areas and bare soils are dependent or independent from each other. The different land cover groups were extracted from a maximum likelihood classified image using masking techniques (Alavi Panah, 1996). The groups are: 1) total vegetation, 2) pistachio, 3) salt crust, and 4) soil materials.

3.4.1. *Results and discussion*

The results of the relationships between the TM thermal and TM reflective bands in different land cover groups of four study areas are shown in table 1. The results indicate that the relationship between the TM6 and other TM bands depends on geographical position, climatic and some other soil and plant conditions. The range of correlation coefficient varies from -0.925 for bare land in the Qom area to +0.81 for vegetated area in the south of Siah-kooh area. The results of the relationship between TM6 and TM reflective bands reveals that the trend of the relationship usually changes at TM4. It means that these relations for the visible spectrum and infrared spectrum is different (table 1).

The results of the relationships between the TM

Table 1 Correlation coefficients between the TM thermal and TM reflective bands on different land cover groups.

Study areas	TM band 1	TM band 2	TM band 3	TM band 4	TM band 5	TM band 7
Qom-bare land	-0.925	-o.919	-0.903	-0.888	-0.418	-0.267
Qom-total area	-0.803	-0.819	-0.800	-0.859	-0.375	-0.200
Qom-vegetated areas	0.321	0.235	0.275	-0.483	0.217	0.535
Abarkooh-bare land	-0.692	-0.716	-0.711	-0.708	-0.686	-0.691
Abarkooh-vegetation	-0.458	-0.512	-o.505	-0.680	-0.488	-0.411
Siah-kooh north-bare	0.145	0.068	0.171	-0.188	0.362	0.372
Siah-kooh north vegetation	0.157	0.175	0.192	0.207	0.368	0.264
Siah-kooh south bare	0.265	0.181	0.165	-0.132	-0.135	0.277
Siah-kooh south vegetation	0.81	0.79	0.81	0.54	0.81	0.80

Table 2. Correlation coefficients between the TM thermal and TM reflective bands on different land cover groups.

Land cover type	TM band 1	TM band 2	TM band 3	TM band 4	TM band 5	TM band 7
Total vegetation	0.81	0.79	0.81	0.54	0.81	0.8
Pistachio	0.75	0.76	0.78	0.72	0.79	0.76
Soil	-0.08	-o,05	-0.07	-0.12	-0.08	-0.07
Salt crust	-0.69	-0.58	-0.65	-0.75	-0.62	-0.67

thermal and TM reflective bands in the Ardakan area with different land cover groups can be categorised into 3 groups: 1) vegetation, b) salt crust, c) soil materials (table 2). The computed correlation coefficients between TM thermal and TM reflective bands 1, 2, 3, 4, 5 and 7 for vegetation class (total vegetation) are 0.81, 0.79, 0.81, 0.54, 0.81, and 0.80 respectively. It was also found that the correlation coefficients between TM thermal and TM reflective bands 1, 2, 3, 4, 5 and 7 for the pistachio class (trees) are 0.75, 0.76, 0.78, 0.72, 0.79 and 0.76 respectively.

These results show that in this vegetated area, TM reflective bands are good indicators of surface temperature. The variation in evaporation is inversely related to radiative surface temperature, thus the TM thermal band as a surrogate of surface temperature shows a negative relationship with NDVI as a surrogate of vegetation density. Gillies and Carlson (1995) found that NDVI is a good indicator of surface radiant temperature. A study of the relationship between NDVI as a surrogate of vegetation amount and the TM thermal band as a surrogate of surface temperature reveals the importance of orchard and agricultural vegetation in lowering their surface temperature by virtue of their association with biomass. Therefore from the negative relationship between the NDVI and TM thermal band we may conclude that the effect of vegetation on lowering the surface temperature can be attributed to the evaporation and/or evapotranspiration. The relationship between NDVI and TM thermal band in the bare soil may be completely different from the result in the vegetated

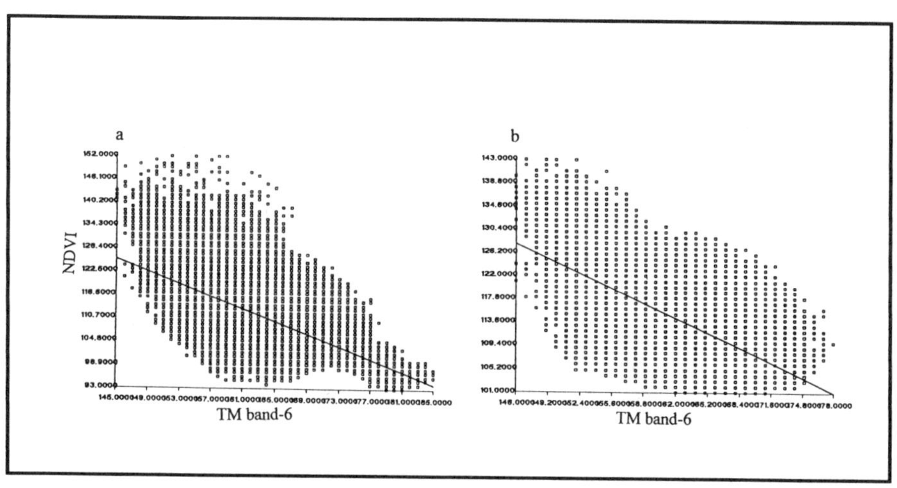

Figure 2. Relationship between TM band-6 and NDVI in the : a) total vegetation and b) pistachio.

area. Norman *et al.* (1995) stated that the amount of information from remote thermal infrared in vegetated areas depends on having a range of NDVI (vegetation amount) and surface temperature in the study area, and it apparently works quite well for the heterogeneous, high-spatial resolution scenes. Negative correlation coefficients between TM thermal band and NDVI for the total vegetation (- 0.69) and pistachio (-0.64) classes were obtained (Figure 2). But the result from the relationship between the TM thermal and TM reflective bands in total vegetation and pistachio classes shows that these two classes are very comparable in all TM bands (except TM band-4). Senesced vegetation is also likely to have a lower radiant surface temperature than bare soil (Jackson, 1982) due to higher aerodynamic surface roughness and higher albedo (Brest and Goward, 1987). Curran (1983) concluded that against a given background of soil or senesced vegetation, VIS reflectance of green vegetation is determined largely by chlorophyll absorption and hence leaf area index. Smith and Choudhary (1991) showed a close relationship between VIS and radiant surface temperature and then they concluded that in a mixed forest and agricultural land use, visible reflectance is a more general index of green vegetation cover than NDVI over South East Australia. A similar conclusion is also indicated for semi-arid and arid shrub lands (Musick, 1984). The inverse relationship between the NDVI and TM thermal band in the total vegetation and pistachio classes reveals the effects of plants on lowering the surface temperature. These results correspond with some investigation relating to the range of foliage temperature at different vigour and/or density. Asrar (1989) states that soil is considerably warmer than the foliage; the difference in temperature could be more than 25°K. The range of leaf temperature can be associated with variations in physiological moisture stress, plant density, soil background, irrigated trees or soil moisture etc. The foliage temperature is warmer when the soil is dry than when the soil is wet (Asrar, 1989). Myers (1970) indicated that the range in leaf temperature is associated with variations in soil salinity from 0.5 to 15 dS/m was 5.4 °C, on June 16 (Fitzgerald, 1974). He also showed the greatest thermal contrast between stressed and non-stressed cotton. The highest positive correlation was found on vegetation and the lowest negative correlation was found on salt crust with dominantly whitish colour in the Ardakan area.

4 CONCLUSION

From the above results we may conclude that the TM thermal and TM reflective bands behave completely differently in different playa margins and different land cover types. It means that the relationship between TM thermal and TM reflective bands greatly depends on the vegetation cover, salt minerals, some other soil surface characteristics and climatic and geographical situations. We may also conclude that temperature is a fundamental thermodynamic quantity that can be used to describe the state of matter. Temperature is an important factor in understanding some physical, chemical and biological systems in the Ardakan area. TM thermal band as surrogate of surface temperature can be useful.

The TM thermal band contains information complementary to the TM reflective bands and this combination of the TM thermal and the TM reflective bands provides a strong tool for land cover classification. The reason that TM thermal band is most often excluded from soil analysis is usually due to its low spatial resolution. But from the obtained result we conclude that in determining the nature of an extensive material, spatial resolution might not be so important, and more detailed spectral information relevant to the physical and chemical composition can be obtained.

REFERENCES

Alavi Panah, S.K., 1997. Study of soil salinity in the Ardakan area (Iran) based upon field observations, remote sensing and a GIS. P. 292, Gent: University of Gent (*Ph. D, thesis*).

Asrar, G., 1989. *Theory and applications of optical remote sensing*. Wiley, New York.

Brest. C.L. and Goward S.N., 1987. Deriving surface albedo measurements from narrow- band satellite data. *International Journal of Remote Sensing*. 8: 351-367.

Carlson, T.N., Perry, E.M. and Schmugge, T.J., 1990, Remote estimation of soil moisture availability and fractional vegetation cover for agricultural fields. *Agric. for. Meteorol*. 52 ; 45-69.

Choudhary, B.J., 1989. Estimating evaporation and carbon assimilation using infrared temperature data: vistas in modelling. *Theory and application of Optical Remote Sensing*, Edited by G. Asrar (New York, John Wiley).

Curran, P.J., 1983. Multispectral remote sensing for the estimation of green leaf area index, *Physiological Transaction of the Royal Society of London*. A. 309: 257-270.

Elvidge,, C.D. 1990. Visible and near infrared reflectance characteristics of dry plant materials. *International Journal of Remote Sensing* 11(10): 1775-1795.

Fitzgerald, E. 1974. Multispectral scanning systems and their potential application to earth- resources survey. *Spectral properties of materials*. Neuilly sur Seine (France) : ESRO. CR-232,

Gillies. R.R. and Carleson, T.N., 1995, Thermal remote sensing of surface soil water content with partial vegetation cover for incorporation into climate models. *Journal of Applied Meteorology*. 34 : 745-756

Mulder, M,, 1987. Remote sensing in soil science. *Development in soil science*, Elsevier, Amsterdam, The Netherlands, 379 p.

Musick H.B., 1984. Assessment of Landsat multispectral scanner spectral indices for monitoring arid rangeland. *IEEE. Transactions on Geoscience and Remote Sensing*. 22, 512-519.

Myers, V.I., 1983. Remote sensing application in agriculture. In: *Manual of Remote Sensing*, Vol. II, (J.E, Esters and G.A.) Thorley, eds.) P. 2111-2228. Falls Church (Virginia) : American Society of Photogrammetry.

Richardson, A. J. and Wiegand, C.L., 1977. Distinguishing vegetation from soil background information. *Photogrammetric Engineering and Remote Sensing*. 43 : 1541-1552.

Jackson, R.D., 1983 b. Adjusting the tasselled-cap brightness and greenness factors for atmospheric path radiance and absorption on a pixel by pixel basis. *International Journal Remote Sensing*, 4(2): 313-323.

Jackson, R.D., 1982. Canopy temperature and crop water stress. Advances in Irrigation, 1. 43-85.

Myers, V.I., 1970. Soil water and plant relations; chapter 6 of *Remote Sensing for Agricultural Purposes*, published National Academy of Science.

Nemani, R.R., and Running, S.W., 1989. Estimating regional surface resistance to evapotranspiration from NDVI ad thermal- IR AVHRR data. *Journal of Applied Meteorology*.

Salisbury, J.W. and D. Aria, D.M., 1992. Infrared (8-14 µm) remote sensing of soil particle size. *Remote Sensing Environment*, 42: 157-165.

Thomas, A.C. *et al*. 1996. Salting particles, playa crusts and dust aerosols at Owens (dry) lake, California. *Earth Surface Processes and Landforrns*, Vol. 21, 621-639.

Thomas. J.R., Meyers, V.I., Heilman, M.D. and Wiegand. C.L. 1966. Factors affecting light reflectance of cotton. In : *Proceedings of the 4th Symposium on Remote Sensing of Environment*, 305-312. Ann Arbor (Michigan) University.

Remote Sensing in the 21st Century: Economic and Environmental Applications, Casanova (ed.)
© 2000 Balkema, Rotterdam, ISBN 90 5809 096 5

Drought monitoring of Spanish forests using Landsat TM satellite images

A.Vázquez, J.M.Cuevas & F.González-Alonso
Laboratorio de Teledetección, CIFOR-INIA, Madrid, Spain
A.Calle & J.L.Casanova
LATUV, Universidad de Valladolid, Spain

ABSTRACT: In this work we combine spatial data from a Digital Forest Map of Spain, the Second National Forest Inventory and TM images to explore the differences in the spectral response of several vegetation types in three years with strong differences in water availability. The results point to the possibility of discrimination between dry and humid years based on the NDVI, although drought conditions affect in different ways to the several vegetation types.

1 INTRODUCTION

During the first half of the present decade, the east part of Spain was severely affected by drought conditions (González-Alonso et al., 1995). The measurement made within the framework of the European Network of Forest Damage showed important defoliation levels in the sampled units located in the southern portion of the Sistema Ibérico (Montoya & López, 1998). This was reflected also in the forest fire incidence registered during that years (Vázquez & Moreno, 1998). Additionally, during the last few years an increasing amount of spatial information relative to the actual forest distribution and characterization is becoming available. This information includes a digital version of a Forest Map of Spain (Ruíz de la Torre, 1990) and the Second National Forest Inventory (IFN-II) carried out from 1986 to 1996 by the National Institute for Nature Conservancy (ICONA).

The objective of this work was to combine spatial data from these sources with Landsat TM images to test for differences in the spectral responses of several vegetation types in tree years with strong differences in water availability.

2 MATERIAL AND METHODS

2.1 *Study area*

The 70 x 60 km study area is located in the southern portion of the Sistema Ibérico (east Spain) between the UTM-30N coordinates 570000 / 4510000 and 640000 / 4459000 and includes parts of Cuenca, Guadalajara and Teruel administrative provinces. Relief is rugged with elevations ranging from ca. 1000 to ca. 1800 m and with most of the territory above 1200 m.

Conifer forests dominate the present vegetation. Main tree species are *Pinus sylvestris* L., *P. nigra* Arnold, *P. pinaster* Aiton, and to a lesser extent *Juniperus sp.* and *Quercus sp.*

2.2 *Satellite data*

We have used three Landsat TM images from 1990 (July 27), 1994 (July 22) and 1996 (July 11). Preprocessing of images included a destriping process and an exoatmospheric reflectance correction process using the ENVI software routines. Geometric correction was based on ground control points (GCP) and first order polynomial equations. 1996 image was corrected against 1:50000 topographic maps (28 GCP and RMSE = 0.87). The other two images were rectified to the already corrected 1996 image. For the year 1990 we used 22 CGP (RMSE = 0.50) and for 1994 also 22 GCP (RMSE = 0.57). The resampling method was cubic convolution and pixel size was maintained at 30 m.

2.3 *Meteorological data*

The study is being carried out in one of the regions most severely affected by the drought that Spain suffered in the first half of 90's. Mean precipitation for the nine hydrologic years (September to August) from 1988-89 to 1996-97 in the three meteorological stations analyzed was 461 mm (Cuenca, 956 m), 385 mm (Tramacastilla, 1254 m) and 354 mm (Teruel,

900 m). Cuenca and Teruel are located close to the study area and Tramacastilla within the study area.

Based on this last station, the cumulative precipitation from the previous hydrologic year to the satellite date in the actual year was 748 mm in 1990, 502 mm in 1994 and 833 in 1996. This implied that 1996 was the most humid year and 1994 the driest with 1990 between both. This pattern was similar in Cuenca but not in Teruel, with precipitation lower before 1996 than before 1990. In the three stations 1994 was the driest year.

2.4 *Sampling procedure.*

The type of vegetation cover was obtained from a digital version (elaborated at 1:50.000 scale) of a Forest Map of Spain. In the present approach we used the main tree species and the proportion of crown cover for these species in each of the original polygons to segregate four vegetation types: (1) High Cover *Pinus sylvestris* (stands dominated by *P. sylvestris* with crown cover greater than 70 %) (Ps1); (2) Medium cover *P. sylvestris* (crown cover between 50 and 70 %) (Ps2); (3) High cover *P. nigra* (stands dominated by *P. nigra* with crown cover greater than 70 %) (Pn1) and (4) Medium cover *P. nigra* (crown cover between 50 and 70 %) (Pn2).

In order to make statistical comparisons among the four conifer forest classes, and to make sure that the sampled point was a forested area, we used the IFN-II 1 x 1 km regular sampling network despite the fact that in the present approach we used only the location of the sampling points. Out of the 1200 IFN-II sampling points located within the study area, there were 500 sampling points within the four classes established according to the Forest Map. These points were distributed as follows: Ps1 = 134, Ps2 = 160, Pn1 = 75 and Pn2 = 131.

2.5 *Analysis procedure*

For each of the 500 sampling points we extracted the mean reflectance value for a 3 x 3 window for the Band 3 (red), Band 4 (near infrared) and for the normalized difference vegetation index (NDVI) based on these bands. We performed comparisons among the four vegetation types for each year and among the three years for each of the vegetation classes. Comparisons were based on non-parametric Mann-Whitney tests.

3 RESULTS

3.1 *Comparisons between years*

In the comparisons between the three years, for each type of vegetation, red band (Fig. 1) provided statistical significant differences between 1996 and the other two years while in the infrared band (Fig. 2) was the year 1990 the different to the others at least in two vegetation types (Ps1 and Pn2).

The NDVI (Fig. 3) follows closely the band 3 patterns and 1996 was statistically different to 1990 and 1994 in the four vegetation types.

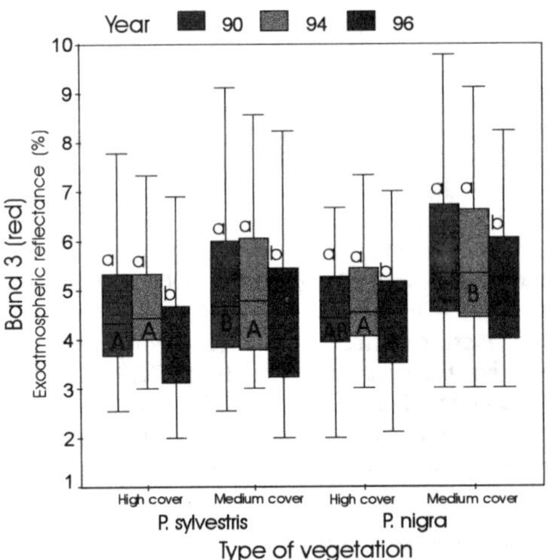

Figure 1. Box-plots showing the Band 3 reflectance value for the three years and the four vegetation types. Each box represent the median and the 25 and 75% percentiles. Vertical lines indicate the 5 and 95% percentiles. For each vegetation type years sharing the same lower case letter are not significantly different, and for each year vegetation types sharing the same upper case letter are not significantly different with (P > 0.05) according to a non-parametric Mann-Whitney test.

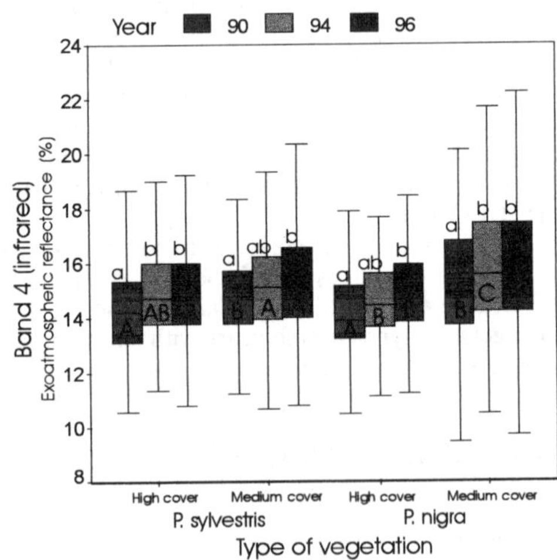

Figure 2. Box-plots showing the Band 4 (infrared) reflectance values for the three years and the four vegetation types. See Figure 1 for explanations.

3.2 *Comparisons between vegetation types*

In the comparisons between vegetation types, independently for each year, we have obtained higher NDVI values for Ps1 although statistical significant differences between this class and the rest of the vegetation types were obtained only for the year 1990 (Fig. 3). The Ps2 type was statistically similar to Ps1 for the years 1994 and 1996 and similar to Pn1 for the year 1990. The Pn1 type was similar to Ps2 in the years 1990 and 1994 and similar to Pn2 in 1996. The Pn2 (lower NDVI values) type was statistically different compared to the rest of the classes in 1990 and 1994 and similar to Pn1 in 1996.

The patterns between vegetation types for the red band (Fig. 1) were quite similar to the ones described for the NDVI: Ps1 and Pn2 were statistically different for the three years but Ps2 and Pn1 depending on the year, could be similar to each other or similar to Ps1.

In the infrared band (Fig. 2) Ps1 and Pn1 (lower reflectance values) behaved similarly in the three years while Pn2, with higher values, was statistically different to the other types in the years 1994 and 1996 and similar to Ps2 in the year 1990.

4 CONCLUSIONS

Although comparison between the three years should be taken with caution, given the lack of atmospheric corrections, the results point to the possibility of discrimination between dry and humid years based on the NDVI values. These differences are less apparent for the red and the near infrared bands. Also, differences in NDVI between dry and humid-years seem to be higher for *P. sylvestris* than for *P. nigra*.

Another preliminary conclusion, based in the different patterns of similitude between the four vegetation types obtained for each of the years, is that drought conditions affect the different vegetation classes in different ways.

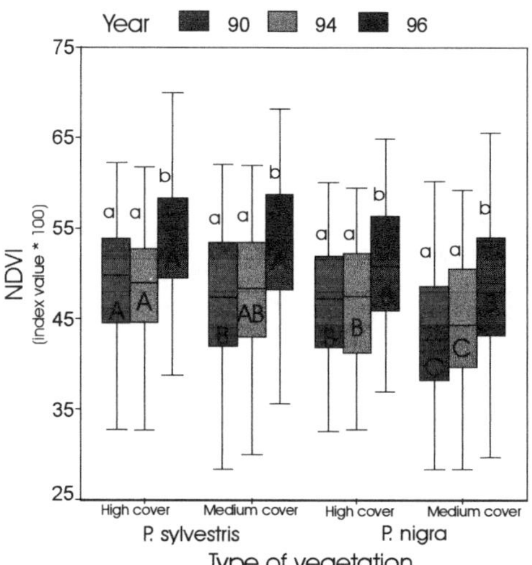

Figure 3. Box-plots showing the NDVI values for the three years and the four vegetation types. See Figure 1 for explanations.

ACKNOWLEDGEMENTS

We thanks the Spanish Ministerio de Agricultura, Pesca y Alimentación (MAPA) for providing finnantial support trought the sectorial research proyect SC96-096.

REFERENCES

González-Alonso F., Cuevas J.M., Casanova J.L., Calle A. & Illera P. 1995. Drought monitoring in Spain using satellite remote sensing. In J. Askne (ed.), *Sensors and Environmental Applications of Remote Sensing*, Proc. 14th EARSel Symposium, Göteborg 1994, A.A. Balkema, Rotterdam.

Ruiz de la Torre, J. 1990. *Mapa Forestal de España, Memoria General*. Instituto Nacional para la Conservación de la Naturaleza, Ministerio de Agricultura, Pesca y Alimentación, Madrid, 191 pp.

Montoya, R. & López, M. (coord.) 1998. *La Red Europea de Seguimiento de Daños a los Bosques (Nivel 1). 1987-1996*. Ministerio de Medio Ambiente, Madrid, 557 pp.

Vázquez, A. & Moreno, J.M. Patterns of lightning-, and people-caused fires in peninsular Spain. *International Journal of Wildland Fire*, 8(2) 103-115.

Remote Sensing in the 21st Century: Economic and Environmental Applications, Casanova (ed.)
© 2000 Balkema, Rotterdam, ISBN 90 5809 096 5

Monitoring of forest change by using multi-temporal satellite data

N. Musaoglu & C.Örmeci
Civil Engineering Faculty, Remote Sensing Division, Istanbul Technical University, Turkey

ABSTRACT: New methods for updating forest information are needed because of the high costs of periodically repeated total field inventories. Continuous updating of forest information is one way of reducing cost. But, quality of the continuously updated information has to be controlled and limited field control resources have to be targeted optimally. In this study forest change was analyzed by using satellites images for the Sarıyer region. In order to evaluate multitemporal data sets 590 rows and 700 columns study area located on the western part of Istanbul has been chosen. For the change analysis, vegetation index and normalized vegetation index algorithms were applied to different date's images. In the final phase, firstly change of the forest area was shown by combination of the different date vegetation index and normalized vegetation index images, secondly change of the coastline was shown by areal expansion.

1 INTRODUCTION

Rich natural resources occupy an important place in the economies of the nations. Determining the natural resources and compiling their inventories and planned use of such resources are among the important parameters in protecting the ecological balance of national economy. The conventional methods of forest data collection and change analysis can not deliver necessary information in a timely and cost effective fashion. To meet the demand for current and accurate data multi temporal remotely sensed images are used as one of the data sources of forest change assessment. Data from the Landsat satellites with its synoptic and regular (16 day and year to year) coverage offer potential for detection and inventory of disturbances and other changes that occur in land use, cover type and cover condition in areas of interest (Sunar at al.1996, Sunar and Musaoglu 1998).

In this study, by applying vegetation index and normalized vegetation index algorithms to the satellite images sensed at different dates, temporal changes in forest lands and coastline were analyzed.

2 STUDY AREA AND DATA CHARACTERISTIC

In order to evaluate multitemporal data sets Sariyer region located on the western part of Istanbul has been chosen (Fig. 1). Sariyer township lies between the coordinates of 28° 61' 25" - 28° 61' 25" east, 41° 16' 37"- 41° 04' 55" north and covers an area of approximately 151 square kilometers.

While the total area of Sariyer was covered with forests in ancient times today only the northwest side is remained as a forest. In the forest areas, the vegetation is formed by natural species of *Fagus oriantalis, Castania sativa, Quercus spp., Carpinus betulus, Tilia argantena* as well as *Pinus nigra, Pinus maritima, Pinus pinea* and other coniferous species that have been planted, which exist pure or mixed in the forest. Apart from this, shrubbery also flourished in the forest lands. Coniferous species grown by planting generally exist in the north east of the area.

In this study, 13 June 1975 dated Landsat MSS, 12 June 1984 and 18 July 1997 dated Landsat TM images were used (Fig. 2).

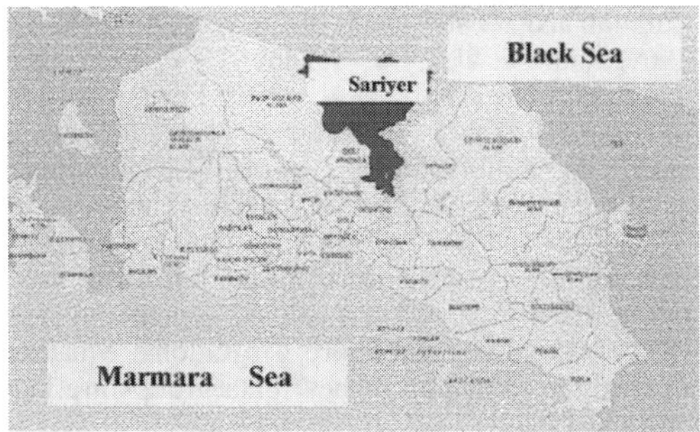

Figure 1. Location of study area. (colour plate, see p. 595).

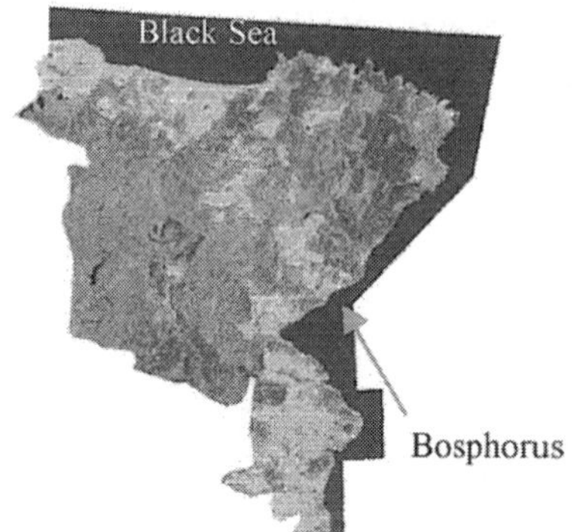

Figure 2. 8 July 1997 dated Landsat TM (Band 7/4/1). (colour plate, see p. 595).

3 METHOD

In the first stage of the study, satellite images were transformed into Universal Transverse Mercator (UTM) coordinate system by using 1/25000-scale standard topographic maps. In selecting the ground control points to be used for transformation, care was shown in homogeneous distribution of the points sharp and net differentiation on the map and satellite image (roads, shores, etc.) and transformation equations were selected as a first-degree polynom.

In the second stage, the change was studied by forming different ratio images by satellite images. In ratio images, being in the shade or in the sun does not affect the image due to the fact that similar objects have same ratio values irrespective of topography, and influence of shade is removed (Chavez and Mitchell 1977, Kowalik et al. 1983). In ratio images spectral reflection curve characteristic gaining more importance by suppressing the reflection characteristics of the objects cause objects with different similar spectral reflection curve slope to appear as similar objects. Different rationing techniques such as vegetation index and normalized vegetation index was developed according to the characteristics of spectral channels of satellite images. These ratio values are being used in studies aimed at establishing the health situation and zonal changes of the vegetation (Jordan 1969, Rause et al. 1973, Spanner 1990, Richardson and Everitt 1992, Lyon et al. 1998). For the purpose of temporal analysis of forest lands in the area of study and non-forest areas, various vegetation index algorithms were applied to satellite images dated 1975, 1984 and 1997. It has been found out that the best results were obtained from the images of vegetation index (VI) and normalized vegetation index (NDVI) (Figure 3 and figure 4). Rational equations used in the formation of new data groups are indicated in Table 1.

Table 1. Ratio equations

Sensor	Equation	
	VI	NDVI
Landsat MSS	B4 / B2	(B4-B2) / (B4+B2)
Landsat TM	B4 / B3	(B4-B3) / (B4+B3)

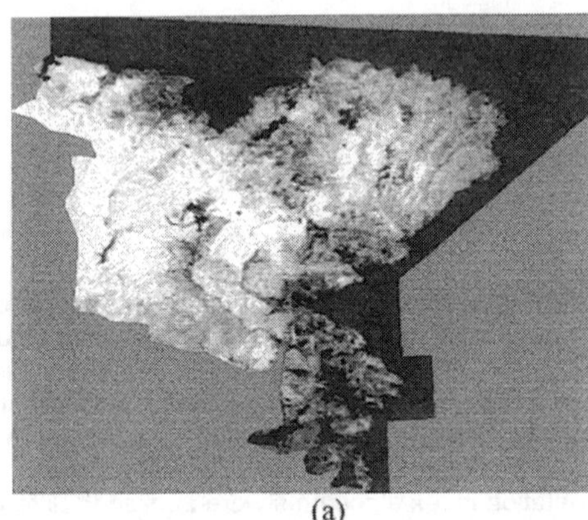

(a)

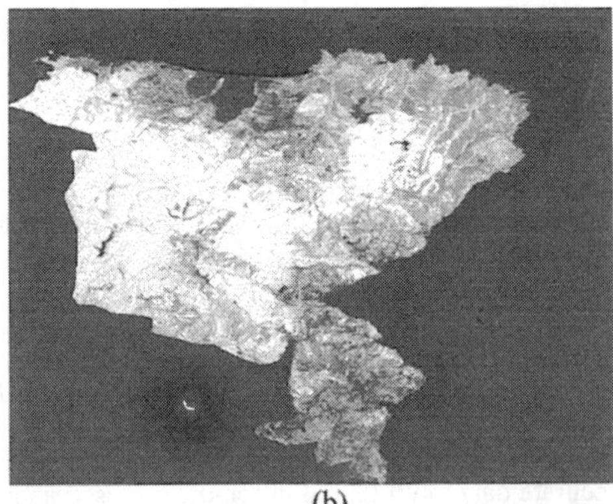

(b)

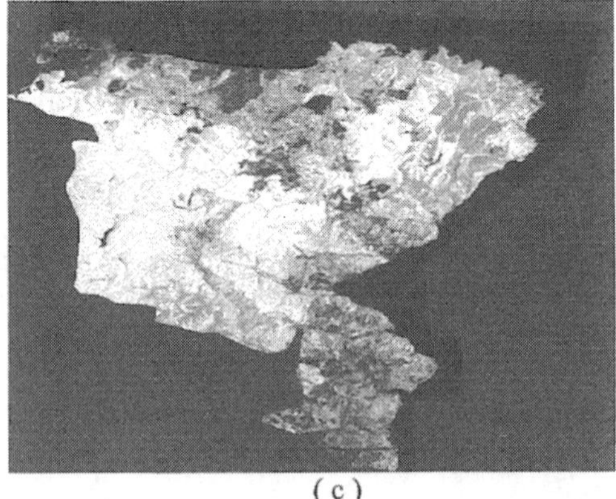

(c)

Figure 3. Vegetation index images a) 1975 b) 1984 c) 1997

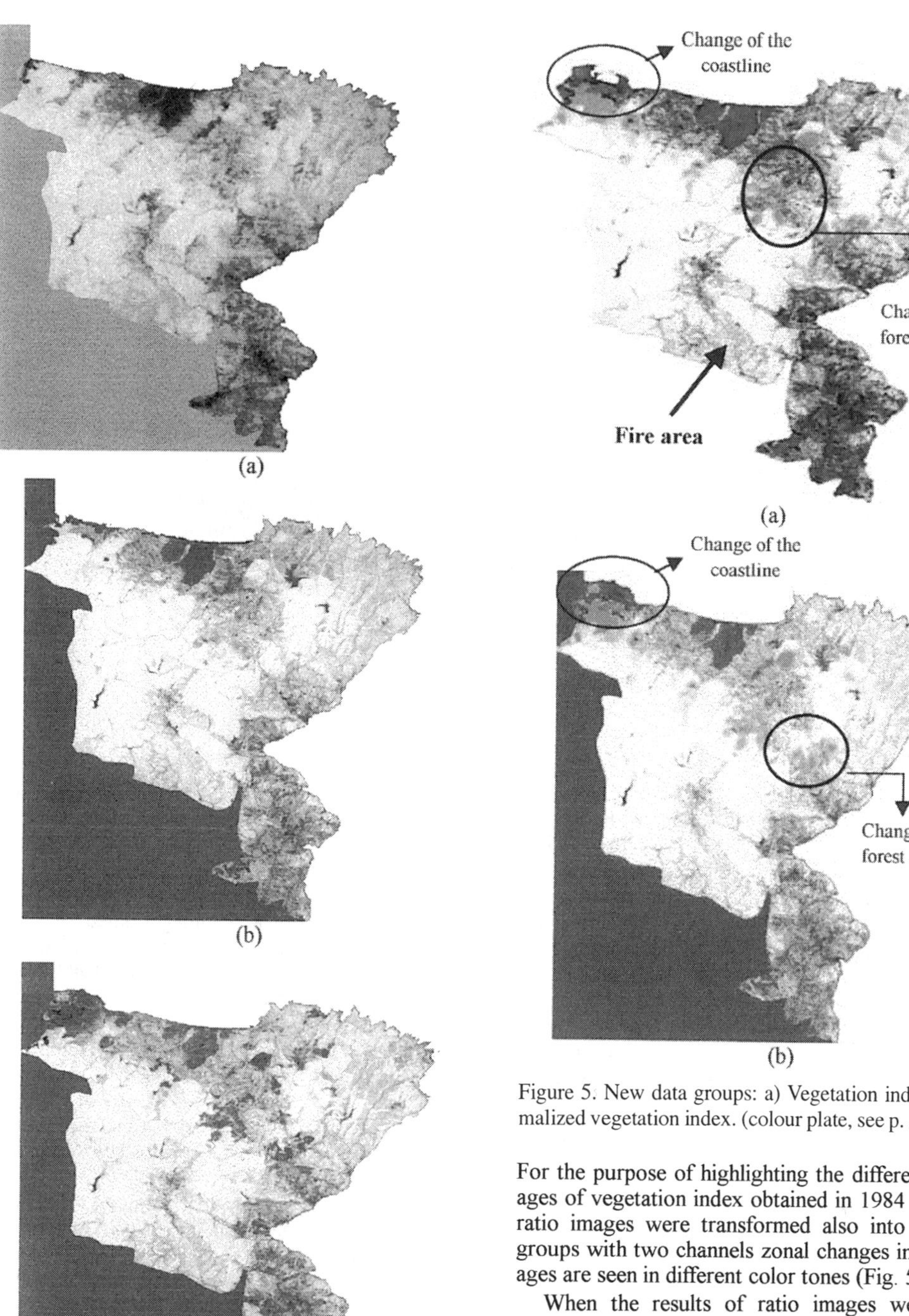

Figure 4. Normalized vegetation index images a) 1975 b) 1984 c) 1997

Figure 5. New data groups: a) Vegetation index, b) Normalized vegetation index. (colour plate, see p. 596).

For the purpose of highlighting the difference in images of vegetation index obtained in 1984 and 1997, ratio images were transformed also into new data groups with two channels zonal changes in these images are seen in different color tones (Fig. 5).

When the results of ratio images were jointly evaluated, it has been determined that temporal changes occurred both in the forest types of the area and the use of land , that the leafed trees being the natural flora of the area were replaced by coniferous trees trough forestation works in time, and that the Black Sea coastline in particular moved towards the sea (Fig. 6).

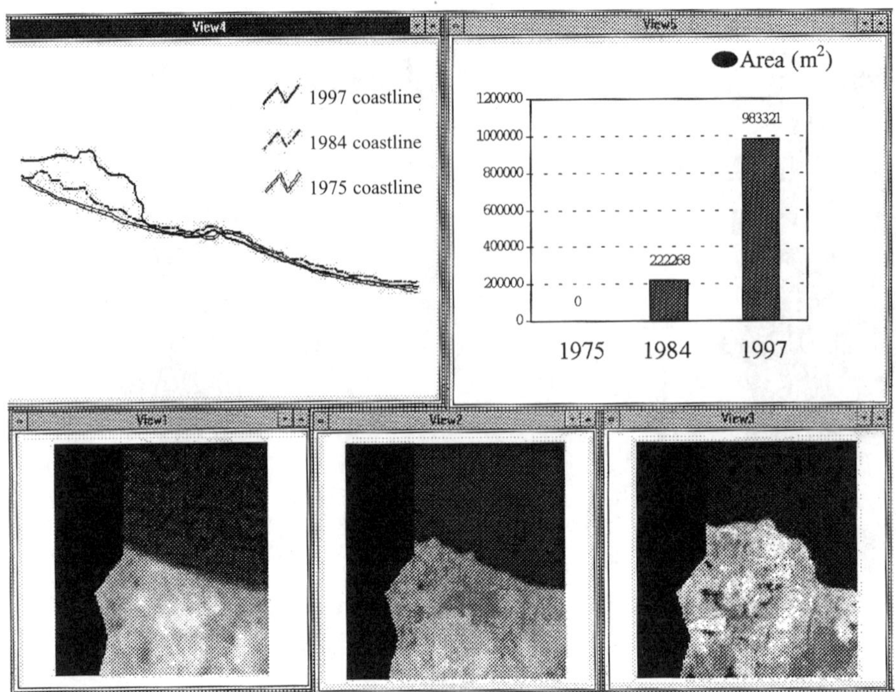

Figure 6: Change of coastline

Surface mining operations in the Black Sea coastline, north of Istanbul city, have changed the natural ecosystems of the area since 1950 (Ormeci and Sunar 1994). Changes in the coastal boundaries can be clearly identified on the original multitemporal satellite images. For the purpose of analyzing the temporal change of the coastline in the area of study, coastline was digitized through satellite image dated 1975, 1984 and 1997 and areal expansion was determined. When the results were evaluated, it has been determined that Black Sea coastline moved towards the sea 222268 m^2 in 1984 and 983321 m^2 in 1997 (Figure 6).

4 RESULTS

Resulting images obtained by means of rationing among the different channels of satellite images were controlled by checks made in the land and by comparing them with the data pertaining to similar area (various maps and photographs).

When the results of the vegetation index of the different dated satellite images are analyzed vegetation area is colored in light grey tones. Non-forest areas, on the other hand, are seen in dark colored tone since they have low reflection values. In the vegetation index results dated 1984 and 1997, the forest areas covered with pine types especially can be easily differentiated in the northeast of the area. In the vegetation index image dated 1975, it is seen that the coniferous trees is scarce and the area is covered with leafed trees which is the natural flora. Although it is the same in the images dated 1975 and 1997, the section lying in the southern part of the area of study, image of the area dated 1984 seems different because of a fire, which occurred in 1997. New forestation works carried out subsequent to the fire in 1997 could not provide full growth of the planted trees thereby forming a difference in the satellite image. As the trees were fully-grown in 1997, the vegetation index image is compatible with that of the one dated 1975. The same difference is seen also in the normalized vegetation index images.

When the normalized vegetation index results are analyzed, it can be seen that the difference between the forest lands and deforested lands in particular is more significant. As the mining sites were opened in the northern part of the area after the year 1984, it is observed that there is a remarkable change in the shoreline when the image dated 1997 is analyzed. In the forest lands located in the northeast part of the area, it is striking that there was destruction of forest lands and empty areas increased. Coniferous trees can be very significantly noticed in the northeast of the area.

5 CONCLUSIONS

In the ratio images obtained from the satellite images dated 1975, 1984 and 1997, it is seen that Sariyer area went through a rapid temporal change and a noticeable decrease occurred in forest lands. Type or extent of change formed by using different image processing techniques can be ascertained from satellite images.

Monitoring the temporal changes of the forest lands which are among the natural resources that are very difficult or sometimes even impossible to retrieve will enable us to find solutions to the potential environment problems on time. Furthermore; accurate, fast and low cost data/information can be obtained in the studies aimed at determining the potentials of forests, monitoring their temporal changes and updating relevant information by using remote sensing data with spectral range fit for the purpose and geometric resolution that are supported from the field check.

REFERENCES

Jordan, C.F., 1969. Derivation of Leaf Area Index from Quality of Light on the Forest Floor, *Ecology*, and Vol: 50, pp.663-666.

Lyon, G.J., Yuan, D., Lunetta, R.S., Elvidge, C.D., 1998. A change Detection Experiment Using Vegetation Index, *Photogrammetric Engineering & Remote Sensing*, Vol: 64, No: 2, pp. 143-150.

Örmeci, C., Sunar, F., 1994. Identification and Multispectral Analysis of Changes of Green Areas and Coastal Zones in The Kilyos- Karaburun Coastline of İstanbul, Turkey, *Second Thematic Conference on Remote Sensing for Marine and Coastal Environments, Proceedings*, pp. I575 -I580, USA.

Rause, J.W., Haas, R.H., Schell, J.A., Deering, D.W., 1973. Monitoring Vegetation Systems in the Great Plains with ERTS, *Third ERTS Symposium*, NASA SP-351, Vol.: 1, pp.309-317.

Richardson, A., Everitt, J., 1992. Using Spectral Vegetation Indices to Estimate Rangeland Productivity, *Geocarto International*, 1: 63-77.

Spanner, M.A., Pierce, L.L., 1990. The Seasonality of AVHRR Data of Temperate Coniferous Forests: Relationship with leaf Area Index, *Remote Sensing of Environment*, 33: 112.

Sunar, F., Ormeci, C., Kaya, S., Musaoglu, N., 1996. Assessment of Multi-Temporal Land Use/Cover Changes Using Remotely Sensed Imagery, A Case Study: Tuzla Region in Istanbul, Turkey, *XVIII ISPRS Congress, Proceedings*, and 9-19 July, Austria.

Sunar, F., Musaoglu N., 1998. Merging Multiresolution Spot P and Landsat TM Data: The Effects and Advantages, *International Journal of Remote Sensing*, vol.19, no 2, pp. 219-224.

Remote Sensing in the 21st Century: Economic and Environmental Applications, Casanova (ed.)
© 2000 Balkema, Rotterdam, ISBN 90 5809 096 5

Integration of NOAA-AVHRR and meteorological data in a GIS – Application for vegetation monitoring in Castilla y León, Spain

P. Illera, J.A. Delgado & A. Fernández Unzueta
Departamento de Física Aplicada I, Remote Sensing Laboratory, E.T.S. de Ingenierías Agrarias, Universidad de Valladolid, Palencia, Spain

A.A. Fernández Manso
Departamento de Ingeniería, E.T.S. de Ingeniería Agraria, Universidad de León, Ponferrada, Spain

ABSTRACT: In this paper we present the design and some results of the GIS developed in our remote sensing laboratory to be used for vegetation monitoring in Spain. The first source of information is Remote Sensing: Meteosat images which allow for the calculation of global solar radiation and NOAA-AVHRR data, that have been employed to calculate the NDVI and surface temperatures. The system also includes air temperature and rainfall recorded by the Spanish Meteorological Institute (INM). Finally, digital vegetation mapping derived from the CORINE land cover program has also been used. The system has been applied for vegetation monitoring following three lines. Firstly, cereal crop yields have been estimated. The second application is for the monitoring of forest areas. Our aim is to characterize the "normal" temporal evolution of the NDVI in order to detect possible changes or problems in near real time. Finally, forest fire danger due to vegetation water stress has been analyzed.

1 INTRODUCTION

The monitoring of the vegetation evolution at the regional scale is a basic operation from the point of view of the management and planning of natural resources. The data from the NOAA's satellites prove to be useful in conducting this type of activity. The satellites provide images that permit the calculation of vegetation indices and the estimation of soil temperature with an approximate spatial resolution of 1 x 1km.

This data can be complemented with other meteorological information with an important influence on vegetation growth such as precipitation, air temperature, solar radiation, relative humidity, wind and evapotranspiration. The Geographical Information Systems (GIS) turn out to be very appropriate to conduct a combined analysis of all the data base information on the territory.

Following this line of thought, we have considered the development of a GIS that integrates Meteosat satellites' data (solar radiation) and NOAA (vegetation indices and soil temperature) with meteorological variables measured on land (air temperature and precipitation) with the aim to conduct studies on the dynamics of the vegetation in the Autonomous Community of Castilla y León (Spain).

2 DATA SET

2.1 *Satellite Data*

An initial source of information were NOAA-AVHRR images captured daily by a HRPT receiver installed in our remote sensing laboratory. The images are calibrated and georeferenced to UTM co-ordinates using standard procedures (Kaufmann & Holben 1993) as well as our own methods (Illera et al. 1996a). In order to eliminate cloudy areas, a threshold technique was used. Ground temperature (Coll 1994) and the Normalized Difference Vegetation Index (NDVI) are obtained. The results presented in this work correspond to NOAA-11 data for 1993 and 1994 (up to September when the satellite operation ceased) and in the early hours of the afternoon, as a result of which the temperatures obtained are close to the maximum daily values. The spatial resolution is 1 x 1 km in the nadir. In order to reduce the amount of information processed as well as to eliminate atmospheric and illumination geometry and observation effects in the NDVI, maximum composite values were used corresponding to 10 day periods.

Meteosat images, also captured and processed in our remote sensing laboratory, have been used to calculate the hourly global solar radiation, with a spatial resolution of around 5 x 7 km, using a statistical model developed and validated by our work team (Illera et al. 1995). As final data we have available the radiation corresponding to 10 day periods georeferenced to UTM coordinates with a 1 x 1 km spatial resolution as with the NDVI-MVC and ground temperature.

2.2 *Meteorological Data*

The study focuses on the Region of Castilla y León for which we had available air temperature and rainfall data gathered by the network of The National Meteorological Institute. Castilla y León covers an area of 94 147 km^2 and is served by a total of 740 stations which gather maximum and minimum temperatures on a daily basis together with total rainfall. The location of the region and the network of stations is shown in Figure 1. In order to be able to superimpose the meteorological information with the satellite data, interpolation procedures have been developed. The temperature is interpolated in accordance with the inverse of the distance and taking into account the elevation obtained through a digital terrain model. A Thiessen polygon method has been used for rainfall. As final data, maps for maximum, minimum temperature as well as accumulated rainfall were obtained, all corresponding to 10 day periods.

2.3 *Digital Mapping*

Mapping developed through the European Economic Community CORINE land cover program (EEA Task Force 1992) has been used. Starting of Arc-Info format vectorial mapping, a new mapping in raster format was developed with a 1 x 1 km resolution. This mapping has 76 different classes, of which 30 correspond to agricultural areas and 35 to forested and wet areas.

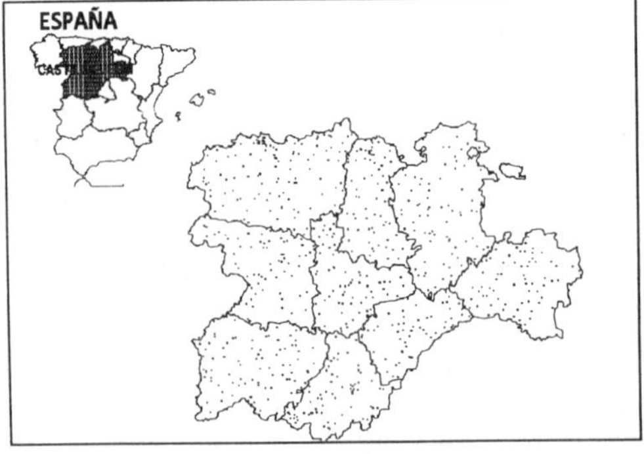

Figure 1. The region of Castilla y León and its meteorological network.

2.4 *Other Data*

In order to develop the various applications of the system, other sources of ground data were available.

Crop yield surveys provided by the Agricultural Department of The Castilla y León Regional Government have been available. These surveys have been carried out by order of the European Union since 1990. They basically consist of a calculation of cultivated area and yield in 700 x 700 m plots (approximately 550), called segments, spread evenly over the whole territory at 10 km intervals.

Fire data information has also been used. It has been provided by the Spanish Forestry Service. We have used only the date, area affected and localisation of each fire within the 10 x 10 km network used by the Forestry Service. Locating fires in this way can lead to inaccuracies when superimposing affected areas on the images.

3 GIS STRUCTURE

All of the information described has been integrated into the Geographical Information System IDRISI. The GIS structure is shown in Figure 2.

4 CROP YIELD CALCULATION

One of the applications developed has been the creation of an empirical agrometeorological model for the calculation of cereal crop yield (Terrab et al. 1997). It should be pointed out that cereal crops (wheat and barley) are dominant in Castilla y León (60% of agricultural land).

The study of crop yield and the evolution of the NDVI has allowed us to find a maximum correlation (r=0.7) between yield and the accumulated sum of the NDVI in the final stage of the phenological cycle (maturation). Following a similar strategy, an analysis has been carried out of the agrometeorological variables available: ground temperature, air temperature, rainfall, solar radiation, Ts-Ta and Ts^2 and an attempt has been made to find the period of accumulation of each variable which most closely correlates to yield data.

Once the optimum accumulated cycles for each variable were obtained (see Table 1) we developed the yield calculation equation by means of a multiple regression linear model, using the stepwise regression technique. During the whole fitting procedure of the model, segments with at least 70% crop use have been included. The final result obtained is shown in the equation (see Table 2):

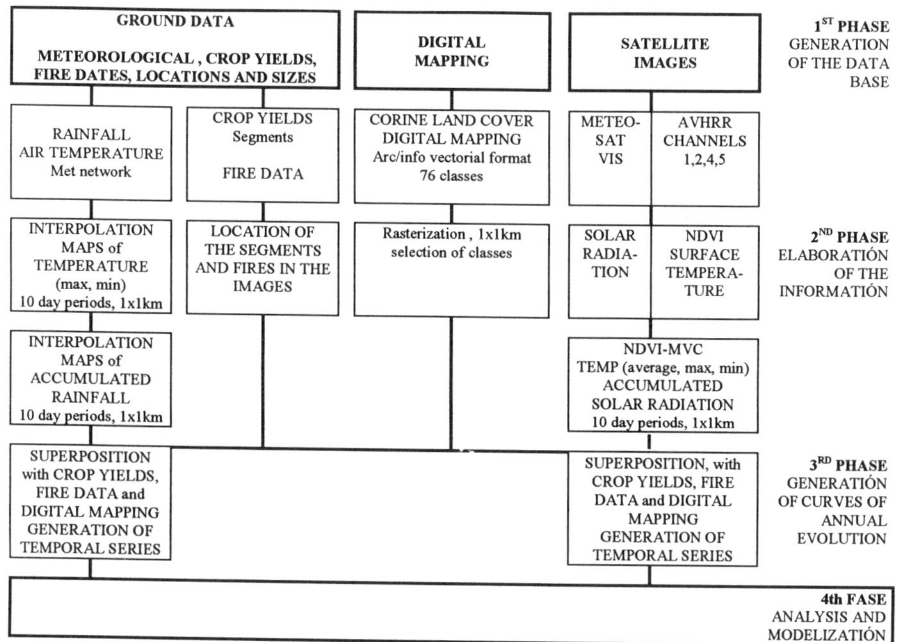

Figure 2. Structure of the GIS.

Table 1. Crop yield estimation. selection of variables.

Correlation	Σ NDVI$_C$ index/day	ΣPr (mm)/day	Σ Ts-Ta (°C)/day	ΣTs^2 (°C)^2/day	ΣR_G (MJ/m^2)/day
Period of summation	01 June ⇓ 31 July	1 March ⇓ 30 June	20 April ⇓ 10 July	20 April ⇓ 10 July	01 May ⇓ 10 July
Yield(kg/ha)	0.75	0.68	-0.66	-0.80	-0.65
Σ NDVI$_C$	1.00	0.04	-0.28	-0.42	-0.38
Σ Pr	*	1.00	-0.68	-0.74	-0.49
Σ Ts-Ta	*	*	1.00	0.78	0.39
ΣTs^2	*	*	*	1.00	0.76

Table 2. Crop yield estimation equation.

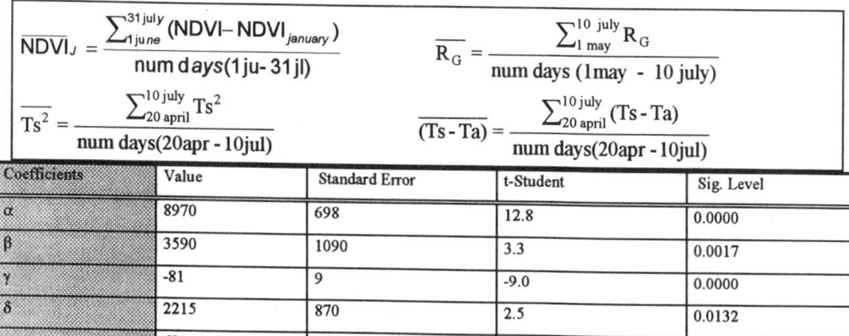

$$\overline{NDVI_J} = \frac{\sum_{1june}^{31july}(NDVI - NDVI_{january})}{\text{num days}(1ju - 31jl)}$$

$$\overline{R_G} = \frac{\sum_{1\,may}^{10\,july} R_G}{\text{num days }(1may - 10\,july)}$$

$$\overline{Ts^2} = \frac{\sum_{20\,april}^{10\,july} Ts^2}{\text{num days}(20apr - 10jul)}$$

$$\overline{(Ts - Ta)} = \frac{\sum_{20\,april}^{10\,july}(Ts - Ta)}{\text{num days}(20apr - 10jul)}$$

Coefficients	Value	Standard Error	t-Student	Sig. Level
α	8970	698	12.8	0.0000
β	3590	1090	3.3	0.0017
γ	-81	9	-9.0	0.0000
δ	2215	870	2.5	0.0132
ε	63	59	1.1	0.0490

$$Yield(kg/ha) = \alpha + \beta \overline{NDVI}_J + \gamma \overline{T_S}^2 + \delta \overline{(T_S - T_a)} + \varepsilon \overline{R_G}$$

The signs of the coefficients in Table 2 show that the meteorological variables introduce in the model information about water stress due to the fact that water is the main limiting factor in the study area.

As a control technique of the model developed we have compared the results obtained for the calculation of crop yield applying the model to all plots with over 10% cereal use and the results obtained from the ground surveys. In Table 3, the good results for the whole of the territory and the acceptable results in the provincial divisions can be seen. As the final equation of the model is available, by means of the application of the GIS we can develop maps for crop yield calculation over the area designated as dry farming land by the CORINE digital mapping.

5 MONITORING FOREST AREAS

As a second application, we have considered the monitoring of forest areas within the region of Castilla y León. Our aim is to characterise the "normal " evolution of the NDVI for the different forest classes present in our digital mapping. However, an analysis of the NDVI values is not enough to carry out a monitoring of the state of the forests, as this index is closely linked to annual climate variation (Cihlar 1993, Schultz and Harper 1993). In order to determine normal evolution, it is useful to have models available which establish the relation with meteorological variables.

In order to do this, we began by analyzing the series of values of air temperature, rainfall and solar radiation. Results are presented corresponding to the classes of pines, deciduous and evergreen forests. A description of the groups and the area they cover in the region is shown in Table 4.

Figure 3 shows the relation between the NDVI and the maximum air temperature corresponding to the group of evergreen forest in 1993. Next to each point, the day of the year to which the data corresponds is indicated. A good lineal relation can be observed from the beginning of the year until day 164. From then on the linear relation is clearly unsatisfactory. Very similar results are obtained for solar radiation. In the case of rainfall (Fig. 4) linear relations with accumulated rainfall from January up to the decade before that studied can be found, although again this ceases to be the case after day 164.

The pine and deciduous groups show a similar behavior although the linear model is valid until day 224 and the deviation of linearity after that point is not so evident. Table 5 shows the results of the linear model for the three types and variables. The

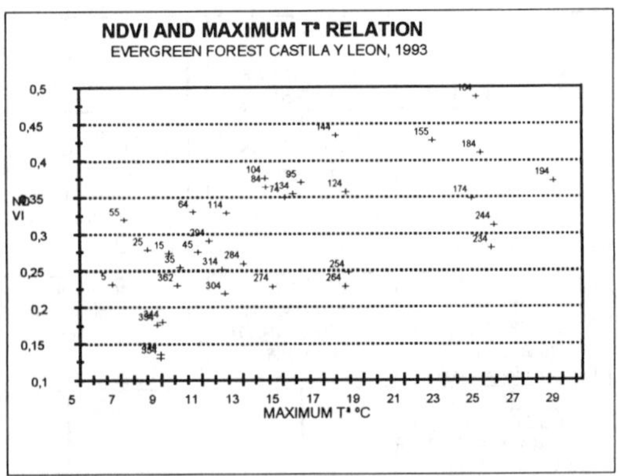

Figure 3. NDVI-MVC and maximum air temperature relation. Evergreen forests, 1993.

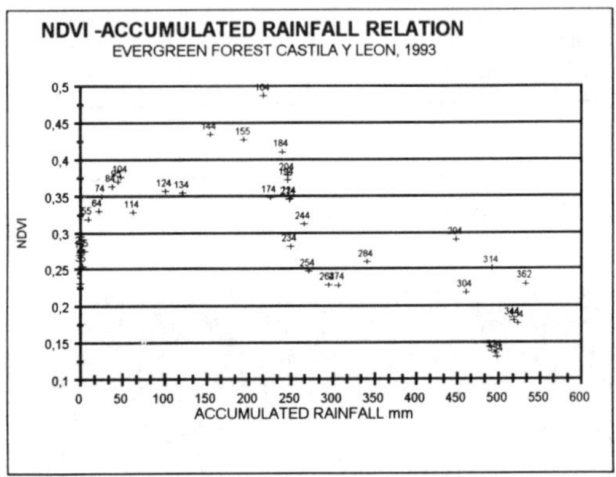

Figure 4. NDVI-MVC and accumulated rainfall relation. Evergreen forests, 1993.

accumulated rainfall is not a good selection to model the whole year. This variable is constant, and thus not adequate, in periods when there is no rainfall. To deal with data of an entire year as a whole, it is necessary to look for a variable capable of reflecting the hydric stress of summer. In 1994 rainfall was very low but the results of the models are similar although the period in which we find good linear relations is more reduced.

The results presented seem to indicate that the NDVI-meteorological variables relation is easier to find in periods of vegetation growth when extreme temperatures do not appear and there is no significant hydric deficit. The best results for deciduous forests, that occupy mountainous or riverside zones, are obtained from this. Regarding pine forests, to say that the class of CORINE cartography is somewhat heterogeneous that is the reason for deciding to conduct a segregation of the different formations: the mountainous or average mountain pine forests, located in the borders of

Table 3. Crop yield estimation. Provincial and regional results.

Province	Year	Official Data (kg/ha)	Estimation model (kg/ha)	Diff. (%)
Avila	94	1800	1900	+5
	93	3600	3300	-8
Burgos	94	3400	3500	+3
	93	4400	4300	-2
León	94	2700	2800	+4
	93	3900	3800	-3
Palencia	94	2900	3200	+9
	93	3400	3600	+6
Salamanca	94	2600	2300	+12
	93	3400	3200	-6
Segovia	94	2100	2300	+10
	93	3400	3200	-6
Soria	94	2000	2200	+10
	93	3900	4000	+3
Valladolid	94	2600	2700	+4
	93	3500	3600	+3
Zamora	94	2400	2200	-9
	93	2800	2500	-11
Castilla -León	94	2600	2700	+4
	93	3600	3500	+2

Table 4. Description of the three forest classes analysed.

CLASSES	Main Species	Total area (ha)
Evergreen forest	*Quercus ilex*	323.470
Deciduous forest	*Fagus sylvatica, Quercus Pyrenaica, Castanea sativa, Populus spp*	381.808
Pines	*Pinus sylvestris, Pinus pinaster Pinus nigra, Pinus halepensis*	753.074

Table 5. NDVI and meteorological variables.

		T max complete year NDVI=A+BT	T max Linear period NDVI=A+BT	Accumulated Rainfall linear period NDVI=A+B.RAIN
Pines	A	0.1475	0.2307	0.2918
	B	0.0088	0.0078	0.0005
	r^2	64.44	74.08	75.47
	Period		224	224
Evergreen Forest	A	0.1725	0.1747	0.2895
	B	0.0063	0.012	0.0008
	r^2	37.98	82.95	76.93
	Period		164	164
Deciduous Forest	A	0.0729	0.1666	0.2664
	B	0.0142	0.014	0.0007
	r^2	76.89	85.12	89.5
	Period		224	224
CT 1	a	0.143	0.145	0.288
65% P. Pinaster	b	0.007	0.010	0.0006
35% P. Pinea	r^2	47.11	71.68	76.40
	Period		164	164
CT 3	a	0.117	0.072	0.245
75% P. Pinaster	b	0.007	0.011	0.0009
25% P. Pinea	r^2	48.44	84.82	69.23
	Period		164	164
CT 5	a	0.147	0.205	0.317
80% P. Sylvestris	b	0.0011	0.009	0.0005
20% P. Pinaster	r^2	71.29	75.80	73.27
	Period		204	204
CT 6	a	0.104	0.154	0.278
60% P. Pinaster	b	0.011	0.010	0.0005
40% other pines	r^2	74.29	84.74	77.38
	Period		204	204
CT 8	a	0.182	0.238	0.328
100% P.Sylvestris	b	0.011	0.009	0.0005
	r^2	75.72	79.37	77.26
	Period		224	224

Castilla y León - P.sylvestris or P.nigra -, from those of the central part, a lot more arid, - P.pinea, P.halepensis or P.pinaster.

This segregation has been carried out by means of The Territorial Biogeoclimatic Classification of Spain CLATERES (Elena-Roselló 1996). The classification, in Arc-Info format vectorial mapping, has been converted to a raster format with a 1 x 1 km resolution. From the 39 classes of our Region a reclassification of 8 main classes has been done with a gradation from zones of smaller altitude (high temperatures, low rainfall) up to the mountainous borders of Castilla y León (colder and more humid).

In Table 5 the results obtained in the case of pines for some of the classes can be seen. As it coul be expected, the best linear models are obtained in mountainous zones classes (P.Sylvestris), being worse the zones dominated by Mediterranean pine forests. For classes 5, 6, 7 and 8 the NDVI-Temperature models for the whole year are not very different from the ones corresponding to the period that we have called "linear".

Finally, as regards current and future work, our aim is to find new climate variables which might prove more significant from an ecophysiological point of view, especially in those aspects related to annual hydric balance with the aim of developing a predictive model for all the climate variations within the study area.

6 FOREST FIRE DANGER ESTIMATION

The last application is the estimation of forest fire danger related to vegetation water stress. The first two procedures are based on the analysis of the NDVI-MVC temporal evolution, its decrease is supposed to be related to an increase in fire danger. As danger indicators the accumulated slope of the evolution (Illera et al. 1996b) and the greenness index (Burgan 1995) are used:

$$AS_n = \sum_{i=1}^{n} \frac{NDVI(t_i) - NDVI(t_{i-1})}{t_i - t_{i-1}}$$

$$G_n = \frac{NDVI(t_n) - NDVI_{MIN}}{NDVI_{MAX} - NDVI_{MIN}}$$

The first one is calculated by using data from spring to the period when the fire occurred and the last one by using data of 1993 and 1994 to determine the maximum and minimum values of the NDVI-MVC for each pixel in the image. The analysis of the indicators shows a behaviour for the different forestry classes which can be seen in Figure 5. They present the highest values during the summer. The differences among different forestry classes seems to be related to the climatology of the area. The superimposition to fire data gives good results as can be seen for G in Table 6. Considering the AS, a danger threshold of AS<=-0.5 leads to 66% of the total fires sample and 53% of the surface burnt detected by the threshold. The results are somewhat worse than those obtained for the greenness, but the AS indicator is calculated using solely data from March to September of only one year, which could be an advantage if NOAA data of a longer series are not available.

The typified values of both indicators with respect to the average value and standard deviation in the area:

$$\frac{Indicator - Indicator\ average}{St\ deviation}$$

could also be useful to estimate the zones in more danger at a given moment (68% of fires and 70% of the surface affected detected to be in danger for Greenness). Finally, both indicators present a close linear relation as can be seen in Figure 6.

Table 6. Greenness index and fire danger.

GREENNESS INDEX (G)				
Year	Total forest fires	Analysed forest fires %	% fires G<=40%	% burnt area G<=40%
1993	1243	86	70	62
1994	2100	83	70	68

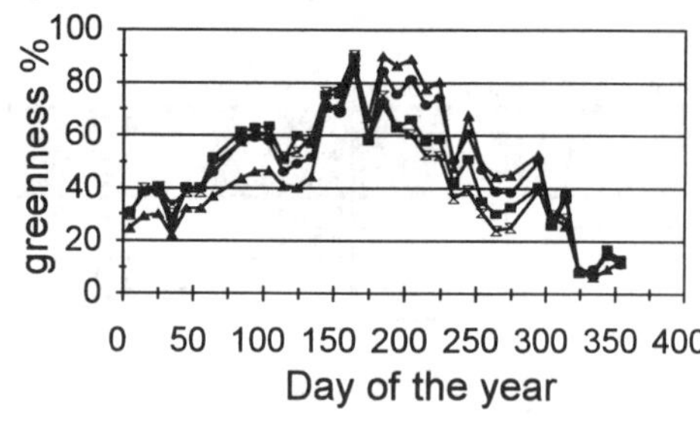

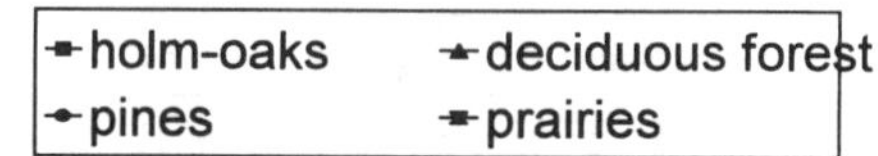

Figure 5. Temporal evolution of the Greenness index in 1993.

Figure 6. Relation between the accumulated slope AS and the greenness G. Fires in 1993.

As regards the third indicator, it is calculated from the surface temperature and NDVI. Both magnitudes present a negative lineal relation, whose slope is believed to be linked to the ratio between sensible and latent heat fluxes, as a result of which it can be used as a measure of hydric stress in forests (Nemani & Running 1989). An increase in the slope, in absolute value, would mean an increase in water stress.

The indicator has been worked out daily in areas of 20 x 20 km only for the year 1994. Its analysis for the different forestry classes (Fig. 7) shows that the values are negative, as it was expected, and drop in the summer, which can be associated with the increase of hydric stress. The majority of the values are between -10 and –30.The evolutions are more or less the same for the distinct covers, although the indicator values are greater for the covers with a greater availability of water. At the time of establishing thresholds we do not get clear results in Castilla y León. If a similar analysis is carried out in regions of the Spanish Mediterranean coast, where large scale forest fires take place, the indicator seems to be useful in these events which could be an estimation of the propagation danger. Finally it has no clear relation with the two indicators described above.

The last indicator is the difference between surface and air temperature, also thought to be related to evapotranspiration or hydric stress. High values of the indicator would mean high stress or fire danger (Vidal et al. 1994). Surface temperature derived from NOAA data of the afternoon orbit is representative of the daily maximum value. Maximum air temperature has also been employed.

In this study average values of T_s and T_a corresponding to ten day periods have been used. The analysis of T_s-T_a for the different forestry classes of the CORINE digital mapping shows maximum stress values during the summer and normal values less than 5 K in Castilla y León. At the time of establishing thresholds of danger, we

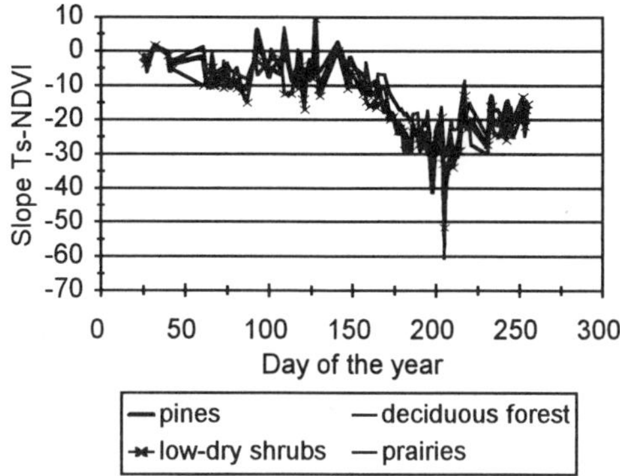

Figure 7. Temporal evolution of the slope between NDVI and Ts in 1994.

again do not obtain clear results. As with the previous indicator, if the study is carried out in Valencia, where large scale forest fires are present we obtain results which indicates that T_s-T_a could also be a good indicator of large scale fires danger. Finally no clear relation with the indicators described above has been found.

7 CONCLUSIONS

A GIS to be used for vegetation monitoring in central Spain has been developed. Meteosat images allow for the calculation of global solar radiation and NOAA-AVHRR data have been employed to derive the NDVI and surface temperatures. The system also includes air temperature and rainfall recorded by the Spanish Meteorological Institute and interpolated to the spatial resolution of the AVHRR images. Finally, digital vegetation mapping derived from the CORINE land cover program has also been used. Data are processed every day and composed in ten day periods NDVI-MVC as well as rainfall amount and maximum, minimum and average temperatures.

The first application is cereal crop yield estimation. A statistical model has been developed by means of NDVI and meteorological data temporal series. The results are quite satisfactory with errors of about 4% at a regional level and 10% at a provincial level. The meteorological variables introduce in the model information about water stress due to the fact that water is the main limiting factor in the study area. Finally, the model allows yields to be estimated using the GIS data which are obtained systematically and at a low cost.

A second application is for the monitoring of forest areas. The series of NDVI data have been analyzed in relation to meteorological variables and linear models with correlation coefficients of around 0.8 have been found. The results seem to indicate that the NDVI-meteorological variables relation is easier to find in periods of vegetation growth when extreme temperatures do not appear and there is no significant hydric deficit. In the arid zones with Mediterranean vegetation new climate variables more significant from an ecophysiological view, especially in those aspects related to annual hydric balance, have to be considered in order to analyze the summer period.

Finally, fire danger due to vegetation water stress has been estimated. Four indicators have been used: greenness (G), accumulated slope of the NDVI temporal evolution (AS), the slope between surface temperature and NDVI (NDVI-T_S) and the difference between surface and air temperature (Ts-Ta). The analysis of the temporal evolution of the indicators identifies the periods of hydric deficit in summer and provides an estimate of its intensity. In the region of Castilla y León, the variability among forest species is something greater and one could consider establishing danger thresholds dependants of the forest species, something that has not been possible due to a lack of data on species affected by fires. The superimposition of fires data and values of the indicators allows us to conclude that the greenness G and the accumulated slope AS seem to be adequate in establishing the danger of fires. Moreover they are considerably related. The results are somewhat better for greenness, but the accumulated slope is calculated with data from only one year, which one could assume it is a good option if there is a lack of temporal series of NOAA data. Regarding Ts-NDVI and Ts-Ta, we have problems to establish thresholds in Castilla y León. If a similar study is carried out in the Mediterranean coast, the results make us believe that both indicators can be related with propagation and be suitable for estimating the danger of large scale fires. Finally, to say that the first three indicators are obtained using only NOAA data and can be calculated in an operational way and in almost real time. In fact, they are sent to the Spanish Forest Service during the period of risk, on the basis of an agreement that has been in effect for the last four years.

8 ACKNOWLEDGEMENTS

The authors wish to thank the Spanish Meteorological Institute, the Agricultural Department of the Regional Government of Castilla y León and the Spanish Forestry Service for having provided the ground data used in this study. Thanks to doctor Elena-Roselló for having provided the CLATERES classification data. Finally, we also would like to thank to the Regional Government of Castilla y León for their financial support and projects which have made this work possible.

REFERENCES

Burgan, R. E. 1995. Use of remotely sensed data for fire danger estimation. *Proc. International Workshop on Remote Sensing and GIS Applications to Forest Fire Management, Alcalá de Henares, Spain, 7-9 September 1995.* EARSeL

Cihlar, J. 1993. A multiyear analysis of the relationship between surface environmental variables and NDVI over the Canadian landmass. *Remote Sensing Reviews* 7: 151-177.

Coll, C. 1994. *An Operational Model to Determine the land Surface Temperature from Satellites.* Doctoral Thesis, Department of Thermodynamics, University of Valencia (in Spanish).

EEA Task Force, 1992. *CORINE Land Cover.* Brochure Prepared as Contribution to European Conference of the International Space Year. Munich, Germany, 30 March - 4 April. 22 pp. 1992.

Elena-Roselló, R. 1996. Un instrumento para la planificación territorial de la reforestación en España : la Clasificación Biogeoclimática Territorial CLATERES. *El campo* 134: 105-118.

Illera, P., Fernández, A. & Pérez, A. 1995. A simple Model for the Calculation of Global Solar radiation using Geostationary Satellite Data. *Atmospheric Research* 39: 79-90.

Illera, P., Delgado, J. A. & Calle, A. 1996a. A Navigation Algorithm for Satellite Images. *International Journal of Remote Sensing* 17: 577-588.

Illera, P., Fernández, A. & Delgado, J. A. 1996b. Temporal evolution of the NDVI as an indicator of forest fire danger. *International Journal of Remote Sensing 17:* 1093-1105.

Kaufmann, Y. J. & Holben, B. N. 1993. Calibration of the AVHRR Visible and Near-IR Bands by Atmospheric Scattering, Ocean Glint and Desert Reflection. *International Journal of Remote Sensing* 14: 21-35.

Nemani, R. R. & Running, S. 1989. Estimation to regional resistance to evapotranspiration from NDVI and thermal-IR AVHRR data. *Journal of Applied Meteorology 28:* 276-284.

Schultz, P. A. & Halpert, M. S. 1993. Global correlation of temperature, NDVI and precipitation. *Advances on Space Research* 5: 227-280.

Terrab, R., Delgado, J. A., Illera, P. & Casanova, J. L. 1997. Combined use of NOAA-AVHRR and Meteosat Images with other Data for the Estimation of Cereal Crop Yield. In *Proc. of the 1998 Meteorological Satellite Data Users' Conference, Brussels, 29th September-3rd October 1997.* EUMETSAT.

Vidal, A., Pinglo, F., Durand, H., Desvaux-Ros, C. & Maillet, A. 1994. Evaluation of temporal fire risk in Mediterranean forests from NOAA thermal IR. *Remote Sensing of Environment* 49: 296-303.

Remote Sensing in the 21st Century: Economic and Environmental Applications, Casanova (ed.)
© 2000 Balkema, Rotterdam, ISBN 90 5809 096 5

Crop spatial characterization in the 18th aquifer, Spain

J.García-Consuegra & F.Abad
Sección de Teledetección y SIG, Departamento de Informática, Universidad de Castilla-La-Mancha, Albacete, Spain

G.Cisneros
Grupo de Tratamiento de Imágenes, Universidad Politécnica de Madrid, Spain

ABSTRACT: The results of a remotely sensed imagery classification depend on a lot of elements (information classes, remote sensed images, conditions of image acquisition, the spectral and spatial resolution etc.) but, mainly, *how* the classification algorithm is trained. In this work, a new pattern characterization is proposed. Its main property is that the spatial variability component is considered in the own pattern; that is to say, the problems inherited from the confidence boundary location of the spectrally homogeneous zones are avoided. The corn crop spatial pattern in the 18th aquifer of the Mancha Oriental (Spain) has been used as the test bed.

1 INTRODUCTION

The results obtained of a remote-sensed image classification process depend on a great number of factors (information classes, remote sensed images, conditions of image acquisition, the spectral and spatial resolution, etc.), but, above all, on the way the classifiers are trained (Mather 1987). Training techniques are typically performed in a supervised, non supervised (clustering) or hybrid way in order to establish the representative spectral characteristics (pattern or range/multidimensional region of validity) for each class. In the supervised training the analyst designates to the classifier the pixel locations of training fields of the various cover types which have informational value or are of interest. Clustering technique has to obtain the characteristics, and sometimes also the number of the classes from the distribution of the pixels in the feature space. Hybrid techniques combine various aspects of the supervised and clustering techniques in order to avoid their weakness. One of these hybrid techniques is described as the Multi-Cluster Blocks. In this method, several small blocks of data are located; each of which contains several cover types and spectral classes. Each data block is individually clustered and then the spectral classes for all cluster areas are combined to form a single data set, which describes the spectral characteristics of all cover types in the entire study area statistically.

For sufficiently large areas, the spatial variability of the characteristics must also be considered (Carbone et al. 1996). Traditionally, spatial variability is introduced by the area subdivision in function of criteria such as bio-climate, terrain elevation, natural regions, soil type and so on (Townshend et al. 1987, Pastor & Broschart 1990). The main deficiency of this method is the spatial confidence limits, that is to say, the boundary location of these spectrally homogeneous zones. The problem that arises is why two close plots (plots 1 and 2 in Figure 1) with the same cover type, in different homogeneous zones, have different patterns whereas two much more distant plots (plots 2 and 3 in Figure 1) have the same pattern. In some cases the boundary location can split a same plot (plot 4 in Figure 1).

The sample design is not the purpose of this work, that is to say, sampling, sample distribution, as well as determination of the sample size, the sample selection procedure, and estimation of the population means, totals, variances, and confidence limits from the sample information. The aim of this research is to study the spatial variability of the corn in the aquifer 18th, if any, as well as to define a methodology to establish a spatial pattern, previous to the classification stage, with the minimum training sites, its assessment and its applicability to other crops.

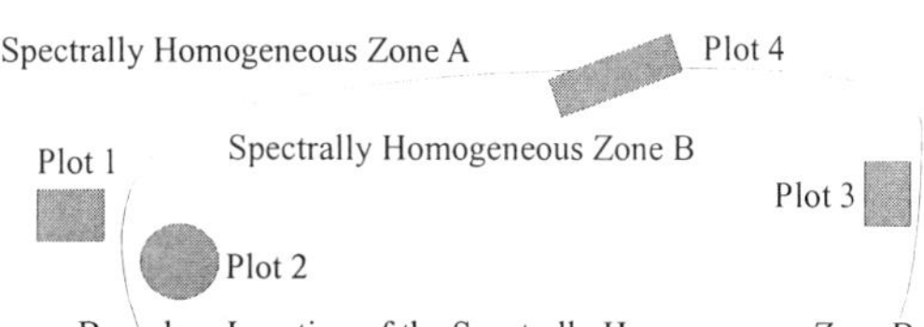

Figure 1. Problems involved in the boundary location of the Spectrally Homogeneous Zone

Figure 2. The location of the 18th Aquifer is mainly between Cuenca and Albacete provinces, Spain.

2 STUDY AREA

The aquifer 18th is located in the south-central portion of Spain (Fig. 2) It extends approximately 120 kilometers in the east-west direction and 118 kilometers in the north-south direction. The aquifer 18th is largely rural and it is specially affected by the dry years.

3 METHODOLOGY

The advantages of homogeneous objects over per pixel classification are already known in the scientific community. So, in a first step the homogeneous objects were obtained using an ECHO (Extraction and Classification of Homogeneous Objects) algorithm. Each homogeneous object has several attributes: the number of pixels, the grey level average, the standard deviation, and its location in UTM.

Due to the aquifer dimensions, as well as the high probability of having a variety of image acquisition conditions, the characterisation with just one pattern (obtained from the supervised training fields) was discarded. The use of Multi-Cluster Blocks or Multispectral Homogeneous Zones was also discarded because of the valid zones limitation namely, the identification of their confident limits in which these patterns are valid.

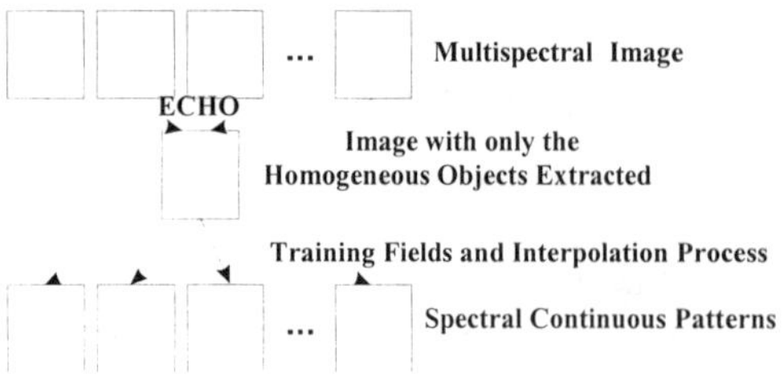

Figure 3. An overview of the methodology to define the Spectrally Continuous Patterns.

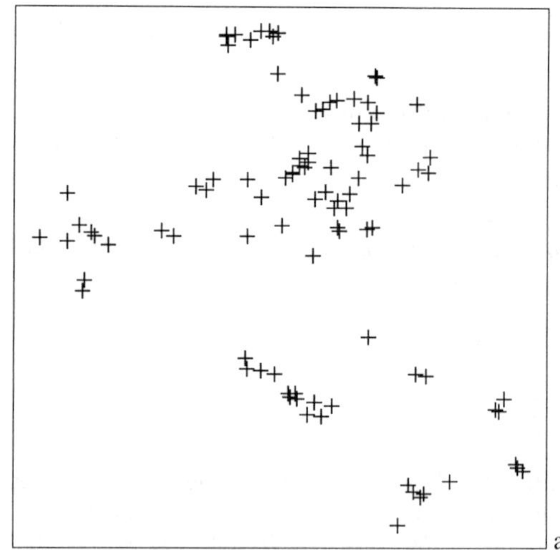

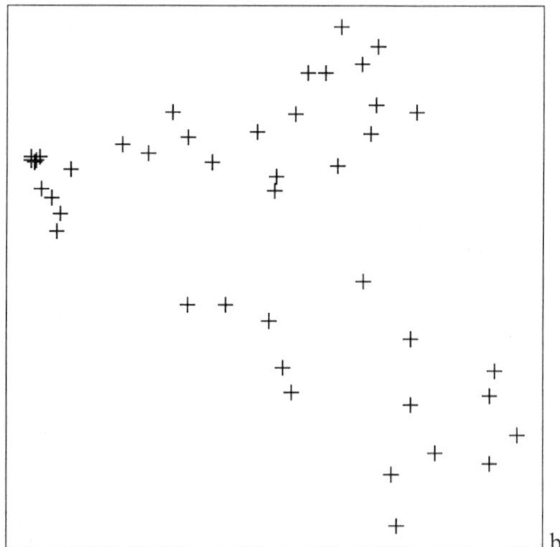

Figure 4. Field reconnaissance distribution training (a) and test (b) samples.

Finally, to avoid the abovementioned problems, a continuous spatial pattern was defined from the interpolation of the homogeneous object values previously identified as corn plots. The field reconnaissance allows us to identify over 149 homogeneous objects covering corn plots, well distributed along the aquifer. This set was divided into two groups: training and test samples (Fig. 4). The first one was used to define the continuous spatial pattern. The kriging method (Burrogh 1986, Cressie 1993) was used to accomplish the interpolation since it can also provide direct estimates of the uncertainty due to the interpolation process. Figure 3 shows an overview of this methodology.

To match the test fields with the spatial pattern, parametric algorithms were prevented by the way in which the spatial pattern is obtained, that is to say, there is no a known probability density function of the corn class for each location (resolution element or pixel) but an estimated centroid. Thus, nonparametric similarity and dissimilarity criteria must be used in a Maximum (minimum) Similarity Decision, when a competition exists between different classes. In this work, we have used the distance object-patterns, at the object coordinates, as the validation of our proposal.

4 DATA AND RESULTS

The remote sensed image used in this work is a Landsat TM image scanned on 15 August 1997. In the following tables are the first order statistics of the original bands, as well as the characteristics of both patterns, Table 1 and 2 respectively. Table 3 shows the mean-square error between real value test plot and both patterns. In Figure 5, the iso-lines for the spectrally continuous patterns are shown.

Table 1. Characteristics of the Bands 3,4, and 5

Band	Average	Std. Deviation	Maximum Value	Minimum Value
3	72.915	25.703	255	16
4	91.612	18.743	255	1
5	156.095	46.003	255	2

Table 2. Global and continuous pattern characteristics.

Band	Global	Continuous Pattern	
	Pattern Centroid	Max/Min Value	Range
3	24.14	29/21	8
4	105.53	86/126	40
5	61.734	55/79	24

Table 3. Mean-square error

Band	Global Pattern	Continuous Pattern
3	0.724	0.968
4	2.376	1.82
5	2.083	2.83

The results show that several plots have an unusual high dissimilarity with the continuous pattern producing an increase in the error rates. This dissimilarity is due to plots with special characteristics. Here, by plots with special characteristics we mean those that have a very different spectral signature (due to different acquisition condition) as a result of which they can be classified in a different way, producing an increase of the classification-error rate. When an analysis of the training field is carried out in order to identify these special plots and to remove them of the training field set, an improvement of the statistics is carried out, specially in the spectrally continuous pattern (Table 4).

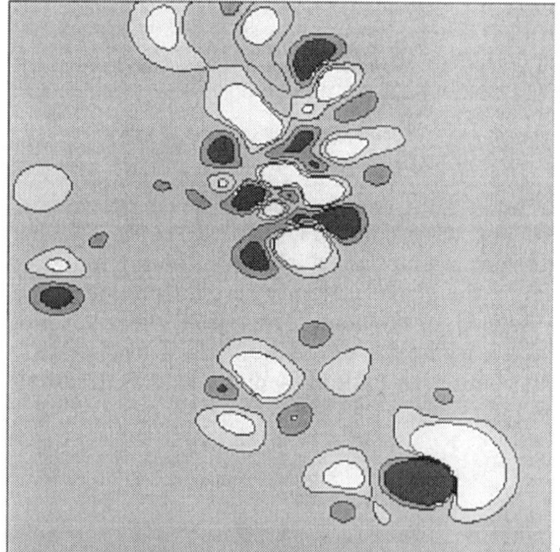

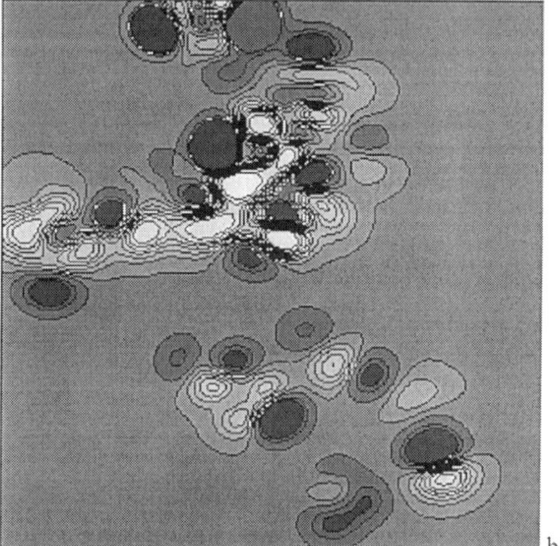

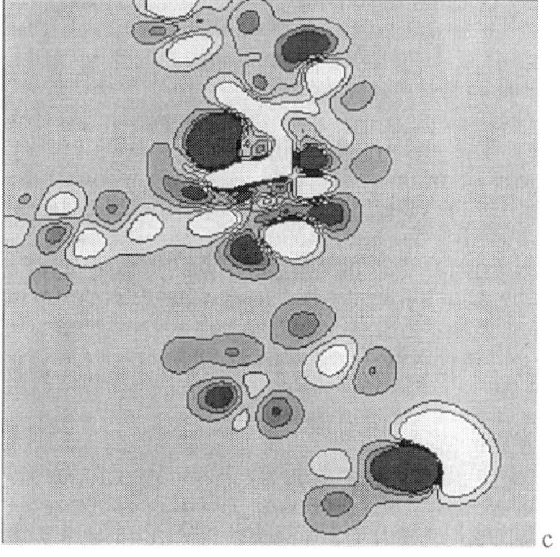

Figure 5. Iso-lines for the continuous patterns in band 3, 4 and 5.

Table 4. Mean-square error after carrying out a training field set analysis.

Band	Global Pattern	Continuous Pattern
3	0.345	0.432
4	1.292	2.180
5	0.905	0.882

On the other hand, our method shows, together those special plots, their location. The highest error made by the global pattern where mainly located in the isolated areas (being away from the main plot clusters), whilst the error rate yielded by the spectrally continuous pattern is distributed between all plots.

5 CONCLUSIONS

Class pattern spatial variability is already known in the scientific community, as are the advantages of homogeneous objects over per pixel classification. The first conclusion is that the corn crop in the 18th aquifer has a significant spatial pattern variation. The results yielded show similar mean-square error to both patterns, but in general, plot to plot, the continuous pattern has a better performance. Several plots have an unusual high dissimilarity with the continuous pattern producing an increase in the error rates. This dissimilarity is due to plots with special characteristics. The proposal methodology identified those special plots and their location. Despite of the results yielded, the differences between both methods will be increased as the study area dimensions are increased.

The use of homogeneous objects entails computational saving in the interpolation and classification process. Furthermore, the use of the homogeneous object centroid is suitable since its main property is the high degree of similarity among the pixels that make it up. Thus, the cluster in the feature space presents a low scattering degree.

This methodology has three important advantages. The first is that, it does not need of any knowledge of the probability density function of the crop. On the other hand, it is very difficult to define a probability density function that varies with space. The second is that the homogeneous-training objects can be determined in a supervised, as we have done, or unsupervised way. And the last advantage is that the homogeneous object classification process is carried out in a Maximum (minimum) Similarity Decision using a non-parametric similarity and dissimilarity criterion.

As might be expected, the spatially continuous pattern yielded the best corn characterisation. Furthermore, the applied methodology can be used with other crops. The necessity of establishing the limits of validity of the statistical patterns obtained by the Multi-Cluster Blocks or Multispectral Homogeneous Zones methods disappears. The space variation obtained is located. Each band defines its own spatial pattern. Thus, just one limit based on homogeneous spectral features is impossible.

An important problem that has arisen is the number and location determination of the training fields. An automatic clustering extraction and analysis algorithm that can be used in each homogeneous object neighbourhood in order to establish a spectral similarity relationship between them may be found in (Buchheim 1989).

In order to introduce the spatial variability, an alternative to (Buchheim 1989) may be to establish the similarity relationship that can be made as an expansion of the region growing algorithm to non-contiguous regions, with the maximum region-region distance parameter as a stopping criterion that limits the match range of a given region with its neighbours. Moreover, it seems logical to think that the crossing of the results obtained by the use of different distances would provide an useful approach to the problem. The dimensions and nature of the study area usually entail a great number of homogeneous objects. Thus, the computational cost in processing them in the abovementioned way is unsuitable. The (RAG–Region Adjacency Graph) approach tries to define structures that establish the vicinity among the homogeneous objects found in an ECHO process. In (Nichol 1990), two algorithms capable of working with a high number of objects in an efficient way are proposed.

This work is the beginning of a search that pretends to study the pattern variability of the crops in the 18th aquifer.

6 FUTURE WORK

Our first aim is to extend this work to other crops and study the discrimination degree introduced. Secondly, to establish its relation or validity with other year patterns, and finally to develop the unsupervised procedure to establish similarity relations as proposed in the conclusions.

7 REFERENCES

Buchheim, M. P. et al. 1989. Semi-automated training field extraction and Analysis for Efficient Digital Image Classification. *Photogrammetric Engineering and Remote Sensing,* vol. 55, no. 9, pp. 1347-1355.

Burrogh, P.A. 1986. *Principles of Geographical Information Systems for Land Resources Assessment.* Oxford, Clarendon Press.

Carbone, G.J. et al. 1996. Application of Remote Sensing and GIS Technologies with Physiological Crop Models. *Photogrammetric Engineering and Remote Sensing*, vol. 62, no. 2, pp. 171-179.

Cressie, N.A.C. 1993, *Statistics for Spatial Data*, John Wiley & Sons.
Mather, P.M. 1987. *Computer Processing of Remotely-Sensed Images*. Edit. John Willey & Sons.
Nichol, D.G. 1990. Region Adjacency analysis of remotely-sensed images. *Int. J. Remote Sensing*, vol. 11, 2089-2101.
Pastor, J. & Broschart M. 1990. The spatial pattern of a northern conifer-hardwood landscape. *Landscape Ecology*, vol. 4, no. 1, pp. 55-68.
Townshend, J.R.G. et al. 1987. Characterization and classification of South American land cover types using satellite data. *Int. J. Remote Sensing*, vol. 8, no. 8, 1189-1207.

Remote Sensing in the 21st Century: Economic and Environmental Applications, Casanova (ed.)
© 2000 Balkema, Rotterdam, ISBN 90 5809 096 5

The use of satellite remote sensing imagery for detection of secondary forests on post-agricultural soils: A case study of Tuchola Forest, Northern Poland

M. Kunz
Institute of Geography, Nicholas Copernicus University, Torun, Poland

A. Nienartowicz & M. Deptula
Institute of Ecology and Environment Protection, Nicholas Copernicus University, Torun, Poland

ABSTRACT: Satellite imagery Landsat TM was used for the evaluation of Normalized Difference Vegetation Index (NDVI) on Zabory Landscape Park. On the basis of satellite imagery it was stated that spatial change of index is correlated with the range of forests in the past presented on old topographic maps and utilisation of the area. To define the influence of the previous land-use on NDVI value the comparison analyses was made on four sites. Two of them on areas which were deforested in the past and then reforested again at the end of 19th and 20th century. Two other sites were localised in areas where forest existence was not interrupted by the temporary agricultural usage of the area. It was stated that NDVI value on pine plantations on the post-agricultural soils is much lower than on pine forests on land that was not agricultural. Obtained index values for forest areas with a different history were compared to the defined in literature Chlorophyll Index values (CI) and states of green biomass in forest and peat bog communities of Zabory Landscape Park and NDVI values for pine plantations in wide geographical range.

1 INTRODUCTION

Secondary forests on the post-agricultural soils cover 1.4 million ha in Poland which is about 16,7% of the whole forest area. As a result of soil degradation by long farming their plant biomass, primary production, chlorophyll index, and species diversity is in most cases lower than in forests restored on clearings shortly after cutting out the previous generation of trees (Barcikowski 1992, Nienartowicz et al. 1998, Wilkon-Michalska et al. 1998, in press). Forests on post-agricultural soils cover over 40% of the area situated in northern Poland Przymuszewo Forest Inspectorate which is part of Zabory Landscape Park. The afforestration of soils used for a time for farming was done here mainly by the Prussian government at the end of 19th and the beginning of 20th century, partly within the framework of the political process „Kulturkampf". It consisted in buying out large numbers of Polish privately owned land estates and setting up state-owned forest cultures. Some farming areas were afforested by landowners, both before World War I and between the two World Wars, when the area under study was part of the restored Polish state. The forest area was extended also shortly after the Second World War, when the big estates were nationalized and reforested in the political action of the Polish government called agricultural reform.

The present composition and pattern of forests in the area under study are closely related to previous human activity. On sandy soils, very much exploited through intensive farming, occurs today dry pine forest belonging in phyto-sociological system to association *Cladonio-Pinetum*. Places, occupied in the past by poor pastures where sheep were grazed and later private forest were planted, are now overgrown with the community *Pinus-Calluna*. In places where forest existence was interrupted only by timber fellings, and the clear-cuttings were soon reforested there are now fresh pine forest *Leucobryo-Pinetum*. Their considerable habitat variability depends to a large extent on the relief, i.e. the height, the slope and the exposure of the terrain. Small fragments of oak-hornbeam and beech forests survived on the sides of river and lake valleys, not fit for cultivation or grazing in place were board-leaved forests have been superseded by pine forests and on brown soils previously used for farming the community *Pinus sylvestris-Deschampsia flexuosa* is now developing.

Measurements done on this area by Barcikowski (1992), and Barcikowski & Michalska (1988) showed that the chlorophyll index on pine plantations on *Leucobryo-Pinetum* habitat is higher that on *Cladonio-Pinetum* habitat which dominates on post-agricultural soils. It refers to mature forests and also all regeneration stages of the forest after clearings. Moreover, on plantations growing on *Leucobryo-Pinetum* habitat the amount of chlorophyll converted to m^2 of phytocenosis is higher than in *Pinus-Calluna* communities, which appeared after reforestation of pastures, and also in *Pinus sylvestris-Deschampsia flexuosa* communities which grew after the introduction of pine-tree to deciduous forest communities, or to fallow lands on more fertile, brown soils. Because of the fact that chlorophyll index in forests on post-agricultural soils is much lower that on pine plantations in fresh forest community *Leucobryo-Pinetum* habitat, and in natural wet forests *Molinio-Pinetum,* and in peat-pine forests *Vaccinio uliginosi-Pinetum*, and on peat bogs, it has been assumed that they will be recognised on satellite imagery after evaluating NDVI index. The aim of this research carried out with the financial support of Polish Research Committee No 6 PO4 F 01313 was to compare spatial distribution and average NDVI values on areas covered by secondary forest on post-agricultural soils and on areas where the existence of forest was interrupted by clearings, and then they were reforested again, in most cases with pine-trees.

2 STUDY AREA

The study area is situated north of the town of Chojnice and west of the town of Brusy (Figure 1).

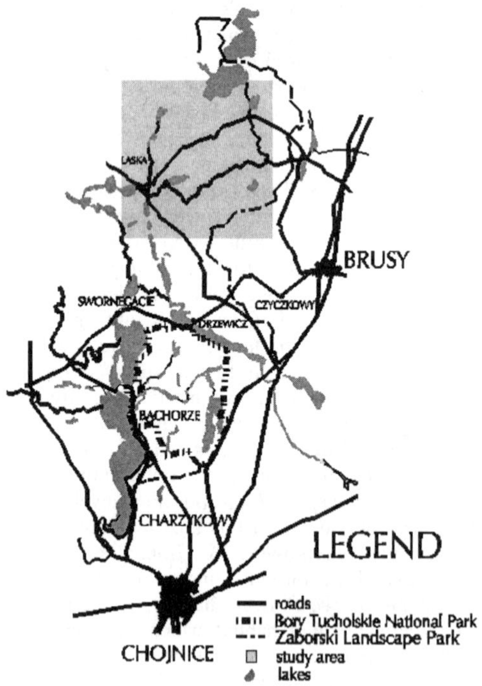

Figure 1. Situation of study area.

Since the latest administrative partition of Poland it has been included in the Pomerania Province. It belongs to the Zabory Landscape Park and in 1996 the National Park Tuchola Forest was set up in the southern part of Zabory Landscape Park.

Vast forest complexes in the area under study are part of Przymuszewo Forest Inspectorate, which is subordinate to the Regional State Forest Management in Torun. Only the forest situated at the north-western edge of the square belong to Osusznica forest Inspectorate, subordinate to the Regional State Forest Management in Szczecinek.

According to Kondracki's (1978) physico-geographical partitioning of Poland, the area under study is part of the macroregion of South-Pomeranian Lakeland and of the mesoregion Charzykowy plain. From the geo-botanical point of view, the study area belongs to the Pomeranian Lakeland Region and Kartuzy district (Szafer & Zarzycki 1972).

The study area lies in the catchment area of the river Zbrzyca and its tributaries, the Kulawa and the Klonecznica. The river Zbrzyca flows through a number of eutrophic lakes and into the river Brda. In the area under study there are also some *lobelia* type lakes and several natural reserves (lakes Nawionek and Laska, the river Kulawa Valley, Stawek Swamp, Channel Lakes).

The settlements are situated mainly in river and lake valleys. The largest villages in that area are those lying on the Zbrzyca: Kaszuba, Rolbik and Laska, as well as Kruszyn situated by lake Kruszyn. Part of the settlements and villages existed there in the second half of 19th century became depopulated and their terrains were later afforested. Such settlements as Zielonki, Lisewo, Polaszek, Piecki, Nowa Laska, Belfort, Kramarska have disappeared. At the end of 19th century the area under study became part of the Prussian Forest Inspectorates Zwangshof and Laska. The former was set up in 1890 following the purchase by the Prussian government of a large land estate. Half of it was fallow land and glades. The Prussian Forest Inspectorate Laska was set earlier. In the mid-nineties of 19th century both inspectorates were about 10 000 ha in area. In the years 1894-1920 the Prussian administration purchased yet another 1788 ha, mainly farming land belonging to the largest land estates in Pomerania. Further extension of forest areas came after 1920, i.e. after incorporation of the area into Polish Republic II and after 1945, following the agricultural reform. Several land estates, situated by lake Kruszyn and in the Zbrzyca valley, near the villages Kruszyn, Kaszuba and Rolbik, were then nationalized. Afforestration of farming land meaning turning cropland into mainly dry pine forest still continues. The afforested areas are mainly small fields, rented until recently to forest workers. The changes of forests range in the area under study are presented in Figure 2.

Most of the forests in the area under study are state-owned. Larger privately owned farms are found by the lake Ksieze, in the nearby Lesno, in Big Chelmy and Male Chelmy, i.e. in the agricultural area called „Brusy island" bordering on the forest complex under study.

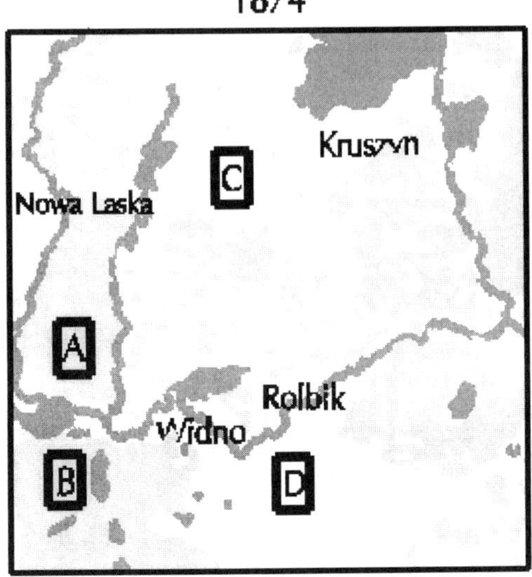

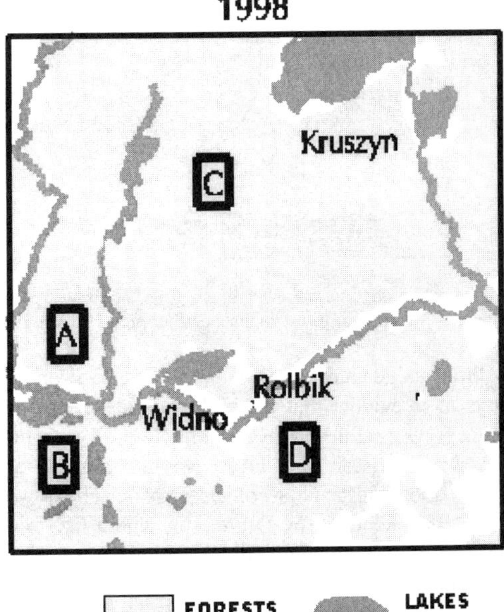

Figure 2 Forest ranges in 1874 and 1998.

Among trees pine-tree dominates on the researched area. On small areas other species dominate, e.g. *Pinus strobus* and *Pinus nigra*. Taking into consideration deciduous trees, stands of trees grow only on the bottom of river peaty valleys and lakes, black alder *Alnus glutinosa*, and beech *Fagus sylvatica* on the slopes of those valleys. The last kind of tree often exists as a understorey in old stands of pine-trees on *Leucobryo-Pinetum*. Other species such as birch-tree *Betula verrucosa*, and lime-tree *Tilia cordata* exist as an addition in stands of pine-trees. Lichens form *Cladonia* predominate in the undergrowth of dry pine forests, and in the undergrowth of fresh pine forests the biggest covered is achieved by chamephytes *Vaccinium myrtillus, Vaccinium vitis-idaea, Calluna vulgaris,* grass *Deschampsia flexuosa,* and mosses *Pleurozium schreberi* and *Dicranum undulatum.* Protected plants exists very often there, e.g. *Lycopodium clavatum, L. complanatum, L. annotinum, Empetrum nigrum*, and even *Linnaea borealis.* Above mentioned species of mosses predominate in pine boskets on *Leucobryo-Pinetum* habitat.

Sokolowski (1965) stated that defined according to classic phyto-sociology method fresh pine forests of the researched area differentiate into four variants: typical, with *Cladina rangiferina*, with *Pteridium aquilinum* and with *Frangula alnus. Cladonio-Pinetum* associations divided into two variants: typical and with *Vaccinium myrtillus.* It proves the existence of important continuum between fresh and dry forests. The reason of phytocenosis differentiation within the association is ground moisture and air humidity, which depend on the eminence and exposition of the area, and distance to lakes.

3 METHODS

Changes in forest range were defined using old topographical maps of 1875 and 1921 and land-use maps of 1894, 1926, 1947, 1953, 1966, 1976, 1988.

To characterise the reflectance of forests on post-agricultural and not post-agricultural soils satellite imagery was used created by Landsat TM on 28.07.1990. With the support of image processing and GIS software three-colour compositions were created and the normalized vegetation index was counted for the whole area of Zabory Landscape Park, according to the following formula:

$$NDVI = [IR - RED] / [IR + RED],$$

where RED and IR are reflectance values equivalent to Thematic Mapper (TM) Bands 3 (RED, 630-690 nm) and 4 (IR, 760-900 nm).

With the support of image processing software the map of spatial variation of NDVI index was prepared for the quarter with a 10 km side covering the area presented on one sheet of Prussian topographic map from 1875 (Figure 2). Spatial variation of NDVI index was also calculated for four areas inside this quarter which differentiate distinctly by the way the land was utilised in the past. Symbols and history of the forests on those areas are as follows:

A) the area situated on the highland on the northern side of Laska lake, where the area was not agriculturally utilised in the past

B) The area situated lower nearby Zmarle and Piecki lakes, where the development of a forest was interrupted only by clearings and not by longer agricultural utilisation

C) The area on the highland between Duze Gluche and Kruszynskie lakes, where agricultural fields existed before it was reforested at the beginning of 20th century.

D) the area where wet pastures existed before reforestation.

The size of each area was about 135 ha, which means each of them covered about 1500 pixels. During the preparation of maps the scope of NDVI index from -1.0 to + 1.0 was divided into 20 classes which had colours used during maps creation.

Using the forest inventory book and information about cultivation activities and achieving wood, the share of forests and other types of plants, the age both structure of stand of trees and forest habitat structure were calculated on comparable areas. After the creation of the spatial NDVI variation map, histograms of index value were prepared, and the average value and standard deviation on each area was calculated.

Differences between the researched areas taking into consideration the value of NDVI index was defined by the numeric classification method. Features of the separate areas was pixels frequency used in the analysis of classes from NDVI scope. As a dissimilarity measure the Euclidean distance was chosen. Dendrogram was made by average variance clustering method (Orlóci 1978). Moreover for characterisation of NDV variability on each area for pixels frequency in histograms Shannon-Wiener's diversity index was used and evenness. In calculations of those indexes and for dendrogram preparation the MVSP program was used (Kovach 1993).

4 RESULTS

On the basis of data analysis in the forest inventory book it was stated that on A area there dominate fresh forests and habitat which transit from fresh to dry (Fig. 3). Forests on that area in most cases belong to the II age class, which means 20-40 years (43%). Forests which are 60-80 years old cover 13%, which are 100-120 years old 10%. On the area of B fresh forest in the only habitat type, and in the age structure of stand of trees forests which are 20-40 years old (29%) dominate. However, there exist older stands of trees on this area. There share of forests which are 100-129 years old is 22%, and which are 120-140 years old 18%. On both areas covered by forests on post-agricultural soils C and D fresh forests dominate. Their share in the structure of habitats amounts to 98% and 100% respectively (Fig. 3). In the age structure forests of II age class, which is 20-40 years, dominate on the C area covering 40%. Forests which are 60-80 years old covered 25%, and 80-100 years old 14%. On the D area the biggest area was covered by 80-100 years old forests is 61%. Years which are 60-80 years old are 7%, and 20-40 years old - 15%.

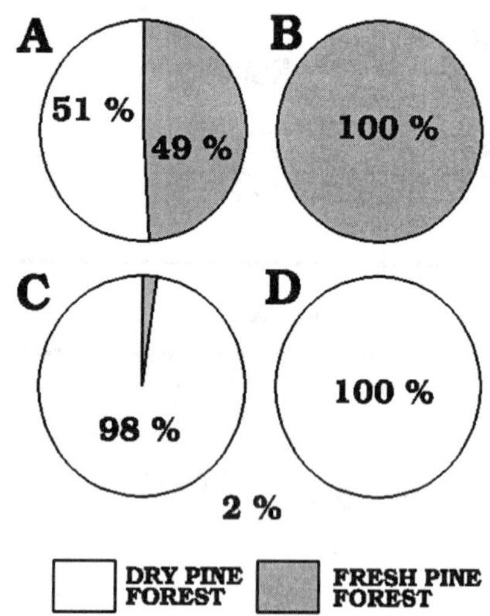

Figure 3. Structure of forest habitats on four plots.

From value analyses of NDVI indexes on separate areas it is evident that the highest value of this parameter was achieved by B area (Figure 4), where the share of fresh pine forest was the highest, and where the oldest stands of trees exist. The lowest value was achieved by D area, on which dry forest dominated. The average NDVI value on those areas amounted to 0.365 and 0.155 respectively. On A area where the existence of forests was not interrupted by agriculture activities NDVI value reached 0.262, and on the second post-agricultural field, e.g. C its value was 0.183.

On the basis of NDVI spatial variation analyses on separate sites it was notified that D area was the most homogeneous, pixels covered 10 classes on them. Its Shannon-Wiener's diversity index amounted to 2.3670, and evenness 0.7126. A and C areas were the most heterogeneous. Their NDVI values belonged to 13 class diversity index 2.7212 and 2.8671 (evenness 0.7489 and 0.7748) respectively. B area pixels covered 12 classes. Its diversity index amounted to 2.8384, and evenness 0.7917.

Separateness of forest area on post agricultural areas from lands with forests not utilised agriculturally is showed on the dendrogram prepared on the basis of pixels frequency in the separate parts of NDVI index. Both pairs of area create separate clusters (Figure 5).

5 CONCLUSION

Results achieved are included into scopes presented for plantations of other kind of pine in different world regions. The highest accordance exists between NDVI values from Zabory Landscape Park and *Pinus densiflora* plantations on the mountain territory of Japan. Lee & Nakane (1997) propose NDVI values from 0.3 to 0.5 for those forests. Gholz et al.(1997) analysing reflectance of *Pinus elliottii var. elliottii* plantations on Florida, USA stated higher values, which are 0.55 - 0.66. Such NDVI values in Zabory Landscape Park are achieved only by the oldest stand of trees on the habitat of a fresh pine forest.

Spatial NDVI diversity is explained very well by the results of chlorophyll index measure and chlorophyll contents in the unit of green biomass on separate phytocenosis types in Zabory Landscape Park.

On the basis of comparisons it was stated that some dependence exists between NDVI values reached in that paper, and results of analysis made using standard method by Barcikowski (1992), and by Barcikowski & Michalska (1988) concerning the amount of chlorophyll in plant communities of the researched area (Figures 6 and 7).

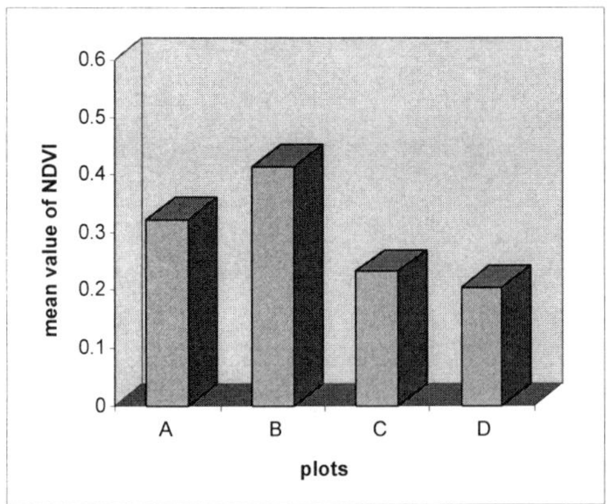

Figure 4. Mean value histogram of NDVI plots A, B, C, D.

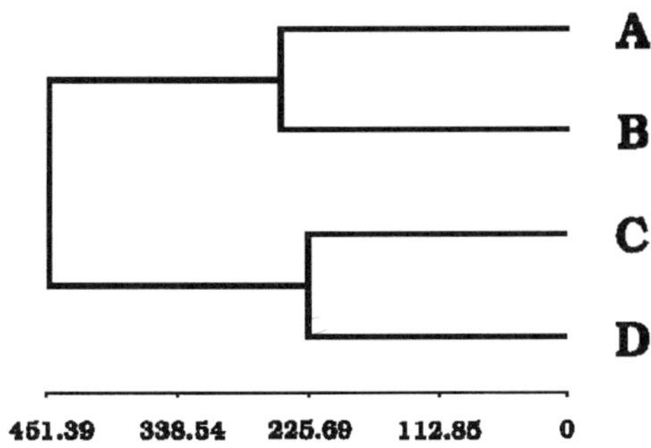

Figure 5. Dissimilarity between plots A, B, C, D.

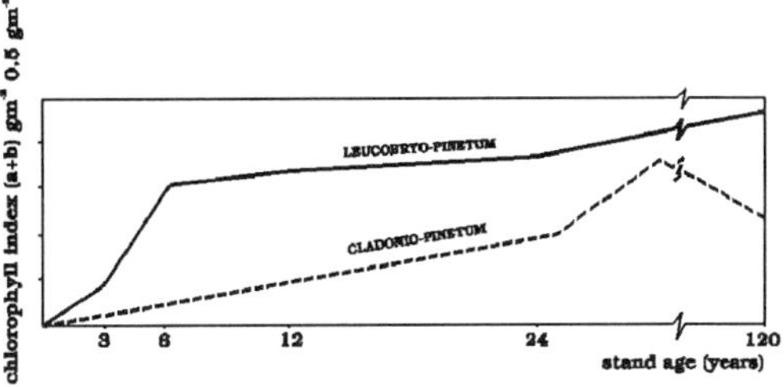

Figure 6. Development of the chlorophyll index in phytocenoses with secondary succession in Bory Tucholskie Forest.

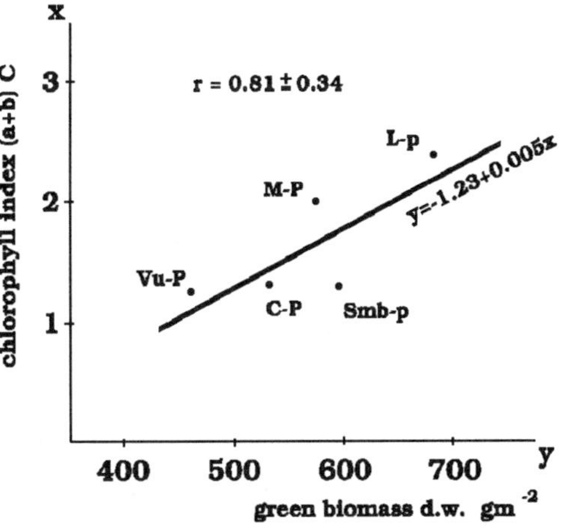

Figure 7. Regression of the chlorophyll index (CI) depending on green biomass in selected associations of Bory Tucholskie Forest (L-p *Leucobryo-Pinetum*, M-P *Molinio-Pinetum*, Vu-P *Vaccinio uliginosi-Pinetum*, Smb-p *Sphagneto magellanici-boreale pinetosum*, C-P *Cladonio-Pinetum*).

The conclusion is that the satellite method can be used to detect areas covered by dry forests, which was agricultural in past times. However, the precise application of satellite methods need making further comparisons of chlorophyll level and NDVI values in dry forests created as a result of natural succession and on post-agricultural fields in fresh forests and *Pinus sylvestris-Deschampsia flexuosa* stand of trees which grow as a result of reforestation of older agricultural fertile soil or as a result of the introduction of pine to deciduous stand of trees.

Substantial age differentiation of tree stands makes the spectral characteristics of plant cover in Zabory Landscape Park more difficult. For precise definition for agricultural influence of soil utilisation on the condition of reforested lands, the analysis of big groups of pixels registered with better resolution, covering the same age plants existing on the same habitat and in places which were utilised in the similar way in the past is advisable.

6 REFERENCES

Barcikowski, A. 1992. Differentiation in the structure and energy flow in phytocenoses with primary and secondary succession. In: R.Bohr, A.Nienartowicz & J.Wilkon-Michalska (eds.) *Some ecological processes of biological systems in North Poland.* N. Copernicus University Press, Torun.

Barcikowski, A. & T. Michalska 1988. *The amount of chlorophyll during secondary succession in pine forest Leucobryo-Pinetum Mat. (in Polish with English summary).* Acta Univ. N. Copernici (Torun) Biol.29:189-198.

Gholz, H.L., Curran, P.J., Kupiec, J.A. & Smith, G.M. 1997. Assessing leaf area and canopy biochemistry of Florida pine plantations using remote sensing. In: Gholz, H.L. Nakane K., Shimoda H. (eds.) *The use of remote sensing in the modelling of forest productivity.* Pp.3-22.. Kulwer Academic Publishers, Dodrecht Boston London.

Konradzki, J. 1978. *Geografia fizyczna Polski* PWN, Warszawa.

Kovach, W.L. 1993. *MVSP - A Multi Variate Statistical Packade for IBM PC's*, version 2.1. Kovach Computing Services. Pentraeth. Wales. U.K.

Lee, N.J. & Nakane, K.1997. Forest vegetation classification and biomass estimation based on Landsat TM data in a Mountain Region of West Japan In: Gholz H.L, Nakane K., Shimoda H. (eds.) *The use of remote sensing in the modelling of forest productivity.* Pp.159-172.. Kulwer Academic Publishers, Dodrecht Boston London.

Orlóci, L. 1978. *Multivariate Analysis of Vegetation Research.* Dr. W. Junk Publishers. The Hague. Boston.

Sokolowski, A. 1965. *Fragmenta Floristica et Geobotanica.* 27:67-79 (in Polish with English summary).

Szafer, W. & Zarzycki K. 1972. *Szata roslinna Polski.* PWN. Warszawa

Wilkon-Michalska, J., Lipnicki, L., Nienartowicz, A. & Deptula, M. 1998. The role of lichens in functioning of pine forests. In: Czyzewska K. (ed.) *Bio-diversity of lichens*, pp.103-121. Wyd. Uniw.Lodz. Lodz.

Wilkon-Michalska, J. & Nienartowicz, A., Kunz, M., Deptula, M.. In press. *Old land-use maps as basis for the interpretaction of the contemporary structure of forest communities in Zabory Landscape Park.* Phytocenosis, Warsaw.

Remote Sensing in the 21st Century: Economic and Environmental Applications, Casanova (ed.)
© 2000 Balkema, Rotterdam, ISBN 90 5809 096 5

The application of aerial photographs for analysis of landscape diversity in Reserve Grzedy, Biebrza National Park, Poland

M. Falencka-Jablonska
Institute of Forest Research, Warsaw, Poland

M. Kunz
Institute of Geography, Nicholas Copernicus University, Torun, Poland

A. Nienartowicz
Institute of Ecology and Environment Protection, Nicholas Copernicus University, Torun, Poland

ABSTRACT: Grzedy reserve which belongs to The World Biosphere Reserve UNESCO and Biebrza National Park grow on the mineral „island" situated among wide-spread peat-swamps of the Biebrza river middle basin. The island is created by the tops of parabolic and oblong dunes, and forests growing there are under the little influence of anthropopression. Temporal sequence of panchromatic and colour aerial photographs presenting this natural phenomena, and real vegetation maps produced by ground exploration and traditional method of cartography have been used for analysis of riches of plant communities changes and their distribution from 1935 to 1997. Present and archival data about the vegetation give the opportunity to compare the amount of information about the forest landscape heterogeneity which has been achieved from the both sources.

1 INTRODUCTION

The estimation and preservation of biological diversity belong to the most important problems of contemporary ecology. They are analysed at the different level of biosphere organisation. Whittaker (1977) writes about resources diversity and landscape diversity changes on the ecological levels of vegetation organisation. The first factor presenting diversity means the species richness or the structure of species domination in the respective types of vegetation community or in the landscape. The second factor defining diversity shows the landscape tapes of habitat changes and populations existing in them along the gradient of ecological factors. It defines the contribution, distribution, and turnover of the vegetation diversity on the bigger area. Biological diversity understood in this way belongs to the problems concerning heterogeneity of landscape (Turner 1987). Habitat types diversity and plant communities existing in their category which exists on the bigger area, Loehle and Wein (1994) describe as landscape habitat diversity.

To define the diversity or heterogeneity of landscape, number of mathematical formulas are used. Shannon-Wiener's diversity index, which is taken from information theory, belongs to the most often used formula. Index value decreases when the number of plant communities in the landscape increases. For the same number of communities index have maximal value if their contribution in covering the analysed area is the same. Among others Romme (1982), Turner (1987), O'Neil et al. (1988), Baker & Cai (1992) and Geoghegan et. al. (1997) recommend this formula to express the heterogeneity of the landscape.

Biological diversity at each level of biosphere organisation, in it also at the level of landscape, depends on natural and anthropogenic factors. Recently the dependence between bio-diversity and technologies used in different sectors of economy has been researched. The analysis results of such dependencies are the basis for optimising human activity in the biosphere. The aim of such analysis is working out models of sustainable management which will preserve species diversity and ecological structures efficiently. It will make the maintenance and development of human population possible.

The landscape is the level of biosphere organisation at which the counteraction to environment degradation, preservation of species diversity, and creation of harmony between the human being and nature is most efficient. Bio-diversity preservation in such an ecological chain is realised among others by steering the structure, size, and space arrangement of ecosystems, and also by introduction of eco-development rules to socio- and technosphere. By careful structure steering it is possible to optimise not only channels of energy flow and matter cycling, but also to preserve existence of very often rare, vanishing, and preserved species of plants and animals. This is why the interest in the theory of landscape ecology and the possibility of analysing data concerning its diversity and heterogeneity from different sources among other by information theory methods and fractal geometry is still increasing (Milne 1988).

The sources of data to define the landscape heterogeneity are cartographic materials prepared on the basis of area exploration and traditional vegetation cartography. Recently, more and more often the diversity and heterogeneity have been defined on the basis of photo-interpretation and satellite imagery. The amount of information about the spatial structure of landscape gained from these sources is various.

The aim of this research is to define the possibility of using multitemporal aerial photographs to characterise states of forest landscape development. The aim of analysis made is also to define the direction of changes concerning landscape heterogeneity within the compass of 30 years. The object of research is the nature phenomenon - forest reserve Grzedy situated among peat-swamps in Biebrza National Park in north-eastern Poland.

The usefulness of aerial photographs is estimated on the basis of difference in the amount of information about the landscape heterogeneity defined in the Shannon-Wiener's formula obtained from the series of aerial imagery and from the source which gives the biggest number of information - the series of real vegetation maps worked out only on the basis of overground exploration. Changes of landscape variety is analysed in relation to economic actions which took place during the last few decades in the area of the research. It is done in conviction that the analyses will support the elaboration of sustainable sylviculture model for this forest complex.

2 STUDY AREA

Grzedy Reserve has been preserved from 1921. It is the part of The World Biosphere Reserve UNESCO and the Biebrza National Park. This protected area which, is included in the RAMSAR convention is in the north-eastern Poland (Figure 1). In respect to administration the area belongs to Podlaska Province, in respect to economy it belongs to forest inspectorate and Rajgród area which covers 6 preserving districts.

Figure 1. Situation of Grzedy reserve in Poland.

Forests which create the complex grows on the mineral island among peat-swamps of the middle basin of the Biebrza river. The island is created by the tops of parabolic and oblong dunes to 24 m over the peat-swamp level high. Diversified sculpture of the area surface and variety of soils give the rich species of plants. On 15801 ha area of this forest complex there are 13 types of forest habitats. In relation to the overall number of 15 types found on Polish lowlands it is as much as 86,7 %.

Isolated situation of Grzedy and the fact that the water level decrease on this area took place only in the first half of 19 century because of the channel building and peat-swamps drainage, were the reason for weak antropopression in this forest. The most intensive man action took place in the southern part of the researched area where the forests on oblong dunes were changed into fields, meadows, and pastures. On the basis of archaeological studies it was estimated that colonisation on the area started about 2000 years ago. During the last few decades the part of fields was afforested. However, from over 50 years no old-fields and grasslands change into forest complexes has been observed.

In the forests of the researched area much of the ground is covered by deciduous and also by coniferous forests. From the broadleaved species the most important is a birch-tree (*Betula verrucosa* and *Betula pubescens*) and a black alder (*Alnus glutinosa*). Smaller area is covered by growing on peat bogs, on the edge of forest complex different species from the group of willows (*Salix*). Pine trees (*Pinus sylvestris*) dominate among the coniferous species. Spruces (*Picea abies*) are reaches less abundance.

Timber exploitation was rather intensive on this area to the first half of 80s. During the last 10 years it decreased. In the middle part of the complex where trees were the most natural, deforestation was much smaller than on the edge of the forest. In the preserved part of Grzedy which covered 8551 ha, in which forests are the most natural, in 1987-1997 1818 m^3 of timber was felled. In the twice smaller Tajno area of 3134 ha during the same period 2814 m^3 of timber was gained.

According to phyto-sociological classification, existing on the big area preserved forests belong to the association of *Ribo nigri-Alnetum* from the class of *Alnetea glutinosae* and association *Circaeo-Alnetum*, from the alliance *Alno-Padion* and the class *Querco-Fagetea*. Dominating in the middle part oak-linden-hornbeam natural forests belong to *Tilio-Carpinetum*. That association is divided into three subassociations: *stachyetosum*, *typicum*, and *calamagrostietosum*. The factor which causes the differentiation is the level of ground water. In south-eastern part of the Grzedy reserve mixed fresh forest are dominating: *Querco roboris-Pinetum* and *Serratulo-Pinetum*, and fresh pine forests with *Peucedano-Pinetum* from the class of *Vaccinio- Piceetea*. Smaller area is covered by wet pine forests *Molinio - Pinetum* and wet, oak-spruce mixed forests (*Querco-Piceetum*), and pine and spruce forests growing on peat bogs *Vaccino uliginosi-Pinetum* and *Sphagno-Piceetum*.

From non-forest areas the largest fields are meadows classified as *Molinietum medioeuropaeum* and *Arrhenatheretum eliatus*. The area of agricultural plantations is very small. An elk (*Alces alces*) has a big influence on plant biomass.

3 METHODS

In the research of landscape heterogeneity of Grzedy reserve multitemporal aerial photographs from 1967, 1979, 1989, and 1997 were used. The description of aerial imagery is presented in Table 1. Information about the diversity of vegetation worked out from photointerpretation analysis were compared with the results of analysis of maps presenting real vegetation created by different authors. Those cartographic materials comprised data about the vegetation condition in 1935, 1956, 1975, 1990, and 1995.

Table 1. Specification of aerial imagery used for the research.

year	type of imagery	scale
1967	panchromatic	1:17700
1979	panchromatic	1:20000
1989	panchromatic	1:17700
1997	colour	1:26000

The analyses of vegetation structure changes visible on aerial photographs and maps was done in Geographical Information Systems (GIS) Bentley/MicroStation program environment. This software comprises subprograms which support two- and three-dimensional projections and realized all typical GIS functions described by Johnson (1990). Its main functions are drawing, projecting, and managing the data base, and processing and cartographic modelling with the support of information which is in data base. Additionally, subprogram MicroStation Descartes was used to scan aerial photos. This program allows for effective projection of raster imagery, for drawing them, transforming, improving, editing, mosaicking, and geographical positioning. Aerial photographs used for analysis were scanned with resolution 1000 dpi (dots per inch), positioned, and rectified. Because of the small changes of field level on the researched area this procedure can be compared to fitting in numerical model of area; as the result orthophotomap is created. Aerial photos transformed in this way were interpreted. At this stage by the visual method on the screen identical areas were separated. The main interpretation features which differentiate those areas were photo-structure and photo-pattern that is to say dimension and arrangement of picture elements and in some cases the colour. The stereoscopy model of aerial photos was sometimes useful for drawing a border between the areas properly. They supported three-dimensional visualisation of analysed area. Then, photo-interpretation key of selected elements covering the studied area was created. On this base separate areas were characterised and named.

Categories of vegetation which were selected on aerial photos, were conditioned by the way the area was explored and by the features of forests.

Taking into consideration quality, aerial imagery from 1967-1997 differentiate very much. Imagery from 1967-1979 were of the lowest quality; only 7 categories of area covering could be found. The best imageries were from 1989-97 where 11 categories could be selected. Higher richness of vegetation categories resulted from the possibility for separation of deciduous and coniferous forests and among these types thin and thick, and young and old forests could be found. However, for the elimination of additional factors influence, which are the quality and conditions for taking photos, on the analyses result of landscape heterogeneity changes direction within the compass of 30 years, on aerial photographs from 1989-97 only those types of vegetation was taken into account whose selection was possible on photographs from 1967 and 1979. Using MicroStation, from cartographic data presenting real vegetation, numeric maps were created for further analyses of landscape structure. Vegetation types distinguished on maps corresponded to syntaxonomic units selected during analysing the vegetation in Grzedy reserve using standard phyto-sociological method proposed by Braun-Blanquet (1964). Syntaxas distinguished on maps corresponded to classification system of vegetation community in Poland elaborated by Matuszkiewicz (1982). Using the opportunity provided by GIS numeric system, the area covered by separate types of vegetation or syntaxa was calculated taking into consideration all those cartographic and photo-interpretative data.

In virtue of the results concerning analysis related to the area covered by separate communities the index of vegetation diversity of Grzedy reserve was calculated. To express the diversity Shannon-Wiener index was used:

$$H = -\sum_{i=1}^{s} p_i \log p_i$$

where:

$p_i = n_i/N$; n_i= area covered by i-th vegetation type or syntaxa (i=1,2,3,...s); s= number of vegetation types or syntaxa in the year of observation; N=area of whole analysed forest complex.

Calculated values of variety indexes were compared with support of histograms. In the analyses of vegetation heterogeneity in Grzedy reserve it was accepted that Shannon-Wiener's diversity index could be discussed as so called Whittaker β-diversity (1977) and could be the measure for landscape pattern or landscape habitat diversity according to the conception of O'Neil et al. (1988), and Loehle & Wein (1994).

Index value was also accepted as the measure of information about landscape heterogeneity supplied by information sources which in analysis made were aerial photos and real vegetation maps.

For the precise definition of differences between separate landscape states, vegetation maps of Grzedy reserve created on the basis of photo-interpretation were compared with the support of numerical taxonomy method. Comparative features of objects was the area covered by separate vegetation types. Measure of differences between objects was Euclidean distance. Agglomerative classification was realised by minimum variance clustering methods (Orlóci 1978).

In calculations and dendrogram preparation numerical taxonomy package MVSP (Kovach 1993) was used.

Table 2. Differentiation of vegetation types on the basis of aerial photographs.

vegetation category	Year				Change
	1967	1979	1989	1997	1967-97
	%	%	%	%	%
thick forests	75.62	67.64	69.48	82.49	+9.08
thin forests	5.34	1.34	5.92	3.78	-29.21
groups of trees	0.51	0.39	2.79	1.10	+115.16
boskets	2.23	7.51	6.44	0.27	-87.70
clearings	7.82	15.40	-	-	-
trunks	1.68	1.68	-	-	-
willow-alder-birch-scrubs	-	-	14.64	11.32	-
agricultural areas and non-forested	6.80	6.05	0.73	1.04	-84.73
Total	100.00	100.00	100.00	100.00	

4 RESULTS

From among possible to select from aerial photographs types of vegetation thick forests dominated in the landscape of Grzedy reserve in 1967. They were on 75% of the researched area. At this time thin forests constituted over 5%, and agriculture areas and other unforested fields about 7% (Table 2). Twelve years later it was observed that thick forests decreased to 68%, and thin forests to 1,3%. It was visible that the area of clearing increased from 7.8% in 1967 to 15.4% in 1979. That fact was connected with the intensive sylviculture, and increased timber exploitation in Poland in 70s.

Strong changes of possible to distinguish vegetation types shares was defined by comparison of photos from 1979 and 1989. In the last period 79% share of thick forests, lack of clearings, and nearly nine times decrease of agricultural areas and other non-forested fields were stated. Moreover, seven times increase in the areas covered by groups of trees was notified. The lack of clearing areas was caused by the recession of Polish economy and introduction of substantial eco-development to forest economy. The increase in the area covered by tree groups and willow-alder-birch scrubs is connected with continuous natural exploitation of forests on peat bogs existing in Grzedy reserve by man.

Analysis of aerial photos from 1997 (Figure 2) revealed that thick forests constituted 82%, willow-birch-alder scrubs over 11%. Comparing those data with the landscape from 1989, insignificant increase

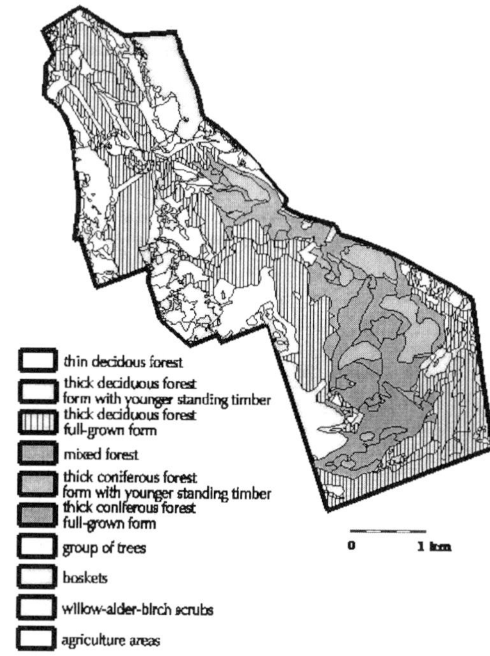

Figure. 2.Interpretation of aerial photographs from 1997.

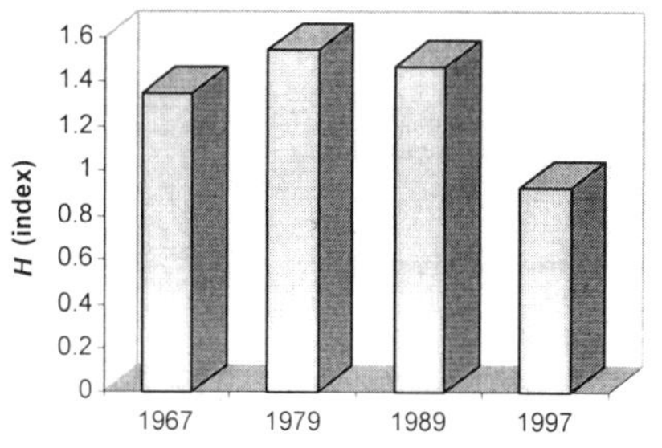

Figure. 3. Pattern diversity changes in Grzedy reserve in 1967-1997 (on the basis of aerial photographs).

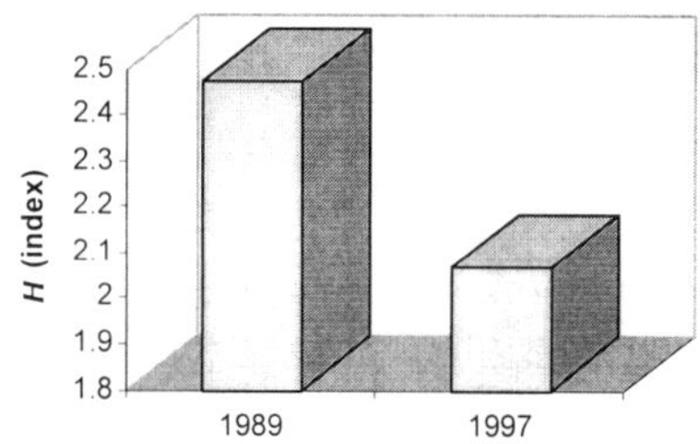

Figure 4. Pattern diversity changes in Grzedy reserve in 1989-1997 with non-reduced data (on the basis of aerial photographs).

in the area of the agricultural and other non-forested fields was notified.

Analysing the changes of area differentiation of vegetation in Grzedy reserve, the comparison of area shares of separate categories in 1967-1997 was also done. Generally the increase in area share of thick forests, group of trees and willow-birch-alder scrubs, and substantial decrease in area share of boskets, clearings and trunks, and agricultural and non-forested fields was observed (Table 2). Heterogeneity index of forest landscape research calculated with the Shannon-Wiener's formula at the beginning showed increase and then decrease in the following years. It reached the biggest value 1.5361 in 1979 which was the period of frequent disturbances caused by clear-cutting. In 1989 index value was a little lower (1.4607), and in 1997 much lower (0.9249) in comparison to maximum value in 1979 (Figure 3).

The usage of entire information gained from aerial photos from 1989 about landscape heterogeneity caused increase of variety index from 1.4607 to 2.4707, which is 1.0100. In case of the photo from 1997 it was increase from 0.9249 to 2.0703 which is 1.1454 (Figure 4).

The value 2.0703 for landscape state was reached taking into consideration 11 vegetation types which were possible to distinguish on aerial pictures.

This index value of diversity is considerably lower than the value 3.2046 calculated on the basis of real vegetation map from 1995 created only on the basis of area exploitation. On this map 43 syntaxas were distinguished. Many of them could be distinguished on the basis of species composition, ground vegetation, which in case of aerial photographs presenting canopy of overstorey would not be possible.

In Poland less intensive wood exploitation at the end of 80s, which influenced heterogeneity of forest landscape, was revealed substantially on dendrogram (Figure 5) presenting the differences in the structure of plant cover in Grzedy reserve.

On dendrogram there are two distinct clusters. The first defines landscape state from 1967 and 1979, the second from 1989 and 1997. Analysing data from Table 2, it can be stated that features differentiating two subgroups were clearings and trunks, and willow-alder-birch scrubs. Trend in diversity index changes defined from real vegetation maps was similar to the one achieved by using multi-temporal aerial photographs. In case of date gained by overground exploration it increased from 1935, and it reached its maximum value in 1990 (Figure 6).

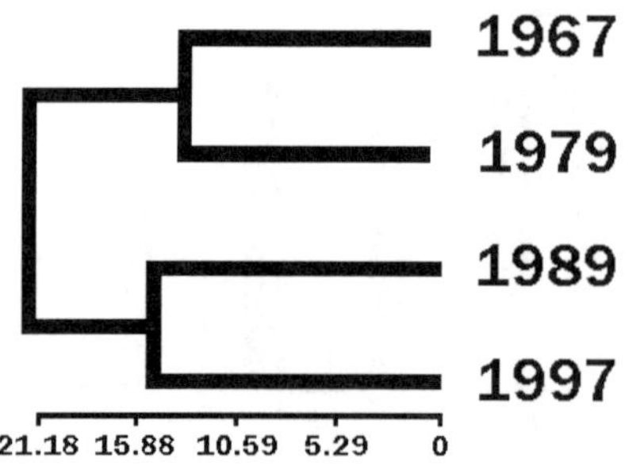

Figure 5. Hierarchical agglomerative classification of separation of pictures from aerial photographs.

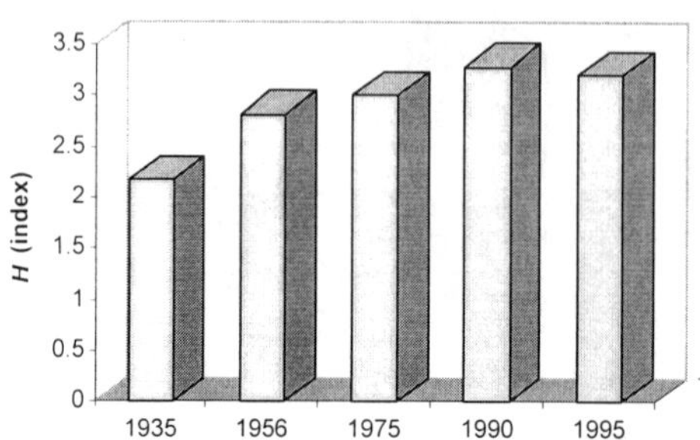

Figure 6. Pattern diversity changes in Grzedy reserve in 1967-1997 (on the basis of traditional cartography).

Likewise in case of aerial photos, in the last period its decrease was visible.

Difference in the course of changes of the index in time between the series of aerial photos and traditional produced vegetation maps refer to the period of the maximum. In case of aerial photographs maximum value was reached in 1979, and in case of real vegetation maps in 1990. Moreover, in case of traditional maps the change of diversity index value between the last two landscape states is not as big as in case of aerial photographs.

The change of diversity index values from maps created only by overground exploration are connected with the processes of succession.

Main direction of changes in vegetation dynamics are progressive changes of alder forests *Ribo nigri-Alnetum* from the class *Alnetea glutinosae* into ash-alder forests from the association *Circaeo-Alnetum* from the class *Querco-Fagetea*, and changes in forests habitat of oak-lime-hornbeam forest into mixed and pine forests from class *Vaccinio-Piceetea*. The reasons for those processes are changes of ground water level and introduction of pine-tree (*Pinus-sylvestris*) in the past into the habitat of deciduous forests. However, on the basis of the series of real vegetation maps it could be stated that during each development state of analysed landscape oak-linden-hornbeam habitats (*Tilio-Carpinetum*) played the most important part.

5 CONCLUSION

Grzedy reserve as a unique natural phenomena has aroused interest in ecologists and in scientists of forestry for a long time. Because of that fact, many maps presenting development states and the structure of its vegetation were created in the past. Today they are the basis for considerations related to the rate and directions of natural ecosystem structures changes. Such data concerning the complexity and succession processes can be the model for the comparison for other forest complexes even more transformed ecologically by human economic activity. Analysis made proved high natural values of Grzedy reserve. There is the great quantity of vegetation types. There are 43 types of associations and subassociations, and more than 10 of habitat types of forest there. From the comparison of mutitemporal aerial photographs and real vegetation maps, apart from the above mentioned facts, the following conclusions may be drawn:

1. most information about species composition and spatial distribution of plant community provides overground area exploration; however, photo-interpretation is a very effective tool which gives opportunity, although with lower precision, to achieve data about the heterogeneity of vegetation from a big area,
2. there exists a big similarity of changes direction in forest landscape during the following years, defined on the basis of aerial photographs and vegetation maps prepared by overground exploration,
3. amount of information gained from aerial photos depends very much from the quality of photographic equipment and weather conditions during taking photos,
4. the highest resolution of colour aerial photographs makes it possible to distinguish more types of vegetation on the landscape than panchromatic,
5. overground exploration provides information about vegetation differentiation at the level of associations and within their compass. Photo-interpretation gives the opportunity to define landscape vegetation distribution at higher levels of syntaxonomic classification, e.g. at the level of alliances, orders, and classes,

6. in delimitation of plant communities on the basis of photo-interpretation, the most important part is played by differences in the composition of tree species. In case of Grzedy reserve big amount of species with wide ecological range e.g. birch-tree and pine-tree in many syntaxonomic units, lowered the diversity level defined from an aerial photograph,
7. economic activities e.g. cutting out trees, reforestation have substantial influence on the level of forest landscape diversity. Increase in diversity by cutting out trees corresponds to among others Franklin & Forman (1987) and van der Maarel's (1993) observations stating that disturbances increase and support diversity,
8. very important sources of information about heterogeneity of landscape are historical maps and archival aerial photographs,
9. obligatory in phytosociology classification systems which at different periods of time (of that field of knowledge) emitted, with different precision, various number of syntaxas, had influence on the value of variety indexes acquired from maps created on the basis of overground exploration,
10. MicroStation program and numerical taxonomy package MVSP are effective tools for processing and presenting data concerning the landscape diversity gained from different sources.

6 REFERENCES

Baker, W.L. & Y. Cai 1992. *The r.le programs for multiscale analysis of landscape strukture using the GRASS geographical information system.* Landscape Ecology 7:291-302.

Braun-Blanquet J. 1964. *Pflanzensoziologie.* Grundzuge der Vegetationskunde. 3. Aufl. Springer. Wien - New York.

Franklin, J.F. & Forman R.T.T. 1987. *Creating landscape patterns by forest cutting: Ecological consequences and principles.* Landscape Ecology 1:5-18.

Geoghegan J, Wainger L.A. & Bockstael N.E. 1997. *Spatial landscape indices in a hedonic framework: an ecological economics analysis using GIS. Ecological Economics 23*:251-264.

Johnson, L.B. 1990. *Analyzing spatial and temporal phenomena using geographical information systems.* Landscape Ecology 4:31-43.

Kovach, W.L. 1993. *MVSP - A Multi Variate Statistical Packade for IBM PC's*, version 2.1. Kovach Computing Services. Pentraeth. Wales. U.K.

Loehle, C. & Wein. G. 1994. *Landscape habitat diversity: a multiscale information theory approach.* Ecol. Modelling 73:11-329.

Matuszkiewicz, W. 1982. *Przewodnik do oznaczania zbiorowisk roslinnych Polski.* PWN, Warsaw.

Milne, B.T. 1988. *Measuring the fractal geometry of landscapes.* Appl. Math. Computat. 27:67-79.

O'Neil, R.V., Krummel J.R., Gardner R.H., Sugihara G., Jackson B., DeAngelis D.L., Milne B.T., Turner M.G., Zygmunt B., Christensen S.W., Dale V.H., Graham R.L., 1988. *Indices of landscape pattern.* Landscape Ecology 1: 153-162.

Orlóci, L. 1978. *Multivariate Analysis of Vegetation Research.* Dr. W. Junk Publishers. The Hague. Boston.

Romme, W.H. 1982. *Fire and landscape diversity subalpine forests of Yellowstone National Park.* Ecol. Monographs 52:199-221.

Turner. T. 1987. *Landscape heterogeneity and disturbance.* Springer-Verlag, New York.

Whittaker, R.H. 1977. *Evolution of species diversity in land communities.* In: M.K. Hecht, W.C. Steere, B. Wallace (ed.), *Evolutionary Biology*, vol. 10. Plenum Publishing Corporation.

van der Maarel, E. 1993. *Some remarks on disturbance and its relations to diversity and stability.* J. Veg. Sci. 4:733-736.

Remote Sensing in the 21st Century: Economic and Environmental Applications, Casanova (ed.)
© 2000 Balkema, Rotterdam, ISBN 90 5809 096 5

Drought monitoring based on thermal and microwave satellite data

K. Dabrowska-Zielinska, A. Ciolkosz, M. Gruszczynska & W. Kowalik
Institute of Geodesy and Cartography, Remote Sensing and Spatial Information Centre, Warsaw, Poland

F. Kogan
NOAA/NESDIS, Satellite Research Laboratory, Washington, D.C., USA

ABSTRACT: The Remote Sensing and Spatial Information Centre - OPOLIS in co-operation with NOAA/NESDIS Satellite Laboratory in Washington and ESA has conducted multidisciplinary studies on impacts of environment on agriculture with application of remote sensing data for several years. Research activities included vegetation monitoring using NOAA/AVHRR as well as high resolution ERS-2.SAR data. Recently, a successful attempt was launched to use low resolution AVHRR data for early drought detection in Poland. NOAA images have been used to derive Temperature Condition Index (TCI) characterising the status of crop development. TCI index represents the relation between actual weekly value of temperature and the temperature, which occurred during both favourable and stress crop growing conditions in the last decade. The differences in TCI values and their spatial distribution for the whole country give information where and when the drought has started and how intensive it was. The research also demonstrated that backscattering coefficient calculated from ERS-1/2.SAR data characterises soil moisture condition and can be used for drought detection. Our research has been conducted for cereals and validated through detailed ground measurements. This paper presents the results obtained for 1992 (drought conditions), 1994 (mean moisture conditions) and 1998 (favourable moisture conditions) for cereals.

1 INTRODUCTION

Drought is one of the major disasters affecting agriculture in Poland. During the drought years the production drops significantly. At the time of the current decade, drought which occurred in 1992 reduced grain yield (24.0 q/ha) for about 20% comparied to 1991 (32.0 q/ha). Therefore, early drought detection and monitoring the area and drought expansion is considered as a very important component in Early Warning System.

In this research of drought detection and monitoring the long term database of low resolution satellite data (AVHRR/NOAA) has been applied. The AVHRR data were collected from the Global Vegetation Index (GVI) data set during the period 1985-1998. The high resolution Landsat TM data were used to differentiate crop and natural vegetation areas, and also to identify areas planted with different crops. This information was used to calculate regional weighted average of AVHRR-based Temperature Condition Index TCI.

The ERS-1/2.SAR.PRI data were obtained from the European Space Agency for PP-PL4, AO2-PL102 and AO3-138 projects.

The different crop growing conditions have been observed in the whole country during the entire period of research. The situation of vegetation and its development conditions were very well represented by remote sensing derived index TCI. Different distribution of this index over time and area indicated that this method would be very useful for crop condition assessment, and drought detection and monitoring. Also, backscattering coefficient calculated from ERS-1/2.SAR data (Laur 1992) has been used to characterize soil moisture conditions under cereals, which occurred in one of the regions localized in central part of Poland.

The results of the research obtained from thermal and microwave data for drought, mean, and favourable vegetation growing conditions (years: 1992, 1994, and 1998 respectively) have been presented in the paper.

2 RESULTS

2.1 *Application of NOAA data*

The research project on crop growing condition assessment and drought detection has been carried out in the framework of co-operation with the NOAA/NESDIS Satellite Laboratory in Washington. This research applied low-resolution data - GVI derived from GAC product. On the basis of GVI Temperature Condition Index (TCI) has been calculated from BT values (Kogan 1997) for each week in the considered period. After smoothing, BT values were normalised to the range of their change in the database as follows:

$$TCI=100*(BT_{max}-BT)/(BT_{max}-BT_{min}) \quad 1$$

where: BT, BT_{max} and BT_{min} are the smoothed weekly BT, its multiyear absolute maximum and minimum respectively derived from NOAA channel four (Ch4) data. TCI values were aggregated to represent each region of Poland through the growing season (Dabrowska-Zielinska et al. 1998). This index changes from 0 to 100 reflecting changes in vegetation growing conditions. The high TCI values indicate low temperature and favourable crop growing conditions while low TCI values indicate dry conditions, unfavourable for development of crops during summer. The research has proved that the low values of TCI in spring represent favourable conditions for vegetation as the brightness temperature of vegetation is close to maximum, which is optimal for this period. Low TCI values during the maximum growth of vegetation (22 week) indicate stress in crop development expressed by high values of brightness temperature close to maximum for this period. Figure 1 shows TCI values from week 18 to 28 (the most important period for development of cereals) for three considered years 1992, 1994, and 1998 for the chosen region localised in central part of Poland.

Figure 1 shows that severe drought occurred in 1992. The high TCI values in spring indicated cold weather and low BT in this part of Poland. The large deficit in water appeared in 20th week of the year (in the middle of May) when TCI values started to diminish from 30 to close to zero. Such unfavourable conditions have resulted in low yield of cereals in the whole country. Different growing conditions occuried in 1998. TCI values were low at the beginning of the growing season that indicated warm and moist growing conditions but from week 21 TCI started to increase, up to high values during

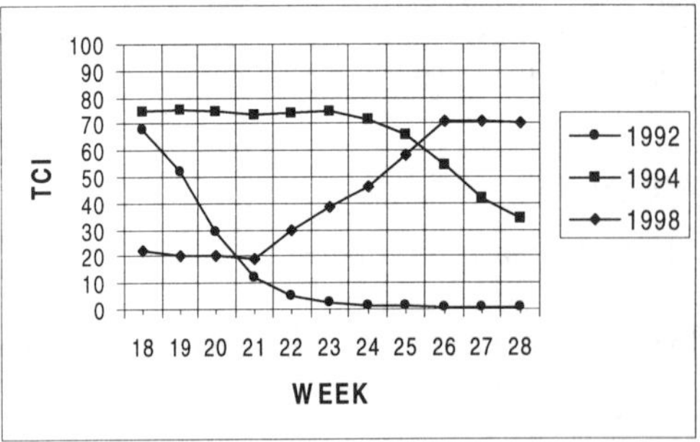

Figure 1. Temporal TCI profiles for chosen region.

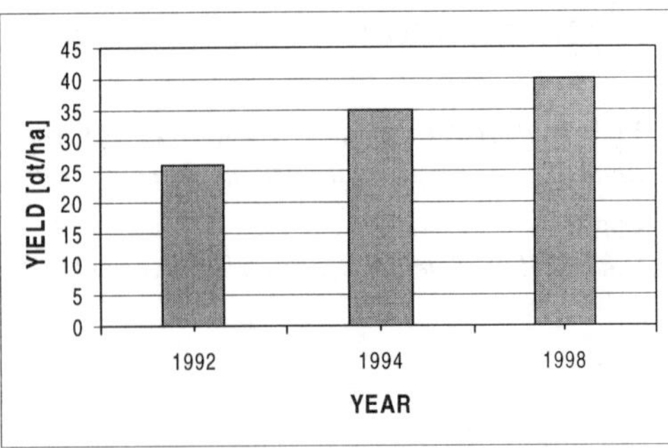

Figure 2. Yield values for cereals for chosen region.

summer. Low values of TCI in spring and high values of TCI in summer yield favourable crop growing conditions giving very high cereal production, one of the best in the last decade. The spring in 1994 was very promising for crop development because of high temperature of crops and appropriate soil water content accumulated during winter. From week 24 of this year (in the middle of June) TCI values started to diminish, up to the value of 30 which resulted that the growing conditions were worse in 1994 than in 1998, but better than in 1992. Considering the crop yield, 1994 has been chosen as an average year from the last decade. Figure 2 presents yield values for cereals for the chosen region for three considered periods: drought in 1992, mean in 1994, and favourable in 1998.

2.2 *Application of ERS-1/2.SAR data*

The images from ERS-1 and ERS-2 satellites covering the western part of Poland taken in 1992–1998 growing seasons have been analysed and backscattering coefficient σ° values have been compared to the parameters, which characterised soil–vegetation conditions.

Throughout each of the growing seasons, and simultaneously to ERS satellites overpasses the measurements of soil and vegetation parameters were carried out at 30 plots of cereals which were represented by point measurements. These parameters were volumetric soil moisture (TDR method), Leaf Area Index (LAI-2000 Plant Canopy Analyzer), vegetation moisture and biomass (laboratory). Also crop type, actual development stage, and growing condition have been monitored.

The impact of ground truth variables on microwave signals expressed by backscattering coefficient σ° has been examined in order to develop the method of soil moisture assessment in agricultural areas covered with cereals.

It was found that the relationship between

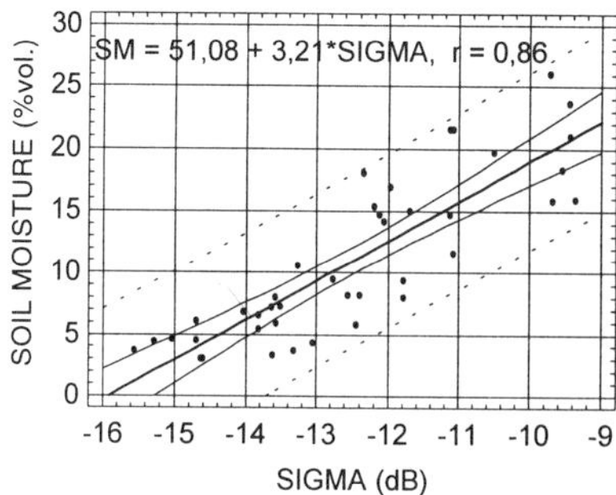

Figure 3. Correlation between backscatter and soil moisture for cereals for LAI>3.

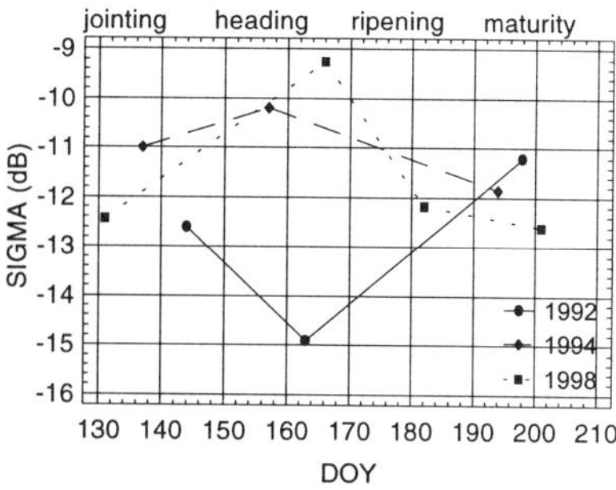

Figure 4. Temporal backscatter profiles for spring wheat.

backscattering coefficient σ° and soil moisture is strongly dependent on vegetation surface roughness. It was assumed that the roughness of vegetation might be characterised by Leaf Area Index (LAI) (Dabrowska-Zielinska et al. 1995). This index characterises the geometry and density of a crop. LAI values changed during development of cereals from low at the beginning of growth, reaching maximum at their heading stage. Thus to assess the soil moisture conditions the roughness of vegetation covering the examined area should be known.

The next step was to create the classes, which characterise the similar vegetation surface roughness. The classes were found within a certain range of LAI values. For each of the LAI classes, which represent similar vegetation surface roughness the relationship between soil moisture and backscattering coefficient σ° has been considered. The best correlation between σ° and soil moisture for cereals has been found for high LAI values (maximum biomass) when crops were in their heading stage (r=0,86). Figure 3 presents this relation for cereals for LAI values grater than 3.

There is a linear relation between soil moisture and σ°, the higher soil moisture the higher backscatter values. For the remaining two LAI classes (<2 and 2-3) correlation coefficient was lower (0,81 and 0,84 respectively). For growing stages from jointing to heading the vegetation moisture was constant equal to 80 – 90%. Then, the vegetation moisture decreased from graining to maturity stage. In each of these stages vegetation moisture values were also constant and not influence backscatter signal.

It is possible to detect and monitor drought conditions over area covered with this type of crop using algorithms developed for cereals (Gruszczynska & Dabrowska-Zielinska 1998). Figure 4 presents temporal backscatter profiles for spring wheat for three various soil moisture conditions: drought in 1992, mean in 1994 and favourable in 1998.

The lowest backscatter values occurred in 1992, and the highest in 1998. Figure 5 and Figure 6 present soil moisture and vegetation growing conditions expressed by LAI for the same years.

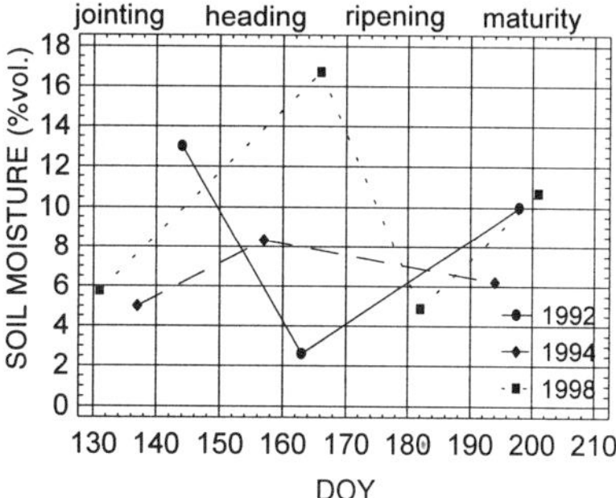

Figure 5. Temporal soil moisture profiles for spring wheat.

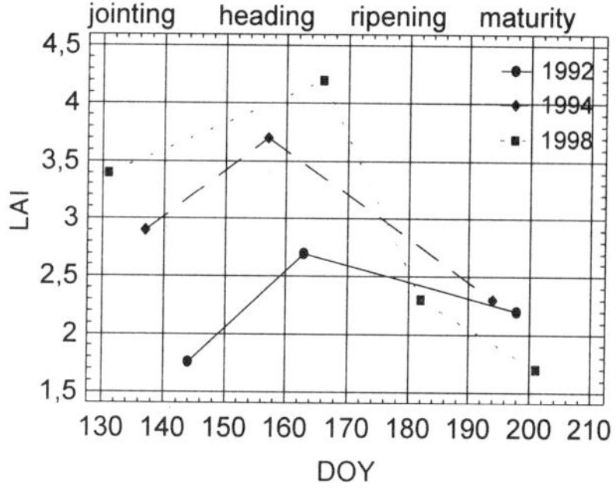

Figure 6. Temporal LAI profiles for spring wheat.

From these two last figures it can be noted that the most unfavourable moisture conditions occurred in 1992 during the heading stage which is the crucial period for cereals development. Such dry conditions caused low LAI values and low yield of cereals. In 1998 the moisture conditions were very good which resulted in high LAI values and high yield of cereals. The year 1994 can be considered as the average year from the last decade regarding crop yield values in Poland. Figure 5 and Figure 6 indicated also that moisture and vegetation conditions in 1994 were better than in 1992 but worse than in 1998. Also, backscattering coefficient values were mean for 1994, the lowest for 1992, and the highest for 1998 during the heading stage. These results proved that radar data reflects soil moisture conditions very well and could be used for drought detection and monitoring over large areas.

3 CONCLUSIONS

Temperature condition index TCI calculated from NOAA and backscattering coefficient calculated from ERS.SAR were considered as good drought indicators. The differences in their values and their spatial distribution give the information where and when the drought has started and how intensive it has been. Clouds very often obscure Poland due to unfavourable weather conditions especially during the growing season. Radar data give the possibility to measure and monitor soil moisture during cloudy conditions. These two methods are very promising for early drought detection and monitoring.

REFERENCES

Dabrowska-Zielinska, K, M. Gruszczynska, K. Moran, M. Janowska & K. Stankiewicz 1995. Visible, infrared and microwave data as a source of information about vegetation status. *Proc. of the Meteorological Satellite Data Users Conference*, Winchester, U.K.

Dabrowska-Zielinska, K., F. Kogan, A. Ciolkosz, M. Gruszczynska, U. Raczka, W. Kowalik and R. Jankowski 1998. New method of drought detection based on NOAA satellites and it's impact on Polish agriculture. *Proc. of the ASPRS-RTI 1998 Annual Conference*, Tampa, USA.

Gruszczynska, M. & K. Dabrowska-Zielinska 1998. Application of microwave images from European Remote Sensing Satellites (ERS-1, ERS-2) for soil moisture estimates. *Journal of Water and Land Development*. No 22:7-17.

Kogan, F.N. 1997. Global drought watch from space. *Bulletin of the American Meteorological Society*. Vol. 78, no 4.

Laur, H. 1992. Derivation of backscattering coefficient in ERS-1.SAR.PRI products. *ESA Bulletin*, ESRIN, Italy.

Remote Sensing in the 21st Century: Economic and Environmental Applications, Casanova (ed.)
© 2000 Balkema, Rotterdam, ISBN 90 5809 096 5

Irrigated crop identification and surface estimation with Landsat TM imagery in La Mancha oriental region, Spain

C. Martínez Beltrán, A. Calera Belmonte, J. R. Ruiz Gallardo & J. Sánchez García
Instituto de Desarrollo Regional, Sección de Teledetección y S.I.G., Universidad de Castilla-La Mancha, Albacete, Spain

ABSTRACT: This paper presents a methodology for supervised classification in a semiarid zone such as La Mancha (Spain) to identify irrigated crops with Landsat TM imagery. The process developed for discrimination is based on the different spectral response and its different temporal evolution according to the crop phenological evolution. The classification procedure has followed multitemporal supervised classification by combining maximum likelihood algorithms, decision-tree criteria and context classifiers, which operate on scenes obtained on different dates. It has produced an improvement when we have introduced other dates, obtaining a global precision from 82.7% to 92.7% for classification accuracy. To consider the level of precision in the estimated crops areas from the classified image, we compare these areas with the calculated areas from the digital rural cadastre, through the intersection of the land use raster map and the digital rural cadastre vector map in a Geographical Information System.

1 INTRODUCTION

In La Mancha (Spain) as in other world regions with semiarid or arid climate, water is a valuable natural resource, whose principal utilized volume is for agriculture. In the study area we estimate a global annual water consumption of 425 hm ^{3}; 400 hm ^{3} of which are for irrigation, and only 25 hm ^{3} for urban supply and industrial uses for a population of 275000 habitants (Santa Olalla et al, 1999). These resources are mainly obtained by the exploitation of subterranean waters through wells, whose piezometer levels have been decreasing over the last 20 years.

Aiming at the sustainability of groundwater use, it is urgent to adopt measures towards the rational exploitation of the aquifer. Therefore it is essential to find the necessary tools for shared integrated water management. The identification of irrigated crops and their spatial distribution on the aquifer will improve management of the available water resources for agriculture.

Remote Sensing has become a very useful tool in projects with classification of vegetal coverage, mainly if we work with large surfaces where fieldwork needs a great amount of economical and human resources. Remote Sensing is integrated in a Geographical Information System when the elementary unit is the cadastrial subplot (Calera et al, 1999). This information layer is the Digital Rural Cadastre (RDC), which enables the spatial identification of crops.

In this study we suggest a multitemporal classification methodology based in Landsat TM images, whose spatial resolution is adapted to the size of the different plot that form our countryside. This methodology is mainly based on irrigated crop discrimination, which allows us to know the crop spatial distribution in each crop season and makes available follow- up.

The most important parameter to estimate groundwater abstraction is the surface occupied by each class of irrigated crop, with different water requirements. The hydric consumption is calculated in the form S_iC_i, where S_i is the crop surface and C_i is the assigned consumption.

Thus we have developed a procedure to compare surfaces estimated both from imagery classification and from RDC. RDC is the best source to obtain estimations of surfaces. The process we carry out will demonstrate the importance of the context classifier.

1.1 *Study area*

This study centers on a wide region of 8000 km^2 in La Mancha, situated on the 08-29 aquifer. The 08-29 hydrologic system is located in the Southeastern part of the Iberian Peninsula, and more specifically in the eastern section of La Mancha, fig.1. In this 08-29 aquifer, there are 100000 hectares of irrigated crops with recent irrigated technology. A pilot zone exists with 32228 ha of surface where fieldwork has been intensified. The main orographic characteristic is a

Figure 1. Aquifer 08-29 in the eastern part of La Mancha (Spain) and the pilot zone.

level plain whose medium height between 650m and 800m.

2 METHODOLOGY APPLIED

In the methodology we have carried out (figure 2), the classification procedure has followed multitemporal supervised classification by combining maximum likelihood algorithms, decision-tree criteria and context classifiers, which operate on scenes obtained on different dates. These dates are 15th April, 24th May, 25th June, 11th July, 12th August and 19th September of 1996.

We have begun with a supervised classification based on the maximum likelihood method with field work in the study area, utilizing only two dates, 24th May of 1996 and 12th August of 1996, because these dates completely cover the study area and because we need one image of spring and another of summer to discriminate the maximum number of crops which belong to both seasons. We have utilized band 5 and the NDVI as a combination of bands 3 and 4.

Although the objective is the discrimination of irrigated crops, when we select training areas to carry out the supervised classification by the maximum likelihood method, we have to consider dry crops also in order to avoid confusion with irrigated crops (Congalton et al, 1998). The different crops to discriminate are shown in figure 3. But with only two dates, there are some crops that can't be discriminated in this classification by the maximum likelihood method, because their phenological development is very similar to others. An example would be garlic, which belongs to Spring irrigated crop type B (RegpriB), because its phenological evolution is very similar to the dry grain phenological development on dates in May and August (which are the dates utilized in the classification) and therefore, its NDVI.

However, the crop class Spring irrigated crop type B (RegriB) will be classified with the application of decision-tree classifiers, introducing new suitable dates, as can be observed in figure 2.

2.1 *Analysis*

From this initial classification we perform an error matrix that is shown in table 1, comparing, pixel by pixel, the classified image with the fieldwork data.

The elementary unit is the cadastrial subplot, because a plot can be divided into several subplots and each one of them is identified with a property register. In this matrix, the statistical index kappa $\kappa = 0.760$ and the classification accuracy reaches 82.7%.

2.2 *Estimation of the principal confusion between crop classes.*

The fieldwork done in Barrax, Tarazona and La Gineta plus the previous error matrix has allowed us to check the classified image, and we can distinguish the crops that produce the principal confusion. Then, a succession of decision-tree classifiers are applied, based on the adequate NDVI to reduce the confusion produced, as much by omission errors as commission errors between the crop classes already classified, and for discriminating other new classes.

2.3 *Application of decision-tree classifiers*

Using the phenological evolution of the different crops and the NDVI temporal evolution of each one of them, the decision-tree classifiers will decrease the confusion between crops by sequentially introducing the NDVI of the most suitable date to discriminate one crop from the rest. We have 6 differ-

Table 1. Error matrix from the supervised classification with the maximum likelihood method with two dates

	Classified image →							
Reference data ↓	Regpri	Regve	Regnoes	Not irrigated	Pixels without classify	Total	Producer's accuracy (%)	Omission error (%)
Regpri	4619	68	907	3334	288	9216	50.1	49.9
Regve	1	10342	164	531	37	11075	93.4	6.6
Regnoes	3	66	5387	15	16	5487	98.2	1.8
Not irrigated	5	17	32	5997	11	6062	98.9	1.1
Total	4628	10493	6490	9887	352			
User's accuracy (%)	99.8	98.6	83.0	60.7				
Commission error (%)	0.2	1.4	17.0	39.3				

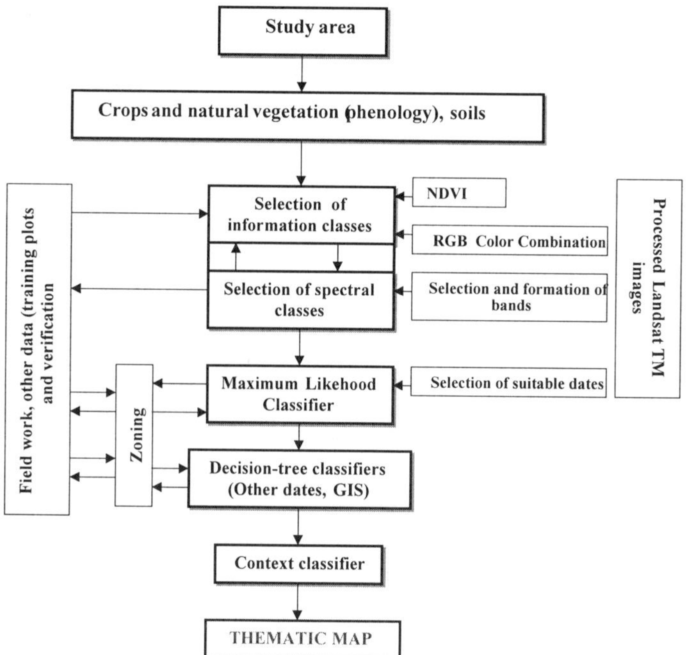

Figure 2. Classification methodology

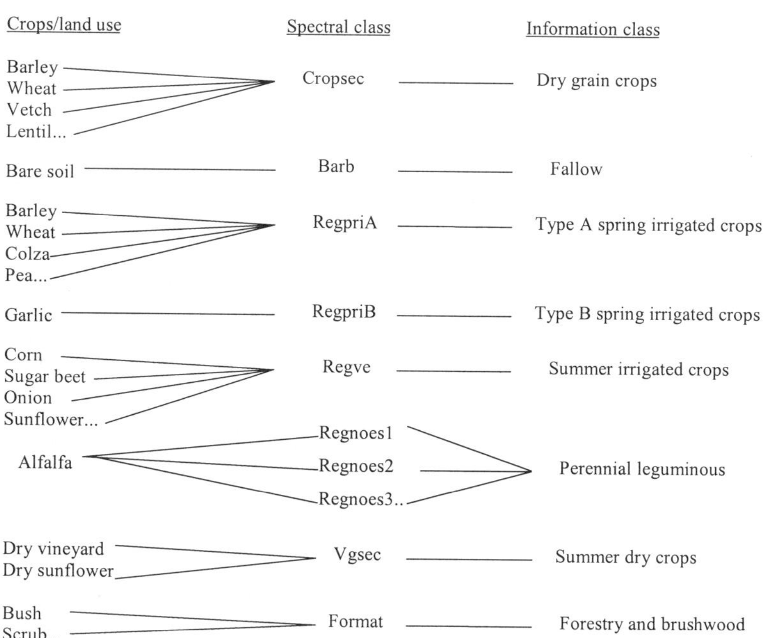

Figure 3. Crops, spectral classes and information classes in the classification

ent dates in 1996 and in each one of the classifiers the needed dates are introduced. The following section illustrates how a decision- tree classifier works.

2.3.1 *An example of a decision-tree classifier.*

This decision- tree classifier discriminates a new class that is garlic, which belongs to RegpriB. At the beginning it was impossible with only two dates, because the phenological states are very similar for RegpriB and Cropsec in May and August, so their NDVI were easily confused too. Due to this, we now introduce the images whose dates are 15th April of 1996 (15Ap96), 24th May of 1996 (24My96) and 25th June of 1996 (25Jn96), because these dates best separate these two crops. These dates, show the greatest difference in their NDVI as shown in figure 4:

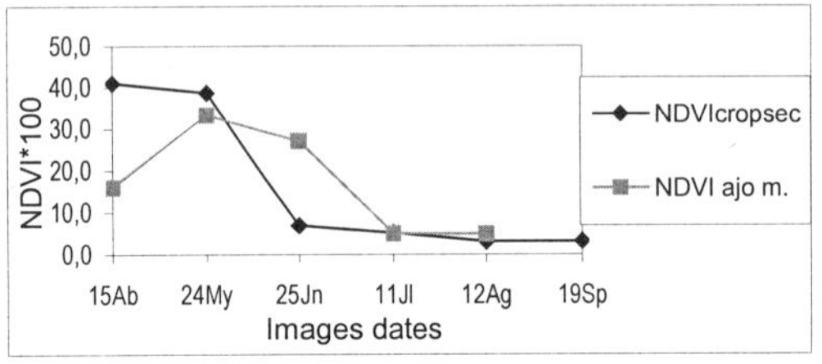

Figure 4. Evolution of NDVI*100 for dry grain (Cropsec) and garlic (RegpriB)

The tree decisions are shown in figure 5:

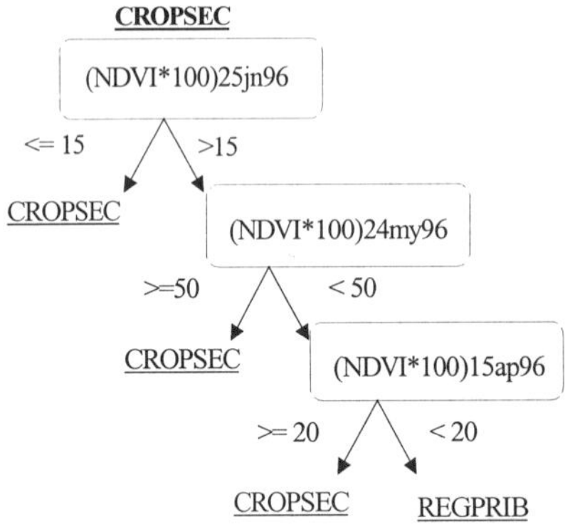

Figure 5. Classifier applied to discriminate as garlic those pixels classified as dry grain

In all decisions shown in figure 5, we have tried to discriminate those pixels that are really garlic and have been classified as dry grain, taking up a succession of decisions according to the NDVI evolution that is shown in figure 4. The first decision is that June NDVI*100 was higher than 15. We have begun with this condition because dry grain (cropsec) is harvested at the end of June, so it has decreased to its minimum NDVI*100 and lower than 15, as is shown in figure 4.

In the following condition, the NDVI*100 of May is utilized, because this is when dry grain reaches its highest development, thus its maximum NDVI. Because of this, those pixels, whose NDVI was superior to 50, will be classified as dry grain, while those pixels that have been confused with dry grain and are really garlic have an NDVI*100 lower in May, as we can observe in figure 4. In the same way, in the NDVI*100 evolution for dry grain, the maximum value reached in May is around 40, but we have utilized a value a little higher because in other zones the dry grain reaches higher values, for example, when it frequently rains. If would be pixels that are dry grain and wouldn't comply with this condition applied for May, they will be perfectly classified in the next decision. This last condition applied is that NDVI*100 in April must be lower than 20 to be garlic. If it is higher than this value, it will be classified as dry grain, as can be observed in the NDVI*100 evolution for these crops.

2.4 *Results validation*

From the last classification obtained, once we had applied all the decision-tree classifiers for the study zone, we performed another error matrix, shown in table 2. In this case, we have obtained the statistical index kappa $\kappa = 0.901$ and the classification accuracy reaches 92.7 %

2.5 *Application of a context classifier.*

In a classification there are color spots, more or less homogeneous, which are sprinkled with pixels that belong to other categories, disturbing the geographical unit and impeding the interpretation results. This is because the majority of assignation criteria don't consider ownership by neighbors, or the complexity of the landscape, which is related to the apparition of edge pixels. To solve these problems, a filter can be applied, in this case a median filter. The last objective is to assign isolated or insignificant pixels, thus allowing a clearer spatial distribution of our categories. For that, this filter compares that class assigned to each pixel with the neighboring class, through a mobil window of 3×3 pixels. The median filter is applied to the classified image to complete the areas without classifying or classified as a different crop, within well known crop classes (Bober M.L. et al, 1996). The median filter is based on replacing the central pixel by the neighboring pixel median. This filter is less sensitive to extreme values and contains fewer modifications of data than the arithmetic mean.

Table 2. Error matrix done after the entire classification process has been applied.

	Image classified →							
Reference data ↓	Regpri	Regve	Regnoes	Not irri-gated	Unclassified Pixels	Total	Productor accuracy (%)	Omission error (%)
Regpri	7855	147	103	1027	84	9216	85.2	14.8
Regve	73	10393	142	453	14	11075	93.8	6.2
Regnoes	37	91	5330	26	3	5487	97.1	2.9
Not irrigated	29	65	8	5951	9	6062	98.2	1.8
Total	7994	10696	5583	7457	110			
User accuracy (%)	98.2	97.2	95.5	79.8				
Commission error (%)	1.8	2.8	4.5	20.2				

3 SURFACE ESTIMATION. INTEGRATION OF A CLASSIFIED IMAGE IN A G.I.S.

The information obtained by Remote Sensing through a classified image, has to be processed in order to expedite its integration into a Geographical Information System (GIS), together with other information layers. These layers, with vectorial or polygon format, have cadastrial identification (cadastral subplots) and also have a geographical location, allowing them to be grouped into exploitations.

This way, a process to extract information from raster layer and to integrate it in a vector layer has been developed. This process has been applied to the crossing of the classified raster map and the digital rural cadastre vector map. The objective of this crossing is to do a comparison between surfaces estimated from cadaster and from classified image.

3.1 *Crossing of the classified raster map and the digital rural cadaster vector map*

The integration of the information obtained by Remote Sensing in a G.I.S is done by the intersection of a polygon layer (the digital rural cadaster) with the classified raster map previously obtained. The elementary unit is the cadastrial subplot and its cadastrial identification is the attribute that allows it to relate with other layers, and to assign to each polygon other attributes like: administrative owner, surface, water source, irrigation systems, etc.

With the superposition of both layers, information can be extracted for each plot and can be identified the crop or crops assigned in the classified image, thus obtaining the number of pixels that geographically intersect with each plot. These operations have been implemented over Arc View, obtaining a spatial database, i.e. a vector layer that integrates the raster information. The process inputs are, besides the irrigated crop map and the polygon layer corresponding to the digital cadastre, a table with the description of the codes assigned to each class in the raster map. The code table is used during the process for designing the database linked to the output layer, so that each one of the created fields in this output layer has an identifier obtained as a function of pixel values in the raster map.

The output is a polygon layer with the same geographical base as the input polygon layer. The difference is the alphanumeric database linked to the output layer, which will be built by those fields in the input layer selected by the user, the fields with the cadastral identification, and also a number of fields equal to the number of classes appearing in the raster map. The information stored in these fields is, for each subplot, the cadastral identification such as polygon, plot and subplot, surface, perimeter and the associated crop classes of the classified image and the surface occupied by them (Calera et al, 1998).

The objective, at the beginning, is to estimate the surface occupied by each one of the irrigated crop classes, and to compare this with the estimations according to the digital rural cadastre. To do this, when the crossing has been carried out, we sequentially consider a number of criteria:

1. To select those subplots with only one crop, which is when a crop occupies 80% or more of the total plot surface. This is to eliminate those subplots with several crop classes.

2. To eliminate those plots that aren't for agrarian use, for example, we have applied the following relation to eliminate paths:

$$P/A > 0.05 \tag{1}$$

where:
A is the polygon surface
P is the polygon perimeter

This relation is applied in order to eliminate paths, because we have empirically proven that the polygon perimeter, which belongs to paths and roads, is greater than the surface. Other polygons, such as rural houses, were eliminated by hand.

In the database, there is another field with the majority crop associated to each subplot, and the surface occupied by this majority crop and by the rest of the crops in the classified image, due to superposition errors (between the digital cadaster and the classified image) and to edge pixels.

Table 3. Surface occupied by each one of the crop classes according to the cadastre and to the classified image

Cadastre surface (ha) ↓		Classified image surface (ha) →						
		Regve	Vgsec	Regnoes	Format	Regpri	Barb	Cropsec
Regve	5403	5120	63	138	14	33	1	28
Vgsec	1305	14	1160	3	48	3	28	49
Regnoes	1439	52	3	1350	5	19	0	7
Format	3888	30	41	11	3750	21	1	34
Regpri	3076	30	24	33	24	2843	2	117
Barb	1251	2	48	0	1	1	1118	76
Cropsec	13508	57	244	18	128	34	113	12909

When we obtain the crossing database, where there are 7 types of majority classes, we have obtained 7 different tables, one for each crop class, which is going to be the majority crop in each plot.

From each table (one for each majority crop), we obtain on the one hand, the sum of plot surfaces according to the digital rural cadastre, and on the other, the surface occupied by each one of the crop classes in the classified image (table 3). The total surface occupied by a crop class according to the digital cadastre and the sum of the plot surfaces occupied by the different crops in the classified image varies slightly. This difference is due to the polygonization of the classified image. For each majority crop, it has represented for each subplot the surface obtained through the digital cadastre on axis y, as opposed to the surface occupied by this crop in the classified image on axis x (for Alfalfa in Figure 6 and for Forestry and brushwood in figure 7). This is because the more usual technique to integrate the results, in this case of the digital cadastre with classified images, is the simple regression estimator (González- Alonso et al, 1993):

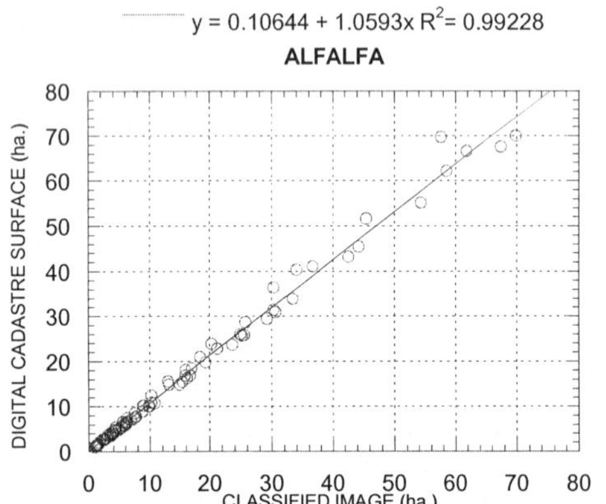

Figure 6. Cadastre surface as opposed to surface calculated from the classified image, of those subplots whose majority class is the Perennial leguminous, Regnoes.

The values of the unit slope would suppose a total concordance between the surface assigned by the digital cadastre and the surface intersected over the classified raster map. The differences with regards to the more suitable situation are due to classification mistakes (superposition and edge pixels). The value of the slope in each one of the graphics indicates the percentage of this crop that is erroneously classified as a different crop. For example in figure 6, for the Perennial leguminous, Regnoes, the linear curve fit slope is 1.0593, which means that there is a 5.9% of this crop classified as another crop. Due to this, we have to consider not only the majority crop but also the rest of the crops in order to make a surface correction.

3.2 *Surface comparison*

We can conclude that it is necessary to consider all classes to correct the surfaces assigned in the classified image, and not only the majority crop class in each subplot. Then, if we consider the hypothesis that surfaces assigned by the digital cadastre is the best approximation to reality that we can obtain, and if we consider that errors due to geometrical superposition between digital rural cadaster and classified image are worthless, we can treat table 3 as a matrix.

From table 3, it is deduced that the area occupied by each crop class can be expressed in relation to the surface obtained by the digital cadastre, because we observe that the sum of each column is the total area occupied by each crop class in the classification.

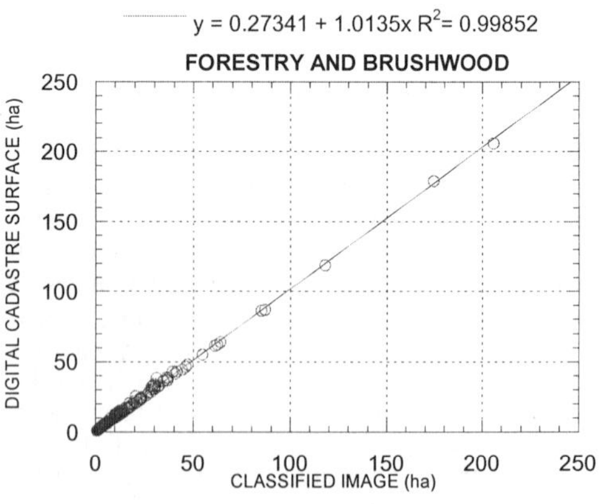

Figure 7. Cadastre surface as opposed to surface calculated from the classified image, of those subplots whose majority class is Forestry and brushwood, Format.

Table 4. Matrix of normalized Cij coefficients

Cadastre surface (ha) ↓	Classified image surface (ha) →						
	Regve	Vgsec	Regnoes	Format	Regpri	Barb	Cropsec
Regve	0.948	0.012	0.026	0.003	0.006	0.000	0.005
Vgsec	0.011	0.889	0.002	0.037	0.003	0.021	0.038
Regnoes	0.036	0.002	0.938	0.003	0.013	0.000	0.005
Format	0.008	0.011	0.003	0.964	0.005	0.000	0.009
Regpri	0.010	0.008	0.011	0.008	0.924	0.000	0.038
Barb	0.002	0.038	0.000	0.001	0.001	0.894	0.061
Cropsec	0.004	0.018	0.001	0.009	0.003	0.008	0.956

Next, table 3 is going to be is normalized (table 4), dividing each row by the "real" surface, i.e. the surface obtained by the digital cadastre, so the total surface of each crop class is reduced to the unit (Hay, 1988).

In table 4, we can observe that for each crop class, there are coefficients Cij (i row, j column), which indicate the surface fraction occupied by each crop in the classification (in the diagonal), and the surface fraction occupied by the rest of the crops, due to classification errors. This matrix formed by the Cij coefficients is called matrix A.

Hay (1989) proposes a method that, from the matrix in table 4, the following relation expressed in the matrix form is obtained:

$$(a,\ b,\ c,\ d,\ .\ .)=(x,\ y,\ z,\ r,\ .\ .)A \qquad (2)$$

where A is the coefficient Cij matrix.

In this relation we have (a, b, c,...), the surfaces of each crop from the classified image, which depend on (x, y, z, ...), that are surfaces occupied by these crops in line with the digital rural cadastre:

$$(x,\ y,\ z,\ r,\ .\ .)=(a,\ b,\ c,\ d,\ .\ .)A^{-1} \qquad (3)$$

By finding inverse matrix of coefficient Cij, $(A)^{-1}$, and replacing it in equation (3), we have (a, b, c,...) which are the crop surfaces obtained by the raster map (classified image), and (x, y, z, ...) which are the surfaces that must be occupied by each one of the crops according to the digital cadastre (Jupp David L.B., 1989).

We are going to apply this methodology in the pilot zone of aquifer 08-29 (which belongs to the classification zone utilized to cross the digital cadastre, in La Gineta and a part of Albacete, and the classified image to obtain the correction matrix). The reason for choosing this pilot zone is because we have an estimation of the real irrigated surface in this zone, supplied by the Junta Central de Regantes de la Mancha Oriental (JCRMO), and we can compare it with the results obtained on applying the surface correction. Considering this pilot zone, in the following table 5, there are the crop class surfaces that we want to correct, which belong to the classification and the results when the surface correction is applied:

Table 5. Comparison of surfaces occupied by crops in the pilot zone. Before and after doing the surface correction, when the median filter is applied over classified image and before applying the surface correction, when a median filter hasn't been applied

	Without median filter	With median filter	
	Without surface correction	Before surface correction	After surface correction
Regve	6259	6424	6561
Vgsec	3887	3648	3624
Regnoes	1646	1727	1637
Format	4647	4118	3997
Regpri	3338	3447	3608
Barb	2209	2421	2534
Cropsec	9809	10354	10245
Unclassified	415	71	71
Total	32210 ha	32210 ha	32277 ha
Total Irrigated crops	11243 ha	11598 ha	11806 ha

When a median filter has been applied, a total surface of 11806 ha of irrigated crops is obtained on applying the surface correction, but without correction we have a total of 11598 ha, this means that, after applying the surface correction, the total surface of irrigated crops increases only 2%.

According to the results obtained in table 5, we can observe that, when a median filter is applied over classification, it is nearer to reality than without applying a median filter. However, the results are slightly modified when a median filter is applied and an inverse matrix is used to correct crop classes surfaces.

4 CONCLUSIONS

A methodology has been designed for supervised classification in a semiarid zone such as La Mancha (Spain) to identify irrigated crops with Landsat TM imagery. The process developed for discrimination is based on the different spectral answer and in the different temporal evolution of this answer according to the crop phenological evolution. The classification procedure has followed multitemporal supervised classification by combining maximum likelihood algorithms, decision-tree criteria and context classifiers, which operate on scenes obtained on different dates.

The results obtained in the supervised classification by the maximum likelihood method have improved through the application of different decision-tree classifiers, according to the confusion observed among crops and introducing new images of more suitable dates for NDVI evolution of each crop. This has improved when we have introduced other dates, obtaining a global precision from 82.7% to 92.7% for classification accuracy.

The fiablility of estimated surfaces from classified image is very important, because the principal parameter to estimate groundwater abstraction is the surface occupied by each class of irrigated crop, with different water requirements.

To consider the level of precision in the estimated crops areas from the classified image, we compare these areas with the calculated areas from the digital rural cadastre, through the intersection of the land use raster map and the digital rural cadastre vector map in a Geographical Information System. When a median filter over the final classification is applied, the surfaces occupied for each one of the crop classes is nearer to the surface according to the digital cadastre, as we can observe in the graphics obtained, and also in the fieldwork data. Because of this, the context classifier is a useful tool to improve surface quantification.

We can consider that a context classifier, when applied to postclassification, is sufficient to improve surface quantification without the necessity of developing a method to correct crop surfaces.

REFERENCES

Bober M.L., Wood D. & Mc Bride R.A. 1996. Use of Digital Image Analysis and GIS to assess regional soil compaction risk. *Photogrammetric Engineering & Remote Sensing*, 62(12). Pp: 1397- 1404

Calera A., Medrano J., Martínez C. & Ruiz J.R 1998. Use of a GIS tool for sustainable management of aquifer 08-29 resources in La Mancha, Spain. *GIS, Technologies and their environmental applications.* Editors: C.A Brebbia and P Pascolo. Pp: 63-72

Calera A., Medrano J., Vela A., & Castaño S. 1999. GIS tools applied to the sustainable management of water resources. Application to the aquifer system 08-29. *Agricultural water management.* Vol 40, nos. 2-3. Pp: 207-220.

Congalton R., Balogh M., Bell C., Green K., Milliken J.A. & Ottman R. 1998. Mapping and monitiring agricultural crops and other land cover in the lower Colorado River basin. *Photogrammetric Engineering & Remote Sensing*, 64(11). Pp: 1107- 1113

González-Alonso F. & Cuevas J..M.. 1993. Vegetation Classification in the Middle Mediterranean area by satellite data. *Int. J. Remote Sensing*, vol 14, no. 6. Pp: 1215- 1219.

Hay A.M. 1988. The derivation of global estimates from a confusion matrix. *Int. J. Remote Sensing*, vol 9, no. 8. Pp: 1395- 1398.

Hay A.M. 1989. Global estimates from confusion matrices; A reply to Jupp. *Int. J. Remote Sensing*, vol 10, no. 9. Pp: 1571- 1573.

Jupp David L.B. 1989. The stability of global estimates from confusion matrices. *Int. J. Remote Sensing*, vol 10, no. 9. Pp: 1563- 1569.

Martín de Santa Olalla F., Brasa A., Fabeiro C., Fernández D. & López H. 1999. Improvement of irrigation management towards the sustainable use of groundwater in Castilla La Mancha, Spain. *Agricultural Water Management.* Vol 40 nos. 2-3. Pp: 195-205.

Remote Sensing in the 21st Century: Economic and Environmental Applications, Casanova (ed.)
 ISBN 90 5809 096 5

The integration of satellite images and GIS for Omerli Water Basin land use changes

C.Goksel
Faculty of Civil Engineering, Remote Sensing Division,Istanbul Technical University, Turkey

H.Turkoglu
Faculty of Architecture, Urban and Regional Planning Department, Istanbul Technical University, Turkey

ABSTRACT: Monitoring land use changes is very important data for urban and regional planning. Remote Sensing Technology can provide a method for acquiring regular, recent information for land use. Geographic Information System with its widespread usage is also a powerful tool for urban and regional planning. The integration of Remote Sensing Technology and GIS will provide maximum data processes in land use planning. Omerli Water Basin is one of the major water dams which supplies 40% of drinking water of the city and approximately covers an area of 620 square kilometers. Population in the water basin has shown a drastic increase since 1985. The rapid urbanization in the area is serious concern for researches, planners and even inhabitants of the city since they are within the boundaries of the water reservoirs of the city. In this study the changes of land use has been examined with remote sensing data based on two different dates (September 1992 and July 1997) Image registration and classification were conducted for digital change analysis. In the classification procedure the Max-likelihood classification algorithm was applied to different data sets.

1 INTRODUCTION

By 2025, the majority of the world's population, around 5 billion people, will be living in urban areas. As cities expand, prime agricultural land and habitats such as forests and waters basins are transformed into land for housing, roads and industry. In developing countries the urban poor, lacking an alternative place to live, often establish informal settlements in ecologically sensitive areas. Urban land use planning can help guide urban development away from vulnerable ecosystems.

Remote sensing technology has been recognized as a useful means of supplying up to date information on activities within the urban environment (Treitz et al., 1992; Ehlers et al., 1990; Forster, 1985). Remote Sensed data may be used to map monitor and estimate the properties of environmental features. Thematic mapping from remotely sensed data is found in numerous fields of study including land-use planning (Foody G. M., 1999).

Remote sensing data can be readily merged with other sources of geocoded information in a GIS. This permits the overlapping of several layers of information with the remotely sensed data, and the application of a virtually unlimited number of forms of data analysis (Lillesand and Keifer, 1987).

The aim of the study to determine land-use changes in an area using with remote sensing data and GIS. Our sample area is Omerli Water Basin. The basin supplies 40% of drinking water of Istanbul. Population in Omerli Water Basin has shown a drastic increase since 1985. There is an urgent need to stop urbanisation process in the area because the damage on natural environment is irreversible.

2 METHODOLOGY

This research has two parts. In the first part of the study is to present natural and building environmental data. The database includes information on the current status of soil characteristics, geology, infrastructure, flora and demographic characteristics. All this data incorporated into the GIS providing for the spatial analysis. The second part of the study based on satellite image processing data. Thematic maps are compared to the merge (Landsat TM and SPOT P) images. Temporal land use changes using satellite data has been realized.

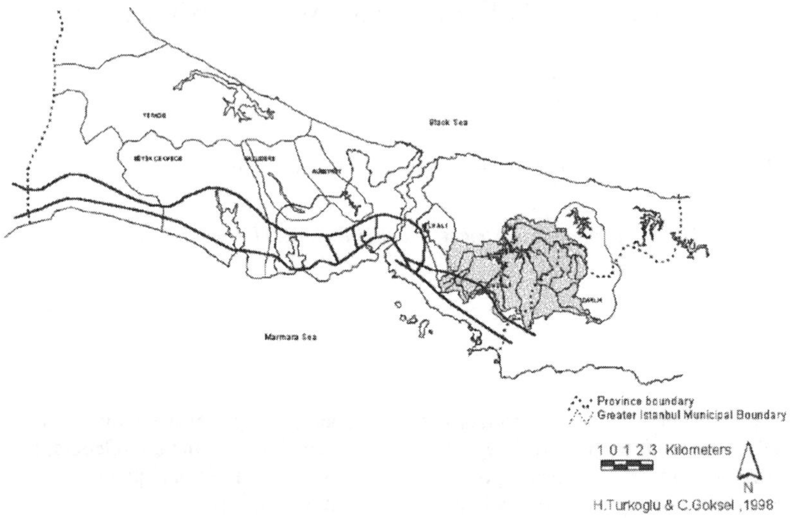

Figure 1. The location of Omerli Water Basin in Istanbul. (colour plate, see p. 596).

2.1 *Data Sources*

In the study many kinds of data were collected from various sources. The satellite images, aerial photographs, standard topographic maps (Scale 1:25000), and various thematic maps, and several photographs have been used as data sources of the study. The characteristics of the satellite images have shown in Table 1.

The satellite data were examined using standard image processing software, ERDAS (v.7.5) and ERDAS IMAGINE, and the results of the digital processing were presented with ARC VIEW (v.3).

The image processing was performed in the Remote Sensing Laboratory, Istanbul Technical University.

Table 1: The characteristics of Satellite Images

Image	Date	Resolution.	Number of bands
SPOT P	13.6.1993	10mx10m	1
Landsat TM	06.09.1992 18.07.1997	30mx30m	7

2.2 *Characteristics of the study area*

2.2.1 *Location*

Drinking water of Istanbul provided by surface water resources through the seven water dams. .Our sample area Omerli Water Basin is located in Anatolian site of Istanbul. (Figure 1).

2.2.2 *Geology*

The watershed covers mainly Paleozoic basement rocks and tertiary cover rocks. In the base of valleys there exist alluvial soil which is not very thick. The reservoirs accumulates surface flow, however small amount of water comes through the underground water. Paleozoic basement rocks are impermeable therefore large amount of underground water accumulation is not expected.

On the other hand tertiary cover rocks which is basically located in the eastern part of the water basin are permeable Figure 2 shows geological characteristics of Omerli region.

2.2.3 Topography and Erosion

Critical slope groups for erosion has been accepted more than 15% slope. In the area the slope group

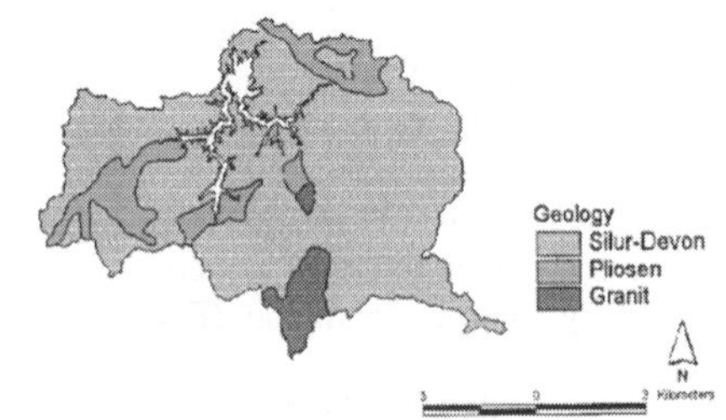

Figure 2. Geological characteristics of the area. (colour plate, see p. 596).

with more than 15% has exceed 40% of total area. In other words especially east and north part of the basin has hilly topography with the slope of more than 15%. Therefore the water basin land is sensitive to erosion. Open areas especially agricultural lands loose the top soil part and its water gradually.

2.2.4 *Flora*

The forest area is approximately 320 square kilometres. Eighteen percent of the forest area is pine trees and eighty two percent of the forest area is leafed trees. Existing forest area is not very efficient in terms of plant materials, therefore it is open to erosion.

2.2.5 *Demography*

Population in the basin has shown a drastic increase since 1985 as shown in Table 2. While the rate of population increase for Istanbul was around 5.3% the same rate for the total water basin were 26.76% for the period of 85-90. There are 24 settlements in Omerli Water Basin. Samandra (Population in 1997: 19524), Sarigazi (22125), Sultanbeyli (82298), Sultan Ciftligi (9747), Yenidogan (4798), Alemdag (6684) and Buyukbakkalkoy (3197) are urbanized areas while other 17 are minor villages scattered in the catchment area (Figure3)

2.2.6 *Land-use*

Industries have scattered and have not been segregated from the generally residential zones (Figure 3) . The military occupies large areas, main parts of which are forest or undeveloped. North part and East part of the basin covered by forest area. In the open areas near the forest there are settlements with the small agricultural land surrounded. In the south part of the basin, urbanised areas, agricultural lands are located.

2.2.7 *Environmental Pollution*

The sewerage system in the area is inadequate. The wastewater is collected in septic tanks from where they are intermittently transferred to their final disposal sites by scavengers. Since the septic tanks are risky for seepage into the ground, there is a possibility of groundwater pollution. Meanwhile at locations close to the rivers, direct seepage into the rivers may also occur.

Industries around the water basin have no treatment plants. Industrial pollution comes from basically metallic industry, plastic industry and animal husbandry. In the area total industrial waste water 337 m^3 per day.

Table 2: Total Population of Omerli Water Basin between 1980-1997

	1980	1985	1990	1997
Population.	22709	36860	60254	257204

There is some pollution due to fertilisers and pesticides used in agriculture in the area. Fertilisers are rich in nitrogen and phosphorous which may cause to the serious problem of eutrophication in receiving waters, nitrate deposition, which is harmful in drinking water, in ground water resources. Pesticides are priority pollutants which are toxic. The most critical part of the area for pollution is the urbanised area which is located south-west part of the area.

2.3 *Image Processing*

The image data sets were geometrically corrected to the Universal Transverse Mercator System (UTM) coordinate system. A first order polynomial transform was calculated to model geometric distortions in the satellite data. Geometric correction has made using twenty ground control points with total root means square (RMS) of 0.49 pixel (14.7 m).

2.3.1 *Classification*

Before the classification, preliminary information was required of the region in order to determine statistical limits. Various data were used to select reliable test areas. Landsat TM data were merged with SPOT P data. These data set are very efficient as a base visual data to analyze land use characteristics.

In the classification procedure the Landsat TM data of 1992 and 1997 are classified into seven classes using the conventional supervised Maximum Likelihood Classification Algorithm. These seven spectral classes were also evaluated as land use categories Figure 4 and Figure 5 shows classification results and statistical results are tabulated in Table 3.

2.3.2 *Accuracy Assessment*

For the accuracy assessment, 500 pixels were randomly selected from the ground-truth coverage for comparison, and error matrix were calculated. For classified images of 1992, overall classification

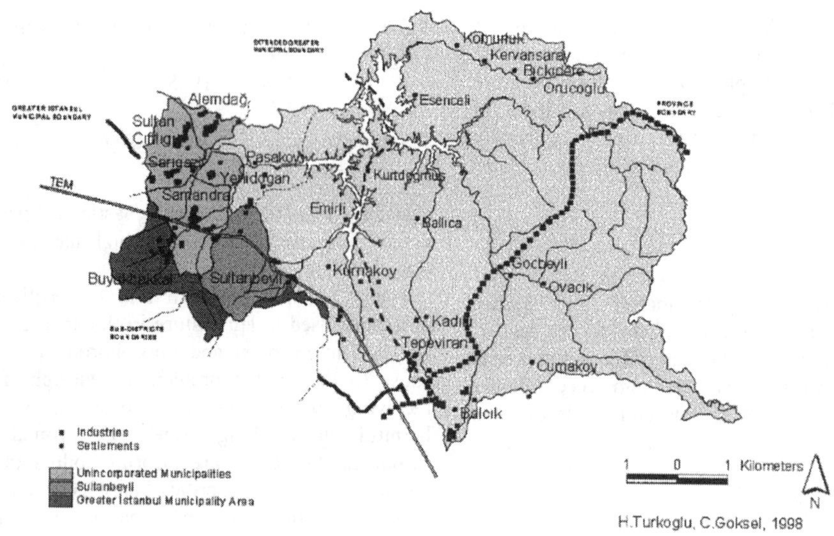

Figure 3. Settlements in Omerli Region. (colour plate, see p. 597).

accuracy for the seven classes was 89.9 %. For the 1997 images, overall classification accuracy for the seven classes was 91.3%.

Table 3:Results of Classified Images

Description	1992 Ha	1997 Ha
Water	1785.33	1825.29
Forest	39169.17	35140.23
Road	360.99	439.86
Industry	164.43	351.63
Settlement	6009.21	6391.26
Wetland	2143.89	1103.13
Bare Soil	11817.99	16198.96
TOTAL	61451.02	61451.02

3. RESULTS

The results of the study show us the classification of satellite image data can be used as land use changes of the Omerli Water Basin. The statistical results of the classification procedure is supported the demographic data. There is a land use change between 1992 and 1997. In the forest area there is 10% of decrease while there is 1% increase in urban areas, 7% increase in open areas. According to the statistical results between 1992 and 1997 instead of creating new settlements, the new residents have settled in old areas creating relatively dense residential areas.

Since the urban development happen mainly on the south western site of the basin where the

Figure 4. Classified image of 1992. (colour plate, see p. 597).

Figure 5. Classified image of 1997. (colour plate, see p. 597).

relatively sensitive environment in terms of natural characteristics the dense urban development should be controlled and the precautions should be taken seriously.

REFERENCES

Ehlers, M. , M. A. Jadkowski, R. R. Howard & D. E. Brostuen, 1990. Application of Spot Data for Regional Growth Analysis and local Planning. *Photogrammetric Engineering & Remote Sensing* v.56, no 2:139-151

Foody, M.G., 1999. The Continuum of classification Fuzziness in Thematic Mapping. *Photogrametric Engineering & Remote Sensing.* v.65, no 4: 443-451

Forster, B.C., 1985. An Examination of some Problems and Solutions in Monitoring Urban Areas from Satellite Platforms. *International Journal of Remote Sensing.* v.65. no 4:443-451

Lillesand, K.M. and R.W. Keifer, 1987. *Remote sensing and Image Interpretation,* Canada: John Wiley and Sons.

Treitz, M. P., P.J. Howarth, & P. Gong. 1992. Application of Satellite and GIS Technologies for Land cover and land use mapping at the Rural Urban Fringe A case Study. *Photogrammetric Engineering & Remote Sensing*, v.58, no 4:439-448

Remote Sensing in the 21st Century: Economic and Environmental Applications, Casanova (ed.)
© 2000 Balkema, Rotterdam, ISBN 90 5809 096 5

Geographical analysis using GIS of agrarian land use in the Province of Valladolid in 1997 with own classification on multiband satellite images

F.J.Tapiador & J.L.Casanova
Remote Sensing Laboratory, University of Valladolid, Spain

ABSTRACT: With the support of several GIS, this work is aimed at studying the characteristics of agrarian land use in the province of Valladolid from a regional geographical analysis viewpoint. Both the methodology as well as the formal approach on which the work is based are innovative within the geographical area of Castile and Leon. The final goal is the objective geographical characterization of the whole province, by means of the explanation of the factors which influence the location of various crops, and the clarification and study of the relations which are established between them as well as the different elements of the physical and antropic environment, laying special emphasis on the interdependencies between the socio-economic characteristics of the population, the historical factors and cultural location.

1 INTRODUCTION

Integrating satellite data and GIS as a means of objectively gathering large amounts of territorial information with a view to its subsequent explanation is not the normal procedure for geographical analysis. It is more common to use field data, which inevitably leads to a delay in performing any kind of work as well as involving a degree of subjectivity. This work aims to demonstrate how satellite support can eliminate such errors without losing either the desirable accuracy in the study or clarity of explanation. From a methodological viewpoint, an explanation is also given of the capacity of the GIS as an efficient tool for the treatment of data and the basis which makes the transposition of the study performed in Valladolid possible to other areas.

Some of the most significant final results of the in depth analysis carried out concern the influence of historical inertia on the location and productive structure of cereal exploitations as well as the tremendous importance of mesoscale and lithologic climate factors in the current development of the Valladolid wine area. This is an extremely important element, not only in the primary but also in the general sectorial economic structure. Other results which have been observed with this set of techniques include the effects of industrialization and the expansion of peri-urban nucleuses on the weak forest canopy of Valladolid.

Finally, it should be pointed out that within the analysis, an important cartographic effort has been made, yielding a series of thematic maps of the area under study which might be used to advance regional study from approaches other than that which has been followed.

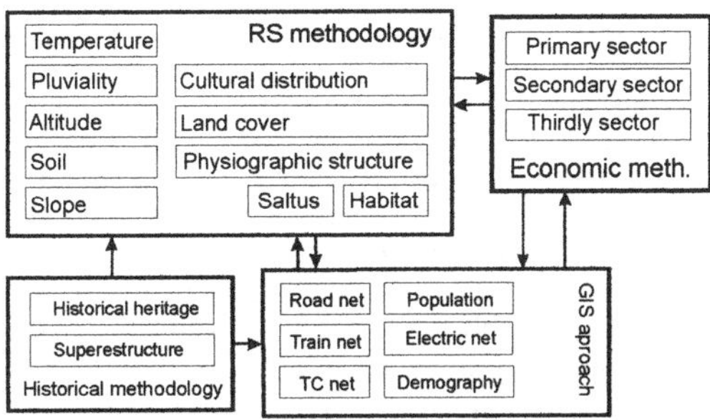

Figure 1. Elements and main factors to considered in the analysis of agrarian land use in the developped countries.

2 OBJETIVES AND METHODOLOGY OF GEOGRAPHIC ANALYSIS

Geography is the science which studies elements that may be individualized on the earth's surface, determining the guidelines for their location, their-inter-relationships and above all their link to man. Within this wide framework, regional studies involve the application of knowledge and general techniques to a limited section of the earth's surface, which may be carried out from a global standpoint -

a regional geographical analysis - or a sectorial analysis, as is the case here. In turn, the methodology used may follow the so-called 'old paradigm', in which a subjective literary discourse is offered on the territory based on experience and the apriorities of the geographer. Alternatively, geography may be considered quantitatively as with any other science, based on contrastable data and hypotheses which may be sustained logically, in other words using explicit mathematical approaches. This approach, which is the one we have opted for and which we will call quantitative geographical analysis, or simply geographical analysis, comprises the following stages:

a) Explanation of experimental data available, both previously processed by other scientists (satellite images, statistics) as well as those obtained by the geographer himself..

b) Basic qualitative model.

c) Development of the preliminary hypotheses on the structure of agrarian land use.

d) Laying out the quantitative model.

The final stage can only be briefly mentioned here and is currently the objective of active research on our part.

3 EXPERIMENTAL DATA

The nucleus of our work consists of the development of mapping and the qualitative approach necessary for the subsequent development of the models which will scientifically explain crop location in the province of Valladolid in 1997. Within the experimental data we can consider as such the satellite images which have been used to carry out land use classification and physiographical analysis. H-resolution and L-resolution sources have been used (Hay, 1986) as a means of covering the microscale and macroscale variables. These were Landsat, IRS-1C, and NOAA images, the latter consisting of a temporal series of 2 years of daily images which have been processed to obtain two magnitudes: ground temperature (split-window algorithm) as well as a of set vegetation indices which were studied in order to choose the most appropriate for the region under study.

Firstly, a thematic classification was performed of the crops using the satellite images with the greatest spatial resolution. Various techniques were used for this: dichotomy, clustering and the analysis of the main components, with several software packages (ENVI, ER-MAPPER, IDRISI) seeking the one which minimized errors. The results, compared with a set of 17 field samples, showed that the most accurate was the classification using neuronal networks, the application of a supervised neuronal network (own software) based on the back-propagation algorithm being the one finally chosen. Classification was then carried out as a base for subsequent analysis, an initial cartographic result being obtained (objective)

Secondly, cross data matching was performed between the classification carried out and the lithologic data base, as it seemed reasonable to assume that one of the most relevant factors in the location of crops in the province of Valladolid was the natural conditions themselves. The software used is our own (LATUV RS SYSTEM). To measure relations a simple statistical model has been used that shows us the dependence of the type of crop on a certain substrate.

Hydric availability and the possible affect of drought periods has been calculated directly with temporal series of NOAA images from which the information corresponding to the Greenness was taken.

Given that the nominal resolution of NOAA images is 1.1 Km, an upscaling of LANDSAT and IRS-1C images was performed following the algorithm proposed by Hay, the objective of which is to preserve a kind of potential throughout the process and which we used as we thought it to be extremely interesting and accurate.

In addition, studies were also carried out of those areas which were difficult to designate in order to determine the temporal evolution of the possible crops in the area with a view to clarifying real land use. A set of graphs were thus obtained that were analyzed qualitatively to clarify unclear areas.

A second kind of data, less reliable, has enabled us to take other factors into account, perhaps no less important but whose quantification is difficult and therefore does not allow us to give them the same relative weight as the previous ones. For the moment they may only be used as qualitative variables. We refer to the structure of property, per-capita rent, the distribution of the population over the territory and the indicators that reflect the industrial fabric of the area as well as the hierarchical relations between population nucleuses. These are values which can only be gauged with official statistics and field surveys, both of which are subject to errors and biases, but which must be taken into account in the analysis. Our own GIS was used (LATUV SIGIM), which has served as a "hidden layer" in the interpretation that obviously cannot only be based on physiographic features.

The richness of a methodology such as the one described lies in this type of integration, as even when the result of the analysis might be misinterpreted at a given moment with the work in the old geographic paradigm, the line of explanation followed of all the data used together with their logical order clearly differentiate them from the latter .

4 LINKING FACTORS WITHIN THE STUDY: HISTORICAL HERITAGE AND EXOGENEOUS FACTORS

It is not possible to carry out a geographic study of this area without taking into account the effects of land concentration, undertaken during the eighties. This policy of the rationalization of exploitations characterizes the physiographic features of the *ager*, and it is possible to appreciate clearly in those images with a greater spatial resolution the defined structure of most crop-orientated plots, which are the most affected by the concentration process. The importance of remote sensing in agricultural analysis can clearly be seen here as verifying such an apparently trivial fact is no easy task without the help of this methodology and is impossible to measure without its help

Another of the fundamental factors that will remain clear throughout the analysis is the result of secular land use over the whole area of the Duero in medieval times. This led to a structure of open fields and uses which, such as the vine, were linked to one very specific fact (monasteries).

Finally, it should be pointed out that Spain's entry into the EEC (EU) led to a series of effects which may not be explained simply by internal factors, effects such as the change in crops in certain areas under CAP subsidies.

These three links, which in the quantitative model would have to be parameterized according to economic and location variables, allow a better understanding of current agrarian land use in the province.

5 ANALYSIS OF AGRARIAN LAND USE IN THE PROVINCE OF VALLADOLID IN 1997

Within the rural environment, there is a classical distinction between *ager* (the cultivated area) and *saltus* (the non-cultivated area). Our work centers on the characterization of the ager. However, this task cannot be carried out without the corresponding reference and study of the interdependencies with the saltus and of course with the *habitat*, the third component of rural areas corresponding to those places where man has established his dwelling. Firstly, and in view of the classification carried out, it can be appreciated how the area corresponding to the saltus is much smaller than the *ager*. This is due to historical consequences of over-exploitation to which the land was subject in this area as a result of its crop orientation and the absence of significant imbalances which have served as a cultural redoubt due to their lack of economic viability in previous times. Practically all of the area except for the small section of the slopes of the barren plains and not always all of them, has been taken advantage of by man, both for agriculturally or for forests. The economic importance of the latter in rural life should not be forgotten. The limited importance of adapted vegetation is reflected in the existence of replanted forest areas, a consequence of the massive plantations of the enlightened policy of the XVIII century that attempted to lessen the effects of deforestation caused by the need for wood in shipbuilding, and in particular the construction of the Spanish fleet.

A satellite view allows three basic areas to be distinguished in the *ager*, faithfully corresponding to the individualized units visible at a ground level. Firstly, the so-called dry cerealist steppe, an area that can be considered as the basis of the *ager* and on which the rest of the units can be explained as discontinuities or enclaves. Secondly, riparian vegetation, with its lineal component, and irrigation crops, scattered over the steppe in the most productive areas for irrigation and/or in which the necessary investment has been made.

As a means of studying the province, a methodology has been selected which consists of choosing a series of areas which we feel characterize individual sections and from the study of which the basic guidelines that characterize the province may be taken.7 areas have been chosen that correspond to the same number of geographically coherent areas covering classic sectors in the literature and in which all the processes worthy of mention may be found.

CONCLUSIONS

Remote sensing is an invaluable tool for regional geographical analysis as it allows the geographer to

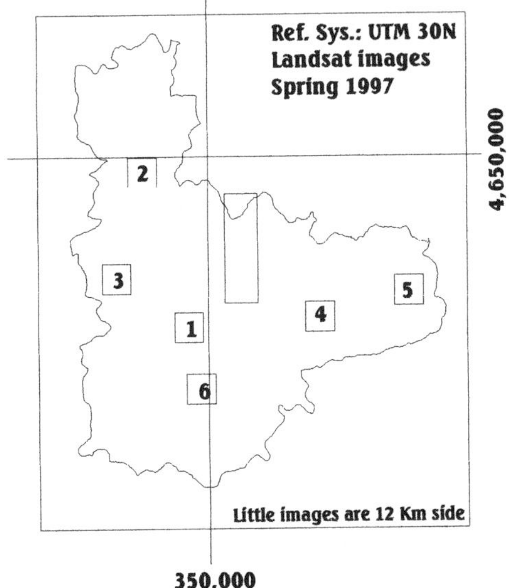

Figure 2. The area of study on the Province of Valladolid

Figure 3. This 5-4-3 image shows agrarian land use in the city of Tordesillas. The effect of land plot concentration is clearly visible in the radial structure top, as is the prevalence of dry land over irrigated land and forest. (colour plate, see p. 598).

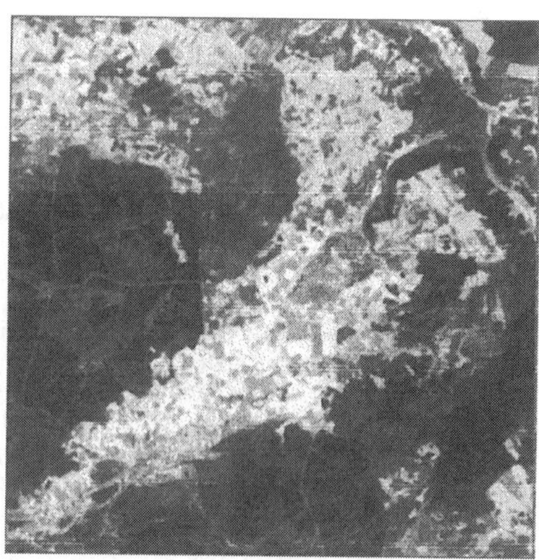

Figure 5. There are only a few spots where forest is the main land cover due to the late re-population in the XVIII century and beyond. These are now leisure and second home areas. (colour plate, see p. 598).

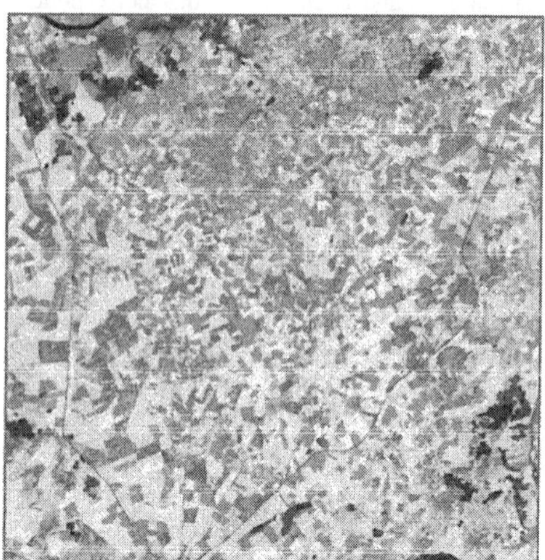

Figure 4. One can see here the small size of the plots in the province as well as other kinds of landscape: Irrigated and dry land alternates with an open field structure, with regular plots in a concentrated village shape. The secondary importance of saltus in this area can be seen. (colour plate, see p. 598).

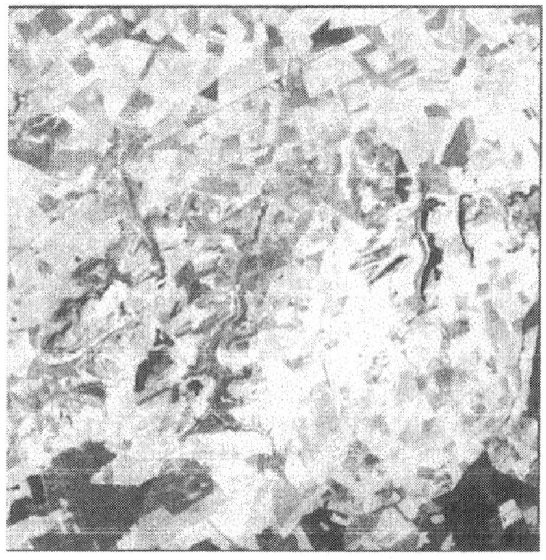

Figure 6. The different substratum between the various levels of plateau does not lead to any special bias in land use, but does in productivity. The structure of property remains similar in both cases. (colour plate, see p. 598).

objectivize his knowledge. From the present study, and references to agrarian land use in the province of Valladolid in 1997, it is possible to draw a series of conclusions that might be classified as simple and complex, depending on how close they are to the hypotheses that could have been put forward before doing any kind of calculation and without the numerical support of GIS. Moreover, the possibility of applying the methodology to other sets of experimental data permits studies to be performed in subsequent years at a minimum cost.

Amongst the simple conclusions, the dependency

Figure 7. The balance between saltus and ager in the province is tilted in favor of ager – plain land, with a historical exploitation of resources. Thus, the primary description can be considered a set of little plots with cerealist orientation. (colour plate, see p. 599).

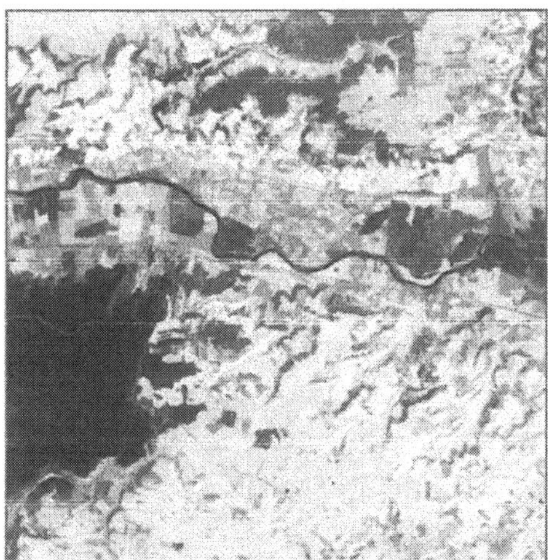

Figure 8. The area of 'Ribera del Duero' wines covers the Duero valley with highly cultivated sectors. The slopes of the bleak plateau remain bare or, in some cases, with reforested Pinus, whereas some woodland is used for second homes, losing its primary function. (colour plate, see p. 599).

on the type of crop as well as the lithologic substratum has been sufficiently described. The degree of dependency, which in our opinion is high taking into account the other factors, indicates an evolved agrarian structure, in which crops are distributed in terms of their natural domains. However, and this is one of the more complex conclusions, the existence

Figure 9. This sector of the river Pisuerga shows the characteristic land use in the bank: Pivot systems for irrigating drought land and other means of transporting water to herbaceous crops such as suger-beet – justifying investment in infrastructure – and wheat when it is amortized, some little greenhouses and market gardens to supply not only the local but regional market. The city of Valladolid city is beneath. (colour plate, see p. 599).

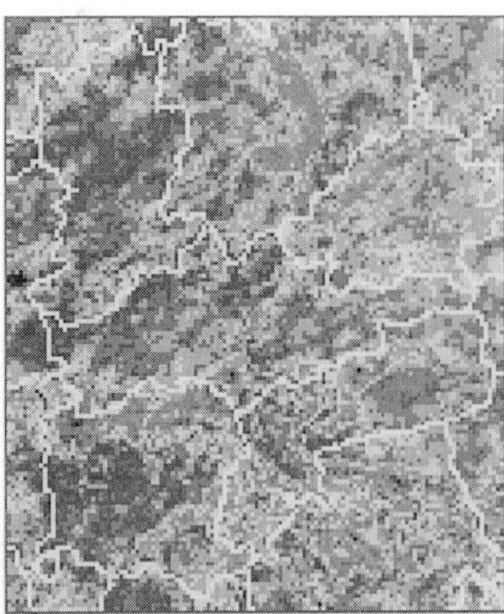

Figure 10. Using NDVI and Greenness with NOAA-AVHRR images, a fairly accurate idea of crop yield can be obtained and related to land physiography, property as well as social and demographic structures. This kind of work can be done with a GIS approach to increase efficiency

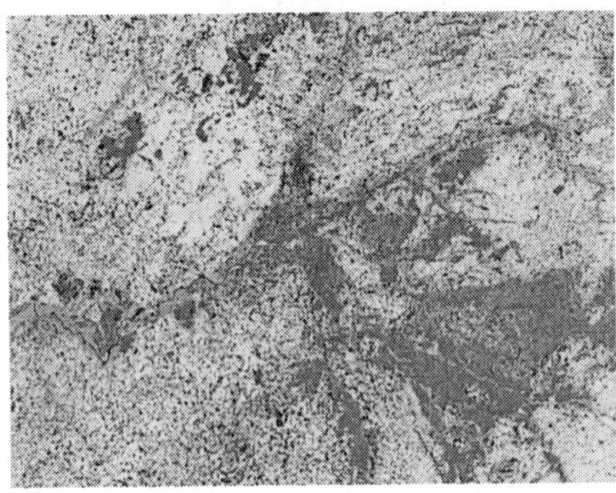

Figure 11. In this (above) classified image (using a back-propagation neural network) centered in Valladolid, we have represented cereal crops (light) herbaceous crops and forest (dark) and other land uses. The shape is clear and shows the dependence of crops on the substratum, slope and historical use.

of crops that occupy areas removed from their natural and expected environment has been seen, a fact explained by economic factors linked to European subsidies.

REFERENCES

Askine, J. (Ed) 1995. Sensors and Environmental Applications of Remote Sensing. *Proceedings of the 14th EARSeL Symposium.* A.A. Balkema.

Atkinson, P. M. & Curran, P. J. 1997. Choosing an Appropiate Spatial Resolution for Remote Sensing Investigations. PE&RS Vol. 63 No. 12.

Dunkel, Z. (Ed) 1995. The use of remote sensing technique in agricultural meteorology practice. COST 77. Workshop.

Elachi, C. 1987. Introduction to the Physics and Techniques of Remote Sensing. Wiley Series in Remote Sensing.

Foody, G. & Curran, P. Ed. 1994. Environmental Remote Sensing from Regional to Global Scales. John Wiley & Sons, Inc.

Goodchild, M.F. & Parks, B. O. & Steyaert, L.T. (Ed) 1993. Environmental Modeling with GIS. Oxford University Press.

Hay, G. J. & Niemann, K.O. & Goodenough D.G. 1997. Spatial Thresholds, Image-Objects, and Upscalling: A multiscale evaluation. Remote Sensing of Environment 62.

Kennedy, P.J. (Ed) 1996. Application of Remote Sensing in European Forest Monitoring (Proceedings)

Lillesand , T. M. & Kiefer, R. W. 1994. Remote Sensing and Image Interpretation. John Wiley & Sons, Inc.

Proceedings of the First International Conference of Geospatial Information in Agriculture and Forestry. 1998. ERIM International, Inc.

Star, J.L. & Estes J.E. & McGwire K.C. (Ed) 1997 Integration of Geographical Information Systems and Remote Sensing. Cambridge University Press.

Worrall, L. 1990. Geographic Information Systems: Developments and Applications. Belhaven Press.

Remote Sensing in the 21st Century: Economic and Environmental Applications, Casanova (ed.)
© 2000 Balkema, Rotterdam, ISBN 90 5809 096 5

Application of remote sensing to the study of oak decline in Aragón, Spain

S.Ait-Bachir & E.Gil Pelegrín
Servicio de Investigación Agro-alimentaria (S.I.A), D.G.A., Zaragoza, Spain

B.Lacaze
CRENAM, Université Jean Monnet, Saint-Etienne, France

ABSTRACT: In the early nineties, a case of oak decline was detected in several parts of southern Aragón (Spain). This phenomenon, locally known as "seca", has received some attention from foresters and ecologists and different explanations have been proposed: severe climatic episodes, stand ageing or fungal attacks.

To analyse the process of oak decline in this region, two Landsat TM-5 images have been used, from 20 August 1993 and 29 September 1996. The first coincides with the period of maximal expression of the decline syndrome. The second, coinciding with an excellent vegetative period, was used as a reference because of the good recovering shown by trees during this year. The study area (400 km2) includes the whole woodlands dominated by Mediterranean and submediterranean oak species, namely *Quercus ilex, Q. coccifera, Q. faginea and Q. pyrenaica*. The comparison of the spectral signature values between the two images, of vegetation index (NDVI) and of green vegetation cover estimated from scaled NDVI has allowed us to confirm that a difference in the stands performance can be clearly detected, showing the critical condition reached by some woodlands during 1993. Although the NDVI values for 1996 are in average higher than in 1993, showing that oak woodland got better through this period, significant differences are found among the different species studied. The syndrome did not reach the same magnitude through the whole area, showing that the "seca" in this region must be analysed as a local phenomenon.

1. INTRODUCTION

During the last two decades, oak woodlands of Spain have been affected by a decline syndrome, locally known as « seca » (drought), which has been observed at tree level, as well as at forest patch level. This evolution is generally reversible, but in some cases the decline process may be intense and can lead to the death of the trees (Montoya, 1992). Several hypothesis have been proposed to explain this phenomenon: pathogens like *Phytophtora cinnamomi* (Brassier, 1992), increased water stress due to climate change (Luisi *et al.*, 1992), or a combination of both causes (Dimitri, 1992; Marçais, 1992).

The present study is a comparative analysis of satellite imagery recorded in 1993 (high level of affectation by decline process) and in 1996 (after recovery of most woodland).

2. STUDY AREA

The study area is located at the southwest part of the Zaragoza province, in Aragon, Spain. The test area covers about 31,000 ha and can be divided into 2 climatic zones: one with sub-mediterranean climate (woodlands dominated by *Quercus ilex* and *Q. coccifera*) and the other with meso-mediterranean climate (woodlands dominated by *Q. faginea* and *Q. pyrenaica*).

Mean annual precipitations are comprised between 400 and 500 mm; a major drought period occurred during the years 1992-94 (cf. Figure 1), which has enhanced the decline process.

3. MATERIAL AND METHODS

A forestry map, available in digital form, has been used to select a set of samples (« polygons ») characterised by dominance of only one oak species (cf. Table 1). For each polygon, additional information is available about canopy height and ecological habitat type.

Two Landsat-TM images have been selected for this study, which were recorded respectively on August 20, 1993 and September 29, 1996.

These 2 images have been geometrically corrected and co-registered; the geographic area covered is defined in UTM geographic projection by Xmin = 473290, Xmax = 694140, Ymin= 4420123, Ymax = 462263.

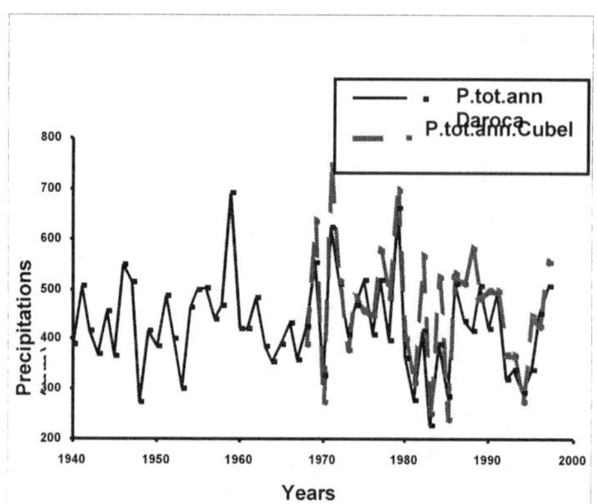

Figure 1: Inter-annual precipitation variations

Table 1: Description of selected woodland polygons in Cubel area.

Dominant species	*Quercus coccifera*	*Q. ilex*	*Q. faginea*	*Q. pyrenaica*
Number of polygons	7	224	37	13
Area (% of total)	2	87	18	3

No attempt of radiometric correction has been done; we used an alternative approach (scaled NDVI) which provides image to image normalisation. NDVI vegetation index is defined as $(DC_4-DC_3)/(DC_4+DC_3)$, where DC_i is the digital count in channel TM_i ; then scaled NDVI is computed as $N = (NDVI- NDVI_0)/(NDVI_{100} - NDVI_0)$, where $NDVI_0$ and $NDVI_{100}$ are respectively the NDVI values observed for bare soil and full canopy cover.

An estimate of green vegetation fractional cover is obtained from the following formulation, taken from Carlson and Ripley (1997):

$$F (\%) \approx 100 N^2.$$

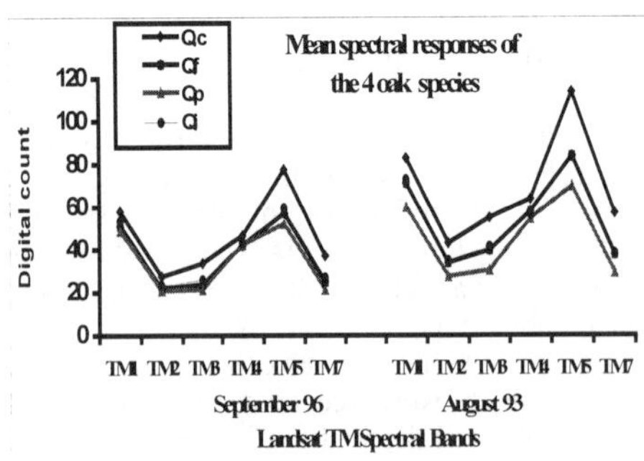

Figure 2: Mean spectral responses of the 4 oak species.

With reference to the analysis of spectral responses of leaves (Elvidge, 1987; Adams and Smith, 1996), we propose 2 spectral indices susceptible to characterise variations in leaf colour and physiological status:

Yellowness index : $(V_{TM2} + V_{TM3})/2$

Brownness/Dehydratation index : $(V_{TM5} + V_{TM7})/2$

where $V_{TMi} = (DC_i \text{ of } 1993/DC_i \text{ of } 1996)$.

In our study, these 2 indices have been tentatively applied to woodland areas.

4. RESULTS AND DISCUSSION

Comparative analysis of spectral responses and of spectral indices of oak species

Mean spectral responses of the four oak species are given at Figure 2 for 1993 and 1996.

At both dates, areas with *Q. coccifera* exhibit higher radiance values in visible channels (TM1 to TM3) and in shortwave infrared (SWIR) channels (TM5 and TM7), which result from lower green biomass amount. For August 1993 the spectral responses of *Q. pyrenaica* remain lower in the visible and SWIR domains, suggesting a better status facing drought conditions.

Then NDVI and other spectral indices have been used to quantify differences between dominant species (Table 2). These values confirm the hierarchy of species facing drought conditions. There are no significant variations between the 3 classes of *Quercus ilex*, suggesting a convergence of spectral responses due to drought effects on tree canopy and understory woodland responses in 1993 (Table 2).

Table 2: Mean values of spectral indices in 1993.

Dominant	Height	Yellowness	Brownness/
Q.coccifera	< 7 m	159.5	150.0
Q. faginea	< 7 m	162.0	148.5
	> 7 m	157.0	149.5
Q. ilex	< 7 m	154.0	141.5
	> 7 m	154.0	140.5
	+ rocks	154.5	141.0
Q. pyrenaica	< 7 m	151.5	139.0
	> 7m	117.5	> 7 129.5

Comparative analysis of estimated green vegetation fraction in 1993 and 1996

Applying the above-mentioned relationship between scaled NDVI and green vegetation cover, we derived the cover maps in 1993 and 1996, using the following classes:

Table 3: Definition of vegetation cover classes.

Class	1	2	3	4	5	6
Green veg. cover (%)	0	0 - 6	6- 26	26 - 46	46 -65	> 65

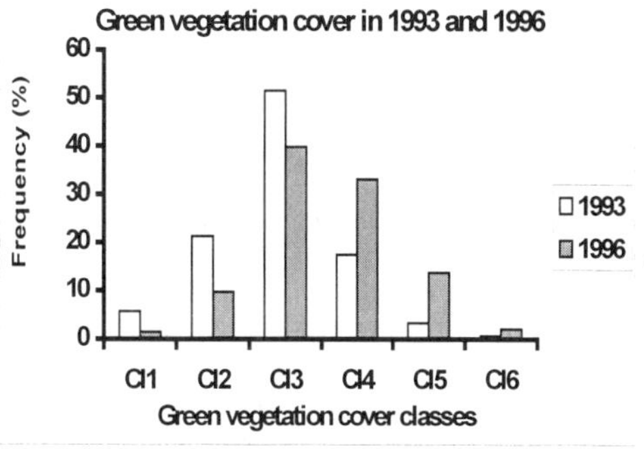

Figure 3: Vegetation cover classes in 1993 and 1996

Results are indicated at Figure 3. The recovery of woodlands is clearly visible: in 1993, about 25% of the polygons exhibit a green vegetation cover less than 6%; in 1996, the lower cover classes are less represented, and the higher classes are significantly increased.

At Figure 4 is displayed the scattergram of green vegetation cover of 1993 vs green vegetation cover of 1996; the same overall trend is obtained (cover of 1996 > cover of 1993), except for some polygons of *Q. pyrenaica* and 2 polygons of *Q. faginea.*

For *Q. coccifera*, there is an increase of cover in 1996, although the estimated green cover remains less than 15%.

For *Q. faginea*, the evolution is not uniform: for low cover polygons in 1993, there is a significant increase in 1996.

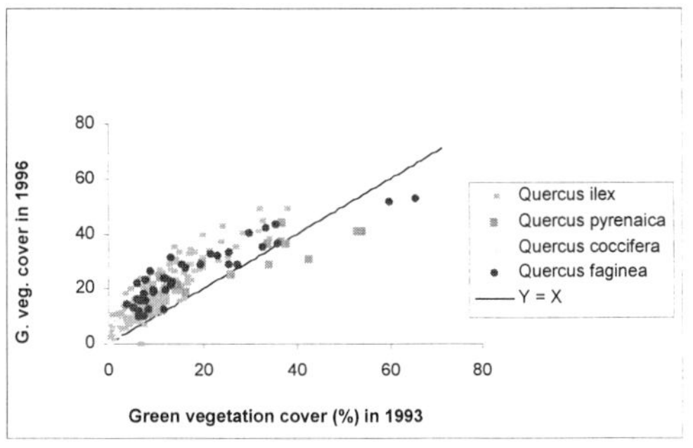

Figure 4: Green vegetation cover evolution between 1993 and 1996.

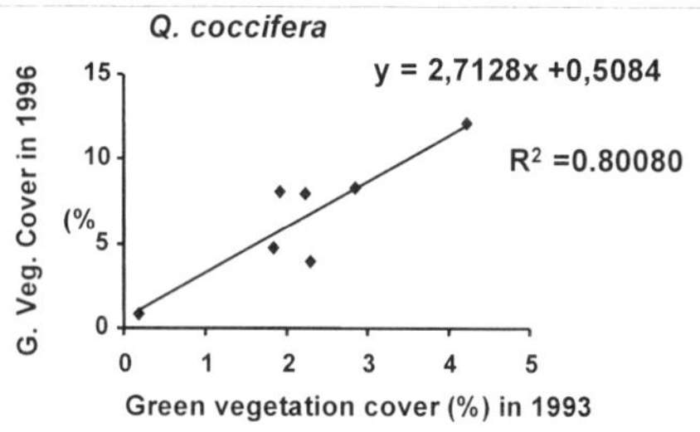

Figure 5: *Quercus coccif*era green vegetation cover comparison between1993 and 1996.

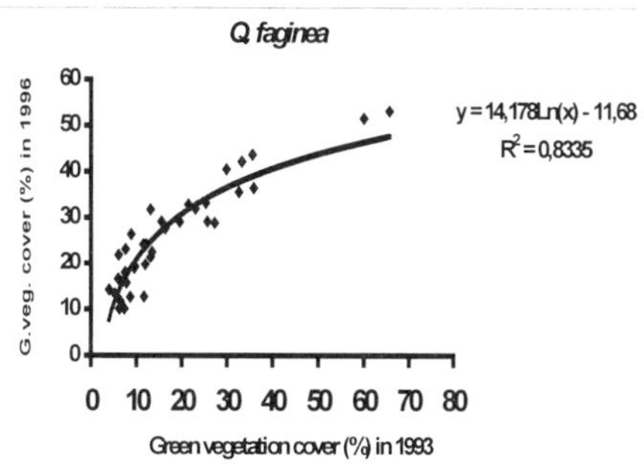

Figure 6: *Quercus faginea* green vegetation cover comparison between 1993 and 1996

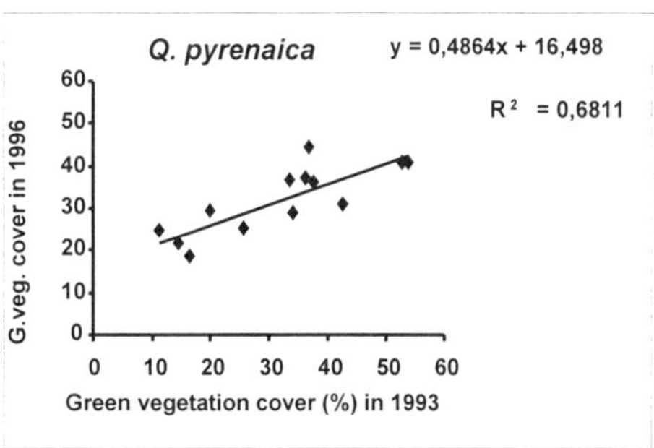

Figure 7: *Quercus pyrenaica* green vegetation cover comparison between 1993 and 1996

However, the increase is less for medium range of 1993 cover (25-40%), and for 2 polygons the 1996 cover is lower than 1993, suggesting a progressive decline, with effects delayed after the peak of drought stress.

For *Q. pyrenaica*, the same phenomenon is observed and there are 6 polygons with lower cover in 1996.

For *Q. ilex*, the observed trend is an increase in green vegetation cover, with some polygons experiencing a very marked increase.

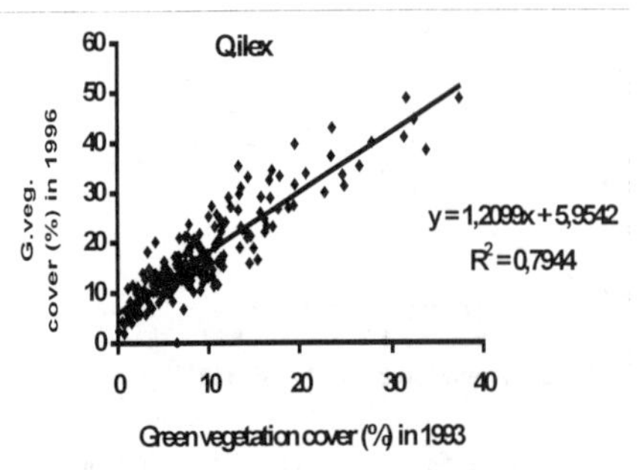

Figure 8: *Quercus ilex* green vegetation cover comparison between 1993 and 1996.

If we simplify the cover information by considering only 3 classes (< 6%, between 6 and 46%, and >46%), then the evolution can be described as follows: 55% of polygon area show an increase in cover, 40% an apparent stability (same vegetation cover class) and only 5% a decrease in vegetation cover, from which two third are polygons with high cover (>45%) in 1993.

5. CONCLUSION

Landsat Thematic Mapper appears as an adequate sensor for woodland monitoring: the spectral responses of forests in drought conditions are quite distinct in visible channels (2 and 3), and in SWIR channels (5 and 7), so that spectral indices can be elaborated to infer greenness, yellowness and brownness/ dehydratation.

From spectral responses and indices, we can identify a hierarchy of dominant species, related with green biomass in 1993: *Q. coccifera* < *Q. ilex* and *Q. faginea* < *Q. pyrenaica*. However, the number of polygons of *Q. coccifera* and *Q. pyrenaica* are too small to generalise the results.

The main result from the multitemporal comparative analysis is that after the peak of drought in 1993, woodland areas generally retrieve a higher green biomass level 3 years later. However, the reverse trend (decreasing cover) is observed for some areas *Q. faginea* and *Q. pyrenaica*, which indicates that a time delay in woodland decline can be observed, and that progressive decline can affect woodland areas which in 1993 appeared as relatively « drought resistant ». This leads to the conclusion that multitemporal satellite data should be obtained on a regular basis (i. e. each year) to monitor forest ecosystem metabolism and resistance to drought stress.

BIBLIOGRAPHY

Auclair, A. N. D., 1992. Extreme winter temperature fluctuation, root and sapwood injury, and oak dieback in central Europe, *Proc. Internat. Congress, "Recent Advances in Studies on Oak Decline"*, Selva di Fasano (Brindisi), Italy, 139-148, Sept. 13-18, 1992.

Brasier, C.M., 1996. *Phytophthora cinnamomi* and oak decline in southern Europe. Environmental constraints including climate change, *Ann Sci For*, 53, 347-358, 1996.

Carlson.N.T and Ripley D.A 1997: On the relation between NDVI, Fractional Vegetation Cover, and Leaf Area Index. *Remote sensing of environment.* N°62, pp 241-252.

Cobos.J.M, Montoya.R et Tuset.J.J 1992: New damages of the Quercus woodlands in Spain. Proc. Internat. Congress, *"Recent Advances in Studies on Oak Decline"*, Selva di Fasano (Brindisi), Italy, Sept. 13-18, 1992.

Dimitri.L 1992: Oak decline in Hessen. Proc. Internat. Congress, *"Recent Advances in Studies on Oak Decline"*, Selva di Fasano (Brindisi), Italy, Sept. 13-18, 1992.

Elvidge, D.E., 1990, Visible and near infrared reflectance characteristics of dry plant materials, *Remote Sensing of Environment*, vol. 11, pp.1775-1795.

Griffin. D. H 1992: Mechanism of disease predisposition by environment stress. *Proc. Internat. Congress, "Recent Advances in Studies on Oak Decline"*, Selva di Fasano (Brindisi), Italy, Sept. 13-18, 1992.

Montoya. J.M 1997: Mortandad de Quercinéas: La perspectiva selvicola y los antecedentes climáticos. La cuestión de Hypoxylon Mediterraneum en el Alcornocal de Mamora (Marruecos). *Reunión de coordinación sobre el decaimiento de las quercinés* (Badajoz 5 y 6 de noviembre de 1997).

Montoya Oliver J.M & Meson García Mª. L 1997: Los factores catalizadores de la "seca de los Quercus". *Reunión de coordinación sobre el decaimiento de las quercinés* (Badajoz 5 y 6 de noviembre de 1997).

Montoya. R 1997 "La seca de encinas y alcornoques" Servicio de protección contra Agentes Nocivos DGCONA. Ministerio de Medio Ambiente. *Reunión de coordinación sobre el decaimiento de las quercinés* (Badajoz 5 y 6 de noviembre de 1997).

Montoya.G 1997: Notas y comentarios sobre la évolución de la seca en España, Departamento de selvicultura. *Reunión de coordinación sobre el decaimiento de las quercinés* (Badajoz 5 y 6 de noviembre de 1997).

Ruiz De La Torre, J. 1990a. Mapa forestal de España E. 1:200.000. *Memoria general. ICONA M° de Agricultura, Pesca y Alimentación.* Madrid 191 pp.

Ruiz De La Torre, J. 1990b. Mapa forestal de España E. 1:200.000. Zarazoga. Hoja 7-4. *ICONA M° de Agricultura, Pesca y Alimentación.* Madrid. 114 pp.

Remote Sensing in the 21st Century: Economic and Environmental Applications, Casanova (ed.)
 ISBN 90 5809 096 5

Modelisation of the NDVI temporal evolution in forestry of Castilla y León, Spain, using climatic data

A.A. Fernández Manso
Departamento de Ingeniería Agraria, E.T.S. de Ingeniería Agraria, Universidad de León, Ponferrada, Spain

P. Illera, J.A. Delgado & A. Fernández Unzueta
Departamento Física Aplicada I, Remote Sensing Laboratory, E.T.S. de Ingenierías Agrarias, Universidad de Valladolid, Palencia, Spain

ABSTRACT: This paper is a description of the methodology applied and results obtained in a study of the relations among the yearly evolution of the NDVI, derived from NOAA-AVHRR images, and meteorological variables. NOAA images are captured daily by the receivers located at our Remote Sensing Laboratory and processed by means of our own software. Air temperature and rainfall recorded by the Spanish Metorological Institute (INM) have been interpolated into the spatial resolution of the AVHRR images. Solar radiation is calculated using Meteosat images, also received at our Laboratory, and a model developed by us. Three forest classes (deciduous forest, evergreen forest and pines) were chosen for 5-year study (1993-1997). These classes have an important presence in Castilla y León and also a well differentiated ecological significance. Models to relate the variables have been developed and an ecophysiological interpretation has been done. Finally, the improvement of the models by means of separating the different climatic zones using the classification CLATERES and also by introducing new variables capable to better measure hydric stress is addressed.

1. INTRODUCTION

The information provided by the NOAA satellites has a great interest in the monitoring of vegetation at a regional level being a useful tool to plan and manage sustainable forest resources. The good temporal resolution of these satellites allows us to consider a continuous monitoring of vegetation changes.

From the combination of red and infrared channels next to the NOAA-AVHRR images one is able to calculate the Normalized Difference of the Vegetation Index (NDVI). The red band is strongly correlated to the chlorophyll content and the controlled infrared by the LAI and the density of green vegetation (Bannari et al. 1995), as a whole they also inform us on the vegetation state and phytomass. Moreover the NDVI is an indicator of hydric stress and of the effect of possible forest plagues or diseases (Sellers 1989). One of the main applications of this index is the vigilance of forest areas since its measurement allows the study of temporal and spatial variations of vegetation distribution. However the study of temporal variation of the NDVI is not sufficient to carry out a monitoring on forest conditions, since this is greatly related to the annual climatic variation. To determine the normal evolution it would be useful to build models that would establish its relationship with climatic variables. This line of work has been followed by other authors who stress the great dependence between the NDVI and the climatology of the zone. The results obtained from this work can be classified according to the climatology of the study areas, in any case in hot zones of the African continent appreciable correlations were found to rainfalls (Hielkema et al 1986, Malo & Nicholson 1990, Davenport & Nicholson 1993), in cold zones the NDVI correlates more to the air (Cihlar 1993) temperature and in warm zones both variables can influence the evolution of the index (Schultz & Halpert 1993). These results are coherent with the ecophysiology of the forests and have been used as a base for the study on its relation in the region of Castilla y León (Spain).

2 MATERIAL AND METHODS

If indeed the NDVI was related to vegetal activity its relation to climatic parameters could be interpreted from the theories that relate climate to vegetation. However, before coming to the interpretation it is necessary to keep in mind a set of factors that have helped in designing the study's methodology .

In the first place, it is necessary to have a good understanding of the study's extensive territory, with a great diversity of forest surface and mesoclimatic variations. This situation has required us to select only the three main forest classes and to stratify, in a second phase, the area in homogenous zones from the physiographic and climatic point of view.

In the second place, complete information must be had on the meterological variables, collected through the Spanish Meteorological Institute's (INM) network of stations (temperature and rainfall) distributed heterogeneously throughout the territory. The strategy to follow considers the generalization of punctual information of the whole study area employing distinct interpolation methods.

In the third place, it is essential to have complete information on the evolution of the NDVI throughout each year within all the years that will be used in the study. The index evolution series was assembled from 37 decennial composites that coincided with other ones from each of the climatic variables selected.

Once the necessary information is obtained, for each forest class the value of the vegetation index and the climatic variables in each of the 37 annual composites for the distinct years of the study through the use of a GIS evolution graphs can be assembled and then be able to proceed to the statistical analysis of data.

To sum up, from the punctual data provided by Castilla y León's Network of Meteorological Stations a series of tracking digital images have been created raster that summarize the values of temperatures every 10 days throughout the territory at a scale where each pixel corresponds to a square kilometre. With the satellite images decennial composites of NDVI's maximum value have been obtained. The climatic information and the NDVI have been extracted for 3 vegetation classes by means of superimposing the digital cartography prepared from Corine Land - Cover. Subsequently, using the Territory Classification of Spain CLATERES, the territory in eight areas has been stratified. The annual evolution curves of the NDVI and climatic variables have been obtained as a final result.

2.1 *Study territory and digital cartography*

The Autonomous Community of Castilla y León is situated in the northern centre of the Iberian Peninsula and occupies 94,147 square kilometres . The relief is defined by a group of high plains (700-1,100 m) in the interior of a vast sedimentary basin whose edges are formed by distinct mountainous systems. This contrasting relief and its geographical situation gives important climatic modifications within a mediterranean climate. These modifications are manifested in the distinct intensity of summer aridity and in the regimentation of temperatures: the aridity is greater and the temperatures lower in a measurement that ascends through the mountainous borders. The regional forest surface is 34.53% of which 53.6% corresponds to coppice formations, 23.8% coniferous and 19.7% leafy.

From the digital cartography in Arc/Info format developed in the Corine Land - Cover Program a new one was generated in a tracking format. Subsequently a thematic group of three great forest classes was produced The data analysed were the average values of a sample of 287,300 ha of perennifolias - mainly *Quercus ilex* which represents 88% of the surface that this species occupies in Castilla y León; 336,500 ha of caducifolias - *Fagus sylvatica, Quercus spp., Castanea satica,* ...- approximately 89% of the regional total; and, 515,700 ha of pine forests which correspond to 68% of the community's total (AFCL, 1997).

The territorial Biogeoclimatic Classification of Spain (CLATERES) (Elena, 1996) divides the Iberian Peninsula into eight Ecoregions and 215 classes. From 39 classes of the 2nd Ecoregion (Duriense) only a reclassification of 8 main classes has been done in function with the proximity within the multifactorial physical slope which appears in the classification.

2.2 *Satellite data*

AVHRR images of channels 1 and 2 have been employed. The Remote Sensing Laboratory of the University of Valladolid, LATUV of the Applied Physics I Department of the University of Valladolid has an NOAA-HRPT receiver from which the digital transmissions by NOAA satellites are captured on a daily basis.

The values of the NDVI are seen influenced by the direction of the reflectance of the earth's surface and the differences of the satellite's spatial resolution in each pixel; moreover, one must consider the disturbances provoked by the atmosphere in the infrared that tends to reduce the index. To minimize these effects, the Maximum Value Composites MVC (Holben, 1986) are used. A ten day period has been selected since it is the closest to the cycle of the satellite.

The Meteosat images are also received by LATUV and data are available every hour. Those of the visible channel are used to calculate the hourly global solar radiation from a statistical model developed and validated (Illera, 1995) The values are obtained for all the Peninsula with a spatial resolution of approximately 5 x 7 km and the model provided errors of 19% at the hourly radiation level and of 9% if an accumulation of the values in a datum of daily radiation is conducted. To reduce the volume of information a process of integration, which uses only data from the hours of 10, 12 and 14, has been designed. Finally, the daily average of global solar radiation corresponding to ten day periods is calculated. As a last step, the Meteosat images are georeferenced to a UTM projection of a 1 x 1 km resolution, to superimpose them on NDVI data and surface temperature.

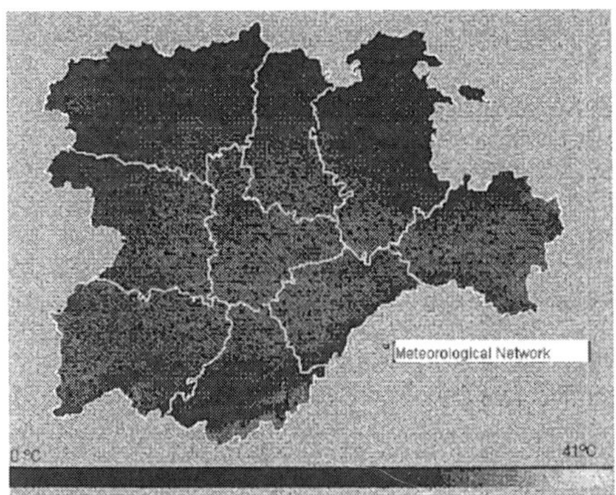

Figure 1. Example of air temperature image Castilla y León, 29-VI-94

2.3 *Preparation of a digital cartography of air temperatures and rainfall according to NOAA scale.*

An interpolation of the thermal values registered at the 240 meteorological stations of the Regional Network has been conducted to obtain a map of temperatures taking into account the relief effect. To achieve that, a Digital Model of Castilla y León's Soil has been used. Each digital level will equal the air temperature in a zone of 1 square kilometre.

Regarding rainfall, from the 740 stations of the INM Network and through a process of Thiessen's polygons images of the whole territory have been obtained. As final data we have maps of distinct temperatures and accumulated rainfall corresponding to ten day periods. The data, like all the others described before, is in a raster format (UTM projection, 1 x 1 km).

3 RESULTS and CONCLUSIONS

3.1 *NDVI and meteorological variables relations in three forest classes*

The results corresponding to 1993 are presented. Data of the NDVI-AVHRR is available, as well as the average, maximums, average of maximums, minimum and average of minimums temperatures, all of them corresponding to ten day periods and covering all the regional territory.

In figure 2 the NDVI-maximum temperature relation can be noticed for the three classes in 1993. In a first analysis a relation approximately lineal can be observed, better in the case of caducifolias -r2 = 0,77 -, that worsens for pine forests -r2 = 0,64 - and clearly deficient for perennifolias -r2 = 0,38 -.

If adjustments are done for squared minimums, the results shown in the first column of table 1 are obtained. The best results are obtained by relating the NDVI to the maximum temperature and for the same ten day period in both variables.

In figure 3, the relation regarding evergreen forest is observed, indicating next to each point the corresponding day of the year. One can see that a good lineal relation exists on day 164. From there, the lineal relation ceases to be satisfactory. The pine forests and caducifolias present a similar deportment, although the lineal relation seems to be valid until day 224 and deviation of the lineal deportment does not turn out to be that evident for the rest of the year. In table 1 the results of regressions corresponding to the lineal period are displayed. The elevated coefficients of correlation is to be stressed.

The results presented seem to indicate that the NDVI-air temperature relation is easier to model in periods of vegetation growth when extreme temperatures do not appear and there is no significant hydric deficit. The best results for caducifolias, that occupy mountainous or riverside zones, are obtained from this. Regarding pine forests, to say that the class of CORINE cartography. is somewhat heterogeneous that is the reason for deciding to conduct a segregation of the different formations: the mountainous or average mountain pine forests - P.sylvestris or P.nigra - from those of the central region, a lot more arid - P.pinea, P.halepensis or P.pinaster.

In the NDVI-rainfall relation, no clear results have been found if the rainfall corresponding to the same ten day period is used. Nor with the previous period. The relations improve when accumulating the rainfall of the previous periods with the maximum value composite of the NDVI. In the figure 3 the rainfall results from January to the said period for caducifolias in 1993 are displayed. The lineal relation is very good until day 224. The rainfall ceases during the summer. In autumn a crescive lineal period seems to exist, although there are few data to conduct an analysis. The pine forests and perennifolias show a similar deportment, although in the last case the rainfall ceases from day 164. In table 1 the results of the regression for the period where the relation is clearly lineal can be seen. The best coefficient of correlation corresponds again to caducifolias.

If the 1994 data is analysed similar results can be found. Moreover the values of the coefficients a and b for the "lineal period" remain sufficiently stable. The difference is that the rainfall was a lot less and the said period is more reduced.

To model the whole year the joint use of both variables can be contemplated, but in figure 3, as can be seen, the accumulated rainfall does not suppose a good selection. This variable is constant, and thus not adequate, in periods when there is no rainfall. To deal with data of an entire year as a whole, it is necessary to look for a variable capable of reflecting

Table 1. NDVI and meteorological variables

		Maximun T^a	Maximun T^a Linear period.	Accumulated Rainfall Linear period.
Pines	a	0,1475	0,2307	0,2918
	b	0,0088	0,0078	0,0005
	r^2	64,44	74,08	75,47
	Periodo		224	224
Evergreen forest	a	0,1725	0,1747	0,2895
	b	0,0063	0,012	0,0008
	r^2	37,98	82,95	76,93
	Periodo		164	164
Deciduous forest	a	0,0729	0,1666	0,2664
	b	0,0142	0,014	0,0007
	r^2	76,89	85,12	94,32
	Periodo		224	224

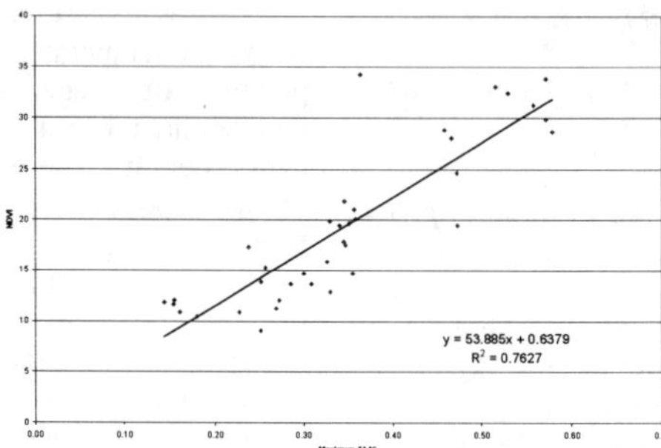

Figure 2. NDVI-maximum air temperature relation for deciduous forest class in 1993.

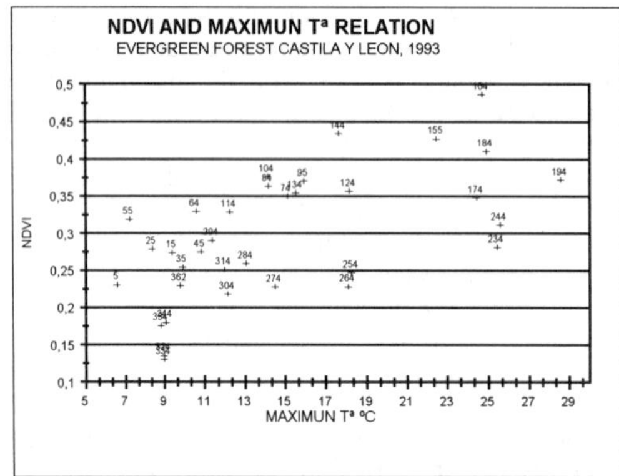

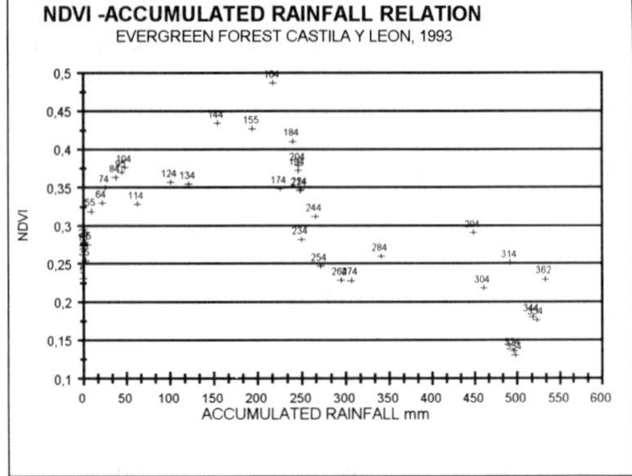

Figure 3. NDVI and maximun temperature relation, NDVI and accumulated rainfall relation

the hydric stress of summer. If only the periods when the NDVI shows a crescive trend are taken, the introduction of both variables will give a good modeling.

3.2 *NDVI and meteorological variables relations for distinct climatic areas.*

Using CLATERES the pine forests have been separated into 8 classes. Once the zones were differentiated, we have proceeded to analyse the relations between NDVI and the meteorological variables following the same methodology as in the previous analyses. As can be proved, in the table 2 and figure 4 a gradation exists from the CT1 in zones of smaller altitude (high temperatures, low rainfall) up to the mountainous border of the CT8 (colder and more humid).

In table 2 the relation between the maximum temperature and the NDVI for the 8 classes of pine forests in 1993 can be seen. The best adjustments are obtained in mountainous zones classes (P.sylvestris), being much worse the zones dominated by mediterranean pine forests.

If an adjustment is carried out by squared minimums, the results displayed in the first column of table 2 are obtained. The best results are also obtained for the NDVI with maximum temperature and for the same ten day period in both variables.

In table 2 it can be observed as well that there is a good lineal relation up to day 164 in the first classes. From there, the lineal relation ceases to be satisfactory. The pine forests in zones with less aridity in summer present a similar deportment, although the lineal relation seems to be valid until day 224. In this table the results of the regressions corresponding to the lineal period are presented. The elevated coefficients of correlation are stressed.

From these results the relation between the NDVI and vegetal activity can be perceived. Forest species are poikilothermal organisms incapable of creating an interior thermal environment, and therefore directly dependent on air temperature.

The results displayed again confirm that the NDVI-air temperature relation is easier to model in periods of vegetation growth when extreme temperatures are not present and no important hydric deficit exists. A future endeavour will be to analyse the temporal series of the NDVI and temperature to check the relation between the vegetation period and the climatic rate, studying how the coordination between the annual variation of temperature and the physiological activity is also conditioned by adapting the climatic history of the area.

The relations improve when accumulating the rainfall of previous periods to the maximum value composite of the NDVI as occurred in the first part of the work. The lineal relation is very good until day 224 in the classes less mediterranean. During the summer rainfall ceases and a drop of the NDVI is produced. In autumn a crescive lineal period seems to exist again.

With regard to water needs, the vascular stoppages can be considered homeohydric. In this case the values corresponding to rainfall, that up to now is the available information, turn out to be sufficiently remote from the actual hydric availability for the plant. Therefore it will be more interesting to find out the true availability of water for vegetal metabolism than the total rainfall. The problem is in defining a simple model of hydric balance that would permit determining the evolution of true availability and thus be able to relate it to the vegetative activity.he exclusively accumulated rainfall allows us to discern two things. In the first place, the moment rainfall ceases and the hydric availability depends on reserves. When water is deficient for plants a series of physiological processes begin which makes the NDVI decline: attemperment of vegetative activity and control of transpiration that can culminate in a detainment of the vegetative activity.

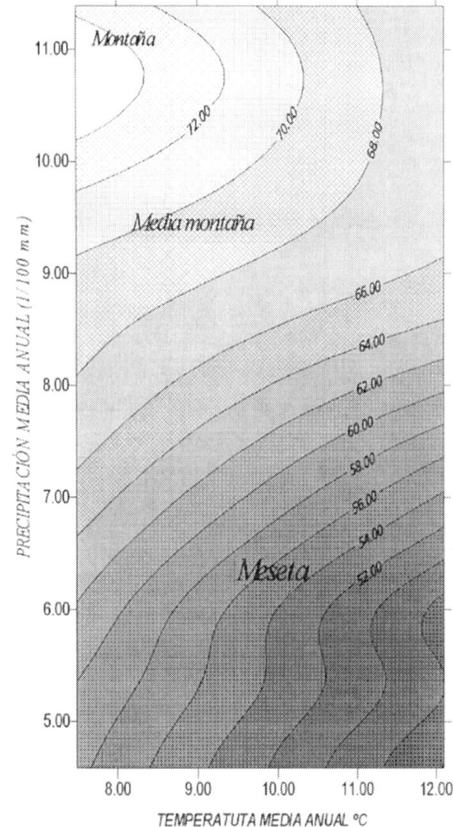

Figure 4. NDVI and maximun temperature (r^2)

If the differences of the plants' deportment are not important when water is abundant, they certainly are when it is deficient, which will condition the interpretation for each of the forest formations. In the second place, the rainfall accumulated to this date when the water contribution ceases does not give an approximation of water that the plant may have as reserve although we know that this will depend on the capacity to retain water from the soil, on the geomorphology and on the annual climatic evolution expressed by true evapotranspiration. Greater accumulated rainfall implies a greater delay in the sudden vegetative stop and a better modeling of the NDVI. In figure 5 one can observe how for the classes of mountain areas CT 8 when rainfall ceases the surface since the month of January it has received 400 mm compared with 200 mm of CT 2 which extends the availability to more advanced dates.

To model the periods with greater hydric stress it is necessary to look for new and more significant variables, especially with aspects related to hydric

Table 2. NDVI and meteorological variables relations for distinct climatic areas

		Maximun T	Maximun T* Linear period	Accumulated Rainfall Linear period.
Clase 1	a	0.143	0.145	0.288
	b	0.007	0.010	0.0006
	r^2	47.11	71.68	76.40
	Period		164	164
CT 2	a	0.133	0.142	0.268
	b	0.006	0.008	0.0006
	r^2	42.77	82.16	54.73
	Period		164	164
CT 3	a	0.117	0.072	0.245
	b	0.007	0.011	0.0009
	r^2	48.44	84.82	69.23
	Period		164	164
CT4	a	0.128	0.176	0.288
	b	0.008	0.008	0.0005
	r^2	59.41	65.13	53.24
	Periodo		174	174
CT 5	a	0.147	0.205	0.317
	b	0.0011	0.009	0.0005
	r^2	71.29	75.80	73.27
	Period		204	204
CT 6	a	0.104	0.154	0.278
	b	0.011	0.010	0.0005
	r^2	74.29	84.74	77.38
	Period		204	204
CT 7	a	0.196	0.192	0.297
	b	0.0097	0.0098	0.0004
	r^2	68.07	71.05	59.36
	Period		214	214
CT 8	a	0.182	0.238	0.328
	b	0.011	0.009	0.0005
	r^2	75.72	79.37	77.26
	Period		224	224

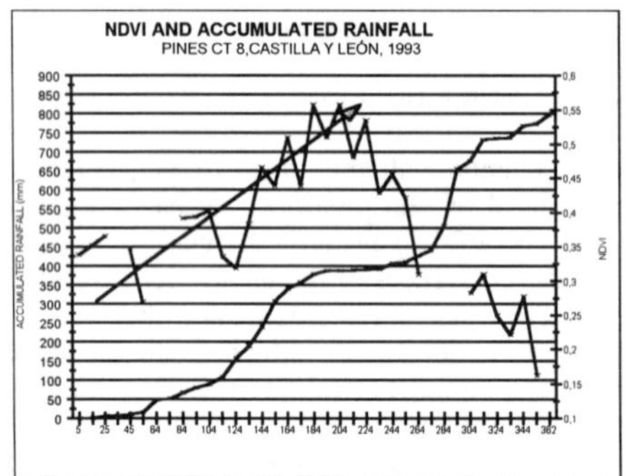

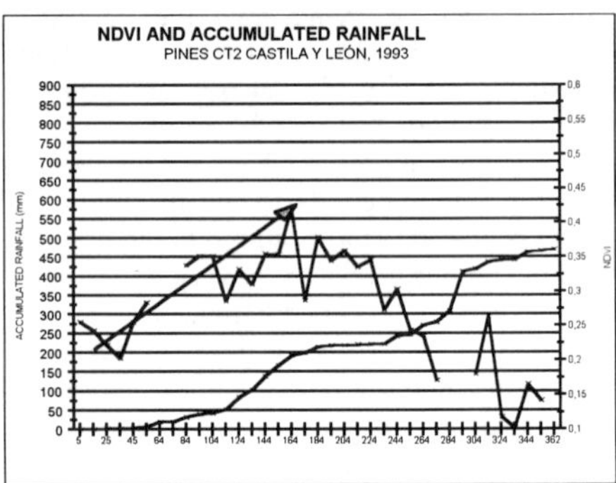

Figure 5 . Rainfall and the hydric availability

balance. This line of work will help in finding adequate models for all the zones of our Autonomous Community.

4 ACKNOWLEDGEMENTS

To the Doctoral Thesis assistance program of the Caja Madrid Foundation. And to C.I.C.Y.T. and Castile and Leon Board for financing the distinct research projects.

5 REFERENCES

Asociación Forestal de Castilla y león. 1997. Resultados del Segundo Inventario Forestal en Castilla y León. *Bolentín Informativo nº 1.*

Bannari, A.D. Morin, F. Bonn, and. Huete, A.R. 1995 *Areview of vegetation indices. Remote Sensing* 13, pp.95-120,

Cihlar, J. 1993. A multiyear analysis of the relationship between surface environmental variables and NDVI over the Canadian landmass. *Remote Sensing Reviews* 7: 151-177.

Davenport. M. L and. Nicholson.. S. E 1993. On the relation between rainfall and the Normalized Difference Vegetation

Index for diverse vegetation types in East Africa. *International Journal of Remote Sensing*. 14: 2369-2389.

Elena-Roselló, R. 1996. Un instrumento para la planificación territorial de la reforestación en España : la Clasificación Biogeoclimática Territorial CLATERES. *El campo* 134: 105-118.

Hielkema , J. U.. Prince, S. D and Astle,.. W. L .1986. Rainfall and vegetation monitoring in the Savanna Zone of the Democratic Republic of Sudan using the NOAA Advanced Very High Resolution Radiometer. *International Journal of Remote Sensing*. 7: 1499-1513. .

Holben B.N 1986. Characteristics of maximun-value composite imagenes from temporal AVHRR data. International Journal of Remoting Sensing, 7: 1417-1434..

Illera, P., Fernández, A. & Pérez, A. 1995. A simple Model for the Calculation of Global Solar radiation using Geostationary Satellite Data. *Atmospheric Research* 39: 79-90

Malo, A. R. and S. E Nicholson. A . 1990. study of rainfall and vegetation dynamics in the African Sahel using normalized difference vegetation index. *Journal of Arids Environments*. 19: 1-24,.

Schultz, P. A. & Halpert, M. S. 1993. Global correlation of temperature, NDVI and precipitation. *Advances on Space Research* 5: 227-280

Sellers. P.J. .1989. Vegetation-canopy spectral reflectance and biophysical prosesses. In *Theory and Applications of Optical Remote Sensing*, (G. asrar, De.), Jhon Wiley, New York, 297 - 335..

2 Meteorology

Remote Sensing in the 21st Century: Economic and Environmental Applications, Casanova (ed.)
© 2000 Balkema, Rotterdam, ISBN 90 5809 096 5

Consequences of spatial resolution transformations for land-use classification

U. Fehrenbach & D. Scherer
MCR Laboratory, University of Basel, Switzerland

ABSTRACT: This paper presents consequences of spatial resolution transformations of input data sets for the classification of areal types and ventilation situations. Areal types are aggregated land-use units determined from areal percentages of pixel classes resulting from maximum-likelihood classifications of Landsat-5 TM and ERS-1 SAR imagery. Spatial resolution of pixel classes is 30 m, while areal percentages were computed for different spatial resolutions between 25 and 1'000 m. Ventilation situations are numerical representatives of the most relevant ventilation conditions, and are classified from land-use and terrain parameters. Areal percentages and digital terrain models with corresponding spatial resolutions were used to derive these input parameters. All input data sets show significant changes for different spatial resolutions effecting spatial distributions and total areas of the resulting areal types and ventilation situations. The paper discusses these effects focussing on the satellite-based data sets. The necessity of adequate spatial resolutions for urban and regional planning applications of remote sensing data could be demonstrated by this study.

1 INTRODUCTION

The main objective of the KABA project (Climate analysis of the region of Basel/Switzerland) (Fehrenbach & Scherer 1998, Parlow et al. 1997, Scherer et al. 1997) was to provide planning authorities with maps containing site-specific and aggregated information on interactions between climate and the heterogeneous spatial structure of the region of Basel at map scales of 1:25'000 and 1:100'000. Due to the size (51 km West-East, 33 km North-South) of the study area and its complex topography with elevations ranging from 250 to 1'200 m a.s.l., new concepts had to be developed to adequately describe settlement structures and climatic conditions in the studied region.

Planning authorities do have a greater interest in the function and composition of urban areas than in single satellite-based land-use types only classified by their electromagnetic surface properties. The concept of areal types, representing functional as well as physiognomic units, was invented to fit the planners needs for more appropriate land-use categories and to reduce problems linked to conventional satellite-based land-use classifications (e.g. mixed pixels).

Ventilation conditions display an important impact on the regional and local climate. They are the result of characteristic combinations of climatic factors which control local and regional wind fields as well as vertical air mass exchange processes near ground (Fehrenbach 1999, Scherer et al. in press). This paper mainly concentrates on the results of the spatial resolution transformations concerning the land-use classification in KABA. In addition, changes of the spatial distribution of ventilation situations are briefly discussed.

2 METHODS

Two already existing land-use data sets had to be merged to cover the whole study area of both cantons Basel-Stadt and Basel-Landschaft. The first one was generated as part of REKLIP (Regional Climate Project) (REKLIP 1995) based on one Landsat-5 TM scene (Scherer et al. 1994). The second data set resulted out of a project named ERSCLiP (ERS-1 Climatology Project) using one Landsat-5 TM and three ERS-1 SAR images (Beha et al. 1996, Parlow et al. 1996a and 1996b, Scherer et al. 1996a). Both land-use classifications were carried out by applying a maximum-likelihood classifier. After merging both images, a set of 15 classes representing urban and rural surface types respectively settlement structures, the so-called pixel classes (see Table 1), formed the basis of the investigations described in this paper. The spatial resolution of this data set is 30 m.
Pixel classes were used to derive spatial distribution of areal types, which are aggregated land-use units

Table 1. Satellite-based pixel classes used in KABA.

No.	Pixel class	Main location
1	Rails	Urban and rural
2	Allotments	Urban peripheral zones
3	High density settlements	Urban
4	Village centers	Rural
5	Building blocks with inner yards	Urban
6	Serial houses	Urban and rural
7	Building complexes	Urban, partially in rural zones
8	Industrial buildings	Urban, partially in rural zones
9	Forests	Rural, partially urban
10	Meadows, pastures and orchards	Rural, partially urban peripheral zones
11	Arable lands	Rural, partially urban peripheral zones
12	Water surfaces	Urban and rural
13	Asphalt and concrete surfaces	Urban and rural
14	Sports fields, parks and village greens	Urban and rural
15	Detached houses	Rural and in urban peripheral zones

Table 2. Areal types used in KABA.

Areal type	Main location
Forestial areas	Rural, partly urban areas
Grasslands, orchards	Rural areas, urban periphery
Waters	Urban and rural areas
Arable lands	Rural areas, urban periphery
Sports fields, parks, and village greens	Urban and rural areas
Horticultural areas	Urban periphery
Extensive railway areas	Urban and larger towns, rural areas
Low density housing	Rural areas, urban periphery
Settlements of mixed structure	Urban areas, town centres
Urban housing	Urban areas
Dense urban housing	Mainly close to city centres
Combined housing and industrial areas	Urban and dedicated rural areas
Commercial and industrial areas	Urban and dedicated rural areas
High-density urban areas	Mainly in the city centres
Extensive traffic areas	Urban and rural areas

determined from areal percentages of pixel classes. Each areal type is defined by a characteristic combination of areal percentages of pixel classes within specific ranges (Fehrenbach 1999, Scherer et al. 1996b, Scherer et al., in press). Areal percentages were computed for grids of varying spatial resolution, i.e. for 25, 100, 250, 500 and 1'000 m.

First, the total area attributed to a certain pixel class located within a window centered around a grid element of the corresponding areal percentage data layer is computed. This area is then divided by the size of the window resulting in the areal percentage. After carrying out this computation for each element of the areal percentage grid and for each pixel class, the spatial distributions of areal percentages are available. The window size was set to one pixel of the resulting areal percentage grid except for the 25 m resolution, where a window size of 100 m was used to avoid numerical problems. Details of the method are presented by Scherer (1998) and by Fehrenbach (1999).

Areal percentage distributions are used as input for a rule-based classifier to determine the spatial distribution of areal types. Rural areal types are mostly delineated by only ensuring the majority of a certain pixel class (e.g. forests are obviously the main pixel class for forestial areas), whereas urban areal types generally demand complex rules combining the areal percentages of several pixel classes with 'AND' or 'OR' conditions. Table 2 shows the resulting 15 areal types. Areal type grid always has the same spatial resolution as the corresponding areal percentage grids used as input in the rule-based classification.

Six typical and distinctive ventilation situations (see Table 3) were delineated for the region of Basel (for a comprehensive description see Fehrenbach 1999 or Scherer et al. 1997). If a ventilation situation is indicated for a certain area, then this reflects a higher probability or an increased relevance of certain atmospheric processes (e.g. hazardous winds) for this area relative to its surroundings. Since topography and land-use are major controls of ventilation conditions, digital terrain model data and surface parameters induced by land-use structures were used as input for a further, rule-based classifier to determine the spatial distribution of ventilation situations. Surface parameters, i.e. aerodynamic roughness length, zero-point displacement and effective building height (total volume of buildings per grid element divided by the size of the grid element), were computed from areal percentage data (Fehrenbach 1999, Scherer 1998). Spatial resolution is the same for the spatially distributed surface parameters as for the areal percentage grids, on which they are based.

Whereas wind-exposed areas and both types of terrain-induced ventilation are extracted only by terrain parameters, areas of undisturbed respectively reduced ventilation are delineated only by land-use parameters. Areas of cold air accumulation need both terrain and land-use parameters for being classified. Resulting ventilation situations were determined for spatial resolutions of 100, 250, 500 and

Table 3. Ventilation situations used in KABA.

Ventilation situation	Brief description
Wind-exposed areas	Higher wind velocities more likely due to exposed topographic situations.
Terrain-induced ventilation paths	Channelled flow due to drainage flows from upslope areas in the vicinity, or by channelling of synoptic winds due to topography.
Terrain-induced ventilation on slopes	Slope winds or drainage flows along gradient lines. Partly channelling of synoptic winds.
Undisturbed ventilation	Undisturbed flow due to small roughness values and weak or missing flow obstacles respectively artificial heat sources.
Reduced ventilation	Modified flow due to higher roughness values and significant flow obstacles.
Cold air accumulation	Reduced wind velocities and enhanced stabilisation of stratification due to cold air accumulation in hollows or by barrier effects at valley narrows or settlement structures.

1'000 m. Therefore, not only the land-use parameters but also terrain parameters had to be computed for different spatial resolutions. The latter was done by aggregating digital elevation data before carrying out the numerical terrain analysis.

3 RESULTS

Table 4 shows average areal percentages (AAP) of each pixel class. AAP is not depending on spatial resolution since the total area occupied by a certain pixel class is only given by the number of corresponding pixels in the 30 m grid. However, 90% values also displayed in table 4 partly show large changes between different spatial resolutions. The 90% value of a certain pixel class represents a threshold for areal percentages not exceeded by 90% of all pixels.

Pixel classes 'forests', 'meadows, pastures and orchards' and 'arable lands' holding large AAP values show decreasing 90% values at coarser spatial resolutions. In total, 90% values are 20 to 30% lower for 1'000 m than for 100 m resolution. This reduction can be explained by the fact that these pixel classes, which typically occur in connected areas, are mixed up with other pixel classes at coarser spatial resolutions. The same holds for asphalt and concrete surfaces respectively for

Table 4. Average areal percentage (AAP) of pixel classes (PC) and 90% values at different resolutions.

PC	AAP (%)	90% value (%)			
		100 m	250 m	500 m	1000 m
1	0.1	0.0	0.0	0.0	0.0
2	0.5	0.0	1.4	1.6	1.8
3	0.1	0.0	0.0	0.0	0.0
4	1.4	0.0	4.3	4.6	4.2
5	0.8	0.0	0.0	0.3	0.8
6	1.9	0.0	3.3	4.7	5.4
7	1.1	0.0	2.8	3.2	3.0
8	1.0	0.0	0.9	1.6	2.4
9	27.9	99.0	87.8	77.7	66.1
10	32.3	87.0	75.0	67.7	61.6
11	20.1	64.0	55.0	49.4	44.3
12	0.9	0.0	0.0	0.7	0.9
13	2.7	9.0	8.1	7.0	6.1
14	2.6	6.0	8.9	9.6	10.0
15	6.4	24.0	20.8	19.5	18.1

Numbers of pixel classes are shown in Table 1.

detached houses, despite their AAP values are much smaller. Village centers and building complexes show only minor fluctuations indicating low sensitivity to spatial resolution transformations. However, most of the urban pixel classes tend to show increased 90% values at coarser resolutions (e.g. allotments, building blocks with inner yards, serial houses, industrial buildings, sports fields, parks and village greens). These pixel classes generally have small AAP values and are spread over a large number of unconnected areas. These patchy areas are frequently aggregated at coarser spatial resolutions resulting in higher 90% values.

Figures 1 and 2 display the spatial distributions of pixels exceeding the 90% value of the corresponding spatial resolution for pixel classes 'serial houses' (Figure 1) and 'arable lands' (Figure 2). The general tendency towards aggregated areas at coarser spatial resolutions can be seen in both cases.

Figures 3 and 4 show total areas of selected urban and rural areal types for the investigated spatial resolutions. The following discussion of the results also includes areal types not presented in these figures.

Between 25 and 100 m, only minor differences occur, since the same window size of 100 m was used in both cases for the computation of areal percentages of the pixel classes. Starting at 250 m, significant shifts in the total area of certain areal types can be seen, particularly if they occupy small areas, i.e. they have low AAP, or display a patchy spatial distribution. Some of them even completely vanish at 1'000 m (e.g. extended traffic areas). A reduction of spatial resolution results in an increase of areas with a mixed residential/commercial structure. Depending on changes in settlement densities at different spatial resolutions, areas of low density housing either turn

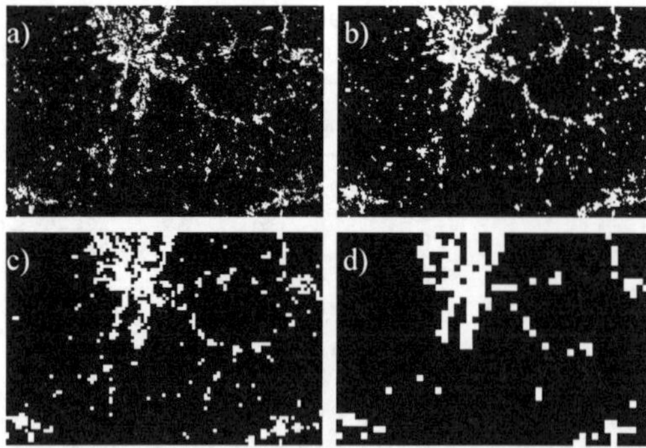

Figure 1. Pixels (white) with areal percentages of serial houses above the 90% value of the corresponding pixel class at different spatial resolutions: a) 100, b) 250, c) 500 and d) 1'000 m.

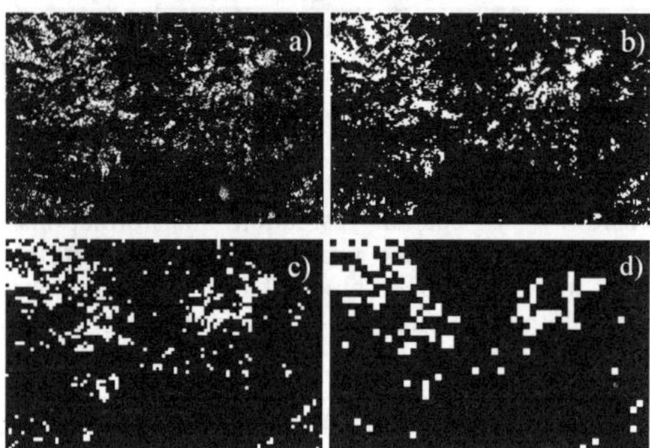

Figure 2. Pixels (white) with areal percentages of arable lands above the 90% value of the corresponding pixel class at different spatial resolutions: a) 100, b) 250, c) 500 and d) 1'000 m.

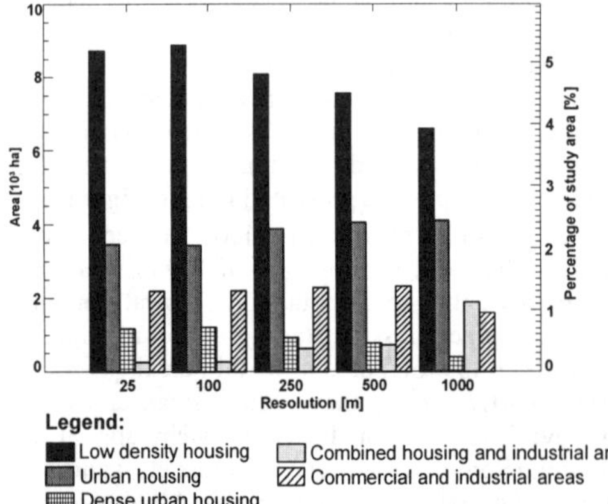

Figure 3. Total area of selected urban areal types at different spatial resolutions.

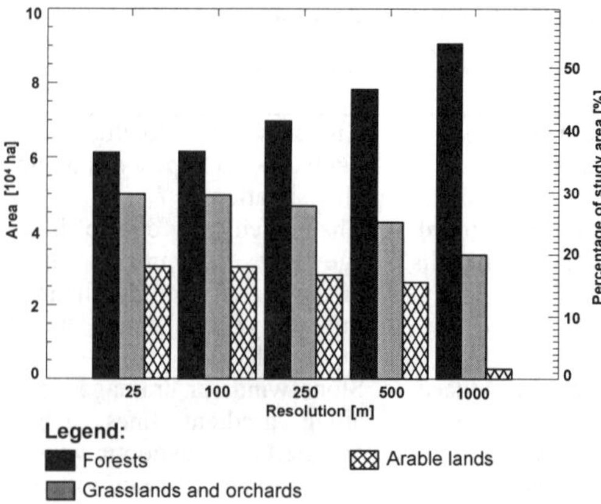

Figure 4. Total area of selected rural areal types at different spatial resolutions.

into areas of urban housing with higher densities, or into sports fields, parks and village greens with lower densities. The total area of dense urban housing areas decreases at coarser resolutions demonstrating the general tendency of larger integration areas to reduce the mean density of highly urbanised regions. Figure 4 implies that forestial areas generally occupy large, connected areas strongly growing in size. Since grasslands and orchards and particularly arable lands display smaller areas their areal extends are reduced at lower spatial resolutions.

In contrast to areal types, even small changes of spatial resolution result in drastic changes of the spatial distribution of ventilation situations. Some general aspects of the spatial distribution especially for the terrain-induced ventilation situations remain stable. However, most of them display an increasing loss of significance at resolutions lower than 100 m. These changes are induced either by coarser patterns of land-use dependent parameters or by oversimplified topographic structures at reduced spatial resolutions.

4 CONCLUSIONS

The dependency of areal percentages on spatial resolution is controlled both by the average areal percentage and by the spatial distribution of the corresponding pixel class. Since this data set is the central element for classifying both areal types and ventilation situations, the influence of spatial resolution transformations on the resulting areal percentages demands more attention in the future. This statement is further strengthened by the fact that highly multispectral or even hyperspectral imagery, which allow a direct determination of areal percentages of elementary surface types, will become more prominent in the next decade.

The concept of areal types derived from areal percentages of pixel classes by a rule-based classifier corresponds much better to the requirements of urban and regional planning authorities than conventional land-use classifications directly based on satellite data. Spatial distributions of areal types are very similar at spatial resolutions of 25 and 100 m, implying that both resolutions show sufficient accuracy for using them for urban and regional planning purposes. For studies that do not require a high degree of spatial accuracy, 250 m resolution may also be considered as adequate. At resolutions of 500 m and less, land-use structures represented by areal types display a significant loss of accuracy, indicating that such coarse spatial resolutions are not applicable to urban and regional planning. It is an open question, whether this problem could be reduced by improving the rule-based classification for areal types.

The results for the ventilation situations indicate that spatial resolutions of less than 100 m can not be recommended for considering ventilation aspects in urban and regional planning. Since land-use parameters are derived from satellite data, this result shows the need for high-resolution satellite systems for this kind of remote sensing research and application.

REFERENCES

Beha, H.-D., E. Parlow & D. Scherer 1996. Land-use classification of the agglomeration of Basel/Switzerland combining multi-sensorial satellite data from Landsat-TM and ERS-1. In E. Parlow (ed.), *Progress in environmental remote sensing research and applications*: 77-82. Rotterdam: Balkema.

Fehrenbach, U. & D. Scherer 1998. *Klimaanalyse der Region Basel - Begleittext zu den Klimaanalyse- und Planungshinweiskarten*. MCR Lab, University of Basel.

Fehrenbach, U. 1999. *Analyse und Bewertung lokal- und regionalklimatisch wirksamer Faktoren in der Region Basel*. Ph.D. thesis, University of Basel.

Parlow, E., H. Gossmann & M. Braun 1996a. *ERSCLIP - 'ERS-1 Climatology Pilot Project' - Endbericht*. Final report for the German Space Agency.

Parlow, E., D. Scherer, H.-D. Beha, H. Gossmann & M. Braun 1996b. Urban climatological parameters derived from multisensor satellite data of ERS-1 and Landsat-TM. *ESA SP-383*: 173-178.

Parlow, E., D. Scherer, U. Fehrenbach & H.-D. Beha 1997. Satellite-based climate analysis of Basel/ Switzerland. *ESA SP-414*: 271-274.

REKLIP (ed.) 1995. *Klimaatlas Oberrhein-Mitte-Süd*. Offenbach, Strassburg, Zürich: IFG, Editions Coprur, vdf Hochschulverlag AG.

Scherer, D. 1998. *Regionale Geosystemanalyse - Theorie und Beispiele*. Habilitation thesis, University of Basel.

Scherer, D., G. Bing, U. Heinzmann, E. Parlow & M. Wüthrich 1994. Problems and applications of a Landsat-TM based land use classification of the trinational region of the Upper Rhine Valley (Germany, France and Switzerland). In R. Vaughan (ed.), *Remote sensing from research to operational applications in the new Europe*: 3-10. Budapest: Springer-Hungarica.

Scherer, D., E. Parlow & H.-D. Beha 1996a. Roughness parameter derivation from ERS-1 and Landsat-TM satellite data for the agglomeration of Basel/Switzerland. In E. Parlow (ed.), *Progress in environmental remote sensing research and applications*: 325-329. Rotterdam: Balkema.

Scherer, D., U. Fehrenbach, E. Parlow & H.-D. Beha 1996b. Determination of aggregated areal types from a Landsat-TM and ERS-1 based land-use classification for the agglomeration of Basel/Switzerland. In E. Parlow (ed.), *Progress in enviromental remote sensing research and applications*: 197-200. Rotterdam: Balkema.

Scherer, D., U. Fehrenbach, E. Parlow & H.-D. Beha 1997. *Klimaanalyse der Region Basel - Technischer Bericht*. MCR Lab, University of Basel.

Scherer, D., U. Fehrenbach, H.-D. Beha & E. Parlow (in press). Improved concepts and methods in analysis and evaluation of the urban climate for optimizing urban planning processes. *Urban Atmosphere/Atmospheric Environment*.

Remote Sensing in the 21st Century: Economic and Environmental Applications, Casanova (ed.)
© 2000 Balkema, Rotterdam, ISBN 90 5809 096 5

Validation of cloud detection algorithms

F. Sospedra, V. Caselles, C. Coll, E. Valor & E. Rubio
Department of Thermodynamics, University of Valencia, Burjassot, Spain

C. di Bella
Instituto de Clima y Agua, Instituto Nacional de Tecnología Agropecuaria, Buenos Aires, Argentina

ABSTRACT: The cloudiness amount retrieved by the algorithm proposed by Sospedra and Caselles (1999) has been compared with the cloudiness observed at ground level during NOAA-14 satellite overpass. The experimental validation took place in four different sites of Argentina (Castelar, Buenos Aires, Capilla del Monte and Puerto Madryn) during two field campaigns (October-November 1997, and May-June 1998). The average difference between modelled and measured values was + 10 %. Best results were obtained for cloud-free and completely covered days. On partially cloudy days, the geometric correction, the evaluation of the exact surface of the observed area, and the possible differences between the observation and the satellite overpass times are the main causes of the increasing differences.

1.- INTRODUCTION

The objective of this work is to describe how the experimental validation of the cloud detection algorithm described in Sospedra and Caselles (1999) has been done. To perform a validation experience with cloud detection algorithms, the first problem found is to obtain adequate data that in, globally distributed, for different seasons of the year, different hours of the day, different cloudiness levels, and different cloud types. This is an important limitation to the objective of having a high number of data. Furthermore, cloudiness is measured in oktas (1 okta = 1 eighth of cloud covered sky) and for this cause, the obtained precision will be restricted. Considering all these facts, it is logical that the number of validation experiments of cloud detection algorithms made in the past has been small. An exception is the work done by Seze et al. (1986) where Meteosat images were compared to the data coming from 124 meteorological observatories of Great Britain and France observed at 12:00 GMT.

In the experience presented in this work, NOAA-AVHRR data were used. The authors decided to carry out a three months field campaign in order to have a reasonable amount of data. The reason was that only one of the four daily NOAA images could be used. In the other three, it was impossible to make adequate cloud observations. Field campaigns were distributed in two different periods of year (late autumn and spring) in order to have the highest number of different meteorological situations. The stations used in the experience were the following: Buenos Aires (34.58 S, 58.48 W), Capilla del Monte (30.85 S, 64.53 W), Castelar (34.60 S, 58.67 W), and Puerto Madryn (42.79 S, 65.01 W). Two of the stations (Castelar and Puerto Madryn) were the basis of the experience with daily measurements while the other two sites were used as examples of a urban area (Buenos Aires) and of the inland of the country (Capilla del Monte).

This work will be divided into three different parts. The first one includes a general description of the validation experiment. The second one includes all the information concerning the results of the experience. Finally, the main conclusions obtained appear in the third part of the work.

2.- VALIDATION EXPERIMENT

The validation experiment described in this work is based on Seze et al. (1986) experience. To work with NOAA imagery caused that only one sky observation per day was available. The total amount of sky observations used in this experience is about eighty-five scenes. A problem to be solved in this experience was the synchronisation between the sky observation and the satellite overpass. To assure a good solution to this issue, we chose only observation areas containing NOAA antennas. The reception of the satellite transmission takes approximately between 10 and 11 minutes. The satellite sends the information to the receiving station sequentially and that the transmission is line by line. This is not an issue of a minor importance. It can be critical in observations where the cloud front is moving at the same time of the satellite overpass. The decision taken was to make the sky observation 5 minutes after the beginning of the satellite except in the Puerto Madryn station where the measurement was made two minutes before because of its particular location (it's located around one thousand kilometres to the south of the rest of the stations).

Two different field campaigns were made in the months of October and November 97 and May and June 98. The objective was to have a higher number of measurements and to have different meteorological situations and different seasons (different water atmospheric profiles). Some of the satellite data were rejected mainly due to two causes: (1) problems with the satellite transmission of the image, and (2) problems in the geometric correction.

As the same time the image was imported, its calibration was performed according the coefficients given by Planet (1988). One of the critical points is the geometric correction. The obtained precision depends on the correct identification of the observed sky in the satellite image. Bannari et al. (1995) grouped the error sources that force to make a good geometric correction into three different categories: (1) Errors coming from the Earth characteristics (motion, curvature and relief), (2) Errors related with the satellite movement (variations in height, velocity and orbit), and (3) Errors related with the sensor (distortion in the images).
In the opinion of Khan et al. (1995), the nearest-neighbour interpolation is the unique method that does not modify the intensity value of the original data but its error is bigger, being around ± 1 pixel. The scenes used in this experience were corrected by using a fourth degree polynomial. To check the quality of the correction made a digitised vector was used, which included the country border and all the state borders. Helped by the coastal line and bearing in mind that most of the rivers were borders between states was not difficult to evaluate the error in the correction. The correction was good as a first approximation but in some zones the error increased up to ± 1 pixel. For this reason the correction was refined in order to have a better identification of the observation area.

To make possible this operation control points were searched in the original image and in one previously georeferenced taken as a model. The majority of points were found in zones where the observations were made in order to have the best accuracy in the observation areas. The error in the geometric correction will acquire more importance, as the observation zone becomes smaller. As a example, in an area of 13 x 13 km^2 and with an error in the geometric correction of ± 1 pixel, the discrepancy between the pixels identified in the satellite image and the true pixels corresponding to the observation zone is around 10 %. This is not important when having large cloudiness but can become essential in partially cloudy situations. This error as it will be seen later in the next section includes the algorithm error and the errors coming from the experience itself.

3.- RESULTS

3.1.- Delimitation of the observation area.

A usual problem that appears when working with large observation areas is that observations tend to overestimate the cloudiness due to the large observation angles (Merrit, 1966). Seze et al. (1986) suggested that the size of the observation area in open fields could arrive to a surface of 100 x 100 km^2. Our experimental zone was open, flat and vast but limited by forest areas. Therefore, our observation area will be smaller to the value previously mentioned.
To evaluate the size of our observation area, a particular image was used. In this image, the end of a cloud front was just in the beginning of the observation area. After having geometrically corrected the image, the cloud was easily identified and then the area was calculated. The obtained size was of approximately 7 x 7 km^2. Taking into account that the error in the geometric correction was around one pixel, all the calculations made in this experience was done for areas of 5 x 5 km^2, 7 x7 km^2 and 9 x 9 km^2. The influence of the size of the

observation area will be low for clear or completely cloudy days. It will be seen later that the model error is independent of the size of the observation area even when partially cloudy situations are analysed.

3.2.- Algorithm application

The inputs to the cloud detection algorithm used are the emissivities and the atmospheric profiles. Ground emissivities are not a critical parameter in the determination of the cloudiness as it was said in Part I of this work (Sospedra et al., 1999). Analysing the observation areas and according to the works of Rubio et al. (1997), the values of 0.980 for the observation point of Castelar and of 0.975 for the rest of areas were chosen. Taking into account that radiosounding data and TOVS information was not available, standard atmospheric profiles were used. The atmospheres chosen were the mid-latitude summer for the campaign of October and November, and the mid-latitude winter for the campaign of May and June (Kneizys et al, 1988).

3.3.- Comparison/Validation

In the field campaigns the most common situation is to have completely cloudy (40.8 % of the analysed situations) or partially cloudy (50.7 %) days. The cloud detection algorithm was applied to almost 100 scenes corresponding to the dates of the field campaigns.

Tables I, II and III show the observed cloudiness, the satellite retrieved cloudiness and the difference between both measurements for all the analysed observation sites. Cloudiness is given in oktas. The algorithm-retrieved cloudiness has been calculated for areas ranging from 5 km x 5 km to 9 km x 9 km as it has been said before. The average of the differences between both measurements has been also included. The first thing it is observed is that this difference does not depend on the area size. The differences range from 0.6 to 1.0 oktas being its weighted average of 0.9 oktas. This result is in accordance with the one given in Part I of this work (Sospedra et al., 1999).

Others facts that can be appreciated in these tables are the following:
- The algorithm is not very sensitive when having very small cloudiness (around 1 okta or less).
- The algorithm shows a good performance when working with completely cloudy pixels.
- As it has been said in Part I of this work, for partially cloudy situations, the influence of the standard atmosphere used is important.

In Table IV, the differences between the observed cloudiness and the retrieved by the model are shown. It can be seen that in approximately the 50% of the situations, there are coincidences between the observed cloudiness and the retrieved by the model. This percentage increases to a 65 % when considering discrepancies until ± 1 okta. Another important fact is that the average difference obtained is the same independently of the size of the observation area.

These results agree with the ones obtained by Seze et al. (1986) that were the following:

- There were obtained a 29 % of coincidences between the meteorological observations and the cloudiness predicted by the used algorithm.
- In a 64 % of the studied situations, the divergence between both quantities was in the range of ± 1 okta (± 12.5 %).
- The highest percentage of success was accomplished in completely covered and completely clear days.

The causes of the biggest divergences found in our experience (approximately a 10-15 % of the studied situations) have been also analysed. These divergences could be explained in approximately two thirds of the situations. From this study, it can be deduced that two are the main causes that lead to these divergences. These are the following:

(1) Errors in geometric correction: In several situations (especially in partially cloudy cases) it can be seen that the divergence between both measurements is caused by an erroneous identification of the pixel. The pixel identified in the image does not correspond to the area where the observation has been made. In some of these situations only going two pixels in north direction the result was appreciably better. Evidently, this error has a lower importance in uniform sky situations, and
(2) Errors in the synchronisation of the observation with the satellite overpass: Some days, situations where the cloud front was moving during the satellite overpass were registered. In these cases, high divergences can be obtained because of in some seconds the cloudiness can substantially change. The sky observed can be different from the image sent by

Table I.- Measured Cloudiness versus Algorithm retrieved cloudiness (in oktas) for Castelar station.

Date	Observed Cloudiness (oktas)	Retrieved Cloudiness (oktas)		
		5x5	7x7	9x9
291097	3	6.8	6.5	6.1
301097	8	8.0	8.0	8.0
311097	6	2.7	2.1	1.9
11197	6	8.0	8.0	8.0
21197	8	8.0	8.0	8.0
31197	3	0.0	0.0	0.0
41197	8	5.6	4.9	4.2
51197	8	8.0	8.0	8.0
61197	5	0.0	0.2	0.5
71197	4	0.8	1.6	2.1
101197	4	7.7	7.5	7.4
121197	8	8.0	8.0	8.0
131197	0	0.0	0.8	0.5
141197	4	0.0	0.0	0.0
110598	3	0.0	0.0	0.0
120598	7	7.0	5.7	5.5
130598	8	8.0	8.0	8.0
140598	8	8.0	8.0	7.5
180598	1	0.0	0.0	0.0
190598	7	7.4	6.7	6.0
200598	8	0.0	0.5	0.7
210598	7	8.0	7.7	7.2
220598	8	8.0	8.0	8.0
230598	8	8.0	8.0	8.0
240598	8	8.0	8.0	8.0
250598	8	8.0	8.0	8.0
260598	8	8.0	8.0	8.0
270598	8	8.0	8.0	8.0
280598	2	0.0	0.0	0.0
290598	8	7.0	6.9	6.7
10698	5	1.3	2.0	2.5
20698	0	0.0	0.0	0.0
30698	1	0.0	0.0	0.0
40698	0	0.0	0.0	0.0
50698	7	2.9	4.1	4.5
50698	8	8.0	8.0	8.0
60698	8	1.9	2.0	2.5
80698	8	8.0	8.0	8.0
90698	3	1.9	1.8	1.8
100698	8	8.0	8.0	8.0
110698	0	0.0	0.0	0.0
120698	5	2.2	1.6	1.5
130698	8	8.0	8.0	8.0
150698	8	8.0	8.0	8.0

Table II.- Measured Cloudiness versus Algorithm retrieved cloudiness (in oktas) for Capilla del Monte and Buenos Aires stations.

Date	Observed Cloudiness (oktas)	Retrieved Cloudiness (oktas)		
		5x5	7x7	9x9
230598	4	2.6	2.1	2.4
240598	8	6.4	7.0	7.4
250598	2	7.7	7.5	7.7
60698	8	0.0	0.0	0.0
70698	8	8.0	8.0	8.0
80698	8	8.0	8.0	7.7
130698	8	8.0	8.0	8.0
140698	8	8.0	8.0	8.0
150698	8	8.0	8.0	8.0

the satellite. The unique solution is to try that the synchronisation was perfect.

These two errors justify part of the divergences but not all. Other are caused by associated errors to the model, as errors in the inputs of the model or errors due to the physical terms not present in the model. Then, the average difference of 0.9 oktas includes the own algorithm errors and the errors associated to the experience design.

4.-CONCLUSIONS

In this paper a methodology to validate cloud detection algorithms has been proposed. The main conclusions obtained in the work can be summarised as follows:

1.- The average error for the experimental validation is 0.9 oktas which is in accordance with the error values obtained in the theoretical simulation included in the first part of this paper.

2.- The cloud detection proposed in Part 1 of this paper achieves a good performance detecting clouds in completely-cloudy and in clear situations. In these cases, a 85 % of coincidences were obtained. In two thirds of the studied situations the value retrieved by the theoretical model and the value coming from the observations is coincident or in the range ± 1 okta.

3.- The knowledge of the atmospheric profile acquires importance specially when working with partially cloudy days. In these days where the interactions ground-atmosphere-cloud are higher, it

Table III.- Measured Cloudiness versus Algorithm retrieved cloudiness (in oktas) for Pto. Madryn station.

Date	**Observed Cloudiness (oktas)**	**Retrieved Cloudiness (oktas)**		
		5x5	**7x7**	**9x9**
110598	8	8.0	8.0	8.0
120598	8	4.8	6.2	6.7
130598	8	8.0	8.0	8.0
140598	1	0.0	0.0	0.0
180598	8	8.0	8.0	8.0
200598	7	8.0	8.0	8.0
210598	7	6.1	5.4	4.9
220598	8	8.0	8.0	8.0
230598	8	8.0	8.0	8.0
240598	8	8.0	8.0	8.0
280598	1	0.0	0.0	0.0
290598	5	0.0	0.0	0.0
310598	0	0.0	0.0	0.0
10698	7	8.0	8.0	8.0
20698	8	8.0	8.0	8.0
30698	8	8.0	8.0	7.7
70698	4	0.0	0.7	1.0
90698	5	3.5	3.1	3.7
100698	0	0.0	0.0	0.0

Table IV.- Differences between the cloudiness retrieved by the model and the observed in the field campaigns.

Cloudiness Difference (oktas)	Average % of situations
0	49
± 1	15
± 2	11
± 3	7.0
Worse than ± 3	18

is essential to have actual atmospheric data or realistic standard profiles.

5.- ACKNOWLEDGEMENTS

We want to thank the collaboration of the "Institute of Climate and Water" of Castelar (Argentina) and in particular of its director, Mr. Cesar Rebella for allowing the use of his centre to perform the validation experiments. The collaboration in the cloud observation of Mr. Roberto de Ruyver, Mr. Patricio Oricchio and Mr. Alejandro Bisigato has also been essential. Finally, we must remember the grants received by Mr. Francisco Sospedra and financed by the Valencian Government. The research was financially supported by the European Community, Environment and Climate Research Program (Contract ENV4-CT95-0166, Climatology and Natural Hazards) and by the Spanish Commission of Science and Technology (project AMB-96-1928-CE).

6.- REFERENCES

Bannari, A., Morin, D., Bénié, G.B. & Bonn F.J. 1995. A theoretical review of different mathematical models of geometric correction applied to remote sensing images, *Remote Sensing Reviews*, 13: 27-47.

Khan, B., Hayes, L.W.B. & Cracknell, A. P. 1995. The effects of higher-order re-sampling on AVHRR data, *International Journal of Remote Sensing*, 16, 1: 147-163.

Kneizys, F. X., Shettle, E.P., Abreu, L.W., Anderson, G.P., Chetwynd, J.H., Gallery, W.O., Selby, J.E.A & Clough, S.A. 1988. *User's guide to Lowtran-7*, Air Force Geophys. Lab. Rep. AFLG-TR-88-0177, Bedford, MA.

Merrit, E, S. 1966. On the reliability and representativeness of sky cover observations, *Journal of Applied Meteorology*, 5, 369.

Rubio, E., Caselles, V. & Badenas, C. 1997. Emissivity measurements of several soils and vegetation types in the 8-14 mm waveband: Analysis of two field methods, *Remote Sensing of Environment*, 59: 490-521.

Seze, G., Drake, F., Desbois, M. & Henderson-Sellers, A. 1986. Total and low cloud amounts over France and Southern Britain in the summer of 1983: comparison of surface-observed and satellite-retrieved values, *International Journal of Remote Sensing*, 7, 8:1031-1050.

Sospedra, F. & Caselles, V. 1999. Retrieving the amount of cloudiness in a NOAA satellite pixel. Part I: Methodology and application, *Journal of Geophysical Research* (submitted).

Remote Sensing in the 21st Century: Economic and Environmental Applications, Casanova (ed.)
© 2000 Balkema, Rotterdam, ISBN 90 5809 096 5

Measurements of CO_2 fluxes in Central Spain using the eddy correlation technique: Preliminary results

M.L. Sánchez, R. Rodríguez, S. Martinez, M. Ozores & M.J. López
Department of Applied Physics, University of Valladolid, Spain

ABSTRACT: This paper presents the results of CO_2 exchange measured in a cerealist steppe in the plateau of Central Spain and describes the main governing factors controlling photosynthesis uptake. The experimental results consisted of soil respiration measured with a soil chamber and net CO_2 and H_2O fluxes measured by means of the eddy correlation technique. From these data, CO_2 assimilation was derived and the influence of air saturation deficit and temperature on the obtained values was examined. Stomatal conductances to water vapour and CO_2 were determined using a "big leaf" model. In an attempt to up-scale the canopy conductance to CO_2 a linear regression fit between these results and the stomatal conductances to water vapour as well as NDVI was performed yielding satisfactory results.

1. INTRODUCTION

At present there is substantial evidence that the accumulation of CO_2 and other greenhouse gases in the atmosphere is contributing to the increasing trend of the temperature observed at a planetary scale. Previous results have shown that the application of forest and agricultural practices play an important role in the global C cycle contributing to sequester or conserve significant quantities of terrestrial C.

In recent years, important land use changes have been carried out in the Region of Castile and Leon in the framework of the EU political agricultural programme. Among these changes, reforestation practices, reduced tillage and crop residue incorporation have been particularly significant.

In 1998, within the framework of a project funded by the Interministerial Commission of Science and Technology, we began to measure CO_2 and H_2O fluxes over barley, one of the most representative land uses of our Region. These measurements form part of a wide programme aimed at quantifying the exchange rates of greenhouse gases in different land uses and providing the adequate tools to up-scale the results measured at surface with those supplied by NOAA satellites at a regional level.

As a part of this programme, the goal of this paper is to present the preliminary experimental results obtained and to describe the models and procedure used to establish a relationship between CO_2 assimilation measured at surface and NDVI (Normalised Difference Vegetation Index). Other issues such as the influence of the main environmental factors on photosynthesis uptake are also presented and discussed.

2. DATA

Experimental measurements were performed in May, June and July 1998. The selected site was a typical cerealist steppe, located on the plateau in Castile and Leon in central Spain. Extreme temperatures ranging from –18 to 42°C, fog in winter and very low moisture (up to 13% in summer) are the main features of the continental climate. Accumulated rainfall is 500 mm per year. CO_2 exchange measurements were performed in May, June and July 1998 and covered most of the growing stage of the vegetation. According to previous studies the roughness length, z_0, is about 1 cm and the representativeness of the fetch required for the evaluation of micrometeorological parameters is adequate in all directions.

The measurements carried out included different temporal and spatial resolution namely, local (soil respiration), canopy (eddy correlation technique) and global (NOAA image satellites) each of which is described below.

2.1. Soil respiration

Soil respiration measurements were performed using a chamber attached to a LI-6400 gas exchange system. Details on similar equipment can be found in Norman (Norman et al., 1992), and here only a brief summary of the experimental procedure will be described. The chamber has a circular section of 41.3 cm^2 and the bottom edge is sharpened so that it can be easily pressed about 1 cm into the soil. The chamber can also be used on thin-walled PVC collars that are sharpened on one end and inserted about 2 cm into the soil. As collars are essential for time series studies, in this paper, they were generally used in most of the measurements. Air from the chamber is drawn to the LI-6400 infrared gas analyser by an internal pump so that CO_2 can be monitored continuously. Measurements can be obtained in about 5 to 15 min depending on the magnitude of flux. The chamber is equipped with a soil temperature probe.

Measurements were carried out in June and July and generally consisted of one or two experiments per day, from 10.00 to 15.00. Three collars were installed in the soil in each experiment and the flux was evaluated by averaging the individual results. Most of the measurements were not concurrent with those measured by the eddy correlation technique, so that a direct comparison between the results of both methods was not possible.

2.2 Eddy Correlation Technique

At this site, a 12m high mast was installed and equipped with slow and fast meteorological instrumentation. Micrometeorological data were recorded at 12m with a Young three-propeller (19Hz) Model 27005 and an AIR fast response temperature sensor (25 Hz) Model FT-1A-T and stored on a PC computer each 0.1s. Micrometeorological parameters were determined by means of the eddy correlation technique using a similar procedure to that described by Rodríguez (Rodríguez 1998). The method is based on spectral analysis followed by a high pass spectral filtering with a cut-off at about 0.01 Hz aimed at removing low frequency oscillations in order to assure the steadiness of the fluctuation measurements of each sample. The data were averaged each 30 min.

The accuracy of micrometeorological parameters using this instrumentation and a sonic (10 Hz) was tested through intercomparison measurements carried out in a forested area. CO_2 and H_2O fluxes were measured with a using a LI-6262 following a procedure similar to that described by Aurela (Aurela et al.1996),. Air was sampled at a flow rate of 5 l/min. The sampling rate was 10 Hz and the data were also stored in a PC.

Table 1. Experimental data measured using the eddy correlation technique

MONTH	PERIOD	DAYS
MAY	20-30th	6
JUNE	10-19th	7
JUNE	20-30th	7
JULY	19-28th	10

Temperature and vertical wind profiles measured with slow response cup anemometers and temperature probes at 1.5, 3 and 9 m were also recorded during most of the experiments and stored in a DASIBI datalogger, and from the continuous stored data 30-minute average values were calculated at each level. Wind direction, and relative humidity completed the available instrumentation. Measurements were carried out over a period of 30 days as shown in the table.

2.3 NOAA Images

From the images supplied by the AVHRR sensor, NDVI was determined following the procedure described by González (González et al., 1995). To minimise the noise of individual data in this paper a 10-day MCV was used and compared with the concurrent mean values of assimilation as is described latter.

3. THEORY

Net carbon exchange comprises a balance between carbon gain through photosynthetic carbon reduction and carbon loss through photorespiratory carbon oxidation and dark respiration. The rate of carbon assimilation depends on plant physiological and environmental variables, such as available energy, water availability and ambient concentrations of several pollutants. It has been largely recognised that healthy plants mainly exchange water vapour, CO_2 and other gases through stomata. Stomata control the flow of water through plants from the soil to the atmosphere. Plants affected by water stress close stomata to save water and as a result assimilation also tends to be reduced. In the literature the exchange of water vapour is quite often parameterised using "a big leaf" model (Baldocchi et al. 1987, Sánchez and Rodriguez, 1997). The surface conductance to water vapour is given by:

$$Gs_{H_2O}^{\ -1} = (Gt^{-1}{}_{H_2O} - (r_a + r_{bH_2O})^{-1}) \quad (1)$$

where $Gt^{-1}{}_{H2O}{}^{-1}$ is the total conductance and in this paper was calculated using the same equation as Tenhunen (Tenhunen et al,1995). r_a is the bulk aerodynamic resistance to the transfer to momentum and r_b is the quasi-laminar boundary resistance:

$$r_b = \frac{2}{u*k}(\frac{S_c}{P_r})^{2/3} \quad (2)$$

where Sc and Pr are the Schmidt and Prandtl numbers.

In the CO_2 case, the net flux is:

$$F = A - R \quad (3)$$

where A and R denote the assimilation and respiration terms. Most of the results of A controlling factors have been obtained at a single leaf-level and up-scaled to the whole canopy using different approaches. Assimilation typically depends on the total conductance to CO_2, ambient CO_2 and intercellular C_i concentrations as well as PPFD and transpiration (Hollinger et al., 1998). In this paper the total conductance to C was determined using a "big leaf" model:

$$A = -Gt_{CO_2}.C(z) \quad (4)$$

where C(z) is the CO_2 concentrations measured at a reference height z, 12m. In this equation we have assumed that the surface concentration was zero. Assimilation was calculated by subtracting an estimate of soil fluxes (see below) to the measured CO_2 eddy fluxes. Important in equation (4) is the assumption that Gt_{CO2} includes all the possible plant physiological governing factors and, consequently, should be regarded as a "surrogate" conductance. With this assumption, surface resistance to CO_2 can be determined with:

$$Gs_{CO_2}^{\ -1} = (\frac{CO_2}{F} - (r_a + r_{bCO_2})^{-1}) \quad (5)$$

4. RESULTS

During our studies, days were generally sunny with temperatures ranging from 5.2°C to 36°C. During the second 10-day period of June, an episode of high temperatures, up to 35°C, together with air saturation deficit, up to 45hPa, occurred resulting in a period of water stress which probably contributed to an acceleration in the senescence of the canopy. Wind speed ranged from 0.1.m/s to 9.1 m/s. To ensure quality of the results all the data recorded when the friction velocity was lower than 0.1 m/s, were rejected.

4.1 Soil Fluxes

In the literature soil temperature is considered to be one of the controlling variables of soil fluxes. Therefore, in this paper a regression fit between soil fluxes and soil temperature was performed and a comparison between the results of both linear and non-linear models was analysed. The best results corresponded to a reciprocal linear fit, as depicted in Figure 1, yielding:

$$F_{soil} = -0.99 + \frac{50.04}{T}; r = 0.80 \quad (6)$$

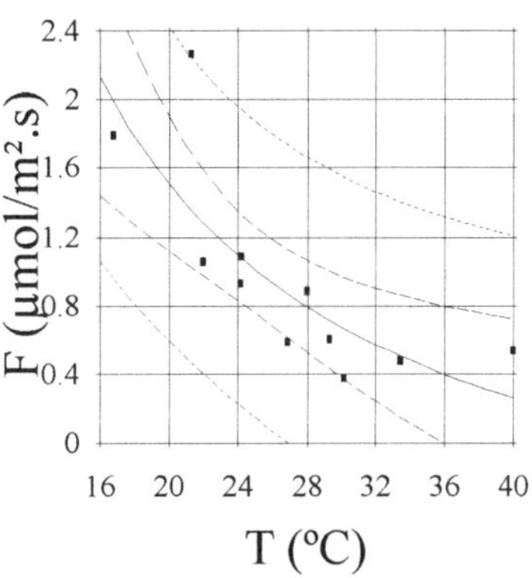

Figure 1. Relationship between Soil Fluxes and Soil Temperature.

4.2 Daily and Seasonal Variation of Net Fluxes

Figures 2 a-c depict the CO_2 daily courses averaged with photosynthesis uptake, began to be observed in the early morning, peaking about noon, from 10.00 to 12.00 and declining from 12.00 to about 18.00. Integrated over 24 h, the canopy was on average a net sink of 0.475± µmol/m^2.s with maximum values in May of –1.35 µmol/m^2.s and minimum ones in July of -0.041 µmol/m^2.s. In June it yielded -0.54 µmol/m^2.s. The results obtained can be considered rather low as compared with other reported values in the literature (Saigusa et al.,1998, Dugas et al,1997)

4.3 Influence of Environmental Factors

To describe the main controlling factors of CO_2 uptake a selection of data was made as follows. Data of eddy fluxes were restricted to the period of time from 10.00 to 15.00 and from these data a mean daily value for each day was calculated. The reason for this choice was that maximum fluxes were generally found during this time interval. In addition they corresponded to the time where soil flux measurements were performed and NOAA satellite images were received. With this approach, and assuming the validity of equation (6) for soil respiration fluxes to the overall data, the mean daily assimilation results were determined by subtracting the soil flux estimates to the mean daily eddy flux. As soil temperature was not routinely measured, based on previous results, these values were estimated by adding 2°C to the air temperature measured at 1.5 m.

Figures 3 and 4 show the relationship between assimilation, the air saturation deficit, D(hPa) and temperature T (°C). Lines represent the fitted model and the 95% significance and prediction levels respectively.

The best fitted model was a linear reciprocal and a linear fit for D and T. The satisfactory correlation coefficients found, 0.86 and 0.87, reveal the great influence of both variables on CO_2 assimilation, with the expected decrease of photosynthesis uptake when D and T increase (Morison,1993).

4.4 Influence of NDVI on CO_2 Assimilation

Up-scaling results from leaves to canopy and from canopy to globe are one of the main priorities in

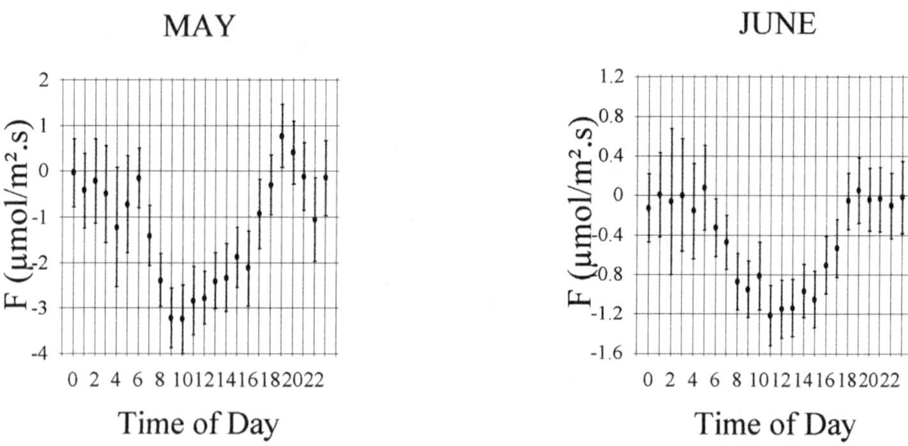

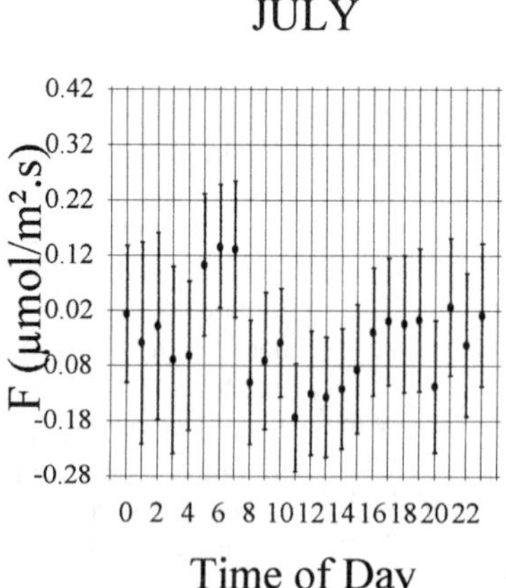

Figure 2 a-c. Daily Patterns of CO_2 Exchange. Dots and Lines represent Mean and 95% LSD Intervals

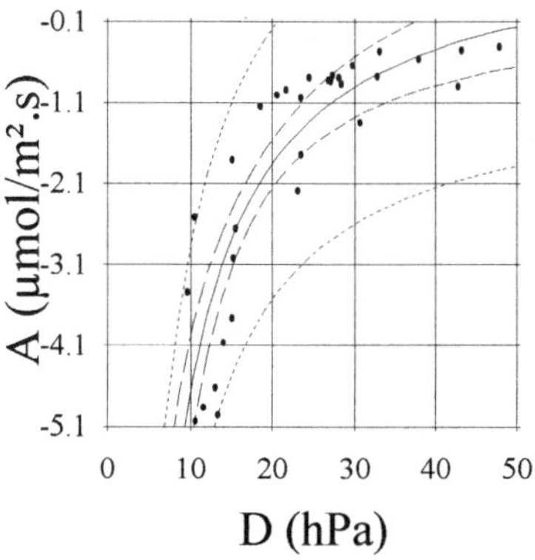

Figure 3. Regression Fit of Assimilation versus D

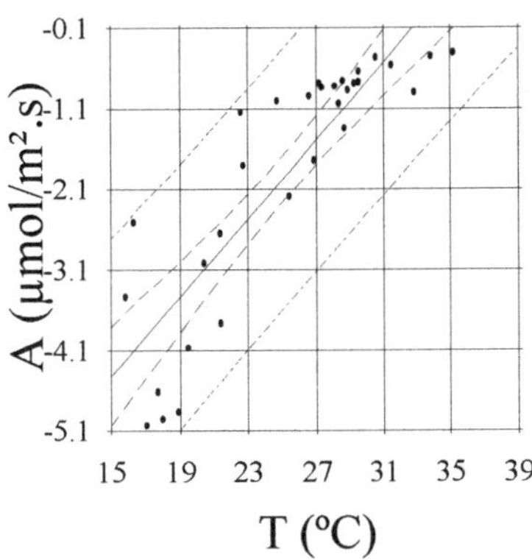

Figure 4. Regression Fit of Assimilation versus T.

climate change issues (Kelliher at. al,1995). Intensive work in this direction has been started quite recently. Among these procedures, remote sensing techniques are attractive and offer some insight into the scaling problem. This is especially true for modelling stomatal conductances.

Stomatal conductance data are generally achieved through direct photosynthesis measurements at a leaf scale or parameterised at a canopy level using single or multilayer models (Baldocchi and Harley,1994, Verhoef and Allen,1998). The simplest ones parameterise the bulk stomatal conductance of different trace gases as a function of NDVI, LAI and stomatal conductance to water vapour ((BAHC, 1993, Erisman et. al.1993). The coefficients involved in the proposed equations are variable depending on land uses, type of vegetation and season of the year (Weseley,1989).

Based on these approaches, and in an attempt to link the information provided by NOAA satellites and the total conductance to CO_2 we have used the stomatal conductance to H_2O and the NDVI MCV of 10-day periods as driving variables. Concurrent eddy CO_2 total conductance was obtained by averaging the mean values of each day (from 10.00 to 15.00) over each 10-day MCV. The best fitted model, depicted in Figure 5, was the following linear regression:

$$\boldsymbol{Gt_{CO_2}} = 0.0021 + 0.23.\boldsymbol{Gs_{H_2O}NDVI}; \boldsymbol{r} = 0.996$$

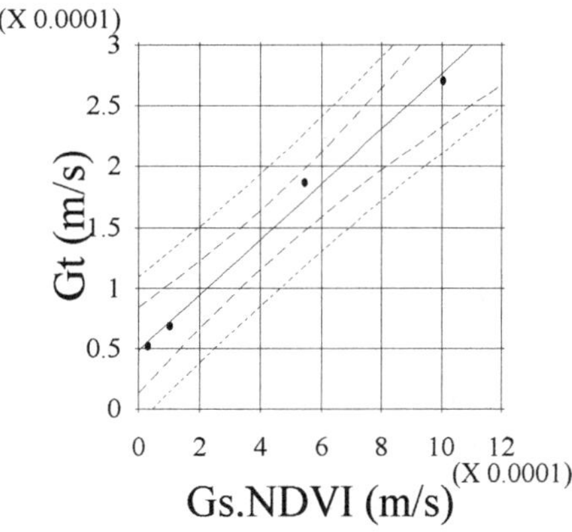

Figure 5. Linear Regression Fit between the Total Conductance to CO_2 to Surface Conductance to H_2O and NDVI.

The satisfactory correlation coefficient found permits us to approximate the photosynthesis rate exchange with:

$$\boldsymbol{A} \approx 0.23\boldsymbol{Gs_{H_2O}NDVI.CO_2}$$

5. CONCLUSIONS

The measurements of CO_2 net fluxes using the eddy correlation technique have shown a strong seasonal variation with maximum values in May, -1.35 $\mu mol/m^2.s$ and minimum ones in July, 0.041 $\mu mol/m^2.s$. Integrated over 24 h, the canopy was on average a net CO_2 sink of 0.48 ± 1.10 $\mu mol/m^2.s$

Soil fluxes measured with a soil chamber have proven the weak contribution of dry soils yielding 0.96 ± 0.56 $\mu mol.m^{-2}.s$. However, the large ratio between soil fluxes and maximum assimilation found (from 10.00 to 15.00), 0.47, reveals the important participation of respiration processes at this site.

The "big leaf model" applied in this paper has proven to be accurate to model stomatal conductance to water vapour and derive the total conductance to CO_2. In this paper the surface conductance was considerably larger than that of r_a and r_b, so that Gt approximated Gs. The important role of stomata as CO_2 sink can thus be inferred.

The satisfactory reciprocal and linear fit of assimilation versus D and T obtained, with a decrease of photosynthesis uptake as T and D increase, again permit us to confirm the important role of stomata as a controlling factor of CO_2 sinks in the atmosphere.

Although further work is necessary to confirm the preliminary results presented in this paper, the satisfactory correlation coefficient between the total conductance measured to CO_2 at surface level and NDVI determined through satellite images offers an interesting future line of research for up-scaling purposes. As a part of this plan additional sampling campaigns at this site as well as in other land uses, such as forests, are currently been organised.

Acknowledgements
This paper has been supported by the CICYT and partially by the Regional Government, to whom the authors wish to express their gratitude. We would also like to acknowledge our gratitude to the Remote Sensing Laboratory of the University of Valladolid for providing us with the NDVI data used in this paper.

REFERENCES

M.Aurela, T. Laurila and J.P. Tuovinen 1996. Measurements of O_3 CO_2 and H_2O fluxes over a Scots pine stand Eastern Finland by the micrometeorological eddy covariance method. Silva Fennica 30 (2-3) 97-108.

BAHC 1993. Global IGBP Change, Report 27. Institute für Meteorology, Freie Universität Berlin.

D. Baldocchi, B.B. Hicks and P. Camara 1987. A canopy stomatal resistance for gaseous deposition to vegetated surfaces. Atmospheric Environment, 21,91-101.

D. Baldocchi and P. Harley 1994. Scaling dioxide carbon and water vapour exchange from leaves to canopy in a deciduous forest: model testing and application. Plant, Cell and Environment,19,234-246.

W.A. Dugas, D.C. Reicosky and J.R. Kiniry 1997. Chamber and micrometeorological measurements of CO_2 and H_2O fluxes fro three C_4 grasses. Agricultural and Forest Meteorology, 83,113-133.

J.W. Erisman, A. Van Pul and P. Wyers 1993. Parameterisation of dry deposition mechanism for the quantification of atmospheric input to ecosystems. Air Pollution Research Report 47,215-233.

F. González, M.Cuevas, J.L. Casanova and P.Illera 1995. Drought monitoring in Spain using satellite remote sensing. Sensors and environmental applications of remote sensing, 87-90.

D.Y. Hollinger, F.M. Kelliher, E.D. Shulze, G. Bauser, A.Arneth, J.N. Byers, J.E. Hunt., T,M. McSeveny, K.I. Kovak, I.Milukova, A. Sogatchev, F. Tatarinov, A. Varlangin, W. Ziegler and N.N Vygodskaya 1998. Forest atmosphere carbon dioxide exchange in eastern Siberia. Agricultural and forest meteorology, 90, 191-306.

F.M. Kelliher, R. Leuning, M.C. Raupach and E.D. Schulze 1995. Maximum conductances for evaporation from global vegetation types. Agricultural and Forest Meteorology, 1-16.

J.I.- Morison 1993. Response of plants to CO_2 under water limited conditions. CO_2 and Biosphere,193-210, Kluwer Academics Publishers.

Norman J.M., R. Garcia and S.B. Verma 1992. Soil surface CO_2 fluxes and the carbon budget of a grassland. Journal of Geophysical Research, 97,D17,18,845-18-853.

M.L. Sánchez and R. Rodíguez 1997. Ozone dry deposition in a semi-arid steppe and in a coniferous forest in Southern Europe. Journal of Air and Waste Management, 47, 792-799.

R. Rodríguez 1998. Determinación de los Parámetros Micrometeorológicos y Flujos de Deposición Seca de Ozono y Aerosoles de Carácter Acido en un Bosque de Conníferas. PhD, Universidad de Valladolid

N.Sougusa, T. Oikawa and S. Liu (1998) Seasonal variations of the CO_2 and H_2O between a grassland and the atmosphere: An experimental study. Agricultural and Forest Meteorology, 89,131

J. Tenhunen, M.Alsheimer, E. Falge, B.heindl, U.Joss, B. Köstner, G. Lischeid, B. Manderscheid, B. Ostendorf, K. Peters, R.Ryel and M.Vedler 1995. Water fluxes in a spruce forest ecosystem: a framework for process study integration. (personnel communication)

A. Verhoef and S.J. Allen 1998. The relative importance of surface and aerodynamic resistances in a multi-source energy-CO_2 model. Physical Chemistry Earth 23, 4,459-463

Wesely M.L. 1989. Parameterization of surface resistances to gaseous dry deposition in regional scale numerical models. Atmospheric Environment,23,1293-1304.

Remote Sensing in the 21st Century: Economic and Environmental Applications, Casanova (ed.)
© 2000 Balkema, Rotterdam, ISBN 90 5809 096 5

Selecting a clear-sky model to accurately map solar radiation from satellite images

G. Rigollier & L. Wald
Groupe Télédétéction et Modélisation, Ecole des Mines de Paris, Sophia Antipolis, France

ABSTRACT: The HelioClim project aims at mapping solar radiation over a wide area that covers Europe, Africa, and Atlantic Ocean. To achieve this purpose, Meteosat images will be processed with the Heliosat method, which turns images into maps of solar radiation at ground level. One of the inputs of this method is the solar radiation under clear sky. The quality of the estimations strongly depends on the accuracy of the radiative transfer model under clear sky. Therefore several clear-sky models are assessed in this paper. It is found that taking into account the turbidity of the atmosphere, the Linke turbidity factor, T_L(AM2), provides better results. The models using T_L(AM2) are then compared. Among them, the « ESRA model », set up in the course of the realisation of the 4th edition of the European Solar Radiation Atlas, fits better than the others. Accordingly, this model is chosen to be used in the Heliosat method.

1 INTRODUCTION

Satellite methods offer an alternative to interpolations of punctual ground measurements, allowing thus to study continuously wide areas. This advantage is well exploited in this study, that only uses satellite images to map solar radiation.

Indeed, the purpose of the HelioClim project, carried out in Ecole des Mines de Paris, is to establish a climatological database of solar radiation over an area as wide as Europe, Africa, and Atlantic Ocean, during the past fourteen years. To achieve such a project, satellite images, especially Meteosat images, will be processed with the Heliosat method, world-wide used.

The original version of the Heliosat method (Cano *et al.*, 1982) has been modified for the HelioClim project, but the principle is still the same.

This method makes use of the digital count of Meteosat images to derive a cloud index, which characterises the transmittance of the atmosphere over the pixel under concern. The atmospheric clearness index is then computed from this cloud index, finally providing global solar radiation.

In order to accurately compute solar radiation, the Heliosat method requires a good radiative transfer model under clear skies. This point is discussed in this paper, where different clear-sky models are first described and then compared.

2 RADIATIVE TRANSFER MODELS UNDER CLEAR SKIES

Radiative transfer models under clear skies give the value of the solar radiation received at ground level in clear-sky conditions.

For some models, the global radiation is split into the beam and the diffuse components, allowing thus a finer description of the radiative transfer.

The inputs are varying from a model to another: the simplest models just take into account the solar elevation, while more detailed ones may include other inputs such as aerosols, water vapour, ozone, elevation of the site, barometric pressure, temperature, or ground albedo, in order to better model the atmospheric transmittance.

The value of clear-sky radiation is needed in the process of the Heliosat method, and the assessment of the solar radiation at ground level from satellite images strongly depends on the accuracy of the radiative transfer model used.

For that reason, we investigate several clear-sky models and study their relevance for the assessment of solar radiation from satellite images.

3 DESCRIPTION OF VARIOUS CLEAR-SKY MODELS

3.1 *Models with solar elevation as unique input*

The first model presented here is the clear-sky model of the WMO (World Meteorological Organization) document 557 (1981, page 120). This model, only acceptable for stations at low altitudes (less than 400 meters), gives the following expression for the global horizontal irradiance under cloudless skies:

$$G_{c(WMO)} = 0.95\, I_0\, \varepsilon\, (\sin \gamma_S) / (1 + 0.2 \operatorname{cosec} \gamma_S) \quad (1)$$

where:
I_0, the solar constant, is equal to 1367 $W.m^{-2}$;
ε is the correction used to allow for the variation of sun-earth distance from its mean value;
γ_s is the solar altitude angle. γ_s is equal to 0° at sunrise and sunset.

The original version of the Heliosat method (Cano *et al.*, 1986) used the model of Bourges (1979) to compute the global irradiance under clear-sky:

$$G_{c\,(Bourges)} = 0.70\, I_0\, \varepsilon\, (\sin \gamma_S)^{1.15} \quad (2)$$

Later, Moussu *et al.* (1989) used the model of Perrin de Brichambaut and Vauge (1982), hereafter noted PdBV, which has the same inputs as the previous models. This one provides larger values than the model of Bourges and is given by:

$$G_{c\,(PdBV)} = 0.81\, I_0\, \varepsilon\, (\sin \gamma_S)^{1.15} \quad (3)$$

These models are very easy to implement. However they do not take into account parameters that influence solar radiation, such as ground albedo or atmospheric turbidity (aerosols, water content, ...) and accordingly yields rough evaluations.

3.2 *Clear-sky models with more inputs*

Other models have more inputs representing the properties of the atmosphere, thus reducing the overall inaccuracy.

The first one is the other clear-sky model proposed in the WMO document 557 (1981, page 124), which uses as another input the atmospheric turbidity. The horizontal global irradiance under clear-sky is given by the following equation:

$$G_{c(WMO)2} = \varepsilon\, (1297 - 57\, T_L)\, \sin \gamma_S^{(36+TL(AM2))/33} \quad (4)$$

where $T_L(AM2)$ is the Linke turbidity factor for an air mass equal to 2. We recall that $T_L(AM2)$ is a function of the optical path for aerosols and water vapour, and that it characterises the transmittance of the blue sky. A value of 1 stands for a dry and clean atmosphere. This WMO model is still simple but it is stated that it is suitable particularly if the solar elevation is higher than 20°-30°.

The other clear-sky models split the global radiation into a beam and a diffuse components. Among them, three clear-sky models have the same equation for the beam radiation: the model of Dumortier (1995), a model developed at the University of Oldenburg (Beyer *et al.*, 1997) based on calculations made using radiative transfer code MODTRAN 3.5 (Kneizys *et al.*, 1996), called the MODTRAN model, and the model developed within the European Solar Radiation Atlas, hereafter noted the ESRA model.

For these models, the horizontal beam radiation under cloudless skies is given by the following equation:

$$B_c = I_0\, \varepsilon\, \sin\gamma_s\, \exp\left(-0.8662\, T_L(AM2)\, m\, \delta_R(m)\right) \quad (5)$$

where:
m is the relative optical air mass;
$\delta_R(m)$ is the integral Rayleigh optical thickness.
The quantity :

$$\exp\left(-0.8662\, T_L(AM2)\, m\, \delta_R(m)\right) \quad (6)$$

represents the beam transmittance of the beam radiation under cloudless skies. The relative optical air mass m expresses the ratio of the optical path length of the solar beam through the atmosphere to the optical path through a standard atmosphere at sea level with the sun at the zenith. As the solar altitude decreases, the relative optical path length increases. The relative optical path length also decreases with increasing station height above the sea level, z. A correction procedure is applied, obtained as the ratio of mean atmospheric pressure, p, at the site elevation, to mean atmospheric pressure at sea level, p_0. This correction is particularly important in mountainous areas. The relative optical air mass has no unit; it is given by Kasten and Young (1989), where γ_s^{true} is in degrees:

$$m(\gamma_s^{true}) = \frac{(p/p_0)}{\sin\gamma_s^{true} + 0.50572(\gamma_s^{true} + 6.07995)^{-1.6364}} \quad (7)$$

with the station height correction given by:

$$p/p_0 = \exp(-z/z_h) \quad (8)$$

where z is the site elevation, z_h is the scale height of the Rayleigh atmosphere near the Earth surface, equal to 8434.5 meters, and γ_s^{true} is the solar altitude angle corrected for refraction.

All the variation of the beam transmittance with air mass is included in the product m δ_R (m). Figure 1 displays the beam irradiance for $p=p_0$ (sea level), and for different values of turbidity factor ($T_L(AM2)$ = 2, 3, 5, 7), as a function of solar elevation.

For the three models, the diffuse irradiance also depends on the Linke turbidity factor. The higher the turbidity, the greater the absorption and scattering of

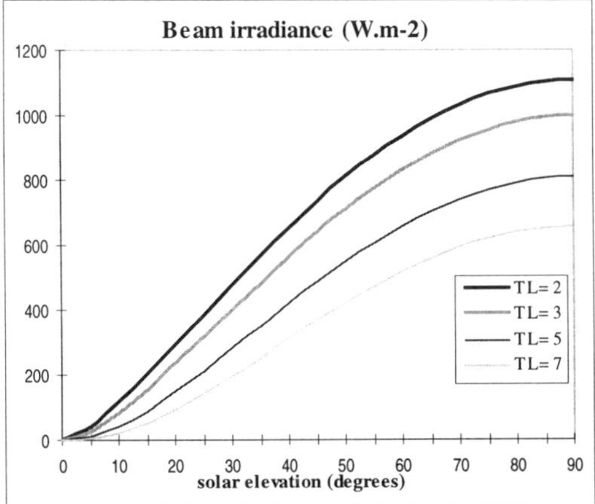

Figure 1. The beam horizontal irradiance for clear sky, B_c.

the solar radiation by the atmosphere, and therefore the higher the diffuse component and the smaller the beam component.

The expressions for the diffuse components are given below.

The diffuse component of the model of Dumortier (1995) is the following:

$$D_c = I_0\, \varepsilon\, (0.0065 + (-0.045 + 0.0646\, T_L) \sin \gamma_S - (-0.014 + 0.0327\, T_L) \sin^2 \gamma_S \quad (9)$$

with the conditions:

$\gamma_S < 70°$

$2{,}5 \le T_L \le 6{,}5$

The MODTRAN model gives the following expression for the diffuse component:

$$D_c = I_0\, \varepsilon\, (a + b\, T_L + c\, T_L^2 + (d + e\, T_L + f\, T_L^2) \sin \gamma_S + (g + h\, T_L + i\, T_L^2) \sin^2 \gamma_S \quad (10)$$

$a = 0.017991$	$d = -0.112593$	$g = -0.019104$
$b = -0.003967$	$e = 0.101826$	$h = -0.022103$
$c = 0.000203$	$f = -0.006220$	$i = 0.003107$

For the ESRA model, the diffuse irradiance is defined as the product of the diffuse transmission at zenith, T_{rd}, and a diffuse angular function, F_d, as following:

$$D_c = I_0\, \varepsilon\, T_{rd}(T_L(AM2))\, F_d(\gamma_s, T_L(AM2)) \quad (11)$$

with the following diffuse transmission at zenith ($\gamma_S = 90°$):

$$T_{rd}(T_L(AM2)) = -1.5843.10^{-2} + 3.0543.10^{-2}\, T_L(AM2) + 3.797.10^{-4}\, [T_L(AM2)]^2 \quad (12)$$

and the following diffuse angular function:

$$F_d(\gamma_s, T_L(AM2)) = A_0 + A_1 \sin(\gamma_s) + A_2 [\sin(\gamma_s)]^2 \quad (13)$$

The coefficients A_0, A_1 and A_2, only depend on the Linke turbidity factor. They are unitless and are given by:

$$\begin{cases} A_0 = 2.6463\,10^{-1} - 6.1581\,10^{-2}\, T_L(AM2) + 3.1408\,10^{-3}\, [T_L(AM2)]^2 \\ A_1 = 2.0402 + 1.8945\,10^{-2}\, T_L(AM2) - 1.1161\,10^{-2}\, [T_L(AM2)]^2 \\ A_2 = -1.3025 + 3.9231\,10^{-2}\, T_L(AM2) + 8.5079\,10^{-3}\, [T_L(AM2)]^2 \end{cases} \quad (14)$$

with a condition on A_0:

$$if\,(A_0 . T_{rd}) < 2.10^{-3}, \quad A_0 = 2.10^{-3} / T_{rd} \quad (15)$$

This condition is required because A_0 yielded negative values for $T_L(AM2)>6$. It was therefore decided to impose this limiting condition to achieve acceptable values at sunrise and sunset.

The diffuse irradiance is represented in figures 2 and 3 for the Dumortier, MODTRAN, and ESRA models for different values of the Linke turbidity factor ($T_L(AM2) = 2$ and $T_L(AM2) = 7$).

Finally, from the previous equations of beam and diffuse irradiances, the horizontal global irradiance under clear-sky for the three models can be deduced from:

$$G_c = B_c + D_c \quad (16)$$

Other models such those described by Perez (1987) or by Iqbal (1983), which not only take into

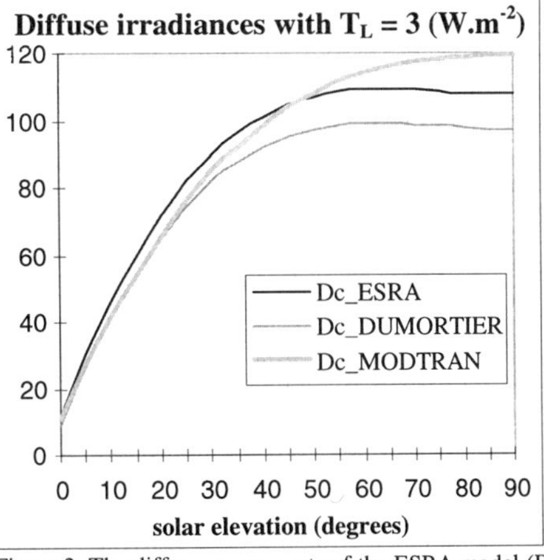

Figure 2. The diffuse components of the ESRA model (D_c ESRA), the DUMORTIER model (D_c Dumortier), and the MODTRAN model (D_c MODTRAN) for $T_L = 3$ at mean sun-earth distance.

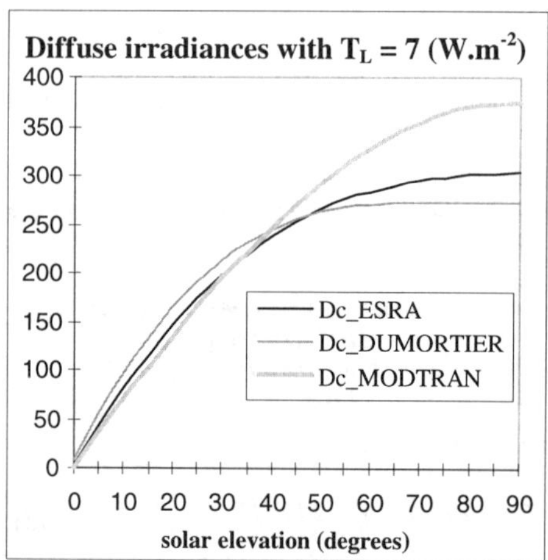

Figure 3. As figure 2, but for $T_L = 7$.

account turbidity, but also other parameters such as ground albedo, ozone, or water vapour, require data that are not always available. In the framework of the HelioClim project, which aims at mapping operationally solar radiation, such models cannot be included. Therefore we are not describing such models in this study. Anyway, some basic comparisons are reported below.

4 COMPARISON BETWEEN DIFFERENT CLEAR-SKY MODELS

The previous models have been analysed by comparison with ground data. The stations of the data set are located in Germany or in Belgium (see table 1). The period of measurement stretches from 1981 to 1990, offering thus ten years of comparison. This data set is extracted from the European Solar Radiation Atlas (ESRA) and offer hourly global and diffuse radiation measurements, except for Uccle (half-hourly global, diffuse, and beam radiations).

The diffuse, or beam component is deduced from the subtraction of the measured component from the global irradiation. Then, using the expression of the beam component given in equation (5), the Linke turbidity factor is deduced from the measurements as following:

$$T_L = - ln(B_c / I_0 \, \varepsilon \sin \gamma_S) / 0.8662 \; \delta_R(m) \, m \qquad (17)$$

This computation provides a "clear-sky ground data set": only are kept measurements corresponding to clear skies, for which the corresponding Linke turbidity factor ranges from 2.0 to 6.5. In fact, fifteen intervals of T_L, partly overlapping each other, were used to distinguish clear from non clear skies: [2.0 - 3.5], [2.5 - 3.5], [3.0 - 3.5], [2.0 - 4.0], [2.5 - 4.0], [3.0 - 4.0], [2.0 - 5.0], [2.5 - 5.0], [3.0 - 5.0], [2.0 - 6.0], [2.5 - 6.0], [3.0 - 6.0], [2.0 - 6.5], [2.5 - 6.5], [3.0 - 6.5].

With this new "clear-sky ground data set", comparisons are made between irradiances estimated with the models previously described and ground measurements. It is assumed that the irradiance of the middle of the acquisition period is equivalent to the hourly or half-hourly irradiation.

The comparisons between irradiances estimated with the different models or measured at the ground stations are summarised as root mean square errors (rms errors) and bias (estimates mean minus observations mean).

A computation has been made to get hourly values from half-hourly ones for Uccle, in order to compare errors obtained from these two kinds of data. The results show similar errors for the assessment of the irradiation on hourly basis than those obtained from half-hourly basis.

4.1 *Comparisons of global radiations estimated from the model of Bourges, Perrin de Brichambaut & Vauge, and WMO*

The models described in section 3.1 have been compared with the clear-sky ground data set (table 1). Results (rms errors and bias) are shown in tables 2 and 3 for the ten years over all stations.

The estimations obtained with the model of Bourges present a high bias whatever the location and the Linke turbidity range. This model systematically under-estimates the clear-sky irradiation. Moreover the accuracy of this model remains poor. The rms error over the ten years ranges from 33 to 58 Wh.m^{-2} for all locations and all T_L ranges. The model of Perrin de Brichambaut & Vauge gives higher radiation values, since the only difference between them is the multiplicative coefficient (0.81 for Perrin de Brichambaut & Vauge, 0.70 for Bourges). Consequently the under-estimation should be corrected. Indeed, the bias is reduced and is even positive indicating now an over-estimation. The rms error is generally lower than the model of Bourges, ranging from 17 to 57 Wh.m^{-2} for the ten years, whatever the turbidity range and whatever the site. However, this model shows a high

Table 1. The data set of ground stations. Period of measurement: from January 1981 to December 1990.

Station Name	Latitude N ; Longitude E	Elevation
Braunschweig(Germany)	52.30 ; 10.45	83 m
Dresden (Germany)	51.12 ; 13.68	246 m
Hamburg (Germany)	53.65 ; 10.12	49 m
Trier (Germany)	49.75 ; 6.67	278 m
Würzburg (Germany)	49.77 ; 9.97	275 m
Weihenstephan (Germany)	48.40 ; 11.70	472 m
Uccle (Belgium)	50.80 ; 4.35	100 m

dependence on the range of T_L used: the higher the turbidities, the higher the errors. The rms errors ranges from 17 to 24 Wh.m^{-2} for T_L between 2.0 and 3.5, and from 48 to 57 Wh.m^{-2} for T_L between 3.0 and 6.5.

To check the validity of the model of PdBV, Moussu et al. (1989) compare it to the clear-sky model described by Iqbal (1983, model C) after the works of Bird and Hulstrom (1981 a, b) for various values of ground albedo, precipitable water thickness, and horizontal visibility. The comparison demonstrates that the shape of the model PdBV is consistent with the model C and that the variation of G_c is well described by the function $(\sin \gamma_S)^{0.15}$. However the magnitude of $G_{c\ (PdBV)}$ suffers from the lack of input parameters, as well as the model of Bourges.

Compared to those models, the first model proposed by WMO, given by equation (1), on the one hand gives lower bias than those obtained with the model of Bourges. On the other hand rms errors remain more stable than those noted with the PdBV model with a change in T_L range. However rms errors are ranging from 26 to 43 Wh.m^{-2}, and the overall accuracy remains poor, also suffering from a lack of inputs.

The results obtained with the second WMO model, see equation (4), are also given in tables 2 and 3. Taking into account the turbidity of the atmosphere really improves the accuracy: the values obtained with this model are close to the irradiances measured at the different stations: the rms error ranges from 16 to 29 Wh.m^{-2} for all locations and all T_L ranges.

Table 2. Global clear-sky radiation in Wh.m^{-2}. Results obtained from the comparison between ground irradiations measured over all stations during ten years and the models of Bourges, Perrin de Brichambaut & Vauge, and WMO, for a T_L ranging between 2.5 and 3.5. The ground mean is equal to 328 Wh.m^{-2}.

	Bourges	PdBV	WMO1	WMO2
mean	290	336	303	333
rmse	53 (16 %)	20 (6 %)	34 (10 %)	20 (6 %)
bias	-38	8	-25	4

Table 3. As table 2 but for a T_L ranging between 2.5 and 6.5. The ground mean is equal to 363 Wh.m^{-2}.

	Bourges	PdBV	WMO1	WMO2
mean	348	402	373	361
rmse	42 (11 %)	51 (14 %)	39 (11 %)	25 (7 %)
bias	-15	40	10	-1

4.2 *Comparison of diffuse radiations estimated from the model of Dumortier, MODTRAN, ESRA, and WMO.*

In the previous section, some models have been tested, showing that the best results are obtained when taking into account the Linke turbidity factor. Another study (Rigollier *et al.*, 1999) investigated other models (Dumortier, MODTRAN, and ESRA) that take into account this Linke turbidity factor. Moreover, these models have another input: they depend on the elevation of the site. Some results of this study are reported below.

For those models, comparisons with ground data have also been made, using several large data sets including the present one. The expression of the beam radiation has been used to deduce the Linke turbidity factor and to choose clear skies data among all. Therefore only the diffuse component has been compared for the models of Dumortier, MODTRAN, and ESRA.

It was possible to investigate the influence of the geographical location on the estimations, since the stations are not only located in Europe, but also in Israel. The results are slightly the same for all sites, it is hence concluded that there is no influence of the ground elevation nor the geographical location on the quality of the estimations.

Moreover, the differences in error observed in 1994 between two remote sites such as Sede Boqer (Israel) and Vaulx-en-Velin (South France) are not higher than those observed between 1981 and 1990 for the different German stations. We conclude that the models are not affected by the climate.

The results computed over ten years show that even if the errors are similar for the three models, the ESRA model always gives the best results for all stations when considering average errors over the ten years. Being the most accurate and most robust model, the ESRA model is consequently recommended by Rigollier *et al.*

We have shown in the previous section that the second model presented in the WMO document (Eq. 4) provides good estimations of the global radiation. When comparing them to the results obtained with the ESRA model, a slight difference is noted, the model of ESRA being fairly more accurate: for the Linke turbidity factor ranging from 2.5 to 3.5 (respectively from 2.5 to 6.5), the rms error is equal to 22 Wh.m^{-2} (resp. 17 Wh.m^{-2}), while the bias is -3 Wh.m^{-2} (resp. -5 Wh.m^{-2}).

Tough this capability is not yet exploited, we feel that further improvements in the Heliosat method may request the knowledge of both the direct and diffuse components for clear skies. While the ESRA model clearly expresses both components, the WMO model presented in equation (4) only gives the global radiation under clear sky and does not have distinct beam and diffuse components. In fact, the WMO document 557 (1981, page 124) presents two distinct beam and diffuse components:

$$B_c = \sin \gamma_S \, [1390 - 31 \, T_L(AM2)] \, \exp[-T_L/(12.6 \sin(\gamma_S+2))] \quad (18)$$

$$D_c = 383 \sin \gamma_S^{[(TL(AM2)+5.7)/30]} \exp[-4/T_L(AM2)] \quad (19)$$

We have compared the WMO beam radiation from equation (18) with that from ESRA for different Linke turbidity factors and solar elevations. This comparison shows that the beam components do not strongly differ.

In the study of Rigollier *et al.*, it has been proved that the ESRA model is accurate: rms errors were found to range from 11 to 35 Wh.m^{-2} for all measurements (different years, different locations) and various Linke turbidity factors for diffuse irradiations up to 250 Wh.m^{-2}. But the comparison between the diffuse components from the WMO and ESRA models shows discrepancies between them, as shown in figure 4. These discrepancies are especially large with high values of Linke turbidity and demonstrate that the WMO model for diffuse component is not accurate enough. These findings proved true when comparing the WMO estimations with ground measurements: the rms error computed for the whole data set (all stations over ten years) for a Linke turbidity factor from 2.5 to 6.5 is equal to 33 Wh.m^{-2} (with a ground mean of 113 Wh.m^{-2}) while the same error for the ESRA model is 22 Wh.m^{-2}.

Besides, as shown in figure 4, the diffuse WMO model does not behave correctly for low solar elevations. Indeed, for a Linke turbidity factor ranging from 2.5 to 6.5, and solar elevations from 5° to 10°, the WMO model widely overestimates the diffuse radiation: the bias over the seven stations of table 1 and during ten years is equal to 27 Wh.m^{-2}, while it is -1 Wh.m^{-2} for ESRA. With the same conditions, the rms error is equal to 34 Wh.m^{-2} and only 18 Wh.m^{-2} for ESRA.

Considering these findings, the diffuse WMO model does not seem to be accurate nor commendable.

Moreover, regarding this WMO diffuse model, two strange facts have attracted our attention:

- Firstly, the values of WMO diffuse radiation computed with equation (19) do not agree with the table XII in page 124 of the same WMO document.
- Secondly, the values obtained while adding up both the beam and the diffuse components from equations (18) and (19), or the values given in table XII, differ from the values of global radiation computed with the equation (4).

Accordingly, we decide to select the ESRA model to compute solar radiation under clear sky. Including this model in the Heliosat method should improve the quality of the estimations of solar radiation from satellite images.

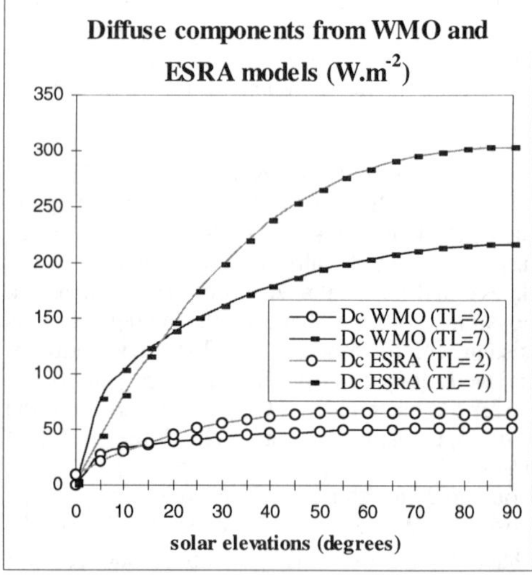

Figure 4. The diffuse components of the WMO (D_c WMO) and ESRA (D_c ESRA) models for Linke turbidity factors of 2 and 7.

5 CONCLUSION

Several clear-sky models have been presented and analysed for the estimation of clear-sky irradiance. The models need different data as inputs. The simplest ones only need the solar elevation, while more detailed ones take into account the transmittance of the atmosphere, by means of the Linke turbidity factor for example. Among them, some models need too many inputs, that cannot be acquired easily. Consequently they cannot be used in the framework of the HelioClim project and have not been studied here.

The purpose of the HelioClim project is to map accurately and operationally solar radiation from satellite images. The results obtained while taking into account the turbidity of the atmosphere are clearly better than the others. After a comparison between these models, that proposed by the European Solar Radiation Atlas (ESRA, 1999) proves to be the most accurate and most robust model as a whole, and is selected for the HelioClim project.

6 ACKNOWLEDGEMENTS

We are indebted to the team of the European Solar Radiation Atlas and the University of Oldenburg for the ground data that allow us to make our comparisons.

7 REFERENCES

Beyer, H.G., Hammer, A., Heinemann, D., Westerhellweg, A., 1997, Estimation of diffuse radiation from Meteosat data. North Sun '97, 7th International Conference on Solar Energy at High Latitudes, Espoo-Otaniemi.

Bourges, G., 1979, Reconstitution des courbes de fréquence cumulées de l'irradiation solaire globale horaire reçue par une surface plane. Report CEE 295-77-ESF of Centre d'Energétique de l'Ecole Nationale Supérieure des Mines de Paris, tome II, Paris, France.

Cano, D., 1982, Etude de l'ennuagement par analyse de séquences d'images de satellite. Application à l'évaluation du rayonnement solaire global au sol. Thèse de Doctorat, Ecole nationale supérieure des télécommunications, Paris, France.

Cano, D., Monget, J.M., Albuisson, M., Guillard, H., Regas, N., and Wald, L., 1986, A method for the determination of the global solar radiation from meteorological satellite data. Solar Energy, 37, 31-39.

ESRA, 1999, *European solar radiation atlas*. Fourth edition, includ. CD-ROM. Edited by J. Greif, K. Scharmer. Scientific advisors: R. Dogniaux, J. K. Page. Authors : L. Wald, M. Albuisson, G. Czeplak, B. Bourges, R. Aguiar, H. Lund, A. Joukoff, U. Terzenbach, H. G. Beyer, E. P. Borisenko. Published for the Commission of the European Communities by Presses de l'Ecole, Ecole des Mines de Paris, Paris, France, 1999.

Dumortier, D., 1995, Modelling global and diffuse horizontal irradiances under cloudless skies with different turbidities. Final report JOU2-CT92-0144, Daylight II. Ecole Nationale des Travaux Publics de l'État, Vaulx-en-Velin, France.

Ielhé, A., Lefèvre, M., Bauer, O., Martinoli, M., and Wald, L., 1997, Meteosat: A valuable tool for agro-meteorology, Final report for the European Commission, Joint Research Center, Ispra, Italy.

Iqbal, M., 1983, An introduction to Solar Radiation (New York: Academic Press), pp. 107-169.

Kneizys, F. X., et al., 1996, The MODTRAN 2/3 Report and LOWTRAN 7 Model. Technical Report, Phillips Laboratory, Geophysics Directorate, Hanscom AFB.

Moussu, G., Diabate, L., Obrecht, D., and Wald, L., 1989, A method for the mapping of the apparent ground brightness using visible images from geostationary satellites, Int. J. Remote Sensing, 10 (7), 1207-1225.

Perez, R., 1987, New simplified version of the Perez diffuse irradiance model for tilted surfaces. Solar Energy, 39(3), 221-231.

Perrin de Brichambaut, C., and Vauge, C., 1982, Le gisement solaire: Evaluation de la ressource énergétique. (Paris: Technique et documentation).

Rigollier, C., Bauer, O., Wald, L., 1999, On the clear-sky model of the ESRA - European Solar Radiation Atlas with respect to the Heliosat method. To be published in *Solar Energy*.

World Meteorological Organization (WMO), 1981. Meteorological aspects of the utilization of solar radiation as an energy source. Annex: World maps of relative global radiation. Technical Note No. 172, WMO-No. 557, Geneva, Switzerland, 298 pp.

Remote Sensing in the 21st Century: Economic and Environmental Applications, Casanova (ed.)
 ISBN 90 5809 096 5

Radiative Bowen ratio to estimate daily fluxes from satellite images

M. Ibáñez & P. Pérez
Departamento de Medio Ambiente i Ciencias del Sòl, Universidad de Lleida, Spain

V. Caselles
Departamento de Termodinámica, Universidad de Valencia, Spain

ABSTRACT : Infrared measurements of canopy temperatures have been extensively used to estimate daily evapotranspiration at regional scale. The aerodynamic and the radiative Bowen ratio formulations give instantaneous evapotranspiration estimates for the daytime period, which can be integrated to obtain a value for daily latent heat flux when daytime detailed surface information is available. This paper proposes a different method for estimating daily evapotranspiration over short well-watered crops with complete canopy from satellite images. The method is based on the radiative Bowen ratio-energy balance method and similarity between the diurnal course of latent heat flux and solar radiation. This regression-based approach requires continuous measurement of air vapour pressure, air temperature, surface radiative temperature and solar radiation during daytime. Daily evapotranspiration is estimated with an error under 15%.

1 INTRODUCTION

Estimating levels of evapotranspiration for agricultural crops is important for adequate water management in semi- arid areas where irrigation is necessary and water is expensive. Many methods exist for estimating instantaneous and daily latent heat flux based on radiative temperature (Kustas et al. 1989, Lhomme et al. 1994). These methods are particularly interesting when their application is feasible at a regional scale thanks to satellite sensors (Moran et al. 1989, Caselles et al. 1992, Kustas et al. 1994).
Most methods for estimating evapotranspiration from radiative temperature use the Monin-Obukhov similarity theory, the limitations of which have already been widely discussed (Sun & Mahrt 1995, Verhoef et al. 1997). Another approach is based on the use of the radiative Bowen ratio (Ibáñez et al. 1998). The radiative Bowen ratio and the aerodynamic Bowen ratio are linearly related over full unstressed crops. The radiative Bowen ratio - energy balance method has proven to be a good tool for obtaining estimates of midday evapotranspiration rates. Errors of around 10% were obtained. It was initially assumed that parameters *a* and *b,* which interrelate the two Bowen ratios, were constant throughout the growing season. However, it was observed that seasonal variations in these parameters could constituted a major drawback of this method.

A number of models have been developed to determine the total amount of water evaporated from a surface during the course of a day. A simple yet accurate way of converting instantaneous evapotranspiration values into daily totals makes use of the similarity between the diurnal course of the latent heat flux and that of other metereological variables. A first approach assumes that evapotranspiration follows the course of solar radiation throughout the daylight period. Jackson et al. (1983) showed that the ratio between total daily solar radiation and a single measurement taken at midday could be approximated by a sine function. This relationship gives rise to a method which gives a strong correlation between measured and estimated daily evapotranspiration.
The objective of this study is to present an original method for estimating daily evapotranspiration over well-watered crops using remote-sensed surface temperature and meteorological data. Close-range instrumentation has been used because it allows the performance of the model to be evaluated with data over different homogeneous crops, free of the complicated effects of the atmosphere and the surface-sensor geometry present in satellite data.

2 METHOD

This original method combines the radiative Bowen ratio-energy balance method and the similarity between the diurnal courses of evapotranspiration and solar radiation to estimate daily evapotranspiration. Daily and daylight latent heat fluxes may be identified and taken into account due to the fact that evapotranspiration during the night is negligible.

The radiative Bowen ratio - energy balance method can be used to calculate instantaneous latent heat flux, λE, by:

$$\lambda E = \frac{Rn - G}{1 + (a + b\beta_r)} \qquad (1)$$

where Rn is the net radiation (Wm^{-2}), G is the soil heat flux (Wm^{-2}), β_r is the radiative Bowen ratio and parameters *a* and *b* depend on crop and weather conditions.

Net radiation can be estimated by:

$$Rn = (1-\alpha)Rs + \sigma\varepsilon(\varepsilon_a T_a^4 - T_r^4) \qquad (2)$$

where α is the albedo, Rs is the incident solar radiation (Wm^{-2}), σ is the Stefan - Boltzman constant ($W\ m^{-2}\ K^{-4}$), ε is the surface emissivity, ε_a is the atmospheric emissivity calculated for clear sky conditions, T_a and T_r are respectively the air and radiative surface temperatures (K).

Soil heat flux can be estimated using its strong correlation with net radiation for full canopy cover crops. The ratio between the two fluxes is around 0.10 for grass (Allen et al., 1994), 0.10 for alfalfa (Clothier et al., 1986) and 0.07 for wheat (Denmead, 1969).

The radiative Bowen ratio, β_r, has been expressed in the form:

$$\beta_r = \gamma \frac{T_r - T_a}{e_s(T_r) - e} \qquad (3)$$

where γ is the psychrometric constant ($kPa\ K^{-1}$), e is the air vapour pressure (kPa) and $e_s(T_r)$ is the saturation vapour pressure at temperature T_r (kPa).

Acceptable instantaneous estimates of evapotranspiration from radiative temperature over unstressed full canopy covers are obtained when parameters *a* and *b* are taken as constants (Ibáñez et al. 1999). However, it must be noted that values for *a* and *b* may vary according to the stage of crop development and prevailing weather conditions.

Daily estimations of evapotranspiration are derived from instantaneous measurements on the basis of the similarity between the diurnal course of the latent heat flux and solar irradiance (Jackson et al. 1983). This result can be expressed by:

$$\lambda E = \frac{Rs}{Rs_d}\lambda E_d \qquad (4)$$

where subindex d refers to daytime accumulated values. Two methods for estimating daytime evapotranspiration are tested in this paper. The first uses equation (1) to obtain estimations of instantaneous latent heat flux. These are converted to daily values by integrating over the daytime hours. Parameters *a* and *b* are considered to be constant for each crop throughout the growing season. Hereafter, this method is referred to as the "seasonal method".

In the second method, equation (1) is combined with expression (4) to obtain:

$$\lambda E_d \frac{Rs}{Rs_d} = \frac{Rn - G}{1 + (a + b\beta_r)} \qquad (5)$$

By rearranging expression (5), it is possible to obtain the following formula:

$$Rn - G = [(1+a)\lambda E_d]\frac{Rs}{Rs_d} + [b\lambda E_d]\frac{\beta_r Rs}{Rs_d} \qquad (6)$$

where $(1+a)\lambda E_d$ and $b\lambda E_d$ are daily constants. Both these factors can be determined through least square regression when continuous values of Rn - G, Rs and β_r are available throughout the daytime period.

Instantaneous Rn - G and β_r measured, or estimated from T_a, T_r and e, together with Rs data, measured throughout the daytime hours, are used to compute $(1+a)\lambda E_d$ and $b\lambda E_d$ applying least square regression in equation (6). Two coefficients and three unknowns are obtained: *a*, *b* and λE_d. The daily evapotranspiration is calculated assuming a given value for parameter *a*. When the crop is short and well irrigated *a* = 0.1 gives good results. This method is useful because it does not require the use of any crop and soil parameters, or wind speed data. The new approach considers *b* to be constant on a daily scale (yet not necessarily throughout the season) and is therefore, hereafter referred to as the "daily method".

3 MATERIALS

Measurements were taken in 1994 and 1997 to validate the proposed method for a number of different homogeneous crops: grass, wheat and alfalfa. Experimental sites were located in the Ebro river basin: a Mediterranean area with semi-continental climatic characteristics. Annual precipitation is less than 350 mm, with minima in

January and July. This semiarid area receives an important quantity of water from the Pyrenees, which makes irrigation possible.
Validation was carried out using Bowen ratio equipment, (Campbell Scientific Inc., Logan ,UK) (Tanner et al. 1987) and a water content reflectometer (CS615, Campbell Sci.) for measuring soil water content. Surface radiative temperature was measured with an infrared thermometer (Model 4000BL, Everest Interscience Inc., Tucson) with an accuracy of ±0.5 K and a field of view of 15°. The data obtained by the infrared thermometer were corrected to take into account the effects of downward longwave radiation and surface emissivity. A pyranometer (Model SP1110, SKYE Instruments) and a two metre high switching anemometer (Vector Instruments, model A100R) were also installed. All measurements were averaged for half hourly or twenty-minute periods.
Irrigation frequencies for the three crops were based on allowable root zone available water depletion, which was maintained below 40% throughout the study period. The crop water stress index, monitored throughout the field tests, never exceeded 0.4.
Grass, wheat and alfalfa were seeded and monitored on plots at three different geographic locations within large irrigated areas of the Ebro valley. Each of the three plots had sandy loam soils.

4 RESULTS AND DISCUSSION

Overcast days were discarded from this study and daylight data were only considered when Rs was larger than 25 Wm^{-2}. The instantaneous measurements of T_a, e and Rs were considered. Continuous radiative surface temperature data were available but T_r was estimated from interpolation of three daylight determinations because this can be the temporal resolution of satellite images. Parameters *a* and *b* were determined each day to study their variability by means of least square regressions of measured Bowen ratio versus calculated radiative Bowen ratio. The coefficient of determination was always higher than 0.90 (13 days for grass, 16 days for wheat and 25 days for alfalfa).
Although parameters *a* and *b* may vary, taking them as constants throughout the growing season provides an operative approximation for estimating instantaneous latent heat flux. The standard deviation of daily computed values presented in Table 1 gives a quantification of parameter variability.
The objective of this paper was limited to presenting the "daily method". Table 2 was computed to test this regression method. The table shows the root mean square error (RMSE) of daily evapotranspiration calculated against a series of other measured values. The performance of the method was studied for different values of parameter *a*. Flux errors were calculated for these values for the three crops: grass, wheat and alfalfa.

Table 1. Mean values and standard deviations of parameters *a* and *b* for the studied data sets.

Crop	$\bar{a}$	$\bar{b}$	σ_a	σ_b
Grass	0.05	2.46	0.05	0.66
Wheat	0.15	1.83	0.05	0.42
Alfalfa	0.04	0.84	0.04	0.20

Table 2.- Root mean square error of daily evapotranspiration estimations from the "daily method", $(RMSE)_D$, with different values for parameter *a* versus daily Bowen ratio - energy balance measured latent heat fluxes for the three data sets.

$(RMSE)_D$ (mm)	$a = 0$	$a = 0.1$	$a = 0.2$
Grass	0.7	0.4	0.6
Wheat	0.9	0.6	0.5
Alfalfa	0.4	0.6	1.1

Tables 1 and 2 are directly related. Lower RMSE (Table 2) were obtained when daily latent heat fluxes were calculated using values for parameter *a* that were closer to the mean computed value (Table 1). Table 3 presents the percentage of error in each case. The smallest mean percentage of errors was found when parameter $a = 0.1$. It was therefore decided to use this value hereafter for applications of the "daily method" on short crops.
The validation of the "daily method" is shown in Table 3. Two conclusions should be drawn:
(a) When the method is applied for a known crop the seasonal constant value of parameter *a* can be estimated and accurate daily values for evapotranspiration can be calculated with a margin of error of around 10%. Taking $a = 0$ for alfalfa, $a = 0.1$ for grass and $a = 0.2$ for wheat, the method gives estimates with 7%, 6% and 10% margins of error, respectively.
(b) When the crop is not known, a value of $a = 0.1$ permits daily evapotranspiration estimates with an error under 15%.
These results show that assuming the value of parameter $a = 0.1$, daily evapotranspiration can be adequately estimated by continuous measurement

Table 3.- Percentage of error in daily evapotranspiration calculations from "daily method" with different values for parameter a for the three data sets.

$(\overline{RMSE})_D / \lambda E_{BR}$ (%)	$a = 0$	$a = 0.1$	$a = 0.2$
Grass	12	7	10
Wheat	22	15	12
Alfalfa	7	10	18

Table 4.- Percentage of error in daily evapotranspiration calculations from the "seasonal method" and "daily method" with $a = 0.1$ for the three data sets.

Crop	$(\overline{RMSE})_S / \lambda E_{BR}$	$(\overline{RMSE})_D / \lambda E_{BR}$
Grass	13	7
Wheat	21	15
Alfalfa	17	10

of T_a, e and Rs and three determinations of T_r even when the phenological stage of the crop and the weather conditions are not fully known.
The "seasonal method" is applied with the mean values for parameters a and b presented in Table 1. A comparison between this "seasonal method" and the "daily method" is presented in table 4. The measured values are better estimated when the "daily method" is applied for the three crops.

5 CONCLUSIONS

The values for parameters a and b expressed in the relation between radiative and aerodynamic Bowen ratio were not constant over the measurement period. Some of the observed variations around the mean value could be attributed to meteorological and phenological scenario. Given that these conditions do not vary substantially during the course of a day, parameters a and b can be considered as daily constants.
The method proposed for calculating daytime evapotranspiration over well-watered full-cover crops based on the radiative Bowen ratio - energy balance method and the similarity between the diurnal course of evapotranspiration and solar radiation gives an acceptable level of relative error. Root mean square errors of less than 15% of the average evapotranspiration measured with Bowen ratio equipment were obtained when the crop was unknown. The margin of error was reduced to 10% when the crop was known and its characteristic value for parameter a was applied. Integrations of daytime instantaneous values of evapotranspiration, taking parameters a and b as seasonal constants, gave larger errors. For applications of the "seasonal method", the crop and its phenological stage should be well known and the parameters a and b relating to these conditions need to have been previously determined.
This "daily method" is particularly interesting because evapotranspiration can be adequately calculated on the basis of continuous daytime measurements of solar radiation, air vapor pressure and air temperature. These meteorological variables are usually known at the regional scale when automatic meteorological stations are available. The satellite images are used to obtain diurnal evolution of radiative surface temperature. This algorithm is useful for remote sensing applications because daily evapotranspiration values are obtained without detailed information about the crop canopy and surface wind speed. The results obtained by applying this methodology are promising for irrigated areas.

REFERENCES

Allen, R.G., M. Smith, L.S. Pereira & A. Perrier 1994. An update for the calculation of reference evapotranspiration. *ICID Bulletin* 43: 35-91.
Caselles, V., J. Delegido, J.A. Sobrino & E. Hurtado 1992. Evaluation of the maximum evapotranspiration over the La Mancha region, Spain, using NOAA AVHRR data. *Int. J. Rem. Sens.* 13: 939-946.
Clothier, B.E., K.L. Clawson, P.J. Pinter, M.S. Moran, R.J. Reginato & R.D. Jackson 1986. Estimation of soil heat flux from net radiation during the growth of alfalfa. *Agric. For. Meteor.* 31: 319-329.
Denmead, O.T. 1969. Comparative micrometeorology of a wheat field and a forest of Pinus Radiata. *Agric. Meteor.* 6: 357-371.
Ibáñez, M., P.J. Pérez, V. Caselles & F. Castellvi 1998. A simple method for estimating the latent heat flux over grass from radiative Bowen ratio. *J. Appl. Meteor.* 37: 386-392.
Ibáñez, M., P.J. Pérez, J.I. Rosell & F. Castellvi 1999. Estimation of the latent heat flux over full canopy covers from radiative temperature. *J. Appl. Meteor.* 38: 423-431.
Jackson, R.D., J.L. Hatfield, R.J. Reginato, S.B. Idso & P.J. Pinter 1983. Estimation of daily evapotranspiration from one-time day measurements. *Agric. Water Mgt.* 7: 351-362.
Kustas, W.P., B.J. Choudhury, M.S. Moran, R.J.

Reginato, R.D. Jackson, L.W. Gay & H.L.Weaver 1989. Determination of sensible heat flux over sparse canopy using thermal infrared data. *Agric. For. Meteor.* 44: 197-216.
Kustas, W.P., E.M. Perry, P.C. Doraiswamy & M.S. Moran. 1994. Using satellite remote sensing to extrapolate evapotranspiration estimates in time and space over a semiarid rangeland basin. *Rem. Sens. Environ.* 49: 275-286.
Lhomme, J.P., B. Monteny & M. Amadou 1994. Estimating sensible heat flux from radiometric temperature over sparse millet. *Agric. For. Meteor.* 68: 77-91.
Moran, M.S., R. Jackson , L.H. Raymond, Ll. W. Gay & P.N. Slater 1989. Mapping surface energy balance components by combining Landsat thematic mapper and ground-based meteorological data. *Rem. Sens. Environ.* 30: 77-87.
Sun, Y. & L. Mahrt 1995. Relationship of surface heat flux to microscale temperature fluctuations. *Bound.-Layer Meteor.* 76: 292-301.
Tanner, B.D., J.P. Greene & G.E.Bingham. 1987. A Bowen-ratio design for long term measurements. *ASAE Paper* No. 87-2503, Am. Soc. Agric. Eng., St. Joseph, MI, 1-6.
Verhoef, A., H.A.R. De Bruin & B.J.J.M. van den Hurk. 1997. Some practical notes on the parameter kB^{-1} for sparse vegetation. *J. Appl. Meteor.* 26: 560 – 572.

Remote Sensing in the 21st Century: Economic and Environmental Applications, Casanova (ed.)
© 2000 Balkema, Rotterdam, ISBN 90 5809 096 5

Investigation of surface characteristics by the synergy of ground observations and NOAA/AVHRR satellite measurements

J. Kerényi & Á. Merza
Hungarian Meteorological Service, Satellite Research Laboratory, Budapest, Hungary

G. Gutman & I. Csiszár
NOAA-NESDIS, Office of Research and Applications, Washington, D.C., USA

ABSTRACT: NOAA/AVHRR Global Area Coverage (GAC) data were combined with conventional ground observations to study physical processes over vegetated surfaces. Ground observations included daily minimum, maximum air temperature, daily mean soil temperature at 2 cm depth, wind speed, global radiation, relative humidity and cloudiness, and daily precipitation from the 1994-1997 period. AVHRR data were taken from a pilot dataset including raw counts of original 4-km resolution GAC data of daily (day- and night) observations over the two 50-km areas in Hungary. These two targets are covering agricultural, forested and marshland surfaces. After calibration and cloud detection 10-day maximum NDVI and surface skin temperature were derived. Surface heat fluxes were characterized by differences between air temperature, surface skin temperature. The above information was used to study seasonal changes in surface conditions. The observed differences over the two targets can be attributed to differences in the dominant vegetation cover.

1 INTRODUCTION

The derivation of large-scale continuous fields of surface characteristics is possible only by the use of high-resolution satellite imagery. Multispectral measurements from the Advanced Very High Resolution Radiometer (AVHRR) on board the operational polar orbiter NOAA satellites provide information on surface reflectivity, land cover (including snow), state and/or amount of vegetation, surface skin temperature, daily temperature range etc. If reliable methodology is developed, inter-seasonal and interannual changes of surface heat exchange processes from time series of AVHRR can be detected. There have been many attempts to use some of this information for the parameterization and measurement of surface energy balance components over limited areas and time periods.

In this study we combine AVHRR-derived information with conventional ground measurements in order to close the surface energy balance equation and describe time series of surface heat fluxes over two target areas with different vegetation cover. It will be shown that vegetation is a governing factor in determining the direction and magnitude of surface latent and sensible heat flow. Evidence will be provided that appropriately preprocessed satellite information is a useful and reliable tool in surface energy budget studies.

2 THE OBSERVATION DATA

The AVHRR/GAC data have been received since July 1993 for two 50x50 km targets at NOAA/NESDIS. One target - surrounding of Szarvas - is agricultural area, the other target - surrounding of Keszthely - is forest and marshland.

Surface data were measured at the meteorological stations in Szarvas and Keszthely. In this study daily maximum and minimum air temperatures (Tmax, Tmin) daily mean soil temperature measured at 2 cm depth (Tsoil), daily global radiation (Gr), daily precipitation, daily wind speed (u), daily mean cloud cover and the daily snow cover data were applied. At the Keszthely ground station the Tsoil value was measured only to the end of June 1997.

3 PREPROCESSING OF DATASET

The brightness temperatures of AVHRR channels 3, 4 and 5 (T3,T4,T5) were derived using the on-board coefficients. For the determination of reflectances in channels 1 (r1) and 2 (r2) the post-launch coefficients (Rao and Chen, 1995, 1996) were used. The determination of channel 3 reflectance followed Stowe et al. (1991) in removing the thermal component based on T4 and T5.

4 PROCESSING OF SATELLITE-GROUND COMBINED DATASET

4.1 *Cloud detection*

To separate the cloud free and cloudy area a cloud detection scheme was used. It is based on T4, r1, Tmax and Tmin values. During the daytime:

for the May-October period the

T4>270 K and Tmax-T4<10 K and r1<15 and std(r1)<1% ,

for the November-April period the

T4>260 K and Tmax-T4<10K and r1<15 and std(r1)<1%

while during nighttime

for the May-October period: T4>270 K and Tmin-T4<10

for the November-April period: T4>260 K and Tmin-T4<10

thresholds were applied.

The NOAA-11 AVHRR data from July 1993- December 1994 were omitted because of the unreliable data due to late observations caused by satellite orbit drift.

After the cloud screening NDVI was calculated daily and LST was derived from a split-window equation including the emissivity (e) correction based on NDVI:

$$LST=T4+2.63*(T4-T5)+1.274+(T4+T5)/2 *(0.156+3.98*(T4-T5)/(T4+T5))*(1-e)/e$$

where $e=1.013+0.0681*\log(NDVI+0.2)$, where 0.2 is added to NDVI to account roughly for the atmospheric effect.

4.2 *Snow cover*

The first part of the work was to investigate the effect of the winter snow on the vegetation development in spring. Snow cover measured at the ground stations was used for the investigation. To derive the snow cover using satellite information, the ratio of channel 1 and of channel 3 albedos was applied. This method was found to clearly identify snow cover with large snow amount. However, further investigation is needed to fully exploit this technique. Surface snow data indicated high snow amount in February 1996. NDVI data from this period indicated a delay in vegetation development, in accordance with results from other sources. Due to anomalous conditions in this year, data from January - May 1996 were not used in investigating the average seasonal variability.

4.3 *Synergy of satellite and ground-based data to characterize surface energy fluxes*

Daily mean surface temperature was calculated from corresponding thermal satellite data by

$$LSTm=0.5*(LST(day)+LST(night)).$$

The 10-day maximum NDVI was derived to assure cloud-free observations. We compared the 10 day maximum NDVI values of the investigated 3 years and we chose the greatest value. The data of this day were used for the investigation of the relationship between the NDVI value and the energy fluxes.

The energy balance equation is

$$R=G+H+LE$$

where R is the net radiation, G is the soil heat flux, H is the sensible heat flux and LE is the latent heat flux.

The sensible heat flux was calculated by the

$$H=\rho*cp*(LST(day)-Tmax)/(ra+rb)$$

equation, where ρ is the air density, cp is the specific heat of air, ra is the air resistance for heat transfer, while rb is the boundary resistance. The values of ra was derived as follows:

$$ra=1/(k*ur)*\ln((z1-d)/z0)$$
$$ur=(k*u)/\ln((z2-d)/z0)$$

where k is the Karman's contant (=0.41), z1 is the height of air temperature measurement (=2 m), z2 is the height of the wind speed measurement (10 m). The zero-plane displacement level (d) and the roughness length (z0) were calculated in the function of the height of the vegetation (h) and the leaf area index(LAI) after Shaw and Pereira(1982):

$$d=1.1*h*\ln(1+0.07*LAI)^{1/4}$$
$$z0=0.29*h(1-d/h)$$

The LAI is derived using the NDVI data after Crosson (1993):

$$LAI=(NDVI+0.2)*4.73-2.08$$

The vegetation height was derived using the NDVI value using equation developed by Gruszczynska (1997):

$$h=2.47*NDVI-0.16$$

The boundary resistance is considered, because the calculation of heat flux needs the aerodynamic temperature (Stewart and Verma, 1992). The relationship between the aerodynamic surface temperature (Tsa) and the radiometric surface temperature (Ts) calculated from the satellite data is:

$$Tsa=Ts+b/(1+b)(Tmax-Ts), \text{ where } b=rb/ra$$
$$rb=B^{-1}/ur$$

The calculation of net radiation based on field measurement (Dávid et al 1990):

$R=0.592[Gr(1-A)]-0.075$ in April;
$R=0.707[Gr(1-A)]-0.638$ in May;
$R=0.717[Gr(1-A)]-0.739$ from June to September;
$R=0.606[Gr(1-A)]-0.41$ in October.

The equation uses the surface albedo (A), which was derived as arithmetic mean of the reflectance of channel 1 and 2.

The surface heat flux was examined using the next equation after Loepmajer (1983):

$G=R*(0.2-0.03*LAI)$

5 RESULTS

Satellite and ground measurements were compared for the two targets. Comparing the ground measurements, higher cloud cover, higher relative humidity and lower global radiation values can be seen at the Keszthely

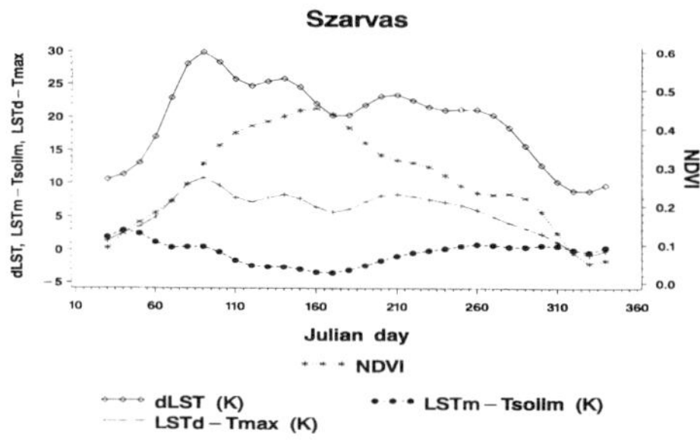

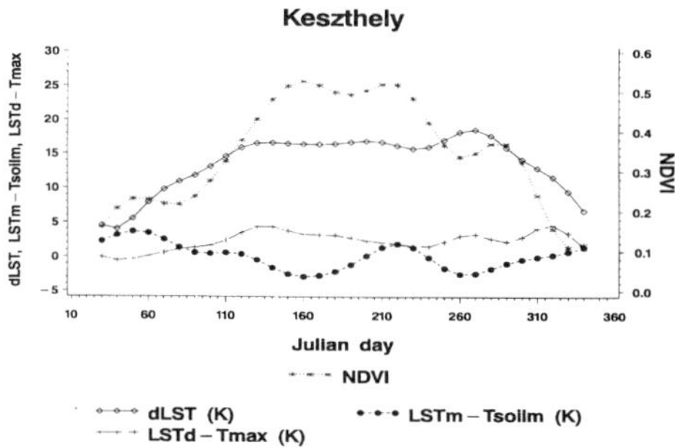

Figure 1. Annual variation of surface thermophysical characteristics over the Szarvas and Keszthely target areas

target, because of the less continental climate and the vicinity of Lake Balaton. Comparing the NDVI values at the two targets we can see (Fig.1.) an earlier greenup in spring at Szarvas target and its reaches the maximum in June, while at the Keszthely target because the different vegetation type the greenup started later, but it decreased only in autumn. At the Szarvas target the difference of LST(day) and Tmax increases rapidly in spring than at the other target.

Surface energy fluxes were evaluated when the NDVI value was larger than 25 %. The calculated energy fluxes for the Keszthely target is illustrated in Figure 2.

Good correlation (0.9) between the NDVI and the latent heat flux can be observed. At the beginning and at the end of the period the latent heat flux increases/decreases rapidly with the NDVI, and when the NDVI is at the maximum the temporal change of LE is smaller.

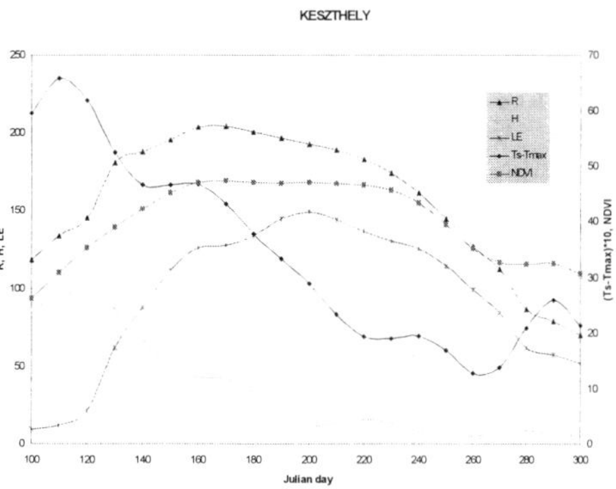

Figure 2. Annual variation of energy fluxes and the NDVI, Ts-Tmax values over the Keszthely target area

6 CONCLUSIONS AND FUTURE PLANS

The synergy of satellite and ground-based observations allows the characterization of surface heat exchange processes over larger areas.
The qualitative evaluation of time series of energy fluxes shows consistence with observed meteorological conditions and vegetation development.
In the future empirical relationships will be sought between NDVI, daily maximum air temperature and the surface thermophysical properties. These relationships will be applied to AVHRR time series within the NOAA/NESDIS GVI (Global Vegetation Index) multiyear multisatellite dataset to investigate if any changes in the variability of energy fluxes occurred during the last decade and whether they are related to global warming phenomenon.

7 ACKNOWLEDGEMENTS

Acknowledgements: This work has been supported by the Hungarian Scientific Research Fund (No 548).

REFERENCES

Crosson,W.L.,Smith, E.A. & Cooper, H.J.: Estimation of surface heat and moisture fluxes over a prairie grassland. 4.Impact of satellite remote sensing of slow canopy variables on performance of a hybrid biosphere model, *J. Geophysical. Res.*, Vol.98.No.D3,4979-4999, March 20, 1993.

Dávid,A.,Takács, O. & Tiringer, C.: Spatial distribution of radiation balance in Hungary on the basis of the data during 1951-1980 period, *Országos Meteorológiai Szolgálat, Kisebb kiadvány*, 66.szám, Budapest 1990.

Gruszczynska,M.,Dabrowska-Zielinska,K., Stankiewicz, K. & Janowska, M.: Remote sensing data applied for monitoring condition of cereals, *Proceedings of the 17th EARSeL symposium on future trends in remote sensing,* Lyngby, Denmark, 17-19 June, 1997, 243-249.

Loepmajer, F.J., 1983: Agrarmeteorologisches Modellzur Berechnung der aktuellen Verdunstung (AMBAV). *Beitraege zur Agrarmeteorologie Nr. 7/83. Deutscher Wetterdienst, Zentrale Agrarmeteorologischie For schungsstelle*

Nagaraja Rao, C.R. & Chen,J.: Inter-satellite calibration linkages for the visible and near-infrared channels of the Advanced Very High Resolution Radiometer on the NOAA-7, -9, and -11 spacecraft, *In. J. Remote Sensing*, 1995, Vol. 16. No. 11, 1931-1942.

Shaw, R.H. & Pereira,A.R., Aerodynamic roughness of a plant canopy: A numerical experiment, *Agricultural Meteorology*, 26, 51-65. 1982.

Stewart,J.B. & Verma,S.B., Comparison of surface fluxes and conductances at two constrasting sites within the FIFE data., *J.Geophys. Res.* 97(D17):18623-18628, 1992.

Stowe,L.L., McClain,E.P., Carey,R., Pellegrino,P., Gutman,G.G.,Davis,P., Long,C. & Hart,S. Global distribution of cloud cover derived from NOAA/AVHRR operational satellite data, *Adv, Space Res.* Vol. 11. No.3, pp (3)51-(3)54, 1991

Remote Sensing in the 21st Century: Economic and Environmental Applications, Casanova (ed.)
© 2000 Balkema, Rotterdam, ISBN 90 5809 096 5

Meteorological radar data processing software

Dejan D. Rancic, Marko S. Smiljanic, Slobodanka Djordjevic-Kajan, Predrag M. Eferica & Predrag D. Vukovic
Faculty of Electronic Engineering, Nis, Yugoslavia

Alexsandar T. Kostic
Military Technical Institute, Belgrade, Yugoslavia

ABSTRACT: This paper describes a real-time software for conventional weather radars. The software is developed to support hail suppression system in Serbia. The system is developed by connecting a specialized data acquisition and processing hardware on a standard PC motherboard. Three additional boards have been installed into the PC: A/D converter for radar signal digitalization, DSP board for real-time signal processing, and A/D, D/A and digital I/O general purpose board that controls the antenna. The architecture of the entire system as well as applied solutions for real-time operation of the system are presented. These includes real-time processing of the radar signal, real-time antenna control, real-time aspect of the meteorological products visualization method and synchronization protocol between different hardware components.

1 INTRODUCTION

Weather radars have been used since the 1950s for qualitative assessment of meteorological phenomena measurable by radars. Research has been directed towards providing short-term forecasting and hydrological data in a clear and unambiguous manner. There are three main applications for this information. The first and most frequent application is for storm warning and hail suppression. The second is short-term forecasting and the third application is in hydrology, involving the measurement of water content reaching the ground and control of the water resources thereafter.

The described system is designed for the first application, but it can be extended for the remaining applications, too. The described software is a part of the Hail Suppression Information System (HASIS) which has been developing in Republic of Serbia (Rancic et al. 1998). The system encompasses a specialized information system, based on Geographical Information System (GIS) technology and object oriented (OO) paradigm (Djordjevic et al. 1996). This system configuration features state-of-the-art radar technology and powerful digital processing. It is enhanced with MS Windows based data processing and communication systems as well as the newest technology graphical user interface. The system acquires the meteorological data of the surrounding atmosphere (about 500km) and presents meteorological information in various radar formats and many other products with map overlay and image zooming, storage, overlay and animation facilities. The system also archives the full data set for subsequent analysis.

In order to track various meteorological phenomena the operator chooses among one of the existing scanning modes, performs scanning in it, observes the reflectivity pattern, decides about a possible mode change and so on. It is interesting to note that for the precise inspection of reflectivity data, user mostly uses manual modes. This is the natural consequence of the fact that the operator can see the reflectivity pattern only in the small area determined approximately by the beam size and the current antenna direction. In other words, the processes of antenna control (where you are looking), data acquisition (what are you looking at) and decision making (should you look this or something else) are all performed by the operator. Having in mind that phenomena under observation exhibit fast, almost unpredictable dynamical nature, it is easy to see that operator must "process" remarkable quantity of information in a relatively short period of time. There are three aspects of the real-time system performances: signal processing, antenna control, and visualization of the meteorological products.

Each aspect will be described in detail, as well as the system architecture. The system is based on inexpensive PC platform running under MS Windows operating system, which provides a user-friendly interface already familiar to the users. Hardware specifics of DSP, A/D and digital I/O boards are connected on PC motherboard impose some constraints and partly determine real-time software structure, especially disposition of its components at proces-

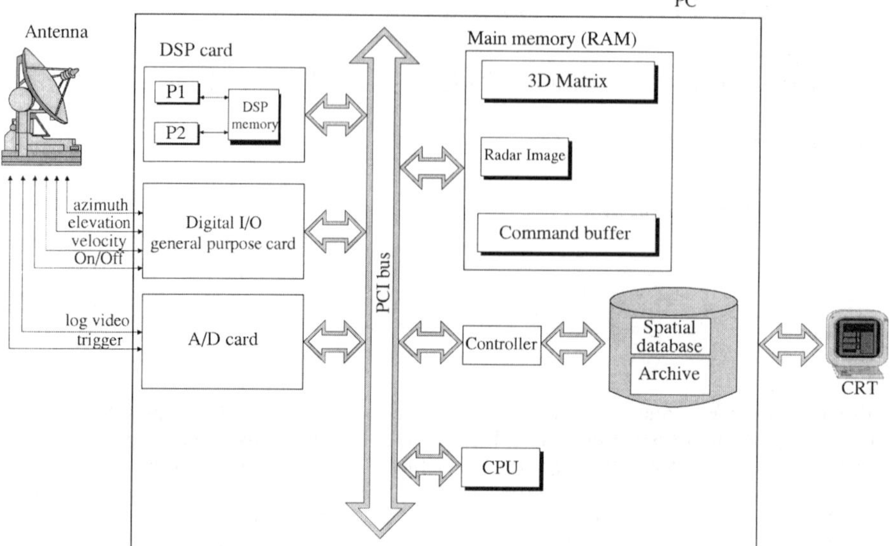

Figure 1. The system architecture.

sors. Also, various tasks performed by the developed software request different priorities and/or service-time delay boundaries, and have a wide range of execution times - characteristics that strongly suggest application of decoupling even parts of the same task which are distributed among two DSPs and host processor. The software is developed using MS Visual C/C++ and assembler.

The remainder of the paper is organized as follows. The second section describes functions and presents the architecture of the system. The third section describes a basic aspects of signal processing. Antenna control subsystem is presented in forth section while the visualization method of meteorological products is given in fifth section. The last section concludes the paper and discusses the system usefulness.

2 SYSTEM ARCHITECTURE

The basic architecture of the system is shown on Figure 1. In order to facilitate real-time radar signal processing and control, main workstation is equipped with three hardware boards:

1. Fast A/D converter (5 MSamples/s, 12 bit), digitizing radar receiver signal,
2. PCI DSP board featuring two TMS 320C40 DSPs, processing digitized radar sweeps, and
3. General purpose A/D, D/A, digital I/O board, for antenna position acquisition and control.

Block acquisition of radar echo samples, initiated by radar trigger signal, is performed by fast A/D converter. The great speed of the A/D converter and the ability of external sampling control allow dynamic change of the resolution for observing radar reflection. Samples are transferred through TMSC40 commport to DSP board and undergo signal conditioning procedure at one of DSPs. Other DSP performs address conversion and transfers, using PCI bus master mode, two blocks holding reflectivity data and their Cartesian addresses in the radar image field to host memory. Host displays acquired image and completely controls antenna motion. Digital I/O general purpose board gathers antenna position parameters (azimuth and elevation angle) and generates the antenna control signal. Video and trigger channels are connected to the A/D board, while antenna azimuth and elevation angle channels as well as the velocity and On/Off ones are connected to the digital I/O general purpose board. Hail suppression methodological activities place the most important constraints (technical requirements) with regard to real-time processing weather echo power and shape of an antenna volume scan. For this purpose, the following functions have been developed in the system:

1. Very precise analysis of the hail cells in the cumulonimbus clouds on the basis of their reflectivity.
2. Time, angle and range averaging of radar echoes from distributed targets (clouds) has been implemented in the system in different degrees, according to modes of operations, through the integration moving window of the adaptable length.
3. Transforming radar coordinates into Cartesian ones and the transfer of data into RAM of the PC.
4. Volume sectorial scans. The limits of the sector are automatically determined by geometrical

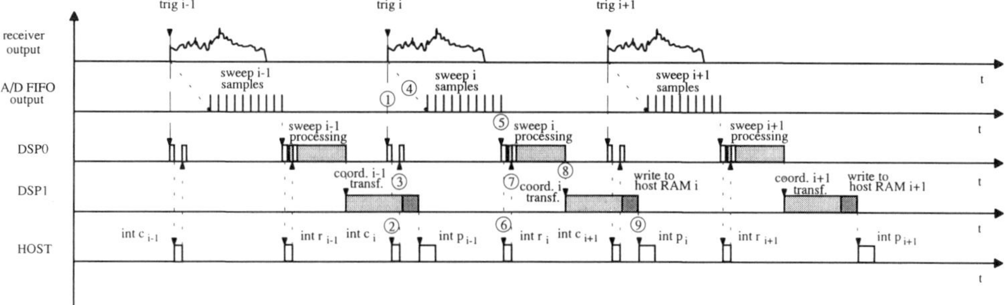

Figure 2. Time sequences of echo signals processing.

characteristics of hail cells at a certain reflectivity level.

5. Displaying of collected data in a chosen format.

6. Hail cells tracking and prediction of their positions.

7. Calculation of the new parameters, constants, look-up tables and their transfer to the DPS board, during the change of the observing mode.

3 SIGNAL PROCESSING

Real time software for signal processing consists of three main parts: DSP0 program, DSP1 program and host processor program. DSP0 and DSP1 programs are downloaded at the beginning of the host application software initialization, using the DSP's system routines. Next phase of initialization commences after DSP programs start-up and its purpose is to establish memory buffer through which radar picture is transferred from DSPs to host. Due to relatively high and constant required data throughput of about 2 MB/s, slow data transferring routines from DSP board device driver are useless, and the only satisfactory solution is to perform direct data transfer to host memory by means of DSP board's PCI controller in bus master mode. After initialization, system starts acquisition and processing of radar echoes (Eferica et al. 1998). Steps involved in processing of one radar pulse echo are outlined in Figure 2.

Radar trigger pulse ($trig_i$) starts RF pulse emission and synchronizes both radar and automated hail suppression subsystem. It starts acquisition of log-receiver signal samples block performed by A/D board as well as acquisition of current antenna position accomplished through steps ① - ③. At the instant ① trigger interrupts DSP0 which in turn issues interrupt to host processor (*int* c_i). During interrupt processing host reads antenna coordinates through digital I/O board and transfers them through *mailbox* register of DSP board's PCI controller. Writing to this register issues interrupt to DSP0 (③) which accepts data and associates them to data belonging to sweep *i*.

A/D converter outputs 12-bit samples to 1024 words long A/D board FIFO register. As FIFO is 16-bit wide, samples are sign-extended. After the FIFO is half-filled, TMS communication port controller starts transmission to one of DSP0 communication ports (④) which lasts as long as FIFO holds more than 512 samples. Hence, FIFO introduces delay of 512 T_S (T_S - sampling period), which is in worst case (for the minimum sampling rate f_{Smin}= 300 kHz) about 1.7 ms. Samples are received by DSP0 commport attached internally to one of six channels of Direct Memory Access (DMA) coprocessor which packs pairs of 16 bit samples in 32-bit words and stores them in one of two input samples buffers. After transmission of preprogrammed number of samples DMA stops and issues interrupt to DSP0 informing it that new sweep data is ready for processing (⑤).

Interrupt Servicing Routine (ISR) interrupts host processor (*int* r_i, ⑥), and, in case DSP0 has finished sweep *i*-1 processing, performs buffer switching, so that DMA receiving buffer and DSP0 processing buffer are exchanged (Fig. 3). At exit, ISR sends signal to main loop and so initiates new cycle of processing. Host ISR servicing *int* r_i resets FIFO and reinitializes A/D board preparing it for new echo acquisition, interrupting DSP0 (⑦) before exiting. Having received this interrupt, DSP0 reinitializes DMA coprocessor.

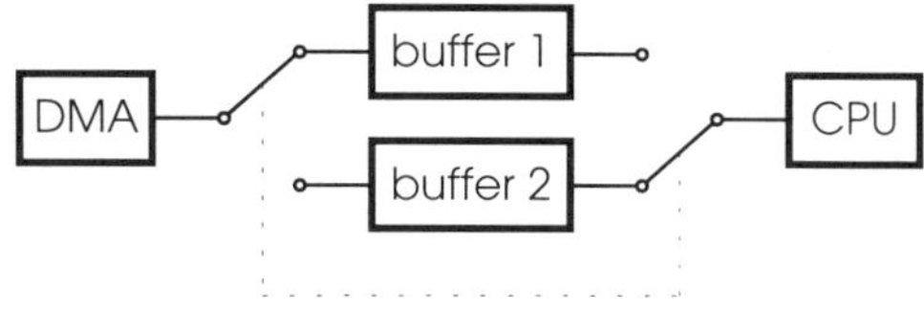

Figure 3. Packed samples receiving buffers.

After unpacking, samples from processing buffer are input to the sweep-to-sweep integration filter whose output is fed to a range integration filter. Its output is compared to threshold obtained according to CFAR algorithm (Kostic et al. 1997). Samples above the threshold are corrected for r^2 and atmos-

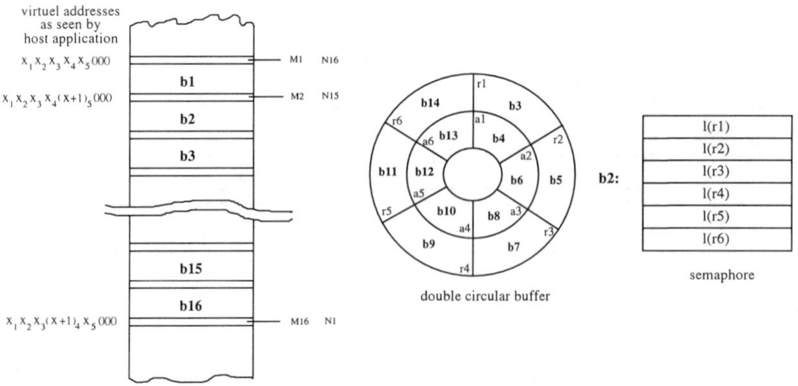

Figure 4. Organization of DSP-host common memory.

pheric attenuation and finally converted from dBm to dBz units. Processed samples are stored in one of the four blocks of circular buffer ($s1 \div s4$), implemented in the DSP global memory (Fig. 4). The length of the data blocks is matched to the number of samples corresponding to sweep maximum resolution. In addition to a length of used block, slot header stored previous samples in circular buffer slot also contains azimuth and elevation of the radar antenna. The next step is a transformation of polar coordinates into Cartesian ones expressed in the form of offsets, i.e. relative addresses of samples in regard to the origin of memory map of radar searched space.

Main loop of the program executing on DSP1 is continuously scanning headers of circular buffer slots and starts coordinate transformation procedure when non-empty slot is observed (⑧). As the transformation is done solely upon data stored in slot header, circular buffer ensures decoupling between processes executing on DSP0 and DSP1. The coordinate transformation procedure output consists of two data blocks: the first, which is string of reflectivity values and the second, holding coordinates of the points in observed space to which these reflectivity values correspond. In general case, reflectivity values block is not merely copy of the circular buffer slot, since, in order to avoid empty spots in radar picture appearing in certain modes of work, coordinate transformation routine also executes elaborate filling algorithm which can assign the same reflectivity to neighboring elementary space cells.

After the coordinates transformation, DSP1, by PCI controller in bus master mode, stores block of reflectivity samples (a_n) and block of addresses (b_n) in one of the six slots, r_n, of double circular buffer implemented in a common space of host RAM (Fig. 4). After storing, DSP 1 writes length of data blocks to semaphore $l(n)$, and interrupts host (*int* p_i ⑨, Fig. 2). Host interrupt service routine transfers reflectivity data to a memory mapped space through indirect indexed addressing with offsets from address block.

Host processor executing algorithms performs further processing of radar picture to get meteorological products. A detailed description of algorithms for the generation of the meteorological products and hail suppression information system is given in fifth section. Finally, let's give a few concluding remarks on real-time software structure for signal processing. Host interrupts *int r* and *int c* are required by DSP board inability to work as bus master in host I/O address space, and had to be implemented in order to control I/O mapped hardware. The reasons for implementation of host interrupt *int p* are manifold: (1) DSP board can address only 32-bit words, and since reflectivity is 8-bit long unacceptable loss of memory space would be incurred; (2) due to philosophy of PCI, favoring block data transfers, and much simpler programming of DSP board's PCI controller chip when data to be transferred have consecutive addresses, transfer speeds are much higher for block transfers compared to random memory accesses; (3) in order to avoid DSP1 executing physical to virtual addresses conversion, calculated addresses are in offset form and actual addresses conversion is automatically done by host processor's Memory Management Unit (MMU).

Introduction of the whole suite of buffers (Figs 3-5) although making software much more complex than it theoretically could be, greatly enhances sys-

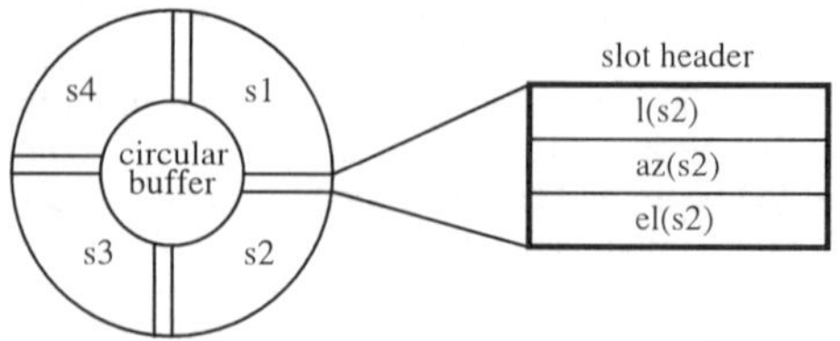

Figure 5. Processed samples circular buffer.

tem robustness and prevent data loss due to unpredicted delays incurred mostly by the host operating system activities running at higher levels of priority, inability of DSP to finish sweep processing by the time next sweep begins, and similar events.

Buffering of data also enables DSPs to use the whole interpulse period of 5 ms for processing although radar echo lasting time for the maximum distance to be observed by this system D_{max}=250 km, is $\tau_{max}= 2D_{max}/c$ = 1.67ms. However, should data loss be inevitable due to prolonged delays, software design guaranties smooth recovery from any point data and/or synchronization loss that can eventually occur.

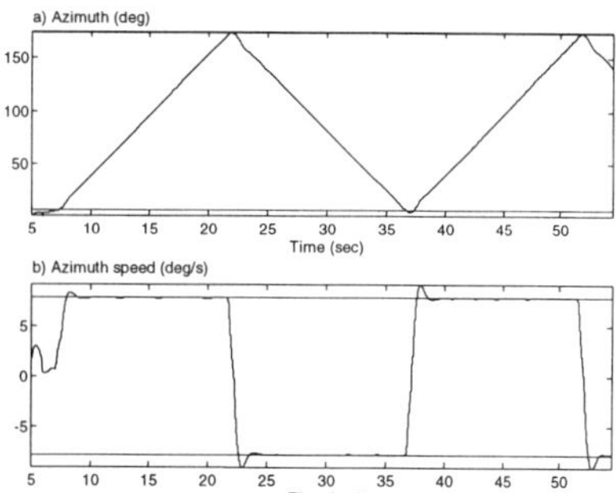

Figure 6. Azimuth speed in ASS regime.

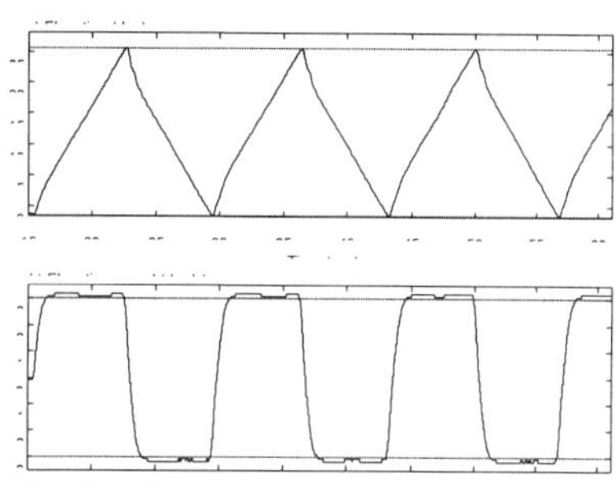
Figure 7. Elevation speed in ESS regime.

4 ANTENNA CONTROL

Design specifications include seven different automatic scanning modes. Four of them are standard PPI (Plan Position Indicator) and RHI (Range, Height Indicator) modes with their sector scanning variants: Azimuth Sector Scanning (ASS) and Elevation Sector Scanning (ESS). The main characteristic of these four modes is that the antenna movement is limited to a plane determined by either constant azimuth (RHI and ESS) or by constant elevation (PPI and ASS). The rest of modes being CAPPI and two volumetric modes - AVS (Azimuth Volumetric Scan) and EVS (Elevation Volumetric Scan), are used for fully automatic reflectivity data acquisition. Design of a control system was carried out at two different levels. At the higher hierarchical level Petri net formalism is used as a simple and efficient tool for modes characterization and design (Murata et al. 1986). On the lower hierarchical level, the direct digital control algorithms were implemented using state space observer techniques and pole placement procedure (Harashima & Kondo 1983) Special attention has been taken in order to avoid the resonance modes associated with the transmission flexibility and suppress the torsion oscillations (Stojic & Vukosavic 1996).

Experienced meteorologists specify overall antenna movement using a number of phases. During each phase, the control system is moving the antenna in a specific way over azimuth and elevation. Using this information it is quite convenient to distinguish between different phases or modes using discrete variables, and to couple the discrete states to different continuous system behaviors. In this way continuous state-plant is supervised by a discrete event system.

The composite system is viewed as a process where a specific task is carried out by changing the structure of the continuous-state plant resulting in a sequence of closed-loop continuous plant behaviors.

Discrete event controller (DEC) selects a continuous state controller (CSC), a reference trajectory and an output function from a set of plant control algorithms, input and output functions respectively. The selected behavior is active until a prescribed plant event or controller event occurs. A continuous state plant (CSP) is described by differential/difference equations. The interface maps signals from the continuous - state domain to symbols in the discrete-state domain and vice versa. The CSP plus the interface are viewed as a Discrete Event Plant (DEP), which is controlled by the (DEC).

Since the velocity signal is not directly measurable and having in mind that chosen control algorithm requires the use of speed (or speed estimate), we designed a full order observer for both azimuth and elevation servo-system. DC motors equipped with large reduction gearboxes as in this case, usually have an output position $\theta(t)$ response to a step input armature voltage $u(t)$ of the form:

$$G(s) = \frac{\theta(s)}{U(s)} = \frac{K}{s(s+b)} \quad (1)$$

The motor coefficients K,b in (1) can be easily determined from the system step response. After this, off-line process identification for both azimuth and elevation servo, discrete time model was calculated, assuming zero order hold and for sample time T_s = 0.005s=$1/f_{PRF}$, with f_{PRF} being radar pulse repetition frequency. Azimuth and elevation observer are further designed and implemented according to (2). As it can be seen from the equation (2) both observers are in standard "position" form without added integral term. Equation (2) also shows that the observers were designed under the assumption that azimuth and elevation servo are decoupled. The experimental results confirm that this assumption can be considered quite valid for most of the modes.

$$\begin{vmatrix}\hat{\theta}(k+1)\\ \hat{\varpi}(k+1)\end{vmatrix} = \begin{vmatrix}1 & E_{12}\\ 0 & E_{21}\end{vmatrix}\begin{vmatrix}\hat{\theta}_1(k)\\ \hat{\omega}_2(k)\end{vmatrix} + \begin{vmatrix}F_1\\ F_2\end{vmatrix}u(k)$$

$$+\begin{vmatrix}L_1\\ L_2\end{vmatrix}(\theta(k)-\hat{\theta}(k)) \qquad (2)$$

Analysis of the technical requirements suggests that every antenna movement phase is characterized by the requirement for trajectory tracking scanning (TTS). TTS is realized using state space feedback with added integral term, leading to the control algorithm of the form:

$$\begin{aligned} u(k) &= K_p(r(k)-\theta(k)) - K_v\hat{\omega}(k) + K_i i(k) \\ i(k) &= i(k-1) + e(r(k)-\theta(k)) \end{aligned} \qquad (3)$$

With defined tracking error $e(k) = r(k) - e(k)$ the dynamics of the closed loop system is now characterized by

$$e(k+1) = e(k) - E_{12}\omega(k) + F_1(K_p e(k) - K_2\hat{\omega}(k) + K_i i(k))$$

$$\omega(k) = E_{22}\omega(k) + F_2(K_p e(k) - K_2\hat{\omega}(k) + K_i i(k))$$

$$i(k+1) = i(k) + e(k) \qquad (4)$$

For the desired aperiodic transient response, characterized by the poles at $z_{1/2} = e^{-\zeta 2\pi f_n t_s}$, with $\zeta = 1$, one can choose undumped natural frequency f_n as the main design parameter. Using the well known pole-placement procedure, the observer, as well as state space feedback gains have been calculated.

During the algorithm implementation it was found that the main obstacle in the extending the servo-loop bandwidth was the mechanical resonance caused by the well known compliance problem. In order to enable the range of stable gains in the presence of flexible coupling we have implemented a simple modification of the position - velocity servo loop according to (Harashima & Kondo 1983). In order to suppress the excitation of resonance modes we apply the following anti-resonance series compensator of the form

$$W_{ar}(z) = \frac{1+z^{-n}}{2}, n = \frac{T_{osc}}{2T_s} \qquad (5)$$

This simple modification in control algorithm resulted in more than doubling the servo system bandwidth.

Experimental results for the ASS PPI and for the ESS RHI regime are shown on Figures 6,7. The proposed control strategy clearly exhibits precise speed regulation without position overshoot on sector limits for both systems.

Experimental results for the AVS and EVS regime are shown on Figures 8,9. Both elevation and azimuth antenna moving is realized using TTS. The proposed control strategy clearly exhibits zero steady state error with little overshoot on sector limits for the azimuth servo.

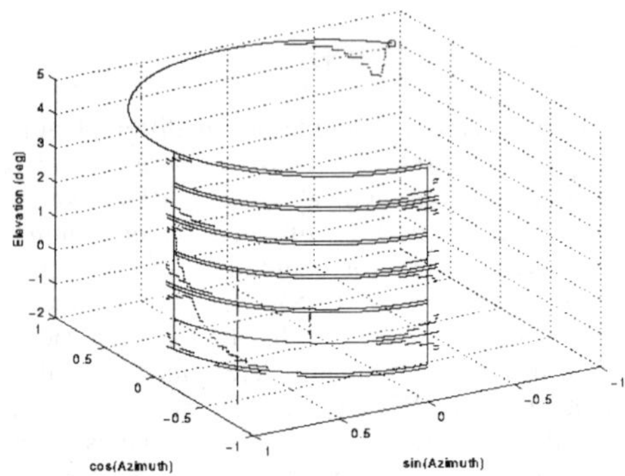

Figure 8. Azimuth and elevation time history in AVS regime.

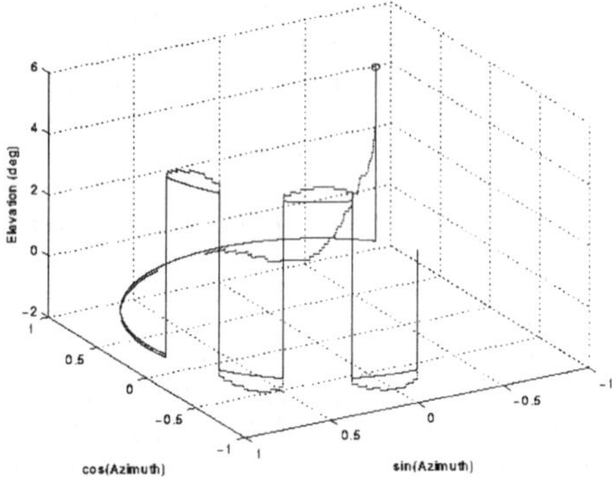

Figure 9. Azimuth and elevation time history in EVS regime.

5 VISUALIZATION METHOD

Real-time visualization of various meteorological data is performed by the host application, which is developed as an MS Windows application and provides a user-friendly interface already familiar to the user and the possibility to interchange data with other MS Windows applications. Visualization is tightly connected with synchronization protocol between different hardware components and this task is also performed by the host application (Rancic et al. 1997). Algorithm of meteorological products visualization and synchronization protocol starts with the selection of working mode (PPI, RHI,...). The next step is determination of antenna and DSP parameters. The host application provides two ways for determination of these parameters. One way is setting parameters using corresponding dialog-boxes provided by the host application. The second way is very sophisticated: the user specifies area of interest on the screen (using mouse) and host application automatically determines all necessary parameters for the best filling of selected area. Some of these parameters are: azimuth and elevation left and right limits, azimuth and elevation step (with 0.1° resolution), range and antenna velocity.

For the communication purposes with DSP application which is running in parallel with the host application, the common memory buffer named Command buffer (Fig. 1) is used. All calculated parameters, constants and look-up tables for antenna and DSP control have been stored to the Command buffer as well as to the memory mapped I/O board for antenna control. When host application finished all necessary computation it sends DSP_START command to the Command buffer. DSP application is waiting for this command and it starts running in parallel with the host application after DSP_START appearance in the Command buffer and process of radar data acquisition and processing begins. Data that have been processed by the DSP are transferred from DSP memory to the two special buffers into main memory (RAM): 3D Matrix and Radar Image. These buffers are used for data exchange between host application and DSP application. 3D Matrix is a special buffer that represents three-dimensional volume around the radar antenna. One-byte elements of the matrix (range 0 - 90) represent reflectivity value in dBz at corresponding spatial points determined by the Cartesian coordinates. This matrix can be saved on disk of main workstation (Archive) for further analyses.

Radar Image is a buffer for real-time display of radar products. In order to fulfill real-time constraints and increase visualization speed, the special color palette has been generated by the host application. The palette is made in such a manner that establishes direct correspondence between radar reflectivity values and palette entries. This enables direct transfer of radar data into video memory. The buffer is transferred 25 times per second directly to the video memory and in this way the display entirely corresponds to the antenna movement. Radar Image, in fact, represents the picture which is displayed. This picture is composed of three components (multilayer architecture, see Fig. 10): radar data geographical data and range markers Radar data are processed data from DSP board and transferred from DSP memory. Also, radar data can be transferred from 3D Matrix if the chosen working mode is not a real-time mode. The second layer consists of geographical data (rivers, towns, etc.) stored in Spatial Database (Rancic et al. 1996) on main workstation disk unit. Range markers with adaptive spatial resolution are part of the third layer. The composed picture is constantly being transferred to the video memory (25 times per second) until the DSP_STOP command is sent to the Command buffer. This command can be sent only on operator's demand.

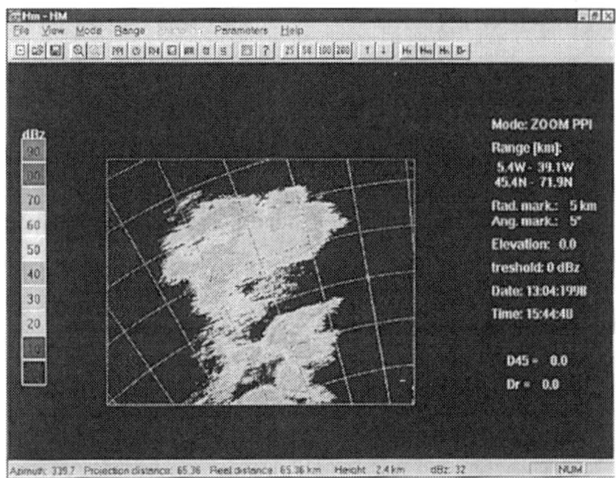

Figure 10. PPI display.

6 CONCLUSIONS

The real-time System for meteorological radars being currently under research and development at Faculty of Electronic Engineering of Nis is presented in this paper. The basic components of the system are described as well as basic software functions that system provides. The chosen object-oriented analysis and design method (Coad & Yourdon 1990, 1991) with object-oriented language for implementation (C++) and the fact that the system is developed for 32-bit MS Windows95 environment, guaranties a high speed and efficiency of the system, as well as its flexibility and expandability. This system is a part of the Hail Suppression Information System that significantly increases the efficiency of hail suppression activities, which are of crucial importance for human goods, especially in critical situations.

The proposed design approach can be extended in two interesting ways. The addition of the reflexivity feedback loop would be the natural hierarchically higher extension to the existing system. This feedback could be used to interpret the reflectivity patterns and focus the reflectivity acquisition to the areas of interest by choosing an appropriate antenna modes. Establishment of the radar network enables even higher control structure that would coordinate the work of several radar sites in the most appropriate way.

REFERENCES

Coad P. & E. Yourdon 1990. *Object-Oriented Analysis*, N. J: Prentice Hall.

Coad P. & E. Yourdon 1991. *Object-Oriented Design*, N. J: Prentice Hall.

Djordjevic-Kajan, S., B. Milovanovic, D. Rancic, A. Kostic, Z. Stanic & L. Stoimenov 1996. Hail Suppression Information System of a Radar Center. *Proceedings of the GIS/LIS '95-96 Central Europe conference, Budapest, Hungary, June 1996:* 102-111.

Eferica, P., P. Vukovic, A. Kostic, B. Milovanovic, D. Rancic, M. Smiljanic & Z. Babic 1998. Weather Radar Signal Processing for Hail Suppression: Radar Products Enhancement and Hybrid Antenna Control Techniques *Proceedings of RADME 98 - Theoretical, Experimental and Operational Aspects of Radarmeteorology, Rome, 9-10 June 1997*: 150-163.

Harashima, F. & S. Kondo 1983. Robust State Observer and Variable Sampling Technique for Performance Improvement of a motor Drive System. *IEEE Trans. Ind. Electronics*, IE-30 (4): 325-329.

Kostic A., Milovanovic B., Eferica P., Vukovic P. & Rancic D. 1997. Weather Radar Signal and Data Processing in Hail Suppression System. *Facta Universitatis, Series: Electronics and Energetics*. 10 (2): 325-340.

Murata, T., N. Komoda, K. Matsumoto & K. Haaruna 1986. A Petri net-based Controller for Flexible and Maintainable Sequence Control and its Applications in Factory Automation. *IEEE Trans. Ind. Electron.*, 33 (1): 1-8.

Rancic, D., M. Smiljanic, S. Djordjevic-Kajan, A. Kostic, P. Eferica, P. Vukovic & Z. Vucinic 1998. Radar Data Processing for Cloud Seeding in Hail Suppression Information System. *Proceedings of RADME 98 - Theoretical, Experimental and Operational Aspects of Radarmeteorology, Rome, 9-10 June 1997*: 137-149.

Rancic, D., S. Djordjevic-Kajan & L. Stoimenov 1996. Spatial Component of Hail Suppression Information System. *Proceedings of XLI ETRAN Conference, Budva, Yugoslavia, June 4-7 1996*. (in Serbian)

Rancic, D., S. Djordjevic-Kajan, A. Kostic, P. Eferica & P. Vukovic 1997. Visualization of Radar Data in Hail Suppression Information System. *Proceedings of Telsiks '97 - 3rd International Conference on Telecommunications in Modern Satellite, Cable and Broadcasting Services, Nis, Yugoslavia, 8-10 October, 1997*: 433-436.

Stojic, M. R. & S. N. Vukosavic 1996. Algorithm for Suppressing the mechanical Resonance in High Performance Servo Drive. *Proceedings of XLI ETRAN Conference, Budva, Yugoslavia, June 4-7 1996*: 405-408. (in Serbian)

Remote Sensing in the 21st Century: Economic and Environmental Applications, Casanova (ed.)
© 2000 Balkema, Rotterdam, ISBN 90 5809 096 5

Evaluation and behavior of the aerosol ε Ångström coefficient from spectral solar irradiance measurements in the Gulf of Cádiz (Huelva, Spain) to assess ocean color atmospheric correction

V.E.Cachorro, R.Vergaz, A. & A.M.De Frutos
GOA-UVA, Grupo de Optica Atmosférica-Universidad de Valladolid & Departamento de Optica y Física Aplicada, Facultad de Ciencias, Valladolid & Departamento de Física Aplicada I, E.T.S.I. Agrarias de Palencia, Spain

ABSTRACT: An operational data base is being developed under Spanish project for the calibration/validation of the ocean color satellite sensor in the regional area of the Gulf of Cádiz. A total characterization of optical aerosol properties: aerosol optical thickness (AOT), particle size distribution and radiative aerosol parameters, etc., is being carried out in order to develop and validate operational atmospheric correction algorithms. Auxiliary data as ozone measurements, pressure and wind speed is also available. Here we present a work applied to the SeaWIFS color ocean sensor in order to address atmospheric correction. Determination and spectral behavior of the Ångström α turbidity parameter and the ε coefficient have been analysed using the aerosol parameters, mainly the aerosol optical depth. This coefficient is the key parameter in the current worldwide SeaWIFS atmospheric correction algorithm. However this algorithm presents great discrepancies in the estimates of pigment concentration in coastal zones (water case 1). As part of calibration/validation procedures, special algorithms are needed adapted for regional coastal zone applications, hence in situ aerosol measurements are required.

1 INTRODUCTION

Remote sensing of ocean color from space provides the means to monitor seawater constituents, like pigment concentration, suspended material, etc.

To derive marine spectral reflectance from top of the atmosphere (TOA) radiance in the visible and near infrared is not easy because of the interference of atmospheric effects, namely, scattering and absorption. Typically, 90% of the TOA signal is composed of photons that not interacted with the water body and, thus, do not contain information on water composition.

The difficulty of the atmospheric correction for color ocean problems was identified with the first space sensor, the CZCS (Coastal Zone Color Scanner, Gordon et al., 1980) and other radiometric space-borne sensor OCTS, POLDER (Polarization and Directionality of Earth Reflectances, Deschamps et al., 1994). The next generation of instruments, such as the SeaWIFS (Sea-viewing Wide Field of view Sensor), MODIS (Moderate resolution Imaging Spectrometer, Ardanuy et al.,1991), MERIS (Moderate resolution Imaging Spectrometer, Rast and Bézy, 1995), will improve the accuracy of ocean and coastal marine reflectance estimates.

Atmospheric correction algorithm attempts to retrieve the water leaving radiance from TOA measured radiance. The most complex problem is the correction of aerosol effects, mainly due to their spectral behavior and their spatial and temporal variability. A priori climatological knowledge of their optical properties is not sufficient to compute accurately their effect on TOA signal.
In situ atmospheric optics measurements are not easy to perform at sea, from a ship or any platform but coastal regional sea sites allow these kind of measurements in an easier way.

This is the aim of this work, in a well-defined regional area in Spain, like the Gulf of Cádiz. Radiation measurements in the visible and near-infrared range allow the determination of the main optical atmospheric parameters needed for atmospheric correction of satellite sensor images in this area.

Furthermore, current atmospheric correction algorithms work well in open ocean areas but not in coastal zones. We are developing an extended project with the objective to perform an operational measurements database for the calibration validation of atmospheric correction algorithm for satellite sensor color application in the area of the Gulf of Cádiz (Spain).

In this work we present the current atmospheric database, mainly addressed to aerosol properties characterization. An analysis of the values and spectral behavior of the Ångström ε coefficient and the α turbidity Ångström is being carried out. The ε coefficient is the physical base of current worldwide SeaWIFS algorithm.

2 ATMOSPHERIC CORRECTION ALGORITHM

2.1 *Physics of the algorithm*

The radiance received by the sensor at the top of the atmosphere (TOA) in a spectral band centered at a wavelength λ, $L_t(\lambda)$ can be composed of the following contributions: $L_{path}(\lambda)$ the radiance generated along the optical path by scattering in the atmosphere. Gordon and Wang (1994) also consider in this term the specular reflection of skylight from the sea surface. $L_g(\lambda)$ is the contribution arising from specular reflection of direct sunlight from the sea surface (sun glitter). $L_{wc}(\lambda)$ is the contribution arising from direct and diffuse downward irradiances reflecting from individual whitecaps on the sea surface. Finally $L_w(\lambda)$ the water leaving radiance, the quantity to be retrieved, which is the contribution of scattering by water body and carries the information of water constituents.

$$L_t=L_{path}+T_dL_g+TL_{wc}+TL_w \quad (1.1)$$

We have removed the dependence on wavelength and angles in L_t for the sake of brevity. T_d is the direct transmittance of the atmosphere and T the total one (direct plus diffuse transmittances) which contains scattering and absorption effects. In that of follow we will consider only the atmospheric effect represented by the L_{path}. More detailed discussion on the other terms can be seen in another paper in this issue (Cachorro et al., 1999c).

For convenience we will work with reflectance ρ instead of radiance, that is the radiance normalized by the solar spectral irradiance outside of the atmosphere Eo

$$\rho=\pi L/\mu_o E_o \quad (1.2)$$

where, μ_o is the cosine of the solar zenith angle θ_o. Therefore equation (1) is reduced to

$$\rho_t= \rho_m + \rho_a +T\rho_w \quad (1.3)$$

Other effects as coupling term between aerosol and molecules interaction (multiple scattering, etc.) are neglected. The contribution of ρ_m can be easily computed by theoretical considerations, by means of the Rayleigh optical thickness and sea level pressure. In the simple scattering approximation ρ_a is given by

$$\rho_{as}= w_o \tau_a P(\varphi)/\mu_o \mu_v \quad (1.4)$$

where ρ_{as} stands for simple scattering, μ_v is the cosine of sensor zenith observation. The aerosol parameters are: w_o is the aerosol simple scattering albedo, τ_a is the aerosol optical depth and $P(\varphi)$ is the aerosol simple scattering function, which depends on the scattering angle φ by means of the two angles of solar position and those of sensor observation.

As the goal of atmospheric correction is to retrieve ρ_w for each pixel at each sensor channels we must determine ρ_{as} at the same level. This can be carried out known the earlier aerosol parameters or by assuming given approaches in order to determine the aerosol type and amount. As we mentioned earlier in situ measurements are difficult and expensive for extended areas, but in the case of regional areas the problem is only at what spatial extension the aerosol parameters must be applied.

In the case of extended areas the aerosol type and amount are eminently variable and in situ measurements are scarce, therefore different approaches must be developed. SeaWIFS algorithm assumes that ρ_w is zero at the two NIR channels (865 and 765 nm) and hence ρ_{as} can be determine for these two bands in the image. However we must also know these reflectances at the other shorter channels. To do this the spectral dependence of ρ_{as} must be known.

On the other hand our knowledge of aerosol models spectral behavior permits to assume a given dependence for ρ_{as}, because in expression (1.4) the simple scattering albedo and Phase function vary weakly on wavelength, therefore the dependence is given basically by the aerosol optical depth. Hence Gordon introduce the atmospheric correction parameter ε, also called the Ångström coefficient as

$$\varepsilon(\lambda_s, \lambda_l)=\rho_{as}(\lambda_s)/\rho_{as}(\lambda_l) \cong \tau_a(\lambda_s)/\tau_a(\lambda_l) \quad (1.5)$$

where we must return to the wavelength symbol to express the spectral dependence. On the other hand it is well known the spectral dependence of the aerosol optical depth expressed by the Ångström law

$$\tau_a(\lambda_s)/\tau_a(\lambda_l) \cong (\lambda_s/\lambda_l)^{-\alpha} \quad (1.6)$$

where α is the Ångström turbidity parameter (Cachorro et al., 1987a, 1987b, 1993). Theoretically, both ε and α depends on the two pairs of wavelengths used to be determine (Cachorro et al., 1987b, 1998b, 1999a). However, the problem is at what spectral extended range a unique value of ε or α can be applied. Gordon has also proposed an exponential dependence in expression (1.6) instead of power law, but both functions give similar results to fit experimental data as we have demonstrated in previous works (Martinez-Lozano et al., 1998; Utrillas et al., 1999). As power law is much used and the parameter α is traditionally used in aerosol studies, we prefer to work with it.

In the SeaWIFS software package (SEADAS) for processing SeaWIFS images the level 2 products generates the atmospheric radiance at 850 nm and the ε coefficient at 765/865 pair of wavelengths. The limitations and uncertainties of the algorithm are obviously involved about the above approaches. In coastal zones the radiance at channels 765 and 865 nm may be non zero and the wavelength dependence of real aerosols not always follow a power (or exponential) law in the extrapolated values of ε or α to other wavelengths, such as the 412 or 440 (1 and 2 channels). Furthermore for greater aerosol optical depth ($\tau a \geq 0.4$) multiple scattering contributions must also be account for. Another remainder problem seems to be related with the absorption of atmospheric aerosols.

We can determine the ε (or α) coefficient for whatever pairs of wavelengths by means of our spectroradiometric measurements covering the spectral bands of SeaWIFS sensor, well using the single scattering approximation (previous the determination of the radiative aerosol parameters) or more easily by the ratio of aerosol optical depths. We are carried both determinations in the developing of the operational database and the aerosol characterization in our area of study. Some of the current results are presented in the next paragraphs.

First work is to determine and analyse the entire above aerosol parameters in order to see how real aerosols follow theoretical approaches, or in other words to validate current algorithm over our area of study. In a second step, this information will be used to develop regional specific atmospheric correction algorithm. Here we present some of the first results.

3 EXPERIMENTAL MEASUREMENTS AND DATA BASE

3.1 *Experimental measurements and data base*

We distinguish two different kind of measurements related with image processing, those we call optical measurements and those of meteorological type. The optical measurements is being carried out by the "Grupo de Optica Atmosférica GOA-UVA" of the University of Valladolid in collaboration with the group of "Estación de Sondeos Atmosféricos El Arenosillo" belonging to INTA (Instituto Nacional de Técnica Aeroespacial). The National Whether Service of Spain provides the meteorological measurements. We must also mention to the CICEM (Centro de Investigación y Cultivo de Especies Marinas, "Agua del Pino") belonging to the "Consejeria de Agricultura y Pesca de la Junta de Andalucia" which carries out sea measurements of optical water properties, pigment concentration, particulate material, etc.

The structure of the data base is not yet well-defined but optical and meteorological data are being structured as separate data. Temperature, pressure and wind are the three basic meteorological data. Cloud cover is also provided by the optical data.

3.2 *Optical atmospheric measurements*

Spectral direct normal irradiance measurements are being performed with the commercial LI-cor LI1800 portable spectroradiometer in the spectral range 300-1100 nm (6 nm of spectral resolution and 1 nm wavelengths sampling). The collimator of the spectroradiometer is moved by an automatic sun-sky tracking system. Specifications about spectroradiometer calibration and characteristics was described in detail elsewhere (Cachorro et al., 1998a). The total error of the irradiance measurements depends on wavelength range but a 5% can be considered as a good estimation for these spectra.

The station is located in a rural site "El Arenosillo" (Huelva) at sea level (37.1°N; 6.7°W) belonging to INTA. This station is the number 213 of Global Ozone Observing System of the WMO with more than 17 years of ozone Dobson observations. The area is of special interest due to the influence of maritime aerosols, African desert dust events and local industrial environment. The period for the optical aerosol measurements extends from August of 1996 to current date. Here we will show analysed data from August 1996 to October of 1997, with more than 407 spectra on 80 clear skies days. Lack of measurements is due to calibration and instrument maintenance.

4 SEAWIFS SENSOR SPECTRAL BAND CHARACTERISTICS

SeaWIFS ocean color sensor are composed of eight channels centered at 412 (yellow substance), 443 (Chlorophyll absorption), 440 (Pigment absorption (Case 2), 510 (Chlorophyll absorption) 555 (Pigments, optical properties, sediments), 670 (Atmospheric correction for CZCS heritage), 765 (Atmospheric correction, aerosol radiance), 865 (Atmospheric correction, aerosol radiance), nm wavelengths. Parenthesis indicates the main features of each band. Bands 1-6 have 20 nm bandwidth and 7 and 8 have 40 nm. The selection of bands is based on considerations about the absorption characteristics of some common in-water optical constituents, extraterrestrial irradiance and spectral transmittance of water vapor, ozone and oxygen transmittance.

Figure 1 shows the spectral features of the eight filters together with the total transmittance of the atmosphere with and without the Rayleigh scattering correction for a typical irradiance spectrum measured by our spectroradiometer. We can observe the position and width of the 8 SeaWIFS bands and how the spectral radiometric response of each filter can be affected by the atmospheric absorption.

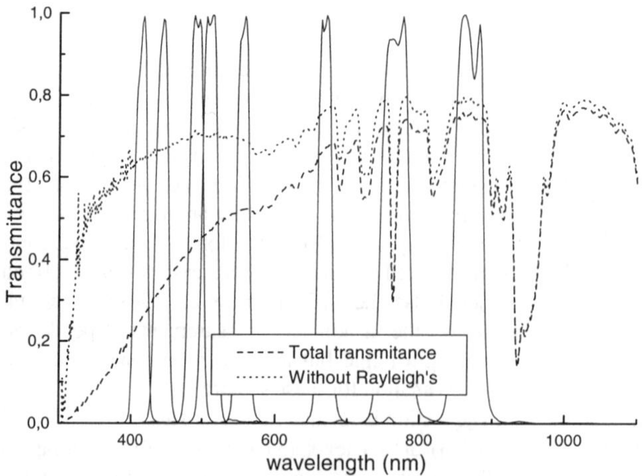

Figure 1. Illustration of SeaWIFS channel transmittance together with the transmittance of typical transmittance spectra of the atmosphere (top curve). Bottom curve is the transmittance after the Rayleigh correction effect.

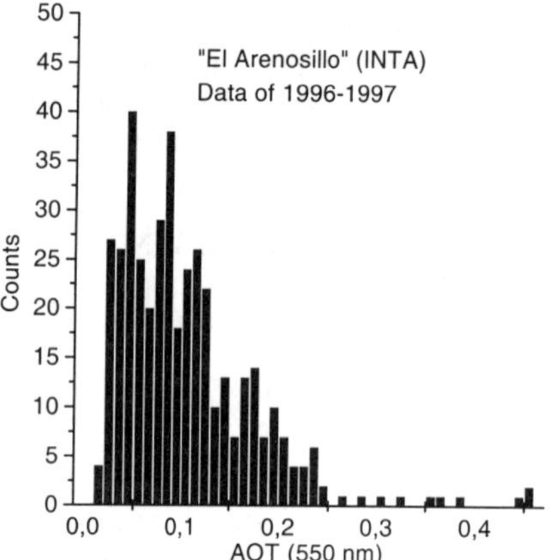

Figure 2. Frequency histogram of the AOT at 550 nm.

5 RESULTS

5.1 *Aerosol optical depth*

The evolution of the aerosol optical depth at 550 nm, considered as the standard turbidity index of the atmosphere, has been analysed in previous works (Cachorro et al., 1998b, 1999a,b). This parameter shows a high variability, ranging from 0.018 to 0.46 with an average value of 0.10 during the 1996-1997 period of measurements. At 765 nm the same average value is obtained.

A first analysis seems to show a flat spectral behavior of the AOT according maritime aerosol characteristics but a more detailed analysis clearly shows two well-defined spectral regions, the NIR one with predominant α negative values, more accentuated for low values of AOT. Figure 2 shows the frequency histogram of these AOT values. This reveals the low turbidity of this area.

Figure 3 shows the determination of the α parameter taking the AOT values given by our spectroradiometer and considering the 7 (765 nm) and 8 (865 nm) SeaWIFS transmittance channels. As can be seen the proportion of negative values is very high.

A determination of the α parameter in the visible spectral range with our spectral AOT values but considering the six SeaWIFS channels was carried out. In this case the mean value is around 1 with more positive values. Both sets of values are represented in figure 4, where we can observe the low correlation existing between these two sets of data.

We have carried out the correlation between the AOT for the different bands of the SeaWIFS channels as SeaWIFS and other radiometric sensors work with band ratios in many applications.

Figure 5 shows the correlations for bands 670/555 and 865/555 respectively. The correlation band to band is very high for close bands but this decreases as much as the bands are separated.

The detailed analysis of the AOT during this period shows the high variability and spectral behavior of real aerosol and the necessity of monitoring for real representation of their features. From the point of view of aerosol studies or atmospheric correction, α or ε parameters are equivalent (see figures 3 and 6). However we will analyse in the next paragraph the latter coefficient.

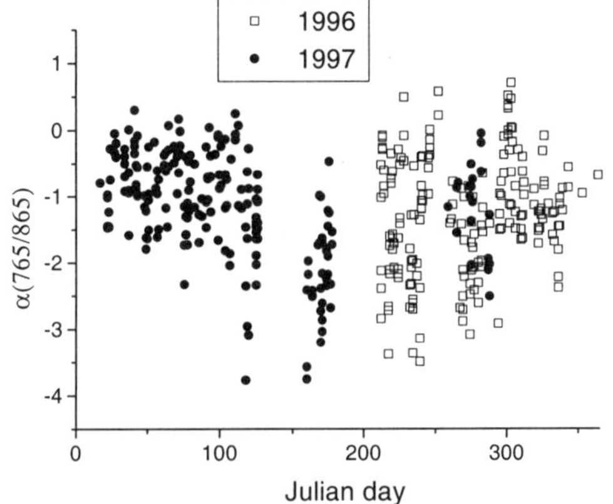

Figure 3. Evolution of the α Ångström parameter determined with the two NIR-INF 7 and 8 SeaWIFS bands of our spectroradiometer during the period 1996-1997.

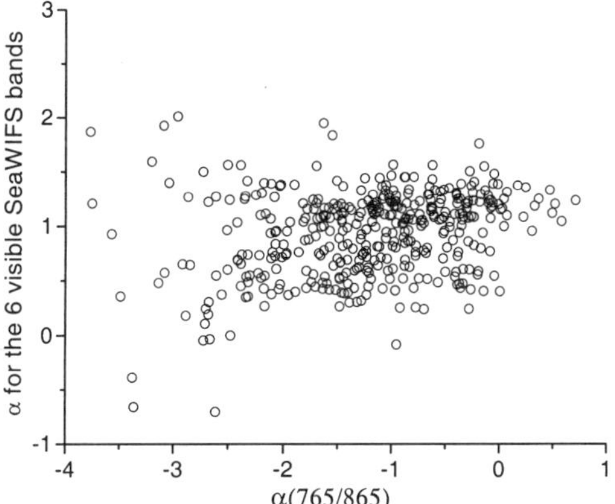

Figure 4. The α Ångström turbidity parameter determined taking the six SeaWIFS bands versus the same parameter determined with the 7 and 8 bands.

5.2 *ε coefficient analysis*

The ε coefficient, which is the key parameter for the atmospheric correction algorithm, has been determined according approach (1.8) for the above mentioned time period. Considering the 7 (765 nm) and 8 (865 nm) SeaWIFS transmittance channels and our spectroradiometic data This parameter is given at level 2 SeaWIFS products. This can be seen in figure 5 where the range of values oscillates between 0.62 and 1.1 with a relatively great variability during the day and day to day. A task of this project is to compare these values with those given by the SeaWIFs processing images during the same period.

For a detailed analysis of this coefficient and in order to study their spectral behavior we have selected different cases corresponding to different turbidity atmospheric conditions. Figure 6 shows 12

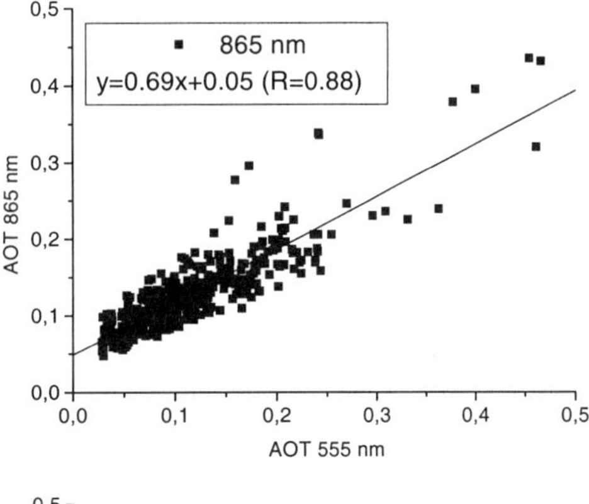

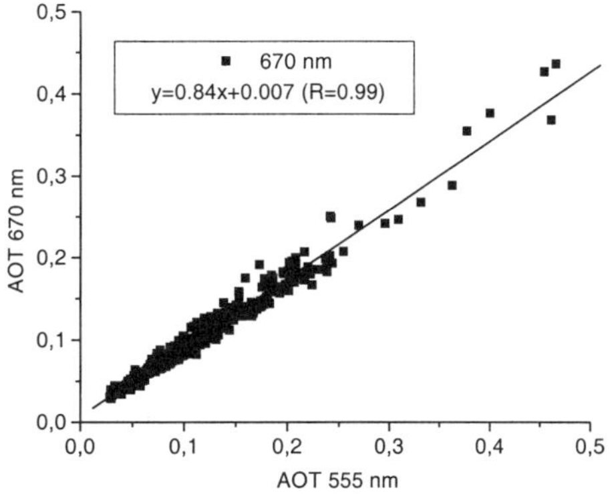

Figure 5. Correlation between the AOT for a) 675 and 555 nm (top) and b) 865 and 555 (bottom) bands of SeaWIFS during the period 1996-1997.

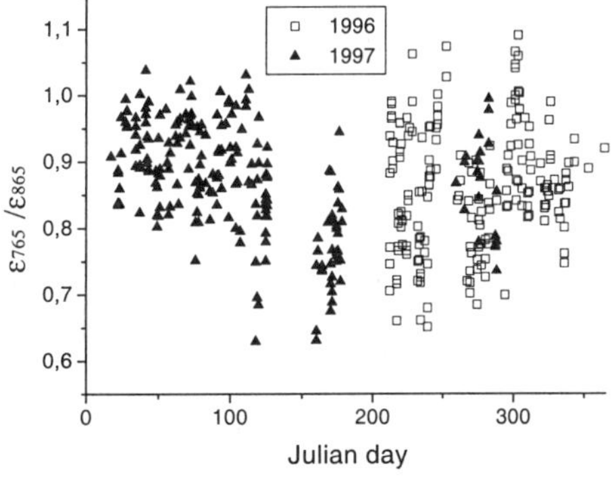

Figure 6. Evolution of the ε(765/865) coefficient determined during the period 1996-1997 (see text).

individual spectra of 12 different days for ε(λ,865) corresponding to the eight SeaWIFS bands. As can be seen the spectral behavior is very different for each spectrum as it is expected according aerosol characteristics. However to fit this spectral behavior to aerosol models according Gordon (1996) algorithm is not an easy task. As can be seen for high ε

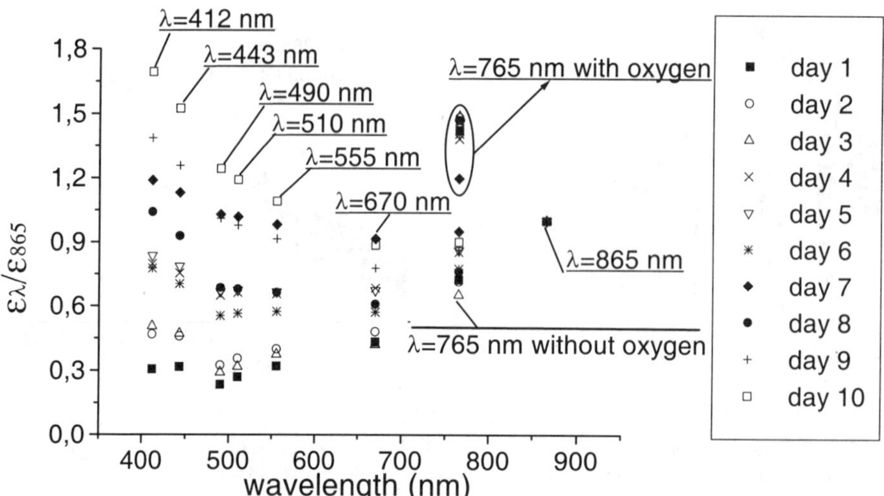

Figure 7. Spectral variation of ε coefficient for 10 real cases of measured aerosol spectra applied to SeaWIFs bands

values the spectral behavior can be modeled very good but for low values of the AOT or low ε values, the spectral features are very complicate and difficult to model (Cachorro & De Frutos, 1996; Schwindling & Deschamps, 1998).

We must note on this figure the values corresponding to band 7 (765 nm). Both set of points, those with and without (top) correction of oxygen absorption (see also figure 1) are depicted. We have followed a simple procedure for this correction similar to that published for water vapor (Cachorro et al., 1998a) which gives same results to that proposed by Wang (1999).

6 CONCLUSIONS

The study carried out shows the high variability and characteristics of the atmospheric aerosol over a given area of special interest for ocean color satellite remote sensing applications. The variability and spectral dependence of the AOT has been analysed based on spectral irradiance measurements carried out during the 1996-1997 period. The AOT data has been specifically applied to SeaWIFS bands and aerosol parameters of SeaWIFS atmospheric correction algorithm, such as the ε and α Ångström parameters has been determined. Thanks to our type of spectrorradiometric measurements, oxygen correction in band 7 is easily accounted for.

The analysis shows the high variability and spectral behavior of real aerosol and the difficulty of its modeling, thus the necessity of real representation of their features.

Preliminary applications of SeaWIFS algorithm to our area of study, the coastal zone in the Gulf of Cádiz presents great discrepancies in the estimates of pigment concentration (water case 1). Therefore improvements and special adapted algorithms are needed for regional coastal applications and hence in situ measurements of aerosol characteristics.

7 ACKNOWLEDGMENTS

This work is partially supported by the CICYT ESP97-1771-CO2-O2. We thank to "Junta de Castilla y León" for supporting the grant of R. Vergaz. We must also thank to the group of "Estación de Sondeos Atmosféricos El Arenosillo" belonging to INTA (Instituto Nacional de Técnica Aeroespacial) in Huelva (Spain), in special to Dr. B. De la Morena and J.M. Vilaplana.

8 REFERENCES

Ardanuy, P.E., Han, D. & Salomonson, V.V. 1991. The moderate resolution imaging spectrometer (MODIS) science and data system requirements. *IEEE Trans. Geosci. Remote Sen.*, 29: 75-82.

Cachorro, V.E., De Frutos, A.M. & Casanova, J.L. 1987a. Determina tion of Ångström turbidity parameters. *Appl. Opt.* 26: 3069-3076.

Cachorro, V.E., Casanova, J.L & De Frutos, A.M. 1987b. The influence of Ångström parameters on calculated direct solar spectral irradiances at high turbidity. *Sol. Energy* 39: 399-407.

Cachorro, V.E., De Frutos, A.M & Gonzales, M.J. 1993. Analysis of the relationships between Junge size distribution and Ångström α turbidity parameters from spectral measure ments of atmospheric aerosol extinction *Atmos.Environ.* 27A,1585-1591.

Cachorro, V.E., & De Frutos, A.M.1996. A revised study of the general Alpha-Junge relationship at solar wavelengths: Aplication to vertical atmospheric aerosol layer. *Atmos. Res.* 39: 113-126.

Cachorro, V.E., Utrillas, M.P. Durán P. Vergaz, R. De Frutos, A.M. & Martinez-Lozano, J.A. 1998a. An study about the-water vapor content determination in the 940 nm band by using using moderated spectral resolution measurements of direct solar irradiance. *Appl. Opt.* 37, 4678-4689.

Cachorro, V.E., Vergaz, R., Vilaplana J.M., De Frutos., A.M., Margarida., R. & De la Morena B. 1998b . Physical and radiative properties of atmospheric aerosols derived from spectral solar extinction measurements in the Golf of Cádiz (Huelva, Spain). *In Proc. Of the 25th Annual European Meeting on Atmospheric Studies by Optical Methods.* Granada. Spain.

Cachorro, V.E., Durán, P., Vergaz, R., & De Frutos, A.M 1999a. Measurements of the atmospheric turbidity of the north-center continental area in Spain: spectral aerosol optical thickness and Ångström turbidity parameters. (unpublished).

Cachorro, R. Vergaz, R J., Vilaplana, J.M., De Frutos A.M., Margarida, R., & De la Morena B. 1999b. Turbidity parameters and radiative properties of atmospheric aerosols derived from spectral solar extinction measurements in the Golf of Cádiz (Huelva, Spain). *In ALPS99: The Contribution of POLDER and New Generation Spaceborne Sensors to Global Change Studies. WK1-p-05. Meribel France.*

Cachorro, V.E., Vergaz, R., Berjón, A., & De Frutos, A.M. 1999c. Developing and testing of an atmospheric correction model for water leaving radiance retrieval in the are of the Gulf of Cádiz based on experimental measurments. *Earsel 19th Symposium.* Valladolid, Spain.

Deschamps, P-Y., Breón, F.M., Leroy, M, Podaire, A, Bricaud, A, Buriez, J.C. & Sèze G. 1994. The POLDER mission: Instrument characteristics and scientific objetives, *IEEE Trans. Geosci. Remote Sen.,* 32: 598-615.

Gordon, H.R , Clark, D.K. Mueller, J.J., & Hovis, W.L. 1980. Phytoplankton pigments derived from the Nimbus-7 CZCS: Initial comparisons with surface measurements. *Science* 210, 63-66.

Gordon, H.R., & Wang, M. 1994. Retrieval of water-leaving radiance and aerosol optical thickness over the oceans with SeaWIFS: a preliminary algorithm. *Appl. Opt.* 33: 442-453.

Gordon, H.R 1996. MODIS normalized water-leaving radiance algorithm theoretical basis document (Mod 18, version 3). Departament of Physis. Univ of Miami. USA.

Martinez-Lozano J.M., Utrillas, M.P., Tena, F. & Cachorro V.E. 1998, Parametrisation of the atmospheric aerosol optical depth using the Ångström power law. *Solar Energy* 63: 303-311.

Schwindling, M & Deschamps, P-Y. 1998. Verification of aerosol models for satellite ocean color remote sensing. *J. Geophys. Res.* 103: 24919-24935.

Rast, M & and Bezy, J.L. 1995. The ESA medium resolution imaging spectrometer (MERIS): Requirements to its mission and performance of its system, in remote Sensing in Action, In P.J. Curran & Y.C. Robertson (ed.). *Proc. Of the 21st Anual Conference of the Remote Sensing Society.* pp. 125-132. Univ. of Southhampton, Southhampton, England.

Utrillas, M.P., Martinez-Lozano J.M., Tena, F., Cachorro, V.E. & Hernandez, S., 1999. Comparison of aerosol optical thickness retrieval from spectroradiometer measurements and from two radiative transfer models. (in press).

Wang., M. 1999. Validation study of the SeaWIFS oxygen A-band absorption correction: comparing the retrieved cloud optical thicknesses from SEAWIFS measurements. *Appl. Opt.* 38: 937-944.

Remote Sensing in the 21st Century: Economic and Environmental Applications, Casanova (ed.)
© 2000 Balkema, Rotterdam, ISBN 90 5809 096 5

Calculation and analysis of erythematic irradiance using experimental data and TOMS ozone data

A. Calle, A. M. Pérez & J. L. Casanova
Remote Sensing Laboratory, University of Valladolid, Spain

ABSTRACT: The current work deals with the concept of erythematic irradiance and analyses experimental values obtained from a measuring station. These experimental values are used to set up a theoretical model which requires as input data the total ozone thickness from the TOMS sensor on board the EP-TOMS satellite. This model will be applied to the generation of erythematic irradiance maps over the whole of the Iberian Peninsula. The distribution of erythematic irradiance data to the population is a priority in summer periods and especially in Mediterranean areas.

1 INTRODUCTION

Control of solar radiation levels is currently of prime importance, above all in Mediterranean countries in which tourist interest or the development of open air leisure activities (mountain, and above all beach) is particularly susceptible to the possible harmful effects caused by solar radiation. The biological effects caused by exposure to solar radiation cover the whole spectrum. However, it is the ultraviolet solar spectrum which entails certain risk due to the harmful effects it can have on most living beings (*Leffell et al., 1996; Hoppe et al., 1996*). Doses above those considered *normal*, lead to serious damage to vegetation systems as a result of the alteration of their chlorophyllous functions. The most widespread effects are a reduction in crops and marine phytoplankton.As regards animals, the damage caused affects various organs, mainly leading to ocular pathologies and loss of immunological system. In the case of humans, it can also be apparent in numerous skin complaints, the most serious of which is cancerous melanoma. Overexposure to this type of radiation also leads to serious alterations in DNA.

These human alterations mainly depend on the characteristics of the individual, in such a way that any study of the risk of exposure would be incomplete if these factors were not taken into account. It is therefore essential to introduce a radiation index for each individual which encompasses the physical conditions of the radiation and the biological type to which it belongs, the so-called *phototype* which will be discussed later.

This risk of overexposure to radiation has increased dramatically in recent years as a result of the constant drop in ozone concentration levels of the total thickness of ozone in the stratosphere (*Stolarsky et al., 1991; Stolarsky et al., 1992; Herman et al., 1993; Varotsos et al., 1993*), which constitutes the main filter for ultraviolet radiation.

2 ERYTHEMATIC IRRADIANCE AND MINIMUM ERYTHEMATIC EXPOSURE DOSE

The biological effects of radiation in the ultraviolet spectrum strongly depend on wavelength. In 1932, the *Comité Internationale de Lumière* presented a division for this area of the spectrum, marking out three regions, A, B y C resulting from the different biological effects they display.

Type A radiation (UV-A) is the nearest ultraviolet spectral region to the visible spectrum and comprises wavelength intervals from λ=400 nm to λ=320 nm. Despite not being absorbed on a large scale by the atmosphere, it is less dangerous as it is less energy-yielding. Its main biological effect is to cause a darkening of the skin.

Type B radiation (UV-B) is the intermediate region of the ultraviolet spectrum, ranging from λ=320 nm to λ=280 nm. This radiation maintains the delicate balance between biological effects which are considered normal. Small increases in the levels received lead to serious damage to the skin and eyes as it is not absorbed on a large scale by the atmosphere and its effects border on the biologically acceptable. The burns it causes are very common as a result of which it is also known as *sunburn radiation.*

Type C radiation (UV-C) is high energy yielding

radiation with more harmful effects such as skin cancers and serious eye complaints. It covers the spectrum between λ=280 nm and λ=200 nm, and is called distant ultraviolet or *germicide radiation.* Practically all of this radiation is absorbed by stratospheric ozone.

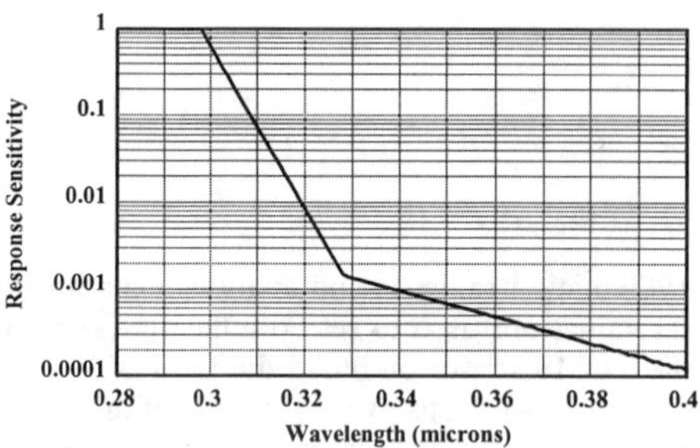

Figure 1. Logarithmic graph of the erythematic action coefficient (McKinlay and Diffey, 1987)

In the analysis of the relation between ultraviolet irradiance and the biological effects caused, the spectral interval corresponds to the λ=[280 nm, 400 nm], wavelength, as lower wavelengths are absorbed. However, the influence of the radiation in this interval is modulated by a coefficient known as the *erythematic action coefficient* which was established by *McKinlay and Diffey (1987).* This spectral coefficient, ε(λ), characterised by the *erythema effect,* is greatest after 0.280 μm and diminishes as the wavelength increases. Total erythematic radiation power is defined by means of the integral of total spectral irradiance weighted by the *erythematic-action coefficient.* This is expressed as:

$$E_{erythem} = \int E(\lambda) \cdot \varepsilon(\lambda)\, d\lambda$$

where $E_{erythem}$ is *erythematic irradiance* and E(λ) is the solar irradiance at wavelength λ at surface level.

Figure 1 represents the *erythematic action coefficient* in a logarithmic scale graph. The spectral values of ε(λ) are given by (*Madronich et al., 1997*):

$$\left\{\begin{array}{ll} \lambda < 0.298\,\mu m & \varepsilon(\lambda) = 1 \\ 0.328\,\mu m < \lambda \leq 0.298\,\mu m & \varepsilon(\lambda) = 10^{0.094*(298-\lambda*1000)} \\ 0.400\,\mu m \leq \lambda \leq 0.328\,\mu m & \varepsilon(\lambda) = 10^{0.015*(139-\lambda*1000)} \\ \lambda > 0.400\,\mu m & \varepsilon(\lambda) = 0 \end{array}\right\}$$

The erythematic action curve is a curve of biological effects such that the value given does not have physical units. The value corresponding to 0.300 μm has been taken as the unit of effectiveness as this is where the maximum erythematic power of ultraviolet radiation is recorded. As can be seen from the graph, an irradiance of I Watts/m^2 at a wavelength of 0.290 or 0.300 μm has an erythematic power equivalent to I*10^4 Watts/m^2 at wavelengths corresponding to the end of the erythematic action spectrum (≈ 0.400 μm).

Erythematic irradiance is still incomplete for determining the harmful effects on individuals as it refers to a value of instantaneous radiation units of power per surface units. Given that the effects of solar radiation occur after exposure they have a cumulative effect, meaning that exposure time needs to be introduced. The *MED* (*Minimum Erythematic Dose*) is therefore defined as the irradiation needed to cause reddening of the skin after a specific period of exposure to radiation. Although reddening of the skin occurs earlier in people with lower skin pigmentation than others with darker pigmentation (in Figure 1. Logarithmic graph of the erythematic action coefficient (McKinlay and Diffey, 1987)other words, it depends on the individual's skin phototype), people with lightly pigmented pale skin are taken as a reference unit. For this phototype, which corresponds to the average central European, the minimum erythematic exposure dose is: 1 MED = 21 mJ/cm^2 = 210 J/m^2.

3 EXPERIMENTAL MEASURING OF ERYTHEMATIC IRRADIANCE

The *Low Atmosphere Research Centre* (CIBA), maintained and controlled by the Department of Applied Physics I at The University of Valladolid has had an erythematic irradiance measuring station since summer 1985. The data obtained by the station have been used for the control and monitoring of the values reached at different times of the year and for the subsequent setting up of a *mono-layer model* which allows the calculations to be extended and an ultraviolet index to be developed which can be used over the rest of the Iberian Peninsula.

The CIBA is set in environmental conditions which are similar to those of a large part of the of the rest of the country, with a Mediterranean type climate with continental features. (*Terán et al., 1987*). It is situated in a rural area whose geographical coordinates are: (Lat, Long): 41° 48' 36'' N, 04° 56' 06'' W; height above sea level: 840 m.

3.1 *Characteristics of the measuring equipment*

Erythematic irradiance has been recorded by an ultraviolet radiation measurer (BIOMETER-501). This is a wide-band radiometer whose response function virtually coincides with the erythematic action curve proposed by *Mckinlay and Diffey (1987)* described earlier. This is a radiometer created from an improved version of the UV measurer developed in the 70s by *Robertson and Berger*, at the University of Temple (USA) which was the first in a global network of ultraviolet radiation operational since 1973. It has been fully operational since the summer of 1995, when the first calibration was carried out at the Izaña observatory in the Canary Islands. It link-up to a computer allows measurements to be taken at 10 minute intervals. The radiometer has a reading range in MEDs in the [0.999] interval with a resolution of 0.01 MED/hour.

The suitable location of the radiometer ensures that measurements are obtained without the interference of any type of contaminating gases, thus resembling clear sky conditions as closely as possible, except for disturbance caused by the occasional presence of clouds.

3.2 *Analysis of the data*

The daily evolution curve of erythematic irradiance coincides with the daily curve of the height of the sun except for a scale factor and reaches a maximum value around 12.00 GMT. This is fairly evident as the zenith solar angle is the most important factor in the values reached by erythematic irradiance. This is why the most important analysis of erythematic irradiance data provided by the radiometer corresponds to the maximum daily values.

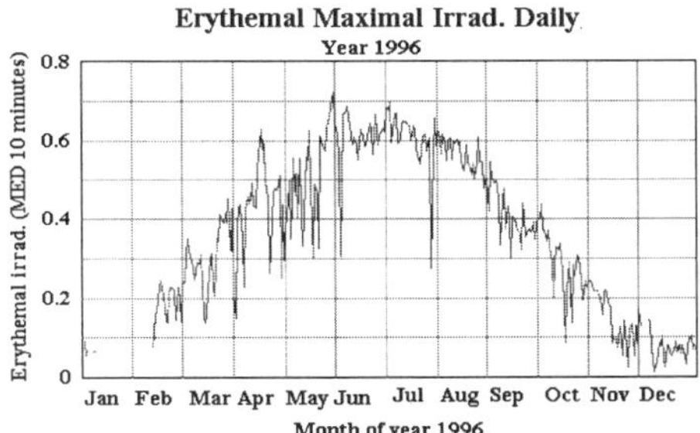

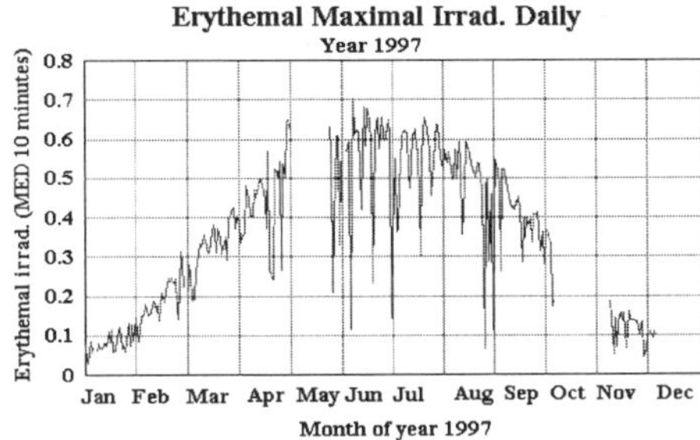

Figures 2 and 3. Maximum daily experimental erythematic irradiance for 1996 and 1997

Figures 2 and 3 contain graphs of the maximum daily erythematic irradiance values for 1996 and 1997 respectively. The values are in MED units per 10 minutes exposure.

As can be seen the spline of the annual curve is parallel to the evolution of the annual solar height. "*Sharp*" drops in irradiance values correspond to cloudy days on which cloud cover reduces irradiance values. This circumstance, despite being characteristic of spring days (for climatological reasons) appeared fairly often in the summer of 1997.

The highest erythematic irradiance values in 1996 were recorded at the end of May and beginning of June, reaching a maximum value of 0.723 MED on May 29th at 12.20 GMT. As regards 1997, the highest values occurred in the middle of June (more in line with the height of the sun than in the previous case), reaching 0.705 MED on June 6th at 12.00 GMT.

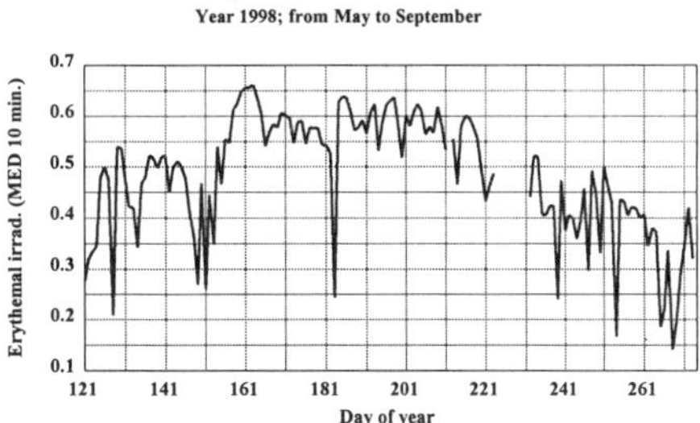

Figure 4. Maximum daily experimental erythematic irradiance for 1998.May-September

Figure 4 shows the experimental data for erythematic irradiance corresponding to the period May 1st to September 30th 1998. The highest value recorded at the measuring station occurred on June 12th when a value of 0.661 MED per 10 minutes was reached. From this graph, the strong variations and deviations from the *standard curve* which occurred during May and September should be underlined.

The decrease in irradiance values due to cloud cover depends on the type and thickness of the clouds (*Bordewijk et al., 1995; Brodeker at al., 1996)*. The most significant drop with regard to the expected values of the spline corresponds to a maximum reduction of 60% in 1996 which occurred in summer. In 1997 the maximum decrease was 80% at the end of spring.

Given that most interest is focused on summer months, for the comparison of values reached in 1996, 1997 and 1998 in this period a comparative histogram of irradiance frequency values has been drawn up. Figure 5 shows these histograms, corresponding to the appearance percentages of the binarised irradiance values in intervals of 0.05 MED for the period June-September.

The results show that in the period analysed the highest values corresponded to 1996. A greater cloud cover in the summer of 1997 led to the existence of unusually low values for this period. 1998 shows a behaviour somewhere between that of 1996 and 1997.

4 SETTING UP OF A MONO LAYER MODEL FOR THE CALCULATION OF ERYTHEMATIC IRRADIANCE

To extend erythematic irradiance measurements to other geographical areas, an attempt has been made to set up a theoretical mono-layer type model for clear skies, in other words considering a homogeneous atmosphere with the same properties as the real atmosphere without cloud cover. The problem dealt with is the lack of experimental data for the characterisation of the atmosphere such as optical aerosol thickness or total ozone data. However, given that in this interval the most important factor is ozone content and geometric conditions, the results obtained have been fairly good as will be seen. We will therefore analyse λ=[0.280µm,0.400µm] wavelengths with a resolution of 0.001µm.

4.1 *Formulation of the model*

The expression which determines the calculation of *spectral solar irradiation* $E(\lambda)$ that reaches the Earth after undergoing absorption and dispersion is the following:

$E(\lambda) = I_{0,\lambda} \cdot F \cdot T_{aer,\lambda} \cdot T_{mol,\lambda} \cdot T_{O3,\lambda} \cdot \cos(\theta) + D_\lambda$

where $I_{0,\lambda}$ is the extraterrestrial spectral solar irradiance at the average Sun-Earth distance. F is the correction factor of the Earth-Sun distance for the period considered, $T_{aer,\lambda}$, $T_{mol,\lambda}$, $T_{O3,\lambda}$ are the transmittances due to the effect of aerosol dispersion, molecular dispersion and absorption by ozone as the main absorbing gas in the ultraviolet spectrum, θ is the solar zenith angle and D_λ is diffuse irradiance.

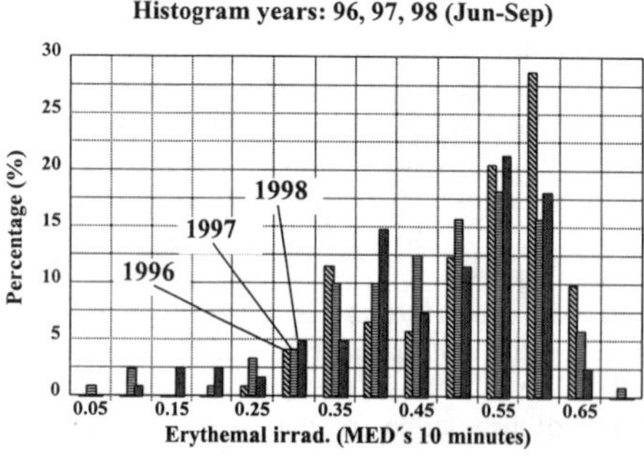

Figure 5. Histogram of erythematic irradiance values (MED 10 min.)June-September, 1996, 1997 and 1998.

As regards the absorption process, the only gas which produces any appreciable absorption in this spectral interval is ozone, and is therefore the only gaseous transmittance in the previous expression. To carry out the calculation of the total transmittance we will consider its definition in which the corresponding optical thicknesses intervene in the form:

$T(\lambda) = T_{mol}(\lambda) \cdot T_{aer}(\lambda) \cdot T_{O3}(\lambda) =$

$\exp[- (\tau_{mol}(\lambda)+\tau_{aer}(\lambda)+\tau_{O3}(\lambda)) \cdot m]$

where "m" is the air mass with the value $m = \sec(\theta)$.

Optical thickness due to molecular dispersion, $\tau_{mol}(\lambda)$, will be calculated from the *Leckner* (*Iqbal, 1983*) approach: $\tau_{mol}(\lambda) = 0.008735 \cdot \lambda^{-4.08}$, where wavelength λ must be expressed in µm.

Optical thickness corresponding to aerosol dispersion is described through the *Angström turbidity formula* (*Kondratyev, 1969*), by the expression:

$\tau_{aer}(\lambda) = \beta \cdot \lambda^{-\alpha}$

β being the parameter related to the density of particles and α the so-called wavelength coefficient linked to the distribution of aerosol size. Due to the lack of aerosol optical thickness data, turbidity coefficients have been obtained from different profiles and periods using the LOWTRAN code version 7 (*Kneizys et al., 1988*).

The results for this spectral interval are:

α= 0.69642; β= 0.24731; Rural aerosol; summer, visibility 23 km; r^2=0.999

α= 0.72706; β= 0.12125; Tropospheric aerosol; summer, visibility 50 km;r^2=0.999

α= 0.72651; β= 0.10410; Tropospheric aerosol; winter; visibility 50 km; r^2=0.999

The previously specified correlation coefficient corresponds to the adjustment provided between the use of aerosol profiles and modelling using the *Angström turbidity* formula.

Optical thickness due to ozone attenuation can be expressed from the mentioned spectral attenuation coefficient (*Cervin et al., 1995; Iqbal, 1983*). Its expression for the atmospheric mono-layer model in which the extinction coefficient is the same in all the atmosphere offering a single wavelength dependency parameter is: $\tau_{O3}(\lambda) = K_\lambda \cdot [O_3]$

with K_λ the absorption coefficient obtained from LOWTRAN-7 for different ozone profiles and with a resolution of 0.001µm, $[O_3]$ total ozone thickness in NTP cm units (*Normal Temperature and Pressure*). Other absorbing gases in the ultraviolet region are ignored as although the values of the efficient absorption section are appreciable, concentration in the atmosphere is low enough not to be taken into account.

For the treatment of diffuse irradiance, we have followed the model proposed by *Bird (1984),* which considers this to be made up of the three basic diffuse processes: molecular, aerosol and Earth albedo respectively:

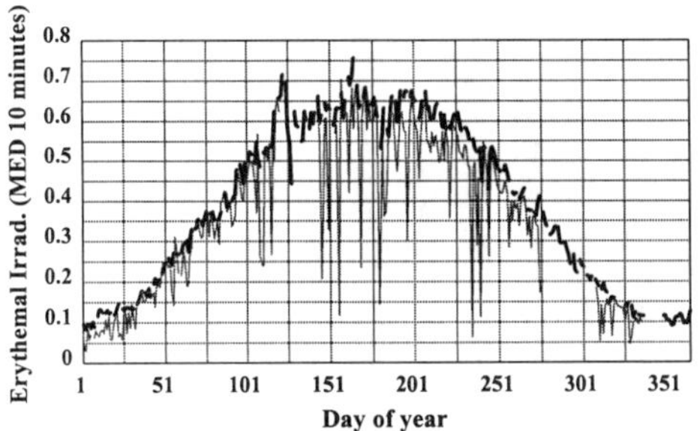

Figure 6. Comparison of maximum daily experimental erythematic irradiance data and those given by the TOMS ozone data model.

$$D_\lambda = I_{d,molecular}(\lambda) + I_{d,aerosols}(\lambda) + I_{d,albedo}(\lambda)$$

Spectral erythematic ultraviolet irradiance is obtained from: $E_{erythem}(\lambda)=E(\lambda)*e(\lambda)$ and the total erythematic irradiance through the integral expression set out in the first section.

4.2 *Comparison: model and experimental data*

In order to verify the results given by the previously mentioned mono-layer model, a calculation has been made of the maximum daily irradiance over the whole of 1997. The expected agreement appears in the spline of the experimental values which appear in Figure 3, in other words, with the expected maximum values, given that the theoretical model offers results for clear skies which are cloud free and rural conditions, with the aim of extending it to the generation of erythematic irradiance maps.

Data from the TOMS sonde on the EP-TOMS satellite, the data from which has been operative since July 25th 1996, have been used as input for total ozone thickness data. These data are available through the internet from the *Godard Space Flight Center* in daily files, with a resolution of 1° in latitude and 1.25° in geographical longitude.

The results obtained after the application of the model to the year 1997 are summed up in Figure 6, in which the experimental data for verifying the spline fit also appear.

The irradiance values have been calculated in MED units per 10 minutes of exposure in order to fit them to those obtained through the radiometer, using standard aerosol profiles for the summer and winter mid-latitude climatological model, with *tropospheric* type aerosol for obtaining clear sky conditions.

The model is obviously not capable of exactly reproducing the sharp alterations caused by the presence of clouds. Its values appear in all cases above the experimental ones. However, it can be seen that in the first half of the year there is good agreement. At the end of the summer period the spline is slightly above the experimental data.

Table 1. Regression parameters

Linear regression: Experimental values vs. Data model	
Slope	*1.01*
Intercept	*0.01*
Correlation coefficient	*0.997*

In order to verify the reliability of the results, an analysis has been carried out of the erythematic irradiance daily evolution curve, using various *clear sky* days corresponding to summer 1997. As in previous sections, the ozone values introduced into the theoretical model are from the TOMS sonde. The comparison of the results shows good agreement, although two discrepancies at different day-time periods should be pointed out.

The results of the theoretical model lead to an overestimate of irradiance at dawn and nightfall, in other words for very high solar zenith angles (between 90° and 85° approximately). It is true, however, that this discrepancy is less significant as it occurs in these circumstances when there is less erythematic irradiance data. The second case occurs at the moment of minimum solar zenith angle, in which the model overestimates the experimental data slightly. The results of the comparison are summed up in Table 1, which shows the data corresponding to a linear regression carried out between the experimental values and the values given by the model. This regression was performed on a total of 370 data sets belonging to situations of totally clear skies. The erythematic irradiance values of these data were situated in the [0.0, 0.66 MED per 10 minutes.] interval.

In the analysis of the daily evolution curves previously mentioned, the greatest discrepancies found between the model and the experimental data on all the days analysed range between 3 % and 8 %.

5 RELATION BETWEEN TOTAL OZONE DEPTH AND ERYTHEMATIC IRRADIANCE

As has already been mentioned, the most important atmospheric factor in the values reached by erythematic irradiance is ozone thickness. The amount of this component determines the delicate balance of the amount of radiation to which an individual can be exposed without suffering any harm.

Figure 7 represents the erythematic irradiance action graphs and the transmittance shown by ozone for a standard amount of total ozone of 330 Dobsons. In the light of this behaviour, the analysis of

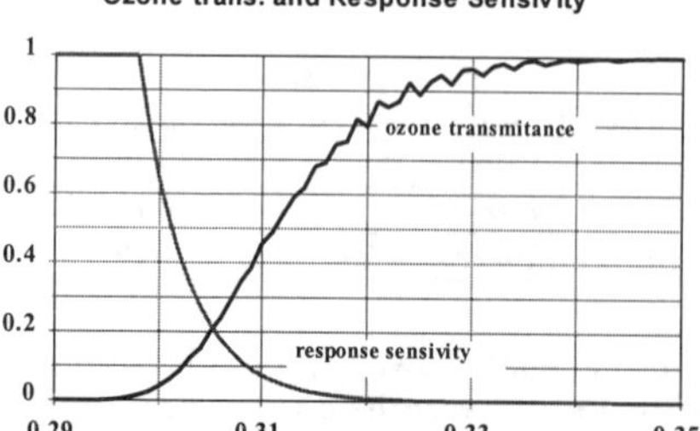

Figure 7. Ozone transmittance and erythematic action.

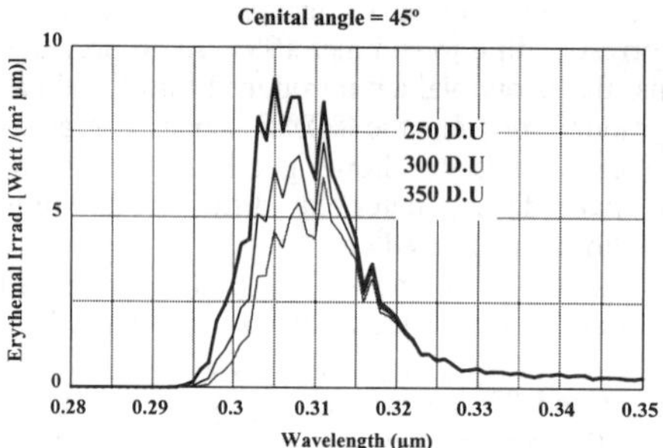

Figure 8. Spectral erythematic irradiance for different total ozone thickness values.

integrated erythematic irradiance is especially interesting when there is variation in the total ozone content. A study has therefore been carried out of the spectral erythematic irradiance for a standard solar zenith angle of 45° for 21 different amounts of total ozone thickness, varying in the [200 D, 400 D] interval in increases of 10 D. Figure 8 contains three graphs representative of the spectral values for three average amounts of ozone: 250, 300 and 350 D respectively.

From the integration of the erythematic irradiance at each point, the relation between this value and total ozone thickness has been obtained. From the integration of these spectral graphs, a relation between the total erythematic irradiance and the total ozone thickness can be obtained. In order to do this, several exponential and potential mathematical expressions have been tried and the best correlation coefficient has been obtained using the following expression:

$$\text{Irrad} = A \cdot [O_3]^{-B}$$

where A=100.65 and B=1.18; $[O_3]$ is the total ozone thickness in Dobsons and Irrad is the integrated erythematic irradiance. The correlation coefficient between both values through the previous expression is *r=0.99997*, with an *average error of* $1.18 \cdot 10^{-4}$.

The coefficients A and B of the previous result are only valid for a solar zenith angle of 45°. This result

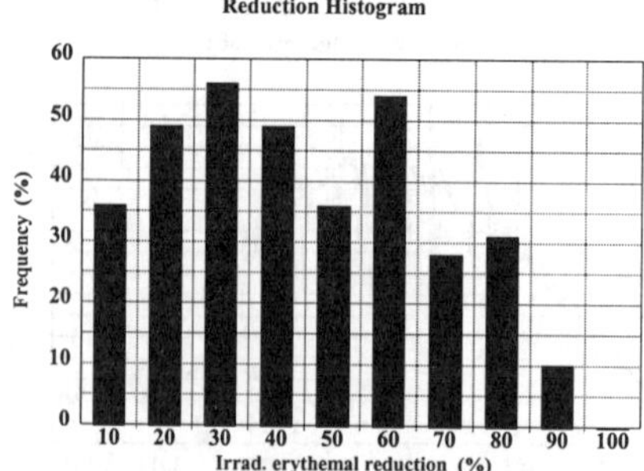

Figure 9. Reduction histogram by clouds cover.

is representative of the strong increase in erythematic irradiance when the amount of ozone varies.

6 REDUCTION OF ERYTHEMATIC IRRADIANCE BY CLOUDS COVER

From the theoretical daily evolution curves analysed and the curves of the experimental values, an analysis of the reduction percentages of erythematic irradiance by clouds has been performed. In order to do this, several days in 1997 which were considered as cloudy were chosen. The selection criteria was performed from a visual analysis of NOAA-14 satellite images, corresponding to channel 2 (near infrared). Images with a greater cloud cover and which ensured cover for thewhole of the day-time period were chosen in order to obtain reduction values for a wide range of different erythematic irradiance values. Figure 9 shows the results.

7 OBTAINING ERYTHEMATIC IRRADIANCE MAPS FROM TOMS OZONE DATA

The distribution of an erythematic irradiance risk index among the population requires the generation of irradiance maps as an easily interpretable aspect. Irradiance maps have therefore been obtained for the whole of Spain using TOMS ozone data, from the EP-TOMS satellite for the whole of summer 1997.

7.1 *Dependence of irradiance on the digital tarrain model*

For working the calculation model, a tropospheric type aerosol model characterised by a visibility of 50 km has been applied over the whole of the country. However, the turbidity coefficients validated for the CIBA measuring station, 840 metres above sea level, are not applicable to the rest of the country due to the different altitude conditions. In order to obtain an irradiance correction equation by height, we have

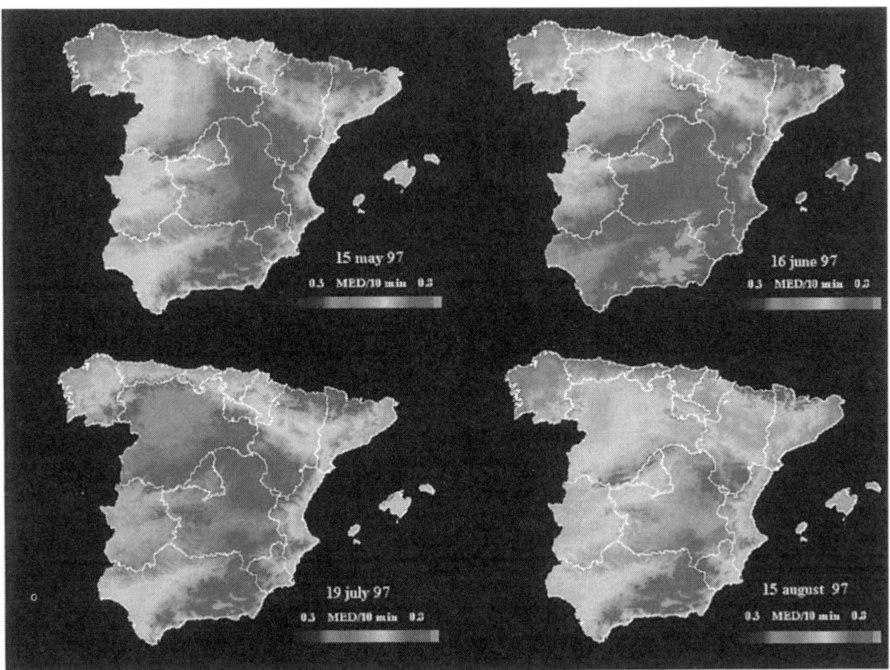

Figure 10: Erythematic Irradiance maps for summer 1997. (colour plate, see p. 600).

used 20 irradiance simulations through the LOWTRAN-7 program at different altitude levels with a separation of 100 metres between sea level and 2000 metres. These simulations have been repeated for different aerosol profiles.

The first point which should be underlined is that the variation in direct irradiance with regard to height is not constant but is lineal with a high correlation coefficient. The results are summed up below, where the increase in irradiance is expressed in percentage terms for each 100 metres further up:

Tropospheric aerosol (vis. 50 km):
$\Delta = h*(-0.36264)+1.652758 \quad r = 0.993$
Continental aerosol (vis. 23 km):
$\Delta = h*(-0.75434)+2.9933714 \quad r = 0.994$
where Δ is the percentage increase per 100 metres increase in height at the height h kilometres above sea level.

7.2 *Resolution of irradiance maps*

In order to generate the maps, TOMS ozone data have been used. These data are located in geographical co-ordinates (longitude, latitude) with a resolution of 1° in latitude and 1.25° in longitude. This resolution is clearly insufficient for the calculation of erythematic irradiance due to terrain fluctuations in the interval mentioned. Pre-processing involved the development of total ozone content maps with a 1 x 1 km resolution, on a UTM projection in time zone 30. The co-ordinates of the ozone maps are:

X = -200000 , 1336000
Y = 3876000 , 4900000

which cover the whole of Spain as can be seen with the exception of the Canary Islands.

The erythematic irradiance calculation model set out earlier is then applied, using the digital terrain model with a previously mentioned resolution and between the same UTM co-ordinates set out.

Figure 10 and Plate ( ) shows the maps for equally spaced days, on which erythematic irradiance has been calculated in MED per 10 minutes exposure for the summer of 1997. The assigned colour palette ranges from 0.3 MED to 0.8 MED. It is important to notice the sharp increase in erythematic irradiance in mountainous areas due to the influence mentioned earlier of the digital terrain model.

8 CONCLUSIONS

An analysis has been presented of the erythematic irradiance experimental data as well as the evolution of its values over the different periods of the year. A mono-layer erythematic irradiance calculation model has also been set up using total ozone thickness values supplied by the TOMS sonde, which adjusts

very well to the spline of the experimental data. This model has been used to obtain the relation between the mentioned irradiance and the
total ozone thickness value.

The analysis of the reduction in erythematic irradiance by clouds has served to characterise the values reached in periods of cloud cover. Finally, the use of a digital terrain model has been applied for the development of erythematic irradiance maps with global coverage of Spain.

These results should be used for the creation of a general exposure risk, including information on the individual's skin type. This relation between radiation and the general phototype, split up into six general categories is laid out in the so-called *ultraviolet radiation index*. The distribution of this index among the population should be a prime concern in summer months; especially in countries such as ours in which tourism and leisure activities are linked to activities which involve prolonged exposure to the sun (leisure activities linked to the beach, mountain)

9 REFERENCES

-BIRD, RICHARD E. (1984). *A simple solar spectral model for direct—normal and diffuse horizontal irradiance.* Solar Energy, vol. 32 No. 4, 1984, pp. 461-471.

-BORDEWIJK, J.A., SLAPER, H., REINEN, H.A.J.M. and SCHLAMANN, E (1995). *Total solar radiation and the influence of clouds and aerosols on the biologically effective UV.* Geophysical Research Letters, vol. 22, No. 16, August, 1995, pp. 2151-2154.

-BRODEKER, G. E., and McKENZIE, R.L. (1996). *An algorithm for inferring surface UV irradiance including cloud effects.* Journal of Applied Meteorology, vol. 35, October 1996, pp. 1860-1877.

-CACHORRO, V.E. y CASANOVA, J.L. (1989). *Puntualizaciones sobre diversos aspectos en la utilización de la fórmula de Angstrom.* Revista de Geofísica, número 45, 1989, pp. 123-130.

-CERVINO, M., CANOSSI, I., GUZZI, R. and TORRICELLA, F. (1995). *Temperature-dependent ozone transmittance in a mono-layer atmospheric model.* International Journal of Remote Sensing, vol. 16, No. 2, 1995, pp. 375-381.

-HERMAN, J.R. and McPETERS, R. (1993) *Ozone depletion at northern and southern latitudes derived from January 1979 to December 1991 Total Ozone Mapping Spectrometer data.* Journal of Geophysical research, vol. 98, no. D7, July 1993, pp 12783-12793

-HOPPE, P., RABE, G., PRAML, G., LINDNER, J., JAKOBI, G. and FRUHMANN, G. (1996). *Are there risk groups in respect to acute effects of environmental ozone ?.* Proceedings of 14th International Congress of Biometeorology, 1-8 September 1996, Ljubljana , Slovenia. Vol. 3, pp. 40-49.

-IQBAL, MUHAMMAD (1983). *An Introduction to Solar Radiation.* Academic Press Canada, 1983.

-KNEIZYS, F.X., ANDERSON, G.P., SHETTLE, E.P., GALLERY, W.O., ABREU, L.W., SELBY, J.E.A., CHETWYND, J.H. and CLOUGH, S.A. (1988). *Users Guide to LOWTRAN 7.* Air Force Geophysics Laboratory, Optical/Infrared technology division. Hanscom AFB, MA 01731. Environmental research papers, number 1010. 16 August 1988

-KONDRATYEV, K. YA. (1969) *Radiation in the Atmosphere.* Academic Press, Inc. International Geophysics Series. New York 1969.

-LEFFELL, DAVID J. and BRASH, DOUGLAS E. (1996). *Radiación solar y cáncer de piel.* Investigación y Ciencia, agosto 1996, pp. 62-68.

-LORENTE, J., REDAÑO, A., CABO, X. De, CODINA, B. y CASAS, C. ( ) *El efecto de la atmósfera en la atenuación del poder eritemático de la radiación solar ultravioleta.* Revista de Meteorología, número 16.

-MADRONICH, S and FLOCKE, S. (1997). *Theoretical estimation of biologically effective UV radiation at the earth's surface.* Solar Ultraviolet Radiation. Ed. Springer pp. 23-48

-McKINLAY, A.F. and DIFFEY, B.L. (1987) *A reference action spectrum for ultraviolet induced erythema in human skin. Human exposure to UV radiation: risk and regulations.* Elsevier Science, pp. 83-87.

-MOLINA, L. T. and MOLINA, M. J. (1986). *Absolute absorption cross-sections of ozone in the 185- to 350-nm wavelength range.* Journal of Geophysical Research, vol. 91, number D13, December, 1986, pp. 14501-14508.

-STOLARSKI, R.S., BLOOMFIELD, P., McPETERS, R.D. and HERMAN, J.R. (1991). *Total ozone trends deduced from Nimbus 7 TOMS data.* Geophysical Research Letters, vol. 18, number 6. pp. 1015-1018.

-STOLARSKI, R.S., BOJKOV, R., BISHOP, L., ZEREFOS, C., STAEHELIN, J. and ZAWODNY, J. (1992). *Measured trends in stratospheric ozone.* Science. Vol. 256, 17 April 1992. pp. 342-349.

-TERAN, M. de, SOLÉ, L. y VILÁ, J. (1987). *Geografía general de España.* Ed. Ariel Geografía, Barcelona 1987.

-VAROTSOS, C. A. and CRACKNELL, A. P. (1993) *Ozone Depletion over Greece as deduced from Nimbus-7 TOMS measurements.* International Journal of Remote Sensing, vol. 14, number 11.1993, pp. 2053-2059.

3 Hydrology

Remote Sensing in the 21st Century: Economic and Environmental Applications, Casanova (ed.)
© 2000 Balkema, Rotterdam, ISBN 90 5809 096 5

Wetland monitoring by the integration of remotely sensed data in a G. I. S. tool

S.Castaño, M.F. Mejuto, A.Vela & A.Quintanilla
Sección de Teledetección y G.I.S., Instituto de Desarrollo Regional, Universidad de Castilla-La Mancha, Albacete, Spain

ABSTRACT: This paper presents a methodology that, by combining remotely sensed data with classic hydrological data, allows the evaluation of land use impacts in the availability of water resources for wetlands. This methodology, developed in the frame of the ASTIMwR project (EU DGXII Project ENV4-ct96-0366), has been implemented in a user friendly G.I.S. software that will help the water manager in monitoring the evolution of wetlands. Two pilot areas have been selected in the Guadiana River Basin, a coastal area at the Guadiana River mouth called O Sapal de Castro Marím, and a continental one called Las Tablas de Daimiel.

1 INTRODUCTION

The ASTIMwR Project ("Application of Space Techniques to the Integrated Management of a river basin Water Resources", ENV4-CT96-0366) is a shared-cost project funded by the Directorate General XII of the European Commission within the Environment and Climate Programme, Area 3.3 (Centre of Earth Observation, CEO) data. This project has developed a user-oriented software that has functionalities tailored to the routine water problems. The main achievement of the project has been to introduce Remote Sensing techniques in the daily work of water managers of the studied areas. Another great advance has been the introduction of a module to monitor the hydrologic evolution of wetlands by means of combining Remote Sensing data with hydrologic-hydrogeological data.

The most typical water-depending environments in a basin are wetlands, which are usually beautiful landscapes of high ecological value. However, water management software seldom includes modules devoted to environmental monitoring.

There are a lot of different kinds of wetlands but the common element to all of them is the presence of water. According to USEPLA (1989) one of the requirements for the legal existence of a wetland is the presence of water in the form of permanent or periodic inundation, or saturation of the soil for a sufficiently long time (at least one week) during the period of vegetation growth. The answers to the questions: "where, when and which waters are present in wetlands?", are the clue to explain the presence of certain types of vegetation and/or soils. So, the hydrologic dynamic is the basic element to define a wetland (Llamas 1991).

Wetlands are the result of the interaction of a lot of factors and this makes wetland casuistic very complex. However, there are some common factors that usually affect to wetlands. In this paper we present a G.I.S. methodology, included in the ASTIMwR project, that can be useful to manage and produce information about wetland hydrologic dynamic in many different situations, using Remote Sensing as the main data source.

2 TEST AREAS

The selection of the test areas has been done regarding the kind of wetlands that are more common in Spain. More than 50% of the Spanish wetlands are linked to aquifers (DGOH 1995). Coastal wetlands are also very important because, while they are less in number, they cover a big surface. According to this, two pilot areas have been selected in the Guadiana river basin: a coastal wetland called O Sapal de Castro Marím, located at the Guadiana river mouth, and a continental wetland called Las Tablas de Daimiel, in the upper basin of the Guadiana river.

2.1 *Las Tablas de Daimiel.*

Las Tablas de Daimiel is a wetland directly linked to an aquifer as many in the Iberian Peninsula. It is located in the confluence of the Guadiana and Gigüela rivers, and it used to be a discharge area of the Mancha Occidental 04.04 aquifer (SGOP 1979), a calcareous aquifer of about 5000 Km^2 and good hydraulic conditions -high transmisivities- (ITGE 1989). It is also fed by the Gigüela river and the Guadiana river. The origin of the water of the Gigüela is mainly superficial runoff while the Guadiana is fed by the aquifer. In the last ten years, Las Tablas has turned into a recharge area from the discharge area that used to be (SGOP 1984; SGOP 1988; López-Camacho et al. 1996). This has been caused by the general lowering of water table levels (SGOP 1991). Las Tablas de Daimiel is a National Park, which implies the maximum legal protection in Spain. Within this park there are a maximum of 1675 hectares susceptible of flooding (Álvarez et al. 1996).

The area has been selected (75,775 hectares) according to its closeness to Las Tablas and to the flow lines of the aquifer, and is extensive enough to be representative of the aquifer (at about 1/6 of the whole aquifer surface). It was studied the part of the Aquifer System 04-04 included in the maps (scale 1:50,000) Daimiel (760) and Villarrubia de los Ojos (737), the area close to Las Tablas de Daimiel. This frame has been also used in previous studies (SGOP 1984) about the influence of the exploitation of subterranean waters in the proximity of Las Tablas de Daimiel (Fig. 1).

2.2 *O Sapal de Castro Marím e Vila Real de Santo António.*

O Sapal de Castro Marím is a coastal wetland located in the Portuguese margin of the Guadiana river mouth (Fig. 2). It has a surface of 2090 hectares. It is an example of a saline wetland where human actions have changed the original landscape of the area. The main environmental values of this wetland are its dominant hypersaline mosaic vegetation (Lousa 1986) and aquatic bird fauna (Farinha & Trinidade 1984). It also has important populations of insects, birds and fish (Xavier 1995). The area was declared Natural Reserve in 1975, which means certain environmental protection even when within the area some human activities still can be developed (salt mining, agriculture...). For the area, there are only a few scientific studies and there exists no cartography for different years.

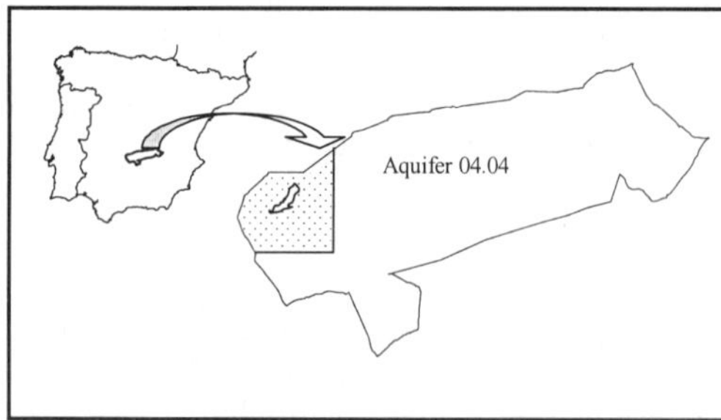

Figure 1. Study area: Las Tablas de Daimiel National Park and the aquifer studied zone.

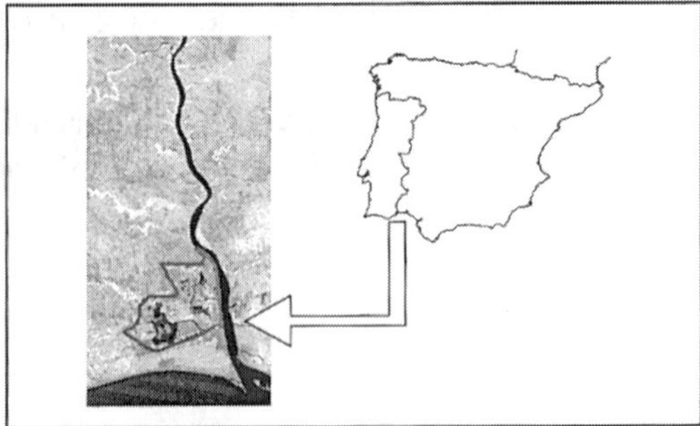

Figure 2. Study area: O Sapal de Castro Marím e Vila Real de Santo António Natural Reserve.

3 OBJECTIVES

The main objective was to obtain results and conclusions about the evolution of the two pilot areas with a methodology, based in E.O. data, easy to extrapolate to other wetlands. The two case studies selected were both very different from the hydrological behavior and the socioeconomic point of view. Nevertheless, both are wetlands that have suffered or are suffering different human actions that can make them disappear.

The major cause of reduction of continental wetland surface in a country like Spain or Portugal is the lack of subterranean water supply, due to the still poor aquifer management. So, it is a primary aim to the relationship existing between the water table and the flooded surface in wetlands. To achieve this aim, the methodology combines the layers of information that are the key to determine the availability of hydrologic resources for these environments (Fig. 3). These are flooded surface, aquifer water table levels and the basic inputs-outputs of water to the wetland-aquifer system (rainfall, pumped volumes and superficial flows).

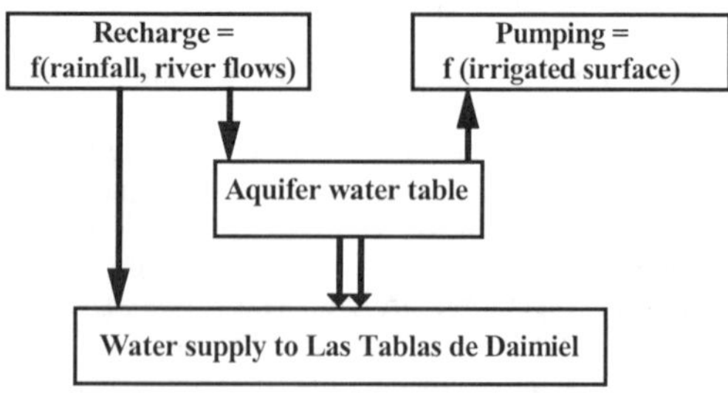

Figure 3. Simplified scheme of the hydrologic dynamic of Las Tablas de Daimiel.

The main causes of degradation of coastal environments are, according to Corre (1976), modifications of topography, irrigation water, fertilisers, land-use changes, changes in the natural cover and farming. Some of these elements, namely those related with land-uses and irrigation, can be monitored by using remote sensing techniques. So, the study in O Sapal of Castro Marím has been focused in the applicability of different Remote Sensing techniques to determinate changes in the vegetation cover, in order to correlate them with mayor changes of the environmental parameters.

4 METHODOLOGY APPLIED.

Several combinations of layers of information have been implemented to assess the evolution of the wetland areas. Of course, the layers of information used for the two test sites were different, in accordance with their dynamics and the needs in each area. However, they cover a lot of topics so that different combinations of the layers used can be extrapolated to other areas.

4.1 *Rainfall and river flow data*

Precipitation is the main entrance of water when working on a basin basis. Indeed, even when considering smaller areas, precipitation data is very useful to evaluate if a year (or a period) is a dry or a wet one. Rainfall is very important for the aquifer recharge, because a low precipitation recharge rate can still provide large volumes of water since it occurs over vast areas. For these reasons the incorporation of rainfall data to the system is essential.

The incorporation of rainfall data to the system is done with raster maps. They can be obtained from interpolations of point data of ground meteorological stations, ground based radar or from satellite images.

Also data of river flows has been implemented into the system as tables and graphics.

Rainfall and river flow data, in addition to the importance that they have for evaluating resources, have been very useful in this project to interpret flooding levels in the wetlands.

4.2 *Evolution of aquifer water table*

Other essential information to monitor continental wetlands is that referred to the aquifers. As explained above, many continental wetlands, and even coastal, are very influenced by the subterranean water dynamics. So, hydrogeological research is a basic tool to understand and to manage wetlands. Basic elements in hydrogeological studies are the water table maps, because the morphology of the water table surface provides basic data about the dynamic of subterranean waters. Moreover, water table maps are useful not only for a qualitative interpretation but for a quantitative interpretation as well (Custodio & Llamas 1983).

These maps were produced as contour maps and afterwards they were converted into raster maps with the use of interpolators.

4.3 *Wetland (flooded surface) evolution*

One of the problems that water managers usually have to face to conserve wetlands is the lack of knowledge of their evolution and dynamics. Often wetlands are poorly mapped and studied as a result of the small interest that they created some years ago. Though, it is very important to have historical records of the flooded surface to study how it has changed.

The objective was to implement a method that automates the detection of water surfaces from Landsat5-TM satellite images. With this method the water manager can obtain some information about wetlands that otherwise it is impossible to obtain when there is no historical record. But it is not only a method to obtain data of the past years, because it can also be a cheap way to monitor the situation nearly in real time.

In the studied area, Las Tablas de Daimiel, field data exists with measurements of flooded surface (PNTD 1998). This has been very useful to calibrate the method and to evaluate its potentialities and limits.

The methodology that was finally used allows a user of the system to classify and monitor water surfaces easily. It produces a map where the water surface is obtained using the following formula for the Landsat5-TM bands (modified from Ángel-Martínez 1995):

$$WaterSurface \Rightarrow \frac{Band4}{Band3} - \frac{Band4}{Band5} \geq 0.4$$

The simplicity of the methodology makes it very useful. However, in cases where the water body is very shallow some pixels can result misclassified. In fact, the water surface seems to be clearly underestimated (fig. 4), being the underestimation bigger when more water is present.

The reason is that the spatial resolution of the Landsat5-TM sensor (30 x 30 m) is not enough to detect neither small water channels nor water under phreatophytic vegetation (i. e. reedbed). This effect is smaller when there is little water, because it is located in the deeper ponds with no vegetation. But for the very wet situations an important amount of water is out of the ponds and under the vegetation.

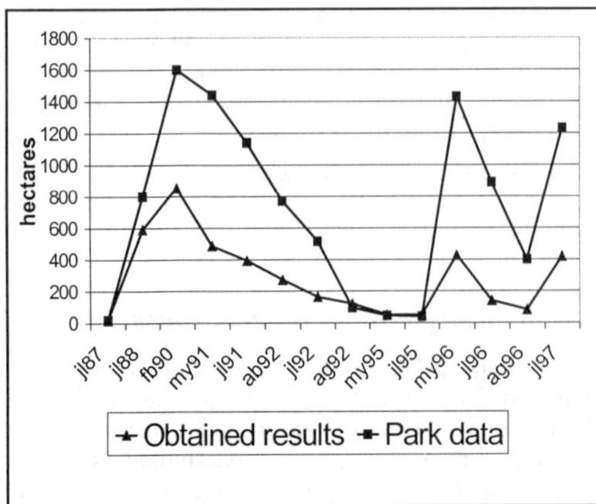

Figure 4: Comparison between values measured in the field and calculated with Remote Sensing data.

Even when these limitations lead to an underestimate of the number of flooded hectares for each date, the methodology is useful to see evolutionary trends (fig. 4). The obtained values show efficiently the evolution of the flooded surface in Las Tablas de Daimiel, and are coherent with values measured in the field, rainfall data and transfers from other basins.

4.4 *Evolution of irrigated surface*

Quantification of water used for irrigation, the most consumptive use in Mediterranean countries (80% of the total water consumption in Spain), is basic for water managers in semiarid environments.

When dealing with big aquifers feeding wetlands and intensely used for agriculture, conventional methods for water extractions quantifying such as water counters installed on each well (also called direct methods) are not affordable by water administrators. Therefore, a low-cost indirect method has to be used. Because of its features, Remote Sensing techniques have proved to be the most suitable for these purposes, since they provide the following advantages:

Wide surface coverage at relatively low costs (comparing with direct methods).

Availability of historical records: Landsat5-TM sensor has been operative since 1984 overflying the surveyed area every 16 days.

Semiarid climates are favorable areas: when summer rainfall is scarce or almost null in drought periods (which occur quite often) the summer irrigated crops, those with the biggest water demands, are easily identified since their spectral response is that of very healthy vegetation within a subdesertic area. Also, cloud coverage is scarce in many dates, what makes it possible to have a choice of various available scenes per studied year.

In order to obtain the best results, discriminating as many irrigation classes as possible, a careful selection of dates was done, taking in account not only the cloud coverage, but also the yearly phenologic evolution of the crops to be classified in the area.

Multitemporal supervised classification has been used to obtain crop maps. The methodology used combines maximum likelihood techniques of classification with decision tree criteria (Friedl & Brodley 1997). These criteria have been adopted by using the Normalized Difference Vegetation Index (NDVI) which is an adequate indicator of crop biophysic parameters, allowing the follow-up of its phenologic evolution (Gilabert et al. 1996). Crop coefficients (Kc) are used for the estimation of the water requirements for each crop (which are related to the phenologic state). Kc are generally empirical ratios of crop evapotranspiration (Et) to some reference Eto, and are calculated for a particular crop at a given grow stage and soil moisture condition (Jensen et al. 1990). This coefficient and the NDVI evolution through the year are closely related (Vela et al. 1998).

The experience acquired in the Mancha region allows us to conclude that the optimal dates to be used in the classification in this area are the following:

Late spring: In the middle days of May, the spring irrigated crops have reached their ripeness, thus their photosynthetic activity and Leaf Area Index (LAI), which are the parameters that condition the spectral response of vegetation in the Normalized Vegetation Index (NDVI), are in their maximum.

Spring-Summer (middle of June): The spring crops have already dried and are ready for harvesting or have been already harvested, while the summer crops have just been sowed, giving both of them a low spectral response.

Summer (end of July or first days of August): The summer-irrigated crops reach their maximum spectral response.

When available, a scene of the middle spring is very useful for better discrimination of drylands and spring-irrigated crops.

The definitive crop map used for the project includes the following three classes:

1 Spring irrigated crops (barley, wheat, garlic, colza, peas...).
2 Summer irrigated crops (onion, sunflower, vegetables, corn, sugar beet...).
3 Non seasonal irrigated crops (alfalfa, irrigated meadows...) and double annual harvest (barley/green beans, peas/corn...).

4.5 *Vegetation index evolution*

The vegetal coverage associated to a wetland area is both its richness and an element to judge the healthiness of the system. Satellite imagery is a powerful tool to monitor the evolution of this vegetation, and thus the wetland evolution.

The two basic methods to monitor the evolution of vegetation by using remote sensing are image classification and vegetation indexes. In the case of irrigation, classified images are a good option since crops are relatively homogeneous covers usually of several hectares (see 4.4). Monitoring the evolution of vegetation indexes can be very useful for natural wetland vegetation, which is usually very heterogeneous and so it is difficult to group in classes.

The methodology based in the analysis of the evolution of values of NDVI has been used for the pilot area of O Sapal de Castro Marím. As it was explained above it is a small coastal area where little information exist about its evolution and dynamics.

For this area NDVI index was calculated for fourteen Landsat5-TM images. It is chosen, whenever possible, a spring image and a summer image for each year, because natural vegetation has the higher values of NDVI in spring and the lower in summer.

The analysis of NDVI values for the whole Castro Marím wetlands is useful only to detect whether major changes in the land coverage have been produced. For areas without significant modification of land uses this kind of analysis of the NDVI gives no results.

To solve this limitation, the evolution of NDVI values has to be studied for smaller areas. In this way, some test plots previously identified in the field are selected (fig. 5). Plots with natural vegetation represent typical assemblages of the area. The main land uses in the Castro Marím wetland area are represented in the plots selected. Also a parcel outside the wetland has been studied. It belongs to a Pine woodland, very stable and apart from areas susceptible of flooding, and it is very useful to compare its evolution with the evolution of the vegetation in O Sapal de Castro Marím.

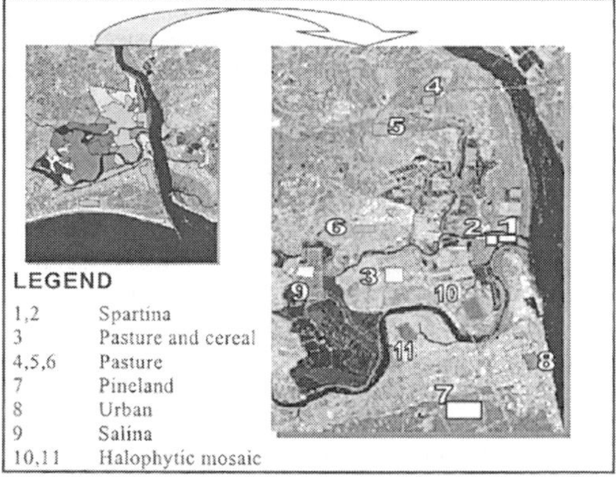

Figure 5. Some of the test plots monitored and the land use map (from 1997 fieldwork) that was the basis to select them.

Once evolutionary data were obtained, they were compared in graphics and tables together with values of other hydrologic parameters that may play an important role in the system dynamics.

5 RESULTS.

Once data included in the system as well as the treatments done of the satellite images has been described, it is going to be explained what kind of results were. The results for both test areas proved themselves usefulness of the methodology. In the following pages the results for each test side are presented.

5.1 *Las Tablas de Daimiel.*

The analysis of data during the studied period allowed to obtain a clear idea of the relationships between irrigation and affections to the aquifer.

Obviously, there is a relationship between rainfall volumes and the water table levels oscillation, as the drought period form 1993 to 1995, when the aquifer reserves reached their minimum). Though, there has been a continued decrease of the water table levels even in the periods when rainfall volume was over the average value, such as the period going from 1985 to 1988 (fig. 6).

The lack of coincidence between rainfall and the aquifer evolution ought to an external factor, that is the water extraction for agriculture.

The results obtained in the classification of irrigated crops (fig. 7) show that the biggest increment of summer-irrigated surface took place during the first half of the 80's (in our quantification for 1978, the surface irrigated in summer was 1400 hectares and increased up to 4300 hectares in 1984). This has resulted in a drastic drawdown of the levels, even when the rainfall volume in this period was higher than the average values.

In the period going from 1993 to 1995 a very strong drought affected the area (the annual rainfall was between 300 and 200 mm, while the average rainfall is around 400 mm). Both spring and summer crops surface decreased, but the levels maintained very low till the humid period starting in 1996 took place.

In this period, most of the farmers changed their crops from summer to spring irrigated because of their lower water needs. In 1990 the irrigated surface in summer had decreased in more than 1500

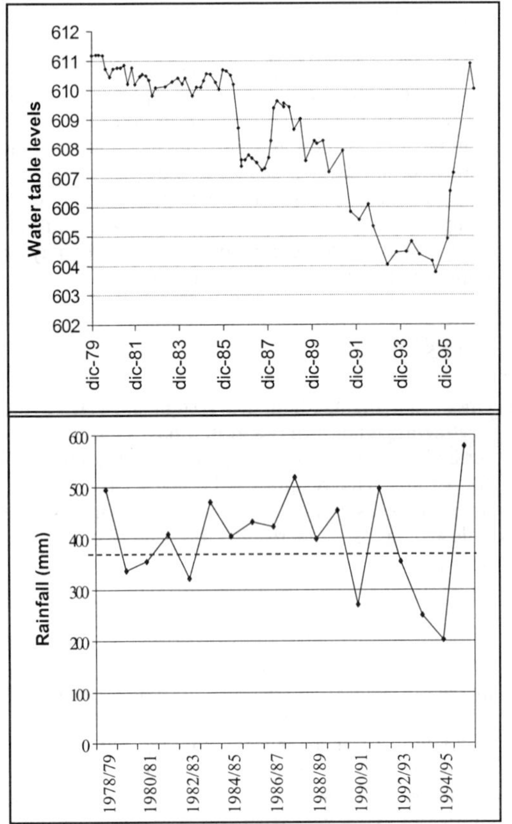

Figure 6. Evolution of the water table levels in a well close to Las Tablas de Daimiel (top) and rainfall volumes for the period 1980 – 1996. The horizontal dotted line in the rainfall graphic is the annual average for a series of 40 years. The graphic shows that the water level decline is not only due to drought.

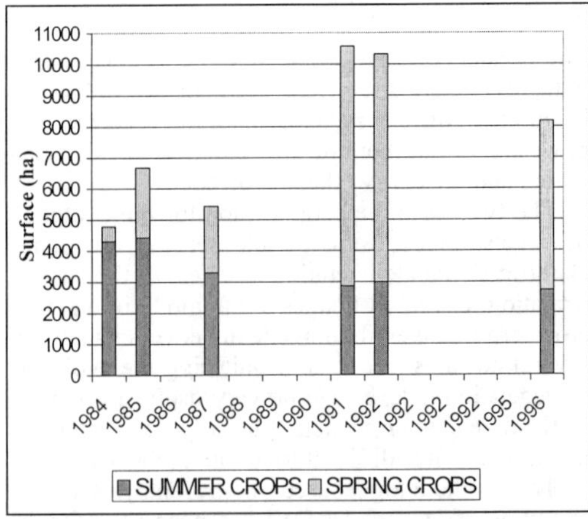

Figure 7. Evolution of summer and spring irrigated crops.

hectares respect to 1985. Thought, the important increase of spring–irrigated crops keeps groundwater table descending, since the water consumption remains almost constant. At this point, it is important to insist in the necessity of three dates (minimum) per year when accomplishing a crop classification. In the case of this aquifer, a classification of irrigated land using only summer scenes should have driven to important errors on the water extraction quantification, because spring crops would not have been detected.

Since 1993, the water administration is carrying out the "Wetland Plan", consisting on an economic aid to those farmers that definitively withdraw the irrigation practices. At this moment, the plan seems to be successful because the spring –irrigated crops have considerably decreased, and the summer crops are in their lower value since the 80's. Nevertheless is estimated that the complete recuperation of the natural wetlands in the area is impossible if the current consumption is maintained.

5.2 *O Sapal de Castro Marím.*

The results obtained for O Sapal de Castro Marím confirmed the interest of the NDVI as a change detection indicator. While in this area image classification, both supervised and unsupervised, was tried and no satisfactory results were obtained due to the natural system complexity, NDVI has been very useful to study the sensibility of the system to certain hydrologic parameters.

From the analysis of the NDVI index for the whole area, it was detected that there have not been major changes in the land use of this area since 1984 until 1997.

From the analysis of the evolution of the NDVI in the test plots it can be concluded that:

Correlation between annual rainfall data and NDVI is low (fig. 8).

Correlation between monthly and twice-monthly rainfall with NDVI is very good, especially for plots with pines, mosaic wetland vegetation (fig. 8) and *Spartina* (a halophytic plant), but it is not always good for pastures, no doubt owing to farming practices. The range of fluctuation of NDVI with rainfall values varies linearly with the seasonal nature of each vegetal cover. For example, for pine woodland, the effect of rainfall can be clearly observed, even when it always keeps NDVI high values. It also can be observed how during the growth of the woodland NDVI increases progressively (1984-90) until the stabilization (1990-97) when it reaches its maximum biomass. On the other hand, NDVI of wetland mosaic vegetation ranges between nearly zero values when rainfall is low for a couple of months to 0.2-0.4 values for wet months.

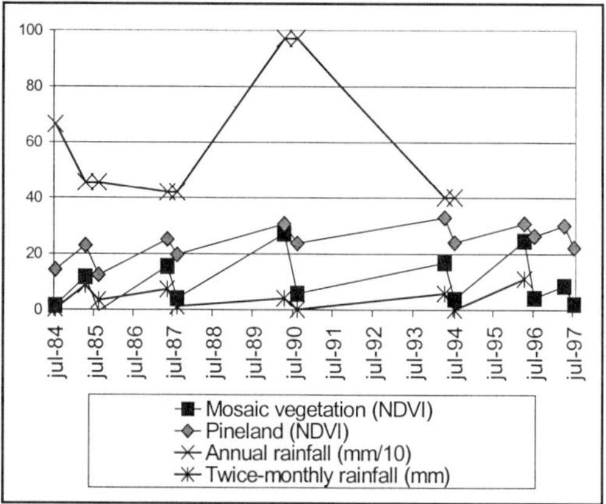

Figure 8. Comparison between rainfall data and NDVI values (from 0 to 100) for two test plots.

There is not a good correlation between the Guadiana river flow data and changes in NDVI. Although higher river flows are coincident with high NDVI values, there is no proportionality between river flow and NDVI. The observed relation is due to the fact that usually wet months in the Guadiana river basin are also wet in the river mouth area where O Sapal is located.

O Sapal de Castro Marím has suffered drastic changes in its landscape due to the human action in this century. Part of its natural values as a bird area is due to human actions, with the creation of pools for salt mining. But there have been no major changes in the land use during the last fifteen years. Changes in the activity of the natural vegetal coverage are directly related to rainfall produced one or two months before the measurement of that activity.

6 CONCLUSIONS

1. Remote Sensing techniques are quite useful and should became a standard in wetland hydrology.
2. Satellite imagery has to be combined with hydrological data, and eventually with other environmental data, to make possible the understanding of the wetland system dynamic.
3. Crop classifications (with Landsat5-TM) are a powerful tool to evaluate irrigated volumes, something critical in areas where wetlands are fed by subterranean waters.
4. The historical record of satellite images allows to study wetlands were no field data has been recorded in the past.

REFERENCES

Álvarez M., Verdugo M. & Cirujano S. 1996. Geografía y Morfometría. In Las Tablas de Daimiel. In *Ecología Acuática y Sociedad*. Álvarez M. & Cirujano S. (Eds.). Madrid: Organismo Autónomo Parques Nacionales.

Ángel-Martínez MªC. 1994. *Aplicación de la Teledetección en la localización de superficie de agua*. Madrid: Centro de Estudios y Experimentación de Obras Públicas (CEDEX).

Corre J.J. 1976. Etude phyto-écologique des milieux littoraux salés en Languedoc et en Camargue. 1. Característiques du milieu. Vie et Milieu XXVO. 2C, pp 175-245.

Custodio E. & Llamas M.R. (1983). *Hidrología subterránea*. Barcelona: Ediciones Omega.

Dirección General de Obras Hidraúlicas (DGOH) 1995. *Libro blanco de la aguas subterráneas. Serie Monografías*. Madrid: Centro de Publicaciones del Ministerio de Obras Públicas Transportes y Medio Ambiente.

Farinha J.C. & Trinidade A. 1984. *Contribuçao para o Inventário e Caracterizaçao de Zonas Húmidas em Portugal Continental*. Lisboa: Instituto da Conservaçao de Natureza.

Frield M.A. & Brodley C.E. 1997. Decision tree classification of land cover from remotely sensed data. *Remote Sensing of Environment*. 61: 399-409.

Gilabert M.A., Gandía S. & Meliá J. 1996. Analyses of spectral-biophysical relationships for a corn canopy. *Remote Sensing of Environment*. 55: 11-20.

Instituto Tecnológico y Geominero de España (ITGE) 1989. *Sistema acuífero nº 23. Mancha Oriental. Serie Manuales de Utilización de Acuíferos*. Madrid: ITGE.

Jensen M.E., Burman R.D. & Allen R.G. (eds.) 1990. *Evapotranspiration and irrigation water requirements. Manuals and Reports on Engineering Practice-No. 70*. New York: American Society of Civil Engineers.

López-Camacho B., Bustamante I., Dorado M. & Arauzo M. 1996. Hidrología. In *Las Tablas de Daimiel. Ecología Acuática y Sociedad*. M. Álvarez & S. Cirujano (Eds.). Madrid: Organismo Autónomo Parques Nacionales.

Lousa M. 1986. *Comunidades halofíticas da Reserva do Castro Marím*. Lisboa: Universidade Técnica de Lisboa.

Llamas M.R. 1991. Groundwater exploitation and conservation of aquatic ecosystems. In *Aquifer overexploitation*. Canary Islands: I.A.H..

Parque Nacional de las Tablas de Daimiel (PNTD) 1998. *Evolución de la superficie encharcada en el Parque Nacional de las Tablas de Daimiel y Volumenes derivados al PNTD desde el Acueducto Tajo-Segura*. Internal Report.

Servicio Geológico de Obras Públicas (SGOP) 1979. *Influencia en Las Tablas de Daimiel de la Extracción de Aguas Subterráneas en la Llanura Manchega*. Internal Report 11/79.

SGOP 1984. *Estudio de la explotación de aguas subterráneas en las proximidades del Parque Nacional de Las Tablas de Daimiel y su influencia sobre el soporte hídrico del ecosistema*. Internal Report.

SGOP 1988. *Evolución de las extracciones y niveles piezométricos en el acuífero de la Llanura Manchega*. Internal Report 06/88.

SGOP 1991. *Evolución de las extracciones y niveles piezométricos en los acuíferos de la Llanura Manchega y del Campo de Montie*l. Internal Report 05/91.

U.S. Environmental Protection Agency 1989. *Federal Manual for Identifying and Delineating Jurisdictional Wetlands*. New York: USEPA.

Vela A., Calera A., Mejuto M.F., Castaño S. & Ruiz J.R. 1998. Application of remote sensing techniques to the evaluation of water consume. In *Remote Sensing, Agriculture and Ecosystems*. Barcelona: SPIE.

Xavier, A. 1995. *Reserva Natural do Sapal de Castro Marím e Vila Real de Santo António*. Vila Real: Parque Natural do Sapal de Castro Marím.

Remote Sensing in the 21st Century: Economic and Environmental Applications, Casanova (ed.)
© 2000 Balkema, Rotterdam, ISBN 90 5809 096 5

The use of Earth Observation for monitoring snowmelt runoff from non-alpine basins

R.G.Caves, O.C.Turpin, C.D.Clark, R.I.Ferguson & S.Quegan
Sheffield Centre for Earth Observation Science (SCEOS), University of Sheffield, UK

B.Johansson
Swedish Meteorological and Hydrological Institute (SMHI), Norrköping, Sweden

ABSTRACT: Methods have previously been developed for mapping snow covered area (SCA) in alpine basins using medium and high resolution optical imagery and for mapping wet SCA using synthetic aperture radar (SAR). This paper discusses the transfer of these methods to higher latitude non-alpine basins. Processing methods are outlined and differences between applying the methods in alpine and higher latitude basins are emphasised. Estimates of SCA derived from both medium and high resolution optical (NOAA AVHRR and Landsat TM) and SAR (ERS) data of a test basin in Northern Sweden are presented and compared. The integration of these SCA estimates into the HBV snowmelt runoff model and the effect on snowmelt runoff predictions are presented.

1 INTRODUCTION

In alpine and high latitude basins where precipitation often falls as snow, information on snowmelt runoff is of interest to hydropower and insurance companies, and environmental agencies involved in managing water resources. Snowmelt runoff can be predicted using models driven by meteorological data (Ferguson 1999), for example the HBV model (Bergström 1992). In principle the performance of such models should be improved by using spatial information on snowpack variables provided by Earth Observation (EO). The HYDALP project is investigating ways of using such EO derived information to improve the performance of snowmelt runoff models in both alpine and higher latitude basins.

The spatial variables of most interest for snowmelt runoff modelling are snow covered area (SCA) since runoff is this times the melt rate, and snow water equivalent (SWE) since this gives total stored water and also controls decline in SCA as melt proceeds. Currently it is not possible to measure SWE from spaceborne sensors at the spatial scales required in basins of even moderate relief, hence good knowledge of SCA over time is vital. Methods already exist for mapping SCA using medium and high resolution optical imagery (Baumgartner & Rango 1995, Dozier & Marks 1987) and for mapping wet SCA using synthetic aperture radar (SAR) (Nagler et al. 1998). Previously the SAR method has only been applied to alpine basins. In this paper all three methods are applied to a higher latitude non-alpine basin and their performance compared.

The methods for deriving SCA from both optical (NOAA AVHRR, Landsat TM) and SAR (ERS) data are outlined in Section 2. Differences between applying the methods in alpine and higher latitude basins are emphasised, including the effects of snow cover variability, cloud cover, topography and repeat coverage patterns. The methods have been tested in the Spey basin in the Scottish Highlands (57° N, 1271 km^2, 198-1288 m) and the Tjaktjajaure basin in Northern Sweden (67° N, 2250 km^2, 450-2090 m). Due to a lack of EO data of the Spey basin during years of good snow cover the methods have mainly been tested on the Tjaktjajaure basin. EO data analysis in the Spey basin has primarily involved field verification of the SAR wet snow mapping method (Caves et al. 1998 & 1999). Time series of SCA estimates derived from ERS, AVHRR and TM data of the Tjaktjajaure basin only, are presented and compared in Section 3. The integration of these SCA estimates into the HBV model and the effect on snowmelt runoff predictions are discussed in Section 4.

2 DERIVATION OF SCA FROM EO DATA

2.1 *ERS SAR*

The SAR wet snow mapping method and its application to an alpine basin are described in Nagler et al. 1998. Here we primarily concentrate on differences in transferring the method to non-alpine basins.

2.1.1 *Theory*

At C-band (5.6 cm), even several metres of dry snow has little effect on the radar backscatter from a surface. The penetration depth of dry snow is of the order of tens of metres and most backscatter is due to scattering from the underlying surface (Nagler 1996, Matzler 1987). By comparison, the presence of even 1 % liquid water by volume reduces the penetration depth to tens of cm. As most losses are due to absorption rather than volume scattering the backscattering coefficient from a surface is substantially reduced when it is covered by wet snow (Nagler 1996). Thus wet snow can be detected in a C-band SAR image by comparing calibrated backscatter values, pixel by pixel, with those in a reference image from a period of no or dry snow cover. This is done by ratioing the two images and then applying a threshold to the resultant ratio image.

2.1.2 *Selection of wet snow detection threshold*

In alpine basins a threshold of less than or equal to -3 dB has been successfully used to detect wet snow in intensity ratio images from the ERS and Radarsat C-band SARs (Nagler et al. 1998).

To check the sensitivity of this threshold in the Tjaktjajaure basin classification was also applied with -2 and -4 dB thresholds. On only one date did the SCA change by more than 5 % when the threshold was varied by 1 dB, up or down. Changes mainly affected the edges of large areas of snow cover and smaller snow patches suggesting that the main area of reduced backscatter due to wet snow is clearly defined. The SCA estimates presented in Section 3 were produced using the -3 dB threshold.

Field verification in the Spey basin indicated that a reduced threshold of -2 dB is more suitable for detecting patchy wet snow cover (Caves et al. 1998 & 1999).

2.1.3 *Image registration and geocoding*

As backscatter is dependent on the local incidence angle the imaging geometry must be exactly the same in the wet snow and reference images. Such repeat pass SAR image pairs are provided by the ERS series of satellites and by Radarsat. Registration of repeat pass images requires only translation. A transform for geocoding ratio images is defined by matching the reference image to a SAR image simulated from a DEM (Guindon & Adair 1992); layover features are used as GCPs. The DEM and geocoded image both have a pixel spacing of 50 m. A RMS geocoding error of less than 1 pixel was achieved using 10 to 12 GCPs.

2.1.4 *Missing coverage*

Areas of missing coverage arise in SAR images due to topographic effects such as layover and radar shadow. However, the intensity ratio method for wet snow detection performs poorly at local incidence angles less than 17° and greater than 78° so we also class such areas as missing coverage (Nagler 1996). The lower bound is due to foreshortening and specular effects at low incidence angles, whilst the upper bound is due to poor signal to noise ratio at grazing angles. All of these effects are range dependent. At near range, where the incidence angle is steeper, layover and foreshortening increase while radar shadow and grazing decrease. For the ERS SAR with its steep look angle, layover and foreshortening are the primary sources of missing coverage while radar shadow and grazing angles are comparatively insignificant.

In high relief alpine terrain missing coverage can affect a considerable portion of an ERS SAR image. For example, between 30 and 40 % of the Zillertal basin (Austria) is affected depending on the range position of the basin within the swath (Nagler et al. 1998). While the amount of missing coverage is less in ERS SAR images of higher latitude basins of more moderate relief, it is still appreciable. For example, between 10 and 20 % of the Spey and Tjaktjajaure basins are affected by missing coverage (Caves et al. 1998 & 1999), although on slopes of specific aspect the area affected can be much greater (> 50 %). These increased amounts often occur at higher elevations where snow cover is likely. Hence missing coverage can severely limit the utility of ERS wet snow mapping even over areas of moderate relief such as the higher latitude basins under study.

It is possible to reduce missing coverage appreciably by combining images of a scene taken from different viewing directions, such as the ascending and descending passes of a spaceborne SAR (Nagler 1996, Piesbergan et al. 1997). For example, the combination of ERS passes reduces missing coverage over the Zillertal and Spey basins to less than 6 and 1 % respectively (Nagler et al. 1998, Caves et al. 1999). However, wet snow cover detection requires that both images be taken close together in time during a period of little change in snow conditions. This is not always possible. At the alpine latitudes (~47° N), where the wet snow detection method has previously been applied, the lag between the two passes is only half a day (Nagler 1996, Piesbergan et al. 1997). However, for the Spey and Tjaktjajaure basins the time lags are respectively 1.5 and 6.5 days. While snow conditions may remain stable over such intervals during a cold period of no snow melt, this is unlikely to occur during the melt periods of interest. Hence, inferences on wet SCA based on ascending/descending pass combinations can be invalid.

The amount of missing coverage can be greatly reduced by using single pass SAR data with less steep incidence. For example, using Radarsat data with a 40° incidence angle missing coverage can be reduced to 10% in the Zillertal basin (Nagler et al. 1998) and to less the 1 % in the Spey and Tjaktjajaure basins (Caves et al. 1998). However, such data was not available for this study. Instead a method was developed for inferring wet snow cover in areas of missing cover. It operates as follows:

1. Using a DEM the image is classified into zones defined by 100 m in elevation and 15° in aspect.
2. The area of wet snow A, and non-wet snow B (dry or no snow excluding missing cover), is calculated for each zone.
3. The wet snow cover proportion is then calculated for each zone using $P = A/(A+B)$.
4. If P is greater than half all areas of missing cover within that zone are classified as wet snow.

The method relies on at least part of any given elevation-aspect zone being unaffected by missing cover. This is the case in the Tjaktjajaure basin where the approach has generated sensible wet SCA maps. The approach has yet to be tested in the Spey basin.

2.1.5 *Inferring dry snow cover*

Following wet snow detection a method is needed to infer dry snow cover at higher elevations. In alpine basins wet snow maps from proceeding years, though from a later period in the melt season and hence showing wet snow at higher elevations, have been used (Nagler et al. 1998). The archive of SAR data of Tjaktjajaure is not sufficiently comprehensive to permit this approach. Instead a rule based approach has been developed for inferring dry snow cover at higher elevations. A pixel is inferred as being dry snow if one of the following is TRUE:

1. It lies at a lower elevation than the mean elevation of wet snow AND lies uphill from a large area of wet snow. A *large area* is defined as 2.5 km^2 at the beginning of the melt season but is decreased to 1 km^2 as snow patches reduce in size during the melt season.
2. It lies at a higher elevation than the mean elevation of wet snow AND lies uphill from wet snow.
3. It lies at least 200 m higher than the mean elevation of wet snow.

This last rule is included to get round the problem of lips on the edges of cirques impeding searching for pixels lying uphill of wet snow. The value of 200 m was set by experimentation.

While this approach circumvents the need for archived data it assumes that snow cover is complete at higher elevations. This is often not the case; for example, exposed ridges are often snow free. By contrast the archived data approach can distinguish between snow covered and snow free areas at higher elevations. However, this does assume that snow cover patterns remain similar from year to year.

2.2 *Landsat TM*

The main processing steps are radiometric correction, classification and geocoding.

2.2.1 *Radiometric correction*

Classification is based on planetary albedo the calculation of which involves:

- Linear conversion of digital numbers to radiance. Pre-flight values were used for the offset (Epema 1990) while the gain was taken from the field calibration campaign closest in time to the image date (Thome et al. 1997).
- Scaling of radiance to planetary albedo. Scaling requires knowledge of the solar elevation angle, the mean solar exoatmospheric irradiance and the earth-sun distance (Epema 1990).

No atmospheric correction is applied.

2.2.2 *Classification*

Snow is differentiated from non-snow by thresholding the ratio of planetary albedo in band 3 (red) to that in band 5 (SWIR) (Rott & Markl 1989)

$$\text{IF } \frac{R_3}{R_5} \geq c_1 \text{ THEN class = SNOW} \tag{1}$$

where c_1 is a pre-set threshold. Rott & Markl used $c_1 = 2.7$ for alpine areas. A subjective assessment of the snow maps indicated that this threshold is slightly too low for Tjaktjajaure. Instead $c_1 = 3.0$ was used. The simple threshold resulted in some areas (e.g. water) with low albedo in band 3 but an even lower albedo in band 5 being classified as snow. To prevent this the rule was adapted to:

$$\text{IF } \frac{R_3}{R_5} \geq c_1 \text{ AND } R_3 > c_2 \text{ THEN class = SNOW} \tag{2}$$

with $c_2 = 0.1$ since the albedo of water is considerably lower than that of snow in the visible region of the electromagnetic spectrum.

2.2.3 *Geocoding*

Images were geocoded using the PCI SMODEL and SORTHO routines. GCPs were selected from a 1:100 000 scale map of the basin scanned at 300 dpi and previously geocoded images. The RMS geocoding error was less than 25 m for all images.

2.3 *NOAA AVHRR*

Snow mapping is based on midday AVHRR images from NOAA PM satellites (NOAA 11 and 14).

2.3.1 *Calibration*

AVHRR images were downloaded from the Dundee receiving station pre-calibrated to planetary albedo (channels 1 and 2) and brightness temperature, BT (channels 3 to 5).

2.3.2 *Geocoding*

Geocoding is based on a modified version of the AVHRR Near Real Time Processing (ANRTP) software (Marçal & Slater 1997). This involves a hybrid of both automatic and manual geocoding:

1. Ephemeris data are used to calculate the pixel coordinates of several hundred points defined in a regular grid over the geographic area of interest (Cracknell 1997). This automatically generates a set of GCPs which are used to define a third order polynomial transform for resampling the image.
2. GCPs are manually selected from the resampled image to determine the residual error and hence to define a first order polynomial transform for correcting the geographic coordinates of the original regular grid of GCPs.
3. The corrected GCPs are used to define a new third order polynomial transform for resampling the original image into the desired projection.

Sub-pixel geocoding accuracy (< 1 km) was achieved using 10 or fewer, manually selected GCPs.

2.3.3 *Classification*

Images are classified into snow, vegetation, cloud and "other" using a hierarchical threshold scheme (Voigt et al 1998); see Table 1. A flag is also set when cirrus cloud is detected. Classification requires channels 1 and 2 to be radiometrically corrected for the Earth-Sun distance and for topography (division by the cosine of the solar incidence angle), and calculation of the normalised difference vegetation index (NDVI) and the reflective portion of channel 3 (Gesell 1989). Table 2 lists the thresholds used for classification. The values for BT_4_max and BT_4_min are set high to account for warm spring melt conditions.

Table 1. Hierarchical threshold scheme for classifying AVHRR data. R_i and BT_i indicate the reflectance and brightness temperature in a channel.

```
IF BT4 ≤ BT4_min OR R3 ≥ R3_max THEN
  class = CLOUD
ELSE
  IF BT4–BT5 ≥ ΔBT45 THEN
    set CIRRUS flag
    R1_min = R1_min + R1_add
  END IF
  IF NDVI > NDVI_min THEN
    class = VEGETATION
  ELSE IF R1 ≥ R1_min AND BT4 ≤ BT4_max THEN
    class = SNOW
  ELSE
    class = OTHER
  END IF
END IF
```

Table 2. Thresholds used for AVHRR classification.

Symbol	Description	Value
R_1_min	Minimum channel 1 albedo for snow.	25 %
R_1_add	Increased value for R_1_min for cirrus.	5 %
R_3_max	Maximum channel 3 albedo for snow.	8 %
BT_4_min	Minimum channel 4 BT for snow.	-10°C
BT_4_max	Maximum channel 4 BT for snow.	15°C
ΔBT_{45}	Minimum channel 4 and 5 BT difference for cirrus.	2°C
NDVI_min	Minimum NDVI for vegetation.	10 %

3 RESULTS OF SCA ESTIMATION

Based on runoff records and available EO data of the Tjaktjajaure basin, the 1992, 1996 and 1998 melt seasons (late May to mid July) were chosen as the most suitable for deriving time series of SCA from EO for use in verifying and updating simulations of snowmelt runoff generated by the HBV model. The acquisition dates of AVHRR, TM and ERS data . during these melt seasons and the derived SCA estimates are plotted in Figure 1. The plots showing "AVHRR (corr.)" are discussed in Section 3.4.

3.1 *ERS SAR*

Within any given ERS 35 day repeat cycle near full coverage of the basin can be obtained from 4 passes, two ascending and two descending. Despite this regular coverage ERS SAR images were acquired for two dates only during the 1992 and 1996 melt seasons. The two images from 1992 are too close in time while the two from 1996 are too separated in time to give a useful indication of snowmelt. However, images were acquired on seven dates during the 1998 melt season. Temperature records were used to select suitable reference images from cold periods during the winter. For SAR the total (wet and dry) SCA is plotted in Figure 1.

3.2 *Landsat TM*

Only images with minimal cloud cover (< 10 %) were selected. Three cloud free images are available for the 1992 melt season and one for the 1996 melt season. No cloud free high resolution optical data are available for the 1998 melt season.

There are no independent data available for accessing the accuracy of the snow maps. However, subjective comparisons of the snow maps and the false colour composites used to generate them indicate a very high level of correspondence. The only exception is snow in cast shadow which is classified as non-snow. The area of snow in cast shadow appears to constitute only a small percent of the image. Nevertheless a classification approach that employs an illumination angle (e.g. Dozier 1989) or topographic correction should circumvent this problem.

3.3 *NOAA AVHRR*

Images showing more than 40 % cloud cover, after classification, were rejected. Snow cover maps were obtained for five images during the 1992 and 1996 melt seasons and nine during the 1998 melt season. The average interval between snow maps ranges from 6 days in 1996 to 8.4 days in 1992. However, gaps of up to 20 days occur. AVHRR has been used to obtain snow maps of alpine basins with a similar frequency (S. Voigt, pers. comm.). AVHRR cloud cover amounts are also indicated on Figure 1. Cloud cover was much more prevalent during 1998 than during 1992 and 1996. Within each sub-basin/elevation zone the SCA has been rescaled to correct for the cloud covered area.

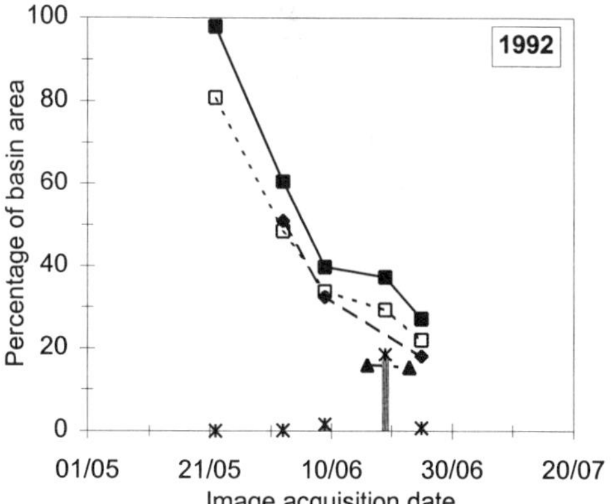

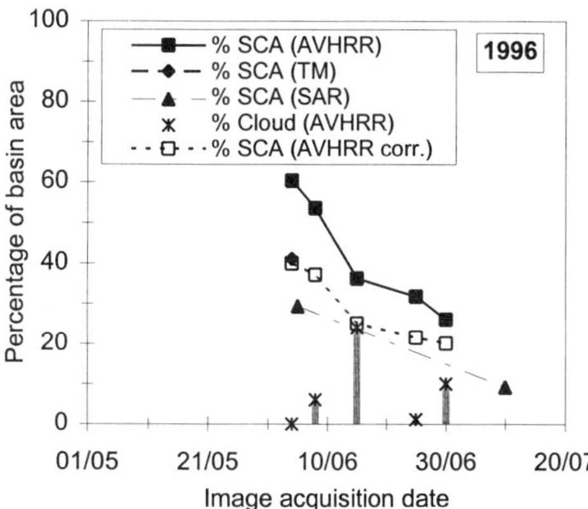

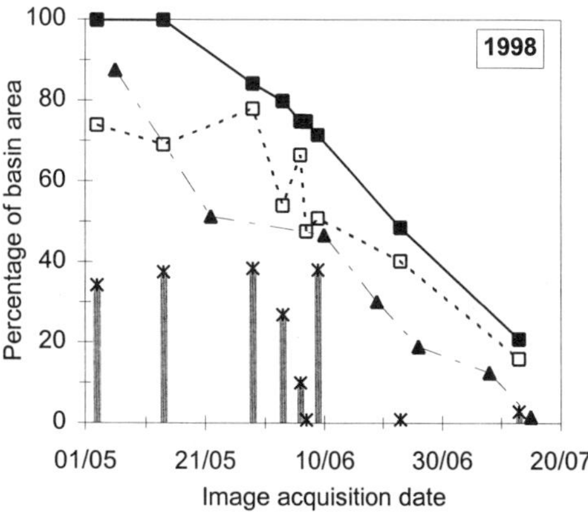

Figure 1. Percentage SCA estimated over the Tjaktjajaure basin using NOAA AVHRR, Landsat TM and ERS SAR data during 1992 (top), 1996 (middle) and 1998 (bottom). The percentage cloud cover in the AVHRR images is also indicated.

In some images 8 % appeared too high a value for R_3_max (Gesell 1989). A value of 5 or 6 % would be more suitable otherwise what appears to be cloud is classified as snow. Areas affected by cirrus were classified in a manner which agreed with surrounding pixels which were not affected by cirrus. Areas classified as snow were often surrounded by pixels classified as "other" indicating mixing of vegetation and snow signatures.

3.4 *Comparison of SCA from different sensors*

While AVHRR, TM and ERS all show a steady decrease in SCA over the melt season in all three years there are considerable differences in the SCA estimated by the sensors. This is most clearly seen on dates with SCA from two or more sensors (e.g. 25 June 1992) but it is also evident from the offset between the SCA plots from the three sensors. AVHRR overestimates SCA as compared to TM, while ERS SAR underestimates SCA as compared to TM. This means that SCA estimates from different sensors cannot be treated as equivalent for purposes of validating or updating modelled SCA.

A detailed intercomparison of the SCA derived from coincident or near coincident (< 3 days) data from the different EO sensors was conducted with the aim of

- Identifying the location of differences in SCA in terms of sub-basin, elevation and aspect.
- Explaining the physical basis for the differences.
- Determining if methods can be defined for correcting the SCA estimated by NOAA AVHRR and ERS SAR to that estimated by Landsat TM.

No ground truth data are available for independently accessing the accuracy of the snow maps. Hence, TM was selected as the most reliable source of "true" SCA due to its high resolution. Intercomparisons were based on the 1992 EO dataset as this is the only year for which TM data are available on more than one date. Coincident AVHRR data are available for all TM acquisition dates (2, 9 and 25 June) but the only near coincident date for ERS SAR data is 23 June. Both snow cover maps and plots of SCA by elevation were compared. The analysis showed that:

- AVHRR overestimates SCA as compared to TM due to AVHRR detecting areas of patchy snow cover as being continuous.
- Agreement between AVHRR and TM derived SCA tends to be best over flat or gently sloping high areas, including glaciers. Agreement is less good over steep slopes and along narrow valley floors where TM derived SCA is patchy.
- As snow cover becomes very patchy at lower elevations it is only detected by TM.
- High exposed ridges which are classified as snow free by TM are classified as snow covered by both AVHRR and SAR. AVHRR is unable to re-

solve such features while the method used to infer dry snow cover in SAR takes no account of them.

- Both TM and AVHRR derived SCA exhibit similar patterns with respect to elevation.
- SAR derived SCA shows quite a different pattern with respect to elevation as compared to both TM and AVHRR derived SCA.
- Differences between the snow maps do not exhibit any systematic dependence on aspect.

3.5 *Correction methods*

3.5.1 *AVHRR*

The intercomparison indicated that the difference between TM and AVHRR derived SCA was primarily one of resolution and that a possible solution lay in relating AVHRR channel 1 albedo to TM SCA. Scatterplots of AVHRR channel 1 albedo against proportional TM SCA (per AVHRR pixel) exhibited a similar linear relationship on all three dates with a correlation coefficient always greater than 86 %. Figure 2 shows the scatterplot for 9 June. The results from all three dates were used to define a linear estimate of proportional SCA from channel 1 albedo. Channel 1 albedo values between 7.7 and 69.1 % are linearly mapped between 0 and 100 % proportional SCA, outside this range values are truncated. The estimate has a standard error of 14%. Areas classified as cloud covered are masked out. However, no correction is made for increased albedo and hence proportional SCA due to cirrus cloud. The method appears to produce sensible results but has yet to be fully verified. One problem is that snow free areas with an albedo higher than the lower cut-off will be classified as having some snow. This correction helps explain why the original AVHRR classification threshold overestimated SCA as the snow detection threshold of 25 % channel 1 albedo equates to only 28 % proportional TM SCA.

The method was used to reclassify all AVHRR images from 1992, 1996 and 1998. The results are plotted in Figure 1 as "AVHRR (corr.)". The correction reduces the AVHRR SCA to values much closer to the TM SCA. However, the corrected estimates no longer show a steady decrease in SCA over the 1998 melt season. Increases could be due to new snow cover, this has yet to be checked. They could also be an artefact of the large amounts of cloud cover (near 40 %) on certain dates and the SCA rescaling used to correct for it, although the fastest change in SCA appears to occur between consecutive days (6, 7 June) with low cloud cover. Further analysis is required of the amounts and spatial distribution of cloud cover in the AVHRR data from 1998.

3.5.2 *SAR*

The difference between TM and SAR is more problematic. Differences are elevation dependent.

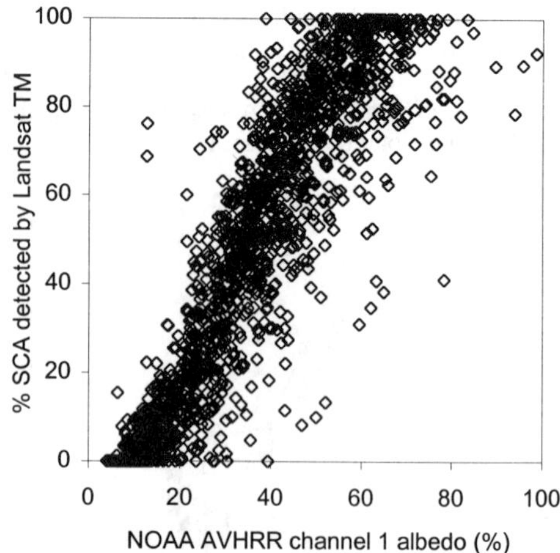

Figure 2. Scatterplot of AVHRR channel 1 albedo against proportional TM SCA on 9 June 1992.

SAR overestimates at higher elevations (at least partially due to the method used to infer dry snow cover) but underestimates at lower elevations (due to failure to detect patchy or shallow snow cover). It is not yet clear how to correct for these differences.

4 HBV UPDATING USING EO DATA

Traditionally the HBV model is updated before a forecast by comparing observed and simulated runoff. If runoff is underestimated, a temperature correction is added to increase the melt rate and vice versa. This procedure is based on the assumption that SCA is simulated correctly but the timing is wrong. The availability of EO derived SCA allows the simulated SCA to be corrected as well. As HBV uses several reservoirs to store precipitation and hence delay the passage of water through the basin to the outlet, SCA updating from EO data must be combined with traditional updating otherwise these other reservoirs are ignored.

An important feature of the HBV model is that it keeps track of the amount of water stored in the snow pack (the SWE), this enables long-range forecasts of runoff to be made. In order to update the model, it is necessary to relate EO derived SCA to the simulated SWE (Turpin et al. 1999). This is done via the simulated SCA which is derived from simulated SWE using a snow distribution sub-model within HBV. The EO updating procedure is based on the assumption that SCA is correctly modelled.

For purposes of modelling the basin is split into sub-basins (4 in the case of Tjaktjajaure), elevation zones (100 m interval) and land cover classes (forest, field, glacier or lake). Hence, the EO derived SCA to be used in updating must be presented accordingly. Each sub-basin is updated separately. Given the EO derived SCA for a sub-basin one searches for the date with matching simulated SCA. The parameters defining the state of the model on the EO acquisition date are then updated with the model parameters on the date with matching SCA. Simulation is then restarted.

Due to the considerable differences in SCA estimated by AVHRR, TM and ERS only one data source can be used for updating at any one time. So far this has only been done for AVHRR. The corrected AVHRR SCA estimates have yet to be used in combination with TM.

In 1992 HBV overestimated spring flood runoff (April to July, inclusive), in 1996 there was a very slight underestimation and in 1998 a clear underestimation. Updating of snow cover on days that AVHRR data were available led to an increased volume error towards the end of the 1992 melt season, similar results to those without updating in 1996 and considerable improvement in 1998 (Table 3). The poorer results after updating for 1992 may be explained by new snowfall shortly before the 25 June 1992 image. The model probably failed to correctly describe the distribution of rain and snow on this occasion. The SCA shown by the image represented a very thin snow pack, but the same percentage SCA could only be recreated in the model by selecting a state from almost two weeks previously. This illustrates one of the weaknesses of the updating procedure.

The updating procedure was also tested on the Landsat TM image from 4 June 1996. As this was a year when the model performed well and the simulated and observed snow cover were similar, a great improvement was not to be expected. Rather there was a poorer discharge fit and a slight increase in the underestimation of June runoff. This could either be explained by an underestimation of the snow pack, or by an underestimation of rainfall.

With more data available, we may find that the HBV model does not always describe the snow distribution within a sub-basin correctly. If the differences between modelled SCA and EO derived SCA are systematic, the model can be recalibrated with new values for the parameters describing snow fall distribution and temperature lapse rate. There is, however, a risk that the snow distribution varies between years depending on the prevailing weather conditions during the accumulation period, in a way not fully reflected by the meteorological data or the model simulations. It is likely that the largest errors in runoff simulation occur in years with an unusual snow distribution, and it might thus be possible to further develop the updating routine to redistribute the snow within a sub-basin if enough data become available.

Table 3. Effect of updating HBV simulated SCA using AVHRR data for 1992, 1996 and 1998.

	1992	1996	1998
Default R^2	0.89	0.93	0.87
Updated R^2	0.85	0.93	0.92
% Change	-4.5	0.0	+5.7

5 DISCUSSION

ERS and AVHRR data are to be used to derive maps of SCA for use in *forecasting* snowmelt runoff during the 1999 spring melt season. This requires the EO data to be transferred to SCEOS, processed and the derived SCA maps transferred to SMHI for input to HBV all within 24 hours of data acquisition by the sensor. During the 1999 forecast period ERS images will be acquired approximately once every ten days and AVHRR images will be acquired when cloud cover permits; once every 7 to 10 days going by previous experience. TM data will not be used for forecasting due to the infrequency of cloud free coverage.

Normally orders for ERS SAR PRI data are processed within two weeks. Hence, a special request for fast data delivery had to be made to ESA. Data are being downloaded to the DLR Neustrelitz receiving station and processed by the German Processing and Archiving Facility (D-PAF). Once processing is complete D-PAF inform SCEOS the data are ready for transfer via FTP. Test transfers of PRI images (130 Mb) have taken between 7 and 12 minutes.

A complete test of near real time processing was conducted on 3/4 February. An ERS SAR image of the basin was acquired on an ascending pass at 20:24 on 3 February. The data was received and processed overnight by D-PAF. SCEOS were informed the image was ready for transfer the next morning. After data transfer generation of the geocoded ratio image took 50 minutes and only required a few minutes operator intervention. Processing was complete one and a half hours after the start of data transfer to SCEOS and fourteen and a half hours after data acquisition, well within the near real time forecasting requirement. Processing of descending pass images should be quicker as they are acquired mid-morning.

As the Dundee receiving station already provides AVHRR images online within less than an hour of acquisition, no special arrangements are required for near real time transfer of the data. Processing of each AVHRR image takes no more than half an hour.

To determine the effect on HBV forecasting of SCA update using different EO data sources, four types of runoff forecast will be generated:

1 Forecasting without EO input.
2 Forecasting with SCA derived from AVHRR (no correction).
3 Forecasting with SCA derived from SAR (no correction).
4 Forecasting with SCA derived from AVHRR with correction to proportional TM SCA.

6 CONCLUSIONS

EO methods for deriving the SCA of alpine basins from high and medium resolution optical (Landsat TM and NOAA AVHRR) and SAR (ERS) data have been transferred for use in higher latitude basins.

While the methods for deriving SCA from TM and AVHRR data only required slight changes in detection thresholds the SAR method required more substantial modification. The main area of wet snow cover was detected using the same threshold as used in alpine basins. However, the method failed to detect areas of patchy snow cover. The more moderate relief in higher latitude basins causes there to be less missing cover than in alpine basins, although the area affected is still appreciable and unlike in alpine basins it cannot be reduced by combining ERS ascending and descending passes. The preferred solution is to use SAR data from less steep incidence. In the absence of such data, elevation and aspect information can be used to infer wet SCA from unaffected areas. Dry snow cover at higher elevations must also be inferred. Inferences based on elevation offer an imperfect solution as they fail to take account of snow free areas due to slope angle or wind.

All three sensors detect a steady decline in SCA over the 1992, 1996 and 1998 melt seasons. However, there are large differences in the SCA derived from the three sensors. The overestimation of SCA by AVHRR relative to TM is primarily due to resolution and the threshold chosen for snow classification. This can be corrected by estimating proportional SCA from the AVHRR channel 1 albedo. However, it is not yet clear how to deal with the differences in SCA derived by TM and ERS. These differences are caused partially by the method used to infer dry snow cover and they vary with elevation and temporally.

So far, time series of SCA derived from AVHRR only have been used to update runoff forecasts produced by the HBV model. The results have been inconclusive. Depending on the year EO update causes performance to increase, remain steady or decrease. Factors affecting the improvement in performance from EO update include how well the model is performing prior to this and new and thin snow cover. The four types of forecasts to be produced during the 1999 melt season, including that using the corrected AVHRR SCA, will hopefully clarify the situation.

ACKNOWLEDGEMENT

This work was conducted as part of HYDALP (Hydrology of Alpine and High Latitude Basins) which is a CEO Shared Cost Action Project of the European Union DGXII 4th Specific Programme for Climate and Environment (Contract number ENV4-CT96-0364).

REFERENCES

Baumgartner, M.F. & Rango, A. 1995. A microcomputer-based alpine snow-cover analysis system (ASCAS). *Photogrammetric Engineering and Remote Sensing*, 61(12): 1475-1486.

Bergström, S. 1992. *The HBV model - its structure and applications*, SMHI RH 4.

Caves, R.G., Hodson, A., Turpin, O.C., Clark, C.D., Ferguson, R.I. & Quegan, S. 1998. Field verification of SAR wet snow mapping in a non-Alpine environment. In Proc. *2nd International Workshop on Retrieval of Bio- and Geophysical Parameters from SAR Data for Land Applications*, 519-526, ESA SP-441.

Caves, R.G., Hodson, A., Turpin, O.C., Clark, C.D., Ferguson, R.I. & Quegan, S. 1999. Application of SAR wet snow mapping in a non-Alpine environment. Submitted to IEEE Trans. Geoscience and Remote Sensing.

Cracknell, A.P. 1997. *The Advanced Very High Resolution Radiometer*, London: Taylor and Francis, p. 140.

Dozier, J. 1989. Spectral signatures of alpine snow cover from Landsat Thematic Mapper. *Remote Sens. Env*, 28: 9-22.

Dozier, J. & Marks, D. 1987. Snow mapping and classification from Landsat TM data. *Annals of Glaciology*, 9: 97-103.

Epema, G.F. 1990. Determination of planetary reflectance for Landsat-5 Thematic-Mapper tapes processed by Earthnet (Italy). *ESA Journal*, 14: 101-108.

Ferguson, R.I. 1999. Snowmelt runoff models. *Progress in Physical Geography*, in press.

Gesell, G. 1989. An algorithm for snow and ice detection using AVHRR data: an extension to the APOLLO software package. *Int. Jnl. Remote Sensing*, 10: 897-905.

Guindon, B. & Adair, M. 1992. Analytic formulation of spaceborne SAR image geocoding and value-added product generation procedures using digital elevation data. *Canadian Journal of Remote Sensing*, 18(1): 2-12.

Marçal, A.R.S. & Slater, M.T. 1997. A system for near real time processing of NOAA-AVHRR data: an application to snow monitoring in Scotland. In Proc. 6th Int. Conf. on Image Processing and its Applications.

Matzler, C. 1987. Applications of the interaction of microwaves with the natural snow cover. *Remote Sensing Reviews*, 2(2): 259-391.

Nagler, T. 1996. *Methods and Analysis of Synthetic Aperture Radar Data from ERS-1 and X-SAR for Snow and Glacier Applications*. PhD Thesis, University of Innsbruck.

Nagler, T., Rott, H. & Glendinning G. 1998. SAR-based snow cover retrieval for runoff modelling, Proc. of *2nd Int. Workshop on Retrieval of Bio- and Geo-physical Parameters from SAR data for Land Applications*, 511-517, ESA SP-441.

Piesbergen, J., Holecz, F., & Haefner, H. 1997. Multi-source snow cover monitoring in Eastern Switzerland, *Proc. of 3rd ERS-1 Symposium*, 871-875, ESA SP-414.

Rott, H. & Markl, G. 1989. Improved snow and glacier monitoring by Landsat Thematic Mapper, In Proc. *Earthnet Pilot Project on Landsat TM application*. 3-12, ESA SP-1102.

Thome, K., Markham, B., Barker, J., Slater, P. & Biggar, S. 1997. Radiometric calibration of Landsat. *Photogrammetric Engineering and Remote Sensing*, 63: 853-858.

Turpin, O.C., Ferguson, R.I. & Johansson, B. 1999. Use of remote sensing to test and update simulated snow cover in hydrological models. *Hydrological Processes*, 13, in press.

Voigt, S., Koch, M. & Baumgartner, M.F. 1998. A multichannel threshold technique for NOAA-AVHRR data to monitor the extent of snow cover in the Swiss Alps. Submitted for *IAHS Publication*.

Remote Sensing in the 21st Century: Economic and Environmental Applications, Casanova (ed.)
© 2000 Balkema, Rotterdam, ISBN 90 5809 096 5

Three years of mapping irrigation water in the Flumen irrigation district, Spain

T. Martín-Ordóñez, Ma. A. Casterad & J. Herrero
Servicio de Investigación Agroalimentaria, Laboratorio de Agronomía y Medio Ambiente (DGA-CSIC), Zaragoza, Spain

ABSTRACT: A methodology to estimate and map irrigation water volumes is presented. The information used was: (i) crop maps obtained by supervised classification of satellite images; (ii) net water requirements obtained by FAO method from meteorological data; (iii) soil maps; and (iv) irrigation efficiency values. From these data we obtained several graphical and numerical documents related to irrigation water in 1993, 1994 and 1996 in the Flumen district (Spain). After our estimations in the three studied years, the irrigation patterns were quite stable both in the whole irrigated district and by morphopedological unit. The method can be applied in other irrigated areas and will be useful for agricultural and environmental management.

1 INTRODUCTION

Agriculture relies on irrigation in central Aragón because of the semiarid climate (Herrero & Snyder 1997), and it is the same in many countries where water is scarce and disputed. Remote sensing can help to account for and manage this strategic resource, and in particular to understand the effects of the application of water to soils, so preventing unwanted effects of irrigation, like seepage, water ponding or salination.

In this way Casterad & Herrero (1998) presented Irrivol, a method involving remote sensing to estimate the irrigation water volume applied in an irrigation district.

In the present article, the land cover map obtained by remote sensing is converted to a map of net water for each of the three years, using Irrivol to assign a water volume to every pixel. These maps are combined with the soil map to obtain irrigation water maps that enable the study of the distribution of water in the landscape. The same information could easily be extracted for soil units, municipalities, or other demarcations.

2 MATERIAL AND METHODS

The studied irrigation district of Flumen (322 km2) is located in Huesca province (Aragón, Spain), and was designed for gravity irrigation, which is still the most common irrigation technique in Flumen. One of the limiting factors for crop production in this district is the salinity of some soils (Herrero & Aragüés 1980).

The methodology to estimate and map the water volumes applied in Flumen in 1993, 1994 and 1996, includes five different steps (Martín-Ordóñez et al. 1998).

2.1 *Obtaining land cover maps by remote sensing*

The maps of crop and other land uses were obtained by multitemporal supervised classification using a maximum likelihood classifier. A total of seven Landsat 5 TM images were processed, two from 1993 (6 March and 12 July), another two from 1994 (28 May and 29 July) and, finally three from 1996 (15 April, 18 June and 20 July).

Before classification the images were corrected by the minimum histogram method (Chavez 1988) for atmospheric effects, and by ground control points for geometry.

The training areas were taken automatically from the ground inventory. This ground truth was also used to evaluate the classification by confusion matrices. After classification, a threshold of 95 percent of confidence level, and a 3 × 3 majority filter were applied.

The treatments were carried out using ERDAS software.

2.2 *Calculating the crops' net water requirements*

The net water requirements (WRn) of the crops were estimated using the FAO Blaney-Criddle methodology (Doorenbos & Pruitt 1977), with meteorological data recorded at the Sariñena weather station.

WRn of rice was not calculated because in Flumen it is cultivated in continuously flooded paddies. The dose of 15,000 m^3 ha^{-1} reported by Tolosa (1990) was adopted as the average water applied by farmers to this crop. The variability in the FAO water requirements was used to introduce the variability between years and months (Casterad & Herrero 1998).

2.3 *Calculating crop hectarages at each morphopedological delineation*

We used a previous soil survey by Rodríguez-Ochoa (1998, pers. comm.). This vectorial cover was rasterised in order to combine it with our land cover maps. The soil map was simplified by considering only the morphopedological units, i.e. by grouping the soil units as shown in Table 1. This resulted in a map containing 257 delineations of the three morphopedological units.

2.4 *Estimating the net water requirement volume by map unit*

The volume of the net crops' water requirements, or "net volume" (Vn) was calculated for every morphopedological unit by multiplying every crop hectarage in the unit by the water requirements of every crop present in the unit:

$$Vn = (H_j \times WRn_j) \times 10^{-6} \quad (1)$$

where Vn = volume (hm^3) of the net water requirement of each morphopedological unit; H_j = j crop hectarage in each unit; and WRn_j = j crop's net water requirement (m^3ha^{-1}).

2.5 *Estimating the irrigation water volume*

The values of irrigation efficiency attributed to every morphopedological unit (Table 1) were taken from field measurements made by Lecina (1998) These values were multiplied by the volumes of Vn to estimate the volumes of irrigation water for each morphopedological unit.

$$Vi = Vn \times Efficiency^{-1} \quad (2)$$

where Vi = volume (hm^3) of irrigation water.

Table 1. Morphopedological units in Flumen with their irrigation efficiency coefficients, and their corresponding soil units.

Morphopedological unit	Efficiency	Soil units
Platforms	0.4	A11, A12, A13
Slopes	0.6	B1, B21
Terraces and Fluvial bottoms	0.8	C1, C2, C3, C4, D1

In the case of rice (see 2.2) we took as Vn and Vi as the same figure, from Tolosa (1990).

3 RESULTS

3.1 *Crop hectarages*

Three land cover maps were obtained by the multitemporal supervised classifications.

The main crops discriminated were alfalfa plus forage, winter cereals (wheat and barley), rice, sunflower and maize. Other less representative land uses discriminated were pines (3 percent of the Flumen area) and other classes (non-agricultural, orchards, fruit and almond trees, etc., that totalled about 10 percent).

About 2 percent of Flumen are non-irrigated enclaves, occupied mainly by winter cereals. Around 30 percent of the district was uncropped.

The principal crops in 1993 and 1994 were alfalfa plus forage, and winter cereals (barley and wheat), which occupied around 40 percent of the district. The combined area of these crops in 1996 was only about 23 percent of the district.

Rice and maize behave in the opposite way. Rice has expanded from 1993 to 1996, almost doubling in area. Traditionally, rice was limited to the saline-sodic soils, but in recent years its high profit has stimulated an increase in its area.

The increase in the area of maize is linked to the decrease in sunflower. The subsidizing of sunflower sowing by the Common Agricultural Policy of European Union resulted in a large increase in sunflower in 1993, replacing maize. In 1994, sunflower and maize tended to equalize, while in 1996 maize increased (14 percent of Flumen), greatly surpassing the area of sunflower (around 3 percent).

In 1996 flax was detected in Flumen, although only 90 ha were located.

The distribution of the crops in the morphopedological units also varied throughout the study.

In the platforms, alfalfa plus forage and winter cereals were the most represented in 1993 and 1994 (52 percent of the platform area). In 1996 maize and alfalfa plus forage were the most represented (19 and 18 percent of the platform area respectively), and winter cereals decreased to 11 percent. Sunflower also decreased from 10 percent in 1993 to 2 percent in 1996.

The most represented crops on the slopes in 1993 were winter cereals (21 percent of the slope area), alfalfa plus forage (19 percent), sunflower (12 percent) and rice (10 percent), and maize was absent. As in the platforms, alfalfa plus forage, winter cereals and sunflower decreased to 13, 10 and 3 percent respectively in 1996, while rice and maize increased (14 and 6 percent respectively).

The same trend occurred in terraces and bottoms. Alfalfa plus forage, winter cereals and sunflower decreased from 1993 to 1996 (from 24 to 16 percent, from 21 to 10 percent, and from 12 to 3 percent of the terraces and bottoms area respectively). While rice and maize increased (from 16 to 23 percent and from 1 to 15 percent respectively).

3.2 *Volumes of irrigation water*

The water volumes were estimated from the main crops discriminated which represented about 90 percent of the cropped area of Flumen.

Different numerical and graphical documents were produced both for the net water requirement volume (Vn) and for irrigation volume (Vi).

The highest Vn and Vi occurred in 1996, opposite to 1993 while in these values were the lowest (Table 2), although differences were slight.

Table 2. Volumes (hm^3) of the estimated net crops' water requirements (Vn) and irrigation water (Vi) in the morphopedological units of Flumen district (Spain) in 1993, 1994 and 1996

	Platforms		Slopes		Terraces & F. Bottoms		Total	
	Vn	Vi	Vn	Vi	Vn	Vi	Vn	Vi
1993	27	64	65	97	38	43	130	204
1994	29	71	68	98	40	45	137	214
1996	27	62	77	104	46	50	150	216

The highest water volumes always appeared on the slopes (52 percent of the area of Flumen), with around 50 percent of the total volumes.

Alfalfa plus forage took around 50 percent of platforms Vi in all three years. However, the second place in the ranking of Vi varied, from winter cereals in 1993 and 1994 to maize in 1996.

Two classes, alfalfa plus forage and rice, were responsible for more than half of the Vi on the slopes.

Most of the Vi in the terraces and bottoms was taken by rice, with the highest being in 1996 with 53 percent of the total Vi in the unit.

Table 3 shows the average values of Vn and Vi per hectare. The averages for Vn were between 3653 and 6149 m^3ha^{-1}, while Vi ranged from 5450 to 9599 m^3ha^{-1}.

Table 3. Mean volumes (m^3ha^{-1}) of the estimated net crops water requirements (Vn) and irrigation water (Vi) in the morphopedological units of Flumen district (Spain) in 1993, 1994 and 1996.

	Platforms		Slopes		Terraces and Bottoms	
	Vn	Vi	Vn	Vi	Vn	Vi
1993	3653	8608	3814	5450	5086	5805
1994	3969	9599	3964	5681	5369	6086
1996	3719	8422	4482	6036	6149	6793

The Vi per hectare was also obtained in every delineation of the three morphopedological units. Figure 1 shows the Vi per hectare of the 257 morphopedological delineations, classified in five ranges.

Almost all the area of the slopes and terraces and fluvial bottoms had Vi per hectare of between 1000 and 10,000 m^3ha^{-1}.

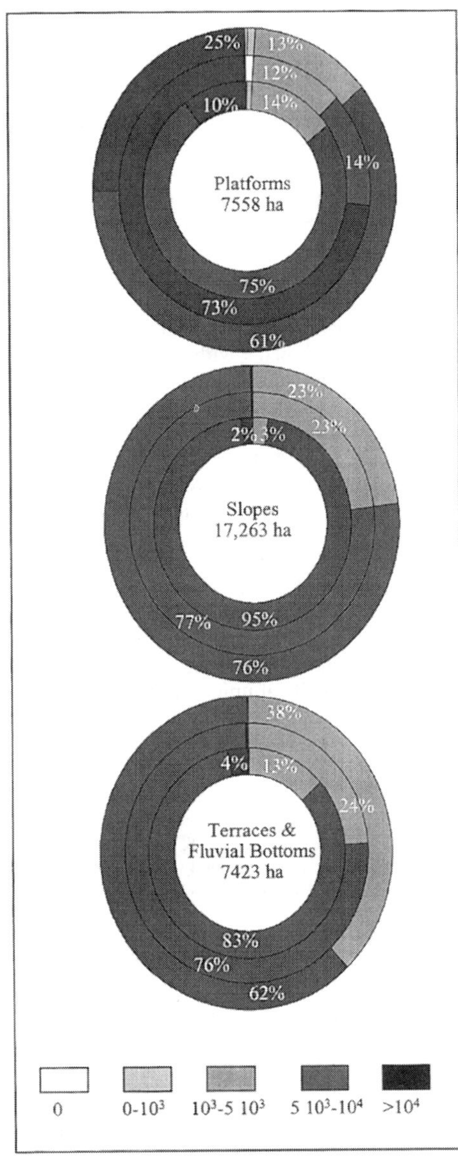

Figure 1. Distribution of the estimated irrigation water (Vi) per hectare in the morphopedological delineations (1993, external ring; 1994 medium ring; and 1996, internal ring).

Platforms were the only morphopedological unit with an important percentage of their surface with Vi per hectare below 10,000 m^3ha^{-1}.

The methodology presented estimates the water parameters related to irrigation, which are associated with every individual pixel. The maps obtained should therefore allow the water use in a specific area to be followed, and changes over time to be monitored. The digital cartography obtained can easily be introduced into a GIS allowing the better management of the irrigation districts.

4 CONCLUSIONS

We have shown the use of Irrivol for mapping the estimated irrigation water in an irrigation district. This information can be easily extracted for soil units, municipalities or other demarcations, and can be used to assess the evolution of water applications throughout the years.

The methodology presented allows knowledge of where the water has been, is being or will be applied within an irrigation district over a season. This information, combined with soil, geomorphic, geological or other maps can help to detect areas prone to seepage, water ponding, salination/desalination, or pollution by agrochemicals. If the data were obtained during the current season, warnings for farmers or for environmental or water agencies would also also be possible.

The estimated irrigation volumes remained stable in the morphopedological units of Flumen district during the three years studied. The above data and results can be used as input in future modelling either of the entire irrigation district or of the landscape or of political units. Similar activities can be undertaken for other irrigated districts.

REFERENCES

Casterad, M. A. & Herrero, J. 1998. Irrivol: A method to estimate the yearly and monthly water applied in an irrigation district. *Water Resources Research* 34(11): 3045-3049.

Chavez, P.S. Jr. 1988. An improved dark-object subtraction technique for atmospheric scattering correction of multiespectral data. *Remote Sensing of Environment* 24: 459-479.

Doorenbos, J. & Pruitt W.O. 1977. *Guidelines for predicting crop water requirements* (Rev.) Irrigation and drainage paper, 24. F.A.O. Rome, 144 pp.

Herrero, J. & Aragüés, R. 1988. Suelos afectados por salinidad en Aragón. *Surcos de Aragón* 9:5-8.

Herrero, J. & Snyder, R.L. 1997. Aridity and irrigation in Aragon, Spain. *Journal of Arid Environments* 35: 535-547.

Lecina, S. 1998. *Estudio de mejora del riego en la zona noroccidental de la acequia de Rufas (regadío de Flumen, Huesca).* Proyecto Final Carrera. ETSIA. University of Lleida. 146 pp. + annexes.

Martín-Ordóñez, T., Casterad, M. A. & Herrero, J. 1998. Cuantificación y cartografía de los volúmenes de agua de riego aplicados en un regadío. *XVI Congreso Nacional de Riegos. Palma de Mallorca, junio 1998. Asociación Española de Riegos y Drenajes. Govern Balear.* pp. 503-510.

Tolosa, M. 1990. *El cultivo del arroz.* Informaciones técnicas DGPA. Diputación General de Aragón. Zaragoza. 8 pp.

Remote Sensing in the 21st Century: Economic and Environmental Applications, Casanova (ed.)
© 2000 Balkema, Rotterdam, ISBN 90 5809 096 5

Satellite monitoring in Amudarya river basin hydrology

V.E.Chub
Central Asian Research Hydrometeorological Institute, Tashkent, Uzbekistan

ABSTRACT: On the middle flow of the Amudarya river are observed significant errors of water balance, which are not possible to explane. At analysis satellite image was found out is bright expressed anomaly in distribution brightness on earthly surface, that has allowed to derive the hypothesis about the strong influence of the ground water flow on water balance in the middle flow of the Amudarya river.

1 INTRODUCTION

In the producing of the Earth surface images by the NOAA satellite pictures the brightness characteristics of the satellite picture are presented the pixels of the specified colour. Using the monitors with high reproducing capacity the image is presented as the brightness distribution in the range of 0-255 units, so it looks like the tonal photo, but with such wide brightness range the eye can not recognize the transition in 1-3 brightness units which sometimes causes the loss of the image informativity.

When using the monitors with the limited brightness interval it is necessary to define the interval borders in such a way that the image can be recognized as geographical feature and, the terrestrial object (water bodies, rivers, etc.) should posses the recognizable contours and outlines.

For the selection of the needed brightness range the following approach was used: the whole brightness range of NOAA-14 satellite images was divided into the intervals in such a way that the same number of pixels was within the range of each interval.

In such case the image becomes recognizable as geographic feature, and its further processing is then possible from the point of view of the correction of the interval values for the further image analysis (calculation of areas, lengths of the objects, etc.)

2 METHODOLOGY

When producing the image of the terrestrial surface the anomaly in the brightness values distribution was found in the middle current of Amudarya river. When the wide brightness range was used for the image producing such phenomenon can not be seen by eye. The preliminary image analysis has shown the existence of the significant ground water flow which is perpendicular to the river current during the year and the width of the brightness anomaly is changed during the year. Such brightness distribution on the image is determined by the differences in the reflectiving: from the terrestrial surface which can be produced by the different vegetation types and depth of the ground water layer.

For the clarifying of the possible reasons of the existence of such anomaly special analysis of the different subject maps contained in the complex research and reference atlas of Uzbekistan (issued in 1982) was carried out. In the atlas the maps of the geological structure and resources of the depth, soil, hydrogeology, vegetation, etc. are presented. None of the maps of the atlas shows the similarity between the anomaly and the other objects.

Evidently, this anomaly is insignificant relatively to the adjacent sectors, that is why it can not be detected by the ground observational facilities, as the difference in the image brightness

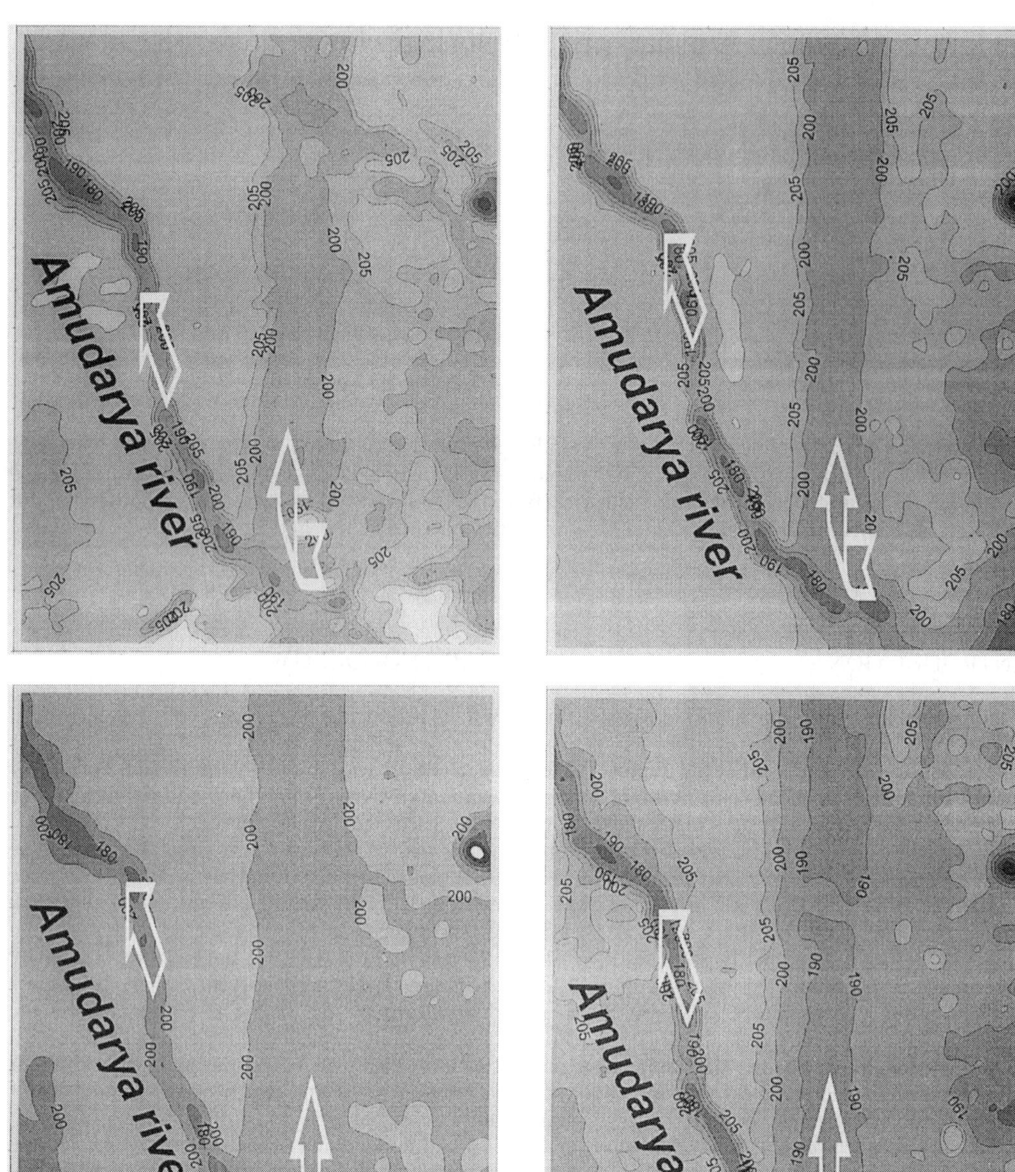

Figure. A satellite image of the Amudarya river basin on the middle flow. Dynamics of the surface brightess is shown. The arrows show the flow direction of river and groundwater.

is not more than 3 units, though it occupies much space. From the other hand, this anomaly is, probable more marked during the last years, for example, in the course of new gas deposits mining in the surroundings of Gazli town, as the channel is directed strictly to this area.

3 RESULT

The detected phenomenon makes it possible to propose a hypothesis about the existence of the close relation between the river and ground water on the specified sector which is necessary to take into account for the water balance calculation. This hypothesis gives the ground for the further investigations in this direction.

The figure successively shows the dynamics of the brightness anomalies (dark tone in the northern direction from the river channel) in the middle Amudarya current by NOAA-14 successive photos, 2-nd channel (3 June, 4 June, 21 June, 29 July 1996).

LITERATURE

Lillesand T.M., Kiefer R.W. 1987.Remote sensing and image interpretation, *John Wiley & Sons,* New York, 170p

Myagkov,S.V. 1995. A method for analysing satellite images.*Processing of the second international workshop*: 299 - 304. Saskatoon, Canada.

Chub V.E. 1996. Direction for application of satellite information to the development of the system for hydrometeorological process surveying in Glavgidromet of the Republic of Uzbekistan.*Tashkent workshop, June, 1996*

Wobber F.J. 1969.Environmental Studies Using Earth Orbital Photography. *Photogrammetria,* vol.24. No3-4, p.107-165.

Remote Sensing in the 21st Century: Economic and Environmental Applications, Casanova (ed.)
© 2000 Balkema, Rotterdam, ISBN 90 5809 096 5

Application of mathematical functions for the monitoring of the small water bodies

V.E.Chub, N.E.Gorelkin & S.V.Myagkov
Central Asian Research Hydrometeorological Institute, Tashkent, Uzbekistan

ABSTRACT: For decision of some tasks of the analysis of the small size objects on satellite image is offered to use the mathematical functions of the matrix transformation. The results of calculations of the analysis of the water surface located on the Amudarya river are presented.

1 INTRODUCTION

One of the main problems of the monitoring of the water bodies using the satellite images is the problem of the identification of the water bodies borders. For the solution of this problem the techniques based on the analysis of histograms of the brightness distribution of the satellite images are often used.

In this work it is proposed to apply mathematical functions for the operations with matrices. It is known that NOAA satellite image present the brightness matrices. At the display of the image as the succession of images onto the monitor screen the contours of the object and the size of its presentation are defined by the parameters of the computer screen. Thus, the scale of the output image and its configuration are determined by the technical characteristics of the equipment.

2 METHOD AND RESULTS

In the case of the construction of isolines by the brightness characteristics many local spots appear which are objective ones because of the integral character of the image producing, but during the visual analysis this adds much diversity of colors which hampers the analysis. In the course of automatic processing the local spots can cause significant errors in the calculation result.

If the values of the brightness matrices are changed before the processing by the smoothing functions, it is possible to smooth not only the contours of the objects, but also the local noise in the image. For the nontransformed image where the geographical distance is changed from point to point in x and y directions the following operator was used:

$$\begin{aligned} U_{j,i} = {} & A^2 U_{j-1,i} + U_{j,i-1} + \\ & + A^2 U_{j,i+1} + U_{j+1,i} - \\ & -2(1-A^2) U_{j,i} \end{aligned} \qquad (1)$$

where: U = value of the pixel brightness, $A = h_y / h_x$, h_x, h_y = side of the rectangle cell.
The result of using of equation (1) on Figure 2 are presented. For the transformed image the equation for calculation is simplified.

Satellite image (NOAA-14) of the water storage's complex of Tuyamuyun water distributor on Amudarya river is presented. It is seen that the lines of the water bodies contours are smooth (Figure 2) in comparison with the non-processed image (Figure 1). Many separate pixels which are not attributed to the water surface, disappeared.

For the more detailed analysis of the different interfaces the equation of the gradient of the brightness field is used:

$$gradU = \lim_{v \to 0} \frac{\oint\limits_{\Sigma} U dS}{v} \quad . \qquad (2).$$

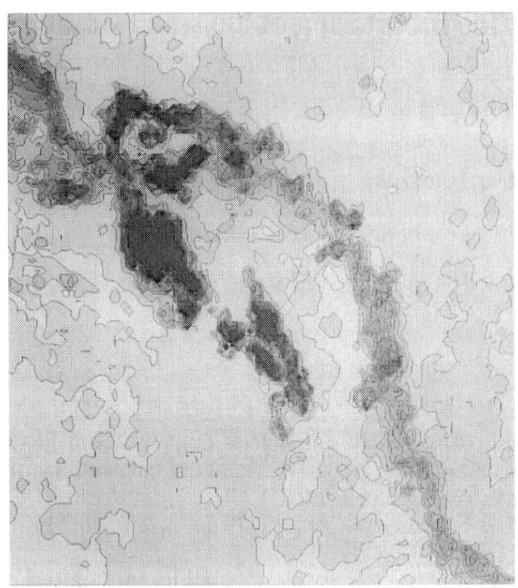

Figure 1. The satellite images of the Tyuamuyun water storages on the Amudarya river is showing. (Numerical matrix of NOAA-14).

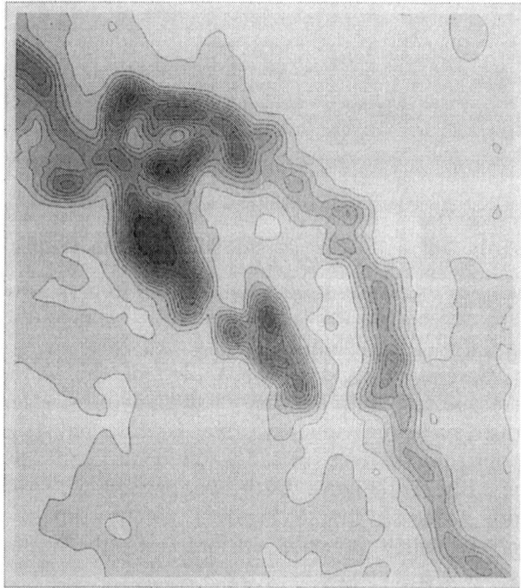

Figure 2. A result after calculation and transformation of numerical matrix at the equation (1) is show.

The result of the application of equation (2) is shown on Figure 3. The disappearance of noise on the sectors with insignificant change of the brightness transfer is especially noticeable in the sectors which are not attributed to the water bodies.

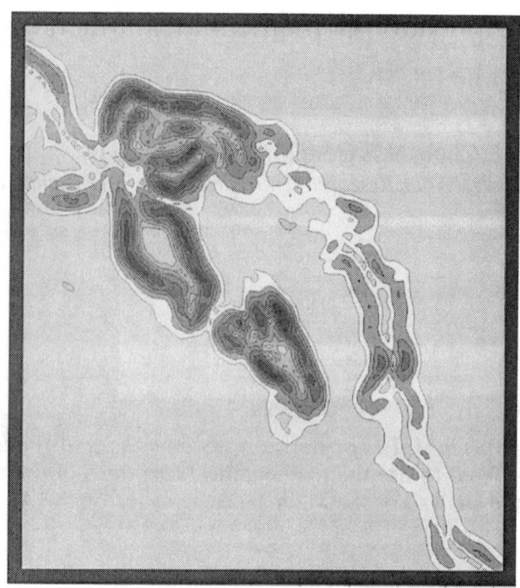

Figure 3. A result after transformation of numerical matrix (1) at the equation (2) are presented.

3 CONCLUSION

Thus, the computation of the area of the prescribed brightness level can be made directly calculating the area of the contour outlined by the smooth line. With this the noise caused by the integral characteristics of the small-scale images, disappear.

LITERATURE

Lillesand T.M., Kiefer R.W. 1987.Remote sensing and image interpretation, *John Wiley & Sons,* New York, 170p

Wobber F.J. 1969.Environmental Studies Using Earth Orbital Photography. *Photogrammetria,* vol.24. No3-4, p.107-165.

4 Methodology

Remote Sensing in the 21st Century: Economic and Environmental Applications, Casanova (ed.)
© 2000 Balkema, Rotterdam, ISBN 90 5809 096 5

Generation and integration of thematic documents from satellite images and ground sensors for environmental analysis and management

S.Ormeño & I.Molina
E.U.I.T. Topográfica, UPM, Polytechnical University of Madrid, Spain

ABSTRACT:

The diversity of variables influencing the environment conditions, and particularly the desertification processes due to soil erosion, implies that a significant number of not controlled variables are involved in the analysis of these different factors. Moreover, the amount and nature of the data that must be considered, makes up that the analysis procedures will not provide obvious conclusions. A large number of efforts have been done in order to model these phenomena, e.g. the USLE equation (Wiskmaier and Smith).

In this current study, a methodology is described in order to derive environmental information from NOAA-AVHRR images, establishing the suitability of the different documents for environmental analysis, mainly for the assessment of surface soil conditions (extremely important in soil erosion processes). In addition, other thematic documents are derived from meteorological ground stations. A preliminary analysis has been performed in order to determine the relationship between this information and the satellite data; then, these results were compared with other thematic maps (available cartography). This work has been carried out in Castilla la Mancha, a Spanish peninsular region with severe desertification problems.

1 INTRODUCTION

The natural medium presents a number of habitability conditions that are limited between some values whose extreme are tolerated from living beings. When these conditions are modified by natural or artificial means exceeding certain values, living beings are accommodated temporally adapting their life cycle, but they can also emigrate towards other areas, returning only when the initial conditions are re-established, or they can even extinguish.

Desertification is known as the decrease of biomass production, and consequently, the productivity in agriculture, cattle raising and forestry. Desertification is preceded by others environmental effects, that are understood as the process of setaside of a territory due among other reasons to the lack of financial resources.

The origin of desertification can be climatic or artificial. The strongest situations of artificial desertification are produced as a consequence of silky soils and soil-loss, this last can be of hydric or eolian origin.

Professor C. Roquero (Polytechnical University of Madrid), has described the main factors influencing soil erosion in a general sense and particularly for Mediterranean soils. This author places deforestation and the new cultivated areas as responsible of the main cases at these latitudes.

In Spain are found clear examples of artificial desertification, some of them are located in the "Sierras of Guadalajara, Soria, Almería and Granada".

In 1960, Wischmeier and Smith presented their well-known USLE equation used for the estimation of soil weight loss by area unit (Tm/Ha) due to water erosion. Since then, several approximations and efforts have been carried out in order to automate the application of this equation. In this context, the current work studies the automation possibilities for modelling erosion processes.

Additionally, it must be also considered as main factor the scale of the cartographic documents, at which this study has been carried out. For this purpose, the following factors have been taken into account:

AVHRR image with ground resolution of 1.1. Km (NOAA-SAA).

882 climatic ground stations for the whole Iberian Peninsula.

A Digital Terrain Model with 1 km resolution.

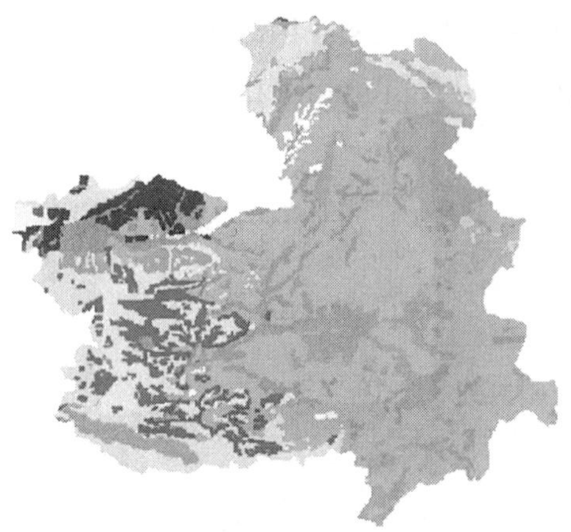

Figure 1. Soil map of Castilla La Mancha. (colour plate, see p. 601).

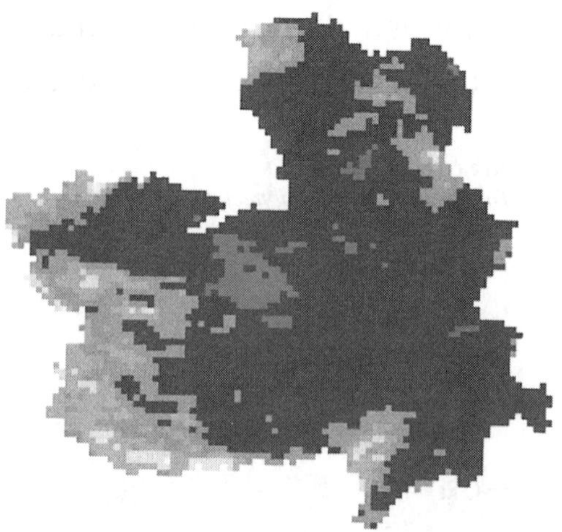

Figure 2. Apparent thermal inertia. (colour plate, see p. 601).

In order to perform the integration of the different documents, the layers involved in this work were georeferenced to a Lambert Conformal Conic projection.

2 COMPILATION OF THE SOILS DOCUMENT

The soil document is derived from the atlas of "Castilla La Mancha" (1/2.500.000), edited by the "Junta de Comunidades" in 1990.

This document shows 13 soil categories obtained from a mixed Baldwin/French Taxonomy. The correspondence with the USDA soil Taxonomy (approx. 1975) is the following:

1. OCHREPT(1)	On Ca++, forestrial browns
2. XERALF(1)	No Calcic browns
3. OCHREPT(2)	On lythics deposits
4. OCHREPT(3)	On SiO2, humid browns
5. OCHREPT(4)	On Ca++, red browns and browns
6. XERALF(2)	On SiO2, Mediterranean reds
7. XERORTHENT(1)	On SiO2
8. AQUENT/FLUVENT	
9. XERORTHENT(1)	On Ca++
10 OTHERS	Lythomorphic
11 XERALF3	On Ca++, Mediterranean reds
12 OCHREPT5	On SiO2, meridian browns
13 PSAMMENT	Sandy soils

3 VARIABLES CONSIDERED FOR THE ANALYSIS

3.1 *Identification of variables*

In this work, six variables have been taken into account in order to characterise the 13 soil classes.

First, a Normalized Difference Vegetation Index (NDVI) has been derived from NOAA-AVHRR data, then a Digital Elevation Model (DEM) of this territory and its corresponding slope map (SLOP) have also been used. The Apparent Thermal Inertia, Potential Evapotraspiration models and an Aridity Index were finally introduced as input variables. These lasts are described in the next sections.

3.2 *Thermal inertia*

Thermal Inertia is defined as the measure of a material response to the variations in temperature. The thermal conductivity (k), the specific heat (c), and material density (D) are directly involved through the following expression:

$$TI = \sqrt{(D*c*k)} \quad \text{in cal cm}^2\ \text{s}^{-1/2}\ {}^{o}\text{C}^{-1} \qquad (1)$$

Since the previous parameters must be measured by contact methods, they can not be consequently determined by remote sensing. Instead, Sabins (1987) suggests to apply the so called Apparent Thermal Inertia (ATI).

The Apparent Thermal Inertia (ATI) is estimated from:

$$ATI = (1 - a) / \Delta T \qquad \text{in \% } {}^{o}\text{C}^{-1} \qquad (2)$$

where "a" is the albedo, and ΔT is the variation of temperature. The albedo and temperature variations are estimated from AVHRR data. First, the albedo is derived applying the calibration values to the first spectral band. The temperature change is obtained from spectral band 5 (11,5 – 12,5 μm), with 12 hours of difference.

The significance of this document lies in its direct relationship with the bedrock properties and therefore with soil groups.

3.3 *ET_P (Potential Evapotranspiration) and ARI (Aridity Index of De Martonne from ground stations)*

In order to apply these models, data gathered from the meteorological ground stations network of the National Meteorologic Institute (Spain) have been used.

The variable value (ET_p) is derived from the method of Thornthwaite and proceeds from the National Agroclimatic Atlas. This variable is expressed in mm of water per year. For this method, a maximum ground water reservoir is considered, which is equivalent to 100 mm. These data are given in total yearly ET_P

The "De Martonne Aridity Index" is defined by the following expression:

$$ARI = P/(T+10) \quad \text{in mm/°C} \qquad (3)$$

Considering total yearly values for the precipitation (P), and average yearly values for the temperatures (T).

4 RECOGNITION OF CLASSES BY MEANS OF STOCHASTICAL CRITERIA

The recognition of soil classes using stochastical methods implies the integration of the previous variables into a classification scheme. Initially a multivariate analysis has been applied. This work has been done analysing the correlation between the different layers (table 1). It can be seen that the splopes document shows a low correlation with the remaining variables, corresponding the highest values to the Et_p.

Then, clustering results are derived from a sampling process using the previous six variables and the 13 standard soil classes (French Taxonomy – Classification).

It can be appreciated (table 2), that some categories can be easily discriminated from the separability analysis: such as C7, C8, and C9, all of them Entisols, and up to some extent C3, Inceptisols.

A sensibility analysis has also been performed in order to demonstrate the influence of the variables

Table 1. Correlation Matrix of the 6 variables

	NDVI	ATI	ETP	ARI	DEM	SLOP
NDVI	1.00	0.95	0.97	0.94	0.94	0.44
ATI	-	1.00	0.97	0.95	0.92	0.43
ETP	-	-	1.00	0.94	0.92	0.41
ARI	-	-	-	1.00	0.95	0.45
DEM	-	-	-	-	1.00	0.45
SLOP	-	-	-	-	-	1.00

Table 2. Results of the multivariate analysis

Number of Samples: 1995 / Number of Samples for each soil class												
C1	C2	C3	C4	C5	C6	C7	C8	C9	C10	C11	C12	C13
44	75	111	159	816	125	24	7	19	164	90	338	22
Upper limit of overall error (%) from the Battacharya distance												
100												
7.73	100											
12.67	10.66	100										
30.04	17.53	17.44	100									
32.87	10.06	14.52	41.51	100								
17.18	13.64	14.74	27.51	23.48	100							
1.07	2.34	0.85	1.6	0.94	12.51	100						
0.23	0	0	0.03	0.12	0.1	0	100					
8.31	10.35	20.83	11.28	7.84	7.87	0.08	0	100				
30.46	8.27	7.67	31.03	37.98	18.79	0.93	0.28	5.24	100			
22.49	4.61	17.24	23.99	26.65	18	1.17	0.03	12.36	21.32	100		
19.05	14.99	8.81	26.56	25.11	41.21	12.85	0.42	5.42	23.52	16.46	100	
18.76	12.34	7.29	23.6	20.13	22.33	1.59	0.01	4.11	17.93	18.87	23.7	100
A posteriori error of classification (from cont. table)./ % of samples correctly classified												
C1	C2	C3	C4	C5	C6	C7	C8	C9	C10	C11	C12	C13
38.63	45.33	87.73	6.28	11.52	36.00	95.83	100	78.94	23.78	57.78	20.12	45.45
Average 48.73 %												

according to the classification result. As significant results, it can be found, that eliminating ET_P, it is possible to discriminate category C11 (one type of Xeralf), though the global reliability can decrease; eliminating the slopes (SLOP), class C2 (another Xeralf type) can also be discriminated.

5 RECOGNITION OF CLASSES BY MEANS OF ARTIFICIAL NEURAL NETWORKS

The Artificial Neural Networks theory provides a suitable methodology for classifying multivariate data. It exists many structures (Net Architecture) for implementing calculation processes that relate an input vector with an output vector; this last will indicate the belonging class of the input vector.

The network model that has been applied, is the Back-Propagation. Here three net architectures (A, B, C) have been tested.

The base structure for architectures A and B is the following:

a) Number of variables 6: dimensionality of input vectors to classify.
b) Number of classes 13: dimensionality of output vectors.
c) Number of samples 1995.

The results of both architectures are given in the following table (table 3).

Table 3. Results of Architectures A and B

Number of Samples: 1995						
Arch. A	nd.11	nd.12	nd.13	nd.14	nd.15	
	6	10	0	0	13	
(I)	(II)	(III)	(IV)	(V)	(VI)	(VII)
0.8	0.8	0.7	3000	0.37	50.17	0.70
Arch. B	nd.11	nd.12	nd.13	nd.14	nd.15	
	6	7	9	11	13	
(I)	(II)	(III)	(IV)	(V)	(VI)	(VII)
0.8	0.8	0.7	3000	0.65	42.06	0.74

The legend for this table is the following:

(I) Sigmoid function slope factor
(II) Sigmoid function displacement factor
(III) Convergence velocity factor
(IV) Number of iterations
(V) Maximum error
(VI) Percentage of truly classified samples
(VII) Sum of the RMSE by iteration

Type B architecture (3 or more intermediate layers) does not provide usually good results in this context. From the different performed tests, the best results were achieved using architecture A: with 30000 iterations, recognition of 50.17% of the samples was reached.

In architecture C, it has been tested the recognition of the mean values of the 13 classes for the 6 considered variables, reaching the results given in table 4.

Table 4. Results of Architecture C

Number of Samples: 13						
Arch. C	nd.11	nd.12	nd.13	nd.14	Nd.15	
	6	10	0	0	13	
(I)	(II)	(III)	(IV)	(V)	(VI)	(VII)
0.8	0.8	0.7	2709	0.006	100	0.02

It can be appreciated that, in order to recognise these 13 classes, the introduction of the mean values for each class allows a 100% of identification.

From these results, it can be stated that the recognition difficulty by means of linear methods is intrinsic to the variability of the samples for each class.

6 CONCLUSIONS

The six analyzed variables do not define completely the soil categories according to a standard classification. Some entisols and inceptisols types can be correctly recognized through the analysis of these variables. To some extent, the recognition difficulty comes out from the high values variability that corresponds to each category.

The ANN methodology can improve the results of the Bayesian algorithms, however it supposes an increase in time and hardware resources.

REFERENCES

Ormeño S. 1999. Software "Ormeño Imagen" for analysis of models and multiband images.

Ormeño S., Roquero C. Bruña M.A., Gómez B. (1998): Cartografía de Suelos y Climatología en Castilla La Mancha. Integración de Imágenes AVHRR en su Análisis. EUITTopográfica, Polytechnical University. of Madrd.

Roozekrans H. 1984.-Mapping of desertification parameters using NOAA-AVHRR data. Meeting of SPIE. Rome.

Roquero C.-1982. "Ciclo irreversible de la erosión en suelos mediterráneos". Polytechnical University. of Madrd..

Sabins Jr. 1987. Remote Sensing, Principles and Interpretation, 2nd Edition. W.H Freeman and Company.

Soufi K 1997.- Estudio comparado de los procesos de desertificación causados por la erosión del suelo en zonas equiparables de España y Siria mediante teledetección. Polytechnical University. of Madrd.

Wischmeier W.H., Smith D.D. 1960. A Universal Soil-Loss Equation to Guide Consevation Farm Planning". 7th Int. Congress of Soil Science. Madison.

Remote Sensing in the 21st Century: Economic and Environmental Applications, Casanova (ed.)
© 2000 Balkema, Rotterdam, ISBN 90 5809 096 5

Accurate geodetic positioning of satellite laser ranging stations using a short arc geometric solution

M.Catalán-M
Real Instituto y Observatorio de la Armada, San Fernando Cadiz, Spain

P.Villares & M.Catalán-PU
Department of Applied Physics, University of Cadiz, Puerto Real, Spain

ABSTRACT: In the European area satellite global orbit accuracy could be improved to the centimetre level and extended towards the Atlantic applying short arc techniques, considering the favourable geometry of the tracking network. To improve the accuracy of the global orbits over the area we have developed an original short arc numeric models. The main features of our technique consist of a detailed "a priori" analysis of the geometric shape of the passes with respect to the laser stations, to detect correlations and to solve out for the orbit components allowing the better accuracies in the local orbit solution.We have checked our short arc solution (SA) by the direct comparison of its results with the tracking data, with the NASA precision orbit ephemeris (POE).The results conclude that our short arc solution presents a local improvement in the precision and accuracy over NASA global orbit.

1-INTRODUCTION

The fundamental problem with tide gauge measurements is that they measure sea level change compared with a fixed crustal reference point, sufficient for measuring some local socioeconomic impacts of sea level rise but of little interest for climate change studies.

This is an important difference, since the long crustal rebound and tectonic uplift rates at many tide gauges are at the same level as the expected signal due to real sea level rise.

In these conditions we need to validate the tide gauge results using, as an independent global measurement techniques, absolute gravimetry and precise space geodetic techniques, GPS and SLR, to resolve the crustal motions at the required accuracy.

Satellite altimetry presently provides the most powerful tool to observe sea level variations globally and synoptically. When an altimetric satellite is tracked and its position computed with some centimetres accuracy, and adequately corrected for a variety of geophysical effects and referenced to the geoid, the altimeter data yield the global dynamic height of sea level associated with barotropic currents with the higher accuracy. Tapley et al. (1992); Zlotnicki et al. (1989).

Satellite altimeter should provide improved measurements of the global sea level change over shorter averaging periods, because of their truly global coverage and direct tie to the geocentre through the satellite tracking stations. However, for altimeter missions such as Geos -3, Seasat, Geosat and ERS-1, errors in the satellite orbit and corrections obscured in the past the sea level rate signals.

The TOPEX-POSEIDON satellite flies in a nearly circular orbit at an average altitude of 1336 km and inclination of 66.05°. It has a period of 112 minutes and a ground track repeat cycle of 9.9156 days, after completing 127 revolutions. The rate of change for the argument of the perigee is near zero to satisfy the frozen repeat orbit criteria. Orbit maintenance are done every three months to keep the orbit ground track repeating within 1km and the mean orbit eccentricity near zero. The rms radial orbit errors for T/P are inside 3- 4 cm due, largely, to improved gravity and non conservative force models. Tapley et al (1992), Marshall et al. (1995).

T/P carries four independent tracking data systems: Satellite Laser Ranging (SLR), Doppler Orbitography and Radiopositioning Integrated by Satellite (DORIS), GPS demonstration receiver and the Tracking and Data Relay Satellite System (TDRSS).

2- THE SHORT ARC TECHNIQUE

According to preceding ideas the use of geometric methods based on the direct laser tracking data, as short arc techniques, has the potentiality of giving an improvement over the accuracy of the global orbit determination in limited geographical regions.

The method has the advantage that does not

require numerical integration and that the computed orbit is directly tied to the position of the tracking stations in the terrestrial reference frame.

Some authors have investigated such methods with close to one centimetre precision result over arcs of about 4000 Km. Sinclair et al (1993); Bonnefond et al (1995), Barlier et al (1992).

A further problem is that the laser station net track, at best, only one third of the available passes, due to the weather and other restrictions, so the number of tracks with good enough tracking for short arc analysis would be small, except in regions with a dense laser tracking stations coverage.

In any case this geometrical approach, with the use of some objective criteria, allows the determination of an improved satellite orbit, without the need of a better dynamic model.

This leads to a better geodetic satellite positioning, with respect to the terrestrial reference frame, over restricted areas with good SLR coverage and the improvement of regional ocean dynamic studies possibilities of the altimetric satellites.

Therefore, the aim of the present work is to present the short arc orbit computation method we have developed, including the criteria, characteristic and error budget, as complete as possible by its cross comparison with other results.

The main features of our technique consist of a detailed "a priori" analysis of the geometrical shape of the passes with respect to the laser stations, to have the best orbit components determination over an orbit of 3000-4000 km. Figure 1.

The method developed is useful not only for Topex-Poseidon orbit, but also for all satellites tracked via SLR and with light modifications for any other technique, GPS, Doris, Prare.

The method is independent of the dynamic equation and so can improve locally the global orbit on relation to the Terresrial Reference Frame, with an accuracy only limited for the geographical position of the tracking stations that could reach, in the future, the 5-mm accuracy considering the developments in the last generation of SLR stations. Degnan (1993).

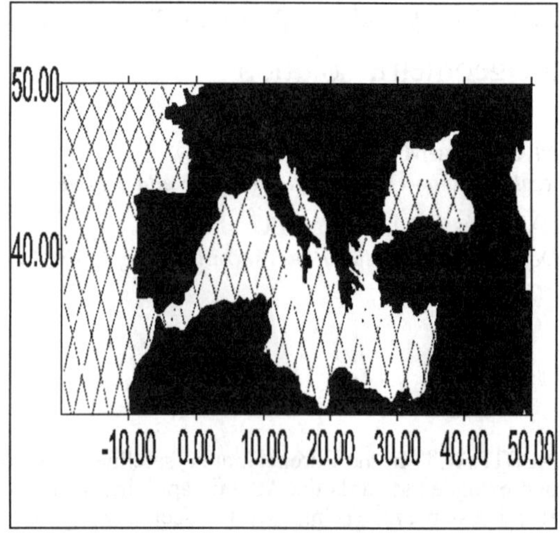

Figure 1: The geographic position of the European Satellite Laser Tracking Station net (SLR) allows to compute, with centimetres accuracy, short arc orbits for altimetric satellites overflying west Mediterranean.

3 -THE DETERMINATION OF SATELLITE ORBIT CORRECTIONS BY SHORT ARC TECHNIQUES.

We have used in our formalism as inputs the Earth Orientation Parameters (EOP) from the International Earth Rotation Service (IERS), the satellite laser ranges from the EUROLAS normal point files and the satellite coordinates and velocity from NASA (POE) ephemeris. Satellite position is treated with a 10 terms Hermite polynomial, to obtain its position and velocity in the Inertial Time of the Date reference frame (ITOD). Stations coordinates corrections have followed the IERS standard recommendations for solid earth tides, tectonic displacements and loading effects. Other corrections applied came from the satellite center of mass - center of phase segment, computed according
with the satellite attitude law. Aberration and the centroid of laser footprint corrections were also taken in account. (Marshall et al. 1995).

Our short arc formalism improves locally POE ephemeris computing the corrections to this input orbit when the satellite overflies the European laser tracking net, typically 10 to 15 minutes, covering an orbit length of about 4000 km. It would be convenient to point out that the corrected orbit is no longer a better solution of the dynamic motion equations sed a kinematic correction adjustment representing local orbit errors fits of the global model, estimated after a least square solution that considers the dense SLR tracking data available in Europe. In our method we have considered two steps. First we made an "a priori " analysis of the geometrical shape of the satellite tracks, to detect possible correlations to get the selection of the variables to be solved.
Second we make an " a posteriori " analysis of our results to validate our short arc solution by comparison with other solutions.

Our formalism starts considering that the satellite position was defined in a Gauss mobile coordinate system by its along track (t), across track (n) and radial
(r) orbit corrections, figure 2.It is possible to model these corrections as simple constants, or better as linear functions of time or with a periodic time dependence variation related with the orbital period.

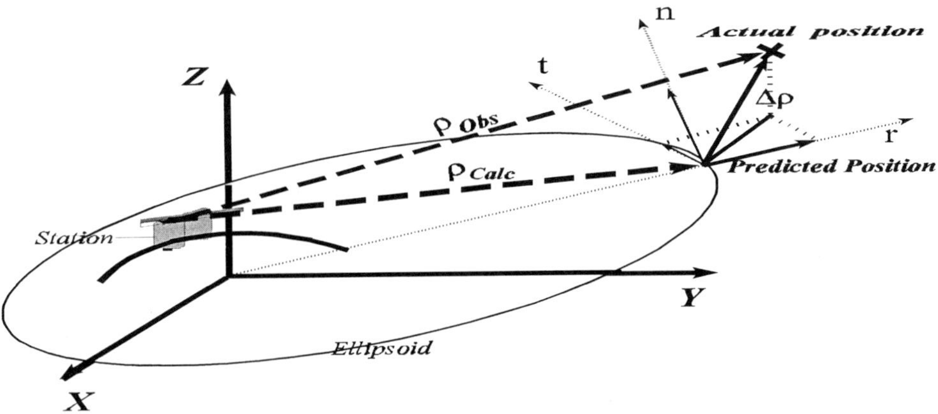

$$\rho_{obs} - \rho_{calc} \approx \frac{\partial \rho_{Calc}}{\partial r} \Delta r + \frac{\partial \rho_{Calc}}{\partial t} \Delta t + \frac{\partial \rho_{Calc}}{\partial n} \Delta n \quad (1)$$

Figure 2: The reference frames of the short arc technique consider that the laser tracking station net is located in the adopted geocentric terrestrial reference frame (X, Y, Z). The satellite position corrections are computed in a Gauss system of coordinates mobile over the satellite orbit (r radial, n across track, t along track). Our short arc correction equations can be reduced to the formula (1) " a priori " error model, as a compromise specially accurate for Topex-Poseidon orbit characteristics. Our formalism computes the correction parameters that produce the best fit for the residual between the computed and observed satellite to station range

The last model considers that the temporal variation of the corrections can be assumed as arbitrary constants plus forced periodic and secular terms. Mismodelling give errors in the periodic terms corrections near the orbital period. (Sinclair 1989, Engelis 1987, Chelton et al.1993, Zanderbergen 1990, Bonnefond et al 1995)

According with preceding ideas in our short arc model we can assume, as a first step, a time linear dependence of the orbital errors in the three track components, takes the time as the interval lasted from the satellite pass over the closest point of approachment (CPA) between the satellite ground path and the tracking station. We note that a high correlation between two of the correction parameters of the orbit, found out from either range or range rate data, will result in some poor estimates of the values of the correlated parameters, implying large standard errors.In the case of the Topex-Poseidon satellite, since the precision of the radial component of the input ephemeris will be always bettter than 5cm, the correction of the radial component can be modelled as a simple bias, neglecting the value of its rate to avoid its high correlation with the rate of the across track component, without perturbing the numerical solution fit. (Sinclair, 1989).

Consequently, the short arc correction equations can be reduced to the following " a priori " error model, as a compromise specially accurate for Topex-Poseidon orbit characteristics:

$$\Delta r = R_o$$

$$\Delta n = N_o + N_1 * (T - t_{ref}) \quad (2)$$

$$\Delta t = T_o + T_1 * (T - t_{ref})$$

Considering our formalism uses a least square solution that computes the correction parameter that produces the best fit for the residue between the computed and observed satellite to station range.

Preceding treatment has the advantage of its simplicity and the disadvantadge that the configuration of the tracking net, relative to the orbit, can degrade the short arc solution because the orbital correction parameters may have small values, or to be highly correlated.

As the accuracy of the geometric solution depends on the relative shape of the network formed by the satellite orbit and the tracking net, we have made a careful study to solve out in our short arc solution, only for the uncorrelated variables.

To analyse all possible correlations cases we have simulate several Topex-Poseidon typical orbits.

As an example Figure 3 present some of the results of our numerical simulation showing that at

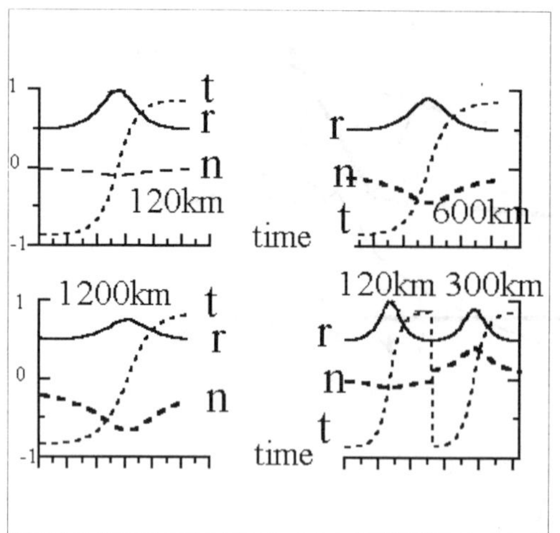

Figure 3: Simulation analysis allows the criteria that could automatically be applied in our numerical simulation analysis, to get the best solution according to the different geometries of the satellite in the tracking net. Lower right figure presents the case of two laser stations located at both sides of the satellite ground path, that track simultaneously the satellite.

distances of the closest point of approach (CPA), between 600 and 1200 km, and for the case of one tracking station. Although the component t is uncorrelated with the others, the r and n are highly correlated, avoiding an accurate solution for the orbit correction parameters.

In the case of two laser stations located at both sides of the satellite ground path, that track simultaneously the satellite, we can see (figure 3, lower right) that although there are a high correlation between the variables r and n from the station at CPA of 300 Km, the low correlation in these variables, for the station at 120 Km CPA, allows to solve out for the correction parameters .

CONCLUSION

Except tracking data no exist a method that could directly measure the absolute orbit accuracy, making the process of assessing orbit quality difficult. Nevertheless several different methods can be used to imply orbit accuracy these include, to compare orbits with independently computed ephemeris, which as an aggregate could provide a reasonable error estimate.For past altimetric missions the altimeter range residuals at the crossover orbits points were widely used as an independent measure to quantify errors in the satellite radial position. However, presently for Topex mesoscale sea surface variation, uncertainties in modelling ocean tides and geoid errors are all large compared with the residual orbit error.

The aim of this section is to present the validation of our short arc solution comparing our orbit (SA) directly with the satellite orbits from NASA (POE) and DGM-E04 for ERS satellites. .Figure 4 presents the comparison of our orbit (SA) with other solutions. The comparison has been made at the

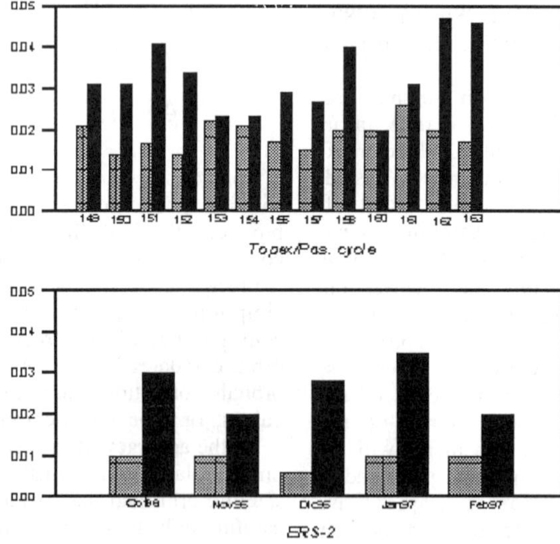

Figure 4: Upper figure presents the std of the residual between the laser station range observations and the station-satellite ranges using the NASA-GSFC global orbit (Black). Same figures presents in red the same information using our short arc solution (OSA, in grey). Lower figure presents the same residual analysis against the Delft global orbit (DGM-E04), in black, and our short arc solution (OSA, in grey)

middle point of the short arc length.

The results of the comparison conclude that our short arc solution (SA) presents a local improvement in the precision and accuracy over NASA and DGM-E04 global orbits.

It would be convenient to note out, that although POE-NASA orbit is accurate better than 4 cm in its radial component, at a global scale, the horizontal components are computed with a lower level of accuracy. According to the locally improved short arc orbit can be used to relocate statistically a tracking laser station, which for any reason does not reach Lageos satellite, in the terrestrial reference frame of the tracking stations used to compute the local short arc orbit.

ACKNOWLEDGMENTS

The authors would like to thank the European Space Agency for its support during the development of the research programm for validation of the ERS satellite altimeter data.

Thanks also to the French agency AVISO and IFREMER for the information received with the Cdrom altimeter data.

Thanks also to the E.U. Research Programms SELF-II and CANIGO for its support.

REFERENCES

Archiving, Validation and Interpretation of Satellite Data in Oceanography , AVISO User handbook ,Merged TOPEX - POSEIDON Products (GDRs),Edition 2,AVISO Altimetrie, 1993

Barlier, F., Bonnefond P., Boudon Y., Exertier, P. 1992 Region alOrbit precision in the Mediterranean Area First *ERS-1Symposium* Cannes 4-6 December.

Bonnefond P.,Exertier P., Schaefer P., Barlier F. 1995 First results of the Topex-Poseidon validation in the Mediterranean area. *Journal Geophy. Research*, vol 100. N° C12, 25365-25382.

Chelton D.B., Schlax M.G. 1993 "Spectral characteristics of the time dependant orbit error in altimeter heigh measurements" *Jour. Geoph. Res.*Vol 98.N° C7, 12579-12600.

Degnan J, 1993 "Milimiter accuracy satellite laser ranging: A review" *Contributions of Space Geodesy to Geodynamics: Technology geodynamic series* vol. 25 pp 133-162 Washington.

Engelis T., 1987 "Radial orbit error reduction and sea surface topography determination using satellite altimetry" *Depart. of Geod. Science. and Surveying.Report 377*, Ohio State University.

Marshall J.A., Zelenski N.P., Klosko S.M., Chinn D.S., Luthcke S.B., Rachlin K.E., Williamson R.G., 1995 "The temporal and spatial characteristics of Topex-Poseidon radial orbit error" *Jour. Geoph. Res* Vol 100, N° C12,25331-25352.

Sellers P., Cross P.A. 1990 "A pseudo short arc technique for precise regional satellite laser ranging baselines" *Manuscripta.Geodaetica*, 15:207-227.

Sinclair A.T. 1989 "The determination of orbit corrections by short-arc techniques, with the application to ERS-1 ". *Manuscripta Geodaetica* 14:238-246.

Sinclair, A.T., Appleby, G.M., 1993 "A short arc method for determination of station coordinates and baselines applied to the Mediterranean Area"*Contributions of Space Geodesy to Geodynamics: Crustal Dynamics Geodynamics* 23,389-396.

Tapley B.D., Schum C.K., Ries J.C., Suter R., Schutz B.E., 1992 Monitoring of changes in global mean sea level using Geosat altimeter Sea level changes: Determination and effects Geophy. *Monograph.69.IUGG Vol.II*, 167-180.

Zlotnicki V., Fu L.-L., Patzert W., 1989 Seasonal variability in global sea level observed with Geosat altimetry *J..Geoph. Res., Vol 94*, N° C12, 17959-17969.

Zanderbergen R.C.A., 1990 Satellite altimeter data processing: from theory to practice *Delft University Press* .

Remote Sensing in the 21st Century: Economic and Environmental Applications, Casanova (ed.)
© 2000 Balkema, Rotterdam, ISBN 90 5809 096 5

Remote sensed image rectification in an automatic way

A. Martínez & F.J. Abad
Sección de Teledetección y SIG, Universidad de Castilla-La Mancha, Albacete, Spain

J. D. García-Consuegra
Departamento de Informática and Sección de Teledetección y SIG, Universidad de Castilla-La Mancha, Albacete, Spain

ABSTRACT: Image registration methods may be used to solve problems such as remote sensed image rectification. In this paper a study is carried out on the possible application of already existent image registration algorithms to automated control point selection and matching. Initially we present existing image registration techniques that broadly fall into two categories: the area-based and the feature-based methods. Feature-based methods are shown as the more suitable for this task. These techniques generally present robustness against stronger geometrical deformations and support multiespectral characteristics. Various experiments are used to show how different algorithms deal with the conditions that typically affect remote sensed images. A methodology for automated image rectification is proposed based on existent registration algorithms and the assumptions derived from experimental results. Due to the flexibility of this methodology, it can be introduced gradually into the remote sensing process, saving time and human resources.

1 INTRODUCTION

Remotely sensed data preprocessing generally refers to geometric and radiometric corrections. Data providers apply different degrees of geometric and radiometric corrections to the raw data. Users normally apply a second preprocessing depending on the data fidelity requirements of the data application such as change detection, classification and remote sensing/geographic information system data integration. This paper is focused on the automation of geometric preprocessing, referenced as image rectification, which has become a more and more important task before image analysis. Remotely sensed data application to agriculture, geology, oceanography, etc. and high spatial sensor resolution resulting from technological developments has lead to algorithms dealing with large amounts of data.

Image rectification is a two-step process. The first step establishes the correspondence between two images and determines the geometric transformation that matches one image with the other. This process is called image registration [1]. In this paper the image to be processed is called a sensed image, which can differ in acquisition time, source sensor and/or view point from a reference image (previously rectified by supervised or orbital methods). A point identified in both images is called a control point. Due to geometric distortions, control point coordinates differ from one image to another. In the second step, using these differences, the rectification algorithm obtains the geometric transformation between images. Generally, a bivariate polynomial function is obtained from control points for estimating new coordinate values of the sensed image. Finally, this transformation is applied to the sensed image, where each pixel in the rectified image is obtained by using an interpolation method over the original image.

The objective of our methodology is to identify control points and their coordinates in both images in an automatic way. These control points can be used alone or joined to other control points obtained in a supervised way in order to calculate the geometric transformation and apply it to the sensed image.

This paper is organized as follows: in the first section a discussion of the existent alternatives for image registration is made, in the second section our methodology is proposed and details are given, in the third section experimental results are presented, and finally the fourth section discusses the conclusions.

2 PRESENT ALTERNATIVES FOR IMAGE RECTIFICATION

2.1 *The geometric transformation step*

In the introduction section we have seen that the rectification process consists of a two-step procedure. Since the second step consists in calculating an approximation function or a surface fitting, it represents a well-known approximation theory problem. The problem for automating this second step is the selection and implementation of the approximation method adapted to the properties of each image each imagery properties (see (Goshtasby 1988) and (Fogel & Tinney 1996) for example). In the last task of this step, the resulting image is generated by means of the approximation function and the interpolation of the original image. This means that digital processing involves the automation of the second step in a relatively simple way.

2.2 *The control point extraction step*

The process of obtaining control points has been traditionally the most expensive task in the rectification process, carried out in a supervised (manual) or semi-automatic manner. Another possible automation of image rectification consists in using orbital models, which unfortunately requires knowledge of sensor status at acquisition time and involves a high implementation complexity and calculation time (Palà and Pons 1995). Control points can be extracted comparing the sensed image against another image, a georeferenced or geocoded image or a map. Image registration arises at this point. Registration techniques have been widely studied and applied where various images of the same scene must be analyzed. They can obtain an image-to-image geometric transformation, extracting control points from it o vice-versa. In those techniques terrain relieve is not taken into account so there is no explicit modeling of terrain effects. Although better results can be achieved using relief (Palà and Pons 1995), the digital elevation model is not always available. Registration techniques can be used satisfactorily when image fidelity in not critical. For some applications, image-to-image rectification using registration may have even less geometric error than rectifying multiple images to a common map projection (Niblack 1986).

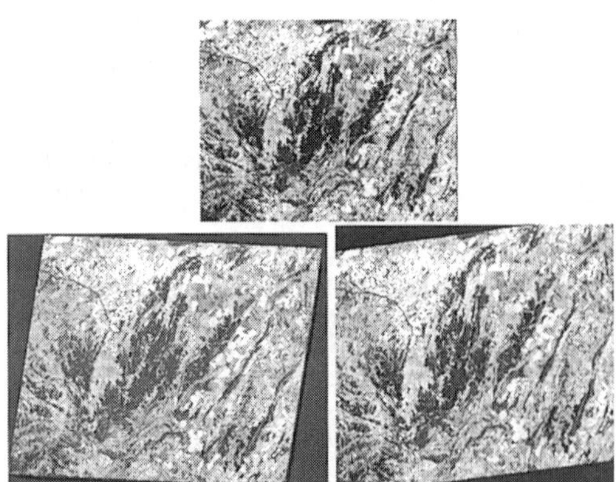

Figure 1. Original image (upper), affine transformation (left) and incorrectly registered result (right) using a correlation approach.

2.3 *Image registration techniques*

Automatic control point extraction generally consists in a two-step process. First, points or significative structures are extracted from each image and two sets of points are obtained. In the second step a correspondence between those points must be achieved in order to generate control points. This step is called image matching or point matching in the relevant literature.

Depending on the method used for point extraction or matching, registration techniques can be classified into two categories: correlation based and symbolic methods (Flusser & Suk 1994).

2.3.1 *Correlation based methods*

These methods are based on a similarity measure between images. Normalized Cross-correlation is the basic statistical approach used for this measure (Dani & Chaudhuri 1995). Other correlation measurements can be applied as in (Van Wie & Stein 1977) based on binary edges or calculated in the Fourier domain as in (Srinivasa & Chatterji 1996). Two dimensional normalized cross-correlation is given by

$$\Phi = \frac{\sum_{(i,j)} a(i+k, j+l) b(i,j)}{\left[\sum_{(i,j)} a^2(i+k, j+l)\right]^{1/2} \left[\sum_{(i,j)} b^2(i,j)\right]^{1/2}} \qquad (1)$$

where $a(i+k, j+l)$ and $b(i,j)$ are pixel values in each image. $(i+k, j+l)$ and (i,j) are all pixel coordinates in the windows where correlation is calculated. A small window of points in the first image is statistically compared with windows of the same size in the second image and the centers of the matched windows are considered control points. As can be seen, a high computational complexity is displayed. Due to the fact that gray-levels of the images are directly used for comparison, sensitivity to gray-level variations of the original images is in evidence so that these methods are not well-adapted to multi-sensor image registration. Correlation is sensible to rotation and scaling.

In Figure 1 an example of the application of the correlation method based on the method described by (Dani & Chaudhuri 1995) shows the inappropriateness of this method for images with general affine transformation. However, it is useful for images that are misaligned by small rigid or affine transformations.

2.3.2 *Symbolic methods*

Symbolic methods do not work directly with gray-level images. Instead, they use scene features such as line intersections, edges and regions. They are well adapted to multisensor or multispectral characteristics since features are used for matching instead of image gray-level and are considered the primary approach in order to register two images whose type of misalignment is unknown (Gottesfeld 1992). Normally, this method supports more geometric transformations models, including common deformations to remote sensing such as horizontal shear or those generated by terrain relief (Brown 1992).

But this technique involves the following conditions:

- Some distinguishable features must appear in both images to be able to obtain a correspondence between them. This is a hard restriction that can not be carried out in images with very few human or natural elements. For example, in Figure 2 a real image is presented where direct application of region-based methods have not obtained sufficient or stable regions to work.
- If the scene contains a high level of repetitive similarity, the extracted features will have similar descriptors that can lead to an erroneous match. For example, an urban or agricultural area can show this problematic. See Figure 3 for the application of a method based on (Flusser & Suk 1994) to an agricultural area that leads to a mismatch.
- In the feature extraction step, the process time can vary depending on the image content and can be very long. Furthermore, as the spatial resolution increases, the sensibility to noise is increased.

Moreover, the application of image registration techniques to large image sizes represents a serious problem (for example, the size (6920 × 5760) of a full frame acquired by the Thematic Mapper sensor on board of Landsat IV and V satellites). A control point matching step is required after feature extraction. A large number of control points, together with the usually high complexity of matching algorithms (Rezende & Lee), involves considerable computing time, increasing mismatch risk because of the greater combination possibilities between points.

Figure 2. An example of Landsat TM scene where features to be extracted are insufficient.

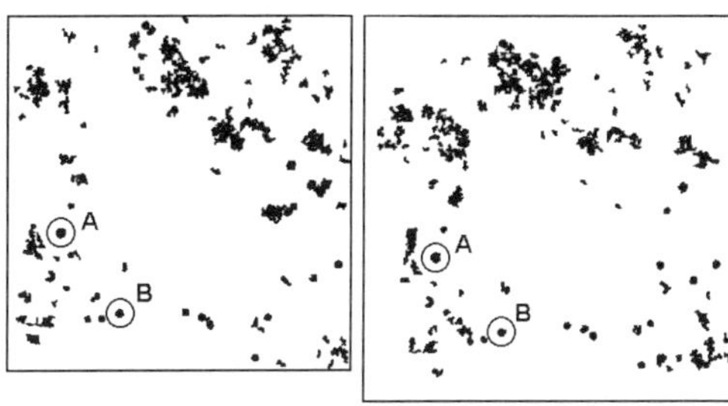

Figure 3. Feature descriptors for regions A and B gives a minimum distance for A and B in reference and test images respectively.

3 PROPOSED METHODOLOGY

Our method has been designed for avoiding problems discussed in the previous section. A mix method has been developed trying to take advantage of each technique type. The method can be resumed in the following steps (see Figure 4):

1 Multi-resolution is applied to the images involved in order to obtain a multi-resolution pyramid.

2 A symbolic or any other method capable of obtaining the geometric transformation between images is applied. This step is carried out at the top of the multi-resolution pyramid so that only an initial approximation is achieved.

3 The initial solution is refined in each lower level of the multi-resolution pyramid. This improvement process is carried out in two ways: cross correlation is used for readjusting existent control points and new ones are extracted in each level using the transformation defined in the upper level.

4 A final set of control points is obtained at the bottom level (original image). The sensed image is finally rectified using these control points.

3.1 *Multi-resolution*

To avoid image size problems, multi-resolution has been used to reduce computation time, keeping the registration accuracy as much as possible. A multi-resolution pyramid is created, making up each new level by averaging values from a 4x4 neighborhood in the previous level (Burt P.J. et al. 1981). A father-son relationship is defined between pixels in adjacent layers. Each pixel in the pyramid is indexed by three variables (i,j,l), where i and j are its row and

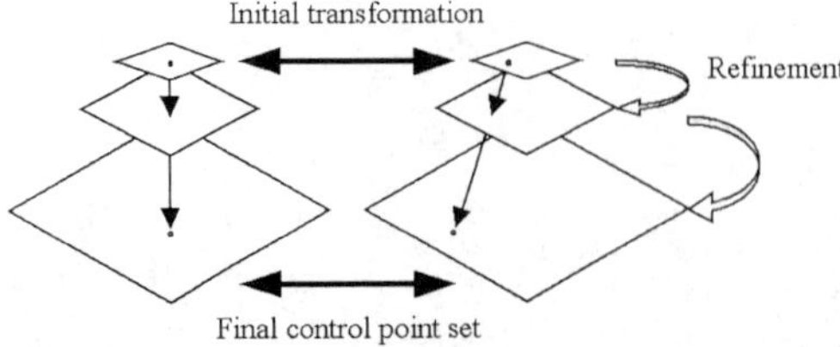

Figure 4. Control points are extracted at the top of the pyramid and refined until final image accuracy is achieved.

column number within that image and l is the level of the image in the pyramid. Therefore, the value of the pixel (i,j,l) is

$$a(i,j,l)=\frac{1}{16}\sum_{x=0}^{3}\sum_{y=0}^{3}a[2i+x2,j+y,l-1] \quad (2)$$

where $a(i,j,l)$ denotes the gray level at a point (i,j) at level l.

The symbolic method applied in the next step takes advantage of noise and image size reduction, showing robustness even when the image does not contain small distinguishable regions at the bottom level of the pyramid as in forest areas. Simpler region detection algorithms can be applied because of resolution reduction smooth textured regions.

3.2 *Initial approximation to the geometric distortion between images*

The feature extraction step supposes the second and more critical stage, which is carried out at the top of this pyramid. At the top, image size is considerably smaller and noise effect is reduced from the bottom to top level of the pyramid, as the averaging process yields blurred layers. This blurring in images can be found contrary to our aim (partial image segmentation), but on the other hand, it is a way of extracting features alone with a high contrast so the features are fewer and more stable.

A symbolic method is used at the top of the pyramid in order to obtain the initial image registration.

First, images are convolved with a LoG operator. Edges are detected at the zero-crossing points, resulting in a binary image. Features are extracted as closed boundary regions, using a region-growing scheme.

Once a set of regions is obtained, feature descriptors are calculated using affine moment invariants (AMI). These are moment-based descriptors of planar shapes, which are invariant under general affine transformation. Full derivation and comprehensive discussion on the properties of invariants can be found in (Flusser & Suk 1993) and a practical application in (Flusser & Suk 1994). If we denote the regions in the sensed image and reference image by S_i (i=1..n) and T_j (j=1..m) respectively, a feature distance d_{ij} between S_i and T_j is obtained from:

$$d_{ij}^2=\sum_{k=1}^{6}\left(AMI^k(S_i)-AMI^k(T_j)\right)^2 \quad (3)$$

where k represents the 6 first moments.

A table of $m \times k$ elements is computed initially containing the k-most similar regions S_i given by equation (3) for each region T_j and not exceeding a pre-set perimeter or area difference thresholds.

Each row in this table represents a possible control point, where centers of gravity are selected as a control point. Each row contains the k best matches ordered by the distance to T_j. A re-sorted version of the table is obtained using a trust index. A quicksort algorithm is employed for ordering table rows taking I_i (4) as the order leader. A simple trust index has been used that tries to measure the column match singularity. If $S_{(i,j)}$ is the region in the row i and column j of the table:

$$I_i=d\left(T_i,S_{(i,1)}\right)-\sum_{j=2}^{k}\frac{1}{2^j}\left(d\left(T_i,S_{(i,1)}\right)-d\left(T_i,S_{(i,j)}\right)\right) \quad (4)$$

Note that in the best case, when $d\left(T_i,S_{(i,j)}\right)$ is near to zero for j=2..k the result tends to zero. Whilst in the worst case, when $d\left(T_i,S_{(i,1)}\right)=d\left(T_i,S_{(i,j)}\right)$ for j=2..k, the result is the feature distance $I_i=d\left(T_i,S_{(i,1)}\right)$.

The algorithm for point matching consists basically in an exhaustive search of correct matches in the image space. This computationally expensive task can be carried out because the image size and region number is reduced at this level of the pyramid, together with the advantages of the sorted table. On the other hand, we made an exhaustive search that can report a correct result even with a limited number of features. A search criterion has been developed that operates as follows:

1. Three control points from the first columns are selected as true matches from the top to the bottom of the table.
2. The affine geometric transformation defined by those three control points is calculated.
3. Control points in which distance from the original control point to calculated projection is less than a threshold value are considered correct matches. If the number of correct matches is greater than a threshold (a preset percent of m), control points can be considered a correct match.

True control points are generally positioned at the top of the table, so in step 1 two real control points are quickly selected and a short search time is usually needed. The existence of three true control points with a minimum feature distance (assigned in the first column) is the only condition for algorithm operability. The algorithm can fall into a mismatch if a low threshold is selected for step 3 and will not be able to find a solution if the threshold is higher than the number of regions common to both images. An estimation of the number of corresponding regions from the total extracted must be carried out for a correct selection of this threshold (Donoho & Huber 1983).

A final refinement process is applied consisting in an iterative process whereby a linear least squares solution is used to compute affine transformations, adding control points that agree with it until no new control points are added.

In the worst case, manual or semi-automatic control point can be extracted when available registration methods can not achieve a correct result. In any case, the human effort required will be minimal.

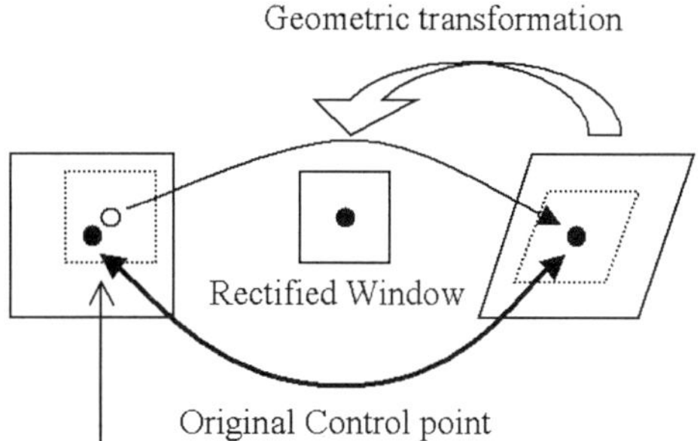

Figure 5. In the refinement process, control points are readjusted using a rectified window correlation.

3.3 *Obtaining a finer solution*

The results of the previous section are a set of control points that establish an approximately correct transformation between sensed and reference image. At this point, a refinement process can be applied in order to obtain accurate results at the bottom of the pyramid, given that a correct match has been calculated at the top level. Our method deals with the idea that control points can be extracted using the geometric transformation defined in the pyramid level immediately above. At the top level, images can have a geometric distortion up to affine degree due to the nature of the symbolic registration algorithm, although more complex deformations are tolerated. At lower levels, control points may define a non-linear geometric distortion. Global polynomials are used in order to calculate an approximation to the image deformation in each level. This approximation does not have to be very accurate because our aim is not the rectification of the image, but the estimation of the control points at each level and their local adjustment. Each control point is recalculated at each level detecting peaks of normalized cross-correlation between the reference image and the rectified image window surrounding control points in the sensed image (see Figure 5). Any control points can be discarded if a sharp correlation peak is not found as discussed in (Althof et al. 1997). This method yields acceptable results and a low computational cost. Other points can be added to the refinement process. For example, a coordinate can be selected randomly and a control point obtained from the general transformation. Then, the local window correlation procedure can be used in order to obtain a new accurate control point.

3.4 *Image rectification*

A function relating coordinates between images is calculated using a global or local approximation method to the geometric transformation defined by a final set of control points. This set can be make up of the automatically obtained control points, others extracted in a supervised way if there are no available rectified images covering all sensed scene, or control points obtained by another procedure. Finally, the sensed image is rectified using this transformation function and the nearest neighborhood rule or other interpolation method.

4 EXPERIMENTAL RESULTS

4.1 *Validation methodology*

In order to validate the accuracy of the methodology, the method must be applied on two images with a known geometric deformation. In our case, two types of tests have been used: reference image and sensed image rectified by using a real control point set extracted by a human operator; and secondly, deformations produced by randomly generated control points. Control points are used in order to be able to obtain the error relative to the control points set obtained by the automatic way. The root mean squared error (RMSE) is obtained between the projection of control points defined by the geometric transformation used in the generation of the test image and the same control point but projected by the geometric transformation using the automatically obtained set of control points. We will have finally a set of points in the sensed image with two different projections in the reference image, from which differences we can calculate an RMSE value.

4.2 *Test with real images*

For the example, a floating quarter scene acquired by the Landsat TM satellite has been used. The sensed image is 2945 rows and 3600 columns. After manually detecting a total number of 50 control points the image was rectified using a second order global polynomial approximation and a RMSE less than one pixel. This rectified image was used as the reference image.

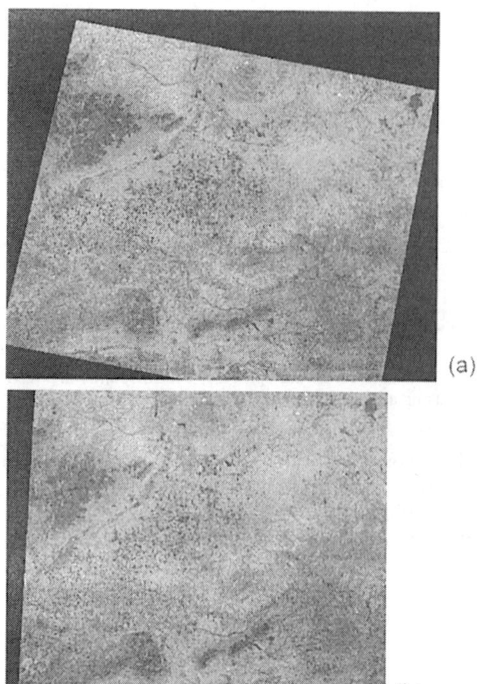

Figure 6. Reference (a) and sensed (b) image (floating quarter scene, Landsat TM) used in the example.

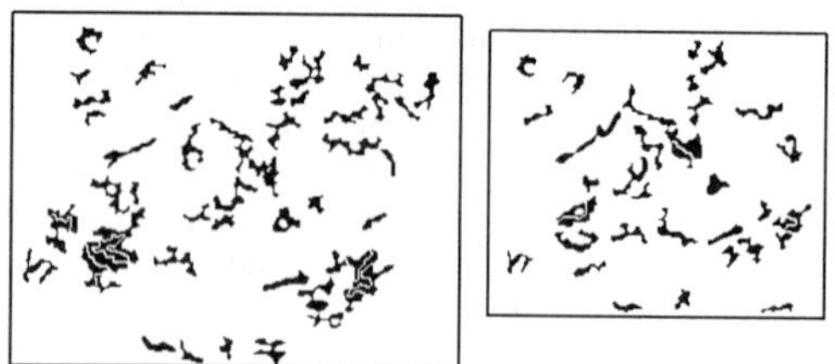

Figure 7. Extracted regions in upper level of the pyramid.

Multi-resolution was applied, creating a 5 level pyramid, until upper level reached a row or column size less than 512 pixels. Top level images were convolved by the LoG operator and a total of 54 and 36 regions were extracted after close boundary detection using the region growing scheme (see Figure 7).

In the image registration step a total of 25 control points were extracted. Using a grid, a total number of 400 points in the sensed imaged were added as control point candidates. 46 control points were discarded during the refinement process due to non-relevant correlation peak detection, so finally 379 control points were extracted by the automatic way.

A final RMSE of 1.68 was obtained from the comparison of original and obtained control point sets.

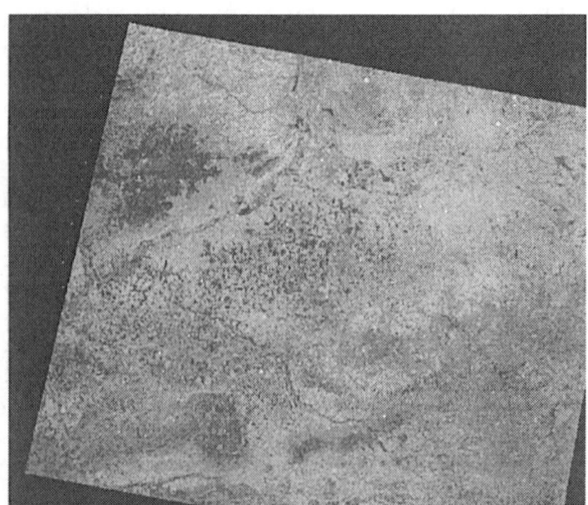

Figure 8. Final rectified image and control points used for the geometric transformation.

4.3 *Test with synthetic images*

For the example, a portion of a Landsat TM image has been used. The sensed image is 600 rows and 600 columns. Control points were generated in order to obtain a rotated image. Each control point is modified adding a random noise so that more complex deformations are produced. The image was rectified using a 3 order global polynomial approximation. This rectified image was used as the reference image.

Multi-resolution was applied, creating a 2 level pyramid, until the upper level reached a row or column size less than 512 pixels. Top level images were convolved by the LoG operator and a total of 124 and 114 regions were extracted after close boundary detection using the region growing scheme (see Figure 10).

In the image registration step a total of 40 control points were extracted. Using a grid, a total number of 400 points in the sensed imaged were added as control point candidates. 207 control points were discarded during the refinement process due to non-relevant correlation peak detection, so finally 239 control points were extracted by the automatic way.

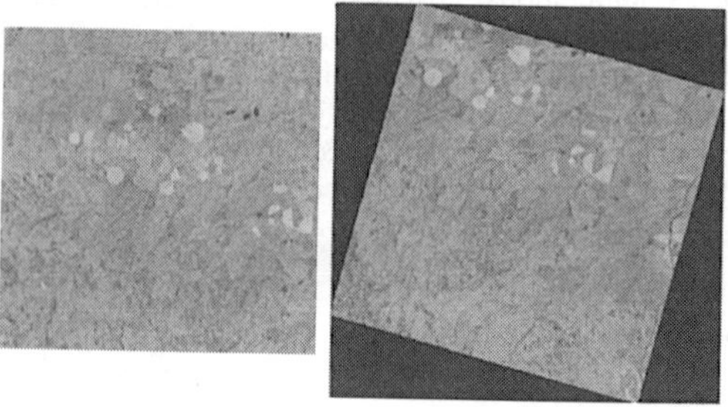

Figure 9. Reference and sensed image used in this test.

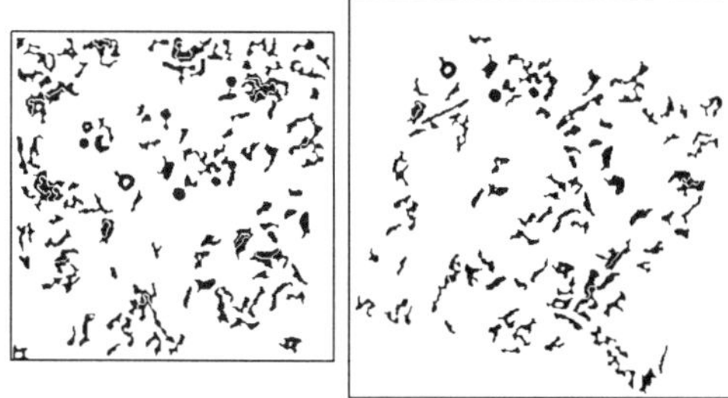

Figure 10. Extracted regions in upper level of the pyramid.

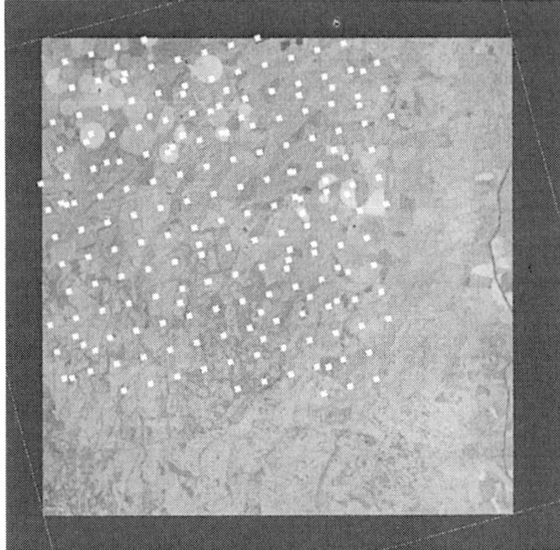

Figure 11. Final rectified image and control points used for the geometric transformation.

A final RMSE of 2.43 was obtained from the comparison of original and obtained control point sets.

5 CONCLUSIONS

Present alternatives for image registration applied to remote sensing image rectification has been presented. Usable techniques have been drawn and a method has been developed taking advantage of each technique. The method has been shown to be scalable and applicable even when the initial registration algorithm fails with minimum human assistance. The refinement process is accurate enough for rectification of images in applications that do not require important precision. All this gives a method that can be introduced flexibly in the rectification process with a notable reduction of operation time.

REFERENCES

Althof R.J, Wind M.G.J. and Dobbins J.T. 1997. A Rigid and Automatic Image Registration Algorithm with Subpixel Accuracy. *IEEE Transactions on Medical Imaging*. vol. 16, no. 3: 308-316

Burt P.J. et al. 1981. Segmentation and estimation of image region properties through co-operative hierarchical computation. *IEEE Transactions on Systems, Man and Cybernetics*. Vol. SMMC-11(12): 802-809.

Dani, P., & Chaudhuri S. 1995. Automated assembling of images: image montage preparation. *Pattern Recognition* Vol 28, No 3: 431-445

Donoho D.L. and Huber P.J. 1983. The notion of breakdown point. In P.J. Bickel, K. Doksum and J.L. Hodges, Jr., editors, A Festschirift for Erich L. Lehman. Wadsworth, Belmont, California: 157-184.

Flusser J. and Suk T. 1993. Pattern recognition by affine moment invariants. *Pattern Recognition*. vol. 26, pp. 167-174

Flusser J. & Suk T. 1994. A Moment-Based Approach to Registration of Images with Affine Geometric Distortion. *IEEE Trans. On Geoscience and Remote Sensing*. Vol 32 no 2 pp 382-387

Flusser J. and Suk T. 1994. A Moment-Based Approach to Registration of Images with Affine Geometric Distortion. *IEEE Transactions on Geoscience and Remote Sensing*. vol. 32, no. 2: 382-387

Fogel, D.N., & Tinney, L. 1996. Image registration using multiquadric functions. Technical Report Number 96-01, National Center for Geographic Information and Analysis, University of California, Santa Barbara.

Goshtasby A. 1988. Registration of images with geometric distortions. *IEEE Trans. Geosci. Remote Sensing*. 26: 60-64

Gottesfeld, L. 1992. A Survey of Image Registration Techniques. *ACM Computing Surveys*. vol. 24, no. 4.

Niblack, W. 1986. An introduction to digital image processing. Englewood Cliffs, N.J. Prentice-Hall International.

Palà V. & Pons X. 1995. Incorporation of Relief in Polynomial-Based Geometric Corrections. *Photogrammetric Engineering & Remote Sensing*. Vol 61, No 7: 935-944

Rezende P.J. & D.T. Lee 1995. Point set pattern matching in *d* dimensions. *Algorithmica*. Vol. 13: 387-404

Srinivasa B. & Chatterji B. 1996. An FFT-Based Technique for Translation, Rotation, an Scale-Invariant Image Registration. *IEEE Transactions on Image Processing*. vol. 5. no. 8. pp.1266-1271.

Van Wie P. & Stein M. 1977. A Landsat digital image rectification system. *IEEE Trans. Geosci. Electro*. Vol GE-15: 130-137.

Remote Sensing in the 21st Century: Economic and Environmental Applications, Casanova (ed.)
© 2000 Balkema, Rotterdam, ISBN 90 5809 096 5

Architecture of a new generation of remote sensing ground segments

Klaus Seidel
Image Science, Remote Sensing, Communication Technology Laboratory ETH, Zürich, Switzerland

Mihai Datcu
German Aerospace Centre, DLR, German Remote Sensing Data Centre, DFD, Oberpfaffenhofen, Wessling, Germany

ABSTRACT: Due to the huge volume and high complexity of remote sensing data, the management of image archives demands high level representation of information. Augmentation of the data with meaning is a key issue for the success of query processes, and has to include the model of user's understanding. The information retrieval process is split up into two steps: objective information extraction and semantic representation. These concepts are applied for the design of a new generation of remote sensing ground segment systems.

1. INTRODUCTION [1]

Modern communication systems are managing huge volumes of data, like images, voice/sound, video, or results of simulations. Applications are in all major fields of science; e.g. medical image and data archives, integrated multimedia systems. However, the field of remote sensing is faced with very high volumes of data and information. The main problems encountered in the design of such systems are data and information representation, coding and understanding, and the adaptation to the user conjecture.

We propose an advanced remote sensing ground segment architecture designed for easier and interactive access of the information and a broader dissemination of remote sensing data. The exchange of information in the system can be understood as *image communication*, i.e. the transfer of information using image content characteristics and visualization techniques.

This article presents the concept developed at the German Aerospace Center, DLR Oberpfaffenhofen in collaboration with the Swiss Federal Institute of Technology, ETH Zurich, for information retrieval from remote sensing (RS) data. The research line has as very pragmatic goals the design of future RS ground segment systems permitting fast distribution and easy accessibility to the data, real and near real-time applications, and to promote the implementation of decision support systems.

1. The text has been derived from a similar presentation at the DFD Nutzerseminar, in München, October 14-16, 1998

2. PRESENT STATUS IN THE FIELD OF IMAGE CONTENT REPRESENTATION

Starting in 1993 the leading groups in the field of query by image content from image archives, were Media Lab at MIT, IBM Almadem and U.C. Santa Barbara (Swanberg and Jain, 1994, Gupta et al., 1991, Pentland et al., 1994).

The first journals dedicated to image content representation and related topics in data mining and information visualization, have been published in the period from 1995-1997:

- a special issue on data bases (IEEE Computer, 1995)
- a special issue on data mining (IEEE Knowledge and Data Engineering, 1996)
- a special issue on digital libraries (IEEE PAMI, 1996)
- a special issue on visualization (IEEE Computer Graphics, 1997)
- a special issue on image data bases (Pattern Recognition, 1997)

In the meantime the growing interest in managing very-large distributed information systems is reflected also in the progress of image archiving technology. It is now difficult to enumerate the groups involved in this field. A measure of the progress is the number of demonstrators available on different web sites. A synopsis can be found i.e. http://www.cs.cmu.edu/afs/cs/project/cil/ftp/html/v-demos.htm.

Almost all of the previously enumerated systems rely on the integration of elements from image recognition in the query system. Not so many groups are involved in investigations of knowledge representation and perception for adaptive query specifications (Cox et al., 1996). Commercial or operational systems are not so frequent and are based on very simple principles (Barber et al., 1994, Finn, 1996).

The operational applications in Remote Sensing data management are still based on "classical" technologies based on inventories and catalogues indexed by geographical location, sensor type or time of acquisition. The table from the web site http//: www.vision.ee.ethz.ch/ lists several of the operational systems.

Recently, IBM in collaboration with NASA designed the "Satellite Image Explorer". The system is based on a multiresolution approach for finding similar areas of coregistered images and for finding similar patterns within the same image. Presently the emphasis is on developing real-life applications of the previously developed techniques. A description of the project can be found on the web site http://www.research.ibm.com/networked_data_systems/9auc/.

Similar systems, both for query by image content and data mining are under development at NEC, NASA-JPL, and NASA-Ames Research Center.

3. DATA AND INFORMATION RETRIEVAL

The next sections present an operational system in exploitation at DFD, and the results of two years of research of DFD in collaboration with ETH Zurich.

3.1 Conventional satellite data ground segments

Conventional satellite data ground segments comprise sensor mode planning and data acquisition systems, data ingestion interfaces, processing capabilities, a catalogue of available data, a data archive, and interfaces for queries and data dissemination. The transfer of data mostly relies on common electronic networks or high capacity tapes. DFD's Intelligent Satellite Data Information System (ISIS), provides data catalogue search and retrieval. In order to support a variety of hardware platforms ISIS operates different interfaces and is available via common communication lines. The graphical interface GISIS provides the most comfortable access to the data of the DFD catalogue offering the following services: distributed catalogue retrieval from more than 200.000 data entries with about 150.000 digital images, supported by a map browser and geographical names; display of data-set footprints on a map; automated network transfer of digital quick-looks (compressed images) for visual inspection; transfer of full resolution data (full or interactively defined subsets) for some data products; guide information about sensors, data products and distribution modalities; on-line ordering; on-line placement of user comments. The system can be accessed from http://isis.dfd.dlr.de/.

3.2 Future ground segments

Future ground segments have to be capable of handling multi-mission data acquired by SAR and optical sensors and have to provide easy user access to support the selection of specific data sets, fuse data, visualize products and compress data for transmission via Internet. In particular, the search for data sets has to support individual queries by data content and detailed application areas as well as capabilities for automated extraction of relevant features and the application oriented representation of results. In the case of high resolution SAR and optical images, we face an enormous volume of data combined with very specific analysis requests posed by the users. In order to reconcile these conflicting aspects we suggest the development of tools and a scalable system for information retrieval (Seidel et al., 1997):

a) query by image content from large image archives,
b) data mining in the set of images obtained as a query response, and
c) scene understanding from the images found to be relevant for the user's application.

The diagram in Fig. 1 presents the system concept.

The three levels of the scalable system are designed to accommodate the huge volume of data containing high complexity structures. They perform a progressive data reduction with simultaneous increase of the extracted image detail.

3.2.1 Query by image content

The goal of query by image content is to enable an operator to *see* into a large volume of high complexity images. The incorporation of content-based query and retrieval techniques into the RS image archives will meet the requirements expressed by today's

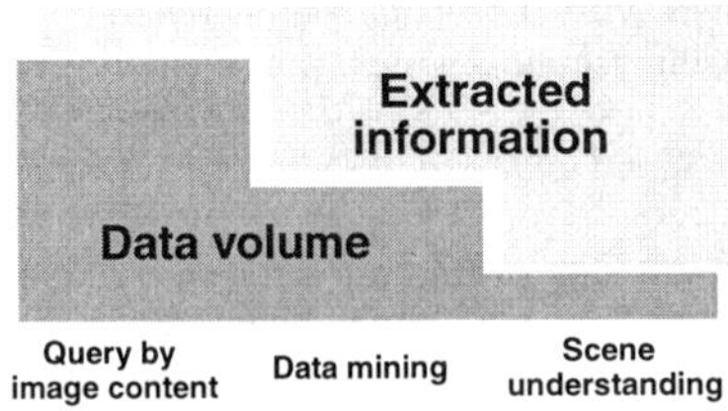

Fig. 1: Diagram of the scalable system for information retrieval from very large image archives. Due to the problem complexity the system scales down the amount of data to be analysed such that the detail of extracted information can be increased.

users and at the same time lay foundations for the efficient handling of the huge data-flow that is expected from the next generation of satellites. Drawing the connection from the image signal to the image content is the key step in providing an intelligent query from large RS data bases, thus making the user able to *see* into the data base before actually retrieving data.

The *design concept* for query by image content in large image archives aggregates three components (Benz et al., 1998, Seidel et al., 1998b, Rehrauer et al., 1997):

i) image content representation and definition of similarity functions,
ii) a multiscale approach with two goals, retrieval at different image scales and fast progressive discarding of non-interesting images,
iii) automatic re-clustering of image archive according to the user query.

The *elaboration of the system* is based on a Bayesian approach for the hierarchic information representation in large image archives (Schröder et al., 1997, Datcu et al., 1998, Schröder et al., 1998, Seidel et al., 1998a):

i) image signal modelling and feature extraction
ii) hierarchic clustering in the joint model&feature space of the image archive,
iii) semantic analysis of the query and user conjecture modelling.

The method aims at obtaining a better selectivity of the query process for different subjective user requests, and it is a basis for the design of optimal multi-dimensional signal representation in large over-the-net distributed archives.

The diagram in Fig. 2 presents the integration of the query by image content subsystem with the architecture of a conventional remote sensing ground system.

At ingestion in the physical archive the image content is extracted and presented in form of signal features and stored in the inventory. The indexing mechanism of the data base management system allows us to cluster together all images containing *similar* content. The *similarity* is defined adaptively depending on the user conjecture, i.e. the hypothesis he uses in a certain application.

Many of the RS image interpretation techniques are oriented to the extraction of radiometric or multispec-

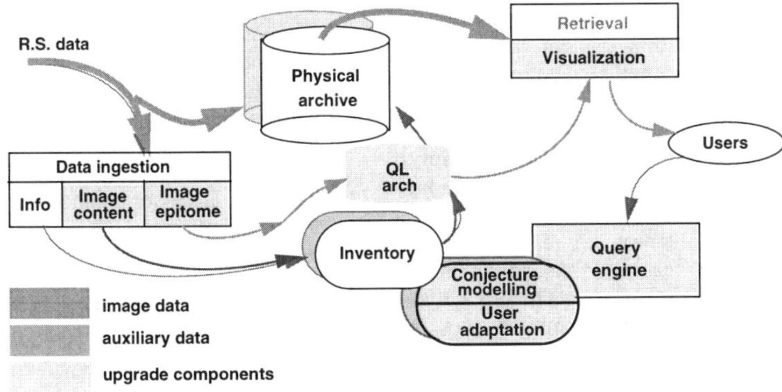

Fig. 2: Integration of the query by image content subsystem with the architecture of a conventional remote sensing ground system. At ingestion in the physical archive the image content is extracted and presented in form of signal features and stored in the inventory. The indexing mechanism of the data base management system allows us to cluster together all images containing *similar* content. The *similarity* is defined adaptively depending on the user conjecture, i.e. the hypothesis he uses in a certain application.

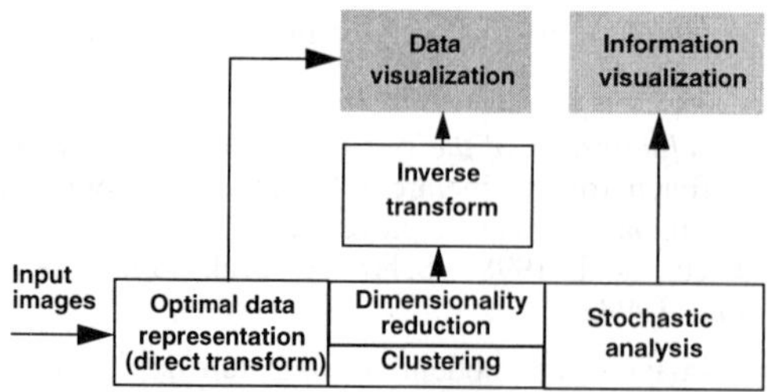

Fig. 3: Block diagram of a data and information mining tool. The tool allows the interactive exploration of the set of images obtained as a query response. Both data and information visualization methods will be used after interactive search of images relevant in a certain application.

tral features. These approaches are sometimes hindered by the limited information available in the observations, e.g. by the small number of spectral channels. Additional information can be extracted using spatial (structural) analysis. We put emphasis on structural information extraction as texture features (Datcu et al., 1998, Schröder et al., 1998).

The concepts have been verified on a test system including query specification using a 7 words vocabulary and based on three types of features: multispectral (Healey and Jain, 1996), texture (Schröder et al., 1997), and image complexity. A demonstrator is available on the web site http//:www.vision.ee.ethz.ch/~rsia/rsia2.html (Seidel et al., 1998c).

3.2.2 Data mining

The expected answer of a query is a set of images presumably containing similar structures. Further refinement of the information content in the now reduced data set, can be obtained by data mining. The goal is to explore the information content of the images and decide which ones are relevant for the user's application. The tool we propose is mainly based on interactive decision using data visualization, e.g. image browsing integrated with image segmentation, or information visualization, e.g. exploration of the cluster structure in the feature space and the interactive design of the decision surfaces. The block diagram of the data and information mining tool is presented in Fig. 3.

The first experiments led to the development of the image epitome: integrated image restoration, feature selection and compression in the wavelet transformed domain. The image epitome is one of the tools for mining in image data sets (Datcu et al., 1997b).
The data and information mining tool will be integrated with the visualization module of the remote sensing ground segment presented in Fig. 2.

3.3 Scene understanding

The last step of highest detail information extraction is scene understanding. Scene understanding is the attempt to extract information about the physical characterization, the structure and geometry of the three-dimensional scene from a two-dimensional image which is explained elsewhere (Datcu, 1996, Datcu et al., 1998). The task of the scene understanding process is to find the scene which best explains the observed data. Scene understanding is an ill-posed inverse problem. The solutions we adopt are model based formulations in the general frame of Bayesian inference. The implemented algorithms allow accurate texture segmentation under adverse noise conditions, very high spatial and radiometric detail extraction from remote sensing images, super-resolution of targets, and surface reconstruction, automatic segmentation and target detection in mountainous areas (Datcu and Holecz, 1993, Datcu et al., 1997a, Walessa et al., 1998).

4. ASSESSMENT OF THE NEW IMAGE INFORMATION EXTRACTION TECHNOLOGY

The classical task in the interpretation of remote sensing data generally assumes that the source of information is just *one* image. The methods applied for information extraction are image enhancement, image segmentation, feature extraction, fitting physical models to the data, etc. A more accurate interpretation of remotely sensed scenes is obtained by synergistically combining signals from different sources, e.g. fusion of data acquired from different sensors or multitemporal analyses. One of the difficulties in many data fusion applications for remote sensing image classification is the identification of the data sets relevant for the user. Here the information retrieval becomes a data and information mining task.

The explosion in sensor technology, both high resolution and frequent repeat passes, results in an increasing number of large multimission remote sensing archives, e.g. DFD stores more than 80 data product types adding up to more than 200.000 images and increasing in 3 years up to 1.500.000 images! Thus, the problem of image content extraction should be reformulated such to take into consideration the *new source of information*: the image archive. The state of the art remote sensing archives are managed by relational data base management systems able to query for the data with no reference to the information content, except the visual browsing using quick-look images. This technology is limited to applications involving a reduced number of images and also the accuracy is restricted by the poor quality of the quick-look images.

The query by image content methods we develop are intended to overcome the informational bottle-neck of classical approaches and also to stimulate the user in finding new scenarios for data interpretation, e.g. *find all images containing cities surrounded by forest.* It is known that the distinction between the perception of information as signals and symbols is generally not dependent on the form in which the information is presented but rather on the conjecture in which it is perceived, i.e. upon the hypotheses and expectations of the user. Thus, the new technology requires a different attitude of the user and also a careful design of an intelligent, adaptive query engine.

5. SUMMARY

The project is a continuation, extension and reinforcement of existing research activities in the fields of data mining, scene understanding and query by image content. We plan to pursue advanced interdisciplinary research in the areas of data visualization, information visualization in very close synergy with latest developments in signal analysis and information theory. We expect a new field of expertise: *visual data mining*, i.e. discovering new, unexpected information in image archives. This is in continuation of activities in the field to query by image content which aims at finding in an image archive the images which contain predefined features (of interest in certain applications).

6. ACKNOWLEDGMENT

The authors would like to thank Gottfried Schwarz for fruitful discussions. In addition, we would like to acknowledge the contributions from Hubert Rehrauer and Michael Schroeder especially for making available the demonstrator. The project is supported by the ETH Research Foundation (project: 41-2637.5).

7. REFERENCES

Barber, R., Flickner, M., Hafner, J., Lee, D., Niblack, W., Petkovic, D., Ashley-J, J., McConnell, T., Ho, J., Jang, J., Berkowitz, D., Yanker, P., Vo, M., Haas, D., Lassig, D., Tate, S., Chang, A., van Houten, P., Chang, J., Petersen, T., Lutrell, D., Snedden, M., Faust, P., Matteucci, C., Rayner, M., Peters, R., Beck, W., and Witsett, J. (1994). Ultimedia manager: Query by image content and its applications. In *Digest of Papers. Spring COMPCON 94*, San Jose, CA, USA. IBM Almaden Res. Center.

Benz, U., Datcu, M., Schwarz, G., and Seidel, K. (1998). Image communication: New features of a remote sensing data ground segment. In *CEOS SAR Workshop, ESTEC.*

Cox, I. J., Miller, M. L., Omohundro, S. M., and Yianilos, P. N. (1996). PicHunter: Bayesian relevance feedback for image retrieval. In *Int. Conf. on Pattern Recognition, Vienna*, volume 3, pages 362–369.

Datcu, M. (1996). Scene understanding from SAR images. In Stein, T. I., editor, *Remote Sensing for a Sustainable Future, IGARSS'96*, volume 1, pages 310–314.

Datcu, M. and Holecz, F. (1993). Generation of synthetic images for radiometric topographic correction of optical imagery. In Chavez, P. S. and Schowengerdt, R. A., editors, *SPIE Proceedings on Recent Advances in Sensors, Radiometric Calibration and Processing of Remotely Sensed Data, Orlando, USA*, volume 1938, pages 260–271. SPIE OE/Aerospace Sensing.

Datcu, M., Luca, D., and Seidel, K. (1997a). An inverse problem approach for the segmentation of the snow cover in satellite images. In Gudmandsen, P., editor, *Future Trends in Remote Sensing, 17th EARSeL Symposium 1997, Lyngby, Denmark*, pages 321–328. A. A. Balkema Rotterdam/Brookfield.

Datcu, M., Schwarz, G., and Walessa, M. (1997b). Wavelet Feature Coding for Quicklook Synthetic Aperture Radar Images – An Image Epitome. In *Proc. of EOS/SPIE, The European Symposium on Aerospace Remote Sensing-Sept'97, London*, volume 3217, pages 58–66.

Datcu, M., Seidel, K., and Walessa, M. (1998). Spatial information retrieval from remote sensing images—Part I: Information theoretical perspective. *IEEE Trans. on Geoscience and Remote Sensing*, 36(5):1431–1445.

Finn, R. (1996). Query by image content. *IBM Research*, (3):22–25.

Gupta, A., Weymouth, T. E., and Jain, R. (1991). An extended object oriented data model for large image data bases. In Günther, O. and Schek, H.-J., editors, *Advances in Spatial Databases, 2nd Symposium SSD'91, Zuerich*, pages 45–61. Springer-Verlag.

Healey, G. and Jain, A. (1996). Retrieving multispectral satellite images using physics-based invariant representations. *IEEE Tr. PAMI*, 18(8):842–848.

IEEE Computer (1995). IEEE Computer - Special Issue on Databases. Journal.

IEEE Computer Graphics (1997). IEEE Computer Graphics and Applications - Special Issue on Visualization. Journal.

IEEE Knowledge and Data Engineering (1996). IEEE Transactions on Knowledge and Data Engineering - Special Issue on Data Mining. Journal.

IEEE PAMI (1996). IEEE Transactions on Pattern Analysis and Machine Intelligence - Special Issue on Digital Libraries. Journal.

Pattern Recognition (1997). Pattern Recognition - Special Issue on Image Databases. Journal.

Pentland, A., Picard, R. W., and Sclaroff, S. (1994). Photobook: Tools for content-based manipulation of image databases. In Niblack, W. and Jain, R. C., editors, *Storage and Retrieval for Image and Video Databases II*, volume 2185, pages 34–47. SPIE.

Rehrauer, H., Seidel, K., and Datcu, M. (1997). Multiscale Markov random fields for large image datasets representation. In Stein, T. I., editor, *A Scientific Vision for Sustainable Development, IGARSS'97*, volume 1, pages 255–257.

Schröder, M., Rehrauer, H., Seidel, K., and Datcu, M. (1998). Spatial information retrieval from remote sensing images—Part II: Gibbs Markov random fields. *IEEE Trans. on Geoscience and Remote Sensing*, 36(5):1446–1455.

Schröder, M., Seidel, K., and Datcu, M. (1997). Gibbs random field models for image content characterization. In Stein, T. I., editor, *A Scientific Vision for Sustainable Development, IGARSS'97*, volume 1, pages 258–260.

Seidel, K., Mastropietro, R., and Datcu, M. (1997). New architectures for remote sensing image archives. In Stein, T. I., editor, *A Scientific Vision for Sustainable Development, IGARSS'97*, volume 1, pages 616–618.

Seidel, K., Schröder, M., Rehrauer, H., and Datcu, M. (1998a). Meta features for remote sensing image content indexing. In *IEEE Intern. Geoscience and Remote Sensing Symposium IGARSS'98*, volume 2, pages 1022–1024.

Seidel, K., Schröder, M., Rehrauer, H., and Datcu, M. (1998b). Query by image content from remote sensing archives. In *IEEE Intern. Geoscience and Remote Sensing Symposium IGARSS'98*, volume 1, pages 393–396.

Seidel, K., Therre, J.-P., Datcu, M., Rehrauer, H., and Schröder, M. (1998c). Advanced query and retrieval techniques for remote sensing image databases (RSIA II and III). Project Description Homepage. http://www.vision.ee.ethz.ch/~rsia/rsia2.html.

Swanberg, D. and Jain, R. (1994). Knowledge guided parsing in video data bases. In *Proceedings of IS&T/SPIE Symposium on Electronic Imaging: Science and Technology*, page 214. IS&T/SPIE.

Walessa, M., Nico, G., and Datcu, M. (1998). Precise information retrieval from SAR and INSAR images of rough terrain: A Bayesian approach. In *SPIE Europto Barcelona '98*, volume 3497. SPIE.

Remote Sensing in the 21st Century: Economic and Environmental Applications, Casanova (ed.)
© 2000 Balkema, Rotterdam, ISBN 90 5809 096 5

'Supervised ART-II' neural network for Landsat TM image classification

Kamal R. Al-Rawi, Consuelo Gonzalo & Estibaliz Martínez
Departamento de Tecnología y Sistema Informática, Facultad de Informática, Universidad Politécnica de Madrid, Spain

ABSTRACT: A new system for automatic classification of Landsat TM, based on Supervised ART-I and Supervised ART-II, has been developed. The sensitivity of the system has been tested for all domains of dynamic parameters: the vigilance parameter ρ [0,1], and the dynamic learning parameter β [0, 1]. This has been achieved due to the simple architecture of Supervised ART-I and Supervised ART-II. The system has been applied for classifying a scene (256 x 240 pixels) of a Landsat TM image. The scene corresponds to the area around the Spanish city "Talavera de la Reina". Training the data with 9000 exemplars, using ρ=0.98 and β=0.50, leads to 85.82% classification accuracy, for the remaining 52,440 pixels, for all the thirteen classes of the scene. The training times are 3.79 and 2.99 minutes for Supervised ART-I and Supervised ART-II, respectively on SUN 4 SPARC Station. Classification accuracy can be increased, using the same training set, by increasing the vigilance parameter and the dynamic learning rate. However, the learning and classifying time will be longer, but still in near real-time.

INTRODUCTION

Mapping land-cover, using remotely sensed data, is a very active area of research, due to the advances in space and computer technology (Benediktsson *et al.* 1990). Conventional classification is usually employed for this task. However, neural networks are heavily used in the last decade. The main advantages of neural networks over conventional classifiers such as Maximum Likelihood Classifier (MLC) are that: 1) They are non-parametric, therefore, the probability distributions for each class are not required. This allows us to introduce ancillary data (slope, topography, aspect, etc.), in addition to the spectral data to the network, which many authors reported that can increase the classification accuracy (Carpenter *et al.* 1997, Benediktsson *et al*, 1990). Moreover, neural networks are more robust when the distribution is not gaussian (Paola and Schowengerdt 1997, Hepner *et al.* 1990). 2) Unlike conventional classifiers, neural networks are able to manage fuzzy classifications (Yool 1998, Paola & Schowengerdt 1997, Wamer & Shank 1997). The numbers in the output represent the strength of the classes membership of the specific input. This is very important when we deal with low spatial resolution. 3) The parallel feature of a neural network allows us to increase the speed of the classification process. This can be done by implementing it on parallel computers (Salu & Tilton 1993, Heerrnann & Khazenie 1992). 4) The flexibility of neural networks for classification improvement (Carpenter *et al.* 1997). 5) It has the ability for establishing arbitrary decision boundary (Paola & Schowengerdt 1995, Tzeng et al. 1994).

Multi-Layer Perceptron (MLP), with back propagation learning, is the most commonly used neural network in the literature to classify remotely sensed data. This is due to the preferable learning approach of the network, which is based on minimising the error between the output of the network and the target value. While some authors have reported that conventional classifiers perform better than MLP (Solaiman & Mouchot 1994, Mulder & Spreeuwers 1991), many authors have reported that MLP perform better than MLC in classifying remotely sensed data (Paola & Schowengerdt 1994, Yoshida & Omatu 1994, Heerman & Khazenie 1992, Hepner *et al.* 1990). The classification performance of MLP can be improved by using ancillary data in addition to the spectral data (Benediktsson *et al.* 1990). However, employing MLP as a classifier causes many problems. The architecture of the network is not fixed. The number of hidden layers and the number of nodes in each hidden layer must be determined by trial and error. This is a very costly process bearing in mind the long training time of the network. In addition to the fact that MLP might fall in a local minima during the training phase. Moreover, MLP might not converge. Using a low learning rate to avoid the convergence problem makes the long training time of the MLP network much longer. (Heermann and Khazenie 1992) suggested using parallel computers to reduce the training time. This reduces the training time but increases the hardware cost.

For classification of a Landsat image, (Carpenter *et al.* 1997) reported that, MLP did not converge, using learning rate=0.6 and momentum rate=0.4, after 212 minutes of training time on a SUN 4 SPARC Station, using 100,000 input presentations. They employed a lower leaming rate to avoid the convergence problem. The training time exceeded 1000 minutes, while the classification accuracy was less than 27%. They reported that Fuzzy ARTMAP (Carpentar *et al.* 1992) perform better classification accuracy than MLP, with lower training time. They reported also that Fuzzy ARTMAP and MLC perform the same level of classifying accuracy. Fuzzy ARTMAP has been employed also by (Mannan *et al* 1998) to classify (512 x 512) pixels, of an image of Linear Imaging Self-scanning Sensor (LISS-II) of Indian Remote Sensing Satellite (IRSIB), for their 13 classes. They reported that Fuzzy ARTMAP performs better than both MLC and MLP in classification accuracy. The average classification for six data sets are 84.7%, 80.3%, and 79.9% for Fuzzy ARTMAP, MLC, and MLP, respectively. They reported that the training times were slightly less than those for MLC, but many times faster than MLP.

Unlike MLP, Fuzzy ARTMAP has a well defined architecture, it always converges, and can tune itself to represent sub-classes by generating a new category node when the committed one has failed to represent an input feature. However, the main drawback to Fuzzy ARTMAP lies in their complex architecture and mathematical complexity. They are constructed from a pair of Fuzzy ART modules connected by a map field. Fortunately, two generations of supervised ART, with simple architecture, have been developed recently. The first one is Supervised ART-I (Al-Rawi 1999), and the second one is Supervised ART-II (Al-Rawi *et al.* 1999a). They are constructed from a single ART module. Therefore, they are faster in training both binary and digital data than Fuzzy ARTMAP. The classification accuracy, for Supervised ART-I and Supervised ART-II, are the same as that for Fuzzy ARTMAP. Supervised ART-I is suitable to deal with a small number of classes and homogeneous data. while Supervised ART-II is suitable to deal with large numbers of classes and non-homogeneous data. However, both of them can perform training and classification in near real time. Due to this feature, Supervised ART-II has been applied for mapping burned areas (Al-Rawi and Casanova 1999), and for fire detection (Al-Rawi *et al.* 1999b), using AVHRR images. when the number of category nodes are few, the training can be done in real time.

This study has been conducted to build an automatic system, based on Supervised ART-I and Supervised ART-II, for mapping land-cover, using a Landsat TM image. The sensitivity of the system will be tested for all the domains of dynamic parameters. This can be done due to the simple architecture of Supervised ART-I and Supervised ART-II. The effect of these parameters on training time and classification accuracy will be addressed.

1 DATA

The data of this work is a scene of (256 x 240) pixels of the Landsat TM image 201/32 EDSAT. The scene corresponds to the area around the Spanish city of "Talavera de la Reina". The spectral values, which are considered here, are band-1 (0.45-0.52) μm, band-2 (0.52-0.60 μm, band-3 (0.63 - 0.69) μm, band-4 (0.76-0.90) μm, band-5 (1.55-1.75) μm, and band-7 (2.08-2.35) μm. The resolution of all these bands is (30 x 30) metres. Band-6 has not been used because it has a different resolution (120 x 120) metres and operates in the infra-red region (10.4-12.5) μm of the spectra. The normalized values [0,1] of the bands, as well as their complements have been simultaneously introduced to the network, together with class code. The class code of each pixel has been determined by a supervised field visit to some of the training areas established during the image classification process with ER-Mapper. Different combinations of dynamic parameters; the vigilance parameter ρ[0,1] and the dynamic learning parameter β [0,1] will be employed. This will let us understand the effect of these parameters on the sensitivity of the system from both the training time and the classification accuracy points of view.

2 LEARNING

2. 1 *Learning with Supervised AR T-I*

I) For each input features, the choice value is computed for each committed category node as follows:

$$T_j = \frac{\sum_{i=1}^{2M}(A_i \wedge w_{ij})}{\alpha + \sum_{i=1}^{2M} w_{ij}}; j = 1...C$$

where C is the total number of committed nodes, w_{ij} is the weight which connects the input component A_i with the committed category node j, and α is the choice value parameter $(\alpha > 0)$.

2) The maximum choice value node among all the committed category nodes are determined;

$$T_J = max\{T_j; j = 1...C\}$$

3) The match value for the winning node is computed;

$$\sum_{i=1}^{2M}(A_i \wedge w_{iJ}) / M > \bar{\rho}$$

The match value is tested against the predetermined vigilance value. If the match value of the winning node is greater than or equal to it and the node passes the class matching, the weights of this node are trained:

$$w_{iJ}^{new} = \beta(A_i \wedge w_{iJ}^{old}) + (1-\beta)w_{iJ}^{old}; i = 1..2M$$

Otherwise a new node with maximum choice value should be redetermined. If the match value is greater than the vigilance parameter, the match value is assigned to the vigilance value. We keep doing this until either, one of the category nodes can represent the current input, or a new node should be committed. In the latter case the input features are assigned to the weights of the newly committed node. The vigilance parameter ρ should be reset to its baseline value ($\bar{\rho}$), before a new input pattern is introduced to the network. For more details about Supervised ART-I see (Al-Rawi 1999).

2.2 *Learning with Supervised ART-II*

1) For each input, the choice value is computed for each committed category node for all stacks as follows:

$$T_{j_k k} = \frac{\sum_{i=1}^{2M}(A_i \wedge w_{ij_k k})}{\alpha + \sum_{i=1}^{2M} w_{ij_k k}}; j_k = 1..C(k); k = 1..L$$

where $w_{ij_k k}$ are the weights, which connect each category node j_k in each stack k with all input nodes i (i=1...2M) where M is the dimension of the input vector A. C(k) is the number of committed nodes in the stack number k. α is the choice value parameter (α>0).

2) The maximum choice value node is determined for each stack. Then the maximum choice value node among those maximums is chosen to represent the current input:

$$T_{J_k} = max\{T_{j_k k}; j_k = 1...C(k)\}$$

$$T_{JK} = max\{T_{J_k}; J_k = J_1...J_L\}$$

3) The match value for the winning node is computed:

$$\sum_{i=1}^{2M}(A_i \wedge w_{iJK}) / M$$

The match value is tested against the predetermined vigilance value. If the match value of the winning node is greater than or equal to it and the node passes the class matching, the weights of this node will be trained:

$$w_{iJK}^{new} = \beta(A_i \wedge w_{iJK}^{old}) + (1-\beta)w_{iJK}^{old}; i = 1..2M$$

Otherwise a new candidate from the same stack, which this current node belongs to must be determined, to represent its stack. For more details about Supervised ART-II, see (Al-Rawi et al. 1999a).

3 PERFORMANCE

3. 1 *Learning performance*

Training with 9000 exemplars for Supervised ART-I and Supervised ART-II has been done using a SUN 4 SPARC Station. The sensitivity of the networks for all the domain of the vigilance parameter and the dynamic learning rate are shown (Figure 1-a,b). Thanks to the simple architectures of both Supervised ART-I and Supervised ART-II, which make the construction of these figures practically possible. The regions of Supervised ART-I and Supervised ART-II, and the number of category nodes have been determined (Figure 1-c,d). The Supervised ART-I training region lies under the heavy line of figure I-c,

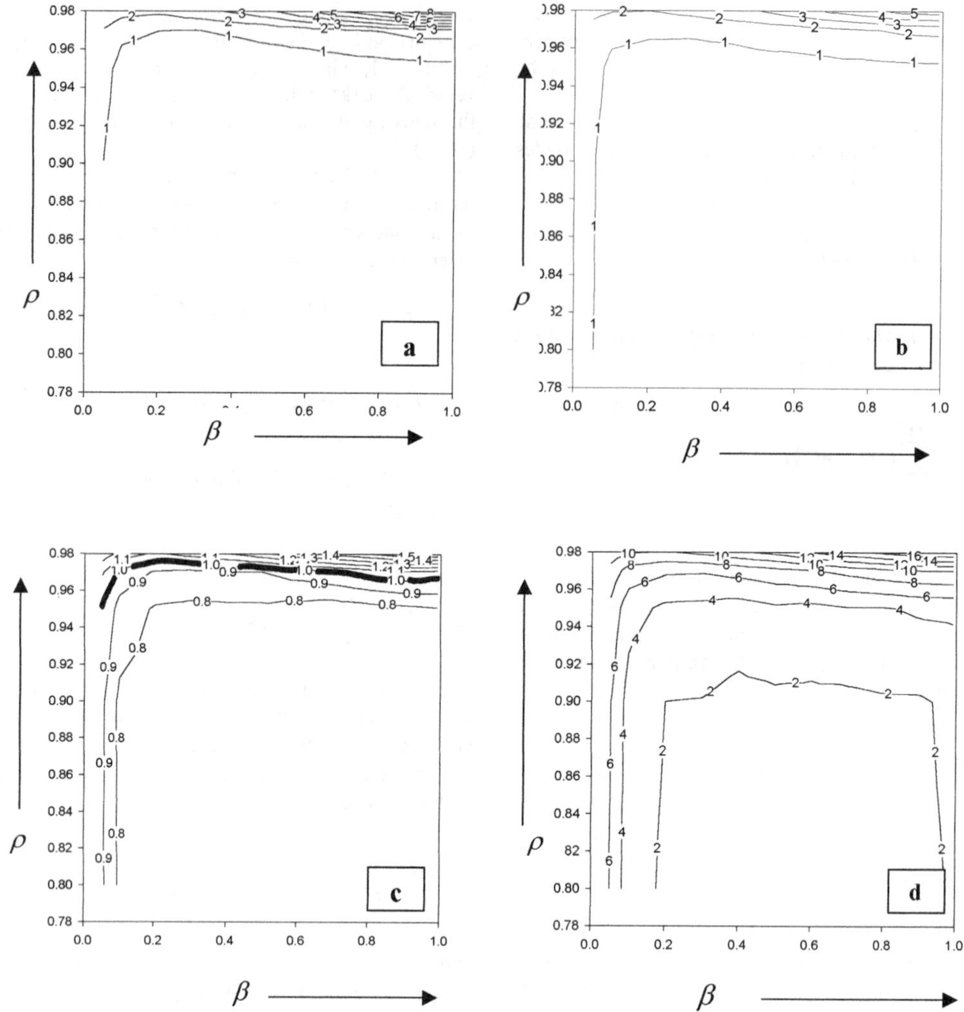

Figure 1: Learning time for Supervised ART-I and Supervised ART-II, in the domain of the vigilance parameter and the dynamic learning rate, using 9000 pixels of the Landsat TM image. a) Learning time in minutes for Supervised ART-I. b) Learning time in minutes for Supervised ART-II. c) The ratio of learning time for Supervised ART-I / Supervised ART-II. Below the heavy line (ratio=1) Supervised ART-I should be employed, otherwise Supervised ART-II should be used. d) Number of category nodes in hundreds.

while the training region for Supervised ART-II lies above it. Supervised ART-II performs better when the number of category nodes exceeds 1000. Otherwise Supervised ART-I should be employed. The training times are in minutes over all the domain of the dynamic parameters Such short learning times make Supervised ART-I and Supervised ART-II very powerful tools for classifying large scale remotely sensed data.

3.2 *Classification performance*

Classification performance for the remaining 52,440 pixels of the image is very good. We have got 85.82% classification accuracy, with classification time around 25 minutes using SUN 4 SPARC Station (Figure 2) and 9.50 minutes using ALPHA Station 500. The values of dynamic parameters for optimum classification have been determined. They have been represented by the heavy line in figure 2-b. As the

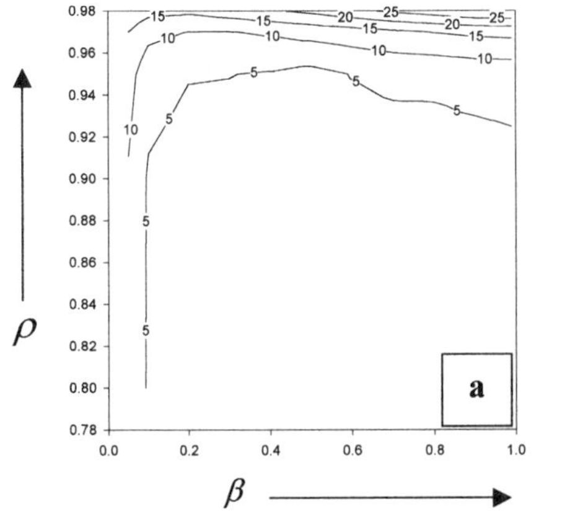

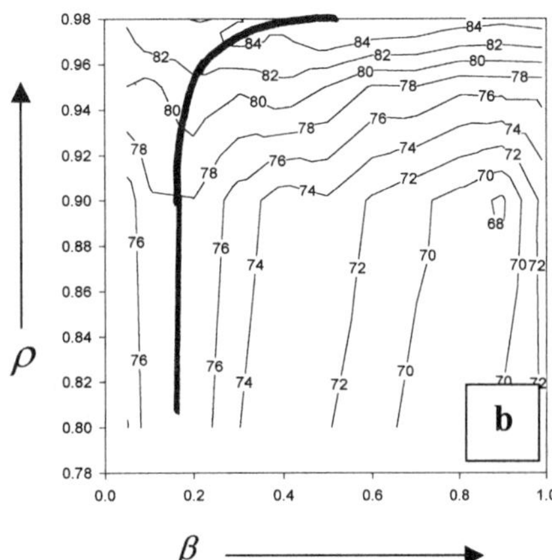

Figure 2: Classification process, in the domain of the vigilance parameter and the dynamic learning rate, for the lifted 52440 pixels of the Landsat image:

a) Classification time in minutes.
b) Classification accuracy. The heavy line represents the best classification. As the vigilance parameter increases more category nodes are generated. The share of each category node of the training exemplar is reduced. The dynamic learning rate should be increased.

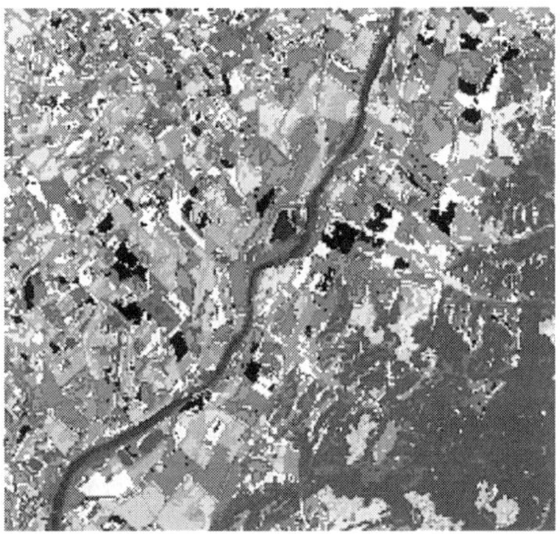

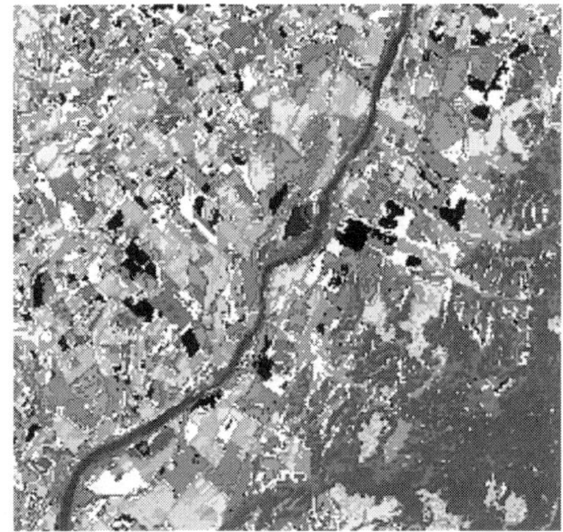

Figure3: The left image represents the real data. The right one is the classified image using Supervised ART-II, with vigilance parameter ρ=0.98, dynamic learning rate β=0.50, and training set of 9000 exemplars. The classification accuracy is 85.82%. (colour plate, see p. 601).

vigilance parameter increases, the number of category nodes increases. Therefore, the dynamic learning rate or the training exemplars should be increased. This is because the number of training exemplars for each category node is decreased.

We used different training sets (6000, 3000, 1000, 600, 200). The optimum classification line of figure 2-b moves to the right with lower curvature as the training set decreases. For the 200 training set it is almost a straight line and moves extremely to the right. This means a large dynamic learning rate should be employed here (close to 1). We used also a 15,000 training set. The optimum classification line (the heavy line in figure 3-b) moves to the left, in this case. This is to decrease the learning rate in order to avoid over learning.

Table 1: Training and classification statistics for Landsat TM image at individual class level. The network has been trained using 9000 exemplars. The classification accuracy for the training set is 94.19%. The classification accuracy for the remaining 52,440 pixels of the image is 85.62%. The vigilance parameter ρ=0.98 and the dynamic learning rate β=0.50 are used. The learning time is 2.99 minutes and the classification time is 20.97 minutes, using SUN 4 SPARC Station. The dark numbers in the right column represent the total number of pixels for each class. The dark numbers in the last row represent the total number of pixels which are assigned by the network to each class.

CLASSE NAME	**TRAINING**				**CLASSIFICATION**		
	Correct pixels	**Total no. of pixels**	**Classification Accuracy %**	**Category Nodes.no**	**Correct pixels**	**Total no. of pixels**	**Classification Accuracy %**
1-Meadow	1449	1475	98.24	189	5470	6094	89.76
2-Mountain	286	310	92.26	63	5320	6923	76.85
3-Fallow land-3	641	652	98.31	140	2514	2795	89.95
4-Irrigated land	561	578	97.06	35	2696	2898	93.03
5- Alfalfa	377	394	95.69	42	1521	1667	91.24
6-Wetland	13	13	100	1	84	88	95.45
7-Forest	34	35	97.14	15	2474	3457	71.56
8 Wheat	507	594	85.35	59	3918	4703	83.31
9-Natural veget-2	1363	1532	88.97	127	7154	8026	89.14
10-Fallow land-1	233	247	94.33	50	2099	2779	75.53
11-Fallow land-2	575	627	91.71	81	2028	2330	87.04
12-Natural veget-1	2416	2519	95.91	450	9128	10059	90.74
13-River	22	24	91.67	3	596	621	95.97
	8477	**9000**	**94.19**	**1255**	**45002**	**52440**	**85.82**

Table 2: The confusion matrix for the classification process for the 52,440 pixels of the Landsat image. The diagonal values represent the number of pixels, which are correctly classified. The non-diagonal values represent the number of missed classified pixels for each class and the class which they are assigned to. The dark numbers in the right column represent the total number of pixels in each class. The dark numbers in the second lowest row represent the number of pixels, which are assigned to each class by the classifier. The light numbers in the lowest row represent the total number of pixels, which are wrongly classified to each class.

		CLASSIFICATION													
		1	**2**	**3**	**4**	**5**	**6**	**7**	**8**	**9**	**10**	**11**	**12**	**13**	
ACTUAL	**1**	5470	16	0	2	95	0	1	39	204	0	0	267	0	**6094**
	2	72	5320	0	0	0	0	328	0	751	105	0	347	0	**6923**
	3	0	0	2514	0	0	0	0	0	0	1	107	173	0	**2795**
	4	2	0	0	2696	11	0	0	118	71	0	0	0	0	**2898**
	5	50	0	0	8	1521	0	0	70	18	0	0	0	0	**1667**
	6	0	0	0	0	0	84	4	0	0	0	0	0	0	**88**
	7	0	197	0	0	0	1	2474	0	762	2	0	0	21	**3457**
	8	90	0	0	421	226	0	0	3918	48	0	0	0	0	**4703**
	9	236	173	0	347	17	0	14	55	7154	0	0	30	0	**8026**
	10	0	81	12	0	0	7	18	0	0	2099	402	160	0	**2779**
	11	0	0	131	0	0	0	0	0	0	88	2028	83	0	**2330**
	12	420	149	104	0	0	0	0	5	86	35	132	9128	0	**10059**
	13	0	0	0	0	0	0	25	0	0	0	0	0	596	**621**
		6340	**5936**	**2761**	**3474**	**1870**	**92**	**2864**	**4205**	**9094**	**2330**	**2669**	**10188**	**617**	
		870	616	347	778	351	8	70	287	1940	231	641	1060	21	

The classification performance for the Landsat TM image is shown in figure 3. The overall classification accuracy, using vigilance parameter $\rho=0.98$ and dynamic learning parameter $\beta=0.50$, is 85.82%. The reference image and the classified image are shown in figure 3-a&b, respectively. For coloured images see the colour plate. Each class is assigned one colour. The classification accuracy for the 13 classes of the image and number of category nodes are: [river 95.97% (3), wet land 95.45% (1), irrigated land 93.03% (35), alfalfa 91.24% (42), natural vegetation-1 90.74% (450), fallow land-3 89.95% (140), meadow 89.76% (189), natural vegetation-2 89.14% (127), fallow land-2 87.04% (81), wheat 83.31% (59), mountains 76.85% (63), fallow land-1 75.53% (50), and forest 71.56% (15)], see table 1. The confusion matrix for this classification is shown in table 2. The diagonal values represent the number of pixels which are classified correctly. The non-diagonal values represent the number of pixels which are wrongly assigned to each class. Natural vegetation-1 and natural vegetation-2 have the large share for mis-classified pixels. They are assigned 1086 and 1940 pixels, respectively. This represents 42.5% of the total mis-classified pixels. Forest contributed 762 pixels to natural vegetation-2. Mountains contributed 347 and 751 pixels to natural vegetation-1 and natural vegetation-2, respectively.

4 CONCLUSIONS

1) The behaviour of both Supervised ART-I and Supervised ART-II, for training remotely sensed data, for all the domains of the dynamic parameters is well understood.

2) Supervised ART-II should be employed when the number of category nodes is in thousands. Otherwise Supervised ART-I performs better. However, Supervised ART-II can be employed here too, since the learning time is very short when the number of category nodes is less than 1000, which is less than a minute.

3) An automatic system for classifying Landsat TM images, with very good classification accuracy, has been developed.

ACKNOWLEDGMENTS

We gratefully thank Professor Jose Luis Casanova, Director of the Remote Sensing Laboratory (LATUV), University of Valladolid for using the facilities of the laboratory. Our thanks also to Professor Santiago Ormeno Villajos, Dept. of Topographic & Cartographic Engineering, University Polytechnic of Madrid, for the Landsat TM image. Finally our thank to Francisco J. Tapiador, Alfredo Romo and El Mostafa Louakfaoui of LATUV for their help in preparing the manuscript.

Remote Sensing in the 21st Century: Economic and Environmental Applications, Casanova (ed.)
© 2000 Balkema, Rotterdam, ISBN 90 5809 096 5

Mode of spectral angular distribution function extraction

A. Nekrassov
Taganrog State University of Radio Engineering, Russia

ABSTRACT: Using the expression representing the normalized radar cross section of the sea surface as a multiplication of the plane slightly rough surface normalized radar cross section and the coefficient of small-scale surface modulation by slopes of large-scale components, the mode for extraction of the spectral angular distribution function in the case when only an azimuth diagram of the normalized radar cross section of the sea surface are known, is stated. The angular distribution function estimation is represented as a ratio of the current averaged azimuth value of the normalized radar cross section to the averaged value of the normalized radar cross section in up-wind direction. Thus, the spectral angular distribution function can be estimated using the airborne scatterometer measurement by circle track flight with inclined one-beam fixed position antenna or by rectilinear track flight with rotating antenna. This mode of the spectral angular distribution function extraction allows to obtain the modulated directional wavenumber spectrum if the modulated directional wavenumber spectrum in up-wind direction is known.

1 INTRODUCTION

The process of the sea roughness has a random character, and the sea roughness spectrum is used to describe one and to calculate sea roughness parameters.

Determination of multi-dimensional spectrum is a complicated problem, and so one- and two-dimensional spectra are used in practice.

In oceanography, the spectra of the sea roughness have been measured using pitch-roll buoys, cloverleaf buoys, arrays of vertical staffs or optical methods (Guissard 1993). Application of radar methods from flying apparatus allows to obtain operational information about the sea roughness spectrum practically from any part of sea or ocean.

The mode of spectral angular distribution function extraction using airborne scatterometer is described in this paper.

2 MODE DESCRIPTION

Normalized radar cross section (NRCS) $\sigma^\circ(\theta,\alpha)$ of a plane slightly rough surface which satisfies the small perturbation assumption is presented by the following expression (Bass & Fuks 1972):

$$\sigma^\circ(\theta,\alpha) = 16\pi k^4 \cos^4\theta S(\kappa,\alpha) \tag{1}$$

where θ = the incident angle, α = the azimuth direction, k = the wave number in air, $S(\kappa,\alpha)$ = the directional wavenumber spectrum, κ = the spatial wavenumber.

Real sea surface can be well represented by a two-scale model in which the surface consists of small patches of slightly rough surface that are tilted by the large amplitude long waves. The NRCS of tilted slightly rough surface is calculated and averaged over the tilts through a probability density function to give the mean value of the sea surface NRCS. The averaged NRCS of the sea surface in polar coordinates $\sigma^\circ_{av}(\theta,\alpha)$ can be presented as follows (Garnaker'yan et al. 1986):

$$\sigma^\circ_{av}(\theta,\alpha) = \int_{-\cot\theta}^{+\infty} W_{\gamma(\alpha)}(\theta,\alpha)\sigma^\circ(\theta',\alpha)\partial\tan\gamma(\alpha) \tag{2}$$

where $W_{\gamma(\alpha)}(\theta,\alpha)$ = the probability density function of the tangent of the slope angles $\gamma(\alpha)$, $\sigma^\circ(\theta',\alpha)$ = the plane rough surface NRCS for observation with the local incident angle:

$$\theta' = |\theta - \gamma(\alpha)| \tag{3}$$

Using (1), (2) and the directional wavenumber spectrum of the form of (Garnaker'yan 1985):

$$S(\kappa,\alpha) = BU\kappa^{-3,5} \tag{4}$$

where B = the parameter, U = the wind speed over sea surface, Garnaker'yan has shown that the averaged NRCS of the sea surface could be represented as a multiplication of the plane slightly rough surface NRCS and the coefficient of modulation of the small-scale surface by slopes of large-scale components (Garnaker'yan et al. 1986):

$$\sigma^{\circ}_{av}(\theta,\alpha) \approx \sigma^{\circ}(\theta,\alpha) k_{\mathrm{mod}}(\sigma_{\gamma}(\alpha),\theta) \quad (5)$$

where $k_{\mathrm{mod}}(\sigma_{\gamma}(\alpha),\theta)$ = the coefficient of modulation depending on the sea surface slope variation $\sigma_{\gamma}(\alpha)$ in azimuth direction α and incident angle θ.

Let the modulated directional wavenumber spectrum $S_{\mathrm{mod}}(\kappa,\alpha)$ be represented as follows (Nekrassov 1995):

$$S_{\mathrm{mod}}(\kappa,\alpha) = S_{\mathrm{mod}}(\kappa,0^{\circ}) Q_{\mathrm{mod}}(\theta,\alpha) \quad (6)$$

where $S_{\mathrm{mod}}(\kappa,0^{\circ})$ = the modulated wavenumber spectrum in up-wind direction $\alpha = 0^{\circ}$, $Q_{\mathrm{mod}}(\theta,\alpha)$ = the spectral angular distribution function. Then using (5) the spectral angular distribution function can be estimated as a ratio of the current averaged azimuth value of the NRCS to the averaged value of the NRCS in up-wind direction $\sigma^{\circ}_{av}(\theta,0^{\circ})$:

$$Q_{\mathrm{mod}}(\theta,\alpha) \approx \frac{\sigma^{\circ}_{av}(\theta,\alpha)}{\sigma^{\circ}_{av}(\theta,0^{\circ})} \quad (7)$$

Thus, the spectral angular distribution function can be estimated using the azimuth NRCS diagram obtained in the airborne scatterometer measurement by circle track flight with inclined one-beam fixed position antenna or by rectilinear track flight with rotating antenna.

3 CONCLUSIONS

1. The spectral angular distribution function can be estimated using the azimuth NRCS diagram obtained in the airborne scatterometer measurement.

2. The proposed mode of the spectral angular distribution function extraction allows to obtain the modulated directional wavenumber spectrum in case when the modulated directional wavenumber spectrum in up-wind direction is known.

3. The obtained result could be used in operational research of sea from flying apparatus.

REFERENCES

Bass, F.G. & I.M. Fuks 1972. *Wave scattering from statistically rough surface*. Moscow: Nauka. *in Russian.*

Garnaker'yan A.A. 1985. *Electromagnetic wave scattering by statistically rough surface*. Taganrog: TRTI. *in Russian.*

Garnaker'yan, A.A., B.I. Orekhov, V.A. Buryak & N.I. Neilo 1986. About anisotropy of microwave radiosignal reflected from the sea surface. *Proceedings of the North-Caucasus Scientific Center of the High School. Technical Sciences* 56(4): 115-118. *in Russian.*

Guissard, A. 1993. Directional spectrum of the sea surface and wind scattering. *Int. J. Remote Sensing* 14(8): 1615-1633.

Nekrassov, A.V. 1995. About microwave signal scattering by the sea surface. In *Scattering of electromagnetic waves – Interdepartmental Thematic Scientific Collected Articles*: 98-103. Taganrog: TRTU. *in Russian.*

Remote Sensing in the 21st Century: Economic and Environmental Applications, Casanova (ed.)
© 2000 Balkema, Rotterdam, ISBN 90 5809 096 5

Correction of the sea wave height measuring results in airborne dual-frequency scatterometer measurement

A. Nekrassov
Taganrog State University of Radio Engineering, Russia

ABSTRACT: The use of the dual-frequency radar having a narrow beam antenna pointed toward the nadir direction allows to measure the root-mean-square height of sea waves. For operational measurement of the sea wave height by microwave dual-frequency scatterometer it could be based on a flying apparatus. Such placing of radar requires to stabilize the antenna location, or to correct the result of measurement in accordance with the antenna location instability arising due to flying apparatus' random evolutions. The mode of the sea wave height determination by squared module of two-frequency correlation coefficient was investigated, and the algorithm for correction of measured sea wave height was proposed. This algorithm allowed to correct the measured sea wave height in accordance with sensing system parameters, altitude, list and tangage of flying apparatus and to increase the altitude of the mode application.

1 INTRODUCTION

The information about sea wave height and sea surface wind vector is required for a pilot to provide hydroplane safe landing on water. The sea wave height information in the area of seaplane station can be obtained using in situ instrumentation. But for hydroplane safe landing on water in open sea or ocean an all-weather airborne instrument for remote measuring of the sea wave height is required. Radar techniques have exhibited good potential for rapid and accurate measurement of sea wave condition over large areas (Jones et al. 1975). There are radar altimeter and dual-frequency scatterometer with nadir-looking antenna beam.

Using radar altimeter the information on the wave height is deduced from the return pulse shape, and typical accuracy of significant wave height measurement is of order ±0.5 m (or 10% if more) for wave heights between 1 and 20 m (Komen et al. 1994). The dual-frequency radar allows to measure squared module of two-frequency correlation coefficient that depends on the root-mean-square height of sea waves which is in interdependence with the significant wave height (Jones et al. 1975).

Unfortunately, the radar altimeter provides insufficient accuracy of measurement of low significant wave heights. The correlation method of the sea wave height measuring requires to use a very narrow spatially stabilized antenna beam to increase the altitude of its application for measurement of significant wave heights H_s between 0.5 and 2.3 m that important for taking a decision about hydroplane safe landing on water (Garnaker'yan & Sosunov 1978).

A way of this problem decision is discussed in the paper.

2 CORRECTION OF THE SEA WAVE HEIGHT MEASURING RESULTS

For operational measurement of the sea wave height by microwave dual-frequency scatterometer having a nadir-looking antenna, it is necessary to stabilize the antenna location, or to correct the result of measurement in accordance with the antenna location instability arising due to flying apparatus' random evolution.

The expression for module of measured two-frequency correlation coefficient $|\rho_m(\Delta f)|$ with taking into account the altitude of flight H_0, the deviation of antenna beam from nadir, and the decorrelation due to antenna pattern has been obtained by Garnaker'yan (Garnaker'yan & Sosunov 1978):

$$|\rho_m(\Delta f)| = \exp\left[-2\left(\frac{\sigma_h 2\pi\Delta f \cos\theta}{c}\right)^2 - \frac{\left(\theta_a^2 R_0 2\pi\Delta f \tan\theta\right)^2}{5.5c^2 K_{\theta_a,R_0}^2}\right] \Big/ K_{\theta_a,R_0} \quad (1)$$

where σ_h = the root-mean-square height of sea waves, Δf = the frequency separation of radiated signals, θ = the incident angle, c = the speed of light, θ_a = the antenna beamwidth, R_0 = the distance between radar and observed surface square middle:

$$R_0 = \frac{H_0}{\cos\theta} \tag{2}$$

K_{θ_a, R_0} = the decorrelation coefficient:

$$K_{\theta_a, R_0} = \sqrt{1 + \left(\frac{\theta_a^2 R_0 2\pi\Delta f}{5.5c} \right)^2} \tag{3}$$

Experimental research has shown the following interdependence of 3% sea wave height $H_{3\%}$ ($H_{3\%} \approx 1.32 H_s$) and the root-mean-square height of sea waves (Davidan et al. 1971):

$$\sigma_h = (0.19 - 0.23) H_{3\%} \tag{4}$$

Increase of the maximum altitude of method by antenna beamwidth narrowing is limited by admissible antenna size for it placing on airplane. Another way to increase the altitude of mode application is the measurement with increased decorrelation magnitude and adequate correction of its influence on measuring result. For instance, for radar having $\theta_a = 1.5°$ and $\Delta f = 64$ MHz the maximum altitude to measure $H_{3\%} = 0.64$ m with $K^2_{\theta_a, R_0} = 1.02$ is 820 m, and one is 2670 m for measurement with $K^2_{\theta_a, R_0} = 1.2$.

Based on considered formulation an algorithm for measuring of the 3% sea wave height has been proposed. The algorithm takes into account parameters of sensing system, altitude, list γ_{fa} and tangage θ_{fa} of flying apparatus. In accordance with it the corrected 3% sea wave height is (Nekrassov 1996):

$$H_{3\%} = \frac{c}{1.688\Delta f} \sqrt{\frac{-0,5}{\cos^2\theta_{\theta\gamma}} \left[\ln\left(\left| \rho_m(\Delta f) \right|^2 K^2_{\theta_a, R_0} \right) + 11\tan^2\theta_{\theta\gamma} \left(1 - K^{-2}_{\theta_a, R_0} \right) \right]} \tag{5}$$

where $\theta_{\theta\gamma}$ = the incident angle deviation from nadir because of flying apparatus list and tangage fluctuation:

$$\theta_{\theta\gamma} = \arctan\left(\sqrt{\tan^2\gamma_{fa} + \tan^2\theta_{fa}} \right) \tag{6}$$

3 CONCLUSIONS

1. Proposed algorithm allows to correct the measured sea wave height in accordance with sensing system parameters, altitude, list and tangage of airplane, and to increase the altitude of the correlation mode application from flying apparatus.

2. The obtained result could be used in oceanographic research and in hydroaviation for creating the airborne equipment for hydroplane safe landing on water.

REFERENCES

Davidan, I.N., V.A. Rozhkov, B.M. Andreyev, L.I. Lopatukhin, Yu. A. & Trapeznikov 1971. Probabilistic characteristics of the sea roughness, methods of their analysis and calculation. *Transactions of GOIN* 57. *in Russian*.

Komen, G.J., L. Cavaleri, M. Donelan, K. Hasselmann, S. Hasselmann & P.A.E.M. Jansen 1994. *Dynamics and modelling of ocean waves*. Cambridge: Cambridge University Press.

Garnaker'yan, A.A. & A.S. Sosunov 1978. *Radiolocation of the sea surface*. Rostov-on-Don: Rostov State University. *in Russian*.

Jones, W.L., W.L. Granthman, L.C. Schroeder, J.W. Johnson, C.T. Swift & J.L. Mitchell 1975. Microwave scattering from the ocean surface. *IEEE Trans. Microwave Theory Tech.* MTT-23(12): 1053-1058.

Nekrassov A.V. 1996. Airborne radar complex for safe landing on a water surface. In *Nature and Human Being: Interaction and Vital Functions Safety – Proceedings of the International Scientific Symposium of Students, Post-Graduate Students and Young Scientists, Taganrog, 5-6 September 1996*: 89-91. Taganrog: TRTU. *in Russian*.

Remote Sensing in the 21st Century: Economic and Environmental Applications, Casanova (ed.)
© 2000 Balkema, Rotterdam, ISBN 90 5809 096 5

Shaping a field of acoustic parametric array in periodic structure with diffuse borders

M.Y.Georgi & N.P.Zagrai
Taganrog State University of Radio Engineering, Russia

ABSTRACT: A possibility a realization of nonlinear acoustic interactions in layered media, for whole remote sensing and studying the ocean inhomogeneities, is considered. Was it earlier supposed that exists a discrete changing the physical parameters of media along or across the direction of spreading a primary field of radiation. Such presentation of real media may be elaborated and extents for more general cases of continuos, smooth (including and periodic) varying the parameters of media, and in the first place along the direction of spreading of nonlinear interacting of field primary pumping waves. Such presentation of media distinctive parameters spatial changing can be interpreted as a stratified structure of medium with diffuse borders. Herewith real structures sub-bottom domains and bottom possible to represent as a layers set with diffuse borders in most cases, i.e. borders, in which properties of contacting medium are changed sufficiently fluent.

1 INTRODUCTION

The (functioning) work of acoustic parametrical array (APA) in an actual medium at nonlinear interaction of elastic waves essentially depends on a structure of nonlinear interaction region (RNA), which represents a medium of propagation a field of primary pumping waves (PPW).

Usually, considering a possibility of a nonlinear acoustic interactions realization in stratified mediums (Georgi & Zagrai 1998), the discrete model medium physical parameters variation along or across direction of primary field propagation is used. Such representation to common cases continuous, smoothness (including periodically) variations of medium parameters, and first of all across of propagation direction of nonlinear interacting field of PPW. Thus the functions of a medium typical basic parameters variation can be represented by an analytically by various manner, including modulation both linear, and nonlinear properties of a medium.

2 THEORETICAL DERIVATION

The real structures of sub-bottom areas and bottom can be presented as a set of layers in most cases with the diffusion boundaries, i.e. boundaries, on which the parameters of contacting mediums vary smoothly.

The inhomogeneity of an elastic medium is caused first of all by density changes of an elastic medium as its physical characteristic, then velocity of propagation of elastic acoustic oscillations and parameters of nonlinearity (first of all a square-law nonlinearity). In the given consideration the following laws of these significant parameters variation are offered:

$$\rho = \rho_0 + \rho_1 \cdot \cos\left(\frac{z}{a}\right), \quad c = c_0 + c_1 \cdot \cos\left(\frac{z}{a}\right),$$

$$\varepsilon = \varepsilon_0 + \varepsilon_1 \cdot \cos\left(\frac{z}{a}\right) \qquad (1)$$

The inverse task, at knowledge of basic regularities of a field secondary radiation shaping, allows carrying out effective diagnostics of a structure and composition of actual layers.

The input equations in the mathematical exposition will be solved by a method of consequent approximations (Novikov et al. 1981) for the definition of secondary waves fields of combinative frequencies (WCF) (particular a wave of difference frequency WDF) nonlinear interaction of initial PPW radiation:

$$\Delta p^2 - \frac{1}{c_0^2} \cdot \frac{\partial^2 p^{(2)}}{\partial t^2} = -\frac{\varepsilon}{c_0^4 \cdot \rho_0} \cdot \frac{\partial^2 \left(p^{(1)}\right)^2}{\partial t^2} \qquad (2)$$

where the solution of the first approximation for two-frequent primary field is put as:

$$p^{(1)} = A_1(x,y)\cdot\Phi_1(z)\cdot\sin\left[\omega_1\cdot\left(t-\frac{z}{c_0}\right)+\varphi\right]+$$

$$+A_2(x,y)\cdot\Phi_2(z)\cdot\sin\left[\omega_2\cdot\left(t-\frac{z}{c_0}\right)\right], \quad (3)$$

and the solution of the second approximation (particularly for a field of a difference frequency wave (DFW) is represented as):

$$p^{(2)} = \frac{1}{2}\cdot A\_(x,y,z)\cdot e^{i\cdot\Omega\cdot t+i\cdot\varphi}$$

$$+ \textit{complex conjugate}, \quad (4)$$

In which the magnitude is a following relation:

$$A\_(x,y,z) = \frac{\varepsilon\cdot\Omega^2}{c_0^4\cdot\rho_0}\cdot\frac{e^{-i\cdot k\cdot R_0}}{4\cdot\pi\cdot R_0}\cdot Dl(z)\cdot Dt(x,y) \quad (5)$$

where $D_l(z)$ = longitudinal aperture factor, defining structure of a secondary field in dependence from distribution of a primary field along direction of its distribution, $D_t(x,y)$ = transversal aperture factor, defining structure of a secondary field in dependence from transversal distribution of a primary field concerning direction of distribution, ω = circular frequency of a field WDF, c_0 = sound speed, ρ_0 = undisturbed denseness of a medium, R_0 = distance to a viewpoint.

The method of immersing thus allows analytically to determine unknown required expressions, the calculation on which in the elementary cases of the components functions can be defined analytically or in complicated case - by a numerical integration obtained in an outcome of a mathematical solution as integrals.

For an inhomogeneous stratified structure were in the RNI of primary pumping waves fields, within the framework of the given model the expression for a longitudinal aperture factor $D_l(z)$ will look like:

$$Dl(z) = \frac{\Omega^2}{4\cdot\pi\cdot R_0}\cdot e^{-i\cdot k\cdot R_0}\int_0^l \frac{\left(\varepsilon_0+\varepsilon_1\cdot\cos\left(\frac{z'}{a}\right)\right)}{\left(\rho_0+\rho_1\cdot\cos\left[\frac{z'}{a}\right]\right)}$$

$$\cdot\frac{1}{\left(c_0+c_1\cdot\cos\left(\frac{z'}{a}\right)\right)^4}\cdot e^{\frac{-i\cdot\omega\cdot 2\cdot\sin^2\left(\frac{\theta}{2}\right)z'}{c_0+c_1\cdot\cos\left(\frac{z'}{a}\right)}}dz' \quad (6)$$

where a = spatial period of a medium properties modification, which takes into account it heterogeneity: $a = c/f = \lambda$ - wavelength, l = length of a layer, f = linear frequency.

After integration of expression (6) on an extension of nonlinear interaction region (RNI) l the following relation of its angular dependence (directivity pattern) from magnitude of a spatial period a of elastic medium physical properties modifications - is obtained:

$$Dl(z) = \frac{\Omega^2}{\rho_0\cdot c_0^4}\cdot\frac{e^{-i\cdot k\cdot R_0}}{4\cdot\pi\cdot R_0}l\cdot e^{-i\cdot k\cdot l\cdot\sin^2\frac{\theta}{2}}\cdot\Bigg[\varepsilon_0+$$

$$+\left(\frac{e^{-i\cdot k\cdot l\cdot\sin^2\frac{\theta}{2}}\cdot\left(4(ak)^2\sin^4\frac{\theta}{2}\right)}{1-4(ak)^2\sin^4\frac{\theta}{2}}\right)\cdot\varepsilon_1\cdot\left[\frac{\sin\left(kl\sin^2\frac{\theta}{2}\right)}{kl\sin^2\frac{\theta}{2}}\right]$$

$$+\frac{\varepsilon_1}{a^2\left(1-4(ak)^2\sin^4\frac{\theta}{2}\right)}\cdot e^{-i\cdot k\cdot l\cdot\sin^2\frac{\theta}{2}}\cdot\left(\frac{\sin\frac{l}{a}}{\frac{l}{a}}\right)$$

$$+\frac{\varepsilon_1 kl\sin^2\frac{\theta}{2}}{\left(1-4(ak)^2\sin^4\frac{\theta}{2}\right)\cdot la^2}\cdot e^{i\left(\frac{\pi}{2}-kl\sin^2\frac{\theta}{2}\right)}\left(\frac{\sin\frac{l}{2a}}{\frac{l}{2a}}\right)^2\Bigg] \quad (7)$$

3 CONCLUSIONS

Based on obtained expressions becomes possible to realize an analysis and ascertainment of a degree and influence of inhomogeneities, smoothly varying in the environment, of physical properties (linear and nonlinear) elastic medium on a secondary field structure. As well as the directness of secondary fields derived in an outcome of nonlinear interactions of (PPW) primary pumping fields.

Thus essential there is an ascertainment of a various influence degree as magnitudes of the additions to a fixed value of considered magnitudes, and their spatial period of a variation in relations with of wavelengths both primary fields, and combinative frequencies waves of secondary fields with the purpose of the most important relations ascertainment, essentially influencing to an APA field shaping.

For this purpose the series of parameters values and their various combinations, overlapping a really probable range of their modifications in actual conditions, were selected. This series also permits to make under the analysis of outcomes common conclusions in established relations. The obtained outcomes represented in Figures 1 – 4. Calculations was performed with following input data $\Omega = 250kHz$, $\rho_0 = 1000kg/m^3$, $c_0 = 1500m/\sec$, $\varepsilon_0 = 5$.

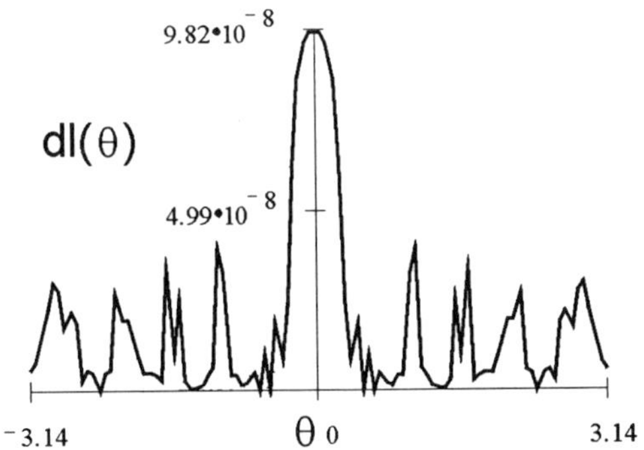

Fig.1 Spatial period of inhomogeneities $a = 4mm$

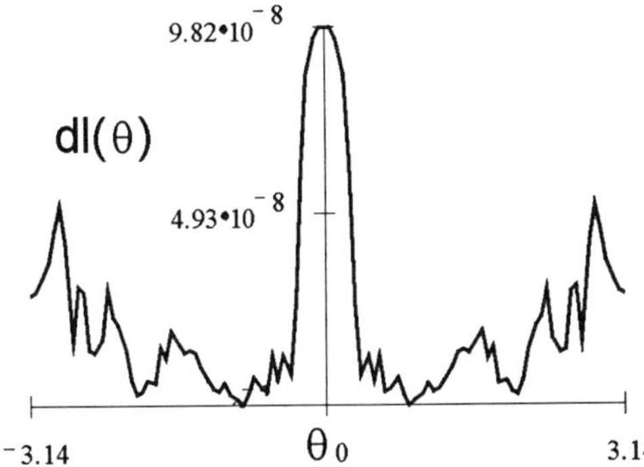

Fig.2 Spatial period of inhomogeneities $a = 20mm$

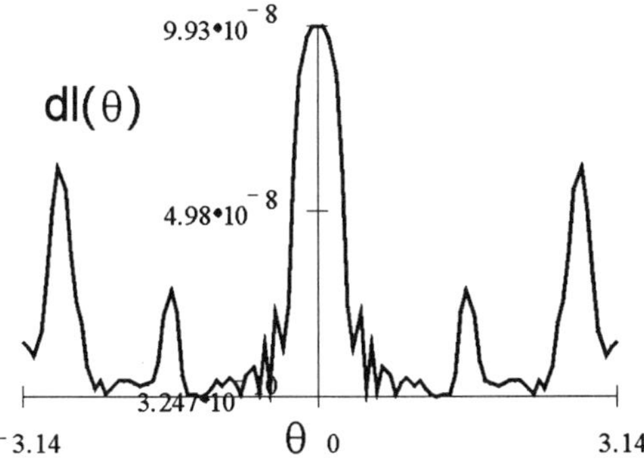

Fig.3 Spatial period of inhomogeneities $a = 40mm$

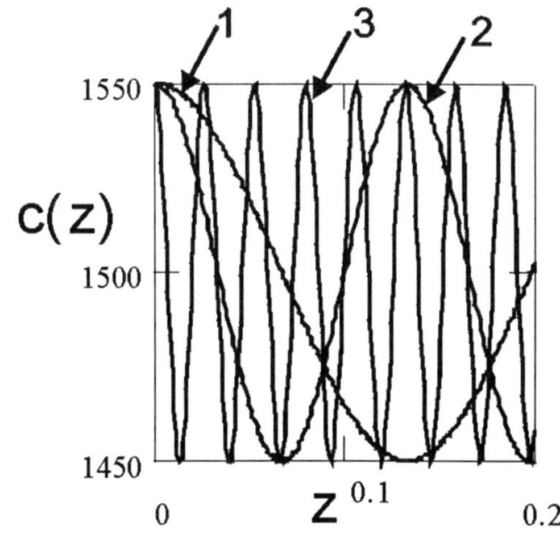

Fig.4 Sound speed profiles for cases represents on Figures 1 - 3

REFERENCES

Georgi M. & N. Zagrai 1998. Layered Structure Influence on Nonlinear Parametric Array Characteristics. *Proceedings of the Fifth International Conference "Remote Sensing for Marine and Coastal Environments,"* vol. II (II-18), San Diego, California, USA.

Novikov B., O. Rudenko. & V. Timoshenko 1981. *Nonlinear Hydro-acoustic* – Leningrad, Sudostroenye.

Remote Sensing in the 21st Century: Economic and Environmental Applications, Casanova (ed.)
© 2000 Balkema, Rotterdam, ISBN 90 5809 096 5

Quality and quantity of ground control points in the orthorectifying process

M.C.Sánchez Trigo & J.V.Giráldez Cervera
Departamento de Agronomía, Universidad de Córdoba, Spain

ABSTRACT: Aerial photography is the basis of Remote Sensing development. Despite the latest scientific and technological advances, which have made possible the use of aerial platforms for remote observation, aerial photography continues to be a commonly used technique. Digital orthophotos to obtain metric information and its capability of integrating new vector data are directly obtained from them. However, the metric information will be available after transforming the conic projection of the photographs into an orthogonal one by the standard orthorectifying process. This process, based on numerical methods, has been improved with the help of new computation facilities. This study is done under the GRASS GIS which has an specific command, "i.ortho.photo", for this purpose. This tool requires scanned photographs, a Digital Elevation Model (DEM) of the area, and several ground control points located in them. The collinearity equations determine the relationship between the pixel co-ordinates and the ground control points. In this study a sensitivity analysis is carried out with the aim of establishing the correlation between the number and location of the ground control points with the root-mean-square-errors.

1 INTRODUCTION

The great developments of Geographical Information Systems (GIS) allow an improved understanding of environmental changes. GRASS (Geographical Resources Analysis Support System) developed by the U.S. Army Construction Engineering Research Laboratory (USA Corps of Engineers 1991) is one system that can easily import scanned aerial photography.

Using aerial photos, a large number of environmental questions can be answered, they are helpful especially in areas where no thematic maps are available. Using them within a GIS a basin can be characterized. Digital photographs are ortho-rectified to obtain digital orthophotos which are considered to be one of the most important parts of GIS databases due to their feasibility of superposing vector data on them to obtain metric information (Hoolihan & Maupin 1993). So, they can be used as a base for digitizing field boundaries and consider variations in vegetation coverage as well as the consequences of human activities, like the cutting of trees or the constructions of new roads.

The GRASS GIS contains a routine, "i.ortho.photo", to perform orthorectification. Digital orthophotos can be the base maps to enhance all analytical aspects and to obtain more detailed information about environment. Therefore, those images can be of great help for future applications.

2 METHODOLOGY

2.1 *Digital Elevation Model*

A Digital Elevation Model (DEM) is an ordered array of numbers representing the spatial distribution of elevations above some arbitrary datum in a landscape (Burrough & McDonnell 1998). DEM are subsets of the Digital Terrain Models (DTM) whose arrays comprise a larger amount of terrain attributes (Moore et al. 1991).

The ideal structure for a DEM may be different if it is used as a structure for a dynamic hydrologic model than if it is used to determine the topographic attributes of a landscape. The most widely used data structures consist of square-grid networks because of their ease of computer implementation and computational efficiency.

The optimal data structure for a DEM depends on its application. The development of a dynamic hydrologic model requires different qualities in the DEM to the allocation of merchandise in the regional food market. The most widely used data structures consist of square grid networks, which are more easy to implement in a computer increasing computational efficiency.

Therefore the DEM chosen for this study was generated digitizing the contour lines of the 1:10.000 scale topographic maps, under the GRASS GIS, with which a 20-meter spatial resolution mesh was obtained (Fig. 1).

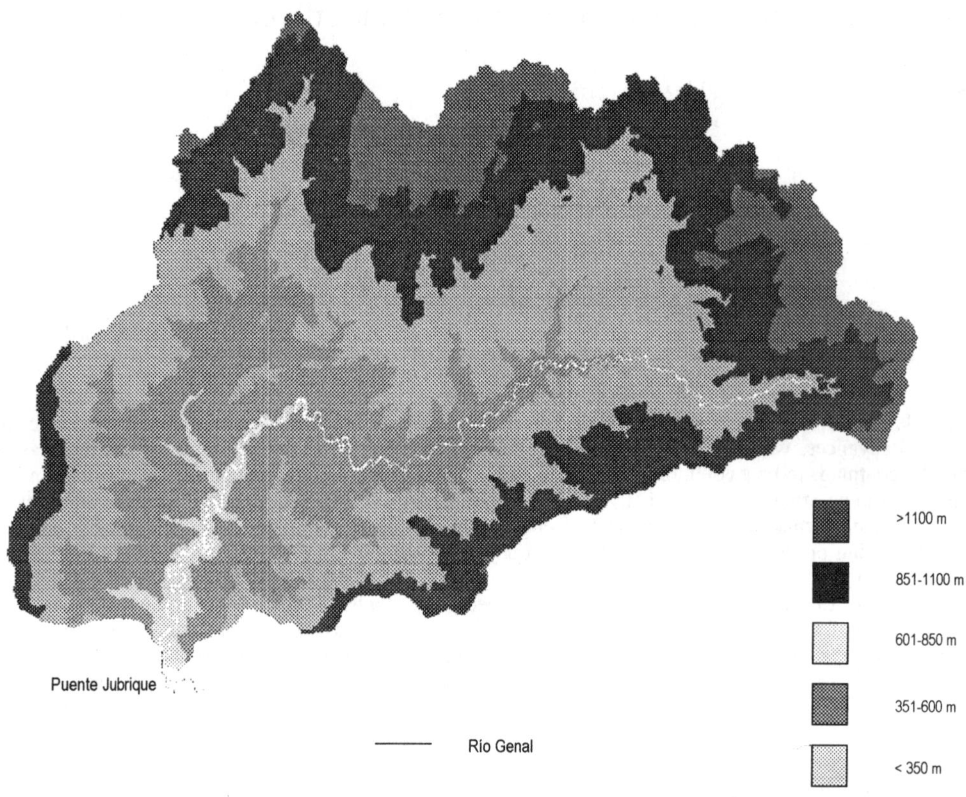

Figure 1. Digital Elevation Model.

Figure 2. Scanned aerial photography imported into GRASS.

Figure 3. Ortho-corrected aerial photograph.

2.2 *Orthorectifying process*

The orthorectifying process converts the conic projection into a central projection system. Aerial photographs are a typical case of conical projection system, where all lines connecting homologue points share a point, the focus (González et al. 1998). The orthorectification transforms those lines linking homologue points in parallel lines, where metric information can be easily retrieved.

Digital orthorectification requires as input, digital files from the aerial photographs, several ground control points (GCP), the Digital Elevation Model (DEM) of the area and the camera calibration data. A routine scanning procedure of the photographs yields the digital files imported later into the GIS GRASS, in this case (Fig. 2).

The digital images are set of elements, pixels, or raster cells. To establish the relationship between the terrain and the aerial photography coordinates, a set of ground control points must be identified in both the photograph and the 1:10.000 scale topographic maps. There must be at least eight easy-identifiable or conspicuous points homogeneously distributed.

Once the ground control points are selected their coordinates are determined. From the plane coordinates of the DEM, the collinearity equations (1, 2), will be used to re-generate the orthophoto points (Welch et al. 1984, González et al. 1998). So, aerial photographs are rectified into a projection like UTM with an ortho-rectification onto a DEM (Fig. 3).

$$x'_i = c_k \frac{a_{11}(X_i - X_0) + a_{21}(Y_i - Y_0) + a_{31}(Z_i - Z_0)}{a_{31}(X_i - X_0) + a_{32}(Y_i - Y_0) + a_{33}(Z_i - Z_0)} \quad (1)$$

$$y'_i = c_k \frac{a_{12}(X_i - X_0) + a_{22}(Y_i - Y_0) + a_{32}(Z_i - Z_0)}{a_{31}(X_i - X_0) + a_{32}(Y_i - Y_0) + a_{33}(Z_i - Z_0)} \quad (2)$$

where X_i, Y_i, Z_i are the ground co-ordinates for the point i, while x'_i, y'_i are the image co-ordinates (i = 1...n). The parameters $a_{\alpha\beta}$ (α=1,..3; β=1,..3) are:

$$a_{11} = \cos(\varphi)\cdot\cos(\chi)$$
$$a_{12} = -\cos(\varphi)\cdot sin(\chi)$$
$$a_{13} = sin(\varphi)$$
$$a_{21} = \cos(\omega)\cdot sin(\chi) + sin(\omega)\cdot sin(\varphi)\cdot\cos(\chi)$$
$$a_{22} = \cos(\omega)\cdot\cos(\chi) - sin(\omega)\cdot sin(\varphi)\cdot sin(\chi)$$
$$a_{23} = -sin(\omega)\cdot\cos(\varphi)$$
$$a_{31} = sin(\omega)\cdot sin(\chi) - \cos(\omega)\cdot sin(\varphi)\cdot\cos(\chi)$$
$$a_{32} = sin(\omega)\cdot\cos(\chi) + \cos(\omega)\cdot sin(\varphi)\cdot sin(\chi)$$
$$a_{33} = \cos(\omega)\cdot\cos(\varphi)$$

with ω, φ, and χ as the camera rotating angles and c_k the camera constant (Wolf 1983).

To check if the GCP are set properly the analysis-function is used (Fig. 4). The root mean square errors (RMS error) are calculated and the bad GCP can be disabled in order to diminish it, so that it is less than the ground resolution (Fig. 5).

The minimum number of GCP for a polynomial transformation depends on the order of the polynom

$$A = \frac{(p+1)(p+2)}{2} \quad (3)$$

where A = number of GCP; p = the order of the polynom.

However as we are working with an ortho-rectifying process a minimum number of 4 points are necessary.

A good rule of thumb is to mark at least 12 to 15 points, which are eventually distributed over the entire imagery group file in order to obtain an accurate transformation parameter for the rectification process. The RMS error may increase with more points added, but the transformation parameter will be more accurate over the entire image.

A RMS error of less than or equal to approximately one resolution unit (pixel) for the image being rectified is generally considered acceptable

3 STUDY AREA

The proposed methodology is applied to the Genal River Basin in Ronda (Málaga), tributary of the Guadiaro River, in Southern Spain, a Mediterranean mountain basin of 160 km^2. The elevation range from 232 to 1429 m. Average annual rainfall is 1000 mm rainfall, with two main rain seasons in autumn and spring (Lorite 1997).

The area is somewhat unstable with frequent landslide events. Landslide scars have been studied by Sánchez et al. (1998).

Though brushwood are the main vegetative coverage, chestnut trees (*Castanea sativa L.*) are been introduced recently increasing the area dedicated to this crop. This change in the vegetation pattern and the intensification of human activity, with tree cutting and opening of new forest roads, has contributed to a degradation of the landscapes, with a higher erosion risk.

So, it is a basin with high-risk erosion potential because of the intensive agricultural activities in chestnut-tree plantations, high slopes and intense rainfall.

4 RESULTS AND DISCUSSION

Twenty two aerial photographs of the study area were ortho-corrected, obtaining different RMS errors for each one. The number of selected GCP varied between 9 and 16 depending on the photograph and the feasibility of reaching appropriate ones (Table 1).

Not all the points were used, only those that gave the minimum error whose number varied between 8 and 13 (Table 1), and forcing the error to be less than one resolution pixel (20 meters).

No correlation was found between RMS errors and the number of selected points as Figures 6-7 show. Nevertheless, RMS errors of the photographs with uniformly distributed GCP, marked with a superindex 1 in Figure 8, were well below 20, lower than those corresponding to heterogeneously distributed points.

It must be pointed out that because of being the selection of points highly subjective, the same person was in charge of doing this job to diminish this bias as much as possible.

LOCATION: genal GROUP: g1064_3_18 MAPSET: ir2lotoi

Analysis of control point registration

	error			image		target		
#	east	north	target	east	north	east	north	elev.
1	-122.7	-39.9	129.1	690.8	-1994.8	302553.0	4051556.0	809.0
2	-27.1	-20.7	34.1	1412.4	-2179.8	303704.0	4051251.0	809.0
3	107.1	-37.0	113.3	696.1	-1157.8	302422.0	4052931.0	423.0
4	-216.4	151.1	264.0	297.6	-830.5	301831.0	4053472.0	621.0
5	57.6	40.0	70.1	1101.6	-1491.7	303149.0	4052297.0	490.0
6	29.1	-38.5	48.2	568.4	-290.5	302190.0	4054525.0	348.0
7	-17.6	43.3	46.8	1434.0	-529.1	303764.0	4053993.0	512.0
8	18.0	-38.9	42.9	1519.6	-2627.4	303824.0	4050576.0	880.0
9	-53.5	-97.1	110.8	1192.4	-2597.2	303356.0	4050648.0	916.0
10	152.9	120.4	194.6	396.7	-2440.4	301944.0	4050752.0	608.0
11	-107.4	-12.0	108.1	2415.7	-946.1	305437.0	4053208.0	527.0
12	95.1	7.0	95.3	2265.0	-1715.4	305020.0	4051967.0	861.0
13	-10.0	-36.1	37.4	1359.6	-681.0	303630.0	4053761.0	471.0
14	19.3	-21.1	28.6	598.7	-267.2	302251.0	4054553.0	350.0

Overall rms error: 114.88

Figure 4. Example of an error report of an image, using all the selected points.

LOCATION: genal GROUP: g1064_3_18 MAPSET: ir2lotoi

Analysis of control point registration

	error			image		target		
#	east	north	target	east	north	east	north	elev.
1	-3.9	-1.4	4.1	690.8	-1994.8	302553.0	4051556.0	809.0
2	-16.0	-7.1	17.5	1412.4	-2179.8	303704.0	4051251.0	809.0
3	5.2	1.0	5.3	696.1	-1157.8	302422.0	4052931.0	423.0
4	14.2	-1.9	14.3	297.6	-830.5	301831.0	4053472.0	621.0
5	not used			1101.6	-1491.7	303149.0	4052297.0	490.0
6	3.9	4.3	5.8	568.4	-290.5	302190.0	4054525.0	348.0
7	0.9	-12.6	12.7	1434.0	-529.1	303764.0	4053993.0	512.0
8	19.8	2.3	19.9	1519.6	-2627.4	303824.0	4050576.0	880.0
9	-11.7	7.3	13.8	1192.4	-2597.2	303356.0	4050648.0	916.0
10	2.7	-19.7	19.9	396.7	-2440.4	301944.0	4050752.0	608.0
11	-2.5	12.9	13.1	2415.7	-946.1	305437.0	4053208.0	527.0
12	-8.5	0.3	8.5	2265.0	-1715.4	305020.0	4051967.0	861.0
13	not used			1359.6	-681.0	303630.0	4053761.0	471.0
14	0.3	15.9	15.9	598.7	-267.2	302251.0	4054553.0	350.0

Overall rms error: 13.63

Figure 5. Example of an error report of an image, without using the "bad points".

According to the number of disabled points 18.9% of them were considered as "bad points" but more than 70% of these were labeled as "edge of crops" while the others were "edge of buildings".

5 CONCLUSIONS

The main advantage of GIS as an environmental managing tool is their ability to integrate elements of different sources of information. In this way complex processes, like the photo orthorectifying may be easily handled through simple operations with the digital data.

With the use of ortho-rectified images geo-referenced information of any source can be overlaid.

Aerial photographs are a very important base for

Table 1. Number of selected and used points for each photograph and their RMS errors.

Photograph	Selected points	RMS error*	Used points	RMS error**
1	9	205.290	8	44.23
2	14	205.290	10	138.50
3	10	201.220	10	155.38
4	13	155.380	11	14.62
5	12	433.770	11	15.01
6	13	99.550	12	11.32
7	12	73.670	10	9.83
8	11	5628.04[1]	8	10.55
9	16	14887.00[1]	13	16.67
10	14	114.880	12	13.63
11	11	74.460	10	10.72
12	13	73.200	11	10.04
13	12	9.760	12	9.76
14	13	28.850	10	15.02
15	12	61.800	9	12.99
16	13	230.730	12	15.55
17	13	66.730	11	17.91
18	12	423.200	11	10.72
19	12	125.050	11	7.97
20	12	68.300	9	20.08
21	12	321.170	10	21.68
22	13	266.080	11	55.29

* RMS error using all the selected points

** RMS error disabling the bad points

[1]Those values were not introduced to obtain the graphics to reduce bias

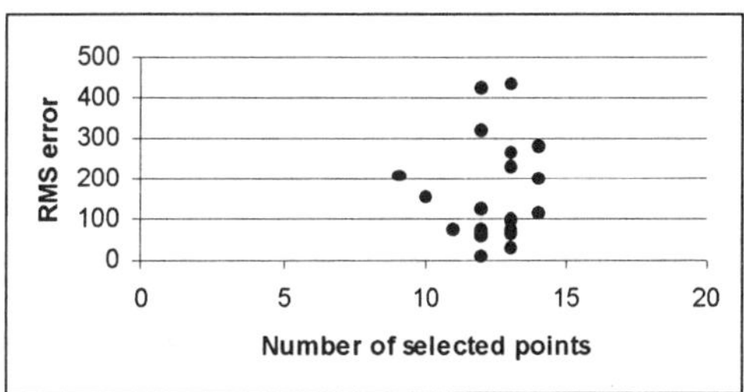

Figure 6. Relationship between the number of selected points and the RMS error.

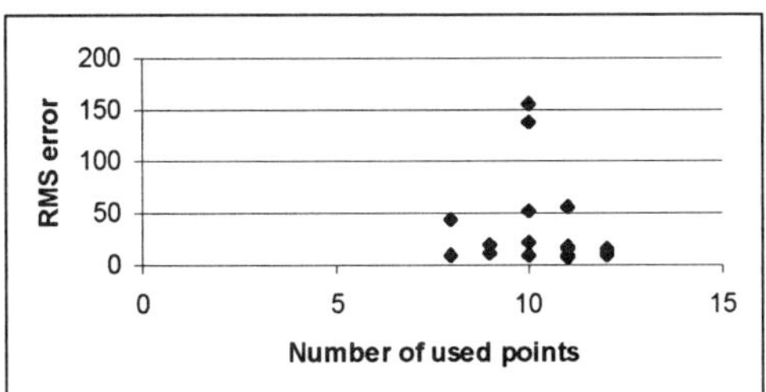

Figure 7. Number of used points and the associated RMS error.

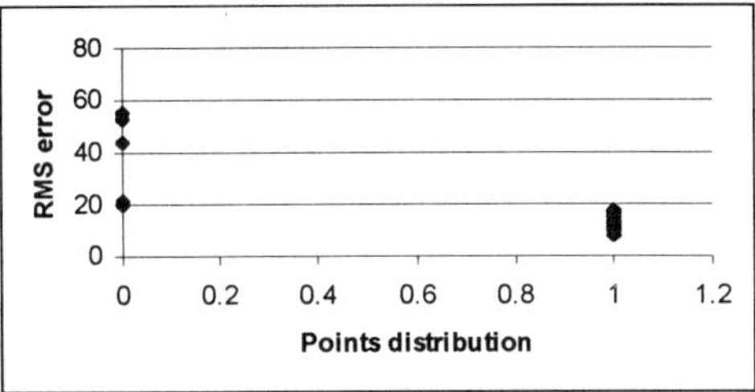

Figure 8. Point distribution and the associated RMS error.

taking decisions for environmental managers and policy makers, mostly if consecutive-year photographs are available.

A homogeneous distribution of GCP leads to a better image transformation than the use of a high number of selected points, so quality does not imply quantity when referring to GCP in orthorectifying photographs.

It is advisable to select permanent points like crossroads or edge of buildings as they introduce fewer errors than those which change within the time as for example edge of cropped areas.

6 ACKNOWLEDGEMENTS

This study is part of the AMB96-1158 research, titled "Estrategias para el control de la erosión en zonas de alta montaña", financed by the Spanish Commission for Science and Technology (CICYT).

7 REFERENCES

Burrough, P.A. & R.A. McDonnell. 1998. *Principles of Geographical Information Systems*. New York: Oxford University Press

González Dugo, M.P., M.C. Sánchez Trigo,V. Ortiz Sánchez, J.L. Tirado Valencia & A. Muñoz Rastrero 1998. Control de Solicitudes de Ayudas por Superficie para la Campaña de Comercialización 1996-1997 (Cosecha de 1996), asistido mediante fotografía aérea en la Comunidad Autónoma de Andalucía. *Proceedings of the IV International Congress of Project Engineering.*: 1445-1452.

Hoolihan, B.D. & J.A. Maupin 1993. Superposition of Vectors on Digital Orthophoto Mosaics Using GRASS. Geographic Information Systems: Proceedings of the Seventh Annual GRASS Users Conference:73-78.

Lorite Torres, I. 1997. *Análisis del modelo hidrológico TOPMODEL y su aplicación a una subcuenca del río Genal.* Trabajo Profesional fin de carrera. Universidad de Córdoba.

Moore, I.D., R.B. Grayson & A.R. Ladson 1991. Digital Terrain modelling: a review of hydrological, geomorphological, and biological applications. *Hydrological Processes 5*:3-30.

Sánchez Trigo, M.C., I. Lorite Torres, P. Montesinos Barrios, M. López Aguilar, J.V. Giráldez Cervera & J.L. Ayuso Muñoz 1998. Análisis de zonas de concentración de flujo subsuperficial en cuencas de montaña mediante

Teledetección y Sistemas de Información Geográfica. . *Proceedings of the IV International Congress of Project Engineering:* 1539-1547.
U.S. Army Corps of Engineers, GRASS4.0 User´s Reference Manual 1991. Army Construction Engineering Research Laboratory. Champaign: Illinois.
Welch, R., T.R. Jordan & A.W. Thomas1984. A photogrammetric technique for measuring soil erosion. *Journal of Soil and Water Conservation,*39: 191-194.
Wolf, P.R. 1983. *Elements of Photogrammetry.* New York: McGraw-Hill.

Remote Sensing in the 21st Century: Economic and Environmental Applications, Casanova (ed.)
© 2000 Balkema, Rotterdam, ISBN 90 5809 096 5

Tool for integration and bi-directional conversion of remote sensing data

A. Arquero, C. Gonzalo, E. Martínez & O. Gordo
Departamento de Arquitectura y Tecnología de Sistemas Informáticos, Facultad de Informática, Universidad Politécnica de Madrid, Spain

ABSTRACT: The aim of the work presented in this paper has been to develop an I/O interface that makes easy the communication between a commercial tool, ERMAPPER 5.5, an application developed to remote sensed images fuzzy classification and some external systems. Some options of this interface are the integration of different kind of information (spectral data, class code, pixel position, etc...) screen displayed by the diverse options of a commercial tool. The integrated information will be represented in one o more files according to user demand; besides, the proposed tool supports formats conversion from the commercial tool to other particular remote sensing applications and the inverse conversion. The tool presents very good performances as: a comfortable work environment, and a high user election flexibility of the format and final configuration data, speed, and access facilities.

1 INTRODUCTION

It is well known that remote sensing data of the earth's surface acquired from either aircraft or spacecraft platforms are readily available in digital format. Even data that initially are not recorded in digital form can be converted into discrete data by use of digitizing equipment. The great advantage of having available data digitally is that computer can process it, using commercial tools. These tools present almost endless applications in remote sensing field: correction, enhancement, analysis, classification... Evidently an important aspect of these tools that should be considered is the import/export image formats provided for them. Usually these formats are largely vendor specific and they are not compatible with other image applications. A possible solution to this problem is to perform several intermediate format changes; but this process could produce the lost of some advantages tool, as for example speed and simplicity. Nevertheless, sometimes, those formats do not allow representing different types of information simultaneously, being this aspect very useful in some areas of remote sensing. In this sense, the aim of the work presented in this paper has been to develop an I/O interface that makes easy the communication between a commercial tool, ERMAPPER 5.5, and an application developed to remote sensed images fuzzy classification (Martínez 1999).

2 TOOL DESCRIPTION

Some initial specifications imposed to the tool design, have been a comfortable work environment, and a high user election flexibility of the format and final configuration data. In order to get the first aspect the tool has been endowed with an user interface window oriented system. About the second, two different libraries have been programmed using the primitives supplied by ERMAPPER and the programming language C.

One library has been implemented for raster image reading (GRL_inter) and other to generate output files (Image_wri). Perhaps the best characteristic of these libraries is that they work with the image stored in the main memory. That provides the tool very good performances such as speed, flexibility and access facilities. Moreover, these libraries present a powerful ability to graphic visualization. All these characteristics have been obtained through the exploitation of commercial tool capabilities.

The design of these libraries is completely general, in the sense that it can use many different images format. That is, only the implementation should be changed, but not the interface. Thus, format changes are very easy, and they only imply the substitution of a dynamic library by other different one. For that dynamic techniques have been used.

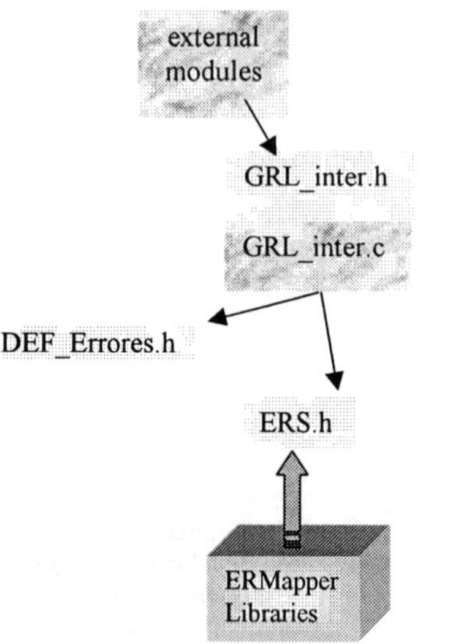

Figure 1. A module scheme of the raster image reading library

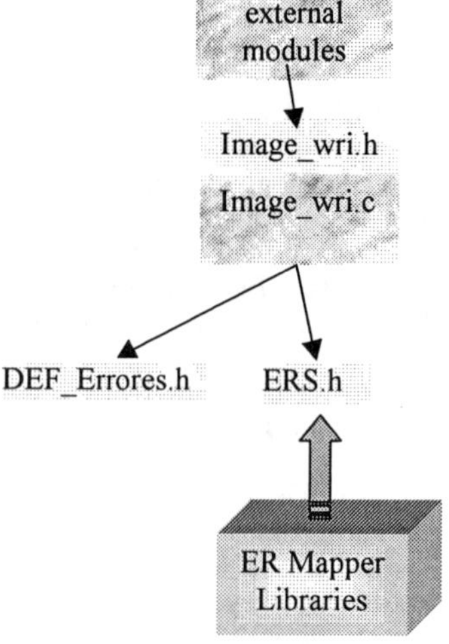

Figure 2. A module scheme of the raster image writing library

2.1 Image reading library

Two files integrate this library: the header (GRL_inter.h), which is common for any format and so it does not change; and the C codes (GRL_inter.c), which are depended on the format. These two files are represented in figure 1. This figure represents a modular scheme of the reading image library interface. Here, ERS.h represents the header file, which allows us to access to ERMAPPER libraries; DEF_Errores.h stores error codes and their meaning; and external modules can be different applications that require to read images from the considered format.

The library contains 18 functions, whose basic tasks are: the opening and closing files, which implies the load and unload of main memory; and information selective extraction from that files (position, spectral information, attribute number, class number, training pattern number).

Also this library allows writing all image information in a text file, which can be used in different applications, as it will be described later.

2.2 Output files library

It has been already mentioned that the output of the general tool is classified remote sensed images. Then, one objective of the functions implemented in the library Image_wri is to represent such information in an image file to be directly visualized and interpreted by ERMAPPER. It should be noted that, in this library, it has been needed to implement different structures that allow to integrate all information associated to a classified image. Figure 2 represents a modular scheme of the writing image library interface. The different modules displayed at the figure have a similar interpretation to the corresponding modules at figure 1. This library contains six functions, which allow one to generate the header image files (*.ers), and the corresponding dataset to be visualized with ERMAPPER. In other words, it is possible to exploit the graphical powerful of this kind of software to visualize image classified by external classifiers at such tool. A flux diagram of the writing process is displayed at figure 3. The diagram is explained by itself. Moreover, the library provides functions that allow building a text file, including all image attributes and interpreted by ERMAPPER.

3 APPLICATIONS AND RESULTS

The developed tool has been used to three particular applications: construction of text files to represent different image attributes in distinct forms; representation of scattergrams in standard spreadsheets; and classified image visualization. The input and output images format has been of kind ERMAPPER. Two different files describe these images: one of them is the header file, which contains general information about the image; and the other one includes spectral information.

The required image data order is sometimes application dependent on. Thus, for example, the training phase of different neural network models require disposing of different arrangement of the image data. Sometimes, it is better to feed the neural network with these data ordered, but other time it is

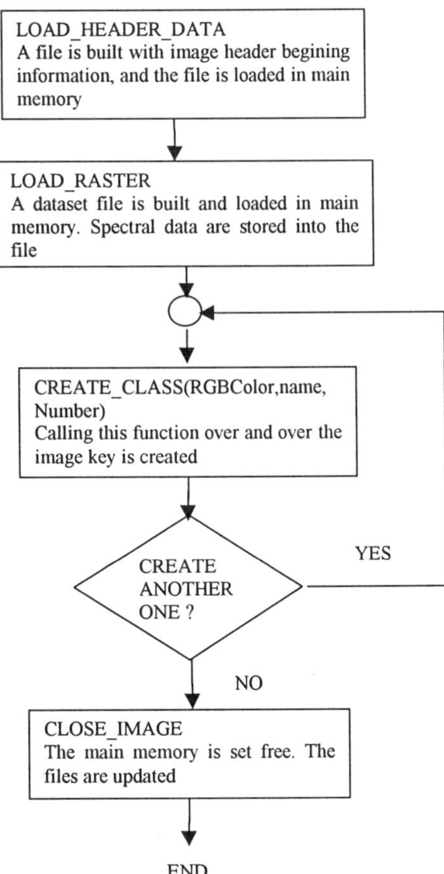

Figure 3. Flux diagram of the image writing process.

0 0 102 58 73 105 116 60 12
0 1 124 70 93 118 159 94 12
0 2 110 53 61 111 167 94 12
0 3 102 57 68 119 134 76 12
0 4 116 66 85 109 180 111 3
0 5 124 68 94 101 188 117 3
0 6 124 68 94 101 188 117 3
0 7 126 71 98 109 202 131 3
0 8 124 63 80 106 196 124 3
0 9 95 43 45 98 133 69 12
0 10 84 38 37 98 79 36 9
0 11 80 37 33 102 71 30 9
0 12 79 36 34 104 67 24 9

Figure 4. Text file segment generated by the image reading library

better that the data go into the net in a random way. Our tool provides both options. It generates a text file with all information (spatial, spectral, ...) associated to each pixel, in the order selected. These re-

NOTA: El pixel superior-izquierdo de la imagen tiene coordenadas :
Numero linea = 0
Numero columna = 0
La primera coordenada identifica el número de linea y la segunda el de columna
EL SIGNIFICADO DE CADA LINEA ES:
<numero_pixel>: (<num_linea>, <num_columna>) <valor_bandaI>, <valor_bandaJ>, ... <texto_region>

0: (27, 23) 88 41 42 114 121 46 [suelo]
1: (27, 24) 85 39 36 119 115 41 [suelo]
2: (28, 23) 86 42 43 113 118 42 [suelo]

Figure 5. Segment of a file, which contain the spatial, spectral and thematic information associated a different pixels.

sults have been used to train some neural network as NEFCLASS (Nauck 1997) and Supervised ART (Al-Rawi 1999). Figure 4 shows a part of one of such files. The first a second column of this kind of files represent the pixel coordinates, the next six columns the corresponding spectral signature and the last column displays the class.

Other option supplied by the tool is the generation of a text file with the information of an image classified by the several classifiers implemented in the global tool. It is well known the interest of disposing in a unique text file the spatial, the spectral and the thematic class information corresponding to a classified image. A segment of one of these files is shown at figure 5. In this figure only information corresponding to three pixels is represented; but all pixels are represented in the whole file. This kind of file has made easy the implementation of an option to translate the training and classified image information to standard spreadsheets. This is very useful, since commercial image processor do not always provide friendly data numerical analysis.

Figure 6 displays a scattergram for band 3 and 4 of the training areas of a particular TM image. This scattergram has been generated through a standard spreadsheets, which has imported the information from a file generated by the tool.

Figure 7 shows an image classified by a fuzzy classifier external to ERMAPPER, translated by the developed tool to ERMAPPER format and visualized through this last software. This option provide facilities to handle through ERMAPPER, images processed outside of it.

4 CONCLUSIONS

In this paper, it is presented an I/O interface tool that makes easy the communication between a commercial tool, ERMAPPER 5.5, an application

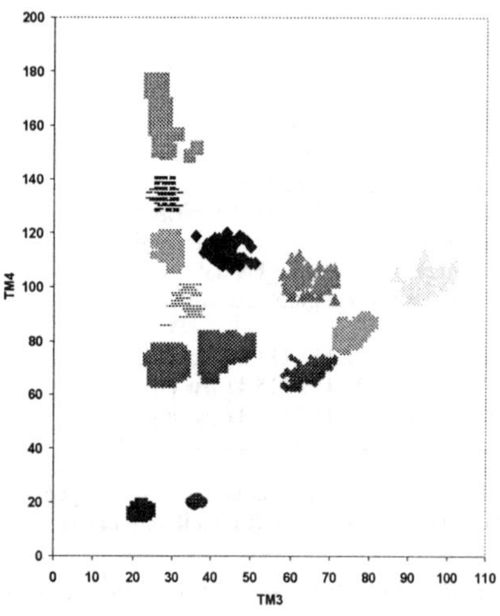

Figure 6. Scattergram for band 3 and 4 of the training areas of a particular TM image, imported by a standard spreadsheets from a file output tool.

developed to remote sensed images fuzzy classification and others possible external systems, like: numerical systems, neural networks, genetic algorithms, etc.

Dynamic techniques have been used to program the two libraries that integrate the developed interface: one for raster image reading (GRL_inter) and other to generate output files (Image_wri). These libraries work with the image stored in the main memory. That provides the tool very good performances such as speed, flexibility and access facilities. Moreover, these libraries present a powerful ability to graphic visualization by exploiting all graphic visualization power given by the commercial tool.

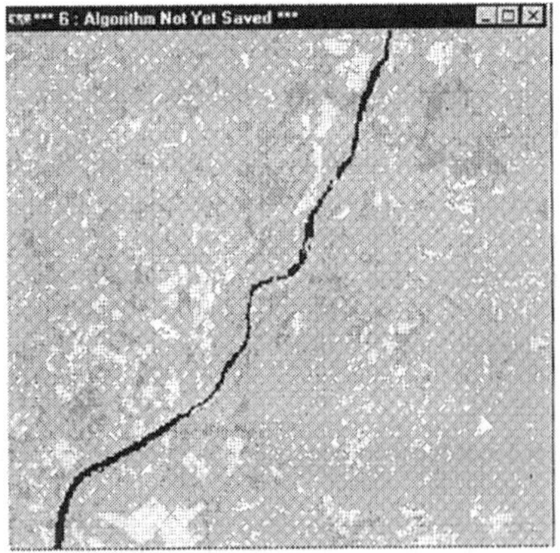

Figure 7: TM image classified by a fuzzy classifier external a ERMAPPER and visualized through it.

The design of these libraries is completely general, in the sense that it can handle many different images format with minimum code changes.

The developed tool allows the integration of different kind of information (spectral data, class code, pixel position, etc...) associated to each pixel image. Such information could be represented in one or more files according to user demand. These files are very useful in some applications like: neural network training, information translation to standard spreadsheets or different representation ways. Besides, this tool should support formats conversion from the commercial tool to other particular remote sensing applications and the inverse conversion, without commercial tool performance loss.

5 ACKNOWLEDGMENTS

The Spanish Advisory Commision for Scientific and Technical Research (CICYT) has supported this work under Grant TIC96-0388.

REFERENCES

ERMAPPER Tutorial. 1995, Earth Resource Mapping Pty. Ltd San Diego.

Nauck et al., 1997, A neuro-fuzzy method to learn classification rules from data, *Fuzzy set and systems,* 89, 277-288.

Al-Rawi, K.R., Gonzalo-Martín C. and Martínez-Izquierdo E., 1999, Supervised ART-II: neural network for Landsat TM image classificaction, *EARSEL 19th Symposium*, 1999.

Remote Sensing in the 21st Century: Economic and Environmental Applications, Casanova (ed.)
© 2000 Balkema, Rotterdam, ISBN 90 5809 096 5

Self-organized criticality in the environment and remotely sensed data

L. N. Vasiliev
Institute of Geography, Moscow, Russia

ABSTRACT: This paper describes one of the most rapidly developing field self-organized criticality in the environment which has attracted immense interest throughout geophysics. Self-organized criticality (SOC) refers to the tendency of the large dynamical systems to organize themselves into a "poised" state far out of equilibrium with propagating avalanches of activity of all sizes. The idea was taken up by (Bak, Tang and Wiesenfeld 1987) who studied the behavior of some models. We outline this concept focussing on particular application of this theory to rainfall. The experiment is built around the rain-gauge measurements at 4 stations in the river basin area of the Ob an Irtysh in the West Siberian Plain. The experience of SOC application to the studies of 49–year rainfall time series in various climatic zones with a resolution of 12 hours, permitted a new understanding of the avalanche-like mechanism of their origin. The critical state can be characterized by the set of exponents and the rainfall obeys the power law behavior.

1 INTRODUCTION

The idea of self-similarity of spatial structures of geosystems has been confirmed in many experiments. Trying not to be too optimistic, we may argue that certain structures demonstrate fractal properties, which make it possible to determine values of scaling exponents and to extrapolate the measurements to minimal sizes. The data about scaling in geosystems, so derived and associated with the components of the latter, characterize only 'frozen' structures leaving open the origin of their generation mechanism. This considerably complicated theoretical justification of the fractal nature of spatial structures and sometimes put to doubt interpretation of the experiments. Indeed, in many cases experimental data fell in the range of self-similarity within 10^2, which could in no way prove the fractality of the structure. The main difficulty such theoretical justification meets is the complexity, diversity of the landscape with its intertangled internal and external relationships and numerous degrees of freedom.

A breakthrough yielding a theoretical approach to explaining dynamic processes in complex systems was made (Bak et al. 1997) when the idea of self-organized criticality was elaborated. This self-organized criticality (SOC) characterizes the tendency of complex dynamic systems to develop till some statistically steady state. The behavior of a system in such a state refers to the tendency to organize itself into a 'poised' state far out of equilibrium with propagating avalanches of all sizes and duration. Numerous SOC phenomena are represented by models of a sand pile, forest fire and evolution. The SOC theory has not yet been fully developed and requires further effort to be applied to geophysical processes in order to answer the question whether natural processes show a common feature typical of the SOC-theory. Probably this paper is one of the first trying to apply the SOC to geographical problems. To take theory and methodological approach into consideration a brief description will be useful to show how the sand pile model implements the main features of self-organized criticality in terms of (Bak et al. 1998).

A pile of sand builds up when sand grains are randomly dropped on a horizontal table. The pile will grow and the slope will increase. Eventually, the slope will reach the critical state and if one more sand grain is added it will slide off and the pile will collapse. The boundary conditions are such that sand can leave the system. The sand pile will grow to a statistically stationary state where the amount of grains added is balanced by the amount of sand falling off the edge. This stationary state is an attractor for the dynamics. The response to the dropping of a single grain of sand can be thought of as a critical avalanche reaction associated with singular behavior. For the sand pile shaped so that its slope is less than critical, the process will be sub-critical and no avalanches would be generated. The critical state is self-organized in the sense that no external tuning is needed to carry the system to this state. This is the principal difference of our situation from the one where the critical point is reached by tuning, say, the temperature in the liquid-gas phase transition, or the Curie point when the ferromagnetic loses its properties. The sand pile ultimately reaches a stationary critical state in which its slope fluctuates about a constant angle of repose, with each new grain capable of inducing an avalanche.

A pile of sand may be considered both literally and metaphorically. Most attractive is a thermodynamic application of this model. Potential energy of sand grains falling down is dissipated in avalanches. Energy/heat accumulation and a locally exceeded threshold trigger avalanches followed by slow relaxation. With the SOC-theory the formation of certain natural systems far from equilibrium may be explained by a process that evolves the system to the state like a dynamic attractor which demonstrates critical properties. The domains where SOC could be applied are not limited to grain substances, still obeying the theory: slowly controlled non-equilibrium systems are being self-organized into a critical state so that the catastrophic events become a 'normal phenomenon' with a well-developed functional dependence.

Sand pile simulation employs the following algorithm: if the height difference between two successive positions along the sand pile exceeds the critical value $z_{i,j} - z_{i+1,j} > \Delta z_c$ the four grains from a site with a height z_i should be redistributed among neighbor-cells

$$z_{i,j} \to z_{i,j} - 4,\ z_{i,j\pm 1} \to z_{i,j\pm 1} + 1,\ z_{i\pm 1,j} \to z_{i\pm 1,j} + 1$$

If employed, this rule leads to a process with the following properties

1. Avalanche dynamics, with the sizes of avalanches statistically distributed in space as well as in time according to the power law with a negative exponent, which makes senceless the use of a characteristic size scale and a characteristic time scale.
2. Self-adjustment to critical state/point. This stationary state is an attractor for the dynamics.
3. Long-duration repose state is interrupted by irregular avalanches.
4. The character of interactions among the neighbors, that is, the microscopic form of grains, does not affect the robust self-organization to the critical state. Only the exponents of the power law dependence may change.

Note that these conclusions rely not only on computer simulation, but also on grain-sorted sand (Nagel 1992) and two kinds of rice, size ratio being different.

2 SELF-ORGANIZED CRITICALITY IN PRECIPITATION

The hydrological cycle and its intensification in the presence of global climate changes are among interesting and useful applications of the SOC theory. We believe that analytically the main property of the hydrological cycle may be described by the SOC model when the energy is being stored in each site, whereas water evaporation enters the atmosphere to be then precipitated as avalanches of rain. Energy dissipation pattern in this avalanche dynamics may help explain the mechanism of behavior of floods and droughts. Note that SOC in the hydrological cycle differs significantly from the situation in the abstract theoretical concept. The idea is that a homogeneous space forms in the sand pile model whereas in a genuine natural system such a space is considerably heterogeneous. That demands that the local critical levels of Δz_c be set up, which are directly relevant to the heterogeneous geographical space.

Precipitation is one of the main components of the hydrological cycle. Scaling of nonlinear behavior is a difficult problem. Clouds microphysics and particle growth, as well as regional and global patterns of temperature, humidity and winds, control the intensity, scales and timing of rainfall at the peak point. Efforts to improve these models are hampered by the lack of realistic data. New analyses of infrared and microwave remotely sensed data are being undertaken to fill the gap in our knowledge of the global distribution of precipitation. The results should establish the necessary link between deterministic and stochastic models of rainfall fields. The relationship between rainfall, runoff and evaporation is highly nonlinear and requires more complex representation schemes.

A large number of papers are well known which emphasize the multifractal character of rainfall timing, however, they do not explore the underlying scaling mechanism, typical of fractal structures (Tessier et al. 1996). The idea of self-organized criticality can be thought of as a paradigm for the ubiquity of spatial and temporal fractals in precipitation distribution. The atmosphere system reaches a stationary state when water vapor added to the system equals, on average, the precipitation leaving the atmosphere in the form of avalanches. The critical state can be characterized by the following set of critical exponents. With s denoting the area involved in an avalanche the distribution of avalanches is described by the power law:

$$N(s) \sim s^{-\beta} \qquad (1)$$

where $N(s)$ is the number of rainfalls over the area s. Similarly

$$N(P) \sim P^{-\gamma} \qquad (2)$$

for rainfall amount in rain-gauge measurements,

$$N(t) \sim t^{-\lambda} \quad (3)$$

for rainfall duration and

$$N(z) \sim z^{-\tau} \quad (4)$$

for time intervals without precipitation.

The avalanche dynamics leads to spatiotemporal complexity, and emerges as a result of extremal dynamics in driven systems. Spatiotemporal complexity is manifested in the formation of fractal structures, the appearance of *1/f* noise and punctuated equilibrium behavior.

3 RESULTS

The experiment is built around the rain-gauge measurements at 4 stations in the river basin area of the Ob an Irtysh in the West Siberian Plain, within 52° and 70° N. Long-term rainfall series cover a 49-year period from January 1936 till December 1984. Rainfall intensity was measured at a time resolution of 12 hours, the minimal recording being 0.1 mm. Strictly speaking, the data do not permit us to determine the amount of precipitation over a given area (i.e., avalanche dimensions), hence we restrict ourselves only to its intensity recorded at the stations.

According to the general SOC theory, the γ, τ and λ exponents were estimated for each series, and the results are shown in Figures 1, 2 and 3. The log-log plots in Figures 1a and 1b show the results for the distribution of the number of events, $N(P)$, against rainfall intensity, P. Two groups of points follow two pretty respectable straight lines with slopes $\gamma_1 = 1.73$ and $\gamma_2 = 0.65$ in Figure 1a and with $\gamma_1 = 1.73$ and $\gamma_2 = 0.76$ in Figure 1b. The fact that the distributions, $N(P)$, in Figures 1a and 1b show two types of power law behavior may be accompanied by an explanation. The events with the largest $N(P)$ values, lying along the straight lines with the slope $\gamma_2 < 1$, occurred among 65 percent rainfall within the interval $P = [0.1, 5\text{mm}]$, which corresponds to about 30 percent of all precipitation there over the 49-year period. There are 35 percent events grouped along the straight lines with slope $\gamma_1 = 1.73$, which corresponds to about 70 percent of all precipitation with $P > 5\text{mm}$.

The events obeying the power law behavior (2), the exponent $\gamma_1 = 1.73$, are influenced by the low time resolution of rainfall recording. Indeed, if precipitation has been recorded for several subsequent 12-hour intervals, in no way should it mean that it was a non-interrupted event. Therefore there could be several rainfalls of smaller intensity within this time interval, and the events should be split up and the new points in the plot should then be shifted left along the $x-$axis. That will increase $N(P)$, the number of events with smaller *P*. Such a redistribution of *P*-value groups will occur as a result of higher time resolution of precipitation recording. The cutoff in the precipitation rate with a 12-hour resolution is present within the range $P < 5\text{mm}$ but it

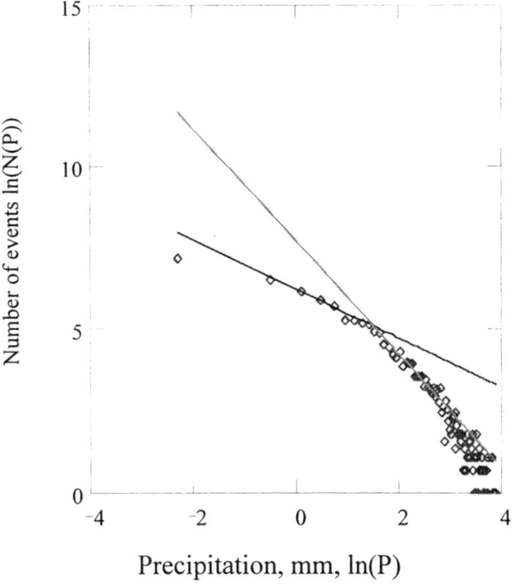

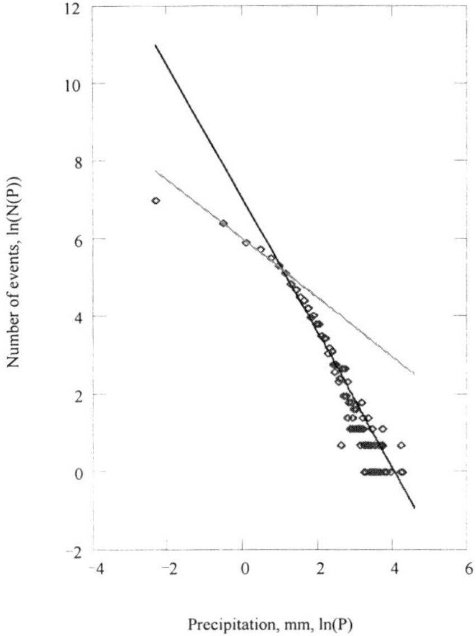

Figure 1. Logarithmic plot of the distribution of precipitation for 49 years (1936 – 1984) for rain gauge measurements: a) Alexandrovskoe: the slopes of the straight lines are $\gamma_1 = 1.73$ and $\gamma_2 = 0.65$; min (*P*) = 0.1mm, max (*P*) = 112 mm. b) Irtyshsk: the slopes of the straight lines are $\gamma_1 = 1.73$ and $\gamma_2 = 0.76$; min (*P*) = 0.1mm, max (*P*) = 76 mm.

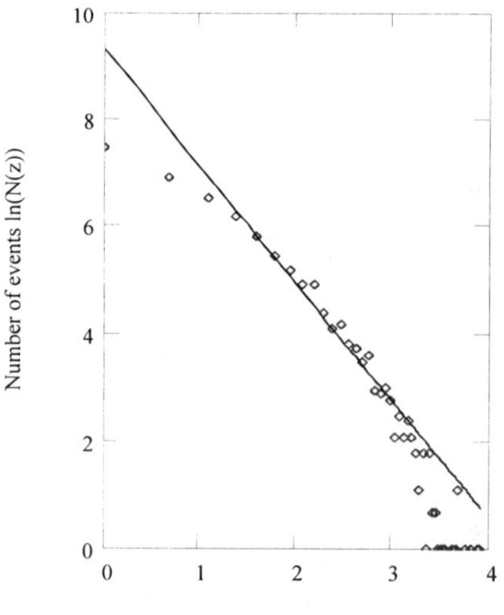

Figure 2. Distribution of timing of dry periods for 49 years for Alexandrovskoe: $\tau = 2.18$, min $(z) = 0.5$ day, max $(z) = 25$ days.

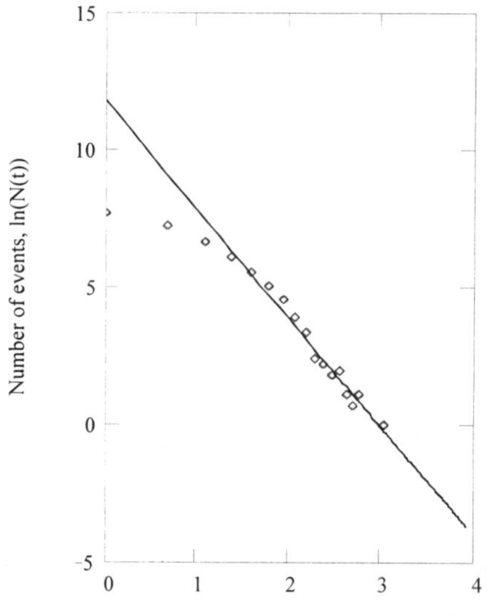

Figure 3. Distribution of timing of rainfall for 49 years for Alexandrovskoe $\tau = 3.87$, min $(t) = 0.5$ day, max $(t) = 10.5$ days.

will decrease due to the higher time resolution. The power law dependence (2) with two exponents γ demonstrates two different SOC behavior patterns of precipitation.

The second parameter of avalanche dynamics in the SOC belongs to the dry time distribution corresponding to a dry period of z–duration. Figure 2 shows $N(z)$- values, the number of cases of a dry period versus its z – duration, the resolution being half a day. The curve is consistent with a straight line, indicating the power law $N(z) \sim z^{-\tau}$. The straight line has a slope $\tau = 2.20$. The crossover in the dry period occurs in the domain of short intervals between rainfalls. At small time intervals the curve deviates from the straight line because the coarse graining effects of the time resolution of rainfall time-series come into play. Figure 3 shows the distribution of rainfall duration. This quantity also obeys a relation of the power law form $N(t) \sim t^{-\lambda}$. The straight line has a slope $\lambda = 3.60$.

In general, SOC signatures identified in the long-term series of precipitation are characterized by three critical power law exponents: γ, τ, λ - the distribution of precipitation at the rain gauge station, the distribution of time intervals without rain and the distribution of rainfall duration. The emergence of the self-organized critical state may connect nonlinear dynamics with the appearance of multifractal structure in rainfall. Scaling of three parameters P, z, t defined by exponents γ, τ, λ and characterizing the scale invariance of the precipitation is one of the argument for calculating the fractal dimension for this process. Standard method will be used for this purpose. The time distribution of events forms a fractal set like the Cantor dust with the range of self-similarity [2, 30 days] and fractal dimensions $D = 0.90 \pm 0.01, D = 0.85 \pm 0.01$. The crossover to $D = 1$ is found at the scale more than 30 days.

A Cantor dust's gap lengths must satisfy $N(Z > z) \sim z^{D}$ in the sense that a gap viewed as a period is without precipitation. Therefore, $\tau = D$ should have been observed, but it is not the case. The power law distribution of the dry periods less than two days is determined by the $\tau > 1$ exponent, however these events amount only to 35 percent. Thus the distribution of the dry periods also exhibits dual scaling behavior of the precipitation within short- and long-term intervals. The experiment indicates that the rainfall is characterized by a set of independent critical exponents and two different time scales of the variables t and z. That means that rainfall dynamics regulates by two independent atmospheric process leading to rainy and dry periods.

The scaling of the time distribution of rainfalls is such that permits its extrapolation up to certain deliberately chosen minimal values (Table 1). Thus, for instance, when the rainfall amount with a time resolution of 1-day was extrapolated to a resolution of ½ day, over a 49-year period the following estimates were derived, the accuracy being 5 percent.

Table 1. Extrapolation from low to high time resolution

Resolution	Measurements	Extrapolation
24 hours	1865	
12 hours	3122	3285
6 hours		5788

This demonstrates how lower resolutions affect the quantities $N(P)$, $N(t)$ and $N(z)$. On the other hand, high-resolution rainfall time series may have the power spectrum which obeys a power law form (Olsson 1995).

Self-organized criticality in precipitation leads – in the case of Cantor dust with unequal parts – to the multifractal distribution of precipitation measure, this precipitation being described by the spectrum of fractal dimensions shown in Figure 4. It implies that rainfalls demonstrate multiscaling behavior over the time scale T:

$$P \sim T^{-\alpha(q)} \quad (5)$$

$$n(P) \sim T^{f(\alpha(q))} \quad (6)$$

where q = moment exponent of a q-th order; $\alpha(q)$ = rainfall amount exponent; and $f(\alpha)$ = fractal dimension. Table 2 summarizes the estimates of SOC exponents for four rain-gauge measurements.

Table 2. The estimates of SOC exponents for rain-gauge measurements

Station	γ_1	γ_2	τ	D	P, mm	$N(E)$
Khanty-Mansiisk	1.63	0.55	2.20	0.908	523±104	5265
Alexan-drovskoe	1.73	0.65	2.18	0.905	487±76	5475
Irtyshsk	1.73	0.76	1.78	0.850	271±67	4042
Kamenny Cape	1.83	0.95	2.38	0.904	381±84	5534

P, mm = average annual precipitation; *N (E)* = number of events with a ½ day-resolution.

Correlation between exponents and fractal dimensions demonstrates their sensitivity to climatic conditions.

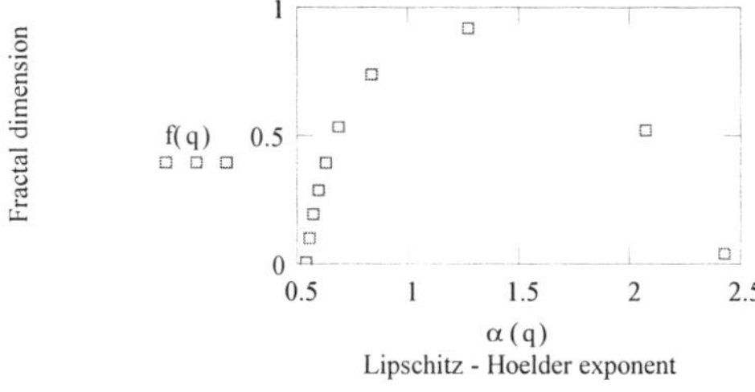

Figure 4. The fractal dimension $f(\alpha)$ of the rainfall time series plotted as a function of $\alpha(q)$ for rain gauge measurement in Alexandrovskoe during 49 years.

4 CONCLUSION

The concept of self-organized criticality combined with the theory of complexity and cellular automation is getting more and more attractive for studying the dynamics of geosystems. The experience of SOC application to the studies of 49–year rainfall series in various climatic zones with a resolution of ½ day, permitted a new understanding of the avalanche-like mechanism of their origin. To fully confirm all aspects of the avalanche dynamics and self-organized criticality in rainfall, microwave and infrared satellite measurements should be employed for estimating the water mass bulk and the area they cover. These satellite data are mainly needed to determine the bulk of the precipitated mass and the process duration for each isolated event and to estimate its 3-D size.

The sand pile model is an idealized homogeneous system, which should be adapted to complicated natural landforms varying their behavior according to external conditions. Hence, one should be careful about the power-law exponents and find each time their geophysical interpretation. The experiment demonstrated that critical conditions leading to SOC appearance, differ in space and time. Further development of these typical features in the model of a sand pile is one of the important trends of practical application of SOC to natural processes.

REFERENCES

Bak, P., C. Tang & K. Wiesenfeld 1987. Self-organized criticality: an explanation of 1/f noise. *Phys. Rev. Lett.* 59: 381-384.

Bak, P., C. Tang & K. Wiesenfeld 1988. Self-organized criticality. *Phys. Rev. A* 38: 364-374.

Nagel, S.,R. 1992. Instabilities in a sandpile. *Rev. Mod. Phys.* 64: 321-325.

Olsson, J. Limits and characteristics of the multifractal behavior of the high-resolution rainfall time series. *Nonliear Proc. Geophys.* 2(1): 23-29.

Tessier, Y., S. Lovejoy, P. Hubert, D. Schertzer & S. Pecknold 1996. Multifractal analysis and modeling of rainfall and river flows and scaling, causal transfer functions. *J. Geophys. Research* 101: 26427-26440.

Remote Sensing in the 21st Century: Economic and Environmental Applications, Casanova (ed.)
© 2000 Balkema, Rotterdam, ISBN 90 5809 096 5

ECEP-ECHO: Extraction and classification of edge pixels in an ECHO approach

J.García-Consuegra & A.Martínez
Sección de Teledetección y SIG, Departamento de Informática, Universidad de Castilla-La Mancha, Albacete, Spain

G.Cisneros
Grupo de Tratamiento de Imágenes, Universidad Politécnica de Madrid, Spain

ABSTRACT: One of the most critical activities of Remote Sensing consists in the classification process. This can be accomplished using the pixel or the homogeneous object as the basic element to be classified. The advantages of homogeneous objects over per pixel classification are already known in the scientific community. On the other hand, a possible source of inaccuracy is the edge pixel classification, due to its mixture nature. In this paper, our interest is focused on refining the homogenous object extraction, and particularly, the precise classification of the edge pixels of those objects.

1 INTRODUCTION

Environmental remote sensing attempts to identify the zones occupied by the discriminated or given types of terrain, and their location, by using multispectral satellite imagery. This process involves labeling each pixel as belonging to a class. Traditionally, the multispectral satellite imagery has been segmented through the classification process. Characteristic feature thresholding or classification can be accomplished by two main techniques: supervised and non-supervised. Many research efforts have been made to reduce or avoid the problems arising from the limitations of the above methods and the nature of satellite imagery. These efforts, despite the promising results produced, continue to use the pixel as the basic independent element to be classified, that is, they ignore, in the majority of cases, the local characteristics given by their position and neighborhood (Deusen 1995).

The ECHO (Extraction and Classification of Homogeneous Objects) approach arose as solution to the “salt and pepper” problem (Kettig & Landgrebe 1975). However, it also provides a solution to a lot of problems in the per pixel classification methods. As a first step, ECHO identifies those areas (homogeneous objects) with the “same” general cover type. Where the term “same” makes reference to the criteria used to establish the similarity among the pixels. In the second and last step, the extracted homogeneous objects are classified, in such a way that all pixel belonging to a same object are labeled in a single classification decision.

ECHO, like those pixel-based classification techniques, shows its main critical point in the so-called "weakness" edge pixels. The classification result is very sensitive to the parameters. So, permissive parameters entail merging between distinct or distinguishable adjacent regions. On the other hand, severe parameters involve an over-segmentation; that is to say, the natural regions in the image are divided into several homogeneous objects.

In this paper, our interest is focused on refining the homogenous object extraction. We are particularly interested in the precise classification of their edge pixels, making at the same time the process more robust against parameter variations. ECEP-ECHO (Extraction and Classification of Edge Pixels in an ECHO approach) stands for the foundations on which the proposal algorithm is based. The process is a variant of the edge detection and region extraction technique integration proposed by Pavlidis, in (Pavlidis & Liow 1990).

In the following sections, the background and outline of the algorithm are described. The next section shows the results of applying this methodology to a small multispectral Landsat TM image. In the last section, our conclusions reached are discussed.

2 METHODOLOGY

In low-level vision, that a cooperation between two techniques for segmentation produces much better results than what would have been obtained by either technique alone has already been demonstrated. In particular, a survey of the region growing and edge detection integration can be found in (Pavlidis

& Liow 1990) whilst, a survey of the multispectral case is done in (Moigne & Tilton 1995).

Region-based methods are based on finding parts of images, which are homogeneous for a given set of properties. We have chosen the region growing technique because it can be used in supervised and unsupervised fashion. At the same time, it allows a wide set of segmentation parameters: statistical texture parameters, fractal measurements, similarity, etc.; and, does not restrict the region shapes to a specific data structure such as a quadtree.

Region-based segmentations always provide closed contour regions and make use of relatively large neighborhoods for decision making. However, the region growing process can lead to three kinds of errors: a) a boundary is not an edge and there are no edges nearby; b) a boundary corresponds to an edge but it does not coincide with it; c) there exist edges with no boundaries near them.

On the other hand, edge-based segmentation approaches track boundaries between regions, based on contrast characteristics, instead of seeking homogeneous regions. Edge-based segmentations provide an accurate location of the region boundaries, but those contours are usually not closed. Furthermore, these methods are often very sensitive to local variations in intensity and are affected by two types of errors: d) there are edge points due to noise; e) edge segments are not closed or not continuous.

The main issues involved in ECEP-ECHO are (Fig. 1): edge detection and region extraction algorithms as well as their integration, edge pixel classification and, finally, region merging.

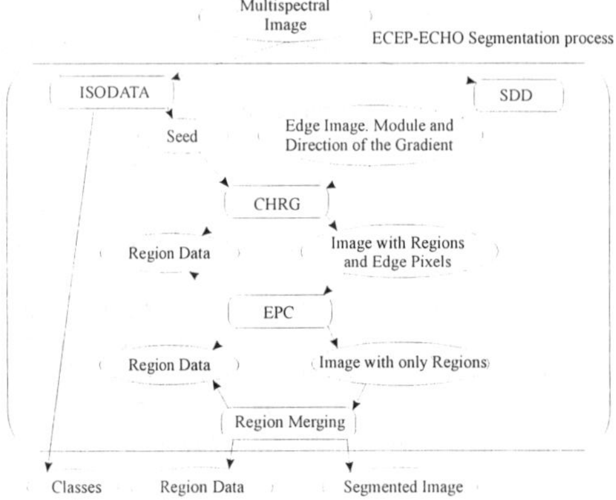

Figure 1. An overview of the texture segmentation algorithm.

2.1 *Edge detection*

Edge detection methods based on the analysis of zero crossings of some second-order differential operator applied to one-band image data have been extensively explored.

Nevertheless, the above-mentioned methods have awakened little interest in the case of multispectral images, probably because of the problems implied by the vectorial character of each of their pixels (Cumani 1991).

In (Cebrián et al. 1993), a study is carried out on the possible application of already existent multiband image edge detectors to the detection of edges in remotely sensed images. Two methods are proposed in this paper, one of which pretends to be optimal for this kind of applications. A one-based image edge detector, based on the second directional derivative (SDD) in the direction of the gradient (Canny 1986, Torre & Poggio 1986), is presented as the best candidate to be extended for the detection of edges in multiband images. The SDD is the most appropriate due to its connection with the natural definition of edges in intensity images, as the loci of maximal gray-level gradient (Cumani 1991).

A combination of the information contained in the different spectra is needed somewhere along the edge detection process, to obtain a single one-band edge image. Depending on the moment in which this combination takes place and how this combination is done, different methods are possible. Four such methods are described in (Cebrián et al. 1993): the feature average method, the multispectral contrast method, the maximal gradient method and the method that combines the edge images of the image bands. Though slow, only the last method is able to preserve all the edges of the feature space. Faster edge detection is only possible by combining the information contained in the different image bands at an early moment in the edge detecting process; but this will make some edges of the image bands undetectable.

The edge detection based on the combination of the edge images of the features establishes that a pixel is a multiband image edge point if, and only if, it is an edge point of any of its bands. This method proceeds as follows:

- Independent application of the SDD monocromatic edge detection method to each image belonging to the feature space.
- Superposition of all edge images obtained from the different spectra (Fig. 5d).
- Reduction of the edge width by a thinning algorithm described next (Fig. 5e).

The application of a Gaussian filter to an image band slightly displaces the edges. The direction of this slight displacement depends, for each edge point, on the luminance function of the image band around the edge point. It is, therefore, understandable, that the superposition of edges of the independently filtered bands will result in variable width lines, where edges of the unfiltered bands coincide. This effect makes multiband image edges unclear. The method to thin the edges to one pixel proposed in (Cebrián et al. 1993) contains the following steps:

- Assign an arbitrary priority to each image band.
- When superposing band edge images, give each pixel a value that identifies the higher priority band, in which this pixel is an edge point. If the pixel is not an edge point of any band, mark it as background.
- Taking in a descending priority order all pixels, which are edge points in some band, check if any of their neighbors is an edge point in a higher priority band. If so, mark the pixel as background.
- Finally, all pixels, which are not marked as background, are given a value that identifies them as definitive edges.

This edge detection method has some drawbacks. The edge detection process has to be entirely repeated for each image band, making it slow. The boundary between two adjacent regions, which are only distinguishable as result of an image combination in the chosen feature space, will not be detected (Fig. 2). The use of the abovementioned, or another, thinning algorithm can misclassify edge pixels.

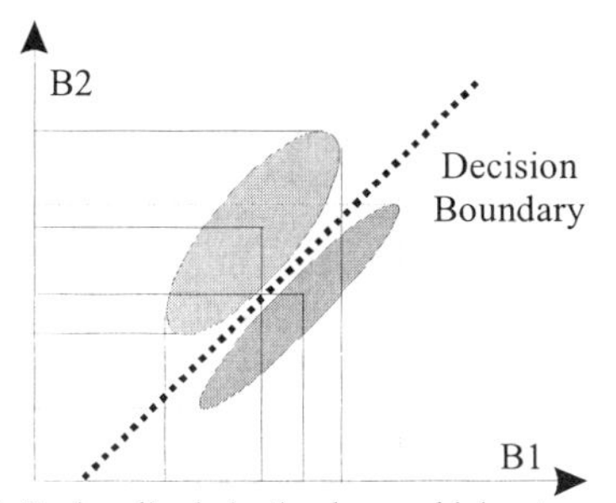

Figure 2. Region discrimination by combining two, or more, bands.

2.2 *Region growing*

Region growing is an iterative process by which regions are merged starting from individual pixels, or another initial segmentation, and growing iteratively until a region encompassing the whole image is reached. A region growing algorithm is characterized by the approaches chosen to solve its critical points:

- How the seed are chosen (if any).
- How the region growing is carried out. Sequential, each region is extracted regardless of the remain regions; or parallel, only the best merging is allowed at each iteration. 4 or 8 connectivity. And, finally, the matching reference used: the spectral centroide of the region or the region-expanding pixel.
- Similarity and stopping criteria

In this paper, a seed set is obtained in an unsupervised way. To accomplish this task the ISODATA clustering algorithm has been modified in order to labeled only those pixels with "enough" similarity degree with the centroid of the spectral classes obtained. This ISODATA version is justified by the fact that those pixels closer to its centroid are the more similar (homogeneous); the rest of pixels, on the other hand, being that present the highest dissimilarity degree. These dissimilarities can be due to artifacts, pixels belonging to a region edge, mixture pixels, different image acquisition conditions, etc.; that is to say, the inappropriate seed pixel.

At each iteration, one or several merges occur. The drawback of sequential growing is the dependence of the resulting segmentation on the order of the merges. On the other hand, if at each iteration only the best merge is authorized (parallel approach), the segmentation is independent of this order but the computation time becomes prohibitive except for small images, even when implemented on a massively parallel processor. The outputs of ISODATA algorithm are a set of patterns of the identified spectral classes as well as a seed image.

In ECEP-ECHO, the sequential technique has been chosen because it is the faster. Its drawbacks have been overcome thanks to the edge data integration in its growing stage as shown in the following point.

Depending on the matching reference the region growing algorithm can be grouped into three categories: Single Linkage Region Growing (SLRG), neighboring pixels whose properties are similar enough to the region-expanding pixel are merged with the region. Centroid Linkage Region Growing (CLRG), in this case the adjacent pixel is matched with some meaningful statistics of the region. Hybrid Linkage Region Growing (HLRG), pixels that are similar are so because their neighborhoods in some special sense are similar. The strength of the SLRG is that boundaries are placed in a spatially accurate way whilst the strength of the CLRG is its ability to place boundaries in weak gradient areas. In ECEP-ECHO the CLRG has been chosen due to its performance with weak edge pixels.

The common similarity criterion used in this kind of techniques is to apply the same user-define similarity threshold to the entire image. This approach lacks in adaptability to local area situations, producing both under and oversegmentation effects. One of the locally adaptable similarity criterion, applied with success in remotely sensed imagery, is the Vector Degree of Match (VDM) similarity (Baraldi & Parmiggiani 1996).

The use of VDM has a great number of advantages, among which its high sensibility with sufficient contrast, in some of the bands. However, VDM has a biased behavior that favors the differences occurring in those bands with a high scattering (standard deviation) in detriment of the rest. This becomes a problem when class discrimination is carried out, as can be gathered from the VDM definition. Thus, as can be seen in Figure 3, the hypervolume of similarity for a given pixel is constant in any dimension (since α is constant), which means VDM takes all differences equally into account. As a solution, a first-order statistical normalization is used (Escribá et al. 1997). This method is based on a linear transformation, the parameters of which are obtained from the mean (μ) and the standard deviation (σ) of both bands.

$$X' = a * X + b, \quad a = \frac{\sigma_{Ref}}{\sigma_X}, \quad b = \mu_{Ref} - (a * \mu_X) \tag{1}$$

where X and X′ are the tonal levels of the band to be normalized and the normalized band, respectively; a and b are the linear transformation parameters; σ_{Ref}, σ_X, μ_{Ref} and μ_X are the mean and standard deviation of the reference band and the band to be normalized, respectively. In order to avoid a loss in the discrimination degree in this transformation, the band with the highest σ must be chosen (Fig. 4).

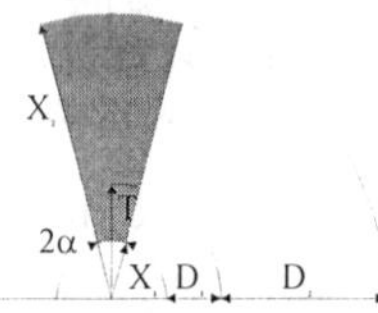

Figure 3. Hyper-volume of similarity for a given pixel (T), in the spectral domain, as defined by the VDM criterion. This hyper-volume is defined by the following polar coordinates: i) the angle 2 α; and ii) the lower and upper bounds of the radius, represented by the moduli of vectors X1 and X2. It is important to observe that D1 < D2. T is not located at the center of the similarity hyper-volume of X.

2.3 *Region growing and edge extraction integration*

In this paper, following the ECHO classification technique, an integration of edge and region data has been carried out in order to take advantage of their relative strengths. A brief survey of edge/region integration techniques can be found in (Le Moigne & Tilton 1995, Pavlidis & Liow 1990). The strength of the region-oriented segmentation is that the regions are extracted by exploiting the spatial information. Its weakness is that segmentation errors often occur at the boundaries of the regions. Furthermore, in the absence of other information, stopping criteria for iterative methods, such as region growing, are often difficult to choose. On the other hand, edge-based segmentation techniques track boundaries between regions, based on contrast characteristics. These methods provide an accurate location of the region boundaries, but the contours are not usually closed, and they are very sensitive to local variations in intensity.

The choice for edge/region integration lies in the timing of this fusion: subsequent to both processings or embedded in the region segmentation. This choice can also be described as integration through post-processing or integration through the definition of new parameters or new decision criteria for the region segmentation. like postprocessing, edge and region information are extracted independently in a preliminary step, and then integrated together (Le Moigne & Tilton 1995, Pavlidis & Liow 1990). When integration is affected through the definition of new region segmentation criteria, the edge information is first extracted, and this information is then used within the region segmentation algorithm (Gambotto 1993).

Here, we chose to perform the integration of edge and region data embedded in the region growing, as a new stopping criterion, at pixel level. In this way, ECEP-ECHO gives a solution to the weak edge pixel besides the sequential region growing processing. Furthermore, in order to enforce the processing of the "weak" edge pixels, the parameters are set up so that an over-segmented image results.

The CHRG procedure, see Figure 1, carries out the edge/region integration. When the image segmentation is accomplished through determining homogeneous regions, then edges are defined as region boundaries. So, the aim of this stage is to classify each pixel as belonging to a region or as edge/boundary (labeled as EDGE).

CHRG has the following inputs: SDD resulting images (module and gradient orientation), those pixels with a module less than, or equal to, a given threshold are labeled as EDGE; the seed images obtained from the modified ISODATA algorithm, and the multispectral image.

In the region growing process, an adjacent pixel will be merged if, and only if, it has enough similarity to the region spectral centroid, measured with the VDM criterion. The region growing process stops when all adjacent pixels to the region is not similar enough or is labeled as EDGE. These pixels represent the *a* and *b* conflictive pixels, in (Pavlidis & Liow 1990). Thus, the misclassification due to this type of errors becomes hereinafter a problem of the Edge Pixel Classification (EPC) process.

As oversegmentation results, the probability of c-type errors (there are edges with no boundaries near them) is reduced significantly. The e-type errors disappear by using the region growing algorithm in the image segmentation.

On the other hand, d-type errors are processed in function of its spatial relationship with the region. So, those pixels affected by noise (artifacts) or small regions representing small acquisition condition variations inside the region (the salt-and-pepper effect) will be merged if: their dimensions are less than a threshold (measured as number of pixels), and their centroid is similar enough to the region centroid. In this way, small regions such as pools are not merged. The final classification of those pixels or small regions that are adjacent to the regions is carried out in the Region Merging stage (Fig. 6a).

2.4 *Edge Pixel Classification (EPC)*

The aim of this stage is to classify the EDGE pixels, as belonging to any adjacent region.

Labeling the edge pixels (yielded in the SDD process) and boundary pixels (obtained in the region growing process) as EDGE, the edge width is increased (Fig. 6a). We can group the approaches to carry out the edge pixel classification into two categories:

- Growing the regions using a SLRG technique.
- Labeled each EDGE pixel as belonging to the region with which has the highest similarity degree.

In the first case, despite boundaries being placed in a spatially accurate way, we have the same problem when a sequential algorithm is taken. However, as the number of unlabeled pixel is reduced, the computation time of a parallel version is also reduced. The second case allows us to take a sequential approach with a high probability of accuracy, since for each EDGE pixel are studied its similarity with all candidate adjacent region. So, ECEP-ECHO chooses this second approach to accomplish this task.

An approach, exclusively based on classifying EDGE pixels as the adjacent region (or limited to those regions in their gradient direction) with high similarity degree, can reduce small regions such as roads or pools. In order to avoid this situation, the EPC process has been carried out in two steps:

- In the first one, EDGE pixels are merged to that region, to which they are similar enough and, are located adjacent: in their gradient direction (when a region is limiting it in each direction); or, in the perpendicular direction to their gradient (when the EDGE pixel is limited by a region and a EDGE pixel). This process is iterative until no new merging is carried out. The remaining EDGE pixels represent the boundary of small regions with local discriminative characteristics.
- In the second steps, those similar EDGE pixels are merged making up regions (Boundary Regions - BR). The common characteristic of this BR is that they are formed by a reduced number of pixels. In this way, the region is the element to be classified.

2.5 *Region Merging*

In order to reduce the number of extracted regions a region merging process is carried out. To that end, any method that calculates the similarity between two regions can be used (Mather 1987). In our implementation, we used the method of region merging by a bi-directional comparison of its histograms (Chang & Li 1994) extended to the multidimensional case. The method establishes, for each region and each spectral band, two limits of acceptance: a maximum and a minimum. These limits are calculated from the spectral distribution of the pixels of the region and from a parameter α provided by the user. Therefore, given the histogram of an spectral band for a region, a number of $(\alpha*size(Ri))/2$ pixels are discarded from both ends of distribution (in this work, we used a value of $\alpha=0.1$). Once the algorithm has calculated the limits of acceptance of all the bands of the regions, the comparison of two regions consists in checking for each band, to check that the average of a region is in the zone of acceptance of the other region and vice versa. In this comparison, several special cases have to be taken into consideration; for example, if there is a large region that is being compared with a smaller one. In this specific case, we only check that the characteristic vector of the smaller region is within the hypervolume of acceptance of the bigger one in the feature space. If two small regions are to be compared, then the non-parametric test of Mann-Whitney (Conover 1980) is used on the characteristic vectors of each region.

3 DATA AND RESULTS

Three-band Landsat V TM images, 128 x 128 pixels in size, showing an agricultural area in Spain (see Fig. 4) were chosen as a test. The month (July) was chosen to facilitate the plots of land extraction and the class differentiability. This area was chosen as a real-world area whose several surface features can be easily recognized by a photo-interpreter.

The following images show the results obtained by the different procedures at each step.

4 CONCLUSIONS

Our proposal solves some problems arising when the VDM is used. The clustering process tries to resolve

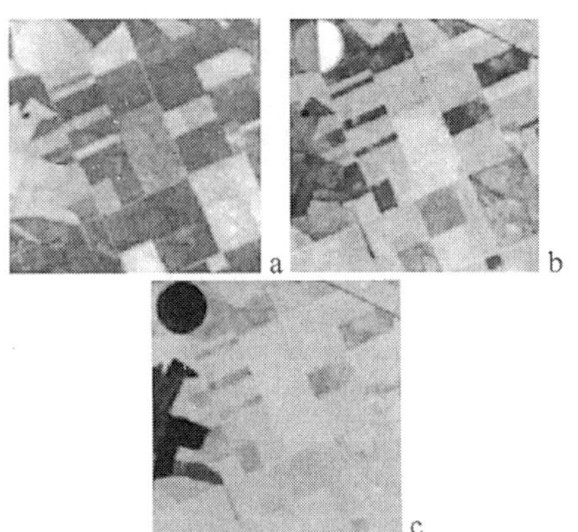

Figure 4. Landsat V TM images, (a) band 5 of May and (b and c) bands 4 and 5 of July.

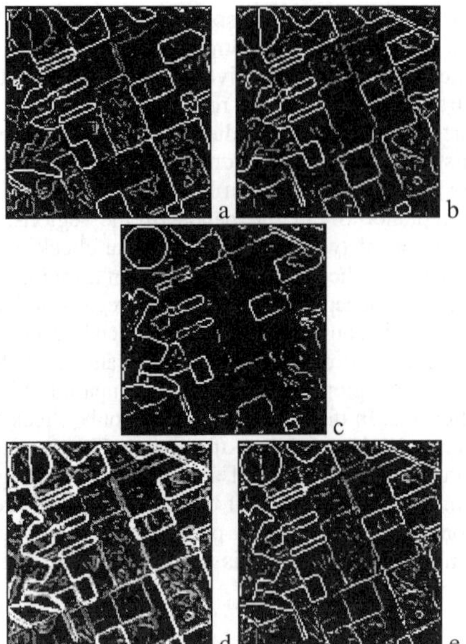

Figure 5. (a, b and c) edge pixels detected from Figure a,b,c after applying the Second Directional Derivative algorithm. (d and e) are the superposed and post-thinned images.

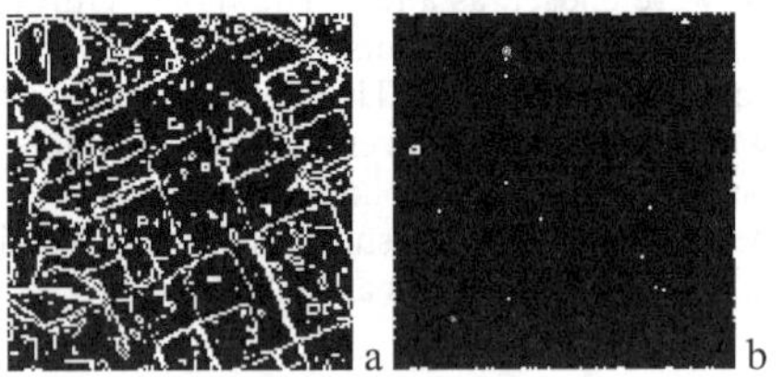

Figure 6. Results yielded by the CHRG and EPC procedures.

Figure 7. Image obtained by the Region Merging algorithm.

the abovementioned region growing constraints. The seed image, generated by the modified ISODATA, is a suitable solution to the problem of initializing the region growing algorithm. On the other hand, the region growing algorithm stops when all the pixels have been labeled.

This algorithm has been applied to an agricultural zone obtaining a high level of accuracy. The methods chosen to carry out each step need few parameters, which cease to be critical when the over-segmentation criterion is assumed. Thus, the whole process can be carried out in an automatic way. The increment in the number of the homogeneous objects can be solved in the merging or in the homogeneous object classification steps.

5 REFERENCES

Baraldi, A. & Parmiggiani, F. 1996. Single Linkage Region Growing Algorithms Based on the Vector Degree of Match. *IEEE Trans. Geoscience and Remote Sensing*, vol. 34, No 1, pp. 137-147.

Canny, J. 1986. A Computational approach to edge detection. IEEE Trans. on Pattern Analysis and Machine Intelligence, Vol. PAMI-8, pp. 679-698.

Cebrián, M. Pérez-Luque, M.J. & Cisneros, G. 1993. Edge Detection Alternatives for Multispectral Remote Sensing Images", *Proc. of the 8th Scandinavian Conference on Image Analysis*, pp. 1047-1056, Troms (Noruega).

Chang, Y. & Li, X. 1994. Adaptive Image Region-Growing. *IEEE Trans. on Image Processing*, vol. 3, no 6, pp. 868-872.

Conover, W.J. 1980. *Practical Nonparametric Statistics*. New TYork: Wiley.

Cumani, A. 1991. Edge detection in multispectral images, *CVGIP: Graphical Models and Image Processing*, vol. 53, no. 1, pp. 40-51.

Deusen, P.C. 1995. Modified Highest Confidence First Classification. *Photogrammetric Engineering and Remote Sensing*, vol. 61, No. 4, pp. 419-425.

Escribá, J.M. García-Consuegra, J. & Castaño, S. 1997. Método de Segmentación para la Generación de Mosaicos con Imágenes del Satélite LANDSAT (TM), *Proc. VII Congreso Nacional de Teledetección*, Santiago de Compostela.

Gambotto, J.P. 1993. A new approach to combining region and edge detection. *Pattern Recognition Lett.*, vol. 4, pp. 869-875.

Kettig, R.L. & Landgrebe, D.A. 1975, Classification of multispectral image data by extraction and classification of homogeneous objects, *Proc. Symposium on Machine Classification of Remotely Sensed Data*, pp. 1-11, West Lafayette.

Mather, P.M. 1987. *Computer Processing of Remotely-Sensed Images*. John Willey & Sons.

Moigne, J.L. & Tilton, J.C. 1995. Refining image segmentation by integration on edge and region data. *IEEE Trans. on geoscience and remote sensing*, vol 33, no. 3, pp. 605-616.

Pavlidis, T. & Liow, Y.T. 1990. Integrating region growing and edge detection. *IEEE Trans. PAMI*, vol. 12, nº 3, 225-233.

Torre, V. & Poggio, T.A. 1986. On edge detection. *IEEE Trans. on Pattern Analysis and Machine Intelligence*, vol. PAMI-8, no. 2, pp. 147-164,1986.

Remote Sensing in the 21st Century: Economic and Environmental Applications, Casanova (ed.)
© 2000 Balkema, Rotterdam, ISBN 90 5809 096 5

An adaptive soft thresholding method for Speckle noise reduction in Synthetic Aperture Radar images

A. Vidal-Pantaleoni, D. Martí & M. Ferrando
Departamento de Comunicaciones, Universidad Politécnica de Valencia, Spain

ABSTRACT: In this paper, a study of different de-speckling methods has lead to the development of a new version of soft thresholding. The new implementation is based on a modified adaptive threshold computation. It uses a logarithmic transformation of the SAR image and a multiresolution representation and it combines the characteristics of additive noise reduction with the localization of information in the coarsest levels of the wavelet transform. However, the main contribution is to adapt the effective threshold to the level of resolution in the wavelet transform and to the local detail. In the study, the performance of this technique has been compared to two well-known methods: Wiener filtering and soft thresholding. The new method has outperformed both algorithms in noise cancellation and in edge preservation over simulated SAR data and ERS-1 SAR images. The edge preservation property has been measured using an edge mask that classifies the error over uniform areas and detail areas.

1 INTRODUCTION

One of the main drawbacks in Synthetic Aperture Radar (SAR) image exploitation and application development is caused by the presence of noise. SAR images contain several sources of noise. In addition to the thermal noise given by electronic equipment, there is another kind of distortion that is typical in SAR systems. It is called Speckle noise, in parallelism with a similar effect that appears in laser imaging (Dainty 1984, Curlander & McDonough 1991). Its origin is the same as Speckle in laser optics or ultrasonic imaging. The relatively narrow bandwidth combined with the surface roughness at the wavelength scale produces a pattern of interference that results in a grainy appearance. Although it is produced by a deterministic effect, Speckle is generally considered a distortion factor and many efforts are made in order to reduce its consequences. Noise affects severely post-processing applications like segmentation and classification. For that reason, it is important to reduce it before feature extraction takes place. Despite of that, there are some situations that take profit of the Speckle contribution like texture exploitation.

There are several approaches towards the problem of Speckle reduction in the literature and many methods have been already developed to improve SAR image quality. A comparison of several speckle reduction procedures is described in Schwarz et al. (1997). In that work, a global assessment considered not only the quality achieved but also the computational effort. In general, Speckle reduction techniques may be classified into two types.

The first approach is devoted to the reduction of the noise variance by averaging. This method is called multilook and it may be implemented by different procedures. One of them is to obtain several acquisitions of the same imaged area by processing a number of antenna sub-apertures or spectrum bands. Then they are added and since the images are statistically independent, the variance is reduced by a factor that is related to the number of sub-images (Curlander & McDonough 1991). This method carries a spatial resolution degradation because full resolution is only obtained when all the spectrum band is processed. Multilook may also be carried out by window averaging. It smoothes the image again at the cost of spatial resolution and poorer results are expected than for sub-aperture multilook processing.

The second technique that is used for speckle reduction is SAR post-processing or image processing. It includes all sorts of local filters and transformations. Some of them are based on statistical locally adapted filters and recently, methods based on multiresolution decomposition have appeared in the literature (Datcu et al. 1996, Zeng & Cumming 1998 and Simard 1998). As a principle, this approach does not affect the original spatial resolution of the image as it occurred in multilooking.

In this paper, a method based on Wavelet decomposition with a new adaptive approach is presented.

It combines the characteristics of additive noise reduction of soft thresholding (Donoho 1995) with the localization of information in the coarsest levels of the Wavelet transform (Mallat 1989), being its main contribution to adapt the effective threshold to the level of decomposition.

2 BASIC THEORY

2.1 *Speckle model in SAR images*

Speckle noise in SAR is produced by the coherent contributions of the terrain scattering points via a complicated interference mechanism (Dainty 1984). However, there have been several attempts to establish a simplified model that could be considered as a good fit of the real Speckle formation process. A good approximation is to consider the Speckle formation as a random walk process walk in the complex domain with many contributions from dephased coherent scattering points. Then, it is modeled with a complex independent stochastic process with exponential distribution and two degrees of freedom (Dainty 1984). Under such assumption, the intensity of the SAR image is Rayleigh distributed. In general, the noise contribution depends on the reflectivity, since bright areas present higher contributions of Speckle noise than dark regions. Consequently, a simplified model considers Speckle contribution as multiplicative noise. This dependence on the data value is a severe drawback when programming a noise reduction algorithm. The well known local statistical filter introduced by Lee (1980) and the adaptive smoothing filter of Kuan et al. (1985) are two approaches that operate directly on the signal dependent noise statistics. Another trend is to transform the image in order to turn the noise process into a signal independent process. That direction may be provided by a simple homomorphic transformation (Arsenault et al. 1976) that approximately turns the Speckle noise into a signal independent standard additive noise (Sadjadi 1987 and Zeng & Cumming 1998). However, the logarithm transformation is not a linear procedure and it would affect the signal statistics and the signal-noise separation as pointed in Sadjadi (1987) and Simard & DeGrandi (1998).

The work of Simard & DeGrandi (1998) chooses as a separation method the normalization by a smooth model. In this work, noise reduction has been carried out after logarithmic transformation following the multiplicative noise Speckle model as a first stage in the chain shown in Figure 1. The signal is modeled as

$$x(t) = \log(sn_s + n_a) \tag{1}$$

where s is the desired reflectivity, n_a is the thermal additive noise and n_s is the Speckle multiplicative model. Since the thermal noise component is negligible compared to n_s, then

$$x(t) = \log(s) + \log(n_s) \tag{2}$$

where $\log(n_s)$ is the additive speckle contribution that is approximately gaussian additive uncorrelated noise (Zeng & Cumming 1998, Arsenault & April 1976).

2.2 *Wavelet representation*

The Wavelet Transform (Daubechies 1988, Mallat 1989) is a recent tool that performs a hierarchical decomposition of the signal space into a nested sequence of approximation spaces. It has been applied successfully to the processing and interpretation of SAR data (Datcu et al. 1996). In the field of Speckle reduction, promising results in the application of Wavelet decomposition have been reported in the literature (Zeng & Cumming 1998, Simard & DeGrandi 1998).

Wavelet transform may be implemented efficiently with a tree filtering scheme that creates a pyramid of lower resolution approximations. The direct and the inverse wavelet transformation of the continuous signal *x(t)* can be expressed as:

$$x_n^m = \int_{-\infty}^{\infty} x(t)\Psi_n^m(t)dt \tag{3}$$

$$x(t) = \sum_n \sum_m x_n^m \Psi_n^m(t) \tag{4}$$

where $\Psi_n^m(t)$ represents a set of basis functions that are dilations and translations of a single function $\Psi(\cdot)$ called mother wavelet

$$\Psi_n^m(t) = 2^{m/2}\Psi(2^m - n) \tag{5}$$

The Wavelet decomposition operates in a two dimensional discrete domain and the implementation of the transformation has been approached through

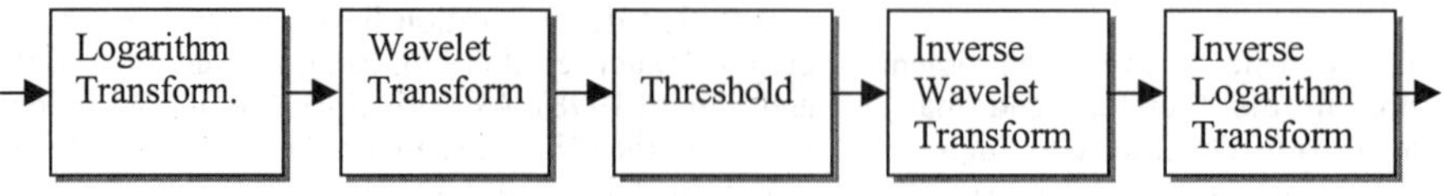

Figure 1. Processing chain in Speckle reduction algorithm.

the efficient tree decomposition described by Mallat (1989). The original image is progressively decomposed by filtering and decimation. The Discrete Wavelet Transform (DWT) has an interesting property of simultaneous space and frequency localization. For that reason, the features are present through the decomposition tree. Therefore, such detail information may be kept and noise can be reduced by applying a threshold detector that does not produce edge distortion. An illustration of the 2-D discrete Wavelet decomposition is shown in Figures 2-3. Figure 2 presents the original image and Figure 3 presents the Wavelet decomposition in three levels using Haar bases.

2.3 *Threshold operation*

Donoho (1995) first proposed a technique that was called soft thresholding and it was applied to reduce additive gaussian noise in a Wavelet decomposed signal. That scheme may be directly applied to a SAR image if an homomorphic operation could turn multiplicative Speckle noise into an additive contribution.

The soft threshold is characterized by a particular implementation of the threshold operator. Figure 1 illustrates both soft and hard thresholding. Hard thrsholding typically yields the minimum mean squared error (MSE), but soft thresholding gets almost the minimax MSE with the property that it achieves a final result as least as smooth as the true solution. As a consequence, in the first case the operation is smoother and gives the best results in SAR imagery (Odegard et al. 1995).

In soft thresholding, the output signal is obtained through the operation described in (6) and hard or classical thresholding is expressed in (7)

$$w' = \begin{cases} sign(w) \cdot (|w| - T) & |w| > T \\ 0 & |w| \leq T \end{cases} \tag{6}$$

$$w' = \begin{cases} w & |w| > T \\ 0 & |w| \leq T \end{cases} \tag{7}$$

Figure 2. Test aerial image.

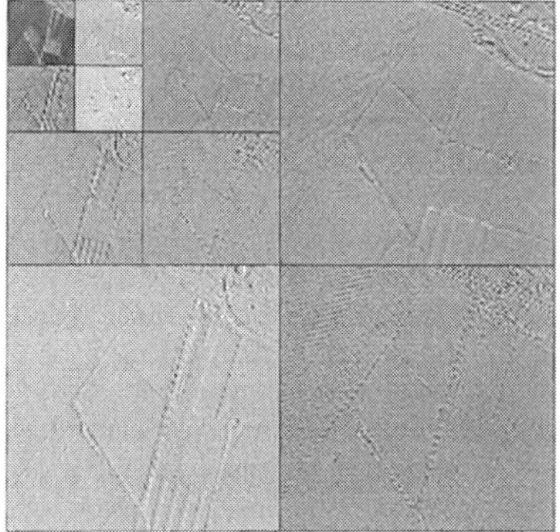

Figure 3. Wavelet decomposition in three levels.

3 ALGORITHM DESIGN

3.1 *Processing stages*

The noise reduction process is summarized in Figure 1. The scheme is the same that was introduced by Donoho (1995) with the logarithmic operation proposed by Odegard et al. (1995). The novelty is introduced in an adaptive version of the threshold implementation.

A logarithmic transformation is applied in order to separate the noise contribution from the desired signal in an additive manner. Other techniques like normalization have been also tested, but the homorphic transformation has achieved the bcst results concerning noise reduction in the tests. Then, the image is decomposed via the DWT using a three level pyramid with Daubechies family of 16th order as described in (Daubechies 1988). This wavelet is

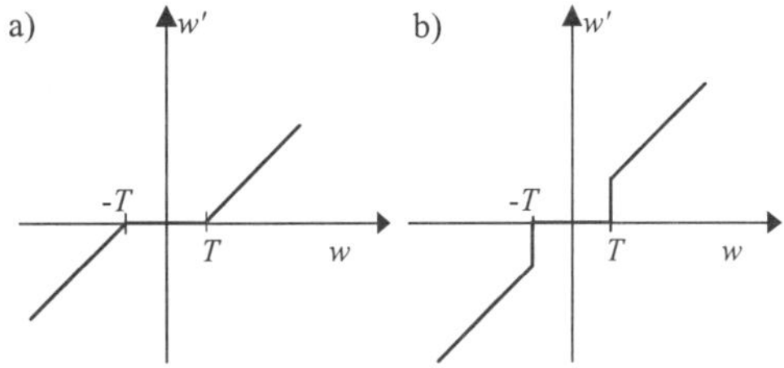

Figure 4. a) Soft thresholding and b) Hard thresholding.

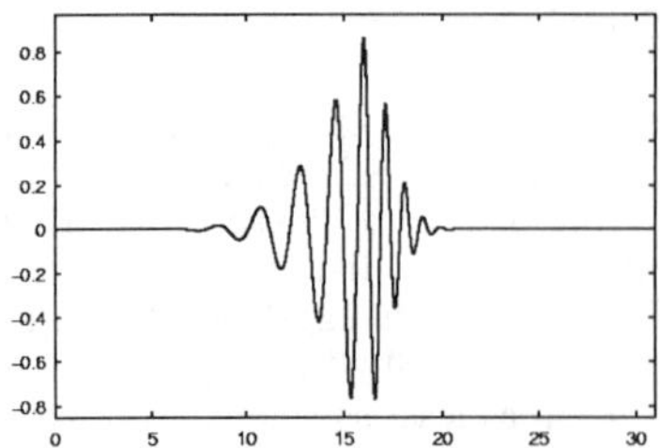

Figure 5. Mother wavelet for Daubechies of order 16.

widely used because it provides compactly supported orthonormal bases. The mother wavelet of this family is shown in Figure 5.

3.2 *Adaptive soft threshold speckle reduction*

One of the key aspects in thresholding in general is the choice of the threshold value. This is the most critical step in the Wavelet shrinkage operation, since a poor adjustment may provide a distorted image or unnoticeable noise reduction. In Odegard et al. (1995) and Wei et al. (1995), the proposed threshold is proportional to the noise variance estimation and it is constant for all Wavelet coefficients except for the low pass image that is not modified since it retains most of the information.

In standard Wavelet shrinkage, the threshold T applied to each wavelet coefficient is computed as

$$T = T_{ef}\sigma \, , \tag{8}$$

where T_{ef} is a design parameter and σ is the noise variance estimate.

The innovation to standard algorithm is a two-fold modified shrinkage operation. First, coefficient processing is carried out independently in each detail sub-image and only the sub-image coefficients are taken into consideration for the computation of the standard deviation. Secondly, the design parameter T_{ef} is not constant but dependent on the level of decomposition according to the following equation that is a modified version of (8)

$$T_{nm} = \gamma^{(n-1)} T_{ef}\sigma \, , \qquad \text{for n=1,...,N} \tag{9}$$

where n is the decomposition level, m is a sub-image index and γ is the SNR adjustment parameter. When n equals zero, no thresholding is applied following the strategy described in Odegard et al. (1995) and Wei et al. (1995) since low pass smooth sub-image retains most of the information and distortion effects would be more severe than possible noise cancellation. The SNR of the transformed image decreases as the level of the decomposition n increases. Meanwhile, the threshold is adapted locally to the resolution level. In summary, this adaptive soft-thresholding noise reduction gets an adjustment not only to the particular statistics of the sub-image (standard deviation) but also to the SNR of the resolution level.

The last stages are inverse Wavelet transformation and inverse logarithmic operation in order to acquire the final SAR image with an improvement in its radiometric quality as described in Figure 1.

4 RESULTS

A simulation has been carried out using an optical aerial image corrupted with multiplicative noise. The original and the noisy image with a SNR of 14.7 dB are shown in Figures 6-7. The selected image presents some detail features like roads and fields and different textures, that are blurred when the noise model is applied to the signal.

In the following results, the performance of the different algorithms is exhaustively analyzed. The evaluation has been done varying the values of the effective soft threshold and the parameter γ that is adjusted according to the image SNR. The results are presented in terms of cancelled mean square error that is equivalent to noise reduction. Therefore, the higher the magnitude of the graph, the best performance is achieved. The noise reduction can be measured globally, but no information about edge and detail preservation would be provided. For that reason, a mask has been designed and used to classify the error cancellation over edges or uniform regions. The mask is provided by a Sobel edge detector and a low pass filter used to smooth the result. So, a normalized edge mask is obtained. The results of the proposed adaptive soft thresholding analysis are shown in Figures 8-10. These graphs show a maximum value that should be the ideal choice for the parameter selection. It can be observed that the method gets an optimum noise cancellation when the effective threshold is equal to 0.1 and γ is equal to 3.75. Therefore, the result using such parameters is presented in Figure 11.

A comparison of the adaptive soft-threshold with other methods has been also carried out. The implemented methods are the Wiener filter (Pascazio & Schrinzi 1996) and soft-thresholding as described by Donoho (1995). Figure 12 presents the de-speckled result with the Wiener method and Figure 13 shows the standard Donoho technique. The best processing parameters have been selected in the results. Figure 14 summarizes the algorithm performance of the three techniques. It shows the edge error cancellation in relation to the achieved uniform error cancellation. The best result is clearly obtained for the proposed adaptive technique and one possible reason for such result is that more information about the image is considered in the noise reduction process.

Figure 6. Original aerial image.

Figure 7. Aerial image corrupted with multiplicative noise.

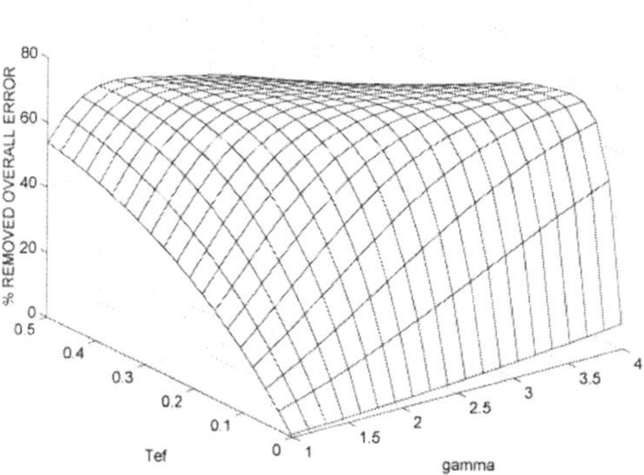

Figure 8. Global noise reduction in adaptive soft thresholding.

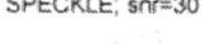

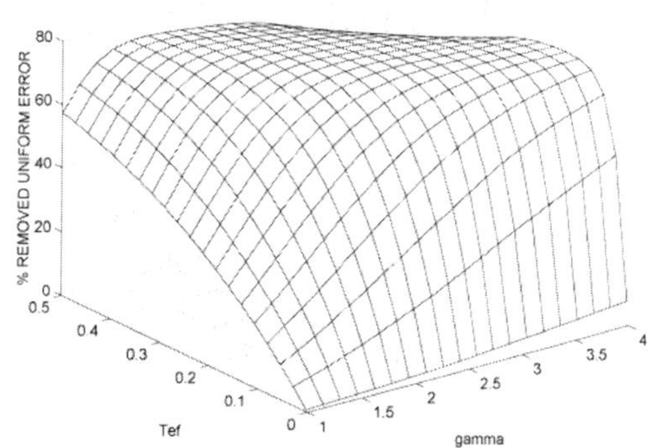

Figure 9. Uniform area noise reduction in adaptive soft thresholding.

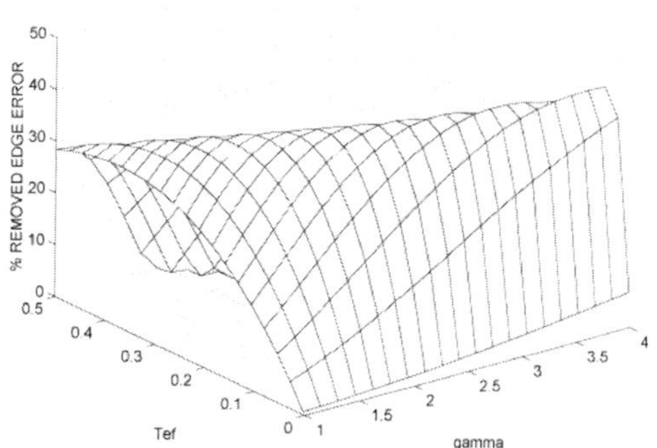

Figure 10. Edge area noise reduction in adaptive soft thresholding

Figure 11. Reconstruction via the adaptive soft thresholding.

Figure 12. Reconstruction with Wiener filter

Figure 13. Reconstruction using soft thresholding.

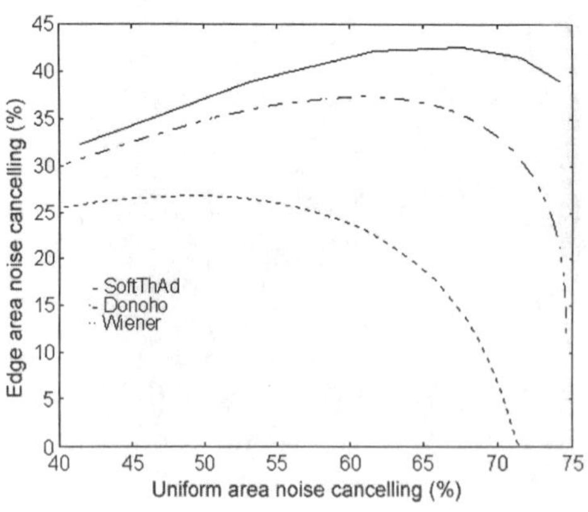

Figure 14. Noise reduction comparison.

In the previous analysis, the visual inspection of Figures 11-13 shows that Wiener produces a smooth result and that standard soft threshold provides a good result that can only be measured in terms of cancelled noise. Once the performance of the new method has been shown on a simulated image, another test using real SAR data has been carried out. The test data is a medium resolution image of the Flevoland region acquired by ERS-1 satellite. This image has been processed using a multilook procedure with 24 looks and a reduced spatial resolution of approximately 160 m (Fig. 15). However, the image presents the typical appearance of Speckle. For comparison purposes, the data has been also processed using a high number of looks (more than 400) and it is shown in Figure 16. The results obtained by the three techniques that are object of comparison are shown in Figures 17-19.

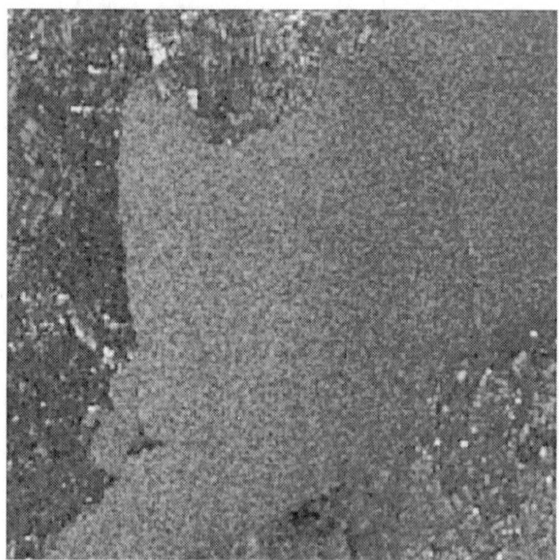

Figure 15. ERS-1 SAR image.

Figure 16. Multilook processed ERS-1 SAR image.

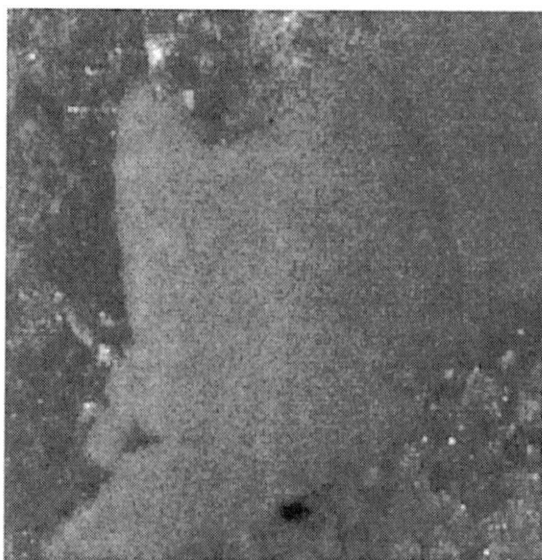

Figure 17. De-speckled SAR image using Wiener filter.

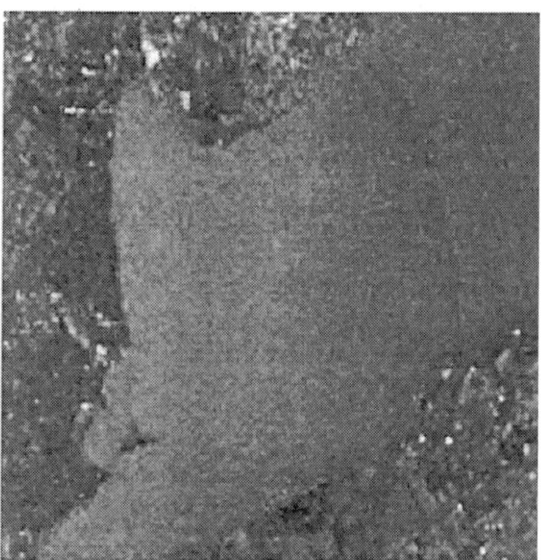

Figure 18. De-speckled SAR image using soft thresholding.

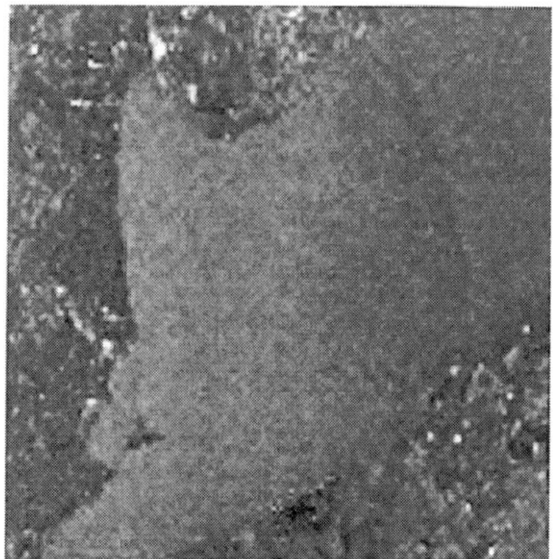

Figure 19. De-speckled SAR image using adaptive soft thresholding.

5 CONCLUSIONS

In this work, we have implemented different techniques that achieve SAR Speckle reduction. The classical Donoho soft threshold denoising and Wiener filtering are compared to a new adaptive implementation of soft thresholding. This new version gives the best performance and it can be shown by the error cancellation analysis. The amount of cancelled error has been measured classifying it into error on uniform areas and edge areas. This gives an assessment not only on the smoothing capacity over distributed targets but also a measure on the detail preservation that is needed. If edge preservation is not taken into account, the method would not assess the lack of spatial resolution that is an undesired effect.

The new version of the soft-thresholding has provided the best results at low computational cost. The test has been performed over real ERS-1 data and over simulated SAR data. The simulated data has been obtained from an aerial image containing many detail features contaminated with multiplicative noise that simulates Speckle characteristics. This experiment enables us to measure the performance of the algorithms with accuracy. As a final conclusion, adaptive soft thresholding has outperformed the other methods in quality results. It gives visually almost the same quality that standard soft thresholding for both simulated and real SAR images, but the detailed measurement of the edge preservation and noise cancellation shows that the new version gives a more accurate result. However, further algorithm developments are left as future research work.

6 ACKNOWLEDGMENTS

The authors want to thank the European Space Agency (Data & Image Processing Section) for the kind provision of ERS-1 raw data.

REFERENCES

Arsenault H. H. & G. April 1976. Properties of Speckle integrated with a finite aperture and logarithmically transformed. *Journal of the Optical Society of America* 66(11):1160-1163.

Curlander J. C. & R. N. McDonough 1991. *Synthetic Aperture Radar Systems and Signal Processing*. New York: Wiley.

Dainty J. C. 1984. *Laser Speckle and Related Phenomena*. New Tork: Springer-Verlag.

Datcu M., D. Luca & K. Seidel 1996. Multiresolution Analysis of SAR Images. *In Proceedings of EUSAR'96*.

Daubechies I. 1988. Orthonormal Bases of Compactly Supported Wavelets. *Communications on Pure Applied Mathematics* 441: 909-996.

Donoho D. L. 1995. De-noising by Soft Thresholding. *IEEE Transactions on Information Theory* 41:613-627.

Kuan D. T., A. A. Sawchuk, T. C. Strand & P. Chavel 1985. Adaptive smoothing filter for images with signal-dependent noise. *IEEE Transactions on Pattern Analysis and Machine Intelligence* PAMI-7(2):165-177.
Lee J. S. 1980. Digital image enhancement and noise filtering by use of local statistics. *IEEE Transactions on Pattern Analysis and Machine Intelligence* PAMI-2:165-168.
Mallat S. G. 1989. A Theory for Multiresolution Signal Decomposition. *IEEE Transactions on Pattern Analysis and Machine Intelligence* PAMI-11:674-693
Odegard J. E., H. Guo, M. Lang, C. S. Burrus, R. O. Wells, L. M. Novak & M. Hiett 1995. Wavelet-based SAR Speckle reduction and image compression. *Proceedings of the SPIE* 2487:259-271.
Pascazio V. & G. Schrinzi 1996. Reduction of multiplicative noise in SAR imaging by Wiener filtering. *In Proceedings of the IGARSS'96* 1577-1579.
Sadjadi F. A. 1987. A perspective on techniques for enhancing speckled imagery. *In Proceedings of the SPIE* 829:158-164.
Schwarz G., M. Walessa & M. Datcu 1997. Speckle reduction in SAR images: techniques and prospects. In Proc of the IGARSS'97 4:2031-2034.
Simard M. & G. DeGrandi 1998. Analysis of Speckle Noise Contribution on Wavelet Decomposition of SAR Images. *IEEE Transactions on Geoscience and Remote Sensing* 36 (6):1953-1962.
Simard M. 1998. Extraction of Information and Speckle Noise Reduction in SAR Images Using the Wavelet Transform. *In Proceedings of the IGARSS'98* 1:4-6.
Wei D., H. Guo H., J. E. Odegard, M Lang & C. S. Burrus 1995. Simultaneous speckle reduction and data compression using best wavelet packet bases with application to synthetic aperture radar (SAR) based ATD/R. *Proceedings of the SPIE* 2491:1131-1141.
Zeng Z. & I. Cumming 1998. Bayesian Speckle noise reduction using the discrete wavelet transform. *In Proceedings of the IGARSS'98* 1:7-9.

Remote Sensing in the 21st Century: Economic and Environmental Applications, Casanova (ed.)
 ISBN 90 5809 096 5

2-D phase unwrapping in Synthetic Aperture Radar interferometry

A.Vidal-Pantaleoni, R.Oviol & M.Ferrando
Departamento de Comunicaciones, Universidad Politécnica de Valencia, Spain

ABSTRACT: Synthetic Aperture Radar Interferometry (SAR) has been identified as one of the most promising fields in Earth Observation Systems. However, SAR Interferometry 2-D Phase Unwrapping remains as a challenging problem, since several sources of error complicate the procedure. This work focuses on Multi-frequency Digital Elevation Model (DEM) extraction. By obtaining data at different frequency bands, more information can be extracted and integrated in a common product and precise DEM products are obtained. The work has been developed in the frame of the SIR-C/X-SAR experiment, taking simulated noisy interferograms for the test of the algorithm. The Multi-frequency approach has been compared to the weighted multi-grid technique implemented with an adaptive multiresolution instantaneous frequency estimator. The result is very promising, since data fusion provide the advantages of each frequency band. Therefore, a very detailed DEM without global error is obtained.

1 INTRODUCTION

Synthetic Aperture Radar (SAR) maps the terrain using an active coherent system. This system has been traditionally exploited in amplitude, but recent developments have discovered the potential of the phase content of SAR products. SAR Interferometry provides the user with accurate Digital Elevation Models (DEMs) of the surface. It is based on the generation of an interferogram out of two or more acquisitions of the same area taken from slightly different look angles.

The phase of the SAR interferogram contains information about the topography, since it is directly proportional to the surface height. However, the phase is obtained in the range $[0, 2\pi]$ due to the corresponding trigonometric inverse operation. Therefore, the phase must be processed using an unavoidable operation called Phase Unwrapping. Unfortunately, the procedure is complicated by the presence of several sources of error and topographic artifacts.

In SAR images, there are various classes of noise. First, thermal additive noise is present. Secondly, speckle noise caused by terrain roughness compared to the radar wavelength appears. In the last case, noise is approximated by a multiplicative model in the intensity image and a statistical uniform distribution in the phase. In the case of low frequencies like L-band, volume scattering introduces error in height computation. Moreover, topographic artifacts like foreshortening, layover and shadow can be present. If topographic height varies very rapidly, then aliasing may appear in the interferogram as explained in Ghiglia & Pritt (1998). All those adverse conditions complicate extremely SAR Interferometry Phase Unwrapping. For that reason, this problem has become a challenge for the scientific community. In the particular case of dealing with a large amount of data to be processed, the issue of getting an automatic Phase Unwrapping algorithm still requires further study.

Phase Unwrapping algorithms are coarsely classified into two categories. The first one is the path following strategy. It identifies phase inconsistencies or residues and tries to integrate the phase without crossing the connecting path of coupled residues. The region growing algorithm is the most outperforming method in this category as indicated in Ghiglia & Pritt (1998). The second class is the Least Squares approach. It derives a smooth phase field by integrating the gradient of the phase observations subject to the Least Squares criteria.

Least-squares algorithms may be weighted or unweighted. That is, they may consider the grade of confidence of the phase value depending on ancillary data for each measurement point. This may be radar brightness or more usually coherence that gives greater weight to those points considered as more accurate according to the level of image correlation. LMS Phase Unwrapping is divided in two stages. The first one is the estimation of the gradient of the interferogram phase. The second step is the integration of that gradient estimate.

In this paper, the multifrequency approach introduced by Lanari et al. (1996) based on the LMS approach has been tested. First, a basic description of the system is given. Then, the multifrequency technique is explained. In the next section, some results are presented. Finally, some conclusions are extracted from the simulations.

2 BASICS OF SAR INTERFEROMETRY

The basic geometry of a SAR system is shown in Figure 1. An active sensor sends microwave pulses to ground and receives the echoes in phase and quadrature. The principle of synthetic aperture enables a much finer resolution than real aperture systems by means of coherent phase processing. Then, a complex SAR image is built up by applying an appropriate processing to the complex raw data in an operation called SAR focusing. The output of this process is a complex SAR image whose phase content was scarcely used until the advent of SAR Interferometry.

In SAR interferometry, two radar images of the same area are acquired from slightly different view angles S_1 and S_2 in order to form a radar interferogram (Fig. 2). The distance between both positions is the baseline B and the range distance is R.

The complex interferogram Z_{int} is obtained multiplying the complex reflectivity at each point of one image by its corresponding conjugate value in the second image, so that the interferogram also preserves useful information about the signal amplitudes I_1 and I_2.

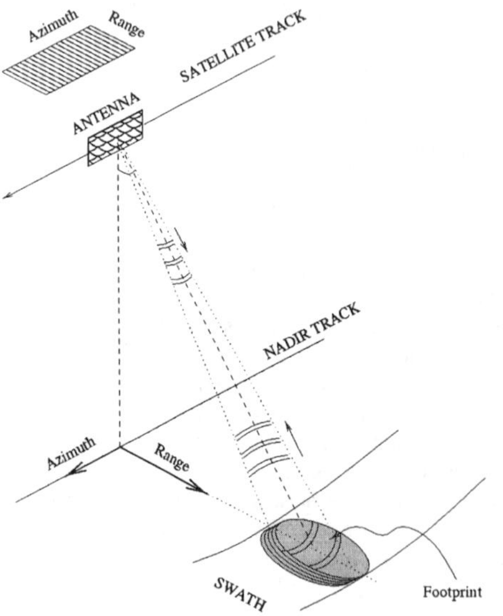

Figure 1. SAR Geometry.

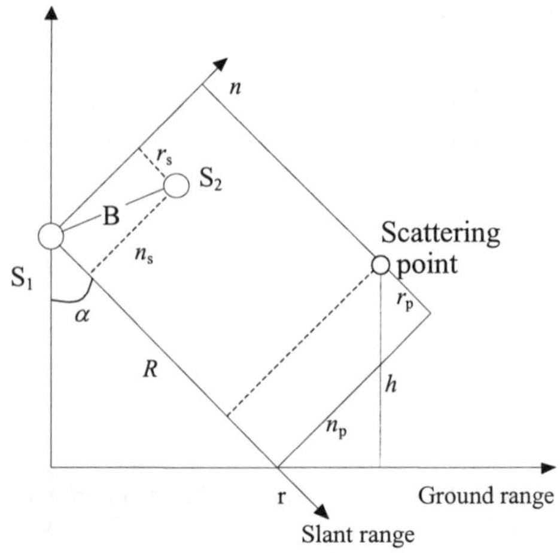

Figure 2. SAR Interferometry geometry.

$$Z_{\text{int}}(x,y) = I_1(x,y) \cdot I_2(x,y) \cdot e^{i(\psi_1(x,y)-\psi_2(x,y))} \tag{1}$$

where x and y represent the pixel indexes in the interferogram map.

$$Z_{\text{int}}(x,y) = I_{\text{int}}(x,y) \cdot e^{i\psi_{\text{int}}(x,y)} \tag{2}$$

where I_{int}, I_1 and I_2 are the interferometric intensities and ψ_{int} the interferometric phase difference.

Following Prati et al. (1996), we derive the relation between the phase ψ_{int} and the terrain elevation h. First, the relation is expressed in terms of slant-range coordinates n and r. An approximation that takes into account that the sensor-target distance is much greater than the range imaged swath is adopted here. In the case of spaceborne SAR, this assumption is usually taken due to the system geometry.

$$\psi_{\text{int}} = \frac{4\pi \cdot n_s \cdot n_p}{\lambda \cdot R} \tag{3}$$

where λ is the radar wavelength, R is the range distance and n_s and n_p are the normal baseline and normal height variation according to Figure 2.

However, the SAR sensor is pointing sideward with a certain mean off-nadir angle α and an undesired phase difference is introduced due to this angle. It may be corrected deriving the expression of the flat terrain contribution that depends on the range position. In this case, we consider a reference slant-range distance R_0 and the actual slant-range position R that depends on the image indexes x and y.

$$\Delta\psi_{ft} = \frac{4\pi \cdot n_s \cdot (R - R_0)}{\lambda \cdot R_0 \cdot \text{tg}(\alpha)} \tag{4}$$

The topographic phase component is then obtained by the subtraction of the flat terrain contribution from the original phase and it is related to the desired height h via the following expression:

$$\psi'_{int} = \psi_{int} - \Delta\psi_{ft} = \frac{4\pi \cdot n_s \cdot h}{\lambda \cdot R_0 \cdot \mathrm{sen}(\alpha)} \tag{5}$$

On the other hand, coherence information gives a detailed measure of phase confidence. This figure is very important to get a correct unwrapped phase because low coherence and noise usually introduce a high amount of error in the unwrapped phase. The complex coherence γ is a measure of the correlation between SAR images and it is given by

$$\gamma = \frac{E(Z_1 \cdot Z_2^*)}{\sqrt{E(|Z_1|^2) \cdot E(|Z_2|^2)}} \tag{6}$$

where Z_1 and Z_2 are the co-registered complex SAR images and $E(\cdot)$ represents the ensemble mean operator. The fact of obtaining high interferometric coherence depends on three primary requirements. First, the baseline should be smaller than the critical one. In second place, the Doppler information of the SAR images should overlap to provide high frequency correlation. In third place, there must exist high temporal correlation. If the acquisitions are carried out at different time slots, then different scattering conditions may affect the interferogram depending on the terrain characteristics, the weather conditions and the frequency band as explained in

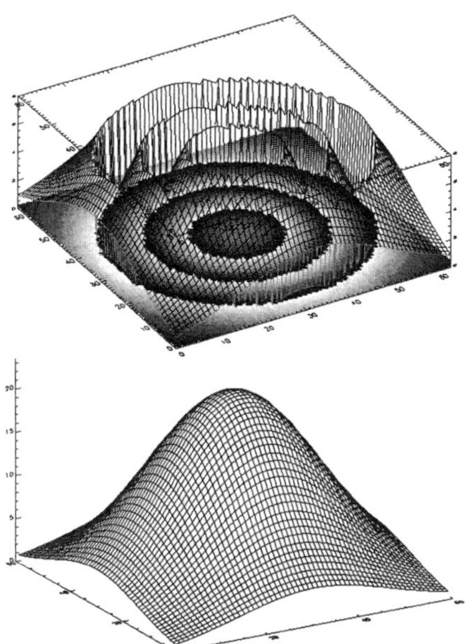

Figure 3. Phase unwrapping of a Gaussian surface.

Table 1 Parameters for X-, C- and L-band.

Band	X	C	L
Wavelength (cm)	3.12	5.67	24.23
Polarisation	VV	VV	VV
PRF (Hz)	1736	1736	1736
Sampling Freq. (MHz)	22.4985	44.9971	44.9971
Azimuth spatial resolution (m)	30	54	54
Slant range spatial resolution (m)	20	17	17

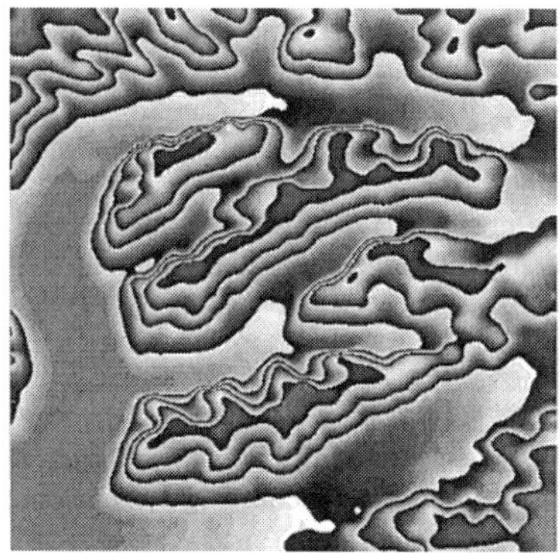

Figure 4. L-band simulated interferogram without noise.

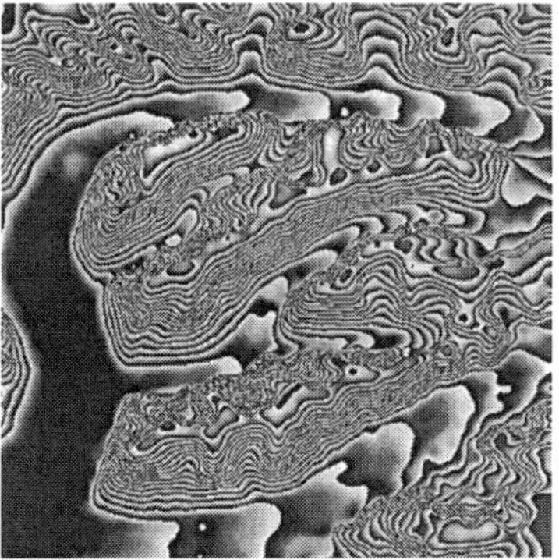

Figure 5. C-band simulated interferogram without noise.

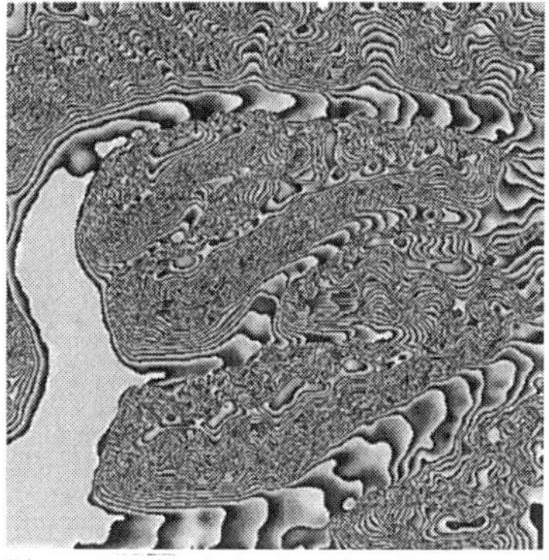

Figure 6. X-band simulated interferogram without noise.

Lanari et al. (1996).

In practical implementations the coherence is approximated by the following expression, as described in Engdahl (1996) and Ghiglia & Pritt (1998)

$$\gamma' = \frac{\sum_n Z_1 \cdot Z_2^* \cdot e^{-i\psi_{\text{int}}}}{\sqrt{\sum_n |Z_1|^2 \cdot \sum_n |Z_2|^2}} \quad (7)$$

where the confidence level of the estimate increases with the size of the window n and the exponential term is introduced to reduce the effect of the topography in the computation of the coherence.

The value that can be measured from the interferometric information is not the phase ψ_{int}, but its principal value ψ_p. It is limited in the range [0, 2π] because of the periodical behavior of the trigonometric functions. The principle value is the sum of the desired ψ_{int} plus an integer number of phase cycles (2π). The problem of 2-D phase unwrapping is therefore to find out that correct integer value. The overall process of 2-D phase unwrapping is shown in Figure 3 for the case of a known Gaussian surface. It is wrapped in the range [0, 2π] and jumps in phase should be cancelled for the proper surface reconstruction. In this case, the procedure is simple and many techniques may be applied to get the unwrapped phase, but in the case of topographic data, the surface itself is difficult to process and undesired noise and artifacts like aliasing are present in the problem. That complicates extremely the phase unwrapping method.

A least square error (LMS) approach for the problem of phase unwrapping may been taken through two methods: direct techniques like Fast Fourier Transform (FFT) and iterative techniques like multigrid. Multigrid has very interesting properties for its flexibility and efficiency concerning computation time. It brings the possibility to use a weighting function in order to overcome the problem of phase inconsistency or residues. An estimate of the phase gradient may be then obtained by the Maximum Likelihood (ML) (Spagnolini 1995) or multiresolution techniques (Davidson & Bamler 1996). The ML method may be carried out with an adaptive method that adjusts the window size and the level of threshold depending on the signal to noise ratio and the number of final outliers as described in Engdahl (1996).

The formulation of the LMS phase gradient integration is extensively explained in Ghiglia & Pritt (1998). The aim of the procedure is the integration of the continuous partial derivatives. Therefore, the input to the process is an estimate of the phase. The minimization of the mean square error between the estimated phase and the true unwrapped phase leads to the resolution of the Poisson Equation. That classical equation can be solved through different techniques. The most efficient is the direct method using Fast Fourier Transform (FFT) or Discrete Cosine Transform (DCT) (Pritt & Shipman 1994). The other option is Multigrid, that is a flexible and efficient LMS estimator. It may incorporate a useful weighting function in the process that usually is related to the interferometric coherence amplitude (Pritt 96 and Ghiglia & Pritt 1998).

3 MULTIFREQUENCY PROCESS

In this work, a technique that combines data from different frequency bands has been studied. The use of different frequency band interferograms of the same region can help in the process of Phase Unwrapping as it was introduced in Lanari et al. (1996). Another approach to the same problem has been described in Papathanassiou & Moreira (1996).

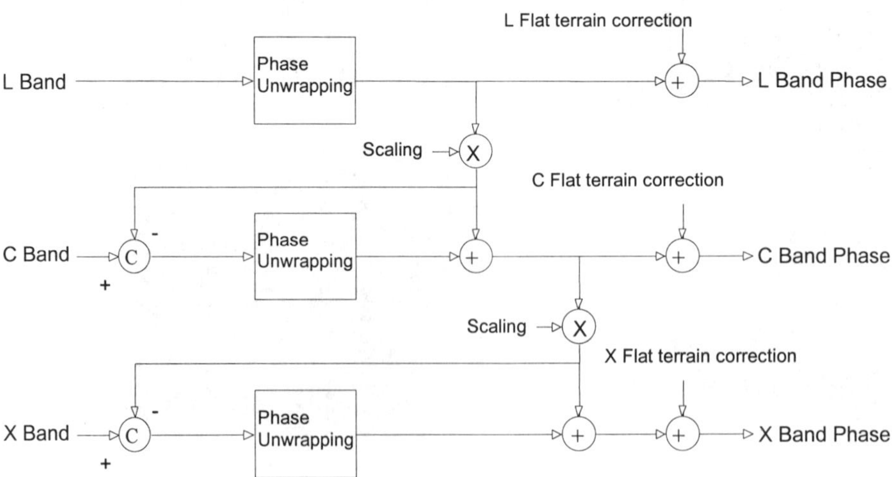

Figure 7. Block diagram of multifrequency process.

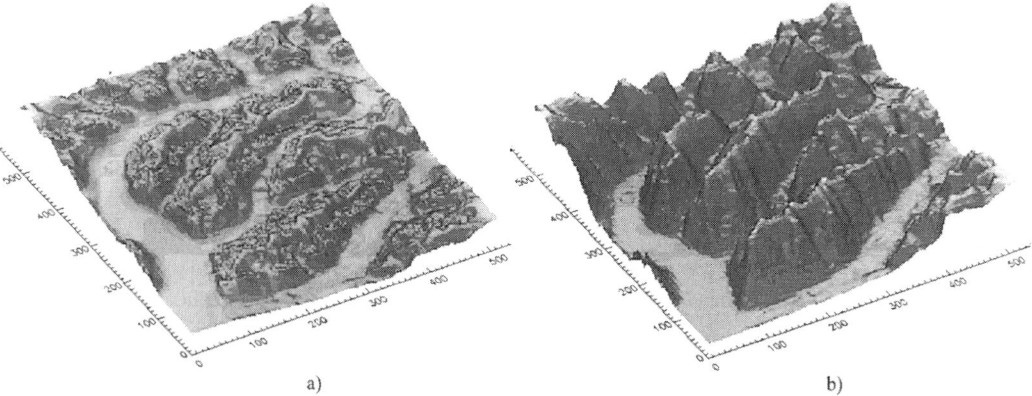

Figure 8. X-band phase reconstruction: a) not assisted and b) assisted.

In this paper, a similar experiment has been implemented and evaluated using simulated data following the approach of Lanari et al. (1996). A direct LMS method (FFT) has been used for the computation of the unwrapped phase for each frequency band. It has been tested over simulated data obtained from a DEM of the glacier Woodfjorddalen (Svalbard, Norway) that is introduced in Eldhuset (1996). The test interferograms have been simulated with the same parameters as the SIR-C/X-SAR experiment. They are shown in Table 1 for the three bands object of study: L, C and X. The simulated interferograms in L-, C- and X-band are shown respectively in Figures 4-6.

Each interferogram presents different characteristics depending on the frequency band. The L-band interferogram is much more robust and less sensitive to the topographic changes. Moreover, its coherence map is usually higher due to the electrical properties of the terrain at L-band. However, height errors are strong compared to the other frequencies as it will be seen below. After obtaining a noisy version of the interferograms, a procedure that combines them is implemented in two stages as described in Lanari et al. (1996). First, an assisted unwrapped interferogram reconstruction that exploits information redundancy is used to get the different unwrapped interferograms. Then, an integration process combines all them in a single product with the best characteristics of each frequency band.

3.1 *Assisted phase reconstruction*

The first step shown in Figure 7 is to obtain the three unwrapped phase maps. In the procedure described in Lanari et al. (1996), the lower band information is used to improve the phase unwrapping process. Taking X-band as an example, the phase reconstruction is drastically enhanced by the subtraction of a phase estimate. Figure 8a shows the phase recon-

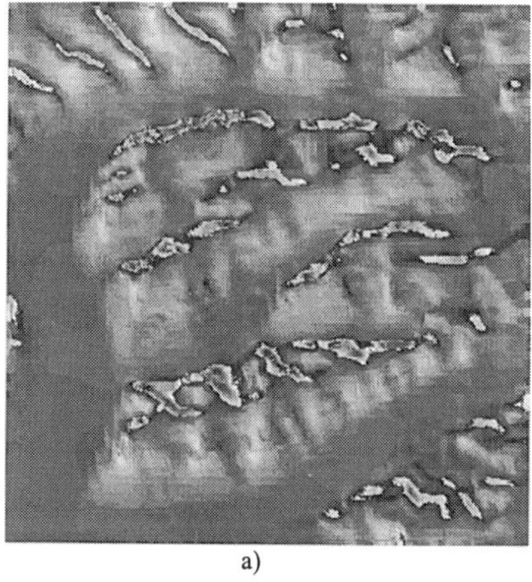

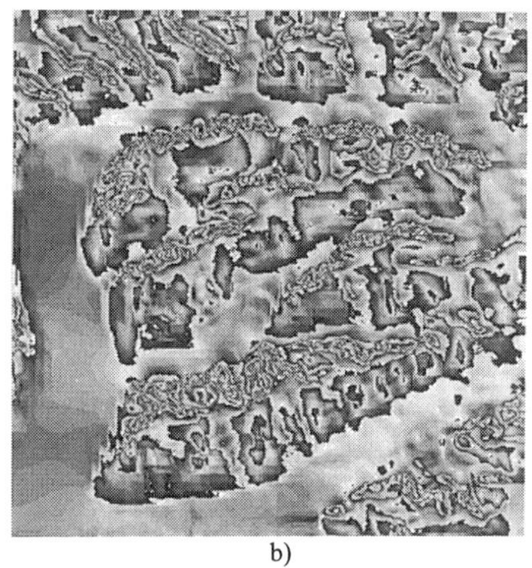

Figure 9. Residual phase in: a) C-band and b) X-band.

struction process for X-band using a direct LMS approach. The method fails completely due mainly to aliasing effects. On the contrary, Figure 8b illustrates the assisted X-band phase unwrapping process that benefits form previous phase knowledge as shown in the diagram of Figure 7.

However, X-band contain very useful detail height information. By subtracting the low pass approximation of L-band interferogram, the residual interferograms show the added information to the smooth model. It is illustrated in Figures 9a and 9b that are the residual C- and X-band interferograms respectively.

3.2 *Fusion of the Absolute Phase*

Previously, the multiple frequency information has been used to unwrap the interferograms. In this subsection, absolute phase fusion is explained according to the method described in Lanari et al. (1996) and summarized in Figures 10-11.

There are three key points that should be taken into account in the fusion stage. The first one is the bias due to unwrapping problems. It is directly related to the lack of coherence or the presence of aliasing. The second aspect is the coherence itself. The confidence level of the absolute phase depends on the coherence of that area and the phase quality is higher in the L-band, since it is not as much dependent on the scene characteristics as it is in X- or C-band. In third place, there is a theoretical limitation in the height confidence that depends on the radar wavelength.

In the case of L-band, height accuracy is much worse than the one offered by the X-band because the standard deviation of the phase error $\delta_{h\varphi}$ is directly proportional to the wavelength λ as described in the following expression

$$\delta_{h\varphi} = \frac{\lambda \cdot R_0 \cdot sin\theta}{4\pi \cdot \cos(\theta - \xi)} \cdot \delta_\varphi \tag{8}$$

where R_0 is the range, θ the incidence angle, B the baseline, ξ the baseline tilt angle and finally δ_φ the standard deviation of the interferometric phase noise.

Following the same scheme, the phase fusion procedure is divided in three steps: bias removal, coherence weighting and wavelength weighting as described in Lanari et al. (1996).

The Kalman filter integrates two inputs and combines them taking the low frequency content of the first signal and the high frequency content of the second input. In order to eliminate the phase bias, the low frequency input is the L-band phase for C-band and the C-band phase for the X-band. The phase with bias I_H can be modeled as the addition of the true phase φ and the bias φ_{bias}. The noisy phase I_L is the addition of the true phase φ and the white noise φ_v, where φ_v and φ_{bias} are statistically independent. The bias can be estimated by low pass filtering the expression shown in Equation 2.

$$I_H - I_L = \varphi_{bias} + \varphi_v \tag{9}$$

The estimated true phase is then obtained by the following expression that subtracts the estimated bias from I_H.

$$\varphi \approx \hat{\varphi} = I_H - \hat{\varphi}_{bias} = \varphi + \varphi_{bias} - \hat{\varphi}_{bias} \tag{10}$$

The next step is coherence weighting. Each phase is directly multiplied by its estimated coherence map γ'_L, γ'_C and γ'_X as described in Lanari et al. (1996) in order to reduce the contribution of low quality phase.

Finally, the last stage is the wavelength scaling that takes into account and corrects the previous coherence weighting. The summation of the three bands is done and it reduces the overall noise

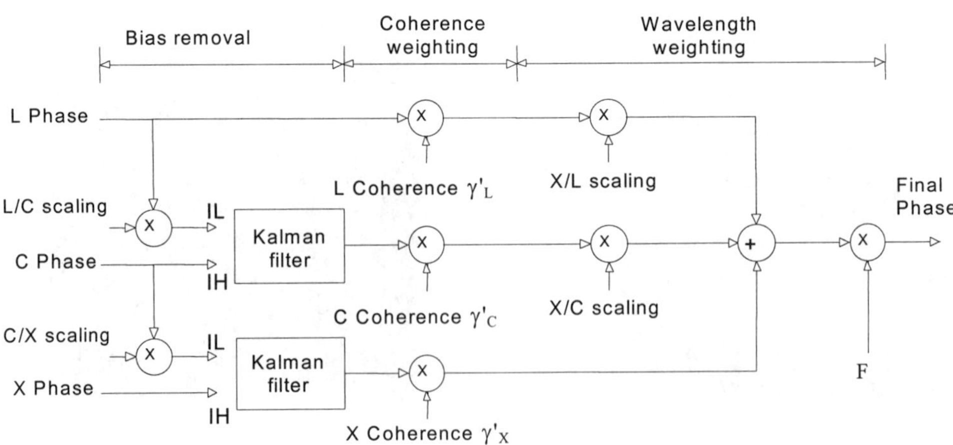

Figure 10. Phase fusion using Kalman filters.

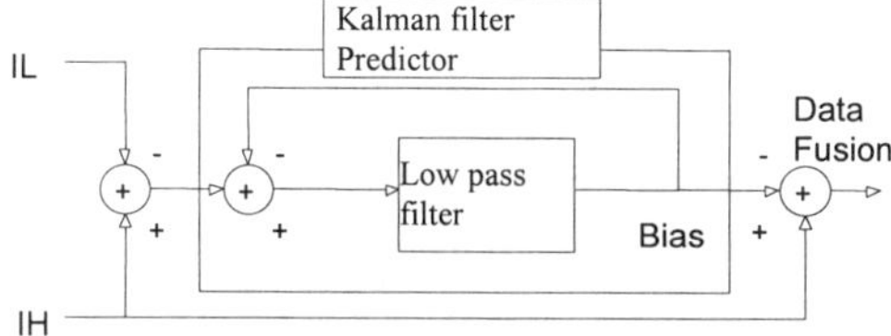

Figure 11. Kalman filter implementation.

because the signals are statistically independent. There is a reduction of the phase variance given by

$$r_f = \frac{\sqrt{1+\left(\dfrac{\lambda_X}{\lambda_C}\right)^4+\left(\dfrac{\lambda_X}{\lambda_L}\right)^4}}{1+\left(\dfrac{\lambda_X}{\lambda_C}\right)^2+\left(\dfrac{\lambda_X}{\lambda_L}\right)^2} \tag{11}$$

where λ_C, λ_L and λ_X are the wavelengths at C-, L- and X-band, respectively. In the case under consideration, this value is equal to 0.79. A correction factor given by F is then applied to normalize the final phase output.

$$F = \frac{1}{\gamma'_X + \gamma'_C\left(\dfrac{\lambda_X}{\lambda_C}\right)^2 + \gamma'_L\left(\dfrac{\lambda_X}{\lambda_L}\right)^2}. \tag{12}$$

The final phase is shown to have the correct global height provided by L-band and the fine detail given by C- and X-band.

4 RESULTS

In this work, the multifrequency process proposed in Lanari et al. (1996) has been implemented in a simulation environment and compared to the result obtained from a single frequency band. The redundancy of data has shown a very improved performance. Multiple band data have not only be exploited for phase fusion, but also to obtain the residual phase maps in the first stage of the procedure.

The final phase reconstruction shows the detailed reconstruction that is typical at high frequencies and the robust global unwrapping given by low frequencies. The noisy interferograms in bands C and X are shown in Figure 12. The L-band phase unwrapping output using the multiresolution intermediate frequency estimator (Spagnolini 1995 and Engdahl 1996) and the multigrid process (Ghiglia & Pritt 1998 and Pritt 1996) is shown in Figure 13a for comparison purposes. The reconstructed result using the multi-frequency process described before is illustrated in Figure 13b. The reconstruction is very accurate and not only it gives a proper global fit which is provided by the L-band reconstruction but it also presents the detailed elevation accuracy given by higher frequencies. The final reconstruction fits with high accuracy to the original elevation information. The result illustrated in Figure 13a fits well to the low frequency surface but it fails in matching the detailed features. On the other hand, the result is much more noisy as it was expected because in the case of one band, there is no noise reduction via averaging. For that reasons, that result is only shown for visual inspection and comparison with the multi-frequency approach.

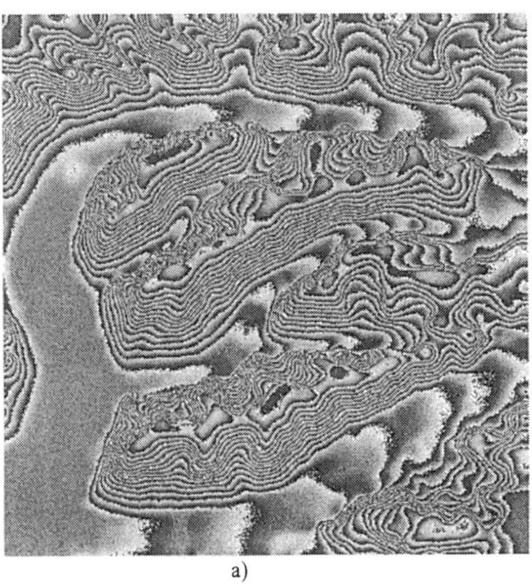

a)

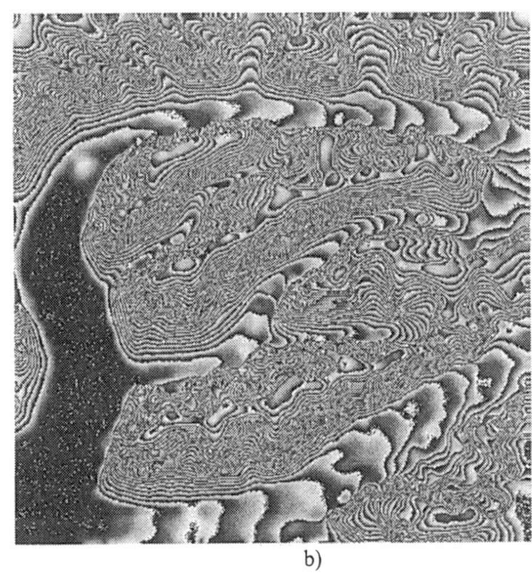

b)

Figure 12 Original noisy interferograms in: a) C-band and b) X-band.

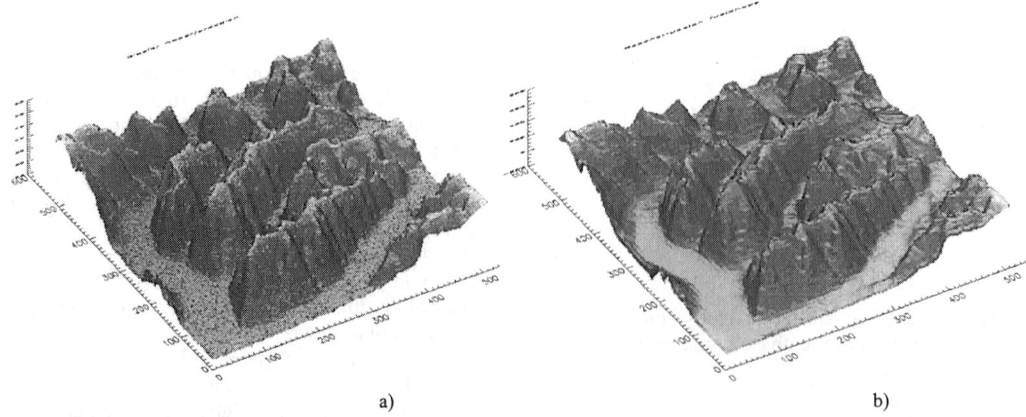

Figure 13. Terrain reconstruction using: a) Noisy L-band interferogram, b) L-, C- and X-band noisy simulated interferograms.

5 CONCLUSIONS

This work has tested the method of Phase Unwrapping based on the use of multiple frequency bands. It has been shown that the result from the fusion of different interferograms gives a robust and detailed output. It combines the high resolution provided by low frequency bands and the robust surface given by higher bands. Moreover, the final noise variance is reduced as independent elevation maps are added in the last stage. As a final conclusion, the integration of information coming from different geometric configurations or frequencies can overcome some of the typical problems faced in the Phase Unwrapping problem. Further work should be done in the combination of this technique with recent developments in Phase Unwrapping like multiresolution processing. A deeper study dealing with real interferometric multi frequency data like the one provided by next SRTM mission would provide a good opportunity to explore this promising method.

6 ACKNOWLEDGMENTS

The authors to want express their gratitude to Estudio Atlas S.A. (IDL and ENVI providers) for establishing a cooperation scheme to carry out this project. They also want to thank the researchers that offer their results via internet for further developments.

REFERENCES

Davidson G. W. & R. Bamler 1996. Robust 2-D Phase Unwrapping based on multiresolution. *In SPIE Proceedings* 2958: 227-237.

Eldhuset K. 1996. First results from ERS tandem INSAR processing on Svalbard. *http://www.geo.unizh.ch/rsl/fringe96 FRINGE'96 ESA Workshop on Applications of ERS SAR Interferometry, Zurich, 30 September to 2 October 1996.*

Engdahl M. 1996. *Phase Unwrapping in SAR Interferometry Using Instantaneous Frequency Estimation.* Msc. Thesis Helsinki University of Technology.

Ghiglia D. C. & M. D. Pritt 1998. *Two-Dimensional Phase Unwrapping: Theory, Algorithms and Software.* New York: Wiley.

Lanari R., G. Fornaro, R. Riccio, M. Migliaccio, K. P. Papathanassiou, J. R. Moreira, M. Schwabissch, M. Dutra, G. Puglisi, G. Franceschetti, & M. Coltelli 1996. Generation of Digital Elevation Models by Using SIR-C/X-SAR Multifrequency Two-Pass Interferometry: The Etna Case Study. *IEEE Transactions on Geoscience and Remote Sensing* 34(5): 1097-1113.

Papathanassiou K. P. & J. R. Moreira 1996. Interferometric analysis of multifrequency and multipolarization SAR data. *In IGARSS'96 Digest* 1227-1229.

Prati, C., F. Rocca, A. Monti Guarnieri & E. Damonti 1990. Seismic migration for SAR Focusing: Interferometrical Applications. *IEEE Transactions on Geoscience and Remote Sensing* 28(4): 627-640.

Pritt M. D. & J. S. Shipman 1994. Least-Squares Two-Dimensional Phase Unwrapping Using FFT's. *IEEE Transactions on Geoscience and Remote Sensing* 32(3): 706-708.

Pritt M. D. 1996. Phase Unwrapping by Means of Multigrid Techniques for Interferometric SAR. *IEEE Transactions on Geoscience and Remote Sensing* 34(3): 728-738.

Spagnolini U. 1995. 2-D Phase unwrapping and instantaneous frequency estimation. *IEEE Transactions on Geoscience and Remote Sensing* 33(3): 579-589.

Remote Sensing in the 21st Century: Economic and Environmental Applications, Casanova (ed.)
© 2000 Balkema, Rotterdam, ISBN 90 5809 096 5

Radiometric characterisation of the olive tree crops

C. Pinilla & F. J. Ariza
Departamento de Ingeniería Cartográfica, Geodésica y Fotogrametría, Universidad de Jaén, Spain

M. Sánchez
Departamento de Ingeniería Gráfica y Sistemas de Información Cartográfica, Universidad de Córdoba, Spain

ABSTRACT: The olive tree crop is a typical example of discreet cultivation that leaves uncovered a large proportion of bare soil. The measured radiance by a spaceborne sensor is the sum of the corresponding to the illuminated fraction soil, shady soil and vegetation. This cause many identification errors, upon not having a pure spectral signature for the crop. In this work the geometric factors that affect to the reflectance of this type of vegetables formations are analysed and it proposes a procedure in order to derive the single vegetable-cover spectral signature.

1 INTRODUCTION

The identification of the land-cover by remote sensing in the solar spectrum is carried out applying statistical techniques to their spectral response. The spectral response of the group of coverings are tried like multivariate aleatory variables, that could be subjected to classifications procedures clustering the response values in function of their means vectors and of their covariance's matrices.

When the observed surface is homogeneous in reference to the geometric resolution of the sensor, the main problem of the classification resides in the confusion between cartographic classes like consequence of their mean vectors similarity.

However, when the sensor resolution is in the same order of magnitude that the change space frequency between the canopy and the background, for example like consequence of a line cultivation, the resulting spectral behaviour is the average between what could be attributed to each one of the surfaces separately. In this case the digital classification obtained presents often many assignment errors.

The olive tree crop is a paradigm of discreet cultivation, whose plantation frame in the Province of Jaén coincides with the available space resolution in commercial sensors. Consequently, the recognition of this cultivation could not be approached by forms or textures identification, but by procedures of statistical classification with multispectral images. Problems of the olive tree crops classification appear here: since the canopy only covers partially the soil and that the covering ratio varies with factors like the frame of plantation, the relative orientation of the surface with the sensor position and the relative position of the observed zone with the scene nadir, the results obtained in classifications of the olive tree crops are frequently not very good. It shows a high percentage of omission errors.

2 OBJECTIVE

The success of the remote sensing based applications depends closely of the capacity to establish a consistent relationship between the reflectance measures made by the sensor and the surface properties (Gilabert 1990). In the study of the vegetal canopies, the reflectance models of the vegetal systems are essential in order to establish that relationship (Goel, 1988). Reflectance models express the interactions between surface elements with the purpose of explaining the relationship between the radiometric measurements and the biophysical characteristics of the scene.

The goal of this work is to contribute with some elements that allow a deeper insight in the radiometric behaviour of the olive tree crops in the Province of Jaén, in order to improve their identification by satellite images.

3 REFLECTANCE GEOMETRIC MODEL

In accordance with Strahler *et al.* (1986), a cultivation of olive tree could be considered as a system formed by discreet elements of a single class (the olive trees) located over a background (soil). Introducing the proposals of Goel (1988) one could distinguish between illuminated zones and shaded zones, for elements both in first plane and in the background. In this way, the olive tree parcel will be examined as a system of four components: directly illuminated trees, shaded trees, directly illuminated soil and shaded soil, since the shaded zones (of trees and soil) will contribute to the total pixel radiance, reflecting diffuse radiation.

Considering a lineal system, the compound reflectance will be equal to the reflectance addition of their components, pondered by their superficial contribution fraction (Goel, 1989):

$$\rho = f_a\rho_a + f_{a'}\rho_{a'} + f_s\rho_s + f_{s'}\rho_{s'} \tag{1}$$

where ρ_a is the illuminated tree reflectance and f_a its fraction; $\rho_{a'}$ is the shaded tree reflectance and $f_{a'}$ its fraction; ρ_s is the illuminated soil reflectance and f_s its fraction; and $\rho_{s'}$ is the shaded soil reflectance and $f_{s'}$ its fraction. In addition it should complete that $f_a + f_{a'} + f_s + f_{s'} = 1$.

The adds $f_a + f_{a'}$ constitutes the fraction of really covered pixel for a vegetal canopy and will be denominated in this work apparent covering coefficient C_a.

3.1 *Determination of the covering coefficient*

Two geometric models are adopted in the plantations of olive tree: the rectangular frame —a concrete case of which are constituted by the square frame—, and the "tresbolillo" frame (Fig. 1).

Graetz & Gentle (1982) simulated bushes by simple geometric bodies like half-sphere located on top of a cylinder. Gilabert (1990) adapted this model to plantations of citric fruits on horizontal surfaces, demonstrating their validity. However, the olive tree morphology advises to liken it rather to a sphere of r radius with their centre located to a h height of the soil (often $h = r$). The importance of the topography in the reflectance measures made on heterogeneous covering surfaces has also been shown (Iaquinta & Fouilloux 1998). For this reason, the model proposed includes the topographical factor and the sensor location.

Considering the rectangular model in first place, a orthogonal covering coefficient C_o could be settled:

$$C_o = \frac{\pi r^2}{m} = \frac{\pi r^2}{ab} \tag{2}$$

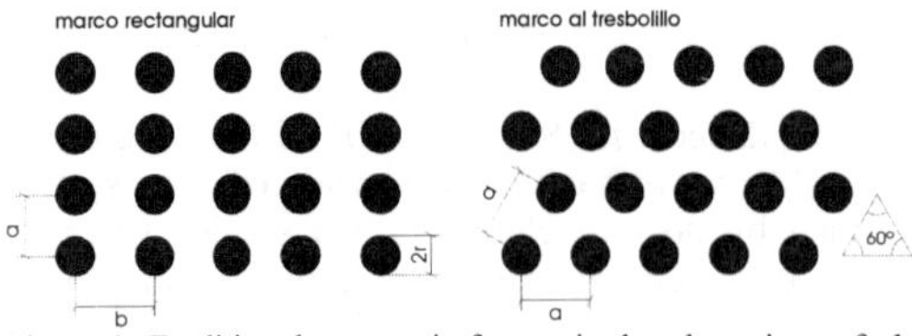

Figure 1. Traditional geometric frames in the plantations of olive tree.

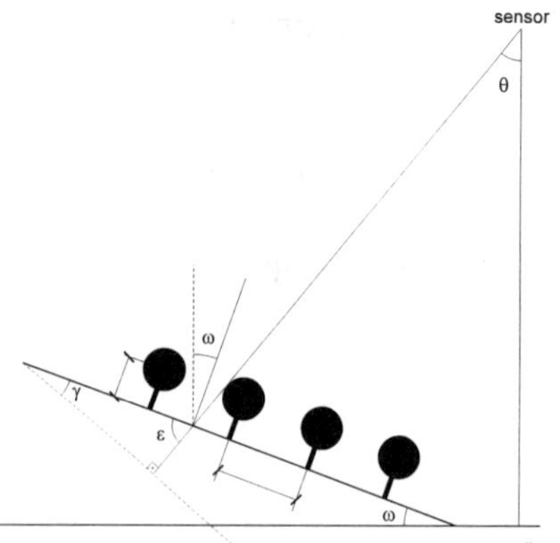

Figure 2. First observation geometric approach.

being r the radius of the top, a the separation between trees of a same line and b the separation between tree lines.

In opposition to the absolute pixel inclination, secondary inclination is the inclination angle according to the perpendicular direction to the satellite trace (turn axis parallel to the track). If one supposes a planar ground pixel with an secondary inclination ω, the trees lines perpendicularly placed to the satellite track and the Earth like plane, a simplified model of canopy fraction for view axis could be found (Fig. 2).

From this model one could obtain the apparent covering factor C_a, which includes the relative orientation of the pixel with the vector sensor-pixel, i.e. the topographical factor:

$$C_a = \frac{C_o}{\cos(\theta - \omega)\cos\phi} \tag{3}$$

where ϕ is the pixel inclination angle according to the parallel direction to the satellite trace, that in the successive it will be called tertiary inclination, and $(\theta - \omega)$ the angle formed by the surface with a perpendicular plane to the visual. If we permit to the tree lines adopt any orientation in relation to the satellite trace (Fig. 3), the exposition would include the most general case, in which value of C_a would continue being the previously expressed.

In order to solve the magnitude of the maximum value of C_a it is necessary to consider the absolute overlap of the top according to the visual due to the terrain slope and to the separation of the trace. One could demonstrate that maximum value is:

$$C_{a\,\text{máx}} = \frac{2r}{(a \,\text{sen}\,\delta + b\cos\delta)\cos\phi} \tag{4}$$

The defined model until now lacks flexibility in its application, because is bases on a very restrictive angular system. It suits generalise the model so that

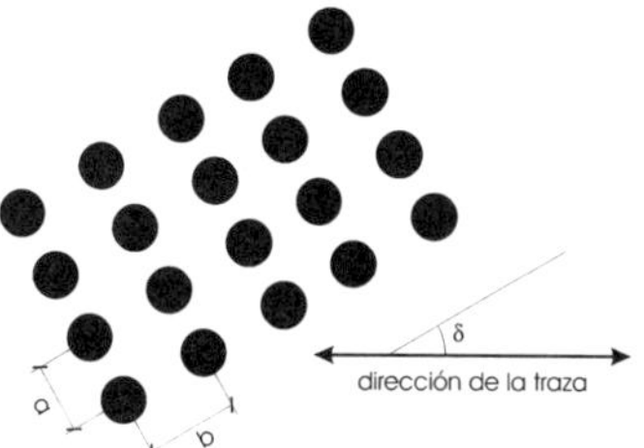

Figure 3. Arbitrary orientation of the olive tree lines with regard to the satellite track.

its application could be automated using a digital model of elevations starting from which the pixel slope and aspect could be derived (Fig. 4).

That's why is necessary define another system of reference polar co-ordinates based of the absolute pixel inclination with regard to the horizontal (α) and in their orientation (κ). The variable change function between the [ω,ϕ] system and the [α,κ] system is:

$$\cos\phi = \frac{\cos\alpha}{\sqrt{\operatorname{sen}^2\alpha\cos^2\kappa + \cos^2\alpha}} \tag{5}$$

$$\operatorname{tg}\omega = \frac{\operatorname{sen}\alpha\operatorname{sen}\kappa}{\cos\alpha} \tag{6}$$

On the other hand, it suits to put the θ angle in function of the pixel co-ordinates (column and row).

If the image is considered geometrically corrected, the habitual format of the image presents two triangular empty regions to left and right (Fig. 5), consequence of having removed the displacement of

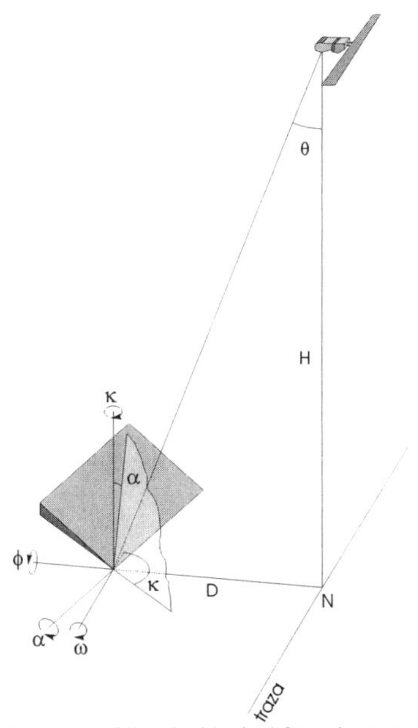

Figure 4. Geometry of the pixel looked from the sensor.

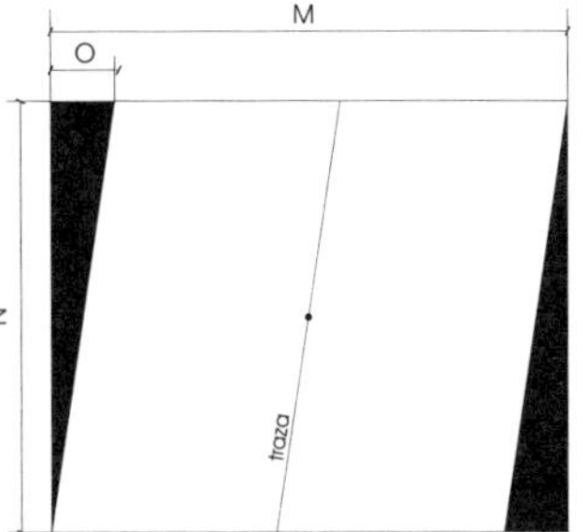

Figure 5. Look of the scene corrected of the Earth rotation effect.

the lines due to the turn of the Earth (Steinwand & Wivell 1993).

Naming O to the maximum number of pixels empties of a row, M to the columns number, N to that of lines of the image, (C, L) to the image-co-ordinates, H to the satellite height (Kramer 1994) and p to the nominal pixel side size in the terrain, the (3) equation remain of the following way:

$$C_a = \frac{\frac{\pi r^2}{ab}\sqrt{\operatorname{sen}^2\alpha\cos^2\kappa + \cos^2\alpha}}{\cos\alpha\cos\left[\operatorname{arctg}\left(p\,\frac{\frac{-O}{N}l + \frac{M+O}{2} - c}{H}\right) - \operatorname{arctg}(\operatorname{tg}\alpha\operatorname{sen}\kappa)\right]} \tag{7}$$

with a maximum approximate value:

$$C_{a\,\text{máx}} = \frac{2r\sqrt{\operatorname{sen}^2\alpha\cos^2\kappa + \cos^2\alpha}}{(a\operatorname{sen}\delta + b\cos\delta)\cos\alpha} \tag{8}$$

The other geometric configuration of the olive tree plantations is the "tresbolillo" frame. For it, the value of C_a has an expression that could be derived easily to the general, since the disposition to the "tresbolillo" only differ geometrically of the rectangular frame in what lines separation is always $b = a$ SQR(3)/2. For this one could write:

$$C_a = \eta \cdot C_o \tag{9}$$

where η is the relative orientation factor of pixel in relation to the sensor, which has a value:

$$\eta = \frac{\sqrt{\operatorname{sen}^2\alpha\cos^2\kappa + \cos^2\alpha}}{\cos\alpha\cos\left[\operatorname{arctg}\left(p\,\frac{\frac{-O}{N}l + \frac{M+O}{2} - c}{H}\right) - \operatorname{arctg}(\operatorname{tg}\alpha\operatorname{sen}\kappa)\right]} \tag{10}$$

On the other hand, the interval of existence of α and κ that interests consider is only that that it allow to make visible the pixel from the sensor, i.e. when the secondary inclination is between θ - π/2 and

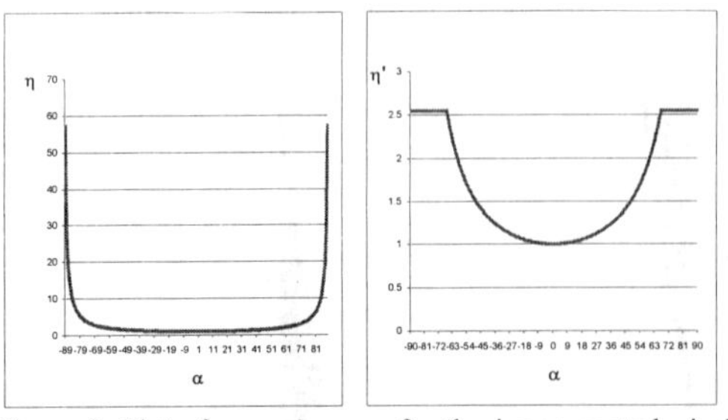

Figure 6. Plot of η y η' vs. α for the image central pixel, $r = 2.5$ m, $m = 10 \times 10$ m^2, $\delta = 0$ and $\kappa = 0$.

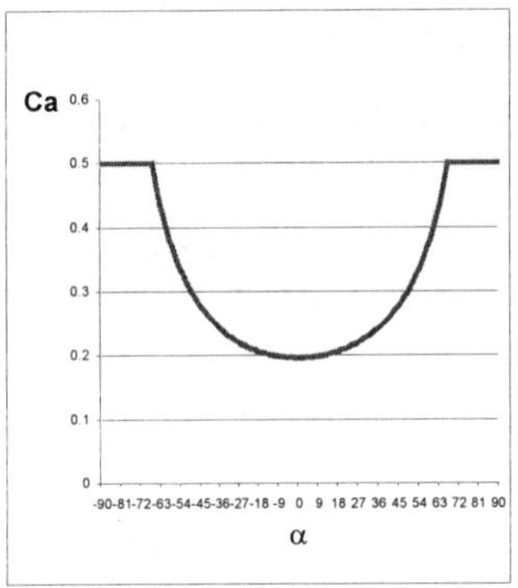

Figure 7. Plot of C_a vs. α for condition from Figure 6.

$\theta + \pi/2$. The η values are enclosed in the interval [0,∝]. However, heigher values of C_a/C_o don't increase the apparent covering coefficient, since this has fixed a limit determined by (8), but rather only increase the overlap produced between the contiguous top. In consequence it doesn't make sense to consider the whole existence interval of η. In this way one settle down the relative orientation truncate factor of pixel in relation to sensor η', equal to η but with a maximum value C_a/C_o.

The Figure 6 shows the behaviour of η and η′ versus the absolute inclination of the central pixel of the image, for a top radius of 2.5 m, a plantation frame of 10 × 10 m^2, and both the lines azimuth and the pixel orientation parallel to the trace ($\delta = 0$ y $\kappa = 0$).
In the given conditions the apparent covering factor represented in the Figure 7 is obtained.

3.2 *Determination of the shaded zones*

The shadow inside the vegetation could be due the lack of direct illumination of the sun or to the interference between plants. In the first case, for the geometry of the scene, parts of the top not illuminated directly for the Sun can be seen from the satellite. It could be carried out an approach to the determination of the shaded fraction, considering the acquisition like practically vertical for all the image pixels. Gilabert (1990) has developed a geometric model of shadowing considering the trees like the juxtaposition of two geometric simple figures: a half-sphere placed on a cylinder

However, starting from the own configuration of the olive, it is more appropriate adjust it to a model based on a tangent sphere to the soil. Being ξ the solar elevation angle, the shaded top zone is:

$$S'' = \frac{1}{2}\pi r^2 (1 - \operatorname{sen} \xi) \tag{11}$$

The contribution of the vegetation component in shade could be despised in comparison with the total scene radiance. Based on this consideration, Gilabert (1990) and García-Haro (1996) have proposed a simplified new three-terms reflectance model:

$$\rho = f_a \rho_a + f_s \rho_s + f_{s'} \rho_{s'} \tag{12}$$

The shaded soil area remaining visible from the sensor is:

$$S' = \frac{\pi r^2}{2}\left[\frac{1 - \operatorname{sen}\xi}{\operatorname{sen}\xi}\right] + \\ + \frac{r^2(1+\operatorname{sen}\xi)}{\operatorname{sen}\xi}\left[\operatorname{arcsen}\frac{\cos\xi}{1+\operatorname{sen}\xi} + \frac{\operatorname{sen}\xi\cos\xi}{(1+\operatorname{sen}\xi)^2}\sqrt{\frac{2(1+\operatorname{sen}\xi)}{\operatorname{sen}\xi}}\right] \tag{13}$$

Particularising this expression for the case of $r = 2$, one could compound the graph of the Figure 8. Being φ the latitude of the place and δ the solar declination for the day of the date, it is verified that $\xi = 90 - \varphi + \delta$, with all angles expressed in degrees. Since $-23°27' < \delta < 23°27'$ and $\varphi > 36°$ for the conditions of the Iberian Peninsula, one could consider that $\xi_{max} \approx 78°$.

If only the S′ values less that the frame area are considered, it is possible to carry out a minimum-quadratic fittings of the (13) equation to simpler functions. They have been tried lineal, polynomials, potentials, logarithmic and exponential fittings and these last two are those that better result has pro-

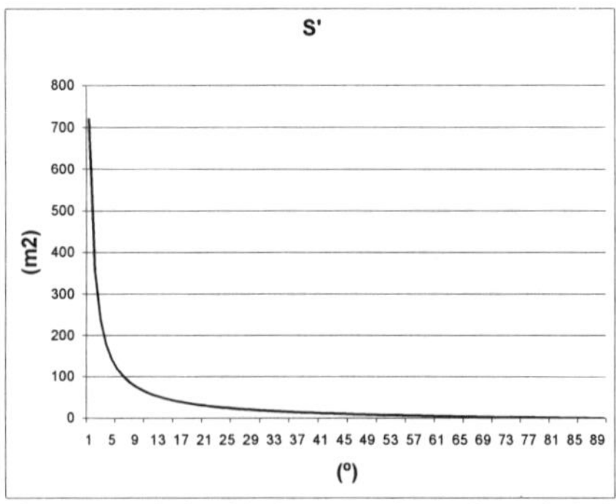

Figure 8. Relationship between the surface of shady soil exposed to the sensor and the solar height ξ.

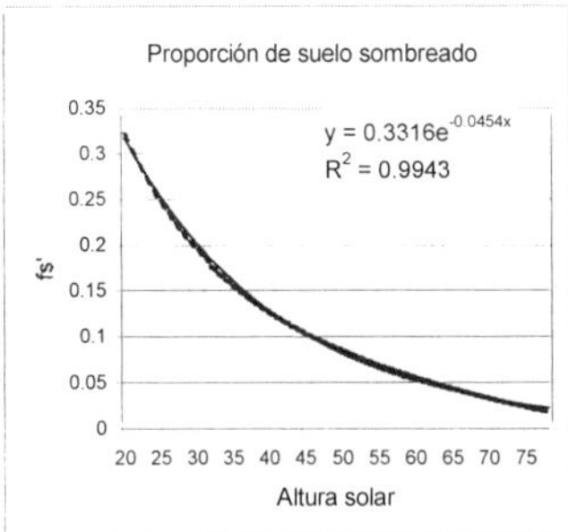

Figure 9. Adjust for $f_{s'}$ between ξ=20° and ξ=78°.

vided. Although it is always possible achieve a very close adaptation with a polynomial function, for the interval in question the function always overcoming the 3rd degree, for what its complexity makes that the substitution of the original function isn't interesting. Therefore, for more simplicity, a exponential adjustment was chosen, being satisfactory the function:

$$f_{s'} = 0{,}6468\, e^{-0{,}0479(\xi-8)} \tag{14}$$

where ξ are expressed in degrees.

Regarding the time to which the satellite goes through the scene (10:10 GMT) it is able to leave very small ξ and delimit the interval of ξ between 20 and 78°. In this case, the adjust becomes:

$$f_{s'} = 0{,}3316\, e^{-0{,}0454(\xi-20)} = k\, e^{-0{,}0454(\xi-20)} \tag{15}$$

whose graph is shown in Figure 9.

The k-parameter is directly proportional to r^2 and m^{-1} (plantation frame). Also, introducing the topographical factor η' in order to adapt better the model to the pixel orientation conditions, it could be write:

$$f_{s'} = 6{,}54\, \eta'\, C_o\, e^{-0{,}0454\xi} \tag{16}$$

3.3 *Single vegetation reflectance*

Now already one is in conditions of rewriting the reflectance function (12) with well-known factors:

$$f_a = \eta' C_o \tag{17}$$

$$f_s = 1 - \eta' C_o \left[1 + 6{,}54\, e^{-0{,}0454\xi}\right] \tag{18}$$

$$f_{s'} = 6{,}54\, \eta'\, C_o\, e^{-0{,}0454\xi} \tag{19}$$

so that the total reflectance is:

$$\rho = \rho_s + \eta' C_o \left[(\rho_a - \rho_s) - 6{,}54\, e^{-0{,}0454\xi}(\rho_s - \rho_{s'})\right] \tag{20}$$

The single vegetation reflectance is

$$\rho_a = \frac{1}{\eta' C_o}\rho + \left(1 - \frac{1}{\eta' C_o}\right)\rho_s + 6{,}54\, e^{-0{,}0454\xi}(\rho_{s'} - \rho_s) \tag{21}$$

Note it that the $1/(\eta' C_o)$ parameter in this expression is derived of the digital terrain model of the zone and

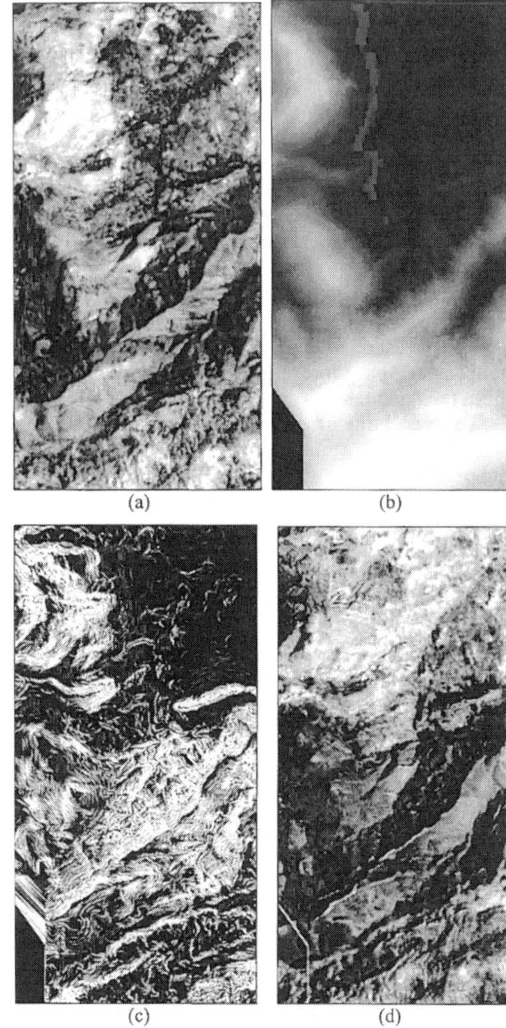

Figure 10. The original band 4 Landsat-TM image (a); DTM (b); η' Factor (c); and band 4 of the single vegetation reflectance image ρ_a (d).

of the knowledge of the plantation frame average, ρ_s can be deduced from image measuring (Baret & Andrieu 1994) and ($\rho_{s'}$ - ρ_s) from the atmospheric scattering knowledge (Kaufman 1989). Therefore both the single vegetation reflectance and compound reflectance ρ (derived on the image DN) remain related directly.

The proposed model has been applied subscene Landsat-TM of 558 samples × 1096 lines belonging to the Albanchez de Úbeda district (Province of Jaén). In the Figure 10 are shown the image, the DTM, the η' factor derived of the model and the single vegetation reflectance image.

At the moment the model is being contrasted using cadastral information and the procedure of identification of olive tree cover is being tuned, since even certain confusion between some classes exists.

4 CONCLUSIONS

The reflectance measured by the TM sensor on a discontinuous of olive tree cultivation can be explained starting from the knowledge of the background reflectance (soil), from the difference of spectral reflectance between illuminated and shaded soil —that can be derived of atmospheric diffuse radiation measures— and from the single vegetation reflectance. Therefore is possible to derive this last from the other data.

The possibility of getting accurately knowledge of the radiation fraction corresponding to one of the subpixel components allow to carry out the opportune image corrections in order to improve the land cover classification.

5 UTILIZED SYMBOLS

a Separation between trees into a same line
b Separation between tree lines
C,L Image co-ordinates for a pixel
C_a Apparent covering coefficient
C_o Orthogonal covering coefficient
D Nadiral distance of the pixel
d Plantation density
f_a Illuminated tree frame fraction
$f_{a'}$ Shady tree fraction frame fraction
f_s Illuminated soil frame fraction
$f_{s'}$ Shady soil frame fraction
h Centre top height
H Satellite height
M Columns number of the image
m Plantation frame (= d^{-1})
N Lines number of the image
O Maximum number of pixels empties of a image row
P Nominal pixel side size in the terrain
r Top radius
S Orthogonal top projection onto soil
S Top tree area
S' Area of shady soil that remains visible for the sensor
S'' Shading top zone
α Pixel inclination with regard to the horizontal
δ Tree line azimuth
ϕ Pixel inclination angle according to the parallel direction to the satellite trace
η Relative orientation factor of pixel in relation at sensor
η' Relative orientation truncate factor of pixel in relation at sensor
κ Orientation pixel
ρ_a Illuminated tree reflectance
$\rho_{a'}$ Shady tree reflectance
ρ_s Illuminated soil reflectance
$\rho_{s'}$ Shady soil reflectance
θ Look visual for a pixel from the satellite
ω Secondary pixel inclination

REFERENCES

Baret, F. & Andrieu, B. 1994. Interet du moyen infrarouge reflectif por caracteriser la vegetation. *Bol. Societé Française pour la Photogrametrie et Teledetection.* 136.

García-Haro, F.J., et al. 1996. Algunas consideraciones derivadas de la simulación de la reflectividad en escenas con vegetación dispersa. *Revista de Teledeteción,* 7: 5-19.

Gilabert, M.A. 1990. *Caracterización radiométrica, en el intervalo solar, de cultivos de cítricos. Modelo geométrico de reflectividad de aplicación en teledetección.* Tesis Doctoral. Universitat de Valencia.

Goel, N.S. 1988. Models of vegetation canopy reflectance and their use in estimation of biophysical parameters from reflectance data. *Remote Sensing Reviews,* vol. 4. Issue I. Harwood Academics Publishers. Londres.

Goel, N.S. 1989. Inversion of canopy reflectance for estimation of biophysical parameters from reflectance data. *Theory and Applications of Optical Remote Sensing.* Ed. G. Asrar. Jonh Wiley & Sons. Nueva York.

Iaquinta, J. & Fouilloux, A. 1998. Influence of the heterogeneity and topography of vegetated land surfaces for remote sensing applications. *International Journal of Remote Sensing.* Vol 19, nº 19. 1711-1723.

Kaufman, Y. J. 1989. The atmospheric effect on remote sensing and its correction. *Theory and Applications of Optical Remote Sensing.* Ed. Asrar, G. John Wiley & Sons. Nueva York.

Kramer, H.J. 1994. *Observation of the Earth and Its Environment. Survey of Misions and Sensors.* Springer-Verlag. Berlín.

Strahler, A.H. et al. 1986. On the nature of models in remote sensing. *Remote Sensing of Environment,* 20:121-139.

Remote Sensing in the 21st Century: Economic and Environmental Applications, Casanova (ed.)
 ISBN 90 5809 096 5

Adjustment of the satellite image to the three-dimensional relief model

L. N. Borovikova, S.V. Myagkov & I. Durdina
Central Asian Research Hydrometeorological Institute, Tashkent, Uzbekistan

ABSTRACT: For identification of satellite image and the earth surface for plain territories the brightness of pixels depends on the surface cover, and then the integrated characteristics of the pixel of display depends on the cover and its reflecting ability. For mountain territories the brightness of the pixels is largely determined not only by type of surface cover, but also by mutual orientation from the sun to hill-slope and from the hill-slope to satellite receiver. The method of snow cover mapping for mountain territories are presented and necessity of use three-dimensional of digital relief model is noted.

1 INTRODUCTION

Aral sea river flow of the 60-90% of the basin is comprised by the snow-melt water. For the forecasting the mountain river flow it is necessary to evaluate seasonal snow reserves on the mountain territory, which is hardly accessible for the carrying out of ground observations.

The use of satellite data is especially important for the assessment of the snow reserves.

Neverless, the problem of the geographic adjustment of satellite images arises, since the mountainous areas are characterized by very rough relief.

That is why the change of the brightness characteristics of satellite image is determined not only by the different surface objects (water, vegetation, snow, etc.), but also the orientation of the sectors and mountain slopes relative to the sun and axis of the satellite photography plays an important role in the formation of the separate sector brightness.

2 METHOD AND RESULTS

In case of the plane adjustment of the satellite image the brightness of the separate sector is defined by the reflectivity which depends on the surface coverage, and in this case the integral characteristics of the image pixel depends on the type of coverage and its reflectivity. For the mountainous surface the brightness is mainly defined not only by the type of the surface coverage, but also by the combined orientation of the sun to the slope and from the slope to the satellite receiver.

It also should be taken into account that some sections are shaded by the mountainous ranges.

The digital relief model should be used for the adjustment of satellite image to mountainous terrain.

The problem of the elaboration of the digital model of the mountainous relief is individual and complex task which is considered in the other sources (Figure 1).

In case of the ordinary superposition of satellite image adjusted by plane to the geographical coordinates onto the altitude matrix of the digital relief model the errors inevitably arise in the calculation of the areas within the brightness intervals by the distribution hystogram. In such case it is recommended to use the number of facts which are known apriori.

For example, it is known that the seasonal snow line on the northern slopes is slightly lower than on the southern ones. This assumption makes it possible to fulfil the adjustment of satellite image with a shift to the northern-southern direction.

For the digital adjustment of satellite image to the three-dimentional relief model the following approach with the application of iterative methods is used.

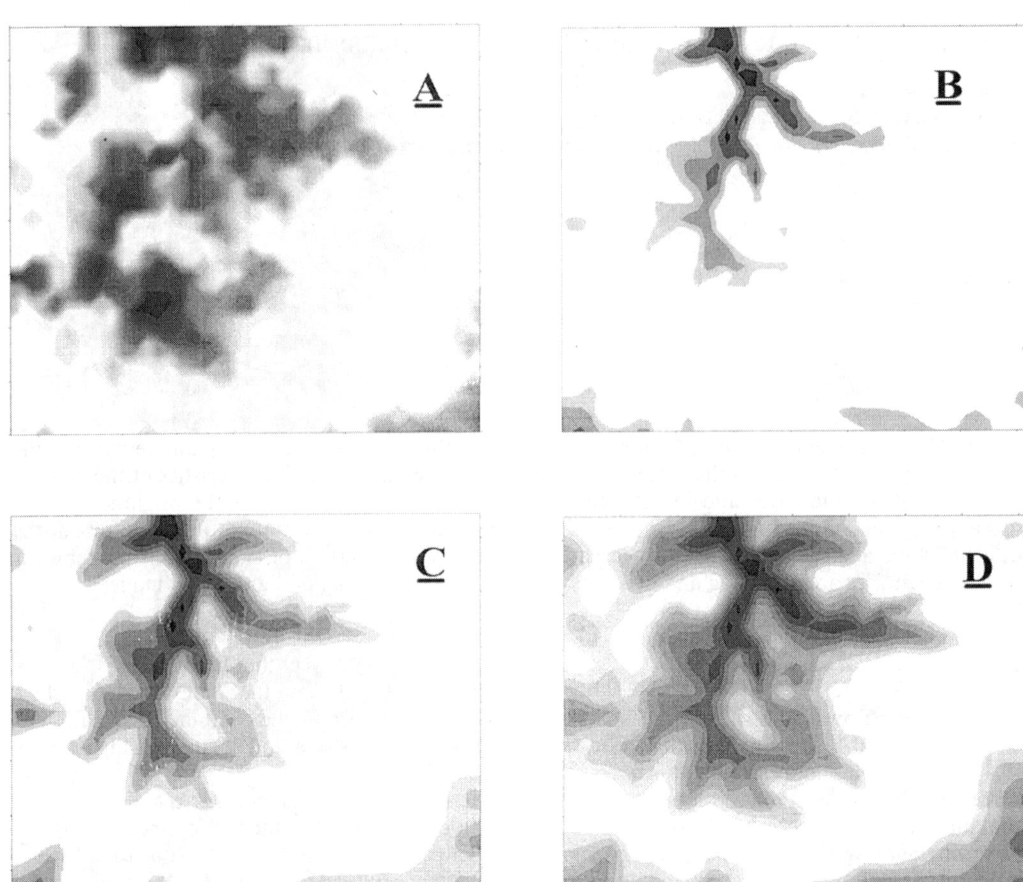

Figure 1. The comparison of the satellite image and image of the snow cover line at relief model calculation. Captions on figures are : "A" is satellite image of NOAA-14 at May 3, 1997, "B" and "C" and "D" is snow area at relief model at different levels of height. "B"=1800 m, "C"=2200 m, "D"=2400 m.

Firstly after the execution of geographical adjustment of the image on the plane and distinguishing the snow cover limits using the method of the isolines construction by the brightness matrix, the iterative calculations using the data of the digital relief model are carried out. The snow covered area shown on satellite image presented in pixels and by the metric system is calculated. For the digital model of the same sector the average value of the snow line is successively prescribed depending on the altitude of the location and the snow covered area is calculated. Then the difference between the snow covered area on the image and produced by the model is calculated. The minimum value of this difference proves the similarity between the mean depth of the snow line on the satellite picture and on the indicated sector of the terrestrial surface.

Secondly with the fixed snow line depth on the digital relief model the overlapping of the picture and model is carried out with step-by-step shift of the picture to south-north-southern direction. The minimum difference between the values of the snow covered areas indicates the closest correspondence between the model and satellite image.

After carrying out of the above mentioned iterations the problem of the estimation of the

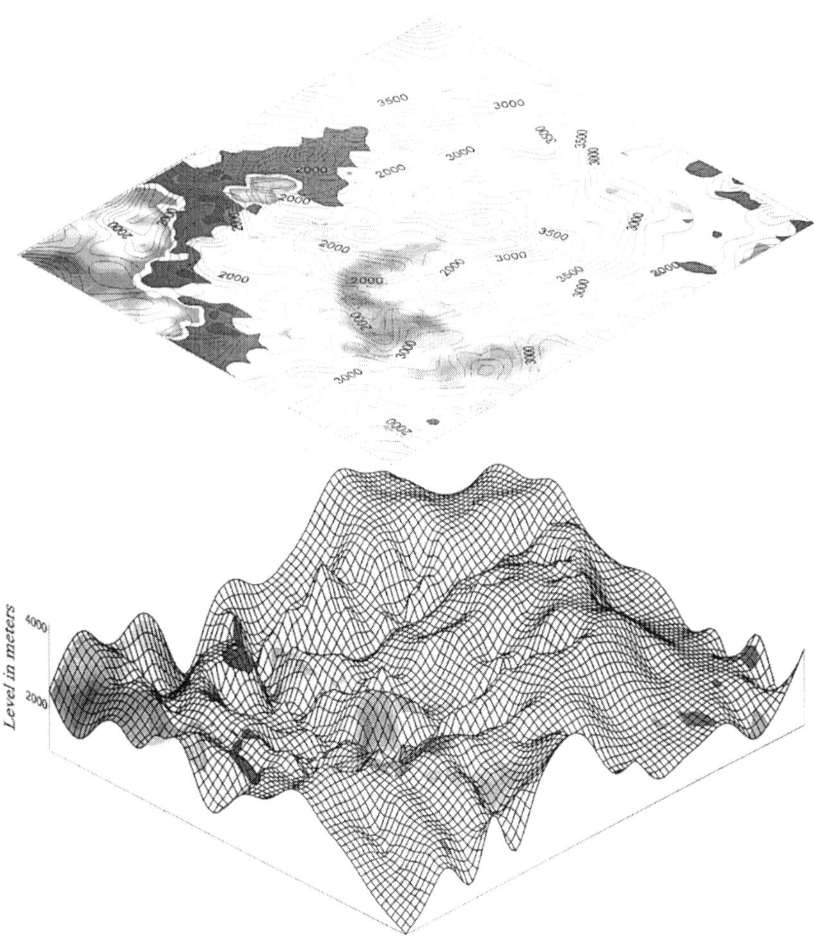

Figure 2. The results of superposition of the relief model and satellite image. More white is more snow.

snow covered areas by the altitude zones can be solved with the least errors. The Figure 2 shows the result of iterational computations and superimposing of NOAA-14 satellite image (channel 2 at May 3, 1997). Onto the three-dimensional model of the relief of the mountain territory of Akbulak river basin (Chatkal river tributary).

The possibility of the snow cover mapping for the mountainous area is snow. The snow cover is indicated by the special tone, while the terrestrial surface by horizontal lines.

Three-dimensional model of mountainous relief is constructed basing on the technique elaborated at SANIGMI.

LITERATURE

Myagkov,S.V. 1995. A method for analysing satellite images.*Processing of the second international workshop*: 299 - 304. Saskatoon, Canada.

Remote Sensing in the 21st Century: Economic and Environmental Applications, Casanova (ed.)
© 2000 Balkema, Rotterdam, ISBN 90 5809 096 5

Textural classification using textural signatures

A. Kourgli & A. Belhadj-Aissa
Laboratoire de Traitement d'Images Institut d'Electronique, Houari Boumedienne University of Sciences and Technology, U.S.T.H.B., Alger, Algeria

ABSTRACT: This paper describes an approach to textural classification of remote sensing images. This study concerns the potential use of high spatial resolution textural features to identify different covers and discriminate between various surface types. These textural features are computed from grey-level co-occurrence matrix statistics at various pixel displacement distances. Textural classification is then performed using textural signatures. The approach is applied to a SPOT multispectral scanner data of the north of Algeria.

1 INTRODUCTION

As a main step in the understanding and description of satellite imagery, texture analysis is one of the most important techniques used in analysis, segmentation and classification of textured images (Kourgli 1998). Texture is simply defined as the relationships between grey levels in neighbouring pixels which contribute to the overall appearance of an image . It represents the first level of spatial properties that can be extracted from a digital image.

In texture analysis, the most important task is to extract texture features which completely embody information about the spatial distribution of grey levels in the image. The two major approaches are : the statistical approach and the structural approach.

The first approach is related to statistical parameters which characterize the stochastic properties of the spatial distribution of neighbouring grey levels of an image. In the second approach, texture is described by the substructures, which are called primitives, their organization and their relationships.

A number of techniques for texture discrimination, texture analyzing and image segmentation have been proposed. See (Haralick 1979), (Gool et al. 1984), (Reed & Hans 1993) and (Kourgli & Behadj-Aissa 1997).

Our interest here, will be in a purely statistical approach for textural classification which is based upon the grey level co-occurrence matrices. This approach has been successfully used on a number of texture analysis problems and these matrices are known to contain enough texture-context information to match a level of human texture perception. The aim is to develop a texture classification algorithm based on grey-level co-occurrence matrices (GLCM).

All statistical texture analysis algorithms share a common problem: providing a features set which measures a visually perceivable quality of texture patterns. The success of the GLCM representation is tied directly to the fidelity with which it captures the structure of the underlying texture. It has been shown to contain texture information, histogram information, shape information and edge information (Julesz 1972). In tests by Conners et al., comparing discrimination abilities of texture analysis tools, the GLCM method performed much better than the grey level run length method, the grey level difference method and the power spectral method (Conners & Harlow 1980).

The GLC matrix is essentially a two dimensional histogram of the number of times that pairs of intensities values occur in a given spatial relationship. It forms a summary of the subpatterns that could be formed by grey level pairs and the frequency with which they occur. This approach been used successfully on number of real-world texture analysis (Haralick 1986).

2 GREY LEVEL CO-OCCURRENCE MATRICES FEATURES

Co-occurrence matrices count how often pairs of grey level pixels, that are separated by a certain distance and lie along a certain direction, occur in a digital image.

They are based on the estimation of second-order joint conditional probability density functions $p(i,j \mid d,\theta)$. Each $p(i,j \mid d,\theta)$, (θ = 0°, 45°, 90°, 135°, 180°, 225°, 270°, 315°) is the probability of going from grey level 'i' to grey level 'j', given that the intersample spacing is 'd' and the direction is given by the angle θ.For a subimage with a window size M×N, $p(i,j \mid d,\theta)$is given by :

$$p(i,j \mid d,\theta) = \text{card}\{ (m,n), (k,l) \in M \times N / \; I(m,n) = i \;\; I(k,l) = j \quad \text{with}(d,\theta) \} \tag{1}$$

These density functions can be written in matrix form

$$\phi(d,\theta) \;\; (\theta = 0°,45°,90°,135°,180°,225°,270°,315°) \tag{2}$$

An element p_{ij} of the cooccurrence matrix $\phi(d,\theta)$ counts how often each pair of grey levels of pixels (i,j) occurs in a window containing N pairs.

$$p_{ij} = p(i,j \mid d,\theta) \;/N$$

If one choose to ignore the distinction between opposite directions; then symmetric probability matrices can be employed, then the GLCM , $S_{0°}(d)$, $S_{45°}(d)$, $S_{90°}(d)$ and $S_{135°}(d)$ can be used where :

$$\begin{aligned} S_\theta(d) &= 1/2\left(\phi(d,\theta) + \phi\,(d,\theta+180°)\right), \\ &= 1/2\left(\phi(d,\theta) + \phi^t\,(d,\theta)\right), \end{aligned} \tag{3}$$

The average of these four matrices is an average matrix $S_a(d)$ which provide an information about the overall repartition of grey levels of the image.

$$Sm(d) = \frac{1}{4}\begin{pmatrix} 1/2\left(\phi(d,0°) + \phi^t\,(d,0°)\right) + 1/2\left(\phi(d,90°) + \phi^t\,(d,90°)\right) \\ +1/2\left(\phi(d,45°) + \phi^t\,(d,45°)\right) + 1/2\left(\phi(d,135°) + \phi^t\,(d,135°)\right) \end{pmatrix} \tag{4}$$

Several measures are computed from the GLC matrix to describe specific textural characteristics of an image. A set feature is defined by Haralick (Haralick an al. 1973) (angular second moment, contrast, correlation, variance, inverse difference moment, sum average, sum variance, sum entropy, entropy, difference variance, difference entropy, information measures of correlation and maximal correlation coefficient), and by Conner (Trivedi and al. 1984) (Inertia, cluster shade, cluster prominence, local homogeneity, energy and entropy).

Commonly used statistics from the GLCM include mean, variance, entropy, energy, local homogeneity, correlation, inertia, cluster shade and cluster prominence. For a subimage containing L grey levels, they are defined as :

2 .1 *Mean (Average)*

$$\mu x = \sum_{i=0}^{L-1} i \sum_{j=0}^{L-1} pij$$

$$\mu y = \sum_{j=0}^{L-1} j \sum_{i=0}^{L-1} pij \tag{5}$$

2 .2 *Variance*

$$\sigma x^2 = \sum_{i=0}^{L-1} (i-\mu x)^2 \sum_{j=0}^{L-1} pij$$

$$\sigma y^2 = \sum_{j=0}^{L-1} (j-\mu y)^2 \sum_{i=0}^{L-1} pij \tag{6}$$

2 .3 *Energy or angular second moment*

$$E = \sum_{i=0}^{L-1} \sum_{j=0}^{L-1} (pij)^2 \tag{7}$$

2 .4 *Entropy*

$$H = \sum_{i=0}^{L-1} \sum_{j=0}^{L-1} pij \;\log(pij) \tag{8}$$

2 .5 *Inertia (Contrast)*

$$I = \sum_{i=0}^{L-1} \sum_{j=0}^{L-1} (i-j)^2 \;\; pij \tag{9}$$

2 .6 *Local homogeneity(inverse difference moment)*

$$L = \sum_{i=0}^{L-1} \sum_{j=0}^{L-1} \frac{1}{1+(i-j)^2} \, pij \tag{10}$$

2 .7 *Correlation*

$$C = \sum_{i=0}^{L-1} \sum_{j=0}^{L-1} \frac{(i-\mu x)(j-\mu y)\, pij}{\sigma x \; \sigma y} \tag{11}$$

2 .8 *Cluster Shade*

$$A = \sum_{i=0}^{L-1} \sum_{j=0}^{L-1} (i+j-\mu x-\mu y)^3 \;\; pij \tag{12}$$

2 .9 *Cluster Prominence*

$$B = \sum_{i=0}^{L-1} \sum_{j=0}^{L-1} (i+j-\mu x-\mu y)^4 \;\; pij \tag{13}$$

In this study, six of the measures above were used. they are : average, variance, inertia, correlation, cluster shade, cluster prominence. It is known that these measures do not gauge all the important texture-context information contained in a GLCM The principal drawback of many texture methods and the GLCM approach is the requirement of large memory and computation time.

3 NEW ALGORITHM FOR COMPUTING GLCM FEATURES

The practical problems encountered in calculating the GLCM are those of memory required an CPU time. Some users deal with the memory problem by limiting the number of grey levels in their data to reduce the size of GLCM (Marceau and al. 1990) or by proposing an alternative to GLCM (Unser 1986) (Peckinpaugh 1991). But these solutions may lead to a loose of texture information.

We propose to use a new algorithm (Kourgli 1997) which permit to compute some of the measures by the use of an improved method based on the computation of different histograms. So that, the double summation is reduced to a simple summation and instead of storing a matrix, we store one, two or three vectors.

From this algorithm six measures of the nine measures commonly used can be computed. They are : mean (1), variance (2), inertia (3), correlation (4), cluster shade (5) and cluster prominence (6) . Table 1 gives the corresponding histogram equations of these measures, for a symmetric matrix.

Table 1. : Texture Features.

	histogram equations
1	$\mu x = \mu y = \sum_{i=0}^{L-1} i \times occ(i)$
2	$\sigma x^2 = \sigma y^2 = \sum_{i=0}^{L-1} (i-\mu x)^2 \; occ(i)$
3	$I = 2 \sum_{i=0}^{L-1} i^2 \; occ(i) \; - 2 \sum_{i=0}^{L-1} i \; occ'(i)$
4	$C = 1/(\sigma x \; \sigma x) \left(\sum_{i=0}^{L-1} i \; occ'(i) - \mu x \; \mu x \right)$
5	$A = 2 \sum_{i=0}^{L-1} \left((i-\mu x)^2 (i - 4\mu x) \; occ(i) + 3(i-\mu x)^2 \; occ'(i) \right)$
6	$B = \sum_{i=0}^{L-1} (i-\mu x)^2 \left(2(i-\mu x)^2 - 4 \mu x (i-\mu x) + 6 \mu x^2 \right) occ(i)$ $+4 \sum_{i=0}^{L-1} (i-\mu x)^2 \; (2(i-\mu x) - 12 \mu x) \; occ'(i) + 6 \sum_{i=0}^{L-1} (i-\mu x)^2 \; occ''(i)$

Where :

$occ(i) = \underset{(MxN)}{Card} \{j\}$ according to $R:(d,\theta)$

$occ'(i) = \sum_{(MxN)} j$ according to $R:(d,\theta)$

$occ''(i) = \sum_{(MxN)} j^2$ according to $R:(d,\theta)$

The others measures : Energy, entropy and local homogeneity represent special cases and the equation can not be written in terms of histograms (Kourgli 1997).

4 TEXTURAL SIGNATURES AND TEXTURAL CLASSIFICATION

Sometimes, multispectral classification schemes are unable to discriminate between different covers. This phenomenon is du to the similarity of spectral signatures in visible and infrared wavelengths. Welch and al 1990 conclude that automatic classification techniques probably need to incorporate the textural information in the scene.

For classification purpose, we use a supervised clustering which consist in partitioning the image into groups according to textural features. So, each measure of the measures above is computed for various pixel displacement distances to constitute a textural signature.

Figure 1 shows variance signatures as functions of pixel distance *d* for three selected samples taken from a SPOT scene of the north of Algeria.

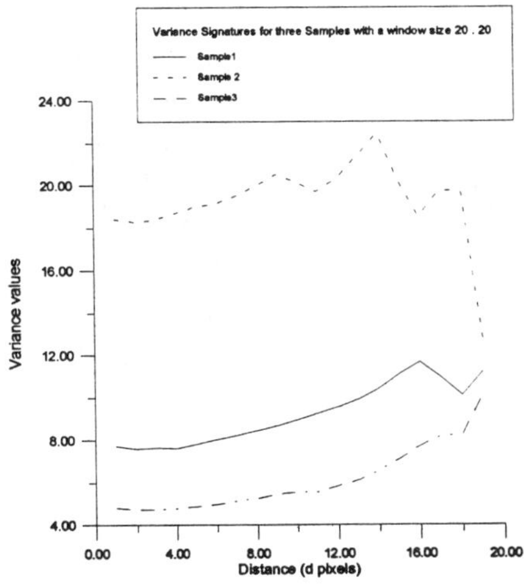

Figure 1. Variance signatures

The procedure of classification is described as follows :

1. 'C' samples with a window size T×T are selected from the image to constitute 'C' classes

1. For each sample, one of the six textural signatures is computed.

$S_i(d)$ for $i=1,2,...,C$ and $d=1,2,...,T-1$

2. The global image is scanned with a window size T×T with a step of one pixel in the row and column directions, then, for each pixel :

- The textural signature S(d) of the moving window is computed.

- The textural feature S(d) is compared to the textural feature of each sample.

$$- D_i = \sum_{d=1}^{T-1} (S(d)- S_i(d))^2 \qquad i=1,2,...,C \quad (14)$$

-Then the pixel concerned is affected to the class which minimize D_i.

The method chosen produces then an image partition according to the textural signature chosen..

We applied these measures to a SPOT scene of an area of the north of Algeria. From the full scene an image window 512 × 512 was extracted (Fig.2). The classification is performed using an optimal moving window size 20 × 20 and ten classes (Fig. 3). For each pixel of the original image, the textural signature is computed according to the moving window. This signature is then compared to the ten classes signature and the concerned pixel is affected to the most similar class using the decision rule.

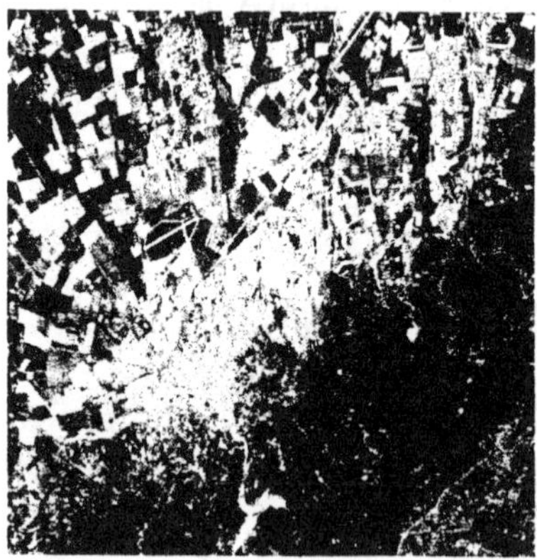

Figure 2. Original Image

Figure 3. Ten selected samples 20 × 20

Figure 4 show the results of textural classification of the original image

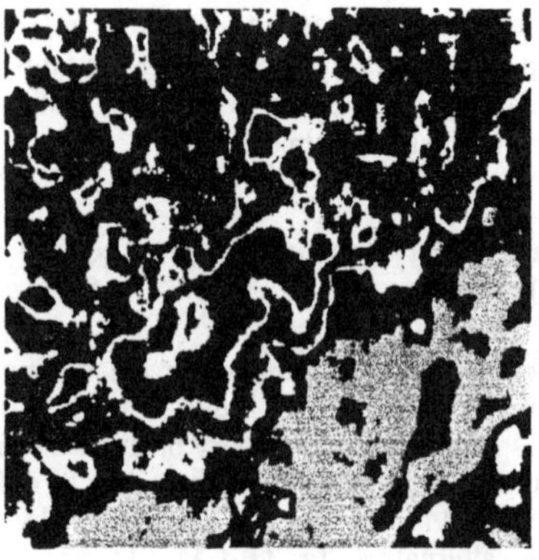

4.1 Image classified by 'mean' signatures

4.2 Image classified by 'variance' signatures

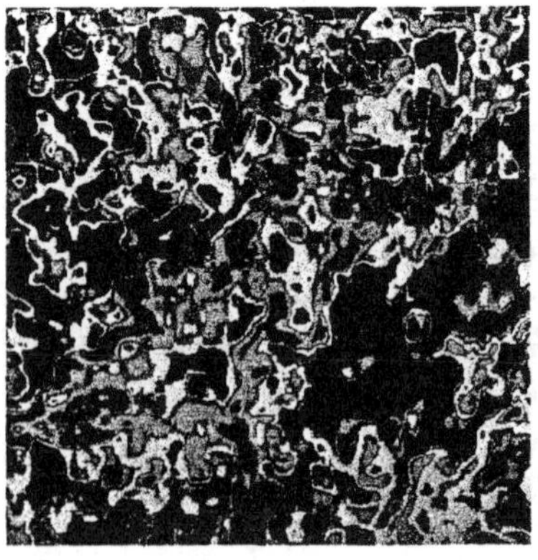

4.3 Image classified by 'correlation' signatures

4.4 Image classified by 'inertia' signatures

4.5 Image classified by 'cluster shade' signatures

Figure 4. Results of textural classification

5 CONCLUSION

This paper has described the approach for classifying textured images. The processing involves the computation of textural features using GLCM.

These results demonstrate the potential usefulness of the textural features. This approach can be easily used to provide simple algorithms for digital image processing. Indeed, the internal spectral variance of classes can be considered as a valuable source of spatial information that can be used as an additional feature in characterizing an identifying land covers.

Open problems that are still under investigation are :

- The unsupervised determination of the optimal window size for processing.
- The interpretation of the images obtained.
- The incorporation of each texture image created ,as an additional band, in spectral feature space for spectral classification purposes to study the possibility of improving classification accuracy.

REFERENCES

Conners, R.W. & C.A.Harlow 1980. A theoretical comparison of texture algorithms. *IEEE trans pattern Anal.Mach.Intell* .PAMI-2:,204-222 .

Gool, L.V. et al. 1985. Texture analysis anno 1983. SURVEY. *Computer Vision Graphics and Image Processing* 29 :336-357.

Haralick, R.M. et al. 1973. Textural features for image classification. *IEEE Trans.Syst.Man.Cybern.*SMC-8:601-621.

Haralik, R.M. 1979. Statistical and structural approaches to texture. *Proc.IEEE.*67:786-804.

Haralick, R.M. 1986. Statistical image texture analysis, *Chapter 11 of Handbook of pattern recognition and image processing.*

Julesz, B. 1972. Visual pattern discrimination . *IRE Trans. Inform. Theory* IT8:84-92.

Kourgli, A. 1997 Analyse de la texture des images satellitaire. *Master Degree U.S.T.H.B. Algiers.*

Kourgli, A. & A.Belhadj-Aissa 1997. Approche structurale de génération d'images de texture. *International Journal of Remote Sensing.* 18. No.17:3611-3627.

Kourgli, A. & A.Belhadj-Aissa 1998. Approche Statistique de Segmentation Texturale d'Images Satellitaires. Int.Conf. 'CNES Statellite-Based : A Tool for the study of the mediterranean', 23-27 November 1998, Tunis.

Marceau, D.J. et al. 1990. Evaluation of the grey-level cooccurence matrix method for land cover classification using SPOT imagery. *IEEE Trans. Geoscience. Rem. Sensing.* 28.N°4:513-518.

Peckinpaugh, S.H. 1991. An improved method for computing grey-level co-occurence matrix based texture measures. *CVGIP : Graphical models and image processing.* 53. N°6:574-580.

Reed, T.R. & J.LM.Hans du buf 1993. A review of recent texture segmentation and feature extraction techniques. *CVGIP:Image Understanding* 57.No.3:359-372.

Trivedi, M.M et al. 1984. Object detection based on grey level cooccurrence. *Computer Graphics an Image Processing* 28:199-219.

Unser, M. 1986. Sum and difference histogram for texture classification. *IEEE Trans.Pattern Anal.Mach.Intell.* PAMI-8. N°1:118-125.

Welch, R.M. et al. 1990. Cloud and surface textural feature in polar regions. *IEEE Trans.Geoscience.Rem.Sensing.* 28. N°4:520-527.

Remote Sensing in the 21st Century: Economic and Environmental Applications, Casanova (ed.)
 ISBN 90 5809 096 5

The MPNA algorithm for atmospheric correction of satellite imagery

M.Ouarzeddine & A.Belhadj-Aissa
Laboratoire de Traitement d'Images, Institut d'Electronique, Houari Boumedienne University of Sciences and Technology, U.S.T.H.B., Alger, Algeria

ABSTRACT: Atmosphere modifies the radiance reflected from the ground and contributes an additive path radiance term. It is necessary then to correct the atmospheric effect to retrieve the surface reflectance. An algorithm is developed for automatic atmospheric correction of satellite imagery of the earth's surface. The algorithm is called the MPNA (Methode des Points Noirs Améliorée). It is solely based on the satellite imagery and published data such as exoatmospheric irradiances. The problem is the estimation of the aerosol optical thickness, since the rayleigh one is generally constant and well known. The algorithm requires that satellite images selected in two suitable bands must contain some dark pixels. These pixels will correspond to dense dark vegetation, water or even shadows caused by topography and will be used to estimate atmospheric aerosol properties.

1 INTRODUCTION

Atmospherically corrected surface reflectance images improve the accuracy of surface classification and earth reflectance data. These images are also a basis for estimating the radiation budget of the earth. The algorithm presented in this study is especially useful when no atmospheric data is available and one has to resort to average climatological atmosphere, e.g. mid-latitude summer atmospheres. If the TM scene contains known reference targets (e.g. lakes) different types of aerosol can be selected.

2 DESCRIPTION OF THE STUDY AREA AND DATA USED

The study area is located in Toscana, Italy. The region is called RADICCONDI. It is a site of 15km^2 approximately. Toscana is situated in the center of north Italy (43.18-43.29°N, 10.95-10.57°E). It has a Mediterranean climate. Figure 1 shows a geographic carte of the test site.

LANDSAT TM images used to test the model are acquired at three different dates. In table 1, information concerning these images are given. Images used are 512x512 pixels dimension. All six reflective TM bands from 0.45µm to 2.35µm were considered for reflective calculation, but only two bands were used in the determination of atmospheric parameters.

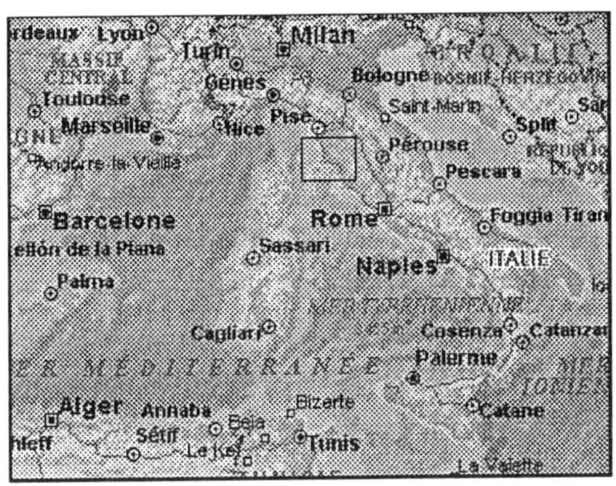

Figure 1. Geographic Carte of the test zone.

Table 1. Information concerning images used in the test of the algorithm.

Image	θ_s(°)	Altitude	date	day d
CS1	64.09	500m	11-11-90	151
CS2	32.18	500m	01-06-90	216
CW1	36.74	500m	06-08-90	312

3 METHODOLOGY

The problem to estimate surface reflectance is the assessment of path radiance. Many authors have shown that atmospheric characteristics can be obtained from space measurements that means satellite images(Isakov et al. 1996)(Wang & Nunez 1992). In this study computation of atmospheric characteristics is based on the dark object method.

4 ATMOSPHERIC CORRECTION METHOD: MPNA

The method called "Methode des Points Noirs Améliorée" is based on the dark object theory (Chavez 1988)(Cracknell 1991). It computes atmospheric parameters using two bands were the absorption is maximal. In fact the computation is focused on atmospheric aerosol characteristics. Molecular ones are well known.

To calculate path radiance, histogram satellite images are used. Since the test site is a vegetation area, we use TM1 and TM3 (Gilabert et al. 1994). Figure 2 shows histogram images for band TM1 and TM3 which correspond to CS2 image.

The DN value corresponding to path radiance is selected directly from the DN frequency histogram of the image. For bands in the visible spectrum there is usually a very sharp increase in the number of pixels at some non zero DN or grey level. This DN value is assumed to be the amount of path radiance in that particular band (Chavez 1988).

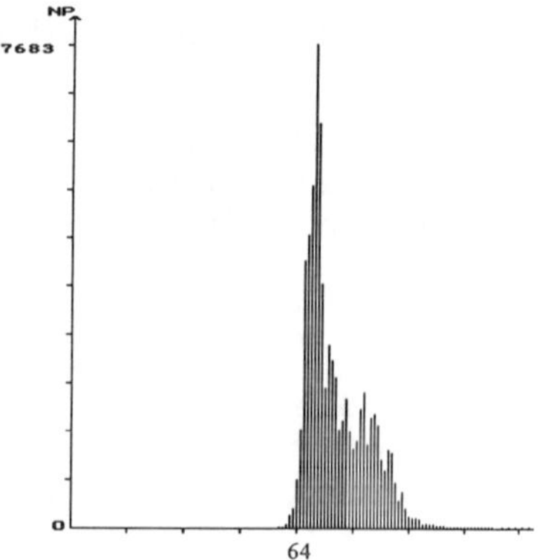

(a)

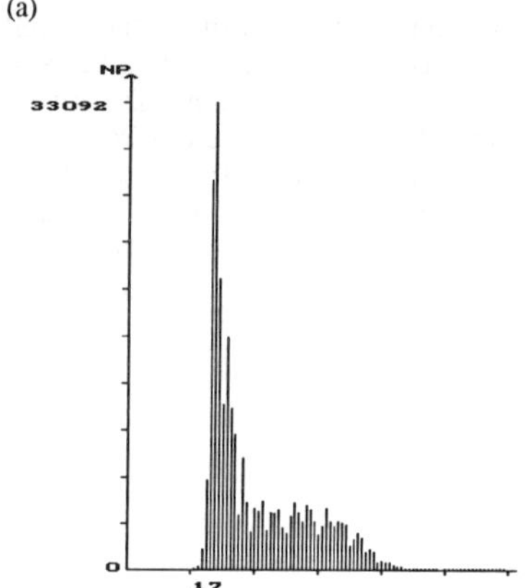

(b)

Figure 2. DN Image histograms for TM1(a) et TM3(b) bands in CS2 image.

5 COMPUTATION OF MODEL PARAMETERS

Remember that satellite radiance measured at satellite level is given by:

$$L_{sat} = L_p + E_g \frac{T^{\uparrow} \rho_{suf}}{\pi(1 - s\rho_{suf})} \tag{1}$$

Where E_g is the global solar irradiance, L_p is the path radiance, $T^{\uparrow}$ is the upward transmittance and ρ_{suf} is the surface reflectance.

If we neglect the environment effect presented by $S\rho_{surf}$ we obtain

$$L_{sat} = L_p + \frac{E_g . T^{\uparrow} \rho_{surf}}{\pi} = L_p + L_{suf} T^{\uparrow} \tag{2}$$

where L_{surf} is the surface radiance. ρ_{suf} is then:

$$\rho_{suf} = \frac{(L_{sat} - L_p)\pi}{T^{\uparrow} E_g} \tag{3}$$

In equation (3) L_p parameter must be evaluated. The other factors can be easily computed using published data [EPE. 90]. For a dark object, L_{surf} is null. Hence we write :

$$L^*_{sat} \cong L_p \tag{4}$$

L^*_{sat} is the radiance which corresponds to the digital count DN* of the dark object determined from the histogram. For the two bands TM1 and TM3 we have:

$$L^*_{sat}(\lambda_1) = A_0(\lambda_1) + A_1(\lambda_1).DN^*(\lambda_1) \tag{5}$$

$$L^*_{sat}(\lambda_3) = A_0(\lambda_3) + A_1(\lambda_3).DN^*(\lambda_3) \tag{6}$$

where A_0 and A_1 are sensor calibration factors.

Since $L^*_{sat} \cong L_p$, we write:

$$L_p(\lambda_1) = A_0(\lambda_1) + A_1(\lambda_1).DN^*(\lambda_1) \quad (7)$$

$$L_p(\lambda_3) = A_0(\lambda_3) + A_1(\lambda_3).DN^*(\lambda_3) \quad (8)$$

Following the single scattering approximation (Hill & Sturm 1991), path radiance is given by:

$$L_p = L_R + L_A \quad (9)$$

where L_R is the RAYLEIGH radiance component and L_A is the aerosol radiance one. L_R is given by:

$$L_R(\lambda) = E_0(\lambda).\cos\theta_s.\frac{P_R(\psi)}{4\pi(\cos\theta_s + \cos\theta_v)} .\left(1 - \exp\left[-\tau_R\left(\frac{1}{\cos\theta_s} + \frac{1}{\cos\theta_v}\right)\right]\right).T_{gaz} \quad (10)$$

L_A is given by:

$$L_A(\lambda) = E_0(\lambda).\cos\theta_s W_0.\frac{P_A'(\psi)}{4\pi(\cos\theta_s + \cos\theta_v)}. \left(1 - \exp\left[-\tau_A\left(\frac{1}{\cos\theta_s} + \frac{1}{\cos\theta_v}\right)\right]\right).T_A \quad (11)$$

$P_R(\psi)$ is the phase function which characterise RAYLEIGH scattering, $P_A'(\psi)$ is the modified phase function which characterise aerosol scattering and θ_v is the viewing angle.

The parameter which characterises aerosols is the optical thickness $\tau_A(\lambda)$. It is given by ANGSTRÖM relation (Iqbal 1983):

$$\tau_A(\lambda) = a\lambda^{-\alpha} \quad (12)$$

a and α are size and distribution aerosols in the atmosphere.

Assuming a linear dependence between $L_A(\lambda)$ and $\tau_A(\lambda)$ (Aranuvachapun 1986), we have:

$$L_A(\lambda) = f(\tau_A(\lambda)) = C.\tau_A(\lambda) \quad (13)$$

$$L_A(\lambda) = \Gamma\lambda^{-\alpha} \quad (14)$$

For TM1 and TM3 bands we have

$$L_A(\lambda_1) = \Gamma\lambda_1^{-\alpha} \ \& \ L_A(\lambda_3) = \Gamma\lambda_3^{-\alpha}$$

Since $L_R(\lambda)$ is well known and using equation (9) $L_A(\lambda)$ for TM1 and TM3 can be derived. Γ and are then calculated.

For a homogeneous atmosphere

$$L_A(\lambda_i) = \Gamma\lambda_i^{-\alpha}.$$

$$\Gamma\lambda_i^{-\alpha} = L_p(\lambda) - L_r(\lambda)$$

5.1 *Computation of global irradiance*

Global solar irradiance $E_g(\lambda)$ is composed by scattering solar irradiance $E_d(\lambda)$ and direct solar irradiance $E_b(\lambda)$. given respectively by:

$$E_b(\lambda) = E_0(\lambda).\cos(\theta_s).T^{\downarrow}(\lambda) \quad (15)$$

$$E_d(\lambda) = E_0(\lambda).\cos\theta_s T_O^{\downarrow}(\lambda). \left\{ \begin{array}{l} (0.5[1 - T_R^{\downarrow}(\lambda)]T_A^{\downarrow}(\lambda)) + \\ (F_c(\theta_s).W_0[1 - T_A^{\downarrow}(\lambda)]T_R^{\downarrow}(\lambda)) \end{array} \right\} \quad (16)$$

$F_c(\theta_s)$ is the factor which characterises the anisotropic MIE scattering.

5.2 *Computation of transmittance*

$T_i^{\uparrow}$ and $T_i^{\downarrow}$ are transmittances in the upward and downward directions for a processus i given as follow:

$$T_i^{\uparrow} = \exp(-\tau_i(\lambda)) \quad (17)$$

$$T_i^{\uparrow} = \exp(-\tau_i(\lambda)/\cos\theta_s) \quad (18)$$

i can be ozone (O), aerosol (A) or molecules (R).

Total transmittance for a process i is given by

$$T_i = T_i^{\uparrow}.T_i^{\downarrow} \quad (19)$$

Total transmittances are then :

$$T^{\uparrow} = \exp(-\tau_O(\lambda) - \tau_A(\lambda) - \tau_R(\lambda)) \quad (20)$$

$$T^{\downarrow} = \exp((-\tau_O(\lambda) - \tau_A(\lambda) - \tau_R(\lambda))/\cos\theta_s) \quad (21)$$

6 RESULTS AND DISCUSSION

6.1 *Direct transmittance*

Direct atmospheric transmittance $T^{\uparrow}(\lambda)$ computed using the MPNA and corresponding to the three scene given in table 1 is shown in figure 3. This transmittance is also calculated using 5S code. The code gives the same values for the three images. All transmittance values given in figure 3 are generally found in the visible and infrared intervals. We remark that direct transmittance varies with wavelength.

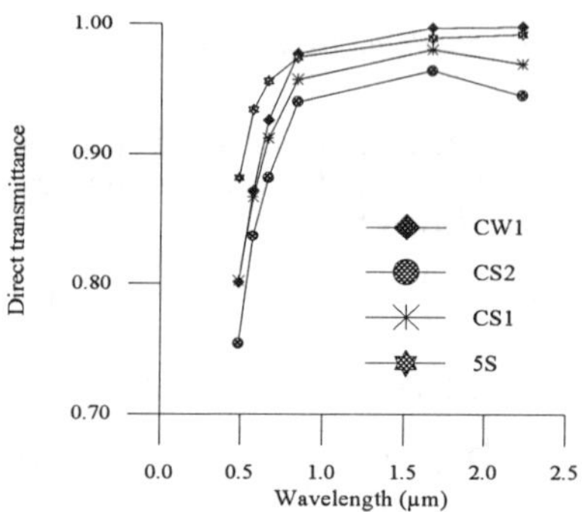

Figure 3. Direct transmittance $T^{\uparrow}(\lambda)$

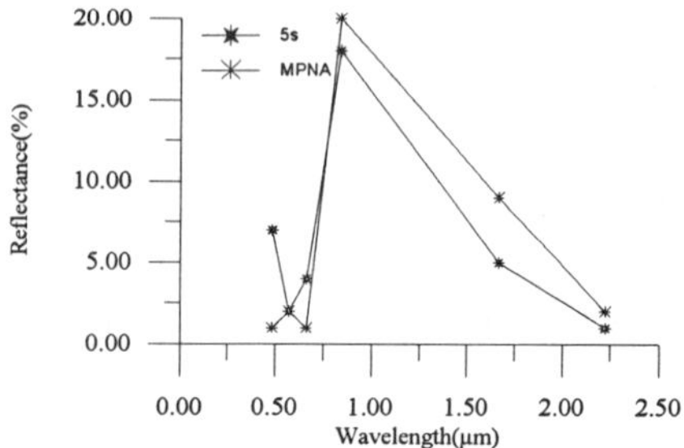

(a)

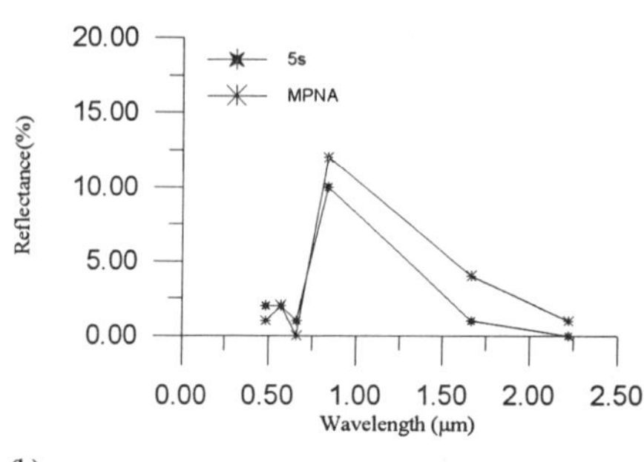

(b)

Figure 4. Comparison of 5S and MPNA results tested on a coniferous vegetation in August (a) and November (b).

6.2 *Irradiances values*

Diffuse, direct and total irradiances corresponding to CS2 image are given in table 2. The contribution of diffuse irradiance to total irradiance in TM5 et TM7 bands is practically negligible compared to direct irradiance. This is in agreement with published results [IQB. 83].

Table 2. Direct, diffuse and total solar irradiance computed by the MPNA for CS2 image.

Band	E_0	E_b	E_d	E_g
TM1	137.20	77.38	16.92	94.30
TM2	162.60	104.46	12.13	116.59
TM3	119.30	81.80	7.40	89.20
TM4	140.60	104.40	4.61	109.01
TM5	49.67	21.60	0.08	21.68
TM7	19.99	8.71	0.02	8.73

6.3 *Comparison with 5S code*

Figure 4 shows a comparison of spectral signatures of coniferous vegetation. Values of reflectance are computed using atmospheric parameters derived by the MPNA method or 5S code.

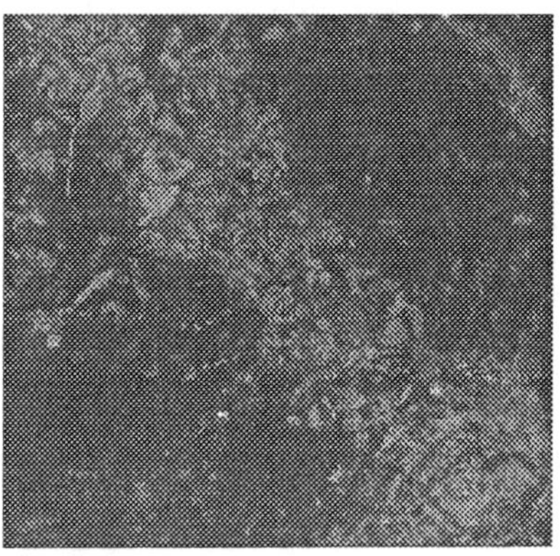
(a)

(b)

Figure 5. Original image (a) and atmospheric corrected image using the MPNA (b) band TM4 (November).

The figure 5 indicates that significant reflectance difference is localised in short wavelength. This difference is caused by variations of atmospheric effects with wavelength, by data introduced to compute atmospheric parameters of 5S code (e.g. Visibility and lambertian surface assumption).

Figure 5.a shows the original TM band 4 image (grey values). Figure 5.b presents the atmospheric corrected relectance image. The corrected image reveals the existence of some details which are not present in the original one.

7 CONCLUSION

A simple method called the MPNA used to retrieve atmospheric effect from satellite images has been presented. Tests have been applied on TM images of different dates, nevertheless this method can be applied to any satellite image which contains dark pixels.

Results obtained using this method were compared to those obtained by 5S code. Difference between values are especially in short wavelength. The accuracy of the atmospheric correction method has been also estimated using spectral signatures taken from non corrected and corrected images of some vegetation. Results can be judged good.

REFERENCES

Aranuvachapun, S. 1986. The atmospheric optical depth spectrum determined from CZCS radiance. *International journal of Remote sensing* : 105-118.

Chavez, P. S. 1988. An improved dark-object substraction Technique for Atmospheric Scattering correction of multispectral Data. *Remote sensing of environnement,* : 459-479.

Chavez, P. S 1989. Radiometric calibration of Landsat TM multispectral Images. *Photogrametric engineering & Remote Sensing*:1285-1294.

Cracknell, A. & L. Hayes 1991: Introduction to remote sensing. *Francis & Taylor, New York.*

Gilabert, M. A., C. Conese & F. Maselli 1994. An atmospheric correction method for the automatic retrieval of surface reflectances from TM images. *International journal of Remote sensing* :2065-2086.

Hill, J.& B. Sturm 1991. Radiometric correction of multitemporel Thermal Mapper data for use in agricultural land cover classification and vegetation monitoring. *International journal of Remote sensing* : 1471-1491

Iqbal, M 1983. An introduction to solar radiation. *New York: Academic Press, INC.*

Isakov, V. Y., R. E. Feind, O. B. Vasilyev & R. M. Welch 1996. Retrieval of aerosol spectral optical thickness from AVIRIS data. *International journal of Remote sensing* : 2165-2184.

Wang, G. & M. Nunez 1992. A new method of estimation path radiance for band ratio applications. *International journal of Remote sensing* :527-539.

Remote Sensing in the 21st Century: Economic and Environmental Applications, Casanova (ed.)

An automatic method to calibrate Meteosat visible images

M. Lefèvre, O. Bauer, A. Iehle & L. Wald
Groupe Télédétection et Modélisation, Ecole des Mines de Paris, Sophia Antipolis, France

ABSTRACT: A method is described for the automatic calibration of the successive operational Meteosat sensors in the visible range since 1985. The autocalibration method is based on a statistical analysis of the images. This approach does not require any information about atmospheric and surface parameters. As a starting point, the absolute vicarious calibration for only one image in the whole time-series is needed, which can be provided by previous published works. Our results are fully consistent with previous studies. Several tests have been performed to validate the method and check its robustness and reliability.

1 INTRODUCTION

For climatological purposes and for any analysis of a time-series, images from satellite should be well-calibrated with respect to each other, in order to ensure that any variation in time is due to change in the signal coming from the observed target, and not from a change in calibration of the observing system. The Earth viewing calibration approach has been recently developed as a backup solution to the possible failure or unreliability of on-board calibration devices. It is based on the knowledge of physical characteristics of some Earth phenomena as well as upon the processing of the digital imagery flowing down from the sensor itself (see e.g., Köpke 1982, 1983; Frouin, Gautier 1987).

Several techniques have been proposed to solve this calibration problem and have been applied to different types of sensors, either for sun-synchronous satellites (e.g. Holben *et al.* 1990) as for geostationary satellites (e.g., Köpke 1982). In the following, we are interested in calibrating the visible channels of geostationary meteorological satellites, and more precisely of the Meteosat satellite-series. Because these satellites have no onboard calibration system, added to the lack of prelaunch calibration, an absolute calibration of the radiometer in orbit for the VIS channel has to be performed to allow for accurate radiance measurements.

2 BACKGROUND

Assuming a linear response of the sensor, the relationship between the emerging radiance from the atmosphere and measured by the sensor, L_{sat}, and the numerical count, NC, is:

$$L_{sat} = \alpha\ (NC - NC0) \quad (1)$$

where α is the calibration coefficient (in $Wm^{-2}sr^{-1}NC^{-1}$) and NC0 is the offset numerical count of the calibration.

Several authors have proposed a method for the calibration of Meteosat data (see a review in Lefèvre *et al.*, 1999). From an operational point of view, standing from the user side, these methods, though accurate, suffer from some drawbacks. The method of Moulin *et al.* (1996) is the most attractive but we found it rather difficult to implement in an operational chain because of:

- the need of radiative transfer computations with a detailed numerical code *i.e.* the knowledge of atmospheric (water vapour, ozone, aerosol optical properties) and surface parameters (e.g., temperature);
- the need of a first test to extract cloud-free pixels from the database of images to create a 12-year dataset of clear sky, 5-day averaged, numeric counts for selected targets (4 targets chosen in desert areas).

Accordingly, we have looked for a method which can be routinely operated in quasi-real time without external inputs. An empirical approach was adopted which makes use of statistical quantities that can be extracted from each image, following the tracks of Asmami, Wald (1993) and Wald (1998) for the AVHRR instruments. Such methods do not have the beauty of methods explicitly simulating the radiative transfer, but they are presently as accurate as the others, far more simple to perform, even in real time, and far more easy to implement.

A new method, called the autocalibration method, has been devised and tested. Actually this method can only perform the calibration of an image relative to another. It follows that to calibrate a series of images, the procedure is the following:

1 use a calibration function found in the literature, and corresponding to a day, noted t_0
2 calibrate the image corresponding to that day (called hereafter the reference image) using Equation 1
3 perform the autocalibration method to calibrate all the other images relative to the reference one.

The first two steps are performed only once.

3 THE METEOSAT SENSORS – THE IMAGES

The Meteosat program, operated by the European Space Agency (ESA) started in 1977. The VIS data used for the calibration were provided by Eumetsat (Darmstadt, Germany), the European organisation for the exploitation of the Meteosat system and distributor of the images. For the study, a special set of Meteosat data was used, called ISCCP-B2. This set of images has been set up in the framework of the International Satellite Cloud Climatology Project (ISCCP), part of the World Climate Research Program (WCRP). This set is derived from the operational Meteosat images, in both visible and thermal infrared bands by a reduction of the number of pixels. The remaining pixels still bear their original values after they have been brought to the spatial size of the infra-red images (5 km at nadir).

Our dataset covers all years since 1985. For that period, five Meteosat satellites and sensors have been operated, at different gain levels. The various Meteosat sensors have different sensitivities due to their difference in the spectral band. Each of the spectral channels can be operated at one of 16 different gain levels. These gain levels are used to obtain the optimum dynamic range for each spectral channel and are adjusted as required (Eumetsat, 1996). These changes in gain affect the sensitivity of a given sensor, and are to be compensated by an adjustment of the operational calibration coefficient.

The spectral response of each VIS channel of each sensor is taken from Eumetsat (1996) and other authors. The overall spectral response of a given radiometer channel is a combination of the optical transmission of the entire optical system and the detector response. The spectral responses S_λ , λ being the wavelenght, have varying shapes, with a maximum transmission between 600 nm and 900 nm. The operational Meteosat satellites have in common a spectral channel at about 700 nm, near the peak of the solar spectrum. Meteosat-1 and -2 have similar spectral response, and are less sensitive than the other sensors below 600 nm. On the opposite, Meteosat-4 and -5 are more sensitive in the first half of the bandwidth. Meteosat-5 is less sensitive than Meteosat-2, -3 and -4 at wavelengths greater than 800 nm.

O'Mongain *et al.* (1983) have emphasised that the calibration depends on the shape of the spectral sensitivity curve of the radiometer. For broad band channels like that of Meteosat sensors, the knowledge of the spectral distribution of the input radiance is mandatory. Values of solar spectral irradiances incident at the top of the atmosphere $I_{0\lambda}$ are those of Neckel, Labs (1984). Then the spectral distribution corresponding to each Meteosat sensor, that is the product of the solar spectral constant with the sensor response (*i.e.* $I_{0\lambda}\ S_\lambda$ in Wm^{-2}) is computed, as well as the total irradiance in the VIS channel for the various Meteosat sensors (*i.e.* integrated from 0.3 µm to 1.1 µm, $\int_{0.3}^{1.1} I_{0\lambda} S_\lambda d\lambda$). More energy flux is available for Meteosat-3, -4 and -5 because the corresponding bandwidths are larger than those of the first two Meteosat sensors.

The effective radiance, L_{sat} (in $Wm^{-2}sr^{-1}$), is the integral of the radiance emerging upward from the atmosphere measured by Meteosat VIS channel, L_λ (in $Wm^{-2}sr^{-1}mm^{-1}$), weighted by its spectral response, S_λ, and is defined as:

$$L_{sat} = \int_0^\infty L_\lambda S_\lambda d\lambda \qquad (2)$$

where L_λ is the spectral radiance at the entrance of the satellite sensor. The spectral radiance L_λ depends on the viewing geometry, *i.e.*, on the solar zenithal angle, on the satellite zenithal (viewing) angle and on the relative azimuthal angle. It is also dependent on the spectral irradiance of the solar constant outside the atmosphere, $I_{0\lambda}$, and on the spectral optical properties of atmosphere and surface. L_{sat} is digitised by the electronics of the sensor onboard the satellite.

A calibration function is often a linear function, with a good approximation (see Eq. 1). It links digital count to reflectances (or albedo) in the visible range. Therefore, two points are necessary to determine this linear function. The farther these points, the more accurate the determination of the straight line. Accordingly, calibration is made by observing a dark target and a bright one. For the visible range, there is no internal calibration. The calibration function has been set prior to launch, and is used throughout the life of the sensor. The pre-flight calibrations are subject to change due to launch constraints and the hostile environment of the sensor: outgasing, deterioration in the optronic system, variation in the spectral filter characteristics, ... They induce changes in the calibration function which are unknown, but should be known for a reliable processing of the images in the visible range.

4 THE AUTOCALIBRATION METHOD

The autocalibration method is based on the analysis of two quantities which are constant in radiance over the time-series. The work of Moulin *et al.* helped us in selecting several time periods during which the calibration parameters were fairly constant. Using our experience gained with the AVHRR sensor, we have searched during these periods, statistical parameters using the fact that in the entire field of view of the Meteosat sensor, the mixed presence of land, ocean, and clouds of different reflectivity over approximately one third of the Earth, whatever the day and time of the year, may lead to the preservation of such statistical quantities with time. In an empirical way, we selected three parameters which are the numerical counts corresponding to a dark target, and to the percentiles 5 % and 80 % of the histogram of the mid-day image. These numerical counts vary in time according to the Meteosat sensor and the date of viewing, but the two quantities that are derived in radiances (and defined in the following) were found constant for the test periods. The analysis of the whole time-series with respect to the calibration proposed by Moulin *et al.* (1996) demonstrated that any drift in these quantities reflect a drift in the calibration of the Meteosat sensor.

For each day *t* for which images are to be calibrated, the three parameters are computed. The numerical count for the dark target, $NC_{t_{dark}}$, is computed using the early image at 0530 UT (slot 11). At that time, most of the portion of the Earth viewed by Meteosat is still in the night. The histogram is built for that night-time part of the image and the first mode gives $NC_{t_{dark}}$.

Then the daytime image at 1130 UT (slot 24) is considered. Only pixels which numerical count is greater than $NC_{t_{dark}}$ are taken into account and the histogram is computed. Percentiles 5 % and 80 % are extracted and this provides the corresponding numerical counts NC_{t5} and NC_{t80}; we recall that NC_{t5} (resp. NC_{t80}) represents the numerical count for which 5 percent (resp. 80 percent) of the surface of the cumulative histogram is reached (percentile 5 %, resp. 80 %). These parameters vary along the time, from one Meteosat sensor to another, from one gain to another for the same sensor, and also from day to day for the same sensor and same gain (Figure 1).

For the day t_0, these three parameters: $NC_{to_{dark}}$, NC_{to5} and Nc_{to80}, can be converted into radiances: $L_{to_{dark}}$, L_{to5} and L_{to80} by the means of Eq. 1:

$$L_t = \alpha_t (NC_t - NC0_t) \quad (3)$$

$$L_{t_{dark}} = \alpha_t (NC_{t_{dark}} - NC0_t) \quad (4)$$

$$L_{to_{dark}} = \alpha_{to} (NC_{to_{dark}} - NC0_{to}) \quad (5)$$

$$L_{to5} = \alpha_{to} (NC_{to5} - NC0_{to}) \quad (6)$$

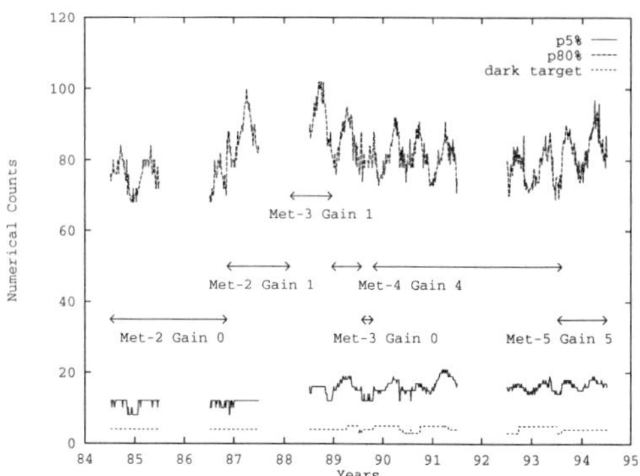

Figure 1. Variation of the percentiles NC_{t5} and NC_{t80} (noted here p5% and p80%) and of the dark target NC_{tdark} for Meteosat VIS channel at 1130 UT and from January 1, 1985 to December 31, 1994. The satellite sensor and the gain are also reported.

$$L_{to80} = \alpha_{to} (Nc_{to80} - NC0_{to}) \quad (7)$$

which leads to :

$$\alpha_{to} = \frac{(L_{to\,80} - L_{to5})}{(NC_{to\,80} - NC_{to\,5})} \quad (8)$$

$$\alpha_t = \frac{(L_{t\,80} - L_{t5})}{(NC_{t\,80} - NC_{t\,5})} \quad (9)$$

$$NCO_t = NC_{t\,dark} - L_{t\,dark}\frac{1}{\alpha_t} \quad (10)$$

The two following quantities are constant with time:

$$\frac{L_{to_{dark}}}{I_{to}} = \frac{L_{t_{dark}}}{I_t} \quad (11)$$

$$\frac{\left(L_{to_{80}} - L_{to_5}\right)}{I_{to}} = \frac{\left(L_{t_{80}} - L_{t_5}\right)}{I_t} \quad (12)$$

where I_t and I_{to} are the incoming irradiances in the VIS radiometer at days *t* and t_0. They are equal to the total irradiances for the VIS sensors under concern above-mentioned, multiplied by the correction of the astronomical distance from the Sun to the Earth for the days *t* or t_0. This ratioing also accommodates for the changes in sensor. After the substitution of Eqs. 9, 10, 11 and 12 into Eq. 3, and some mathematical handling, the radiance L_t at any time *t* and for each numerical count NC_t is:

$$L_t = \frac{(L_{to80} - L_{to5})}{(NC_{t80} - NC_{t5})}\frac{I_t}{I_{to}}\left(NC_t - NC_{t_{dark}}\right) + L_{to_{dark}}\frac{I_t}{I_{to}} \quad (13)$$

5 RESULTS

The autocalibration method has been applied to the Meteosat dataset spanning for the above-mentioned period. The results have been compared to those of Moulin *et al.* who analysed a period similar to ours: 1983-1994. Pixels have thus been selected in the same locations as these authors did, that is four sites in desertic areas. For these pixels, numerical counts have been extracted from each Meteosat image taken at 1130 UT (slot 24) and converted into radiances with the autocalibration method on the one hand,

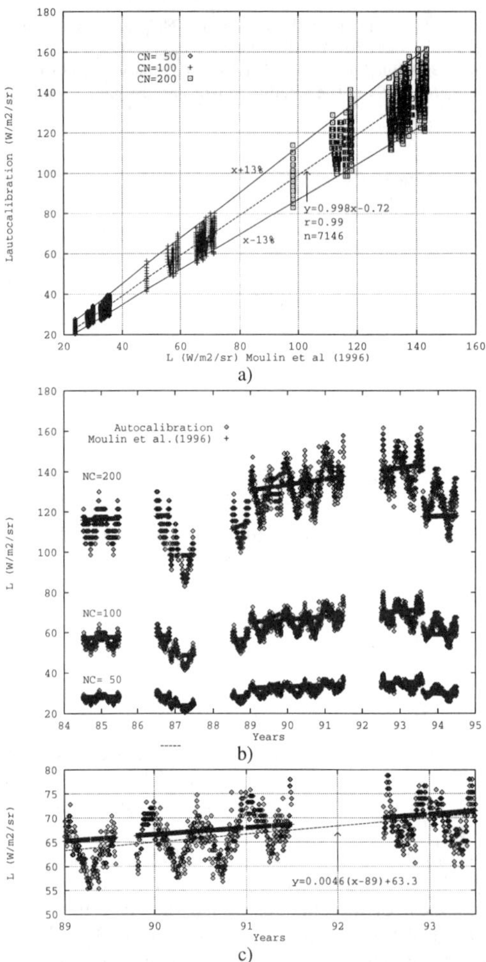

Figure 2. Autocalibration is initialised with Köpke's (1983) coefficients on 01/01/89. Radiance calculations are made for numerical counts NC 50, 100 and 200. ***a)*** Comparison with Moulin *et al.* results. ***b)*** Radiances for both methods. The juxtaposition of the crosses "+" makes the Moulin *et al.* results appearing as thick lines. ***c)*** Magnification of the drawing of ***b)*** for NC=100 during Meteosat-4 period. Also shown is the regression line for the autocalibration results as a function of the year *x* (in floating values).

and with the coefficients published by Moulin *et al.* on the other hand.

As said above, the autocalibration method requires an absolute calibration for the radiance on one image at least. We have tested several dates for the day t_0 with different coefficients of calibration taken from known calibration coefficients calculated by vicarious calibration (Moulin *et al.*, 1996) for June 1983 to December 1994, and from aircraft experiment: Köpke (1983) for the end of 1981, or Kriebel, Amann (1993) for August - September 1989.

All our results are consistent with Moulin *et al.* (1996) results for the same time series from 1985 to 1994. The correlogram between both sets of results shows a slope close to 1 and a correlation coefficient of 0.91. This coefficient is 0.95 for the four sites used by Moulin *et al.*

From the analysis of the time-series, it is obvious that whatever the method is, the fluctuation in calibration within a year is quite large: approx. 15 W m^{-2} sr^{-1} in radiance for NC = 100, that is approx. 23 percent of the mean radiance. It follows that a formula based upon a linear regression such as that proposed by Moulin *et al.* and others does not provide the best results, as demonstrated by the low correlation coefficient between the proposed formula and the samples (0.51). We conclude that calibration coefficients should be assessed on a daily basis preferrably.

Figure 2 displays the results of one of the multiple tests. The initialisation is made with the coefficients of Köpke (1983) set up for January 1, 1989. The autocalibration coefficients as well as the Moulin *et al.* Coefficients are applied to the digital counts NC 50, 100 and 200. The relative rms error claimed by Moulin *et al.* is 13 %, and is reported in this figure.

Figure 3 displays the results for the initialisation made with the coefficients of Kriebel, Amman (1993) set up for September 9, 1989, and for the Tunisian desert.

6 CONCLUSION

A method of autocalibration of Meteosat B2 VIS channel has been developed, based on the analysis of changes in statistical quantities extracted in each image.

We have tested different dates for the absolute reference, as well as different known calibration coefficients calculated from vicarious calibration and from aircraft experiment. All our results are consistent with Moulin *et al.* (1996) results for the same time series from 1985 to 1994. It was also shown that a calibration law based upon a linear function of the day is not providing the best results, though very easy to handle. The autocalibration method performs similarly or better than such formulae. Several tests

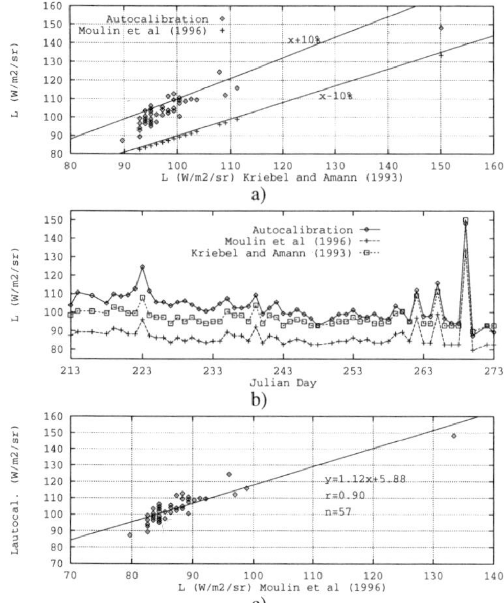

Figure 3. Autocalibration initialised with Kriebel and Amann's (1993) coefficients on 05/09/89 for the Tunisian desert. The radiances are computed for a rectangular shaped sensitivity of the radiometer during August-September 1989. ***a)*** Comparison between autocalibration and Kriebel and Amann. ***b)*** Radiances from 01/08/89 to 30/09/89 (Meteosat-4) for the three calibration laws. ***c)*** Comparison between autocalibration and Moulin *et al.*

have been performed to validate the method and check its robustness and reliability.

Its main advantage is its very easy implementation as it does not require any radiative transfer procedure nor any specific targets to be extracted like in ocean or desert. One single date for the absolute calibration is sufficient, and moreover this single date can be any day of the time series. The present approach provides reliable calibration without external knowledge and/or human operator. It makes it simple to calibrate rapidly a large time series of images.

The autocalibration method can be applied to Meteosat images acquired before 1985 and after 1994. The authors think that it should be also applicable to other geostationary satellites having spectral bands similar to Meteosat, such as GOMS, GMS, Insat, and the first GOES series, but no check has been performed so far.

A work is undertaken at Ecole des Mines de Paris for the calibration of the Meteosat images, starting January 1985 up to now. The calibration coefficients should be made available on the Web site: www-helioserve.cma.fr.

ACKNOWLEDGEMENTS

The help of the Joint Research Centre (Ispra, Italy) is acknowledged: it indirectly initiated that study which also benefited from the collection of several years of Meteosat images made under a contract with JRC within the MARS programme, and held at Ecole des Mines de Paris. Very fruitful discussions have been held with Eumetsat on several aspects: we particularly thank Richard Francis and the archives Department MARF.

REFERENCES

Asmami, M. & L. Wald 1993. A method for an accurate on-flight calibration of AVHRR data for vegetation index calculation. In *Remote sensing for monitoring the changing environment of Europe*: 53-57. Rotterdam: Balkema.

Eumetsat 1996. Annex to the Meteosat-5 calibration report. Meteorological Products Extraction Facility (MPEF), March 1996. Eumetsat, Darmstadt, Germany, 25 pp.

Frouin, R. & C. Gautier 1987. Calibration of NOAA-7 AVHRR, GOES-5 and GOES-6 VISSR/VAS solar channels. *Remote Sensing of Environment* 22 (1): 73-101.

Köpke, P. 1982. Vicarious satellite calibration in the solar spectral range by means of calculated radiances and its application to Meteosat. *J. of Appl. Optics* 21(15): 2845-2854.

Köpke, P. 1983. Calibration of the Vis-channel of Meteosat-2. *Advances in Space Research* 2(6): 93-96.

Kriebel, K. T. & V. Amann 1993. Vicarious calibration of the Meteosat visible channel. *Journal of Atmospheric and Oceanic Technology* 10(2): 225-232.

Lefèvre, M., O. Bauer, A. Iehle, L. Wald 1999. An automatic method for the calibration of time-series of Meteosat images. To be published by *Int J. of Remote Sensing*.

Moulin, C., C.E. Lambert, J. Poitou & F. Dulac 1996. Long term calibration of the Meteosat solar (VIS) channel using desert and ocean targets. *International Journal of Remote Sensing*, 17(6): 1183-1200.

Neckel, H. & D. Labs 1984. The solar radiation between 3300 and 12500 Å. *Solar Physics* 90: 205-258.

O'Mongain, E., M. Goodman, M. Reynolds & C.B. Connor 1983. High-accuracy calibration of optical imaging instruments in space. *ESA Journal*, 7(3): 299-309.

Wald, L. 1998. In-flight interband calibration of the AVHRR data by a cloud-viewing technique. In P. Gudmansen (ed.) *Future Trends in Remote Sensing*: 453-459. Rotterdam: Balkema.

Remote Sensing in the 21st Century: Economic and Environmental Applications, Casanova (ed.)
© 2000 Balkema, Rotterdam, ISBN 90 5809 096 5

Education and research in remote sensing in Algeria

Y. Smara & A. Belhadj-Aissa
Image Processing Laboratory (IPL), Electronic Institute, Houari Boumedienne University of Sciences and Technology, U.S.T.H.B., Alger, Algeria

ABSTRACT : The remote sensing technology transfer in Algeria began in the mid 1970's. Later some national agencies funded projects that included the application of this technology. Generally speaking, these projects had two distinctive aspects. Firstly, in both fieldwork and visual interpretation were carried out in our country. Secondly, digital image processing was carried out in the developed countries.
During the early 1980's, the Image Processing Laboratory has produced software and hardware designed for digital image processing. Around this time, some institutions started to acquire their image processing facilities, specially for teaching and research purposes.
In Education, the Image Processing and Remote Sensing course is intended for post graduate level of education and for professionals from different disciplines who may require the use of the remote sensing technology in their specific field of knowledge. The duration of the course is one academic year. The aim of the course is to provide a thorough training in the skills necessary for the application of remote sensing. The research in the university is organised by projects and in the IPL laboratory methodologies and applications in Image Processing and Remote Sensing are developed.
Although Algeria is a developing country, it is considered that in relation to remote sensing technology of North African regions it is in a strategic position. Given that, according to some ecologist organisations, Algeria bears one of highest desertification rates, a lot of effort is being put into the use of this technology.
Given the present availability of remote sensing in the country, the application of this technology in cartography and other disciplines is obviously within sight. In the near future optical and radar remotely sensed imagery will be used to produce natural inventories that would lead to a better management of the natural resources of the country.

1. INTRODUCTION

Remote sensing technology is a relatively new discipline compared to other related disciplines dealing with the capture of geographical information. All remote sensing processes are carried out using computer hardware and software.

Remote sensing may be viewed as a study, a source or a technique. It is a study of its own in that it has a set of principles to inculcate, matter to examine, collate, store, expand and retrieve ; a mode of transmission ; attitude formation and skills proper to it. As a study, it has a scientific basis in physics, atmospheric chemistry and biological processes. To understand what it deals with or brings out, however, one needs to apply a multidisciplinary approach.

When remote sensing is viewed as a source of data, it is inter-disciplinary. The data given can be studied and treated in different subjects and fields, for example in Geography, geology, biology and range management. Each subject or discipline takes what concerns it from the images. There is an advantage here, that in one subject taking what it wants from an image, the subject benefits from the understanding of others. So remote sensing encourages inter-dependence of disciplines.

The third view of remote sensing is that it is a technique, a means of obtaining data of distant places, features, activities, phenomena or processes. Once one has access to remotely sensed data, there are fine skills to use under the technique, for example overlaying of images on transparencies and the use of computer for combining elements and factors to give desired results.

A brief account is given of the state of remote sensing in Algeria. The evolution of remote sensing education and research in the Image Processing Laboratory (IPL) is very related to the evolution of remote sensing in the country. The IPL laboratory has noticeably improved the education and research activities on remote sensing and image processing.

In this respect, special attention has been given to the role of the University in the transfer and application of this technology. Also, some conclusions are stated pertaining to the future of remote sensing in the country.

2. SOME GENERAL ASPECTS OF REMOTE SENSING TECHNOLOGY IN ALGERIA

The remote sensing technology transfer in Algeria began in the mid 1970's. Initially some researchers were trained in the developed countries. Later some national agencies funded projects that included the application of this technology. Generally speaking, these projects had two distinctive aspects. First, both of field work and the visual interpretation were carried out in the country. Secondly, the digital image processing was carried out in the developed countries.

During the early 1980s, The Image Processing Laboratory has produced software and hardware designed for digital image processing [1][2]. Around this time, some institutions started to acquire their own image processing facilities, specially for teaching and research purposes.

In 1988, the Image Processing Laboratory (I.P.L) of the USTHB university in liaison with the National Institute of Cartography (I.N.C) and the National Centre of Space Techniques (C.N.T.S), the two main organisations in Remote Sensing and Cartography in ALGERIA, organised some local seminars in remote sensing. After these seminars, interest increased substantially among geophysicists for the use of remote sensing techniques but economical constraints limited the organisation and development of a working group.

At the same time, we started a course in the Electronic Institute of the USTHB university in remote sensing focused essentially on the topographical aspects with SPOT satellite data. In effect, since this date, the INC and the IPL with the collaboration of IGN (France) started to develop projects supported in digital image analysis of SPOT HRV scenes acquired over some geographical locations in Algeria.

Although remote sensing technology transfer may represent many benefits for our regions, there are critical factors that make its routine application difficult. Some of these factors include a lack of economic resources, a relative absence of trained personnel, little aid from the developed world and some inefficient bureaucracy in the country.

In 1990, the first annual course is organised by the Office of Outer Space Affairs, through the United Nations Programme on Space Applications and hosted by the Department of Physical Geography of the Stockholm University. This first course is directed at educators from Africa.

3.THE UNITED NATIONS TRAINING COURSE AND ITS IMPACT AT LOCAL LEVEL

The main objective of the United Nations training course is to develop the practical knowledge and skills of educators from developing countries in remote sensing technology and to equip them with the skills needed to introduce the discipline into the educational curricula of universities and institutes in their home countries.

This course allowed us to initiate and organise public lectures in our university, a long with seminars and workshops with other institutions. Discussions are made with institutions able to use remote sensing techniques to transfer our awareness in this field. In our laboratory, we developed techniques on PC stations because these institutions can use it for their applications.

3.1. Education in Remote Sensing

The Image Processing and Remote Sensing course is intended for post graduate level education and for professionals from different disciplines who may require the use of remote sensing technology in their specific field of knowledge. The duration of the course is one academic year. The aim of the course is to provide a thorough training in the skills necessary for the application of remote sensing. The course is organised in two parts, the first one covers the general aspects of the technology. The second part of the course is devoted to the study of the digital image processing tasks. This part includes several practical sessions that span from the simple display of images to image classification. The image processor used in the course is LATIS (Logiciel d'Analyse et de Traitement d'Images Satellitaires) [3][4]. The Laboratory has some images acquired over Algeria by the sensors on board the Landsat and SPOT and ERS-1 satellites. These images are stored in a digital form in both 1600 bpi computer compatible tapes and double and high density floppy diskettes.

The course is divided into fifteen units as follows:

1. remote sensing : a general overview,
2. physical principles,
3. geometric considerations,
4. technology for data acquisition,
5. technology for data processing,

6. active remote sensing,
7. thermal remote sensing,
8. digital analysis
9. image processing methods,
10. image classification,
11. statistics and remote sensing,
12. project design,
13. digital cartography,
14. geographical information systems,
15. remote sensing applications.

We added notions such as Photo-interpretation, techniques on Radar Imagery, multivariate statistical analysis, evaluation of classifications in comparison with ground data.

Practical sessions in Photo-interpretation , Image processing, Demonstrations on G.I.S. over regions such as SWEDEN and La LOIRE (France) have been carried out.

For the development of the course the laboratory has several recent text books, and articles of international journals on the subject. The image processing tasks covered in the course enable the student to apply the basic operations to extract useful information using any image processing software. The course is assessed based on written examinations, practical exercises and a final examination. The normal entry requirement for this course is a third year in the fields of mathematics, physics, engineering, geography or any related science.

Secondly, there are 16 researchers in the IPL preparing Masters thesis or Doctorates. Between 1990-1998, 7 Masters thesis and 2 Doctorates were finalised. The themes of these works are as follows :

Masters thesis

- Image Processing system with full screen interactive menu (Nov. 89).
- Scene description and pattern recognition system (Nov. 89).
- Images filtering and restoration (Nov. 89).
- Images Segmentation to regions by edge detection process. A comparative study (Jun. 91).
- Automatic determination of control points and Image-Image rectification (MAY 92).
- Image- Map geometric rectification process (MAY 92).
- Satellite image analysis and classification process. Application to characterisation of surface states (JUN. 94).
- Automatic determination of training points and Non-parametric Poisson model using estimates of the density applied to satellite image classification (SEP. 97).
- Satellite images texture analysis (SEP. 97)
- Atmospheric radiometric correction of optical satellite images (JUN. 98).

Doctorates

- Contribution to satellite images analysis and processing system. Application to thematic cartography (APR. 98).
- Contribution to satellite images analysis and processing system. Evaluation of correspondence terrain-classified images and multisources radar SAR and optical images integration (APR.98).

In the other part, members of the laboratory have had the opportunity to participate at two long-term fellowships (one year) for in-depth training offered by the European Space Agency within the framework of the United Nations Space Applications Programme :

■ Remote Sensing Instrumentation, tenable at the European Space Research and Technology Centre (ESTEC), Noordwijk, The Netherlands, year 1997.

■ Remote Sensing Information systems, tenable at ESRIN, Frascati, Italy, year 1997 [11].

These two fellowships allowed to improve education and research in remote sensing techniques and contribute to the development of remote sensing capacity of the laboratory.

3.2. Software development and research activities

Research interests of the staff of the laboratory include :

a) Software development.

b) The use of Landsat and SPOT imagery to solve local problems.

c) Use of microwave data in Remote sensing investigations.

3.2.1. Software development at the IPL

When the IPL laboratory was established in 1980, digital image processing capabilities were not available at the university and the Laboratory had to produce software and hardware designed around MC 6800 microcomputers for digital image processing. In 1986, the laboratory had two microcomputers, equipped with a high resolution monitor and a MATROX Video card. It also had an Intel 80386 processor and an Intel 80387 math co-processor. This Centre was assigned a memory capacity on a mini-disk equivalent to 40MB. Initially, public domain software was used, but given the constraints associated with this sort of software the writing of suitable programmes was decided. The selected programming language was the C language, because of its great capabilities. The chosen name for the programme package to be developed was SOWAR (Images in Arabic words)[6]

At present, in 1998 , LATIS software has replaced the initial SOWAR software. The remote sensing group is working on the development of more programmes and also on the improvement of the software already developed.

LATIS allows the user to perform the following tasks, among others :

Displaying of digital images
Colour composites
Geometric correction with G.C.Ps
Least squares adjustment
Images rectification
Atmospheric corrections of satellite data
Radiometric enhancement
Normalised Difference Vegetation Index
Supervised Bayesian and contextual classification
Supervised Poisonian classification
Confusion Matrix generation for classified images
Kappa statistic computation
Radar imagery
Speckle filtering
Segmentation of radar SAR images
Fusion of multi-sources images

3.2.2. Projects and current research in Remote Sensing

The research in the university is organised by projects and in the IPL laboratory methodologies and applications in Image Processing and Remote Sensing are developed. Projects and Current research are in the field of desertification of semi-arid areas, urban areas, coastal pollution and earthquake effects assessment by differential interferometry, among others.

Firstly, the laboratory has defined a project concerning digital image processing, data analysis of optical and radar remotely sensed data for the characterisation, the identification and the monitoring of soils surfaces states. The duration if this project is three years (1993-1996).

Secondly, the laboratory has defined a new project concerning data analysis of optical and radar remotely sensed data for the identification and the monitoring of urban and vegetative areas of the " Gouvernorat du Grand Alger ". This project concerns ten researchers. The duration of this project is three years (1996-1999).

Thirdly, the laboratory takes part in a project which consists of resources assessment and monitoring in the south of the Saharian Atlas of ALGERIA, between Laghouat and Bechar (the distance between these two towns is about 800 Km). This area is both an arid and a semi-arid region. This project is made in three parts :

- Inventory on the region with Landsat images and field work.
- Diachronic study on the region in order to evaluate the evolution of desertification .
- Carry out a map of soil occupation of this region .

This project has been carried out with collaboration of a French laboratory of Physical Geography of the university of PARIS VII . The duration of this project is four years (1991-1995)

Fourthly, the laboratory has had an opportunity to carry out, with the European Space Agency (ESA/ESRIN), a scientific project in Radar Imagery on the same region of the previous project. This project, from September 1993 to December 1994, involved integration of optical and Radar remotely sensed data with numerical methods. In this context, researchers of the laboratory have had scientific staying to carry out this project in December 1993 and June 1994 at the ESA/ESRIN Frascati in Italy [7].

For the future, the Laboratory proposed a project related to the ESA Announcement of Opportunity for the ENVISAT data exploitation. If the project is accepted, it shall start in the year 2000 to 2004.

The IPL laboratory is, from 1991, an observer member of the European Association of Remote Sensing Laboratories (EARSeL). The Laboratory is seeking national and international collaboration to establish a remote sensing database for the Mediterranean region with a view to future projects (coastal pollution, urban areas, etc.) with the neighbouring countries. between 1990-1997 more of thirty five papers and communications have been presented in several aspects of Image Processing and Remote Sensing.

3.2.3. Seminars and workshops

Thanks to the collaboration of the INC institute and the CNTS Centre, we initiated and organised, together, two workshops in February 1991 and January 1993 focused, primarily on the topographical aspects and the study of the arid regions with the contribution of the OSS (Observatoire du Sahara et du Sahel) agency[8][9].

In other parts, the laboratory organised, in March 1998, conferences on Remote Sensing with other institutions such as the Polytechnic Military Institute and the " Gouvernorat du Grand Alger ". The goal of these conferences is to convince decision makers to promote and use Remote Sensing techniques [10].

In the USTHB University, the IPL organises conferences in Remote Sensing for students of graduate level, and this takes place twice a year since 1992. For these conferences, the lecturer can use several means such as slides, video tapes, etc., and the conference closes by a debate with the students.

In the near future (1998-1999) the IPL laboratory plans to organise with the collaboration of the Earth Sciences Institute of the USTHB University a conference monthly concerning remote sensing and G.I.S techniques and applications.

4. CONCLUSION

Although Algeria is a developing country, it is considered that in relation to remote sensing technology of the North African regions it is in a strategic position. Given that, according to some ecologist organisations, Algeria bears one of highest desertification rates, a lot of efforts are being put to use this technology.

Many institutions concerned with environment of studies now have their remote sensing office and start to train their technical staff in the remote sensing technologies and GIS.

One of the most important efforts is done in higher education. In effect, the USTHB University is using considerable efforts to promote the new technologies. From these efforts, the university is updating the different courses and producing knowledge about what is the best approach to apply these technologies. In this respect, the fact that now the image processing capabilities are available at the Image Processing Laboratory is particularly remarkable.

Given the present availability of remote sensing in the country, the application of this technology in cartography and other disciplines is obviously within sight. In the near future optical and radar remotely sensed imagery will be used to produce natural inventories that would lead to a better management of the natural resources of the country.

5. REFERENCES

[1] Y. SMARA, 1985. *Etude et réalisation d'un système de traitement d'images ARTIS 1984.Traitement et visualisation.* J.T.E.A.,Tunis, TUNISIE, Mai 1985.

[2] A. BELHADJ-AISSA, 1985. *Conception et réalisation d'un système d'acquisition et de traitement d'images Application à la localisation d'objet par caméra vidéo.* J.T.E.A.,Tunis,TUNISIE, Mai 1985.

[3] Y. SMARA, 1998. *Contribution au système d'analyse et de traitement d'images satellitaires L.A.T.I.S. Evaluation de correspondance terrain - image classifiée et intégration d'images multisources optiques et radar S.A.R.* Thèse de Doctorat d'Etat en traitement d'images et télédétection, Avril 1998, U.S.T.H.B, Alger.

[4] A. BELHADJ-AISSA, 1998. *Contribution au système d'analyse et de traitement d'images satellitaires L.A.T.I.S. Analyse spatiale-spectrale d'images. Application à la cartographie thématique.* Thèse de Doctorat d'Etat en traitement d'images et télédétection, Avril 1998, U.S.T.H.B, Alger .

[5] NATIONS UNIES, ECA / FAO / UNITAR / UNEP / RCSSMRS, 1991. *Rapport de l'atelier regional sur la télédétection et les systèmes d'information géographique à l'intention des décideurs africains.*26-30 Aout 1991, Nairobi, KENYA.

[6] M. BELHADJ-AISSA, Y. SMARA, 1989. *Mise au point d'un logiciel de traitement numérique d'images géré par un menu interactif plein écran.* 2eme Colloque Maghrébin M.N.I., Nov 1989, Rabat, MAROC.

[7] Y. SMARA, A.BELHADJ-AISSA, B. SANSAL, J. LICHTENEGGER & A. BOUZE-NOUNE, 1997. *Multisource ERS-1 and optical data for vegetal cover assessment and monitoring in a semi arid region of Algeria.* Thirth ERS symposium, 17-21 march 1997, Florence, ITALY.

[8] Y. SMARA, 1991. *Introduction à l'imagerie radar .Présentation du satellite E.R.S.1.* Journées d'études sur la cartographie et la télédetection , Institut National de Cartographie (I.N.C), Alger, 11-12 Février 1991.

[9] A.BELHADJ-AISSA, Y. SMARA, 1993. *Etude du processus de désertification sur le Piémont Sud de l'Atlas Saharien à l'aide d'images satellitaires.* Séminaire du Centre National des Techniques Spatiales (CNTS) "Télédétection Afrique du Nord", Arzew-ALGERIE Janvier 1993.

[10] Y.SMARA, 1997. *Imagerie radar et intégration de données satellitaires multisources.* Conference on " parole et image ", Ecole Militaire Polytechnique (EMP), Alger, 28 Décembre 1997.

[11] M. BELHADJ-AISSA, 1997. UN/ESA fellowship report. ESA/ ESRIN, Frascati, ITALY, Déc. 1997.

[12] J.A. RICHARDS, 1993. Remote Sensing Digital Image Analysis, an introduction. *Springer Verlag Editions, second édition, 1993.*

5 Forest fires

Remote Sensing in the 21st Century: Economic and Environmental Applications, Casanova (ed.)
© 2000 Balkema, Rotterdam, ISBN 90 5809 096 5

Forest and fire monitoring with NOAA AVHRR imagery in the Central African Republic

I. El Ghazzawi
ICL Central Europe, London, UK

ABSTRACT: Each year, millions of tons of forests and savannah all over the world are destroyed. Animal and plant species disappear as a result of deforestation and fires caused by nature and human activity. This drastic reduction in forests has had an adverse impact on the delicate global ecosystem. Using NOAA AHVRR imagery, manipulated by ERDAS Imagine software, a study has been completed to identify, assess and differentiate between forests that remain intact and forests that are on fire, or damaged areas in the Central African Republic. Vector and raster data were overlaid to provide a simple, but effective, Geographical Information System.

1 INTRODUCTION

1.1 *Description of the study area*

The Central African Republic occupies an area of approximately 622,980 km^2 and is located in the centre of Africa (latitude 7°00'N, longitude 21°00'E fall in the centre of the country). The Central African Republic is landlocked and borders Cameroon (from the West), Chad (from the North), Democratic Republic of the Congo and Republic of the Congo (from the South); and shares borders with Sudan (in the East). Further background information is necessary to understand the characteristics if the area. Figure 1 below, shows a map of the Central African Republic.

1.2 *The threat of forest fires*

Deserts cover almost one third of the Earth's land, whereas only a fifth has a fertile covering of trees. Desertification and deforestation are both on the increase. These two facts are not coincidental, man's vicious attack on trees greatly contributes to the creation of deserts. At the current rate of destruction around the world, an area of forest about the size of the United Kingdom is disappearing every year. It has been calculated that this amounts to a forest the size of a football pitch being felled every second.

Forest fires, both natural and man-inflicted, destroy vast areas of vegetation and trees every year. Natural causes triggering fires are principally lightning and spontaneous combustion of dry vegetation. Man-inflicted fires are usually to burn down trees in order to make way for land cultivation, using the ash as a neutral fertiliser; sometimes it is to make way for a new highway or road. Whatever the reason, these fires have powerful impacts on the wildlife, the global ecosystem and the surrounding human population. Prevention and early warning are the only means of reducing the impacts and costs incurred by fire damage.

1.3 *Objectives of the study*

The main objective of the study is to offer international organisations, developers, planners and conservationists a tool to aid in combating deforestation in the area. Identifying, assessing and analysing the biomass burning using coarse resolution meteorological satellite data covering the Central African Republic has accomplished this.

2 ENVIRONMENTAL CONDITIONS

2.1 *Physical Conditions - Geography*

1) Climate – a typical tropical climate; with hot, dry winters and mild to hot, wet summers
2) Terrain - vast, flat to rolling, monotonous plateau; scattered hills in north-east and south-west of the country
3) Natural Resources - diamonds, uranium, timber, gold, oil
4) Land use – the following classification best summarises the land use (estimated in 1993):

- Arable land : 3 %
- Permanent crops : 0%
- Permanent pastures : 5%
- Forests and woodland : 75%
- Other : 17%

5) Natural hazards - hot, dry, dusty harmattan winds affect northern areas; river floods are common
6) Environmental current issues - tap water is not potable; poaching has diminished its reputation as one of the last great wildlife refuges; desertification and deforestation

2.2 *Socio-economic conditions - People*

At the last estimate (July 1998), the population of the Central African Republic was found to be 3,375,771. Of this population 6% (in 1993) are unemployed. Major industrial activities include : diamond mining, sawmills, breweries, textiles, footwear, and assembly of bicycles and motorcycles.

Ethnic classifications in the country are : Baya 34%, Banda 27%, Sara 10%, Mandjia 21%, Mboum 4%, M'Baka 4%, Europeans 6,500 (including 3,600 French). As far as religions are concerned : indigenous beliefs 24%, Protestant 25%, Roman Catholic 25%, Muslim 15%, other 11% . French is the official language, but Sangho (lingua franca and national language), Arabic and Swahili are widely used.

Subsistence agriculture, together with forestry, remains the backbone of the economy of the Central African Republic, with more than 70% of the population living in outlying areas. The agricultural sector generates half of the GDP. Timber has accounted for about 16% of export earnings and the diamond industry for nearly 54%. Major agriculture products are : cotton, coffee, tobacco, manioc (tapioca), yams, millet, corn, bananas and (as mentioned above) timber.

Important constraints to economic development include the country's landlocked position, a poor transportation system, a largely unskilled work force, and a legacy of misdirected macroeconomic policies.

3 PROJECT MATERIAL

3.1 *Equipment and data*

- SUN UNIX Workstations, running ERDAS Imagine (Version 8.2)
- IBM PCs for desktop publishing
- NOAA AVHRR image acquired on 27th January 1993
- NOAA AVHRR image acquired on 14th February 1993
- Vector data produced digitised from local maps producing the following files :
  - **ROADS** containing the road transportation infrastructure
  - **RIVERS** showing all rivers and waterbodies in the area
  - **CTRY** depicting the country and surrounding international boundaries
  - **PROTPY** a digitised map of protected areas in the Central African Republic

3.2 *ERDAS Imagine*

The Earth Resource Data Analysis System, or ERDAS, is a powerful commercially available «off-the-shelf» satellite image processing and GIS package, that allows the user a single comprehensive view of the landscape, containing all relevant geographical, geological, environmental and topographic data. ERDAS Imagine is in a modular form, i.e. each module specialising in a certain specific group (or sets of) tasks. For example, there is a module for classification, a module for spatial modelling, etc. ERDAS is a world leader in geographic imaging and GIS software for commercial, governmental and military applications. Since 1978, ERDAS has pioneered the development of powerful, state-of-the-art decision support systems in which satellite data, aerial photography and GIS can be brought together and exploited for the benefit of the application.

3.3 *NOAA AVHRR*

The Advanced Very High Resolution Radiometer (AVHRR) on board the NOAA TIROS satellite is a good monitoring tool. The satellite provides data four times a day. Each scene covers an area of about 3000 x 6000 km. Data is not expensive and is readily available from the United States Geological Survey (USGS).

4 METHOD AND ANALYSIS

4.1 *Data Preparation*

A supervised classification produced a file containing 3 classes:

1) A forest class, which represents areas of forest cover of 70% or over
2) A fragmented forest class, which represents areas of between 40 - 70% forest cover
3) The non - forest class, corresponding to all areas under 40% forest cover

It should be noted that these files all have 1 km pixels. This file is called **CAFOREST**.

A file called **FIRES** containing a summary of a fire detection algorithm run on three other images acquired around 15 February 1995. If a fire is detected on any one of the three days, a pixel is highlighted. This produces a classified image of which there are only two classes : the fires, and a background class.

Subsetted images of three Central African reserves have been produced: they are of Dzanga Ndoki National Park, Dzanga Sagla Special Reserve and Vassalho - Polo Strict Nature Reserve. The names of these files are listed :

Dzanga Park & Reserve	Vassalho - Polo
DZFOREST	**VFPROT**
DZPROT	**VFFIRES**
DZ93045C	**VF93045C**

4.2 *Image Processing*

The first image to be looked as was that of Jan 27 1993. The colours of pixels are interpreted in the following way:

1) yellow, stubble or dry vegetation - red, warm bare soil
2) green, densely vegetated areas - white, small patches of cloud
3) intense red, indicates an active - grey/black, burnt forest areas
4) blue/purple, indicates smoke plumes

Large areas of smoke and several active fires are to be seen in the eastern parts of the image. The bi-directional effects across the swath of the image increase in an Eastern direction. This is assumed to be caused by a somewhat brighter hotspot canopy effect. This depends on azimuth angle, attitude of the satellite sensor, and type of vegetation canopy.

The vector data may be overlaid onto the satellite image. It must be noted that this cannot be done, unless the images (or data) is geo-referenced and rectified to the same map projection. They are all in Plate Carree projection (simple Lat./Lon.) with a resolution of 0.01 degrees (for the satellite data only), which corresponds to a ground resolution of approximately 1 km at or near the Equator. The accuracy assessment of the rectification process may take place here, when the rivers are overlaid, they may be compared with those on the satellite image.

It was found that the rectification procedure that has been used, is quite acceptable. It has an overall root mean square (rms) error of 0.7 pixels. This is not unusual, since most rectification processes introduce a certain element of error, usually due to human errors in ground control selection, digitising of rivers, roads, etc.

The second image is viewed (dating Feb 14 1993). The linear features of fragmented forest become more apparent (in lighter shades of green), and when the **ROADS** file is overlaid, the fragmented forests always seem to coincide with a road. This is of course, because of the clearing and burning of forests to make these roads. This affects the surrounding area of the road and causes degradation of the forest near to it.

The small areas of deforestation that are surrounded by largely forested areas, appear to red on NOAA AVHRR false colour composite imagery. In this image three such regions have been found:

- in the extreme South of the Central African Republic, towards the East
- in the extreme South of the Central African Republic, towards the West
- in the extreme North of Zaire

When the country borders (**CTRY**) was overlaid on top of the **CAFOREST** file, it is noticed that the classification has shown a fragmented forest area above the Zaire - Central African Republic South Eastern borders, but below, in a very abrupt manner, there is only non-forest areas. This is interpreted in that different countries use different classification techniques, for example a densely vegetated area in Central Africa may be considered to be fragmented forest in Zaire.

As for when the **ROADS** are overlaid, the linear patterns of forest degradation always seem to be where there is a main road. This is not strange. Trees must be felled to make way for the construction of these roads, leaving and environmentally damaged area (for example, fumes from the passing vehicles will most surely have its negative effect on the surviving vegetation).

There are only a couple of areas that not seem to follow this pattern, at the co-ordinates 14°35'E, 4°09'N there is a road that does not fall directly on the fragmented forest area, directly next to it. This is most probably an error; either when digitising the roads, or an error in the rectification of the satellite data. another area that does not follow the general road - fragmented forest relationship is at 21°86'E, 2°55'N, where there is a fragmented forest, but no apparent road. This may be true, but it could have been an error in the digitising of the roads. There appears to be a couple of other linear features that are not associated with levels of forest fragmentation and degradation, but they are only few and are small in length. It is always better to be familiar with the area, in order to have a complete grasp of the area, however usually this is not the case.

The file **PROTPY**, the file containing the borders of the protected areas is overlaid. Only two of the protected areas still remain to be classified as heavily forested (i.e. with a forest cover of 70% or over); these are Dzanga Sagla Special Reserve and Dzanga

Ndoki National Park, in the extreme South West of the country.

As for the relationship between dense and fragmented forest patterns to the network of roads, rivers and country boundaries, we find that there is direct relationship with respect to roads (where there is fragmented forest, there is a road), the rivers have no relationship, and the country borders also have no relationship (except for the different classification systems used by different governments, mentioned above).

The **FIRES** image is viewed. The fires do not all look alike (in shape and size). It is obvious from this image that the size and shape of the forest fire is dependent on the location of the fire. The fire's behaviour varies according to the type of vegetation that is being burnt.

The Central African Republic is divided into three vegetation zones. Each of these vegetation zones have a certain characteristic:

(a) Sahelian zone - desert area where there are no forests (and of course no fires)
(b) Sudanian zone - savannah and short grasses only. Fires are expected to spread relatively easier in the prevailing wind direction at the time of the fire.
(c) Guinea - congolia / sudania zone - characterized by forests, where fires also easily spread.
(d) Guinea - congolian zone - characterized by dense, wet tropical forests. Fires do not burn easily here because of the wet and rainy climate and the dense, lush, healthy vegetation on the surface of the ground.

When the **PROTPY** file was overlayed onto the **FIRES** image, the following areas seem to have forest fires active inside their boundaries:

- in the South of the Zemango Faunal Reserve
- in all parts of the Koukourou - Bangui Faunal Reserve
- in all parts of the Barmingui - Bangoran
- in the South of the Gribingui - Barmingui Faunal Reserve

After the pixels containing fires in the **FIRES** were enhanced (given a higher value), they were overlaid onto the forests file (**CAFOREST**). There were no fire pixels overlapping with forest pixels. This is not unexpected, the cause of this phenomena is, as mentioned above, the fact that fires cannot spread in dense, wet vegetation, it is naturally put out (if it even starts). The areas where fires spread easier, are the areas containing relatively sparser, drier and unhealthy vegetation.

Using the files that are subsetted from the larger image, the Dzanga National Park and Reserve, the percentage of forest cover remaining within their boundaries may be calculated. This is found to be 97%, with evidence of encroachment from the North.

As for the Vassalho - Polo protected region (operated within a buffer zone of less protected area surrounding the most valuable or fragile part of the protected area), the **FIRES** image overlayed onto this image, will indicate the amount of fires surrounding the Strict Nature Reserve. The largest amounts of fires occur in Barmingui - Bangoran, with Gribingui - Barmingui Faunal Reserve next, followed by the least amount of fires in Koukourou - Bangui Faunal Reserve, an no fires in the Vassalho - Polo Strict Nature Reserve.

The Vassalfo - Polo protected region may be digitally processed to find how close the fires are to it. The following table illustrates the buffer depth and the fires contained within this buffer (for example within a distance of 7 pixels away only one fire was found, although there were 6 found in the 6 pixel buffer zone) :

Buffer Zone	Fires	Buffer Zone	Fires	Buffer Zone	Fires
1	1	10	5	19	1
2	1	11	1	20	1
3	2	12	3	21	5
4	1	13	2	22	1
5	2	14	2	23	0
6	3	15	1	24	1
7	1	16	4	25	5
8	5	17	7		
9	5	18	1		

5 CONCLUSIONS AND RECOMMENDATIONS

As a conclusion to the report, it may be safe to say that the NOAA AVHRR sensor, although being of such coarse resolution, has proven to be quite suitable for the identification of intact forests, degraded forests and active/in-active fires in the natural vegetation.

The fire patterns depend on the type of vegetation that is over the area and does not depend on national borders (although forest classification systems and methods applied by governments may vary across borders).

The recommendations for this region must come in the form of an action plan, that will include :

- Policies, Laws and Regulations are to be introduced to prevent the burning of forests for the construction of roads and grazing land

- Regular ground surveys for the monitoring of the forests in the Central African Republic
- Institutional Strengthening and Development of Human Resources whereby the local authorities and management that are to enforce the policies, regulations and laws. They must be trained to carry out their tasks, this should be viewed as a prerequisite for stable agricultural and other relevant economic productivity in the Central African Republic
- Repetitive high resolution satellite coverage of the forests to detect, monitor and measure changes
- Environmental Activities, the trend should be to focus on the decrease of the environmental degradation, the improvement of overall environmental conditions, and the development of the sound scientific basis needed to enhance the understanding of the complex and critical problems in the region, and to derive resolutions to them
- The creation of a series of environmental sensitivity maps leading to an integrated environmental geographic database for the Republic
- Public Awareness and the development of environmental awareness programs is essential for the development and directing of public will and energy to assist and accept the policies, laws, regulations and measures that are to be implemented by the local authorities
- The use of Global Positioning Systems for the ground surveys, ground truth data and classification verification

REFERENCES

CIA. 1997. *World Fact Book* Central Intelligence Agency of the United States of America.

Curran, P.J. 1988. *Principles of Remote Sensing* London: Longman.

ERDAS. 1994. *ERDAS Imagine 8.1 Field Guide* Atlanta: Erdas Incorporated.

ESA ESRIN. 1998. *Earth Watching Anthology* Italy: ESA ESRIN

ESRI. 1993. *Understanding GIS: The ARC/INFO Method* England: Longman Scientific & Technical.

Lillesand, T.M. & R.W. Keifer. 1987. *Remote Sensing and Image Interpretation* London: John Wiley and Sons.

Remote Sensing in the 21st Century: Economic and Environmental Applications, Casanova (ed.)
© 2000 Balkema, Rotterdam, ISBN 90 5809 096 5

Building-up a Burned Area Monitoring System based on Earth Observation data

J.U. Paúl, M.R. Caetano & T. Santos
Centro Nacional de Informação Geográfica, Oeiras, Portugal

ABSTRACT: In Portugal, fires are a major problem, consuming, in average, 100 000 ha/year. A large percentage of these fires - 43% (DGF estimations) - are due to arson, which most of the times is related to land use change purposes. As an attempting to reduce the amount of intentional fires, the Portuguese legislation includes a set of laws that regulate the landuse change in areas devastated by forest fires. However, the control of these areas on the field is very expensive and time consuming, resulting in an ineffective procedure. In this study we propose a more efficient and cost effective procedure to control the compliance of the legislation on burnt forest areas. This procedure is based on a system, BAMS (Burned Areas Monitoring System), built on a GIS that uses multi-temporal dataset of satellite images for detecting burned areas, and for characterizing land use before and after fire occurrence.

1 INTRODUCTION

Vegetation fires are a global phenomenon that has several implications in the way humans interact with the surrounding environment. In fire prone regions of the planet, fire must always be considered as an imbedded element of terrestrial ecosystems, controlling its functions and dynamics. Ones decisions for acting in these ecosystems at local and regional levels must not overlook this natural factor since it imposes particular constraints. The most obvious effect of fire is the drastic change in land-cover that it produces in a short period of time and the extent of landscape that can be affected by it. However, this landscape change can trigger other subsequent land-cover modifications, and frequently the post-fire vegetation cover is different from the pre-fire one. Understanding how fire regimes contribute to land-cover change, is an essential information for defining adequate land management politics that can guarantee a sustainable and wealthy development of often less-favoured areas. For achieving this level of knowledge, monitoring programs of fire activity and land-cover change and are a basic steppingstone. In this paper we present the conceptualisation of a burned areas monitoring system based on EO data to be tested in Portugal.

In Portugal an average of 100,000ha burns every year, representing 3% of the country forest area (DGF data) and about 1% of the continental surface. The magnitude of burned areas extents and its relevance by comparison with other European countries reveals the importance that must be assign to this problem. In Portugal vegetation fires are the most important cause of land-cover change for what landscape planning decisions have to address this issue.

The main reasons for these impressive numbers in Portugal are related with two causes: high risk of fire spread and high risk of fire ignition. The high forest fire spread risk is due to the typical Mediterranean climate characteristics, the demographic trends in rural areas during the last 30 years (Almeida and Moura, 1992; Rego, 1992), and also to inadequate forest management practices. The large number of forest fires (ignitions) is due essentially to negligent human behaviour (20%) and to arson (43%). The latest, a criminal action severely punished by law, is often considered to be partly related to land-cover change motives. As an attempt to reduce the amount of fires and their serious effect on the landscape, the Portuguese legislation includes a set of laws that regulate the land-cover change in areas devastated by forest fires. The control of legislation compliance is however very expensive and time consuming, resulting in an ineffective procedure since the field checking actions have to be carried on for a large number of fire perimeters. Therefore, an automated system to assist in this task is needed.

CNIG is working on the development of a prototype that makes use of EO data and geographical information systems to address this problem. The system on development, BAMS (Burned Area Monitoring System), will provide the means to easily determine the burned areas locations, the land-cover before and after fire occurrence, and access its legal compliance. On a broader context, regional analysis of trends can be performed contributing to estimate the fire role as a driving force of land-cover change. One of the innovative aspects of such a system, is the use of a yearly based multi-temporal dataset of EO data that allows the building up of an updated spatial database of land-cover changes within burned areas.

This task is being developed under the framework of the APERTURE project (ENV4-CT97-0437), which is supported by the Commission of the European Union under its programme on Environment and Climate. This project aims at promoting the use of EO data in judicial and extra-judicial proceedings concerning the implementation of environmental law. For that, four case studies are being analysed in different European countries. In each case study, the use of EO data as a mean of evidence is being investigated, not only in technical terms, but also in respect to the legal and administrative constraints of each member state.

2 STUDY AREA

The system on development is being tested for an area of 2240km^2 (fourteen 1:25000 topographic map sheets) (Figure 1). This area, located in central Portugal, has been severely affected by vegetation fires in the last decade, with an annual average of 10000ha/year. The dominant vegetation are maritime pine (*Pinus pinaster* Ait.) and Tasmanian bluegum (*Eucalyptus globulus* Lab.) along with large patches of Mediterranean shrublands. Land-cover has significantly changed from pine stands towards eucalyptus ones, a more profitable forest species, which has been subject to legal constraints regarding new plantations.

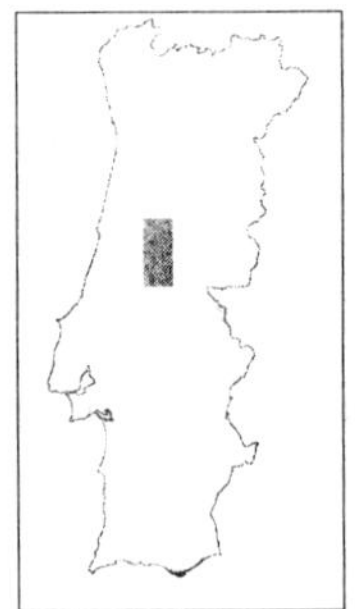

Figure 1 – Study area.

3 BAMS APROACH

BAMS is conceived as a decision support tool. Its prime objective is to integrate in a single system all the information needed to document, analyse and understand the land-cover changes occurring within burned areas, over time. BAMS uses information obtained from EO data as its main input. This information is then analysed within the system to generate products that can fulfil the end-user needs. Two distinct levels of data analysis are therefore considered: image-processing techniques to produce information for feeding the system and GIS operations for automatic analysis of spatial data for generating decision support information.

The system is conceived to provide three basic capabilities of data analysis: mapping, statistics and alarm. The mapping capabilities refers to the possibility of automatically generating maps of burned areas and pre and post-fire land-cover, on users request on spatial location, area attributes, time period or other features included in the databases. The statistics component describes the burned areas and land-cover in numerical terms based on similar user requests. The alarm component refers to the system capability of using information contained in the database to detect the areas were a predefined set of legal restrictions may not have been followed by land managers. The alarm capability is the central piece of the whole system, for what we will describe it in greater detail.

As previously referred, the Portuguese legislation includes a set of laws that regulates the land-cover change in burned areas. In Table 1 we summarise these laws and respective regulated actions (exclusively the actions that are under DGF responsibility and also adequate for spatial analysis are mentioned)

Table 1 – Portuguese Decree-Laws that regulate the actions that occur in burnt forest areas.

DECREE-LAW	ACTIONS
DL139/88	Reforestation of burnt forest-areas is mandatory within 2 years after the fire occurrence.
DL327/90	Forbiddance of land-use changes within 10 years after the fire occurrence, unless ministerial authorization is granted.

In BAMS this set of legal requirements on land-cover in burned areas is used to spatially detect the areas where there is a law infringement. In Figure 2 we show some examples of these law infringements. The detection of areas where the law is not being followed is based on simple operations of spatial

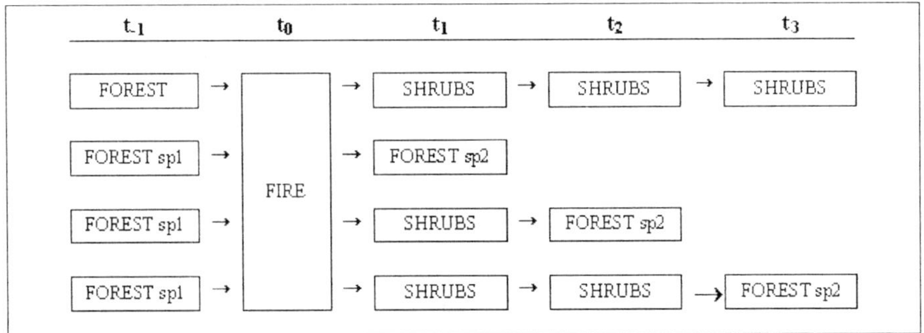

Figure 2. Illegal land-cover transitions in burned areas.

analysis that are automatically produced by the system.

4 DATA DESCRIPTION

In the elaboration of BAMS prototype a period of 9 years (1990-1998) is been considered for methodological development. To develop this system three sets of spatial data are essential: burned area maps, pre-fire and post land-covers.

Burned areas mapping has been performed in Portugal as a regular procedure. Since 1989, the Portuguese Forest Service (*Direcção-Geral das Florestas*) finances the production of this information based on Landsat5-TM data. This information is available for the entire country and contains all the burned areas larger than 5ha since 1995, while for previous years the minimum cartographic unit was set to 15 ha. The mapping is done for the whole fire season, and so individual fires can not be distinguished when they overlap. This information source is provided in paper format (1:100,000) and in digital files. The scales at which these data are produced impose some limitation for use in high detail cartography of fire perimeters. Nevertheless, we believe that this dataset is an excellent starting point for the development of BAMS, and that the availability of the upcoming satellites (Landsat7 and Ikonos, for instance) will certainly allow the production of cartography with increased spatial accuracy for future use.

Another key dataset for the BAMS are the pre-fire land-cover maps. In Portugal high detail land-cover cartography is produced on an approximately 10 years basis by visual interpretation of aerial photographs. This information is available at 1:25000 scale with a minimum cartographic unit of 1ha. For BAMS purposes it is important that this land-cover information is updated to the time of fire occurrence. For that, we are using change detection techniques based on Landsat5-TM imagery. A yearly based multi-temporal dataset of Landsat imagery has been assembled for this purpose. We expect to detect the most important land-cover changes occurred, like clear-cuts or new plantations, within the areas that burned.

For the post-fire land-cover, the same multi-temporal dataset of Landsat imagery is used. In this case change detection is applied in a yearly base to allow the monitoring of changes occurring within burned areas through time.

To assist in image processing procedures, and for analysis of outputs, some ancillary data is also used in BAMS implementation. In Table 2 we list the most important ones.

5 GENERATION OF PRE AND POST-FIRE LAND COVER MAPS USING SATELLITE DATA

The updating of pre-fire land-cover maps for burned areas is done by application of change detection techniques based on Landsat5-TM data. The reference data is the land-cover map of 1990, produced by visual interpretation of aerial-photographs. Change detection analysis is also applied to identify the post-fire land-cover modifications. Several techniques for change detection are known. In this project the ones that are being tested are image differencing (Singh, 1989; Lillesand *et al.*, 1994); spectral mixture analysis (Smith *et al.*, 1990); change vector analysis (Lambin *et al*, 1994); and principal compo-

Table 2. Ancillary data

THEME	FORMAT	DATE
CORINE Land-Cover	Digital	1988
Ortho-photos	Digital	1995
Aerial photographs	Analogic	1990
Digital Terrain Model	Dig./Anal.	-
Water streams	Dig./Anal.	-
Road network	Dig./Anal.	-

nent analysis (Byrne *et al.*, 1980; Richards, 1984).

Our goal is to identify which technique is more adequate for generating the required maps. For pre-fire land-cover maps, we expect to detect the most important land-cover changes occurred, like clear-cuts or new plantations, within the areas that have burned. Similarly, for the post-fire situation we expect to detect the areas where aforestation has occurred or not, and ideally the species used. Aforestation can be assessed based on the detection of soil mobilization activities, which usually produces a change in spectral signals. The magnitude of change in radiometric signals is the basis for all change detection analysis, since remote sensing can assess only land-cover changes that produce a significant spectral modification. Spectral change detectability is sensor-dependent, for what the Landsat-TM data will only provide us the means for studying the most critical changes. However, this limitation might be overcome by the upcoming of new satellite/sensors having an increased spectral and spatial resolution.

Other important aspects in change detection analysis are the need for a precise co-registration of satellite images and the use of radiometric normalization techniques between images acquired in different dates. For the latter case, we are using a temporal image normalization technique based on image data obtained over areas of constant spectral signatures, like water bodies or sand surfaces.

6 CONCLUSIONS

Monitoring of land-cover changes occurring in burned areas is an essential task for understanding how fire activity contributes to landscape modification. Earth Observation data are a very important data source for this task, especially if the information resulting from image processing can be integrated with other spatial data in a GIS environment. This information assembled in a user-oriented system, like the one purposed in this paper can then contribute to various decisions that have to be taken for managing natural resources. BAMS is a pioneer integration of GIS technologies and remote-sensing data for an effective implementation of legislation on the management of burned areas at national level. The success of the implementation of such a system may lead to similar systems for other types of environmental legislation.

7 REFERENCES

Almeida, A. M. S., and P. V. S. Moura. 1992. The relationship of forest fires to agro-forestry and socio-economic parameters in Portugal. *International Journal of Wildland Fire* 2: 37-40.

Byrne, G. F., Crapper, P. F., and Mayo, K. K., 1980. Monitoring land-cover by principal components analysis of multitemporal Landsat data. *Remote Sensing of Environment*, **10**, 175-184.

Lambin, E. F., and Strahler, A. H., 1994. Change-vector analysis in multitemporal space: a tool to detect and categorize land-cover change processes using high temporal-resolution satellite data. *Remote Sensing of Environment*, **48**, 231-244.

Lillesand, T. M., Kieffer, R. W., 1994. *Remote Sensing and Image Interpretation*. 3rd editionWiley&Sons

Lourenço, L., and B. Gonçalves. Cartography of forest fires recorded in Central Portugal from 1983 to 1989. *Proceedings of the First International Conference on Forest Fire Research*, A.08.1-A.08.16. Coimbra, Portugal.

Richards, J.A., 1984, Thematic mapping from multitemporal image data using the principal components transformation. *Remote Sensing of Environment*, **16**, 35-46.

Rego, F. 1992. Land use changes and wildfires. *Responses of Forest Ecosystems to Environmental changes*. Editors A. Teller, P. Mathy, and J. Jeffers, 367-373. London: Elsevier Applied Science.

Singh, A., 1989, Digital change detection techniques using remotely sensed data, *International Journal of Remote Sensing*, **10**, 989-1003.

Smith, M. O., Ustin, S. L., Adams, J. B. and Gillespie, A. R., 1990. Vegetation in deserts: I. A regional measure of abundance from multispectral images, *Remote Sensing of Environment* 31, 1-26.

Remote Sensing in the 21st Century: Economic and Environmental Applications, Casanova (ed.)
© 2000 Balkema, Rotterdam, ISBN 90 5809 096 5

Remote sensing estimation of fuel moisture content

E.Chuvieco, P.J.Vaughan, D.Riaño & D.Cocero
Department of Geography, University of Alcalá, Spain

ABSTRACT: Fuel moisture content (FMC) for Mediterranean species was measured and correlated with laboratory radiometry, NOAA-AVHRR and Landsat-TM images. The best variable related to FMC was the quotient between the Normalized Difference Vegetation Index and Surface Temperature (NDVI/ST). Two other indices derived from Landsat's SWIR bands also showed good correlation. Two multiple regression equations to derive a synthetic FMC for grassland and shrubs are presented, with r^2 values of 0.61.

1 INTRODUCTION

The estimation of plant water content is critical for various applications such as predicting agricultural yields, appraising regional evapotranspiration and determining fire risk conditions. In order to apply remote sensing techniques to carry out this estimation it is first necessary to study in detail the effects of water on leaf and plant reflectance.

In the context of fire danger rating, Fuel moisture content (FMC) is estimated from meteorological danger indices, which try to account for the effect of weather variations on water availability for plants. The relation of weather variables (air temperature, relative humidity, wind, etc.) with FMC has been proved with some dead fuels since they are very dependent on the atmospheric conditions. However, there is little experience regarding live fuels, which are more related to water availability in the soil and to their own physiological mechanisms of minimising water loss. Consequently, the estimation of FMC from meteorological indices may be rather imprecise.

The other conventional method to estimate FMC would be the direct measurement of FMC in the field, which is very costly and time consuming. Therefore, satellite imagery may be a reasonable alternative to derive FMC information. Satellite data cover the whole territory and are directly related to the vegetation status. However, the application of satellite data to FMC can only be used once the relationship between this variable and the spectral characteristics of vegetation covers has been derived. For instance, it ought to prove that FMC changes the way plants reflect or emit electromagnetic energy, with enough clarity to make it distinguishable from other sources of spectral variation (leaf area index, soil background, atmospheric disturbances, etc.). The issue is therefore, whether or not current satellite systems provide the necessary spatial, spectral and temporal resolution for the operational use of these data in real fire management.

A series of laboratory experiments were carried out to better understand the spectral relationship of one particular species to FMC changes in controlled situations. Together with these experiments, data from high and low-resolution sensors have been used to compare and link both sources of information. A brief description of the experiments carried out follows.

2 FUEL MOISTURE CONTENT ESTIMATION

2.1 *Analysis of previous experiences*

Remote Sensing of plant water content has been attempted using mainly optical and thermal wavelengths, although microwave data are also promising in this field (Gogineni et al., 1991). Most of the experiences were performed in laboratory analyses, and commonly at leaf level. These studies have shown that leaf water content is most clearly related to short-wave infrared reflectance (SWIR), ranging from 1.4 to 2.0 μm (Bowman, 1989; Cohen, 1991; Hunt et al., 1987; Jackson and Ezra, 1985; Tucker, 1980). It is well known that these wavelengths present a high absorption of water, and therefore SWIR reflectance is negatively related to leaf water content.

Regarding near-infrared (NIR) reflectance (0.8 - 1.1 μm), conclusions obtained from different studies do not completely agree. Some authors find a small increment as the leaf dries (Bowman, 1989; Hunt and Rock, 1989; Hunt et al., 1987; Thomas et al., 1971), which may be caused by the increase in the refractive index of the mesophyll layer when water is replaced by air. Other authors do not find a significant change in NIR reflectance for leaf drying (Carter, 1991), while some others measure a decrease in reflectance (Westman and Price, 1988). This reduction may be caused by the indirect effects of leaf dryness, such as the decrease in leaf area index or shadowing due to leaf curling (Jackson and Ezra, 1985; Westman and Price, 1988).

Finally, the visible domain does not appear to have enough sensitivity to estimate leaf water content, at least in absolute terms. On the one hand, the chlorophyll reduction after drying increases reflectance, while on the other hand, the decline in water decreases it. Some authors, though, found that the red band is sensitive to water content (Jackson and Ezra, 1985; Ripple, 1986), while others did not observe significant changes (Bowman, 1989; Thomas et al., 1971).

When these laboratory studies are applied to remote sensing image analysis, models to extrapolate from leaf to canopy's reflectance are required. Some authors emphasise the difficulties in estimating plant water content, since they consider that the contribution of leaf water to the whole canopy reflectance is too weak in comparison with other factors: leaf geometry, shadows, soils, etc. (Carter, 1991; Cohen, 1991). However, other authors have found high a correlation between leaf water content and canopy reflectance (Cibula et al., 1992; Hunt and Rock, 1989; Hunt et al., 1987; Westman and Price, 1988). Their conclusions were based on hand-held spectroradiometry and not on actual satellite images.

Experiences with satellite data are somewhat scarce. The most consistent results were found when estimating temporal evolution of FMC for grassland (Chladil and Nunez, 1995; Paltridge and Barber, 1988; Paltridge and Mitchell, 1990), whereas FMC for shrubland showed weaker relationships with satellite data (Alonso et al., 1996; Deshayes et al., 1998).

An alternative to determining canopy water content is the analysis of plant thermal dynamics. When the plant is well watered the increase in air temperature causes evapotranspiration to rise also, which will increase latent heat loss and reduce sensitive heat loss, and consequently, leaf temperature will decrease. On the contrary, when the plant dries, transpiration is reduced and consequently latent heat loss is reduced and sensitive heat loss increases. As a result of this relation, the difference between air and surface temperature is related to evapotranspiration, and indirectly to water content and water stress. Several indices have been proposed based on this difference and additional weather variables (net radiation, air humidity, wind data, etc.). The most currently used are the Stress Degree Day (SDD) (Jackson, 1986), the Crop Water Stress index (CWSI) (Jackson et al., 1981) and the Water Deficit Index (WDI) (Moran et al., 1994). The WDI has been successfully tested as a predictor of fire danger with both NOAA-AVHRR and Landsat-TM data (Vidal and Devaux-Ros, 1995; Vidal et al., 1994).

2.2 *Laboratory analysis*

2.2.1 *Methods*

Since satellite data are acquired in natural conditions, it is complicated to isolate the main factors of spectral variation for a single plot. In our case, we intended to derive the water content information on the spectral signal, which we assumed would be fairly weak, compared with other factors (green cover, soil background, plant geometry, atmospheric conditions, etc.), as reviewed in the previous section. Therefore, several experiments were planned in very controlled situations, in which the satellite signal could be simulated.

The experiments were based on the reflectance measurements taken with a GER 2100 spectroradiometer, that shows continuous spectra of 140 bands in the range of 0.4 to 2.5μm. The spectral resolution of this apparatus is 0.01μm in the visible (VIS) and near infrared (NIR) (up to 1μm) and of 0.024μm in the range from 1 to 2.5 μm. Although the radiometer is fairly accurate, sporadic noise was detected in some bands of the optical mid-infrared, which is why, following a common practice in spectrometry, some post-processing was applied to the raw measurements.

The species collected for this case was rock-rose (*Cistus ladanifer*), a highly representative Mediterranean shrub. The leaves were placed on a tray, which had a layer of black charcoal tracing paper, which gave a very low reflectance. On it, the leaves were arranged in four layers, with the adaxial (upper leaf-side) facing upwards as we would find it in the

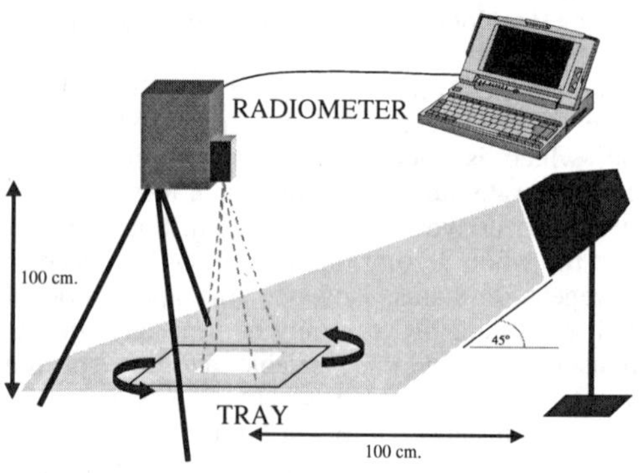

Figure 1. Experimental design for spectral measurements

field. The radiometer's field of vision (IFOV) was always within the tray area. Before each measurement, a rigorously calibrated reference target (Spectralon Reflectance Target, Sample I.D.: SRT-99-050) was used to calibrate the equipment. Samples were oven dried at regular intervals between measurements. The moisture content analysed each time was determined using the following expression:

$$FMC(\%) = \left(\frac{W_f - W_d}{W_d} \right) * 100 \quad (1)$$

where W_f is the fresh weight of the leaves (the one measured each time it was taken out of the oven) and W_d is the dry weight (the one obtained at the end of the process). The result is expressed as a percentage of the dry weight.

2.2.2 *Results*

The results are shown in figure 2. It can be observed that in the VIS there are hardly any reflectance variations as the leaves dry. However, the reflectance in the NIR is reduced, apparently due to a double effect. First, by the drying of the leaves themselves. Second, as a result of the curling and shrinking of the leaves, the leaf area decreases allowing gaps of the tray itself to show up, thus altering the reflectance signal. This causes reflectance values to decrease more than expected.

Throughout the experiment the initial leaf arrangement remained unchanged, to avoid altering the geometry of the sample. Consequently, it was necessary to get a better picture of what the reflectance curve should have looked like, had there been no interference from the tray itself. The scattered curled leaves from the last measurement taken (by then the leaves were completely dry), were bunched up to an area slightly greater than the radiometer's IFOV, with the object of leaving no gaps between the leaves to avoid the tray surface to show.
In this case the reflectance curve resembled to a far greater degree what was expected, especially in the NIR and SWIR. In the NIR, the curve did not drop as drastically when moisture contents decreased. In fact, it was well above the last five measurements, which goes to show the importance of the tray surface reflectance in the last set of measurements. As for the SWIR, the curve of the bunched up rock-rose has a similar tendency. However, previous reflectance curves, regardless the percentage of FMC, seemed to converge into just one curve. The bunched up rock-rose curve is especially relevant here not only for being different from the rest but also due to the promising magnitude of such reflectance differences.

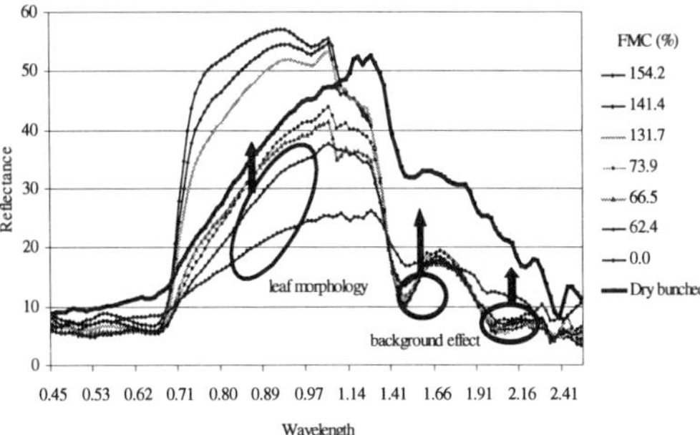

Figure 2. Reflectance curves for different FMC in rock-rose.

Figure 2 shows how the loss of reflectance in NIR and SWIR is influenced by the decrease in the leaf cover. As the leaves dry the leaf area decreases (scattered rock-rose) therefore the signal is a mixture of the reflectance of the rock-rose and the tray surface itself.

The experiment has therefore clarified several aspects of laboratory spectroradiometer procedures, especially regarding the importance of the leaf arrangement, leaf area and its underlying surface in relation to quantifying the resulting reflectance curves. The results narrow down the bands to look out for in future experiments, to quite an accurate range of bands. It has also given a good idea of the weight of each of these variables and serves as a useful and reliable reference for future work. In the SWIR, the reflectance increases less than expected, again due to the tray effect although in this case the result is the opposite.

3 SATELLITE DATA INTERPRETATION

3.1. Low-resolution AVHRR data

The Advanced Very High Resolution Radiometer (AVHRR) on board the NOAA satellite series has been the most commonly used sensor in fire danger estimation, since it provides an adequate temporal resolution (12 hours) for operational danger management. However, it presents several difficulties for FMC estimation, especially its low spatial resolution (1.21 km² at nadir), poor radiometric quality, and lack of a channel in the SWIR, which is the most sensitive band for water content assessment. In this work the afternoon pass of NOAA-14 satellite has been used to estimate FMC, since it is close to the period of maximum temperature and therefore shows most dangerous conditions from a fire management point of view.

Within the European Union Megafires project, two years of satellite data were compared with field measurements of FMC in three Mediterranean countries (Spain, France and Greece). The following variables were derived from AVHRR images: Normalized Difference Vegetation Index (NDVI), Soil Adjusted Vegetation Index (SAVI); Global Environmental Monitoring Index (GEMI); Surface Temperature (ST); Difference between Surface Temperature and Air Temperature (ST-AT); and the quotient between the NDVI and ST (NDVI/ST). Relative Greenness (RGRE) and Absolute Greenness (AGRE) (Burgan and Hartford, 1993) were also computed for comparison among the different plots. Validation of the satellite-derived indices was based on the measurements of foliage moisture content of different species carried out during 2 years on four networks of test land plots situated in three Mediterranean countries. From a methodological point of view, NOAA-AVHRR-derived indices proved to be useful for monitoring FMC as significant correlations have been obtained on the four test land plots (Chuvieco et al., 1999; Deshayes et al., 1998). Depending on the test land plot, different indices obtained the best results: for example, ST and ST-linked indices (ST – AT, NDVI/ST) for Cabañeros (Spain), but ARND and multitemporal-NDVI ones for the Les Maures area (France). It should be clarified that the former used spring and summer data, while the latter only summer measurements and consequently, a much narrower range of FMC values was found. The object of using daily or maximum value composites (MVC) was more precisely studied at Les Maures. The analysis with daily images has, however, suffered from the small number of cloudless days, due to the unusually rainy '96 and '97 summers. The analysis with MVC data, that, to a certain extent, provides clear data during cloudy periods, has shown good potential, with significant correlations, especially when considering several land plots together. Some aspects, brought to light by the project, such as the problem of "change of scale" between the size of field plots (100x100 m) and the size of NOAA pixel (1x1 km) or even the NOAA 3x3-pixel window (3x3 km), should now be investigated.

From an operational point of view, the Megafires project showed that NOAA-AVHRR images contain information that could be used to monitor and provide some mapping of FMC. This would give information on a parameter that is considered important by fire fighting authorities, but that up to now could not be monitored conveniently. This parameter when assessed from space, together with other parameters such as meteorological ones, could be used for an improved estimation of short-term fire risk.

One step further in the Megafires project was to derive a synthetic estimation of FMC from AVHRR data, in order to propose a single index that could be used operationally for FMC estimation. For doing this, several correlation analyses were performed: both considering spring and summer data or just the latter. These tests were focused on the study area of Cabañeros National Park, in Central Spain (200 km south of Madrid). Field sampling was carried out in this area every 8 days from early April to the end of September, from 1996 up to the present. The study area includes very good examples of Mediterranean type vegetation. Plots were located in the central valley of the Park, which has very gentle slopes. Species collected were grassland (3 plots of several types of herbaceous plants), shrubs (2 plots of *Cistus ladanifer, Erica australis, Erica umbelata and Rosmarinus officinalis*), and trees (*Quercus faginea* and *Quercus pyrenaica*), all of which are very common in Mediterranean areas. Three samples of FMC were taken for each species and plot.

Average FMC values of grassland for the three plots sampled on the field were computed. Similarly, average FMC shrub values were computed for the two plots sampled. *Cistus, Rosmarinus* and *Erica* samples were averaged as a whole, in order to derive a single value for the FMC estimation of shrub. The tendencies observed may be summarised as follows:

- Correlation of satellite data and FMC is consistent for grassland when spring and summer trends are included. The most significant variables found were NDVI/ST, RGRE and AGRE, with r^2 values higher than 0.84 (although the first variable accounts for 90 % of the variance).
- Correlation of satellite data and FMC for shrubs is lower, but still significant, with r^2 values of 0.66. The most significant variables in this case are NDVI/ST, NDVI and ARND.
- A set of three satellite variables provides adequate multiple regression values with both grassland and shrubland these being, NDVI/ST, RGRE and ST. For the spring and summer period, they account for 0.83 of the total variance in grassland and 0.51 in shrubland.
- If we only consider the summer time (mid June to the end of September), much lower correlations are found, since the range of FMC values is also much lower. The most significant associations were found for RGRE in the case of grassland, and NDVI/ST for shrubland. The r^2 values decrease down to 0.4 for grassland and 0.23 for shrubland.

Consequently with the previous results, two lines of application for generating a synthetic fire danger index that could be used operationally are proposed. The first one is based on a direct estimation of FMC from satellite data, while the second on an indirect method, which would derive FMC from correlations of satellite data and meteorological danger indices. Both methods are based on the analysis of NOAA-14 afternoon passes (14.30 h. approximately).

3.1.1. Using images from spring season.

Although most fire managers are focused on summer conditions, which are the riskiest, to include spring data in the estimation of FMC provides a wider range of FMC conditions, and makes it possible to better analyse relative changes in FMC among the species. After the analysis of field sampled FMC and satellite-derived indices for the spring and summer season of 1996 and 1997 in the Cabañeros study area, several simple and multiple regression analysis were performed. As said before, different agreements were found for grassland and shrub species. However, in order to propose a synthetic index that could be used operationally for the forest managers, a final regression analysis was applied after mixing together grassland and shrub samples. Consequently, a synthetic fuel moisture content (SFMC) values may be estimated as follows:

SFMC=-106.95 + 14.873*NDVI/TS + 7.181*AGRE - 1.505*RGRE (2)

A determination coefficient (r^2) of 0.619 was found for the samples collected on the field. A scattered plot of the observed versus predicted value of FMC is shown in figure 3.

The satellite variables would be computed as follows:

NDVI / ST = ((NDVI + 1)*125) / ST (in ° C) (3)

AGRE = (100*(NDVI-0.05))/0.66 (4)

RGRE = (100*(NDVI-NDVImin))/(NDVImax-NDVImin) (5)

where NDVImax and NDVImin, are the maximum and minimum NDVI values of that pixel.

AGRE and RGRE are closely related to NDVI, but they are more easily compared in time and space. A previous suggestion to use these indices in operational fire risk estimation was included in the Fire Potential Index (PFI) by Burgan et al. (1998).

SMFC values are estimations of FMC (considering mixed the values of grass and shrub species), and therefore are measured in percentage of dry weight. Coefficients should be tuned up to different vegetation types. The ones proposed are averaged over grassland and shrubland over two years of field measurements.

SFMC values might be converted into danger classes after comparing them with the maximum and minimum values found for a historical series of data. Once the limits of SFMC variation for each pixel were found, a relative danger index (RDI) may be computed as:

$$RDI = \frac{SFMC_i - SFMC_{min}}{SFMC_{max} - SFMC_{min}} * 100 \quad (6)$$

Then, a quantification of danger may be proposed as follows:

Extreme danger: Below 10 %

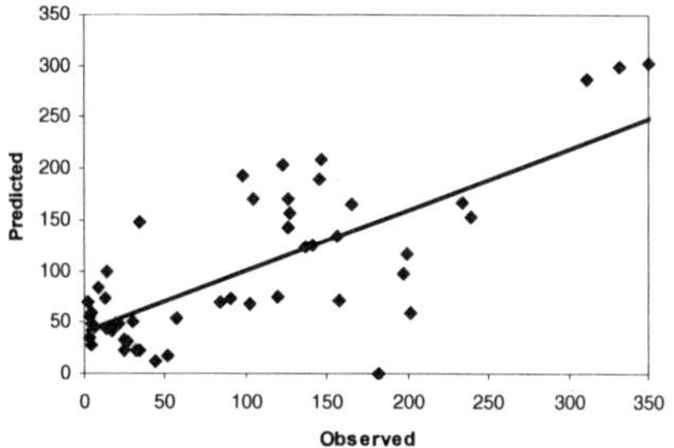

Figure 3. Predicted versus Observed FMC values of shrub speciesand grassland (Spring and Summer data)

High-danger: 10-20 %
Medium-danger: 20-50 %
Low-danger: Above 50 %.

3.1.2. Without images from the spring season.

In this case, only images from the middle of June to the end of September would be available. In this case, the model would become:

SFMC = -4.587 + 2.387*NDVI/TS - 4.062*AGRE + 4.029*RGRE (7)

In this case the r^2 value is slightly reduced (0.613).

The satellite variables would be generated similarly as in 3.1.1. The maximum and minimum SFMC values to compute the RDI would be derived from the analysis of 1998 images. Danger levels would be also computed as in 3.1.1.

3.2 *High-resolution Landsat-TM data*

In spite of its high cost and its 16 day temporal resolution, which is rather low for operational fire danger estimation (daily predictions are usually required by fire managers), Landsat-TM data analysis has two advantages over NOAA-AVHRR images. Firstly, it provides images from the SWIR band (band 5 and 7), which is more clearly related to water content than visible or NIR bands. Secondly, it reduces the uncertainty in correlating field measurements and satellite data, since Landsat's higher spatial resolution makes it much easier to locate target plots than in AVHRR images.

Three Landsat-TM images were acquired for this project. They correspond to dates in which field measurements were available: April 16, July 21 and September 23, all of them in 1997. Besides calculating the indices mentioned, in the case of NOAA-AVHRR images, two other indices were computed: the NDII5 and the NDII7. Both are similar to the NDVI but instead of using band 4 and 3 of the sensor, corresponding to the NIR and Red, they use bands 4-5 and 4-7, respectively, being the 5 and 7 in the SWIR). Also the ratio between the NDII5 and the ST and the Water Deficit Index (WDI) were calculated. Correlations were consistent with expected tendencies, showing the vegetation indices based on the SWIR bands the highest sensitivity to FMC (Chuvieco et al., 1999).

For the most significant correlation, several fitting models were computed to explore the presence of non-linear trends, but since there were few data available no definitive conclusions could be reached.

4 COMPARISON OF LABORATORY AND SATELLITE ESTIMATIONS OF FMC

In an effort to combine results from the three data sources analysed in the previous sections, the following table may be useful for a preliminary assessment of the differences in FMC estimation depending on which source is used. Table 1. shows correlation coefficients between FMC obtained in the 1997 field campaign, and a selection of vegetation indices derived from laboratory and satellite data.

Taking only Landsat TM and laboratory data it is clear that the species chosen for the laboratory experiment (*Cistus ladanifer*), appears to have similar correlation values to those corresponding to average shrubland, in both NDII5, NDII7 and NDVI. This discards the possibility of not having chosen a sufficiently representative species of the study area, in spite of the differences between Cistus and shrub AVHRR correlation with NDVI.

In the case of NDVI/ST, the higher spatial resolution the Landsat TM provides, compared to the NOAA-AVHRR, does not seem to be an advantage for correlation with the Cistus l. AVHRR data correlation with NDVI/ST improves considerably with respect to NDVI especially in the case of grassland. This may prove to be a valuable index to be used with this sensor in the future.

Table 1. Pearson R values computed between Landsat-TM, AVHRR and Laboratory experiment variables and FMC of grassland, average shrub and Cistus ladanifer.

Data source	Species	Vegetation Indices			
		NDII5	NDII7	NDVI	NDVI/ ST
Lab.exper	*Cistus ladanifer*	0.98	0.96	0.92	
AVHRR	*Cistus ladanifer*			0.49	0.62
Landsat TM	*Cistus ladanifer*	0.88	0.90	0.72	0.68
AVHRR	Shrubs (average)			0.49	0.58
Landsat TM	Shrubs (average)	0.90	0.92	0.74	0.69
AVHRR	Grassland			0.56	0.79
Landsat TM	Grassland	0.96	0.94	0.96	0.97

5 CONCLUSIONS

The possibility of comparing FMC results from several sources of information is now more feasible. The use of standard and derived vegetation indices can link information taken from fieldwork and both high and low resolution sensors.

The new NOAA 15, which includes a band in the SWIR, ought to help these comparative studies further, especially for significant indices such as the NDII5 and NDII7 derived from Landsat TM.

Meanwhile, the NOAA-AVHRR remains the most operational data source for fire danger management, in spite of its drawbacks (mainly noise). The most suitable variable for FMC appraisal, that can be obtained from the AVHRR, is the NDVI/ST.

ACKNOWLEDGEMENTS

Funding for this research was obtained through the Control-Fire-Sat and Inflame projects, funded under the European Union's DG-XII, Climate and Environmental Program.

REFERENCES

Alonso, M., A. Camarasa, E. Chuvieco, D. Cocero, I. Kyun, M.P. Martín & F.J. Salas 1996, Estimating temporal dynamics of fuel moisture content of Mediterraneam species from NOAA-AVHRR data. *EARSEL Advances in Remote Sensing*. 4(4): 9-24.

Bowman, W.D. 1989, The relationship between leaf water status, gas exchange, and spectral reflectance in cotton leaves. *Remote Sensing of Environment*. 30: 249-255.

Burgan, R.E. & R.A. Hartford 1993. Monitoring Vegetation Greenness with Satellite Data. GTR INT-297, USDA Forest Service, Ogden, Utah.

Carter, G.A. 1991, Primary and secondary effects of water content on the spectral reflectance of leaves. *American Journal of Botany*. 78: 916-924.

Chladil, M.A. & M. Nunez 1995, Assessing grassland moisture and biomass in Tasmania. The application of remote sensing and empirical models for a cloudy environment. *International Journal of Wildland Fire*. 5: 165-171.

Chuvieco, E., M. Deshayes, N. Stach, D. Cocero & D. Riaño, 1999.Short-term fire risk: foliage moisture content estimation from satellite data. In E. Chuvieco (Editor): *Remote Sensing of Large Wildfires in the European Mediterranean Basin*: 17-38. Berlin: Springer-Verlag.

Cibula, W.G., E.F. Zetka & R. D.L 1992, Response of Thematic Mapper bands to plant water

stress. *International Journal Remote Sensing*. 13: 1869-80.
Cohen, W.B. 1991, Temporal versus spatial variation in leaf reflectance under changing water stress conditions. *International Journal of Remote Sensing*. 12: 1865-1876.
Deshayes, M., E. Chuvieco, D. Cocero, M. Karteris, N. Koutsias & N. Stach, 1998. Evaluation of Different NOAA-AVHRR derived indices for fuel moisture content estimation: interest for short-term fire risk assessment. In: D.X. Viegas (Editor), III International Conference on Forest Fire Research - 14th Conference on Fire and Forest Meteorology. ADAI, Coimbra, pp. 1149-1167.
Gogineni, S., J. Ampe & A. Budihardjo 1991, Radar Estimates of Soil Moisture Over the Konza Prairie. *International Journal of Remote Sensing*. 12: 2425-2432.
Hunt, E.R. & B.N. Rock 1989, Detection of changes in leaf water content using near and middle-infrared reflectances. *Remote Sensing of Environment*. 30: 43-54.
Hunt, E.R., B.N. Rock & P.S. Nobel 1987, Measurement of leaf relative water content by infrared reflectance. *Remote Sensing of Environment*. 22: 429-435.
Jackson, R.D. 1986, Remote sensing of biotic and abiotic plant stress. *Ann. Rev. Phytopathology*. 24: 265-287.
Jackson, R.D. & C.E. Ezra 1985, Spectral response of cotton to suddenly induced water stress. *International Journal Remote Sensing*. 6: 177-185.
Jackson, R.D., S.B. Idso, R.J. Reginato & P.J. Pinter 1981, Canopy temperature as a crop water stress indicator. *Water Resources Research*. 17: 1133-1138.
Moran, M.S., T.R. Clarke, Y. Inoue & A. Vidal 1994, Estimating crop water deficit using the relation between surface-air temperature and spectral vegetation index. *Remote Sensing of Environment*. 49: 246-263.
Paltridge, G.W. & J. Barber 1988, Monitoring grassland dryness and fire potential in Australia with NOAA/AVHRR data. *Remote Sensing of Environment*. 25: 381-394.
Paltridge, G.W. & R.M. Mitchell 1990, Atmospheric and viewing angle correction of vegetation indices and grassland fuel moisture content derived from NOAA/AVHRR. *Remote Sensing of Environment*. 31: 121-135.
Ripple, W.J. 1986, Spectral reflectance relationships to leaf water stress. *Photogrammetric Engineering and Remote Sensing*. 52: 1669-1675.
Thomas, J.R., L.N. Namken, G.F. Oerther & R.G. Brown 1971, Estimating leaf water content by reflectance measurements. *Agronomy Journal*. 63: 845-847.
Tucker, C.J. 1980, Remote sensing of leaf water content in the near infrared. *Remote Sensing of Environment*. 10: 23-32.
Vidal, A. & C. Devaux-Ros 1995, Evaluating forest fire hazard with a Landsat TM derived water stress index. *Agricultural and Forest Meteorology*. 77: 207-224.
Vidal, A., F. Pinglo, H. Durand, C. Devaux-Ros & A. Maillet 1994, Evaluation of a temporal fire risk index in Mediterranean forest from NOAA thermal IR. *Remote Sensing of Environment*. 49: 296-303.
Westman, W.E. & C.V. Price 1988, Spectral changes in conifers subjected to air pollution and water stress: experimental studies. *IEEE Transactions on Geoscience and Remote Sensing*. 26: 11-20.

Remote Sensing in the 21st Century: Economic and Environmental Applications, Casanova (ed.)
 ISBN 90 5809 096 5

Ground-based infrared characterization of fires in the 3-5 μm spectral window

A.J.de Castro, J.M.Aranda, S.Briz & F.López
Departamento de Física, Universidad Carlos III de Madrid, Leganés, Spain

ABSTRACT: During the last years, new techniques have been applied to the detection of forest fires. Satellite-based infrared remote sensing appears to be one of the most promising tools to detect and monitor forest fires. Problems appear related to signal saturation, false alarms, and a high pixel size (about 1 km at nadir for the AVHRR instrument aboard the NOAA satellites). Several ground-based monitoring systems have been implemented by adapting infrared cameras to mobile platforms; but these systems also present similar problems, mainly signal saturation and false alarms. In order to design an adequate sensor system, it is very important to have a complete ground-based characterization of the emission properties of the fires. Ground-based remote sensing equipment has been employed to measure the infrared emission of controlled forest fires. This characterization has been used as the basis to define a bicolor analysis in the 3-5 μm spectral region. This analysis would permit to separate signals coming from real fires from those due to solar reflectances and solar irradiance scattered by the atmosphere. The proposed algorithm has been verified by using different simulated infrared scenarios.

1 INTRODUCTION

In the last years, a tendency to a continuous increase of the number of fires over the whole Earth surface has been observed. The burning of large areas of rain forests, mainly in Amazonia, has been the subject of great concern, but this is by no means the only environmental problem posed by forest fires. In particular, in the Mediterranean area forest fires are one of the main factors of environmental degradation. As many other environmental problems, forest fire has become global as the main reservoirs of oxygen sources and carbon dioxide sinks are dangerously decreasing.

On the other hand, forest fire fighting is currently, in many countries, an exclusive duty of local or regional authorities. In this scheme, management of all works, including fire detection, is conducted at a local level. All currently early detection is ground-based, and the watchtower is the main element regardless of the instrument used, from binoculars to sophisticated IR automatic cameras.

Nevertheless, the global dimension reached by the forest fire problem is forcing a new approach to forest fire management. In fact, a large scale or global approach is foreseen, probably overlapping the actual ground-based systems but changing deeply its current characteristics. This detection approach would combine global satellite-based support, with local, ground-based, specific stations linked to the extinguishing tasks. The integration of different information by means of sensor fusion techniques will be increasingly important, as it offers important support to fire management decisions.

In this scheme, IR sensors should play an important role, both in ground and satellite-based platforms. IR sensors can detect fires at very large distances, even when visibility is impeded by smoke or clouds.

The main problem related to IR remote sensing of fires by automatic systems, whatever the platform, is a good identification of fire in order to avoid a high false alarm rate. In this work we propose to use the spectral and eventually spatial characterization of fire in order to do this.

2 IR CHARACTERISTICS OF FIRE: THE ROLE OF SPECTRAL DATA

In the previously described scheme, it is essential to have an accurate spectral characterization of the fire IR emission. The main goals for IR systems are to combine both a high detection range and a very low false alarms rate, at a moderate cost, and, ideally, with low maintenance.

Cost can be determinant in ground-based systems, whereas low maintenance is one of the critical points in satellite-borne systems. Both factors are directly related to the sensor complexity. From this point of view, some features would be preferred, depending on the platform:

Ground-based systems
- No cooling down (or if needed, only thermoelectric cooling)
- Low electronic complexity in the sensor
- Few sensor elements in the focal plane
- Low maintenance and low power consumption
- Low cost per sensor

Satellite-borne systems
- No maintenance, low power consumption
- High number of elements in the focal plane
- High sensibility

Both platforms
- No moving parts (staring sensor)

In satellite-based systems, one of the characteristics of improved fire sensors would be an early detection capability. That means sensors with a high sensitivity, as the intensity of fires in this stage is extremely low (some hundreds of W/Sr in the 3-5 μm spectral window). But, in general, a satellite based system must also provide other outputs; in particular, fire monitoring as it becomes a great fire. Here, signal saturation is one of the main problems in accurate and high sensitivity sensors.

Whatever the characteristics of the sensor, the use of spectral analysis in the detection process greatly improves the sensor performance, mainly the false alarm rate. This is a key point in automatic fire sensors where confidence in the declared alarms and in their classification in terms of importance or danger is decisive for fire management.

Spectral analysis is specially important when spatial resolution is poor. As an example, subpixel detection can be possible when at least two different spectral bands are used (Dozier 1981).

3 IR FIRE INDEX CONCEPT

Systems offering spectral resolution can vary from a few bands (multispectral sensors) to a large number (hyperspectral sensors) (Website 1999). Although hyperspectral measurements are necessary for compositional analysis where many components are presented (Clark & Roush 1984, Kruse et al. 1995), only a few channels would be needed to detect a mixture of two or three gases (one channel for each gas and one reference channel) (López et al. 1992, López et al. 1994, Meléndez et al. 1995). If we assume that the typical IR fire fingerprint is due to CO_2 emission, whatever the temperature, it would be possible to detect fire and to distinguish it from other IR sources with a sensor having only two channels. These would be obtained by the use of two filters.

This concept has been proved in a previous paper using two measuring bands, located respectively in the medium (mid) and thermal IR bands. In reference (de Castro et al. 1998), the IR fire index (FI) concept has been defined. The two bands used there were, one in the mid IR varying from 4.15 μm (2410 cm^{-1}) to 5 μm (2000 cm^{-1}). The second band used for fire index definition was taken in the long wavelengths, from 9.1μm (1100 cm^{-1}) to 12.5 μm (800 cm^{-1}). The ability of the FI to distinguish among the fire states, from initial stages to the final ones is proved. The index permits to distinguish in an easy way unburned terrain (FI ~ -1) from a high fire intensity (FI = 0.7).

In this paper we deal with a new proposal for a similar index to arrive at a clear differentiation between signals associated to fire and non-fire scenarios, also including differentiation of burned terrain. The proposed method is similar to the one cited before. In order to use one only detector, it is necessary to find a new IR fire index with the discrimination bands working in the mid IR (3 to 5 microns).

3.1 *A Fire Index for the 3-5 μm band.*

For given electrooptical characteristics, namely the optics aperture and FOV, the specific detectivity (or noise equivalent power-NEP), the acquisition time and the number of elements in the focal plane, there are two constraints imposed in a fire detection system: the range and the false alarm rate (FAR). Both constraints are always related: the higher the range the higher the false alarm rate. The main problem to obtain a fire detecting non-imaging system concerns the ability of the system to avoid false alarms. Usually, a false alarm is defined by the system when, whatever the origin of the emission, the signal in the sensor output is high enough to cross over a threshold value previously defined.

An imaging system can also use other tools like spatial analysis (if the image has enough resolution) to detect a false alarm. This is not the case for the system described before, where spatial resolution does not exist (non-imaging) or is poor (fire size is a fraction of the pixel area). In all cases it is well known that spectral resolution will improve the detection figure of the sensor.

The system will incorporate spectral resolution by means of interference filters located before the detector. Although some possibilities exist to do this with no moving parts in the system, as in the spectral retina concept (López et al. 1992, López et al. 1994) or in other developments related to integrated optics, (Frank 1997) the general method in compact systems is to move a filter wheel with the corresponding interference filters arranged in such a way as to pass alternatively before the detector. A synchronism mechanism allows the signal readout when the field of view of the detector is completely covered by the specific filter.

The wheel method for wavelength selection tends to complicate the arrangement when the filter number increases. For this reason we will try to select the optimum filter collection, depending on the required detection characteristics.

Assuming a signal process to avoid false alarms coming from slow variations of background, clutter or scattering of sun emission from the atmosphere the main cause of false alarms in an automatic unmanned IR fire detecting system is the spurious and relatively rapid rising of irradiance on the detector. This is the case of sun appearing after a cloud shadow, and specular or diffuse reflection of sun from rocks or other objects.

In order to design an adequate sensor system, it is very important to have a complete ground-based characterization of the emission properties of the fires. Measurements of well defined fires have been performed during a campaign, in order to obtain spectral characterization of signatures associated to the CO_2 and H_2O emission. Measurements have been performed by using a Fourier transform-based (FTIR) spectroradiometer (de Castro 1998). Different forest fuels have been used for the burns: *Eucalyptus globulus* slash, straw and *Ulex europaeus* shrub. These fuels can be considered as primary carriers of fire, according to the NFFL propagation model. As an example, figure 1 shows the measured spectral irradiance corresponding to the eucalyptus slash burn.

At wave numbers between 2080 and 2440 cm^{-1} (4.1-4.8 μm) the flame shows its characteristic structure caused by CO_2 emission band. The spectral distribution of this important emission band, observed at distances higher than a few meters, appears divided into two bands, known respectively as blue and red spike. The longer the observation distance the wider and deeper the absorption band of atmospheric CO_2 and the smaller the irradiance impinging on the system produced by CO_2 emission band of fire. Nevertheless, both peaks are always present if enough resolution is attained in the spectroradiometric measurements, and these features can be used in order to distinguish fire emission from others.

In order to compare the fire emission with other IR sources candidates to give false alarms, the following scenarios have

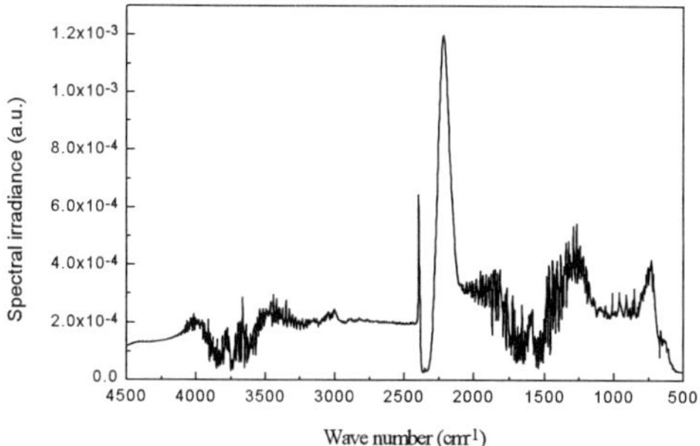

Figure 1: Spectral irradiance corresponding to the combustion of *Eucalyptus globulus* slash.

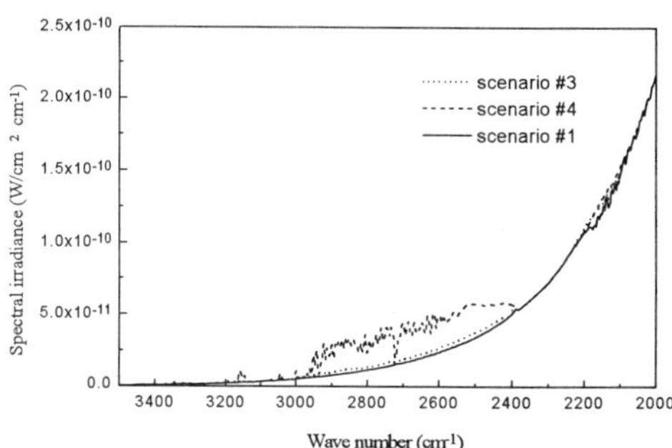

Figure 2: Spectral irradiance calculated for different scenarios in the 2000-3500 cm^{-1} range.

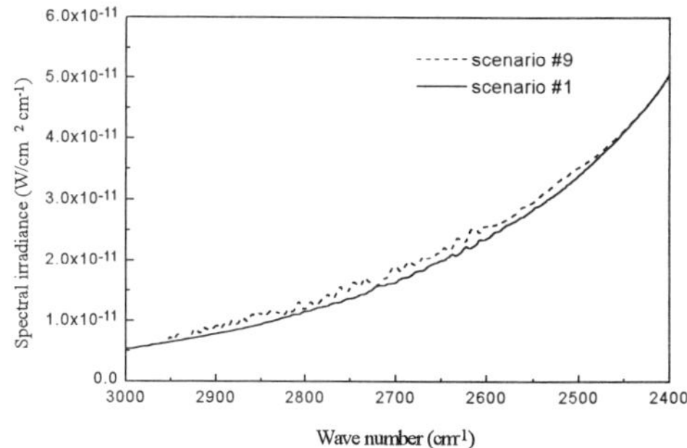

Figure 3: Spectral irradiance calculated for scenarios # 1 and 9 in the 2400-3000 cm-1 range. Differences between both spectra are due to solar radiance scattered by the atmosphere

been simulated. An IR sensor in the 3-5 μm spectral window is located in an observation point 1 km high, with a zenith angle of 101°. A value of 1.2° is assumed for the detector FOV. For these conditions, terrain situated at ~ 5 km is into the FOV of the detector. Atmospheric and solar conditions, as well as surface albedo, have been modeled by using the PcModwin3 code, a PC version of the MODTRAN atmospheric model.

Different irradiance scenarios have been simulated calculating data from radiance measurements of different emission sources. Scenario #1 corresponds to the atmospheric emission (*Midlatitude Summer* atmospheric model), and it will be con-

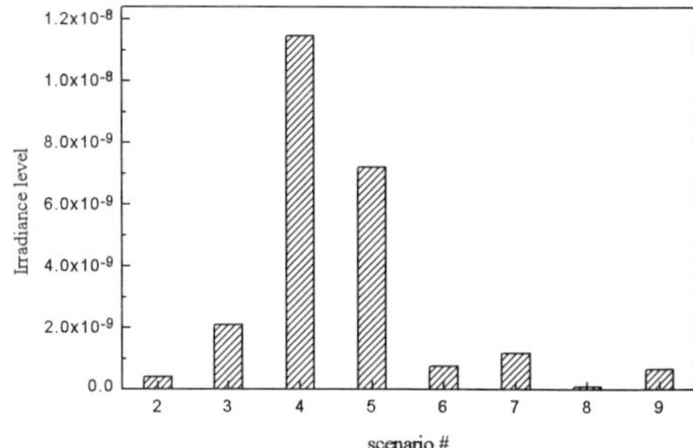

Figure 4: Calculated values of the irradiance measured by filter *f1* under different scenarios.

sidered as the reference measurement of the system. Scenario #2 corresponds to the irradiance of a burning area of 1 m^2 superimposed to scenario #1. This irradiance is calculated by using the experimental spectrum of figure 1. Scenarios #3 and #7 correspond also to incipient fires, and they are calculated from experimental burns of different fuels following the same procedure as scenario #2. Scenarios # 6 and 8 correspond to emission of recently burnt terrain (temperatures less than 200°C). Scenarios # 4, 5 and 9 represent different background conditions including solar radiance scattered for the atmosphere and solar reflections at the surface obtained at different daytime. Figure 2 and 3 illustrate the irradiance spectra calculated for some of these scenarios in the mid IR range.

3.2 *Spectral analysis: one unique filter*

Wide band filter

This is the most common filter used to work in the 3-5 μm spectral window. This is, for instance, the default filter in all the IR cameras working in this region.

Figure 4 shows the integrated irradiance calculated by using a rectangular filter (*f1*) in the 2100-3000 cm^{-1} (3.33-4.76 μm) spectral range. From this figure we can conclude that, with only one wide filter like *f1*, background irradiance can be much higher than irradiance corresponding to the fire scenarios. The irradiance of burnt terrain is indistinguishable from the rest of the cases. One wide band filter for the whole window and large range (5 km or more), would give many false alarms in the sensor and would be inadequate. In fact, it has been tested that, in some cases, irradiance on the detector caused by backgrounds can be a hundred times higher than the irradiance of small fires.

Very narrow-band filter

In some specific cases related to the detection of incipient fires at short ranges (≤1 km) using detectors having a very good detectivity, it is possible to use only one filter to detect them. This filter would be a narrow bandpass filter centered at the CO_2 red spike, with bandwidth less than 50 cm^{-1}@ 2250 cm^{-1}. This filter permits the use of about 10 % of the impinging irradiance in the detector for short distances. This figure rapidly deteriorates when range increases, because of the loss of contrast of the red spike, absorbed by atmospheric CO_2.

3.3 *Spectral analysis: two filters*

In this work we propose the use of two spectral filters and an algorithm to separate alarms produced by sudden variations of background from those coming from real fires.

The first filter would be filter f_1, defined in §3.2. The second filter, f_2, is a band pass filter covering from 2100 to 2410 cm^{-1} (4.15-4.76 μm).

Provided that the system, after conditioning and filtering, gives an electrical signal proportional to the irradiance on the detector, and that each filter would give one independent signal, we can deal of a system with two channels. Channel 1, associated to the filter f_1, gives a signal s_1 .The second channel would give a signal s_2, corresponding to the same scenario seen through the filter f_2. Both s_1 and s_2 signals are normalized to the bandwith of the corresponding filters.

The first operation the system will do is the subtraction in both channels of reference background (BG) irradiance. Scenario #1 is used to simulate this reference value. For a detector with a high field of view (1.2°) this irradiance is 6×10^{-8} W/cm^2.

In order to optimize the range, we propose the system to be in a permanent observation state using filter f_1. In this way the system will improve the detection of small signals wherever they came from, fires of backgrounds. Immediately after one alarm signal has been detected in the first channel, the system would change to observe the same scenario through the second filter f_2.

After the measurement of the possible fire alarm, the system will have got two signals, s_1 and s_2, both referenced to BG in each band and normalized to the bandwidth.

With these two quantities we propose to define a fire index in the mid IR as follows,

$$MFI = \frac{s_1 - s_2}{s_2} \tag{1}$$

This index represents the ratio between the irradiance in the shorter wavelength region (3- 4.1 μm) and the irradiance in the spectral region corresponding to the CO_2 red spike emission. The greater the fire index, the lesser the probability of signal to correspond to a true fire.

Figure 5 represents this fire index for the eight scenarios cited before. From the analysis of this figure it can be derived that all BG are very well separated from real fires. MFI value is always greater than 5 for backgrounds, while MFI< 1.5 for fires. Recent burnt lands corresponding to scenarios #6 and 8 can also be separated.

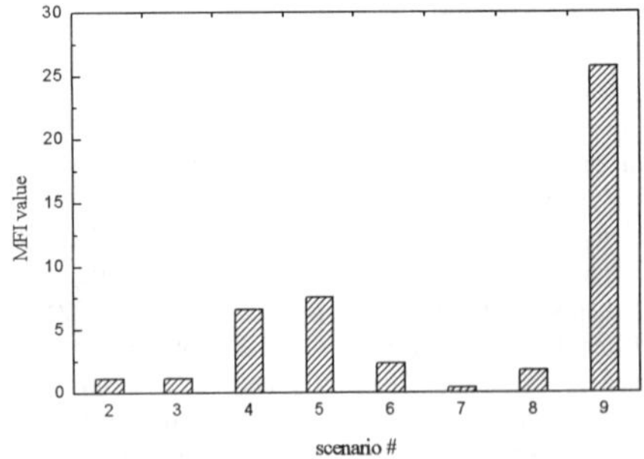

Figure 5: Calculated values of the MFI under different scenarios.

The table below summarizes a clear view of the MFI behaviour for the three cases cited above

Table I: Alarm limits of MFI for different scenarios in the sensor FOV.

Mid Fire Index	Type of alarm
MFI > 5	Background
1.5 < MFI < 5	Recent burnt land (T < 200 °C)
MFI < 1.5	Fire alarm

From table I it can be derived that, for real fires, the averaged irradiance in the shorter band (3.3-4.1 μm) is less than twice the irradiance in the larger band (4.1-4.8 μm), being more than 5 times for backgrounds, whatever the irradiance they give. Recent burnt areas give intermediate values.

4 CONCLUSIONS

An algorithm to reduce false alarms in a surveillance IR ground-based system has been proposed to be used in the 3-5 μm band. The algorithm is based on the fire index concept introduced in this work. This fire index is defined from a bicolor spectral analysis, obtained by using two filters. The fire index permits to separate the contributions coming from fires that those coming from backgrounds. This index can separate the contribution of recent burnt surfaces from the rest of IR sources too. Different scenarios (atmospheric backgrounds, fires and burnt terrains) have been simulated to test the efficiency of this index. Fire scenarios have been simulated from spectroradiometric characterization of the IR emission of controlled burns. Spatial analysis of fires and backgrounds, is now been carried out by using a spectral IR imager, to verify by a different method the ability of the mid IR fire index proposed.

5 ACKNOWLEDGMENTS

The authors wish to acknowledge the Spanish *Comisión Interministerial de Ciencia y Tecnología* (CICYT) for financial support under project no. TIC97-0790-C03-01. One of the authors (S.B.) acknowledges the Spanish *Ministerio de Educación y Cultura* for financial support.

6 REFERENCES

Clark, R.N. & Roush, T.L. 1984. Reflectance spectroscopy:Quantitative analysis techniques for remote sensing applications. *Journal of Geophysical Research* 89: 6329-6340

de Castro, A.J., Aranda, J.M., Briz, S., Meléndez, J. & López, F. 1998. Ground-based infrared remote sensing of forest fires. *SPIE Proc.* 3499, 416-426

Dozier, J. 1981. A method for satellite identification of surface temperature fields at subpixel resolution. *Remote Sensing of Environment* 11: 221-229.

Frank, M., Schallenberg, U.B. & Kaiser, N. 1997. Micropatterned multilayer dielectric filters with two spectral characteristics. *Opt. Eng.* 36:1220-1224

Kruse, F.A., Lefkoff, A.B. & Dietz, J.B. 1995.Expert system-based mineral mapping in northern Death Valley, California/Nevada using the Airborne Visible/Infrared Imaging Spectrometer (AVIRIS). *Remote Sensing of Environment* 44: 309-336

López, F., de Frutos, J., González, A.M. & Navarro, A. 1992. Integrated optical filters for infrared electrooptical gas sensors. *Sensors and Actuators B* 6: 170-175

López F., de Castro, A.J., Meléndez, J. & Meneses, J. 1994. Design of a new sensor of multiple gases based on the integration of interference filters SPIE Proc. 2253: 872-880

Meléndez, J, de Castro, A.J., López, F. & Meneses, J. 1995 Spectrally selective gas cell for electrooptical infrared compact gas sensor. Sensors and Actuators A 46-47:417-421.

Website http://makalu.jpl.nasa.gov/aviris.html

Remote Sensing in the 21st Century: Economic and Environmental Applications, Casanova (ed.)
© 2000 Balkema, Rotterdam, ISBN 90 5809 096 5

Analysis of the environmental factors in forest wildfires in Aznalcóllar and Sierra Bermeja

A. Sanz Redondo
Departamento de Ingeniería Cartográfica, Geodésica y Fotogrametría, Universidad de Castilla La Mancha, Ciudad Real, Spain

J.V. Giráldez Cervera
Departamento de Agronomía, Universidad de Córdoba, Spain

ABSTRACT: Different natural factors make the Mediterranean ecosystem specially vulnerable to forest wildfires: the weather, with extremely long dry summers; the strong winds, which dry the soil and vegetation and spread the fire once it has begun; the abrupt topography and finally, the human factor. In late spring, May 30 to June 4 in 1995, a wildfire broke burning an area of 1800 ha called "El Madroñalejo" (Sevilla). This area is characterized for his rough topography, with deep narrow valleys. One month and half later, an area of 1500 ha of forest was burnt in Sierra Bermeja, (Málaga). The development of temperature and vegetation index maps from the daily NOAA-AVHRR sensor databases, and the synoptical meteorological maps analysis, enable the study of increase in temperature and the NDVI decrease of the previous days and during the fire, as well as the climatic conditions of those dates.

1 INTRODUCTION

Fight against forest fires means a high cost of materials and human support, so the evaluation risk is very important to prevent and to put out them, what let us accommodate the supports and to increase their efficiency.

The combustible as well as the oxygen and ignition energy are just the three components of the fire triangle. Any fire will begin if one of them is not present. In addition, there are several factors which help to the beginning and spreading of wildfires, but they are not direct responsible for it, although they influence in the risk of the occurrence of this event.

The fire risk is related with causality and it is usually separated between human factor and natural factors (e.g. lightning). The fire danger refers to meteorological conditions, temperature and air humidity, speed and wind direction, vegetation water content, etc. In this paper, the term "risk index" refers to both concepts.

Figure 1. Situation map.

In late spring, May 30 to June 4 in 1995, a wildfire broke burning an area of 1800 ha called "El Madroñalejo", in Aznalcóllar (Sevilla). This area is characterized for his rough topography, with deep narrow valleys. One month and half later, an area of 1500 ha of forest was burnt in Sierra Bermeja, (Málaga). In the Figure 1, on see the situation map.

2 RISK INDEX DEFINING FACTORS ANALYSIS

Different natural factors make the Mediterranean ecosystem specially vulnerable to forest wildfires: the weather, with extremely long dry summers; the strong winds, which dry the soil and vegetation and spread the fire once it has begun; the abrupt topography and finally, the human factor with its traditional agricultural practices as the burning of stubble.

Meteorology conditions not only the state of the combustible but also the time when the fire spread risk increases, and its direction and velocity (Hernández et al. 1995).

The previous factors are not the main cause of the fire beginning but they directly help to its spreading. Because of that, burning areas of more than 1 ha are not easily found if the meteorological conditions and vegetation state are not advantageous to the fire (Dubois & Vidal 1995).

2.1 *NDVI decreasing*

The vegetation water stress estimation has been commonly analyzed from multi-temporal NDVI images. So, this index is considered to be a good vegetation health index. Moreover, NDVI decrements are related with plants greenness and vigor, which is fully connected with vegetation water content (Goward & Hope 1989).

Vegetation water content is a difficult estimation parameter due to its scarcely spectral variation, in comparison with other environmental factors (Cohen 1991). However, the spectral characterization of vegetation water stress is possible if its temporal evolution is studied, and if the contrast between death and alive vegetation is done.

Several studies have shown the highly correlation between spectral vegetation indexes, calculated from AVHRR sensor data, and physiological variables as the greenness biomass, leaf area index, evapotranspiration and photosintetically active radiation absorbed by the plant. The vast majority of those studies found significant correlation between NDVI and vegetation tendency (Cihlar et al. 1991), but those relationships were more important in short time intervals, like one or two weeks (Deblonde & Cihlar 1993).

2.2 *Temperature increases*

The temperature is an important factor in fire studies. High surface temperature values indicate the water stress of vegetation easily burning.

The temporal evolution of this temperatures can be studied from NOAA satellite images daily, but this temperature obtained from satellite is not the real soil temperature, due to the atmospheric influence over the radiation in its path to the sensor. As the natural surfaces do not behave as a blackbody, this measures must be also corrected by emissivity. The automated process obtaining surface temperature can be seen in Sanz 1999 in this symposium.

2.3 *Meteorology: wind*

Synoptical scale forecasts are widely precise in a time interval of 48 to 72 hours, so if we try to relate the meteorological conditions at this scale and the wild fire frequency, the risk of fire will be known previously.

Hernández et al. 1995, found two meteorological condition classes (Fig. 2)

The first situation, type I, corresponds to a low thermal at surface level at the Peninsula centered, that is to say, a low pressure area originated by a solar heating over soil. Type II corresponds to a storm, situated in northeastern Spain, with a cold front associated, which is by western. This situation follows to strong winds which help fire spreading.

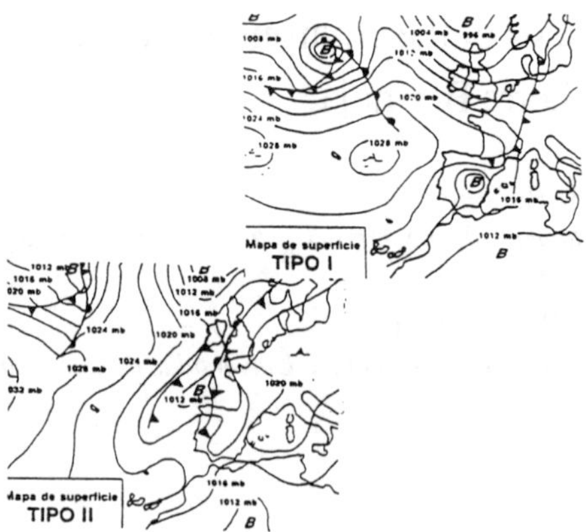

Figure 2. Meteorological conditions synoptical scale

2.4 *Relief factor*

Fires begin in deep valleys, where dry vegetation accumulates, spreading upwards following the maximum slope contour lines. If valleys are narrow, fire can easily go from one hillside to another and they can perform as a smokestack in which valley winds can reach high speed, spreading fire flames. Fluvial vegetation is other index of the way the front fire will take.

3 RESULTS AND DISCUSSIONS

3.1 *NDVI decreasing*

While using remote sensing techniques to measure a crop radiance, the satellite signal is the result of the vegetation and its soil radiance combination. Both of them reflectivity are affected by internal and external crop factors which mask the real tree and/or crop reflectivity values (Gandía 1991).

The external factors are the surface of observed area, the light and the angles under which the observation is done (Gross et al. 1988), as well as the atmospherical conditions when the image is captured.

The internal factors are the proportion of the different parts of the tree, leaves, trunk and fruits, the health of it and the water and nutrients content in the soil, and their chemical composition.

In tables 1 to 2 are the NDVI values for Aznalcóllar and Sierra Bermeja fires, before and during the fires. According to them the affected area has been delimited and it can be observed how the NDVI decreases while the fire days is nearer and in the burning pixels. Once the fire is extinguished, this index keeps approximately constant because vegetation is dead.

In these tables it must be pointed out that NDVI values vary over 0.05, which is the maximum value for the leaves of the tree crown, that we should

Table 1: NDVI decreasing evolution in Aznalcóllar fire

ΔNDVI May 27 to 29 in 1995

0.034	0.022	-0.01	0.003	0.034	0.051	0.092	0.06
0.021	0.059	0.059	0.032	0.031	0.033	0.053	0.05
0.025	0.057	0.073	0.028	0.008	0.006	0.038	0.04
0.038	0.038	0.054	0.035	0.004	0.008	0.032	0.04
0.080	0.085	0.054	0.029	0.010	0.004	0.013	0.04
0.029	0.047	0.032	0.043	0.059	0.039	0.009	0.03
0.033	0.077	0.025	0.032	0.011	0.065	0.040	0.01

ΔNDVI May 29 to 31 in 1995

-0.05	-0.04	0.00	0.00	-0.04	-0.05	-0.07	-0.06
-0.05	-0.09	-0.09	-0.04	-0.02	-0.06	-0.07	-0.05
-0.02	-0.18	-0.16	-0.09	-0.04	-0.10	-0.08	-0.04
-0.29	-0.24	-0.24	-0.26	-0.19	-0.15	-0.08	-0.05
-0.28	-0.28	-0.24	-0.24	-0.24	-0.18	-0.06	-0.05
-0.16	-0.20	-0.19	-0.16	-0.20	-0.19	-0.04	-0.01
-0.12	-0.16	-0.17	-0.15	-0.12	-0.11	-0.07	0.00

ΔNDVI May 31 to June 1 in 1995

-0.01	-0.03	-0.05	-0.01	0.03	0.04	0.02	-0.01
-0.11	-0.09	-0.1	-0.12	-0.11	-0.02	0.05	0.04
-0.09	-0.09	-0.06	-0.1	-0.11	0.02	0.05	0.03
0.01	0.01	0.019	0.04	0.04	0.08	0.06	0.02
0.00	0.04	0.041	0.04	0.04	0.05	0.03	0.03
0.00	0.08	0.12	0.07	0.04	0.06	0.00	-0.01
0.04	0.04	0.09	0.12	0.07	0.05	0.03	-0.02

ΔNDVI June 1 to 2 in 1995

-0.1	-0.1	-0.1	-0.13	-0.15	-0.15	-0.14	-0.1
0.03	-0.05	-0.11	-0.09	-0.08	-0.1	-0.20	-0.14
0.18	0.13	-0.03	-0.07	-0.03	-0.07	-0.13	-0.12
0.21	0.12	0.04	-0.04	-0.03	-0.09	-0.13	-0.10
0.16	0.13	0.077	0.02	-0.05	-0.07	-0.15	-0.12
0.11	0.07	0.01	-0.01	-0.03	-0.11	-0.13	-0.11
0.02	-0.01	-0.02	-0.02	-0.07	-0.1	-0.1	-0.07

ΔNDVI June 2 to 3 in 1995

0.01	-0.01	0.01	0.06	0.04	0.01	0.01	0.01
0.03	0.07	0.04	0.00	-0.02	-0.04	-0.01	-0.02
0.02	0.07	0.06	0.03	-0.02	-0.03	-0.03	0.00
0.01	0.03	0.07	0.09	-0.01	-0.01	0.01	-0.02
-0.02	-0.01	0.04	0.09	0.08	-0.01	0.00	-0.03
-0.03	-0.02	-0.01	0.06	0.06	0.05	0.00	0.01
0.02	0.01	-0.04	0.00	0.03	-0.02	-0.02	0.01

ΔNDVI June 3 to 4 in 1995

-0.02	0.00	0.013	-0.01	0.04	0.07	0.07	0.06
-0.07	-0.06	0.03	0.04	0.04	0.02	0.12	0.12
-0.06	-0.13	-0.08	-0.03	0.01	0.03	0.10	0.06
-0.01	-0.03	-0.08	-0.1	-0.02	0.01	0.05	0.07
0.05	0.01	-0.06	-0.1	-0.08	0.00	0.07	0.09
0.08	0.04	0.01	-0.06	-0.1	-0.05	0.03	0.03
-0.02	0.01	0.05	-0.03	-0.03	0.04	0.07	0.02

Table 2: NDVI decreasing evolution in Sierra Bermeja fire

ΔNDVI July 3 to 4 in 1995

-0.055	-0.095	-0.206	-0.335	-0.375
-0.060	-0.058	-0.209	-0.347	-0.361
-0.108	-0.150	-0.247	-0.353	-0.355
-0.185	-0.321	-0.348	-0.394	-0.373
-0.142	-0.181	-0.315	-0.428	-0.301
-0.150	-0.087	-0.117	-0.168	-0.175
-0.082	-0.066	-0.046	-0.058	-0.102
-0.055	-0.093	-0.11	-0.139	-0.158
-0.042	-0.066	-0.145	-0.245	-0.286

ΔNDVI July 4 to 11 in 1995

-0.030	0.027	0.170	0.303	0.352
-0.019	0.004	0.149	0.292	0.330
0.023	0.092	0.192	0.320	0.319
0.103	0.247	0.304	0.358	0.339
0.096	0.145	0.258	0.387	0.268
0.096	0.054	0.071	0.126	0.129
0.093	0.047	0.055	0.035	0.108
0.074	0.099	0.126	0.094	0.167
0.010	0.067	0.148	0.214	0.272

ΔNDVI July 11 to 12 in 1995

0.156	0.116	0.085	0.082	0.195
0.105	0.063	0.092	0.088	0.158
0.103	0.015	0.031	0.078	0.110
0.111	0.107	0.144	0.119	0.057
0.061	0.147	0.173	0.041	0.016
0.151	0.169	0.080	-0.010	0.004
0.160	0.154	-0.006	0.041	0.028
0.243	0.164	0.067	0.129	0.025
0.246	0.133	0.101	0.112	0.042

ΔNDVI July 12 to 14 in 1995

0.024	0.027	0.016	0.031	0.002
0.075	0.084	0.033	0.031	0.007
0.053	0.097	0.121	0.090	0.087
0.063	0.025	0.033	0.054	0.071
0.097	0.073	0.008	0.037	0.017
0.024	0.102	0.113	0.033	-0.077
0.060	0.097	0.150	0.073	-0.037
0.036	0.102	0.092	0.036	0.040
0.028	0.112	0.061	0.035	0.040

measure with a reflectivity radiometer in field during successive days for calculating the NDVI.

3.2 *Temperature increasing*

Once we have obtained the temperature maps (Sanz 1999) the temperatures increment are in Tables 3-4, where thermal evolution of the days previous to fire and up to its extinction can be observed. According to them the affected area has been delimited in these Tables.

Table 3: Temperature increasing in Aznalcóllar fire

ΔT May 27 to 29 in 1995

13.6	13.8	15.04	16.76	16.80	15.65	13.53	13.5
13.8	13.9	14.73	16.02	16.25	15.90	15.26	14.5
14.4	13.9	13.90	15.62	16.97	16.83	15.97	15.4
14.6	14.6	14.66	15.41	17.27	17.16	16.01	15.7
13.5	12.4	13.27	14.93	16.59	16.76	16.11	16.1
14.4	14.2	15.35	14.92	15.09	15.97	16.30	16.6
16.8	16.3	17.00	17.07	16.34	16.09	16.16	16.7

ΔT May 29 to 31 in 1995

-1.19	-1.15	-1.34	-2.48	-2.19	-1.67	-0.72	-0.88
-2.39	-2.54	-2.15	-4.09	-4.13	0.62	1.18	-1.18
-5.73	-4.44	-3.20	-3.84	1.96	3.74	1.88	-1.32
-7.71	-7.33	0.044	9.63	8.55	9.77	2.92	-1.73
-7.47	-7.62	-1.36	7.75	11.22	13.67	3.44	-2.02
-7.92	-8.11	-7.46	2.80	8.19	5.92	2.53	-1.81
-10.5	-10.6	-10.6	-9.81	-4.45	0.18	-0.54	-3.24

ΔT May 31 to June 1 in 1995

1.33	3.99	3.82	3.48	2.23	1.29	0.81	1.52
-1.56	8.76	15.65	14.10	6.99	1.33	-1.05	0.23
-0.70	1.71	16.05	20.08	7.20	-0.11	-1.43	0.04
1.17	4.39	5.67	3.13	-0.02	-6.62	-2.84	0.69
1.58	5.97	5.32	0.65	-2.04	-8.15	-1.93	1.39
0.89	3.64	7.62	2.26	-0.19	0.41	-0.33	1.75
3.17	3.96	6.36	9.23	6.50	2.16	1.73	3.49

ΔT June 1 to 2 in 1995

21.6	23.3	11.37	7.83	10-32	14.42	14.84	8.01
22.4	26.1	13.37	12.22	14.61	20.62	21.15	12.5
14.5	21.8	14.09	9.47	11.75	17.63	14.77	11.7
13.0	12.9	10.84	10.35	18.03	24.58	17.46	10.6
19.6	11.5	5.88	7.26	14.77	21.82	20.33	16.9
24.3	17.6	9.43	7.10	9.77	14.02	15.63	19.7
21.2	21.7	16.58	10.89	10.40	10.83	7.59	7.87

ΔT June 2 to 3 in 1995

-14.6	-16.4	7.10	15.45	9.10	9.88	13.76	7.14
-15.2	-28.8	-16.3	-6.02	3.26	-2.29	2.92	5.69
-4.96	-16.6	-22.2	-14.7	-4.88	-5.38	5.95	10.05
5.37	-1.55	-11.7	-16.7	-20.2	-15.4	4.37	11.26
3.88	7.73	3.74	-8.68	-15.9	-13.7	-6.80	-4.51
-3.16	3.08	1.70	-0.27	-6.89	-9.39	-10.3	-14.6
-6.03	0.59	-3.52	-2.08	-2.88	-3.30	2.14	-1.71

ΔT June 3 to 4 in 1995

8.94	3.23	-8.19	-9.74	-6.59	-14.9	-24.2	-9.55
14.2	14.7	8.55	3.37	-0.90	-4.80	-19.2	-12.1
7.30	14.7	17.41	13.1	5.50	-6.31	-13.3	-12.4
-7.88	2.61	14.98	15.7	14.21	-0.53	-14.9	-12.7
-14.0	-8.41	2.67	13.4	16.40	4.09	-5.78	-3.62
-8.69	-12.5	-3.87	4.31	9.42	2.64	4.08	5.14
1.87	-9.15	-1.35	1.69	-0.06	-3.69	-2.96	2.89

ΔT June 4 to 5 in 1995

-26.3	-22.7	-22.8	-27.2	-26.1	-21.0	-14.7	-16.2
-26.2	-26.9	-28.4	-30.8	-31.9	-26.9	-15.5	-14.9
-19.9	-25.6	-28.9	-32.9	-32.8	-19.9	-19.0	-20.1
-14.6	-19.0	-26.7	-28.9	-30.1	-22.3	-19.1	-20.7
-14.7	-18.0	-23.8	-27.3	-30.2	-25.5	-20.2	-20.9
-17.8	-14.5	-17.6	-25.1	-26.9	-19.4	-20.4	-22.0
-24.1	-19.7	-20.8	-22.7	-19.5	-14.4	-16.9	-20.4

Table 4: Temperature increasing in Sierra Bermeja fire

ΔT July 3 to 4 in 1995

-0.921	-1.482	-3.326	-4.271
-0.988	-1.476	-2.716	-3.695
-1.789	-1.627	-1.427	-1.299
-1.915	-2.355	-1.641	0.129
-1.094	-1.835	-2.316	-0.039
0.749	1.096	1.198	2.280
3.798	4.422	2.971	2.189
3.803	4.196	3.801	1.449

ΔT July 4 to 11 in 1995

2.547	3.442	4.020	5.190
2.099	2.860	3.935	4.947
2.424	2.698	3.264	4.047
3.385	3.899	3.128	3.832
3.88	4.319	2.945	1.787
2.916	2.037	-0.521	-2.301
-0.342	-0.745	-2.619	-4-480
-1.655	-1.248	-2.101	-4.276

ΔT July 11 to 12 in 1995

2.828	3.029	3.927	3.211
3.762	2.916	2.000	1.298
2.165	2.586	3.192	1.416
0.941	0.757	3.474	2.612
-0.160	0.550	3.525	3.990
0.748	0.674	2.646	5.517
2.670	2.331	3.497	7.317
4.877	4.071	3.844	8.533

ΔT July 12 to 14 in 1995

8.689	8.465	6.625	5.578
6.857	7.177	7.149	6.901
7.727	6.759	5.095	6.911
8.619	7.889	4.951	5.404
9.508	8.088	5.858	5.396
7.904	8.358	6.743	2.711
6.130	7.154	5.773	1.177
5.415	7.462	7.060	2.391

It would have been very useful being able to calculate the differences between soil temperatures from satellite images and air temperature from mete-

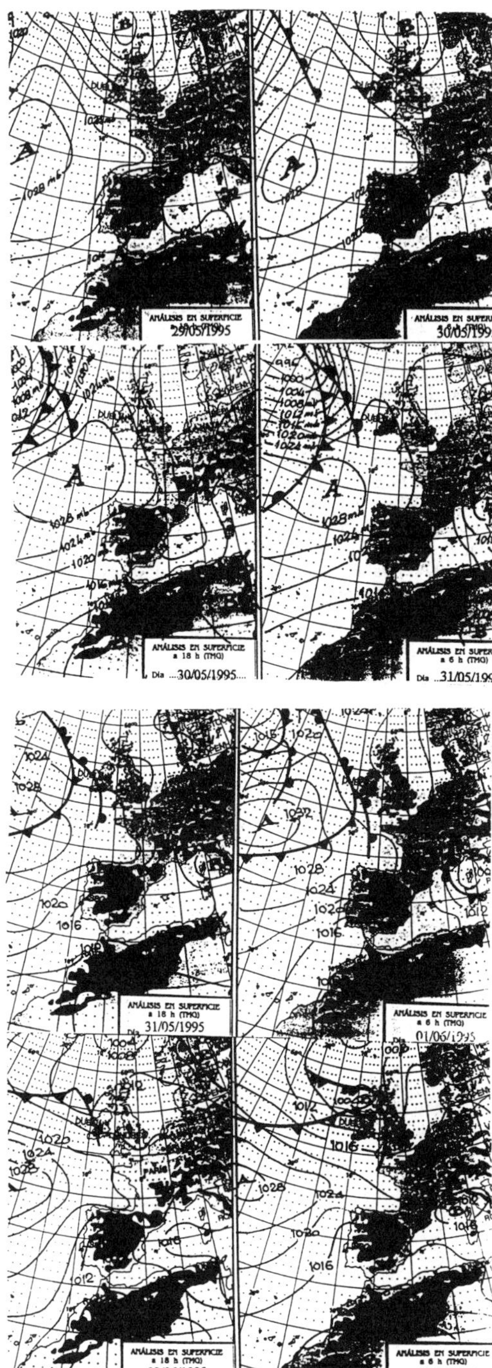

Figure 3. Synoptical scale meteorological maps

orological stations, as this parameter is strongly correlated with water stress indexes for long enough time intervals (Vidal et al. 1994). However, meteorological stations geographical situation, far from the study site and its functioning, not always automated, make difficult this relationship.

The vast multi-temporal images group for Aznalcollar fire, let us make a daily following of thermal evolution, where an increment of 45 degrees can be observed from the first days to latest, and a decrement of 27 degrees after the extinction.

This interval is not so notorious in Sierra Bermeja fire, where temperature increment is not higher than 20 degrees. In this case, the group of available images (there were not images from 4th to 11th of July, and the day were vegetation was burning) was captured at 8 a.m. while the images for Aznalcollar fire correspond to 10 a.m. in previous days of the fire and at 2 p.m. during the fire, what justify the high temperature values in Aznalcollar in contrast with temperature values in Sierra Bermeja. What´s more, those differences in temperature can be explained if we take into account that Sierra Bermeja fire began in an east-oriented area where no vegetation exist. A temperature increase is foreseen when the fire starts to follow a stream, in a young cork oak re-afforested plantation, what could not be verified because of the lack of images.

These temperature maps must not be compared between one fire and another, but their individual evolution, as geographical characteristics, proximity to sea, the climate, vegetation type and its distribution, determine those values.

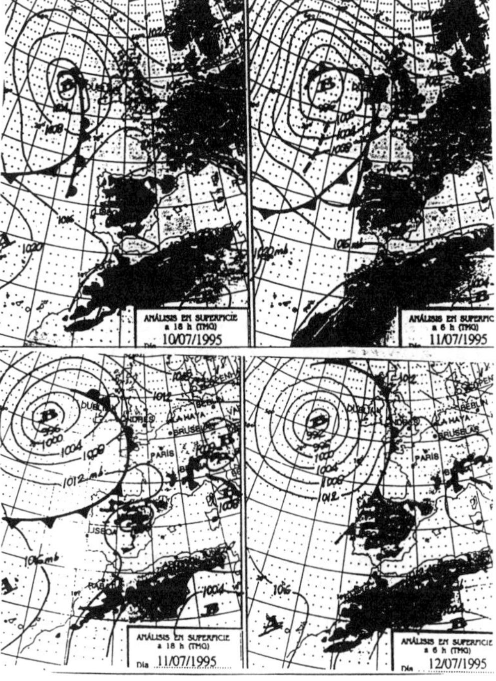

Figure 4. Synoptical scale meteorological maps

3.3 *Meteorology: wind*

During the previous days to Aznalcollar fire (Fig. 3-4) we can observe a similar situation to Type I (Fig. 2) with high temperatures and low relative humidity, which dry the combustible and leave it ready for ignition. Once the fire started storms centered in the Peninsula with eastwards winds help fire spreading.

The meteorological conditions in Sierra Bermeja fire do not correspond to any of the previously described situations, but in some areas and due to temperature differences within several topographic conditions, local winds increase general winds effects leading to thermals. In this case, the fire started in the deepest part of the hill and east-oriented, where air is convection-warmed and goes up, creating ascending violent hillside winds.

3.4 *Relief factor*

The topography in Aznalcollar is undulated in the ascending southeast-northeast direction. In this direction the landscape is hillside with hillock and hills of different slopes. This relief is abrupt and cut-off with deep narrow valleys with rivers and streams crossing them. The lowest height are 50 m over sea level and 100 m are rapidly reached, and even 500 m. The slopes vary between 10 and 20%, except in some areas were the values are higher.

Sierra Bermeja is a mountainous area with even higher slopes than those in Aznalcollar and abrupt topography.

4 CONCLUSIONS

From NOAA images we can calculate temperature and NDVI daily values, which are very useful for knowing the vegetation status. Consequently, we can know its ignition capability and foresee a fire occurrence.

With the synoptical maps with a 48 to 72 hours precision, it can be known the meteorological conditions and the type of dominant winds.

The combined action between slopes and winds let us know the spreading speed better than considering those factors individually. The front spreading fire direction can be estimated as the vectorial sum of fire and wind directions.

5 REFERENCES

Cihlar, J.L., Laurent, St. & Dyer, J.A.; 1991. Relation between the Normalized Difference Vegetation Index and ecological variables. *Rem. Sens. Environm.* **35**:279-298.

Cohen, W.B. 1991. Response of vegetation indices to changes in three measures of leaf water stress. *Photogram. Eng. Rem. Sens.*, **57**:195-202.

Deblonde, G. & Cihlar, J.L. 1993. A multiyear analysis of the relationship between surface environmental variables and NDVI over the Canadian landmass. *Rem. Sens. Reviews*, 7:151-177.

Desbois, N. & Vidal, A. 1995. La Télédétection dans la prévision des incendies de forêts. *Ingenieries- EAT*, **1**:21-29.

Gandía, S. 1991. La teledetección en el seguimiento de las superficies cultivadas. *La Teledetección en el seguimiento de los fenómenos naturales. Recursos renovables: Agricultura.* Curso de postgrado. Depto. de Teledetección, Universidad de Valencia.

Goward, S.N. & Hope, A.S. 1989. Evapotranspiration from combined reflected solar and emitted terrestrial radiation: preliminary FIFE results from AVHRR data. *Ad. Space Res.* **9**:239-249.

Gross, M.F., Hardisky, M.A. & Klemas, V. 1988. Effects of solar angle of reflectance from wetland vegetation. *Rem. Sens. Environ.* **26**:195-212.

Hernández, E.; Díaz, J.; Cana, L.C. & Martín, I.J. 1995. Incendios forestales y meteorología de montaña. *Mundo Científico*, **155**:254-258.

Sanz, A. & Giráldez, J.V. 1999. Temperature maps obtained from satellite images for a forest wildfire temporal evolution analysis. In press.

Vidal A., Pinglo, F., Durand, H., Devaux-Ros, C. & Maillet, A. 1994. Evaluation of temporal fire risk index in mediterranean forest from AVHRR thermal IR. *Rem. Sens. Environ.* 49:296-303.

Remote Sensing in the 21st Century: Economic and Environmental Applications, Casanova (ed.)
© 2000 Balkema, Rotterdam, ISBN 90 5809 096 5

Temperature maps obtained from satellite images for a forest wildfire temporal evolution analysis

A. Sanz Redondo
Departamento de Ingeniería Cartográfica, Geodésica y Fotogrametría, Universidad de Castilla La Mancha, Spain

J.V. Giráldez Cervera
Departamento de Agronomía, Universidad de Córdoba, Spain

ABSTRACT: The conversion of Digital Counts (DC) to surface temperature, for Thermal Infrared wavelenghts in TIROS-NOAA satellites, requires three steps: 1) calibration; 2) conversion of radiance to true temperature; 3) soil temperature after atmospheric and emissivity correction. The transformation of DC to radiance of the scene depends on the blackbody internal temperature and the representative cold space view sensed with the platinum resistance thermomethers. Header words 18, 19 and 20 each contain those measures and header words 23 to 52 provide ten internal target data each for Infra Red channels 3, 4 and 5. The conversion to brightness temperature from radiance is performed using the inverse of Planck's radiation equation. The non-linearity in the calibration is accounted for through the addition of a correction term to the brightness temperature of scene. From true brightness temperature, the split-window algorithm, determines the surface temperature. This methodology is applied to a group of images dated before, during and after a fire in Aznalcóllar (Spain) for a temporal evolution analysis.

1 INTRODUCTION

The forest wildfire high risk in Spain is due to human, ecological, meteorological and topographical factors. If a wildfire starts in an abrupt, risky area, the use of several variables, measured in meteorological stations, which are generally far from the burning area, are very sparse for good risk estimation indices. So, Remote Sensing is considered to be a good technique to support that daily information (NOAA images) of extensive forest areas.

The high spatial, spectral and temporal resolution required by forest wildfire studies justify the use of AVHRR/2 sensor, in NOAA satellites, to determine soil temperature.

The soil temperature calculated from terrestrial radiation measures between the intervals of wavelength (3.7 and 4.1μm) and (10.5-12.5 μm), is not the real temperature, due to the atmospheric influence over the radiation in its path to the sensor. As the natural surfaces do not behave as a blackbody, this measures must be also corrected by emissivity.

In late spring, May 30 to June 4, a wildfire broke burning an area of 1800 ha in Aznalcóllar (Sevilla). The temperature maps obtained from NOAA images dated before, during and after the fire will permit to analyse the temperature increase evolution in those days.

2 METHODOLOGY

The conversion of Digital Counts (DC) to surface temperature, for Thermal Infrared wavelenghts in TIROS-NOAA satellites, requires three steps: 1) calibration; 2) conversion of radiance to true temperature and 3) soil temperature after atmospheric and emissivity correction.

2.1 *Calibration*

The pre-launch calibration relates the AVHRR's output, in digital counts, to the radiance of the scene.

Prelaunch calibrations of the infrared and microwave channels are carried out in a thermal chamber to minimize absorption of radiation in the path between the source and the radiometer and to simulate conditions in space. The radiometer sequentially views the warm calibrated laboratory blackbody, a blackbody cooled (aproximately 77 K) representing the cold space view, and its own internal blackbodies. Temperatures of all blackbodies are sensed with platinum resistance thermometers (PRT's).

The pre-launch characteristics of radiometers, such as their signal-to-noise ratios, stability, linearity of response and sensivity, do not remain unchanged with the time, that's why it is on board sensor calibration, when the image is acquired, is necessary.

To characterize the calibration when the sensor is in orbit, the only data available are those acquired when the AVHRR views space and the internal blackbody. This gives two points on the calibration curve, sufficient to determine the slope and intercept of the linear aproximation applied to determine the scene radiances:

$$N=M*X+I \qquad (1)$$

where N is the radiance observed, X the digital count, M the slope and I the intercept.

$$M_i=\frac{N_{Ti}-N_{SPi}}{\overline{X}_{Ti}-\overline{X}_{SPi}} \qquad i=3,4,5 \qquad (2)$$

where N_{SP} is the radiance of deep space, N_T is the radiance when the instrument views its internal radiance calibration target, and X_{SP} and X_T are the mean counts associated with several observations of space and the internal target, respectively. The number of observations in each case is sufficient to effectively eliminate the residual variances in N_T and X_T. The intercept (I) is calculated for each channel from the equation:

$$I= N_{SP}-M*X_{SP} \qquad (3)$$

The information required to produce AVHRR IR channel calibration coefficients is located in the 103-word HRPT header. The 30 words of internal target data (header words 23-52) provide 10 samples each for IR channels 3, 4 and 5, whose mean value is called X_{Ti}. The 50 words of space view data (header words 53-102) provide 10 samples each for all five AVHRR channels, whose mean value is called X_{SPi}. (Lauritson et al. 1988).

2.1.1. *Internal blackbody radiance calculate.*

To calculate the internal blackbody radiance, it is first necessary to compute the target temperature. The conversion of PRT mean counts to temperature uses the following:

$$T_j=\sum_{j=0}^{4}a_{ij}*\overline{X}_i^j \qquad (4)$$

where X_i is the mean count for PRTj, i = 0, 1, 2, 3, 4; aij are the coeficients of the conversion algorithm; and T_j is the temperature of the internal blackbody calculated from PRT_j.

They are three values of X_i for each PRT. This values are located in the 18, 19 and 20 header words. This values are repeted each four frames (one of each of the four PRT´s) from reference value.

The table I shows a practical example to the model image (05/27/1995).

Table 1. Values for each PRT

	18	19	20	Xi
REF	1	2	2	-
PRT1	256	257	256	257.3
PRT2	268	268	268	268
PRT3	235	235	236	235.3
PRT4	234	233	233	233.3
REF	1	2	2	-

The reference values are easily identified because are less than 10.

The internal blackbody average temperature is:

$$\overline{T}=\sum_{i=1}^{4}b_iT_i \qquad (5)$$

where b_i is the weighting factor of each PRT. The conversion of the average temperature to radiance units is performed using Planck´s equation:

$$\overline{N}_T=C_1v^3\Big/\left[\exp\left(C_2v\big/T\right)-1\right] \qquad (6)$$

v is the central wave number of the channel (cm^{-1}) and C_1 and C_2 are constants (C_1=1.1910659 x 10^{-5} $mW/(m^2srcm^{-4})$ and C_2=1.438833 cmK).

2.2. *Determination of true brightness temperature*

The conversion of the radiance to brightness temperature is performed using the inverse Planck´s equation.

$$T(E)=\frac{C_2v}{\ln\left(1+\frac{C_1v^3}{E}\right)} \qquad (7)$$

The non lineality in the calibration is accounted for through the addition of a correction term to the brightness temperature of the scene. The appropiate correction term is determined by double interpolation in a double input table: the brightness temperature of scene and the internal blackbody temperature.

The Figure 1 shows the designed diagram to automatice the process. Each input image pixel has a brightness temperature value between 245 and 320K. The conditional function I (Figure 1) classifies, to 5 K intervals, these values to classes 0, 1, 2, ...8. Each new values, correspond the a_i and b_i coefficients of lineal interpolation equation $a_{ij}x+b_{ij}$; each scene brightness temperature calculated intervals (j=0, 1,

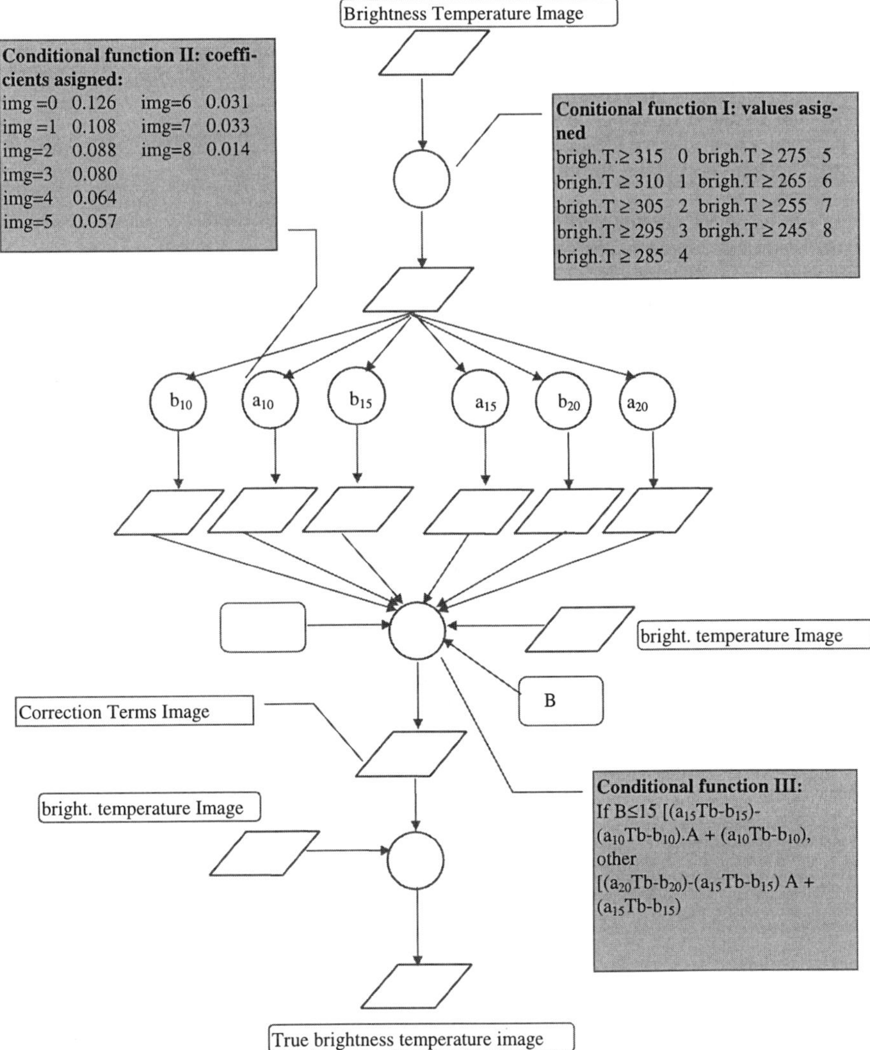

Figure 1: Diagram to calculate the true brightness temperature image.

Table 2: Interpolation equation coefficients.

Scene T	Internal blackbody temperature K channel 4							
	10°C		15°C		20°C		25°C	
°K	a_{10}	b_{10}	a_{15}	b_{15}	a_{20}	b_{20}	a_{25}	b_{25}
320-315	0.126	37.11	0.098	28.48	0.11	32.93	0.096	28.8
315-310	0.108	31.44	0.09	25.96	0.088	26	0.09	26.9
310-305	0.088	25.24	0.104	30.3	0.096	28.48	0.092	27.5
305-295	0.08	22.8	0.089	25.72	0.067	19.63	0.068	20.2
295-285	0.064	18.08	0.076	21.89	0.065	19.05	0.054	16.1
285-275	0.057	16.08	0.061	17.62	0.053	15.62	0.049	14.7
275-265	0.031	8.925	0.013	4.415	0.014	4.9	0.013	4.76
265-255	0.033	9.455	0.023	7.065	0.034	10.2	0.027	8.47
255-245	0.014	4.61	0.02	6.3	0.005	2.805	0.003	2.35

2,...8) for each three or four internal blackbody temperatures (i=10, 15, 20 and 25) satellite NOAA9 and 12 respectivelly. The conditional function IIi assigns these values to classes before called.

The table 2 shows the coefficients values to channel 4 of the example image (05/27/1995).

The other tables corresponding to NOAA9, 10 and 12 satellites to channels 4 and 5, are published (Sanz 1997).

The next equation interpolates to the several values internal blackbody temperatures:

$$\left[(a_i\,Tb + b_i) - (a_{i-1}\,Tb + b_{i-1})\right]A + \left[a_{i-1}\,Tb + b_{i-1}\right] \quad (8)$$

where A is a positional coefficient of the internal blackbody temperature in the interval [i-1,i], divided into 50 parts and B is the internal blackbody temperature value between the limits [10, 15] or [15, 20].

The diagram conditional function III calculats the corrected terms by pixel which will be added to the brightness temperature image to obtain the true brightness temperature one.

From the launching of NOAA13 satellite and the following ones, the calibration is different. The conversion of the DC to radiancia is performed by using equation (1), but the slope (M) and interception (I) coefficients are calculated using the deep space radiance particulary for each channel and different from zero.

$$R_{lineal} = M*X+I \tag{9}$$

The non-lineality correction is performed by calculating a corrected radiance from lineal radiance. The corrected coefficients are tabulated (Kidwell, 1995):

$$RAD = A*R_{lineal} + B*R_{lineal}^2 + C \tag{10}$$

The conversion of this radiance to true brightness temperature is performed using inversa Planck's equation.

2.2 *Determination of soil temperature*

The surface temperature is calculated from split-window algorithm proposed by Coll & al. 1994:

$$\begin{aligned} &T = T_4 + A(T_4 - T_5) + 0.51 + B(\varepsilon) \\ &A = 1 + 0.58((T_4 - T_5) \\ &B(\varepsilon) = \alpha(1-\varepsilon) - \beta\Delta\varepsilon \end{aligned} \tag{11}$$

where ε is the semiaddition and $\Delta\varepsilon$ is the difference between ε_4 and ε_5. The term $B(\varepsilon)$ requires to know the water vapour content (W) in the atmosphere to calculate the coefficients α and β; and the emissivity in the channels 4 and 5.

To evaluate α and β, Caselles et al. 1993, using the LOWTRAN7 model (Kneizys et al., 1988) applied to a temperature and relativity humidity group of vertical profiles, from different standard atmosphere and a radio-sounding group registered by the Meteorology National Institute. The result is a graphic where α and β are easily calculated when the W is known.

The surface emissivity is calculated using Caselles & Sobrino, 1989 equation:

$$\begin{aligned} &\varepsilon = \varepsilon_0 + d\varepsilon = ai + b \\ &a = \frac{\varepsilon_v - \varepsilon_s}{i_v - i_s} \\ &b = \frac{\varepsilon_s(i_v + di) - \varepsilon_v(i_s + di)}{i_v - i_s} + d\varepsilon \end{aligned} \tag{12}$$

$$di = i\text{-}is\text{-}Pv(iv\text{-}is) \tag{13}$$

$$P_v = \frac{(r_{2s} - r_{1s}) - i(r_{2s} + r_{1s})}{i(r_{2v} + r_{1v} - r_{2s} - r_{1s}) + (r_{1v} - r_{2v} + r_{2s} - r_{1s})} \tag{14}$$

where ε_v and ε_s are the vegetation and soil emissivity estimated by referenced values or radiometer measured; i_v and i_s are the vegetation and soil index of a pure pixel; P_v is the percentage of vegetation in a pixel; r_{iv} and r_{is} are the vegetation and soil reflectivities in channel i; and $d\varepsilon$ is a reflection term.

The ε_v and ε_s values are estimated from Salisbury (1992) studies depending on vegetation and soil classes present in the study area, and the $d\varepsilon$ term is calculated from Valor's (1994) experiments.

The Figures 2 and 3 show the used models to calculate the channel 4 and 5 emissivities and soil temperature.

3 RESULTS AND DISCUSION

Those models are applied into a group of images dated from May 27 to June 5 and the results are shown in Figures 4, and 5. In these maps, the fire contour is drawn to show the clear temperature increase of afected area with respect the surrounding areas and to throw the high values of burning pixels into relief.

The estimated temperature has not been contrasted with true values because there are not ground data due to problems in accessing during and after the fire, because of not being able to correlate them with meteorological stations data, usually far from the affected areas.

Data from radio-sounding are valuable for a 100 km distance, and from them on the meteorological measured variables do not characterize the atmosphere profile. Despite of exceeding this value between Gibraltar and Aznalcóllar, those data have been used as they are the nearest to our study area.

The emissivity vegetation and soil values should indicate pure pixels, but because of the spatial NOAA resolution this condition is not followed. Moreover, different vegetation species emissivity was averaged without considering each ones percentage in the pixels.

To finish, a standard aerosols model was applied

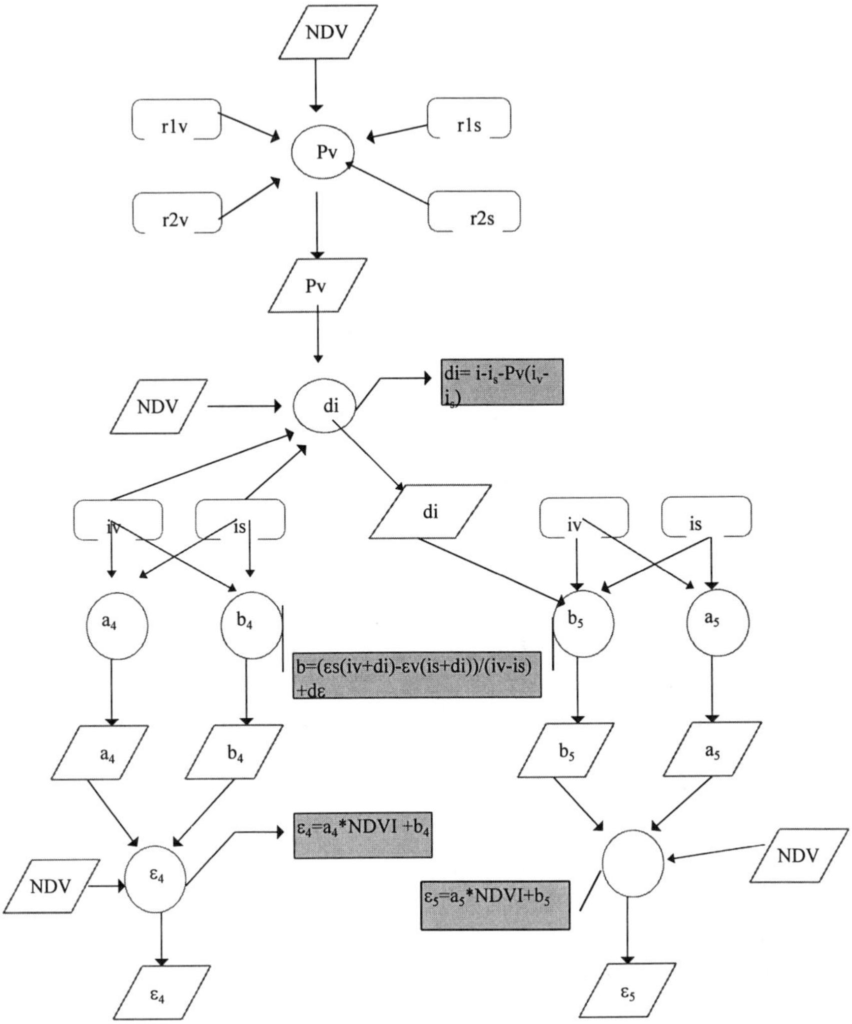

NDV

r1v

r1s

Pv

r2v

r2s

Pv

NDV

di

$di= i-i_s-Pv(i_v-i_s)$

iv

is

di

iv

is

a_4

b_4

b_5

a_5

$b=(\varepsilon s(iv+di)-\varepsilon v(is+di))/(iv-is)+d\varepsilon$

a_4

b_4

b_5

a_5

$\varepsilon_4=a_4*NDVI+b_4$

NDV

ε_4

NDV

$\varepsilon_5=a_5*NDVI+b_5$

ε_4

ε_5

Figure 2: Diagram to calculate channels 4 and 5 emissivities

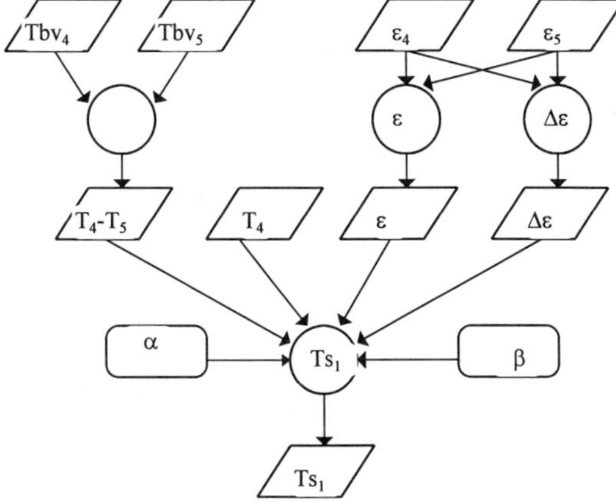

Figure 3: Diagram to calculate the soil temperature

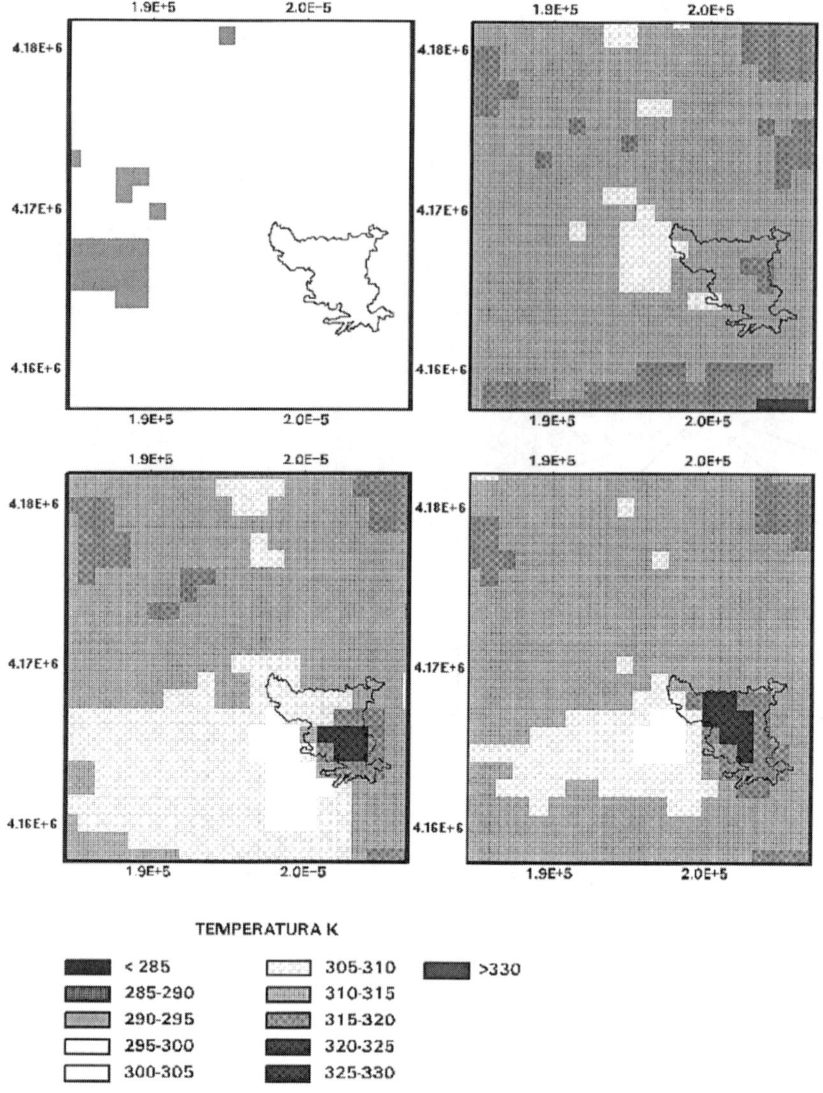

Figure 4: Soil temperature images corresponding the fire occurred in Aznalcóllar (from top to bottom and left to rigth May 27, 29 and 31 and June 1 in 1995).

instead of a specific one according to the atmosphere conditions during the fire.

All those limits make temperature maps are just approximated but useful for our purpose as we are not looking for exact values but their evolution before, during and after the fire.

4 CONCLUSIONS

The scarce number of meteorological stations and their failures, mainly during mid-day, make correlation between measured and satellites temperatures difficult.

At present days there are several atmospheric correction models to forecast temperature with less than 1 degree errors if the surface emissivity and wet atmosphere are known. Those models need data day-time and night-time radio-sounding, aerosols information, meteorological maps and use-land maps to know the vegetation distribution.

Having an automated process for determining soil temperature to study itd evolution, is more important than exact values of this temperature in order to

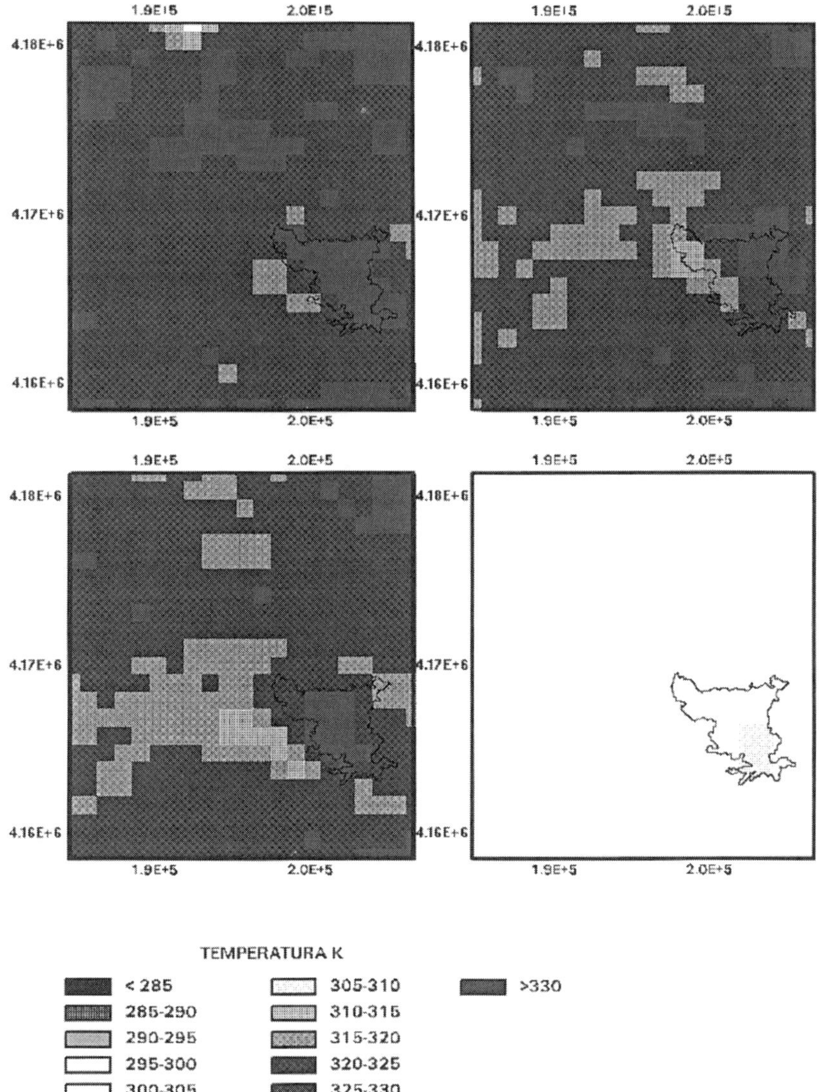

Figure 5: Soil temperature images corresponding the fire occurred in Aznalcóllar (from top to bottom and left to rigth June 2, 3, 4 and 5 in 1995).

forecast high risk burning areas.

In our study case, temperature increases gradually before the fire days and up to it. To be precise, the previous day temperature increased 10 degrees.

5 REFERENCES

Caselles, V. & Sobrino, J.A. 1989. Determination of frosts in orange groves from NOAA-9 AVHRR data. *Rem. Sens. Environ.*, 29:135-146.

Caselles, V. , Coll, C. & Valor, E. 1993. Un método operati vo de determinación de la temperatura de la superficie del suelo. *Revista de Teledetección* 2: 19-26.

Kidwell, K. B. 1995. *NOAA Polar Orbiter data users gui de.*NOAA Information Service National Climatic Data Center. Washington, D.C.

Kneizys, F.X., Shettle, E.P., Abreu, L.W., Anderson, G.P., Chetwynd, J.H., Gallery, W.O., Selby, J.E.A. & Clough, S.A. 1988. *Users Guide to LOWTRAN 7*. Technical Re port AFGL-TR-88-0177, Optical Infrared Technology Di vision, U.S. Air Force Geophysics Laboratory.

Lauritson, L., Nelson, G.J. & Porto, F.W. 1988. Data extrac tion and calibration of TIROS-N/NOAA, radiometers. *NOAA Technical Memorandum NESS 107.* National

Ocea nic and Atmospheric Administration, U.S. Dpto. of Commerce. Washington.
Salisbury, J.W. 1992. Emissivity of terrestrial materials in the 8-14 μm atmospheric window. *Rem. Sens. Environ.*; 42:83- 106.
Sanz, A. 1997. *Teledetección en incendios forestales*. Tesis doctoral. Universidad de Córdoba, España.

Remote Sensing in the 21st Century: Economic and Environmental Applications, Casanova (ed.)
© 2000 Balkema, Rotterdam, ISBN 90 5809 096 5

Automatic mapping of large scale forest fires in Spain using Landsat-TM images

E. Quintano, J.A. Delgado, A. Fernández & P. Illera
Department of Applied Physics I, Remote Sensing Laboratory, University of Valladolid, LATUV, Spain

ABSTRACT: The use of Remote Sensing allows for the estimation and the cartography of areas affected by forest fires. For the inclusion of a cell in the burnt category, visual inspection techniques or soil data have been used to obtaining therefore a dependent value of the area and the study date. In this endeavour two techniques are applied on three Landsat-TM images that allow for the designation of threshold change in a statistically objective manner not dependent on neither the location nor the time of occurrence of the fire. These techniques, previously developed by the Department on NOAA-AVHRR images consist of the application of the differences method and the regression method on previous and subsequent NDVI values in the course of the fire. Secondly, the level of damage caused by the fires was estimated.

1 INTRODUCTION

Around 50,000 fires are produced in the Mediterranean area each year which affect about a million ha. Only in 1994, 184,193 ha burnt in Spain, of which 50% were fires greater than 10,000 ha. The effort demonstrated by the Official Services in prevention work, detection and extinction has been outstanding, applying the most varied and sophisticated techniques. Nevertheless, the preparation of fire statistics still presents some important gaps, despite the fact that Spain was the first country to devise it in 1968.

Estimating the area affected by large forest fires has improved with the incorporation of the GPS technique, although the unburnt interior areas are excluded from the total calculation and a flight is required for each one of the fires taking place.

The same thing does not occur with the quantification of the level of damages which, currently, is achieved by means of considering the effect of factors like the slope, the density of plant cover, etc.

The employment of remote sensing allows to overcome these two inconveniences and determine the affected area by the fire and the damages caused, as has been already shown by various authors, employing not only images taken by the NOAA's AVHRR sensor (Malingreau et al., 1985; Cahoon et al., 1992; Kennedy et al., 1994), but as well as by the Landsat's TM sensor (Kinnaird, A., 1986).

Firstly, various techniques of change detection have been dealt with in the cartography of the burnt areas, some of them start from the use of a single channel where the contrast of the burnt vegetation with the rest of the plant cover is more obvious. (Bruzzone and Serpico, 1997). The images classification methods, both supervised and unsupervised have been, likewise, used by means of the variation in the number of cells which each class undergoes after the fire. For the study on forest fires, detection has been usually based on the following bands: TM3, TM4, TM5 (Setzer et al., 1994; Pereira et al., 1995). Other authors have preferred the use of the KT diagram (Patterson and Yool, 1998) or the principal components analysis (PCA) thanks to its ability to emphasise the burnt areas (Siljestrom et al., 1995). However, Siljestrom did not obtain the expected results, achieving greater precision with the difference or the ratio between the date before and after the fire. The method of the difference between NDVI values before and after the fire has offered satisfactory results as demonstrated in various works based not only on the information obtained from NOAA (Kasischke et al., 1993) but as well as from Landsat (Viedma and Chuvieco, 1997).

The main problem found in the distinct techniques applied to the cartography of burnt areas is the determination of the change detection threshold. The inclusion or exclusion of a cell in the burnt category is usually carried out by means of ground information in a supervised classification or as well in the same images during a recurrent process from which a threshold with a low level of noise is selected as adequate.

On this current work, two algorithms of change

detection have been used. The first is based on the aforementioned method of the NDVI values difference and the second on the regression technique that allows to build a theoretical NDVI. The inclusion of a cell in the burnt category is conducted using statistical criteria, that is why the changes threshold is independent from the rest of the components in the image. These techniques were put forward by the NOAA's AVHRR sensor, obtaining good results (Fernandez, 1997).

In the second section, the quantification of the damage level was obtained through the definition of the Severity Index, based on the normalised difference of the NDVI values before and after the fire. This index was proposed in a previous work aimed at determining the danger of hydric erosion after the the fire (Ruiz Gallardo, 1997).

2 MATERIAL

To conduct this work, various images taken by the Landsat's TM sensor during the summer of 1994 were used. The area covered by the study comprises the south of the province of Castellón and the north and centre zone of the province of Valencia. The reason for selecting this area is due, on one hand, to the magnitude of its fires which substantially exceeded 500 ha and, on the other hand, it allowed comparison of previous works initiated by the same department. The events on which the study focused on were large scale forest fires situated in the following areas: (1) Alcublas (10-August), (2) Altura (10-August), (3) Loriguilla-Calles (2 and 22-July), (4 and 5) Requena (5 and 31-July), (6) Millares (4-July).

To attain the proposed objectives three Landsat images taken by the TM sensor throughout the summer period were used. The actual dates were : May 28, June 29, August 16. During this period 208 fires took place (the majority of which did not exceed 1,000 ha.), which, according to official statistics, destroyed 86,261 ha.

The spectral information finally employed, was the existing one in the TM3, TM4 bands with which the NDVI was calculated for each of the three dates, and in the TM7 which made the radiometric correction of the images possible.

Being a multitemporal study it was essential tackling a correction phase that would permit, on one hand, to improve the radiometric quality of the images and, on the other hand, the superimposition of the same. To eliminate the atmospheric effect, the "Correction of the Histogram by its minimal values" (Chavez, 1988) method was used, which is based on the strong absorption which characterizes areas of water or in shade that should show ND values close to zero in absence of atmospheric effect. To make the superimposition of the images possible, in the first place, the solar inclination angle was corrected by the cosine method and, in the second place, each one of the three images was geometrically corrected through the selection of 77 control points. In both cases, the image used for reference was the one from June 29. The reason for this high number of points is found in the amplitude of the altitudinal range of the study area (200-1300 m.).

3 METHODOLOGY

3.1 *Estimation of the area affected by the fire*

The two methods proposed for the demarcation of the area affected by the fire start from the marked decrease that the vegetative activity undergoes during a fire which produces NDVI values close to zero. The NDVI calculation is, therefore, the starting point for estimating the burnt surface. Once this phase was concluded, the two algorithms of change detection were applied on each pair of images.

3.1.1 *Differences method*

It is supported by obtaining the image resulting from the difference between the NDVI before and after the fire. Thus the pixels that correspond to burnt zones will be those where the decrement of the NDVI has been greater and displays higher values in the image resultant from the difference. The steps to delimit the affected areas were:

1 Calculation of the average (μ_{B-A}) and the typical deviation (σ_{B-A}) of the image resulting from the difference of the NDVI values before and after the fire.
2 Determining the threshold of detection of changes by way of the algorithm:

$$NDVI_B - NDVI_A > \mu_{B-A} + 2\sigma_{B-A} \quad (1)$$

3.1.2 *Regression method*

This method considers that one pixel has been affected by the fire if its NDVI value is very inferior to the obtained through regression for this date.

The steps followed were:

1 Calculation of the regression equation starting from the NDVI values before and after the fire.
2 Obtainment of the NDVI values likely present of not having produced any fire.
3 Estimation of the pixels affected by the fire, through the relation:

$$NDVI_{After} < NDVI_{After\ regression} - 2S \quad (2)$$

S being the error of regression.

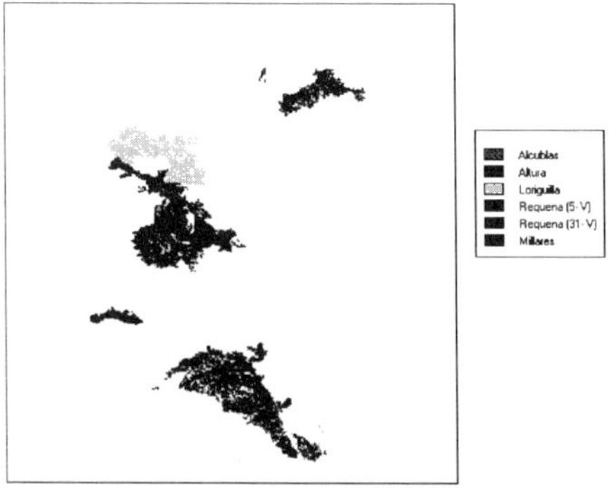

Figure 1. Areas affected by fire according to the differences method.

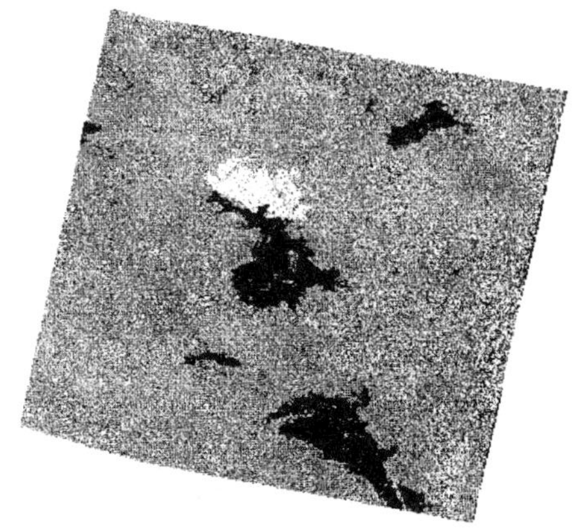

Figure 3.Image of Severity Index for the May to August period

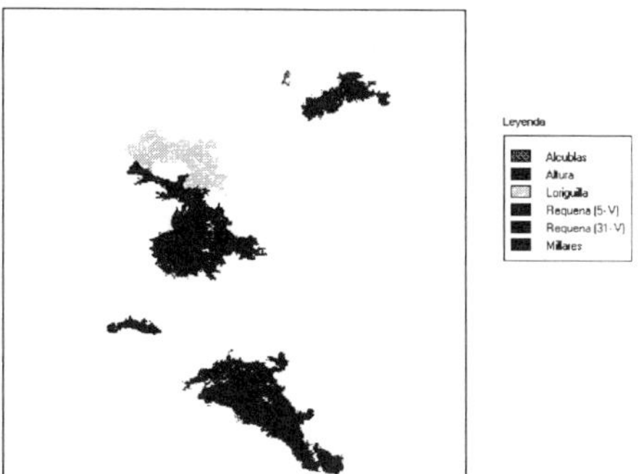

Figure 2. Areas affected by fire according to the regression method.

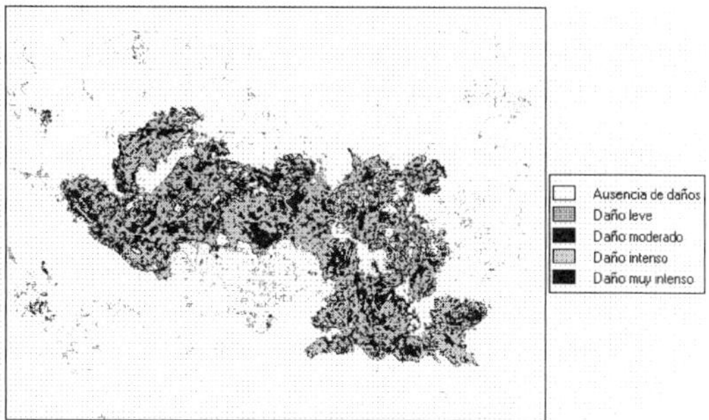

Figure 4. Categories of Damage Intensity for Loriguilla fire.

3.2 *Quantification of the level of damages*

To determine the level of damages associated with each fire, the normalised difference of the NDVI values before and after the disaster were used. The normalisation of said expression tried to facilitate the subsequent process of classification in distinct levels of damage. The general expression of the applied "Severity Index" was the following:

$$IS = \frac{(NDVI_{BEFORE} - NDVI_{AFTER})}{(NDVI_{BEFORE} + NDVI_{AFTER})} \qquad (3)$$

The value of this index in absence of any variation is equal to zero. A deviation in this behaviour supposes the presence of more striking changes the further they are from the origin. Nevertheless, the natural variation of the plant cover causes the value associated with the category of the absence of damages to be put slightly forward of the zero value.

The steps followed to determine the level of damages were:

1 Obtainment of the Severity Index images (IS) through the expression (3) for each of the periods considered.(Figure 3)
2 Elimination of the burnt areas in each image to avoid that the strong decline experienced by the NDVI in the same, would indicate major and minor damages of those really existing.
3 Calculation of the average value (x) and the typical deviation (σ) of the distinct IS images.
4 Classification of the distinct damage categories, considering that one cell had been seen damaged by the fire when its value surpassed the interval defined by (x + σ). This interval was selected to limit the variation considered as normal that the NDVI experiences from one date to the other.(Figure 4) The final division remained as follows:

- Absence of damages: (x + σ)
- Slight damage: (x + 2σ)
- Moderate damage: (x + 3σ)
- Intense damage: (x + 4σ)
- Very intense damage: rest of values

4 RESULTS

ICONA (Institute for Nature Conservation) calculates the surface covered by a fire using as a limit the fire's outer boundary, without taking into account the inner areas which may have been left intact. Through the data gathered by the Landsat's TM sensor it is possible to delimit these zones and exclude them from the total calculation. For this reason, these two situations have been distinguished in the final estimation of the area affected by the fire.(Figures 1 and 2).

When the affected area was estimated from the outer perimeter of the fire, in general, errors in the order of 4% by the differences method and 7% by the regression method were reached, taking always the ICONA data as a base. Nevertheless, in three of the fires these levels were substantially exceeded, reaching levels close to 30% or even 50%. (Table 1). These deviations appeared to be associated with two of the fires, (1) and (3), jointly detected, which supports the difficulty in assuming the estimation of the affected area in fires occurring in neighbouring areas by whichever of the two algorithms.

The results obtained got worse, picking up errors in the order of 15% for the differences method and 6% for the regression one, excluding from the total calculation the unaffected inner areas. (Table 2).

When the verification was carried out using as a base the surfaces taken from the GPS technique, better approximations were obtained, especially in the case of the Millares fire. (Tables 3 and 4).

Table 1.- Burnt areas taking into account the outer perimeter.

ID	Burnt area			Relative error	
	ICONA	Differenc	Regress	Differenc	Regress
	(ha)	(ha)	(ha)		
1	268+256	250,4	367,9	52,2	29,8
2	7000	6491,0	7454,9	7,3	-6,5
3	8310,3	11342,5	11908,2	-36,5	-43,3
4	24769,9	25599,7	26248,6	-3,4	-6,0
5	1650	2148,8	2148,8	-30,2	-30,2
6	25930	25439,9	28409,5	1,9	-9,6

(1)Alcublas (10- August), (2) Altura (10- August), (3) Loriguilla- Calles (2 y 22- July), (4 y 5) Requena (5 y 31- July), (6) Millares (4- July).

Table 2.- Burnt areas taking into account the unaffected inner zones.

ID	Burnt area			Relative error	
	ICONA	Differenc	Regress	Differenc	Regress.
	(ha)	(ha)	(ha)		
1	268+256	250,4	366,4	52,2	30,1
2	7000	5986,1	7352,9	14,5	-5,0
3	8310,3	10699,4	114258,2	-28,7	-37,5
4	24769,9	22623,0	25848,5	8,7	-4,4
5	1650	2124,1	2124,1	-28,7	-28,7
6	25930	23477,4	27995,2	9,5	-8,0

Table 3.- Burnt areas taking into account the unaffected inner zones.

ID	Burnt area			Relative error	
	GPS	Differenc	Regress	Differenc	Regress
	(ha)	(ha)	(ha)		
5	1714,3	2124,1	2124,1	23,9	23,9
6	25033,7	23477,4	27995,2	6,22	-11,8

Table 4.- Burnt areas taking into account the outer perimeter.

ID	Burnt area			Relative error	
	GPS	Differenc	Regress	Differenc	Regress
	(ha.)	(ha.)	(ha.)		
5	1714,23	2148,8	2148,8	-25,4	-25,4
6	25033,67	25439,9	28409,5	-1,6	-13,5

Table 5.- Burnt areas according to the different satellites.

ID	ICONA	NOAA		Landsat	
		Differenc	Regress	Differenc	Regress
	(ha)	(ha)	(ha)	(ha)	(ha)
2	7000	6200	5700	6491,0	7454,9
5	24770	23600	24000	25599,7	26248,6
6	25930	27200	27200	25439,8	28409,5

Table 6.- Errors in the estimation of burnt areas

ID	ICONA	NOAA (%)		Landsat (%)	
		Differenc	Regress	Differenc	Regress
	(ha)	(%)	(%)	(%)	(%)
2	7000	11,4	18,6	7,3	-6,5
5	24770	4,7	3,1	-3,4	-6,0
6	25930	-4,9	-4,9	1,9	-9,6

The comparison between the results obtained from Landsat and those procured by Fernandez through the NOAA, did not allow to reach a definite conclusion given the reduced number of fires on which it was based. However, an improvement of the results could be observed utilising Landsat data, especially, for the differences method, and a minor dispersion of the errors committed. (Table 6).

Apart from the study about large scale forest fires, on a second point, the location in the image of the whole fires occurred during the study period, reflected in the official statistics, was dealt with. The inaccuracy of the said geographic location in the official data impeded obtaining the results hoped for. All the fires that affected more than 50 ha were perfectly identified even though on two occasions the window had to be moved towards the sides. The location of the events where the burnt surface did not reach this level was not possible due to its limited significance and the low level precision reached at in the official statistics in the location of the initial focus. The noise level present in the images, especially in the one obtained by regression made the identification of the foci that burned difficult.

With respect to the quantification of the level of damages, from the visual inspection of the images it can be concluded that the cells appearing with a higher level of damage intensity correspond, in its majority, to the those affected by the fire. Likewise, dispersed pixels exist in the image with slight damage, which in a first approximation do not correspond to burnt surfaces, particularly during the period of May and June. In the interior of the areas that burned, pixels appear with absence of damages associated with flooded surfaces or border zones, which put again into prominence the inadequate precision obtained by official statistics.

5 CONCLUSIONS

Once the detailed operations for delimiting the burnt areas were conducted, the following conclusions could be extracted:

With the algorithms developed by Fernandez Unzueta, all the fires greater than 50 ha were easily detected, even though the noise level obtained by the regression method was higher.

The information gathered by the Landsat's TM sensor makes it able to outline, with a high degree of detail, the precise limits of the fire, delimiting the unburnt inner areas such as reservoirs or communication lines, obtained as affected by official statistics. This capacity to exclude inner areas unaffected by the course of the fire, resulted in attaining greater deviations with respect to areas obtained in Forest Fires Reports.

Regarding the estimation of affected areas, given the reduced number of accidents that were able to be studied, it cannot be inferred, with certainty, which of the two methods obtains better approximations, nor which of the sensors turned out to be more accurate. To draw a conclusion of this magnitude, a larger size sample should be relied on.

With respect to determining the damages level, it was verified that the proposed index associated, as it was expected, the categories of greater damage intensity in the areas affected by the fire, and that there were inner zones that remained intact.

6 ACKNOWLEDGEMENTS

The authors wish to thank the Institute for Nature Conservation for having provided the ground data used in this study.

REFERENCES

Bruzonne, L. and Serpico, S.B. 1997. Detection of changes in remotely- sensed images by the selective use of multi-spectral information. *Int. J. Remote Sens.* 18, no.18: 3883-3888.

Cahoon, D. R. Jr., Stocks, B.J., Levine, J.S., Cofer III, W.R. and Chung, C.C. 1992. Evaluation of a technique for satellite- derived area estimation of forest fires. *Journal of Geophysical Research* 97 no.D4: 3805- 3814.

Chávez, P.S.Jr. 1988. An improved dark- object subtraction technique for atmospheric scattering correction of multispectral data. *Remote Sens. Environ.* 24: 459- 479.

Fernández, A., Illera, P., and Casanova, J.L. 1997. Automatic Mapping of surfaces affected by forest fires in Spain using AVHRR NDVI composite image data. *Remote Sens. Environ.* 60: 153- 162.

Kasischke, E.S., French, H. F., Harrel, P., Christensen, JR, N.L., Ustin, S. L., and Barry, D. 1993. Monitoring of wildfires in boreal forest using large area AVHRR NDVI a composite image data. *Remote Sens. Environ.* 45: 61- 71.

Kasischke, E.S. and French, N.H. 1995. Locating and estimating the areal extent of wildfires in Alaskan boreal forest using composite data. *Remote Sens. Environ.* 51: 263- 275.

Kennedy, P.J., Belward, A.S., and Grégorie, J.M. 1994. An improved approach to fire monitoring in West Africa using AVHRR data. *Int. J. Remote Sens.* 15: 2235- 2255.

Kinnaird, A. 1986. The use of remote sensing in mapping and monitoring vegetational change associated with bushfire events in Eastern Australia. *Geocarto International* 1.

Malingreau, J.P., Stephens, G., and Fellows, L. 1985. Remote sensing of forest fires: Kalimantan and North Borneo in 1982- 83. *Ambio* 14: 314-321.

Pereira JR, A.C. and Setzer, A.W.1996. Comparison of fire detection in savannas using AVHRR's channel 3 and TM images. *Int. J. Remote Sens.* 17, no.10: 1925- 1937.

Patterson, M.W. and Yool, S.R. 1997. Mapping fire- induced vegetation mortality using Thematic Mapper data. *Remote Sens. Environ.* 65: 132- 142.

Ruiz, J.R., Castaño S: and Calera A. 1997. Aplicaciones de la teledetección y S.I.G. a la determinación del riesgo de erosión en zonas incendiadas. *Congreso de IRATI.*

Siljeström, P. and Moreno, A. 1995. Monitoring burnt areas by principal components analysis of multi- temporal TM data. *Int. J. Remote Sens.* 16, no.9: 1577- 1587.

Setzer, A.W., Pereira, M.C.,and Pereira A.C.1994. Satellite studies of biomass burning in Amazonia- Some practical aspects. *Remote Sensing Reviews.* 10: 91- 103.

Viedma, O. and Chuvieco, E. 1997. Modeling rates of ecosystem recovery after fires using Landsat TM data. *Remote Sens. Environ.* 61: 383- 398.

Remote Sensing in the 21st Century: Economic and Environmental Applications, Casanova (ed.)
© 2000 Balkema, Rotterdam, ISBN 90 5809 096 5

An operational system for the forest fire risk assessment by means of NOAA images

A.Calle, J.L.Casanova & A.Romo
Remote Sensing Laboratory, University of Valladolid, Spain

F.González-Alonso
INIA, Ministry of Agriculture, Madrid, Spain

ABSTRACT: The model we show in order to obtain a Forest Fire Risk Assessment combines two kinds of analysis involving NDVI and surface temperature. The University of Valladolid Remote Sensing Laboratory (LATUV) is equipped with a NOAA image receiver, and the risk pattern is practically automated; in this way after receiving the satellite image on a daily basis and applying the above-mentioned algorithms, detailed maps are produced with information relating to local distribution over the whole country, with Forest Fire Risk levels.This information is very useful to the Fire Prevention Services of the Autonomous Communities during the summer season, when the incidence of forest fires is high. This Risk Assessment is also determined for Portugal, Italy and the mediterranean areas of France.

1 INTRODUCTION

In Spain, as in all other mediterranean countries, forest fires are one of the main concerns or the *Nature Protection Services*, as a result of the large extent affected every year. Near-drought conditions favour large-scale fires. Between 1880 and 1980 more than half the years were considered to be "dry", and between 1980 and 1992 eight years were deemed "very dry" (*González-Alonso et al., 1994, 1995, 1997*). Communities in the north-west of Spain suffer a large number of small fires, which are more common just before summer. However, the mediterranean Communities such as Valencia and Catalonia are the most affected ones due to large-scale fires. In these Communities, in 1994, fire destroyed more than 190,000 hectares and five fires affected areas of over 15,000 ha. (*Illera et al., 1995*).

Because of their orbital and spectral characteristics, the NOAA series satellites are very suitable for obtaining forest fire risk assessment in real time over large areas. These satellites have a polar orbit and go around the planet in 102 minutes, flying over the same point, at an altitude of 800 kms., twice daily and with a spatial resolution of 1.1 km. at the nadir. In the case of the Iberian Peninsula this takes place at midday (around 13:00 G.M.T.) and in the early morning. This daily periodicity together with a wide width coverage of about 3.000 km. makes these satellites particularly suitable for the continuous monitoring of processes relating to the earth's surface and its vegetation, such as the prevention of forest fires.

The AVHRR sensor of NOAA satellites is equipped with 5 spectral bands, of which two, visible and near infrared (1 and 2), may be combined to obtain an NDVI (*Normalized Difference Vegetation Index*) magnitude; this provides very accurate information concerning the photosynthetic state of the vegetation, and, consequently, its hardiness. Two other bands in thermal infrared (4 and 5) are used to calculate the *surface temperature*, T_s.

The University of Valladolid Remote Sensing Laboratory (LATUV), in collaboration with the Remote Sensing Area of the National Institute of Agrarian Research (INIA), has for several years been working on fire risk indices, incorporating more data and new algorithms (*Illera et al., 1995*) in order to produce a trust worthy real time risk index which can be sent straight away to those organizations responsible for measures to control vulnerable areas.

For this purpose LATUV is equipped with a NOAA receiver and has developed the software necessary for processing the image. LATUV also has a World Wide Web page where risk mapping is updated daily.

2 ALGORITHM TO DETERMINE FIRE RISK

The most important question about forest fire risk calculation is: what does remote sensing measure ?. the answer is the risk of spread, because the risk of spread is directly linked to the appearance of large fires. Although few in number they devastate large areas. The statistic is that in Mediterranean countries almost 90 % of burnt surface results from fires of over 1000 Ha. In this way, the model used to determine the risk index applies an algorithm which combines the analysis of two types of indicators. The first analyses vegetation degradation for that same day with reference to the state of the vegetation, and the second analyses the relation between the vegetation and surface temperature; this relation correlates with the humidity content. These indicators are analysed below.

2.1 *Vegetation condition*

As was mentioned earlier, the photosynthetic state of the vegetation is determined by means of the NDVI (Normalized Difference Vegetation Index), using the reflectances of spectral bands 1 and 2 of the AVHRR sensor. The NDVI is defined as:

$$NDVI = (R_{inf} - R_{red}) / (R_{inf} + R_{red})$$

where R_{red} and R_{inf} are the reflectance values of channels 1 and 2, corresponding to 0.68 μm and 0.72 μm, respectively.

In order to analyse vegetation degradation for the same day on each pixel, it is necessary to compare the given NDVI value with previous data. We have used time series of daily values for two months before the particular day. For this purpose NDVI images have been used and maximum value NDVI-MVC compounds have been produced every 10 days; in this way, every one day is compared with 6 NDVI-MVC images. The data from maximum value compounds eliminate the *noise* caused by unfiltered faint clouds in image processing and compensate the effects of orbital heterogeneity.

Once the previous time series has been obtained, a *parabolic adjustment* is made in each "i" pixel in order to produce a relation of the type:

$$NDVI_i = a_i \cdot t^2 + b_i \cdot t + c_i$$

where $NDVI_i$ is the NDVI regression value of pixel "i" and t is the variable of the past-time day. The a_i, b_i and c_i coefficients are the specific regression coefficients of this pixel. In this way, the *expected* vegetation evolution of the pixel for that particular day, $NDVI_i$ expected, will be given by extrapolating this curve to the today point in time. It should be added that the compound maximum value images are generated on a daily basis, with daily displacement of the components of each 10-days compound. For each pixel i, vegetation degradation VD_i is represented by the value of the difference between the real value obtained on the image of the day in question, $NDVI_{i,current}$ and the extrapolated value, $NDVI_{i,expected}$. That is:
$VD_i = (NDVI_{i,current} - NDVI_{i,expected})$.

This difference is normalized in respect on average provincial differences and its value is analysed in terms of standard provincial deviation.

2.2 *The hydric stress level*

So as to characterize the hydric stress of vegetation, some authors have suggested establishing the linear relation between surface temperature and the NDVI value (*Nemani et al., 1993*). These authors also point out that this linear relation between the two magnitudes is different for the various types of crops. For this reason, it is a good indicator for establishing fire risk levels because the slope of this relation is a indicator of the real evapotranspiration of the vegetation. Acording to *Nemani, et al. 1993*, for dry conditions as the NDVI in a pixel increases, the surface temperature decrease. If the top layer of soil is wetted by rainfall, much of the absorbed solar radiation is consumed in evaporation and the surface temperature provide no distinction between soils and leaves. Values of this slope Ts/NDVI are negative, in the most of cases, because when NDVI increase, surface temperature decrease, near air temperature.

In order to establish the relation between NDVI and surface temperature, the whole of the satellite image coverage has been divided into cells measuring 20x20 km^2. In addition, a linear regression adjustment between the points for each cell has been calculated, giving:

$$Ts = m \cdot NDVI + n$$

It should be mentioned that the points used in the regression have been filtered by means of a reordering process in accordance with a criterion which selects for a particular NDVI the maximum Ts values. This is done in order to extract points with faint clouds which went undetected in initial image processing, as well as to eliminate points which are possibly *contaminated* by cloud shadows. In this way, of the 400 pixels which make up a cell belonging to a geographically corrected NOAA-AVHRR image, filtering picks out around 60 points, which, in addition, increase the correlation coefficient of the above-mentioned linear relation.

Figure 1 shows the Ts vs. NDVI relation of a generic cell measuring 20x20 km. after the application of the point selection filter.

The parameter characterizing the Ts vs. NDVI relation is the regression slope, m, which has negative values in practically all situations. Positive values near zero have been encountered in situations of high humidity and for areas with a high percentage of vegetation, as is the case in northern regions of the Iberian Peninsula around the beginning of springtime. The slope shows very high negative values when the hydric stress of the vegetation in the cell increases.

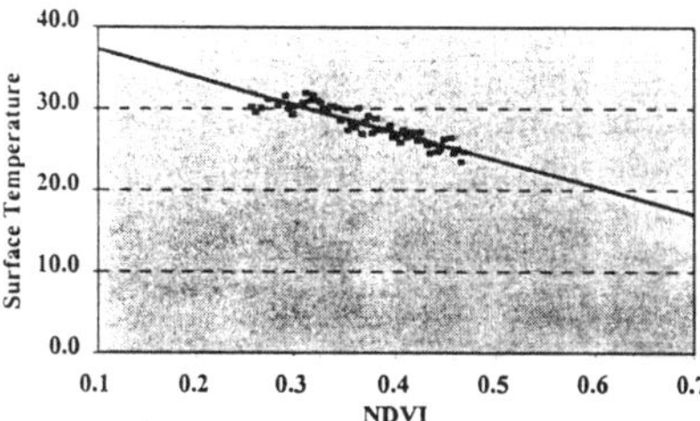

Figure 1. Linear regression between surface temperature and NDVI in a 20x20 km cell, after filtering data.

Figure 2 shows a image of Spain with digital levels corresponding to slope of linear regressions beetwen surface temperature and NDVI in 20x20 km^2 cells.

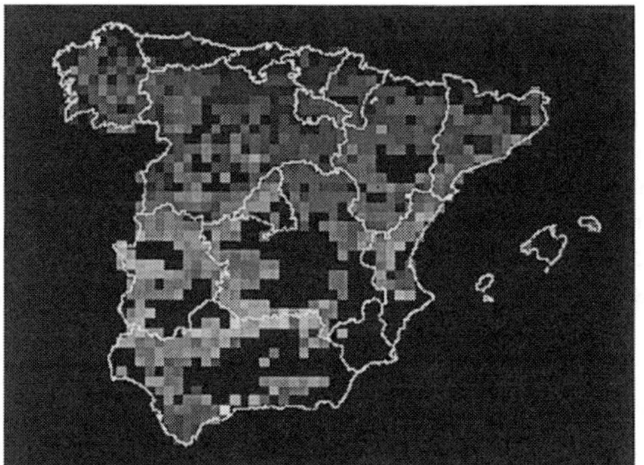

Figure 2. Slopes of linear regression between surface temperature and NDVI in 20x20 km cells throughout Spain country.

3 METHODOLOGY

Since 1993 HRPT format NOAA-AVHRR images have been received at The University of Valladolid Remote Sensing Laboratory.The methodology used to obtain the forest fire risk based on the two indicators mentioned (vegetation condition and hydric stress level) has the following steps:

a) Radiometric and atmospheric correction of channels 1 and 2 of the AVHRR sensor.
b) Daily calculation of the Normalised Difference Vegetation Index (NDVI) from channels 1 and 2 of the AVHRR sensor.
c) Daily calculation of the surface temperature using the algorithm of *Coll et al., 1994*.
d) Geometric correction of NDVI images (*Illera et al., 1996*) and application of clouds mask.
e) Calculation of the composite ten days image corresponding to the maximum value of the NDVI in the ten days period considered. These images are the MVC (maximun value composite) and for each day, the seven MVC previous are calculated moving on one day.
f) Obtaining residual image of vegetation condition; that is: for each pixel the value VD_i, using the algoritm mentioned in the 2.1 paragraph.
g) Obtaining slope image of hydric stress level; that is, for each 20x20 km^2 cell, the m value, using the algorithm mentioned in the 2.2 paragraph.

4 RESULTS: FOREST FIRE RISK MAPS

Despite the fact that all calculations in the process are made with a spatial determination of 1 pixel, our objective has been to obtain trustworthy risk maps for 10x10 km., cells, in concordance with the distribution of fire-fighting staff belonging to the Spanish Nature Conservation Service.

4.1 *Forest cover filter*

The first process undertaken in the analysis of vegetation degradation is to check whether the cell to be analysed may be considered a *forest zone*. For this purpose, we have a land-use map, CORINE, previously degraded to NOAA-AVHRR spatial determination. Of the 64 land uses of CORINE map we have chosen those relating to forests, and we have decided to consider a forest cell that in which at least 30% of its area is occupied by woodland. Map CORINE used for Europe have a resolución of 42 land uses.

A lesser degree of forestation would imply the analysis of areas not likely to burn. In this case (for example, cereal areas during the summer which have become stubble) the risk could be a high one.

4.2 *Quantifying risk assessment*

The first indicator analysed is that relating to vegetation degradation. Here four levels of the $VD_i=(NDVI_{i,current}-NDVI_{i,expected})$ magnitude for each pixel analysed (section 2.1) have been established; these levels are:

Pixel i have risk level L_j

if $VD_i < VD_{average} - A_j \cdot f \cdot \sigma$

with $A_4 = 3.0$; $A_3 = 2.5$; $A_2 = 2.0$; $A_1 = 1.0$;

where $VD_{average}$ is the provincial average of the VD magnitude, f is normalization factor and is the standard deviation of the provincial VD. As may be appreciated, these levels can be associated with the corresponding risk levels, so that N_4 is the number of pixels with level L_4 (an extreme risk situation), N_3 is the number of pixels with level of risk L_3 (a high risk situation), N_2 and level L_2 (a moderate risk) and N_1 and L_1 (a low risk). However, it should not be expected that all the pixels of a given cell are at level 4 for this cell to be considered one of extreme risk; a few pixels at this level are enough for the rest of the cell to be in danger.

In order to establish a criterion with regard to risk determination in a given cell, we have decided on a threshold value of N=10% of a determined level in the cell under study and, of course, we have given priority to the highest risks. The risk criterion regarding vegetation degradation may be resumed as follows:

IF $\sum N_4 \geq N \Rightarrow$ Cell risk = L_4
ELSE IF $\sum N_4 + \sum N_3 \geq N \Rightarrow$ Cell risk = L_3
ELSE IF $\sum N_4 + \sum N_3 + \sum N_2 \geq N \Rightarrow$ Cell risk = L_2
ELSE IF $\sum N_4 + \sum N_3 + \sum N_2 + \sum N_1 \geq N \Rightarrow$ Cell risk = L_1
ELSE $\Rightarrow$ Cell risk = L_0

The criterion for the second indicator, the Ts vs. NDVI slope, m, has been determined as a result of calculations carried out over at least 2 years of study. During this period, m values were determined at different times of the year and for different regions of the Iberian Peninsula. The 4 risk levels have the following m values:

$L_1 \Rightarrow m < -30$; $L_2 \Rightarrow m < -50$;
$L_3 \Rightarrow m < -70$; $L_4 \Rightarrow m < -90$;

The final risk for a 10x10 km^2 cell is given by the average of the two indicators, with priority being given to the vegetation degradation indicator; the cloud filter is more restrictive for the second indicator and occasionally a partially cloudy 20x20 km. cell may not have any regression points when the selection filter is applied; in this case, the risk assigned is the first indicator.

4.3 *The wind factor*

It is very difficult to introduce the wind factor in the algorithm of forest fire risk. Dependence between risk and wind direction is variable with differents places and climatologic situations. But it is easier to analyze the wind speed. The main problem is to get wind data in real time.

To use wind data in real time we obtained these values from TOVS (*Tiros Operational Vertical Sounder*), with HIRS (*High Resolution Infrared Sounder)*, MSU (*Microwave Sounder Unit)* and SSU (*Stratospheric Sounder Unit*) sensors.

Wind data from TOVS images, are obtained using ITPP (*International TOVS Processing Package*) code. These values are available in a spatial resolution of about 100x100 km^2, locating each pixel in geographical coordinates and ITPP provide wind speed and wind direction.

To use these values in our algoritm, we need to build wind field in whole country and in a better spatial resolution using an interpolation model with the inverse of square distance to interpolation point. Figure 3 shows a wind field for a particular day and its meteorological map for comparison.

Wind fields are approximates and they have not perturbations from orographic effects. But there are very good results to provide wind speed.

When wind speed is lower than 3 m/s, the wind factor does not affect to forest fire risk. Between 3 and 8 m/s forest fire risk increase in a linear way (within a scale from 0 to 10, wind factor increase forest fire risk from 0 to 3. Faster than 8 m/s wind factor shows saturation of increase.

4.4 *Integrating risk assessment with territorial data*

Once Risk Index maps have been produced, they are to be sent by fax to the pertinent organizations in each autonomous community. As a result, their presentation should be as accurate as possible and they should be easy to interpret. In this way, the previously described cell network is superimposed on regional maps containing the demarcation of agrarian sectors. As an example, Figure 4 contains daily maps for the

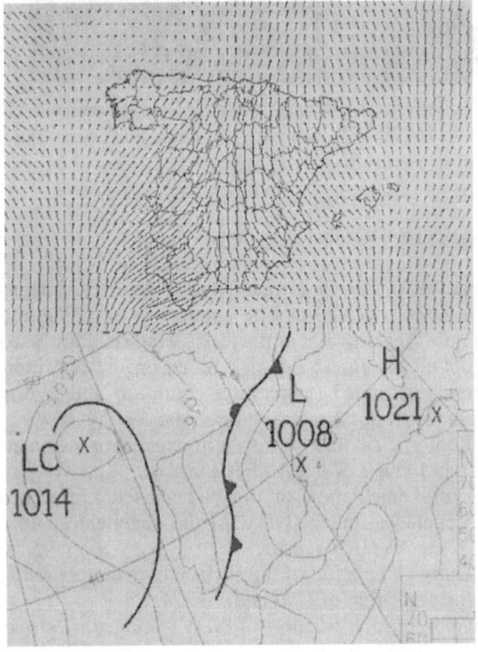

Figure 3. Wind field obtained from NOAA-TOVS and interpolation model; figure shows meteorological map.

autonomous community of Catalonia. Its presentation corresponds exactly to the information sent by fax to all autonomous communities in Spain, and the risks are distributed among the levels: extreme risk, high risk, moderate risk, low risk and without risk. The legend is the following: number 4: cell with exterme risk, 3: cell with high risk, 2: moderate risk, 1: low risk and 0: no risk.The symbol "+" corresponds to forest cells not analysed due to cloud cover, and cells with no symbol are cells without forests.

4.5 *Extension to others countries*

The image from the NOAA-AVHRR sensor picked up daily by LATUV has a width coverage extending from Italy to Portugal (totally inclusive). Depending on the orbital conditions, it is possible to pick up Greece at least 3 or 4 times a week. As regards latitude, the scene goes from the south of Morocco to the British Isles. In addition, our laboratory has a large data base on European land-use distribution thanks to Europe data provided by CORINE, with a resolution of 250 metres. Consequently, we are in a position to produce Risk Assessment maps in all countries sensitive to the problem of forest fires, with the most important characteristic necessary for ensuring effectiveness, that of real time operation.

At the moment, LATUV, besides improving its risk calculation models by introducing new data (digital land model, data from meteorological facilities, etc.) is carrying out risk mapping also for the whole of Portugal and the mediterranean regions of France, where every year forest fires devastate large forested areas. Here analysis is being made of four continental regions: Languedoc Rousillon, Provence-Alpes Cote D'Azur, Rhone-Alpes and the island of Corsica.

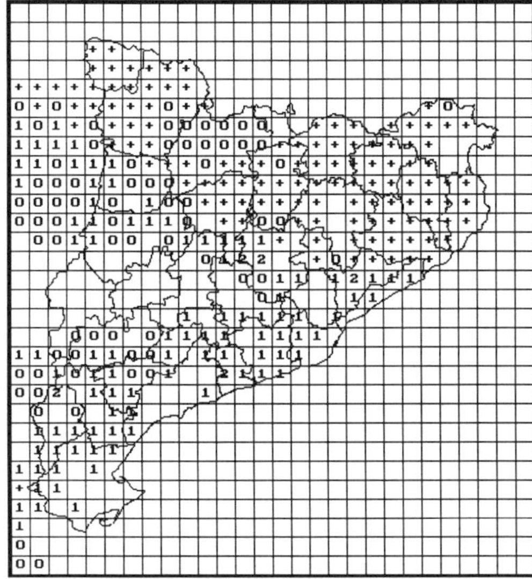

Figure 4. Results: Forest fire risk map for a autonomous community: Catalonia.

5 CONCLUSIONS

A computational model is shown for forest fire risk obtained by means of two indicators: on the one hand, the state of degradation of the vegetation obtained daily via earlier images taken up to two months before, and, on the other, the relation between surface temperature and NDVI, which describes the hydric stress of the vegetation.

The wind speed obtained from NOAA-TOVS in real time is used in the algorithm increasing forest fire risk.

The model presented is currently in operation over all of Spain's autonomous communities, with distribution to each of these taking place on a daily basis throughout the period when most fires occur, namely, march until october. This model is also being used in Portugal, Italy and the mediterranean regions of France. LATUV aims to extend its real time calculation to other mediterranean countries, in particular Greece.

The use of complex systems based on artificial intelligence is the most appropriate method for dealing with all the different factors which influence risk mapping.

6 ACKNOWLEDGEMENTS

This work has been carried out within the colaboration agreement beetwen Remote Sensing Laboratory of the University of Valladolid (LATUV) and the Dirección General de Protección del Medio Natural del MIMA.

7 REFERENCES

-Coll, C., Caselles, V., Sobrino, J.A. and Valor, E. 1994. On the atmospheric dependence of the split-window equation for land surface temperature. International Journal of Remote Sensing, 15, 105-102.

-González Alonso, F., Cuevas, J.M., Casanova, J.L., Calle A. and Illera, P. 1994. Aplicaciones de las Imágenes NOAA AVHRR en el Seguimiento de los Grandes Incendios Forestales en España en el Verano de 1994. Montes, Vol. 38, 1994, pp 26.

-González Alonso, F., Cuevas, J.M., Casanova, J.L., Calle, A. and Illera, P. 1995. Drought Monitoring in Spain Using Satellite Remote Sensing. Sensors and environmental applications of Remote Sensing. Ed. A.. Balkema, Rotterdam, Brookfield 1995. pp 87-90.

-González Alonso, F., Cuevas, J.M., Casanova, J.L., Calle A. and Illera, P. 1997. A forest fire risk assessment using NOAA AVHRR images in the Valencia area, eastern Spain International Journal of Remote Sensing. Vol.18, Nº 10, pp 2201-2207.
-Illera, P., Delgado, J.A. and Calle, A. 1996. A navigation algorithm for satellite images. International Journal of Remote Sensing, 17, 577-588.
-Illera, P, Fernández, A., Calle, A. and Casanova, J.L.. 1995. Evaluation of Forest Fire Danger in Spain by Means of NOAA AVHRR Images. Remote Sensing and Gis Applications to Forest Fire Management EARSeL: Advances in Remote Sensing. Vol. 4, Nº4, November 1996; pp 33-43
-Nemani, R., Pierce, L. and Running S. 1993. Developing satellite-derived estamates of moisture status. Journal of applied meteorology. Vol. 32, pp 548-556.

6 Yield estimation

Remote Sensing in the 21st Century: Economic and Environmental Applications, Casanova (ed.)
© 2000 Balkema, Rotterdam, ISBN 90 5809 096 5

Estimation of cereal yields combining crop growth models and remotely sensed vegetation indices

B.J. Denore & J.L. Meilán – *Independent Consultants, Ibersat, S.A., Valencia, Spain*
J.M. Williams & M. Cole – *Logica UK Limited, UK*
K.J. de Koeijer & J.J. Colls – *University of Nottingham, UK*
C. Trigueros – *Junta de Castilla y León, Spain*

ABSTRACT: A cereal yield estimation system has been developed that incorporates a crop growth model and satellite vegetation indices to estimate areas and yields of wheat and barley at agricultural district level, every month throughout the growing season. The agro-meteorological model simulates biomass accumulation through numerical integration with a daily time-step. Growth is taken as the minimum of water limited growth, calculated from potential water uptake rates and the amount of available water in the rooting zone, and light limited growth, calculated from the fraction of photosynthetically active radiation intercepted by the leaves. The link between this yield model and information from satellite images is founded on the assumption that there is a relationship between the vegetation index derived from the image data and the green leaf area index used in the model. To find the best fit between the model and the image data an optimisation process is used.

1 INTRODUCTION

Traditional predictions of crop yields are either based on sampling surveys or visual inspections of a few sites. The former can be very time consuming, expensive and inaccurate except for the large regions for which the area frame surveys are designed. Visual inspections are highly subjective, though possibly accurate for the specific fields visited. In Castilla y León, Spain's principal producer of wheat and barley, the relevant policy and management unit is increasingly the agricultural district. Such areas are poorly represented in large sampling surveys yet too large for visual inspections to be practical.

This paper presents a methodology that has been used to produce more objective predictions of yields for agricultural districts by combining estimates of crop development and dry matter partitioning provided by an agro-meteorological model with estimates of crop dry matter production from remote sensing data. Within the cereal yield estimation system the crop growth model assumes, in the early stages of crop growth, that crops experience average weather conditions for the region. As the season proceeds real weather data are used to reduce uncertainty of the forecast yields. Also, as crop cover information becomes available from satellite images the numeric values of the parameters used in the model are systematically adjusted to obtain closer agreement between the green leaf area index (GAI) calculated by the model and the vegetation index derived from remotely sensed data.

Sensitivity analysis with 1997 and 1998 data (IRS-LISS, SPOT-HRV, IRS-WiFS) has demonstrated the method's soundness and the importance of 1) the model's parameters and 2) the correct determination of the vegetation index. Further refinements are being made on the basis of 1999 observations.

1.1 *Cereal production in Castilla y León*

Castilla y León is the largest of the 17 Autonomous Communities in Spain and one of the largest regions in the European Union. The region is Spain's primary producer of winter and spring cereals, (in 1994 over 55% of the arable land of the region was dedicated to grain cereals) and agriculture represents the entire primary sector in which 13% of the working population is directly employed (1993 data). It is not surprising therefore that the regional government, the Junta de Castilla y León, is extremely interested in the reliability and accuracy of cereal crop predictions because of the economic and social impact that variations in production imply. The location of the autonomous community of Castilla y León is shown on the map of Spain in Figure 1. Currently yield predictions and field surface estimations are determined in this area by statistical methods, with high errors. This is an expensive procedure that involves field visits and farmers interviews. Further, the results are only reliable for large areas (province level) and the predicted values are dependent on subjective evaluations made by different individuals, which increases their uncertainty.

1.2 *The Cereal YES Project*

The Cereal YES project is supported by the European Commission Centre for Earth Observation (CEO), which is a European initiative to encourage the wider use of information generated by satellites which are observing the planet Earth, its climate, environmental conditions and resources. The project has as its main objective to investigate the use of crop growth models combined with meteorological data and time series of satellite images as an alternative cost-effective method to produce better field and yield predictions for agricultural districts. This methodology has been adapted to cereal models and the resulting system is a tool to assist local policy makers. The target cereals to be monitored are wheat and barley, the predominant cereals of the region, covering 70% of the total cultivated area (1991). The agricultural district of Tierra de Campos has been used for developing the methodology and testing the system.

2 CROP GROWTH MODEL

Agro-meteorological crop models generally describe capture and conversion of resources into dry matter. Under optimal conditions the only limiting resource is solar radiation which determines the potential growth. Conditions for crop growth are rarely optimal, however, and more often low nutrient levels or water shortages affect growth rates. The availability of water and nutrients, therefore, determine the attainable growth (whether water- or nutrient- limited), and this may then be further reduced by pests, diseases, pollution, physical damage by storms or flooding and so on. Available resources and growth reducing factors together determine the actual growth (Lövenstein *et al.*, 1992). The complexity of any such model increases as more (growth limiting or growth reducing) factors are considered.

In Tierra de Campos, variations in cereal yields are known to be highly correlated with rainfall figures. As a result, the availability of water will often be the limiting factor responsible for much of the yield variability in time and space. The crop growth model therefore had to include a module describing water limited crop growth in parallel with a light limited growth module.

The model was designed at the University of Nottingham and has been developed specifically for the region of interest (i.e. Castilla y León). A brief overview of the model and the principles underpinning its design is presented here. A fuller description of the model is to be published elsewhere.

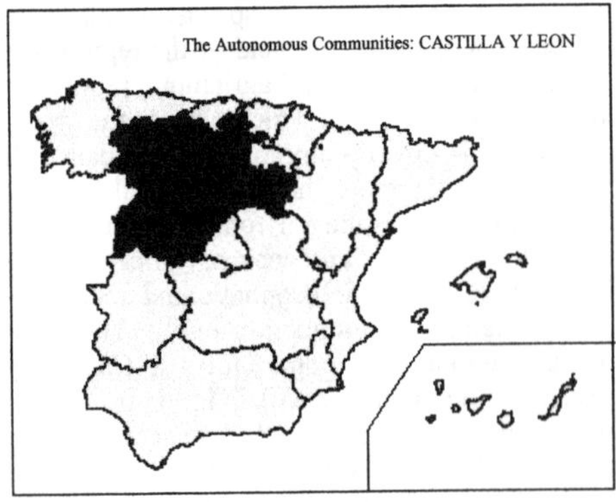

Figure 1 The Autonomous Community of Castilla y León.

2.1 *Model design*

The specific attributes of the model were dictated by the need to produce cereal yield figures for an agricultural district as accurately as possible. The approach adopted was to start from a basic potential growth model of minimal complexity and then identify the main factors influencing the yield variability between seasons (e.g. rainfall) and also the yield variability between agro-ecological zones. Based upon this analysis only the most relevant factors were included within the model helping to minimise its complexity.

The relative homogeneity in soil type, climate and farming practice across the region make it reasonable to treat the entire region as a single agro-ecological zone. However it is necessary to distinguish between irrigated and non-irrigated areas as this has a significant influence on the soil moisture conditions.

A customised light interception and utilisation model was developed for the Cereal YES system by combining a leaf area prediction model for winter wheat with a light interception and water uptake model following the principles of the PARCH model for Sorghum (as developed by Bradley and Crout, 1993).

Both components of the model are driven by weather data. Temperature, rainfall, humidity and solar radiation data are acquired in digital form in near real-time from a local "complete" weather station, while rainfall data are also compiled with an average two week delay from six manual stations in the Tierra de Campos agricultural district. Thirty year long term averages of the data from these stations were taken as the expected weather till the end of the season.

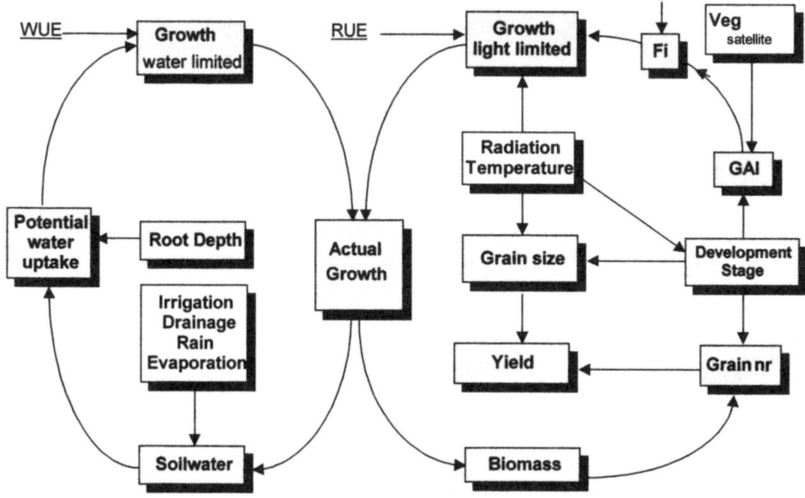

Figure 2 Overview of the crop growth model. The water balance for growth under water limited conditions is represented in the left part of this figure. The right part of the figure represents the light limited growth module.

2.2 *Biomass production*

Biomass production is simulated through numerical integration with a daily time-step. On each day, the resources of light and water, derived from the weather data, are "captured" and "converted" into assimilated dry matter. Depending upon the availability of these resources and the crop's ability to sequester them, growth is considered as either light limited or water limited. Water limited growth (WLG) is calculated from potential water uptake rates and the amount of available water in the rooting zone. Light limited growth (LLG) is calculated from incoming photosynthetically active radiation (PAR) and the fraction of this radiation which is intercepted by the leaves. LLG and WLG are then compared and the minimum of the two is taken as the actual growth.

2.3 *Crop development*

Crop development (phenology) is mainly temperature driven and divides the crop cycle into two main stages; the vegetative stage before anthesis and the reproductive (grain filling) stage between anthesis and maturity. Additionally day length has a minor influence on the rate of development in all cereals and winter cereals require vernalisation, a period with low temperatures required to achieve optimum rates of development later in the season. The phenology module used in Cereal YES is similar to the SUCROS1 phenology module (Goudriaan et al., 1994). Crop development stage is the driving variable for the development of the green leaf area, and thus fully determines the duration and strongly affects the rate of crop growth. The simulated grain weight at time of crop maturity is taken as the estimated yield (section 2.9).

2.4 *Water limited growth*

The water limited growth module considers three soil layers (see Figure 3). The change in water content of each layer is calculated separately from information related to precipitation, soil evaporation, drainage and water uptake by roots.

Drainage (D) from one layer to the next occurs when the net flow of water into a layer is more than the amount required for that layer to reach field capacity (FC). The surplus water then drains to the layer below.

In the top layer (layer 1), soil evaporation takes place when soil moisture is above the permanent wilting point (PWP). The soil evaporation rate has a linear relationship with canopy cover and moisture status of layer 1. On a bare soil at field capacity, soil evaporation is set to 0.9 times the rate of Class A pan evaporation. The top layer has been defined to extend to a depth of 130 mm.

Soil evaporation will thus be high after a rainfall event when the topsoil is moistened. When irrigation and/or rainfall cease, soil evaporation will gradually decline until the top layer has dried out and contact with lower layers is lost. No capillary transport is considered, so soil evaporation does not directly affect deeper soil layers. The effect of shading by the canopy is to reduce direct radiation on the soil and is simulated according to Ritchie (1972).

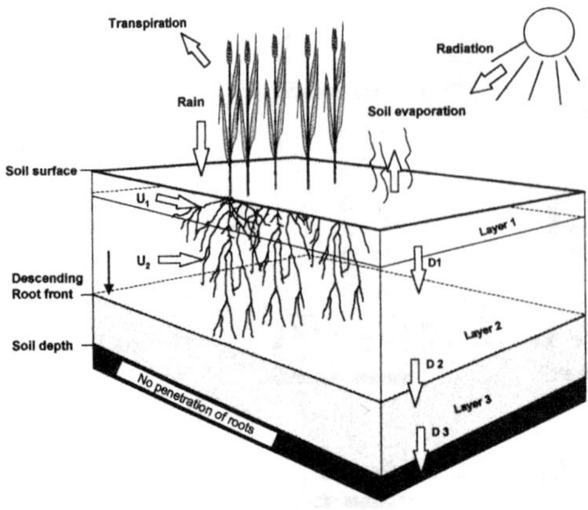

Figure 3 In and outward water flows for the different Layers considered in the soil water balance model.

Layer 2 extends to the root front (RF). Roots do not extract water below the RF, which moves down the soil profile as a result of root extension. The thickness of layer 2 increases until the roots reach the end of the soil profile (i.e. soil depth). The soil profile can be limited by a layer of rock or a layer that otherwise is not penetrated by roots or alternatively the roots can simply reach a maximum rooting depth. The latter is the case in Tierra de Campos where the clay soils are several meters deep.

Layer 3 is the layer between the root front and the maximum rooting depth, and is potentially accessible to the roots. Layer 3 (and the water in layer 3) is gradually transferred into layer 2 as the root front advances.

2.5 *Potential water extraction*

A maximum relative uptake rate XT_{max} (mm of water/mm of soil) is defined for a rooted soil at field capacity (FC). The potential water uptake rate in a certain layer is then calculated from the available soil water in that layer.

The total potential water uptake U_{pot} is equal to the amount of water that the crop can potentially transpire. The contribution of layers 1 and 2 to U_{pot} is calculated separately because differences in soil moisture between these layers are considered.

2.6 *Transpiration equivalent*

Water limited growth is often calculated as the product of potential water uptake and a crop specific constant (water use efficiency - WUE). Here the transpiration equivalent according to Azam-Ali et al. (1994) is employed. WUE is not treated as a crop specific constant but a variable that responds negatively to changes in atmospheric saturation deficit.

The total amount of water extracted from the rooting zone (layer 1 and layer 2) equals the rate of transpiration U. The relative contributions of layer 1 and layer 2 towards the crops water supply, are assumed to be proportional to the potential contributions.

2.7 *Green leaf area index (GAI)*

The green leaf area module uses an empirical mathematical function to predict leaf area. This function is initially set to fit a best estimate of the green leaf area curve and can later be adjusted using vegetation indices derived from remote sensing data (see below). The approach to modeling leaf area is very similar to that adopted by Werker and Jaggard (1997) for sugar beet yield estimation.

GAI is modeled as following the shape of the normal probability density function characterised by a mean μ, a variance σ and green area duration (GAD; see Figure 4). The parameters μ, σ and GAD, represent the timing of the canopy area peak with respect to emergence, the leaf area duration around the peak and the total area under the curve respectively.

2.8 *Light limited growth (LLG)*

The potential light limited growth rate is calculated from the principle of light interception and utilisation first described by Gallagher and Biscoe (1978) and Monteith (1977):

$$LLG = fi * PAR * \varepsilon \qquad (1)$$

Where: fi = Fraction PAR intercepted by green leaves
PAR = Photosynthetic active radiation above canopy ($MJm^{-2}d^{-1}$)
ε = Radiation use efficiency (gMJ^{-1})

The driving variable for crop growth is the photosynthetically active radiation (PAR) which is

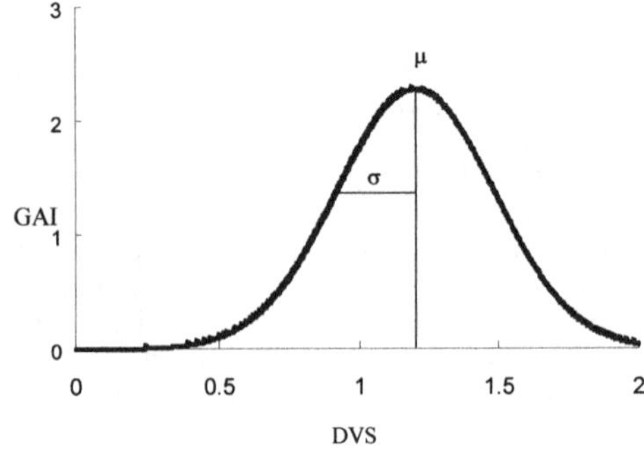

Figure 4 Relationship between development stage and green leaf area index for winter wheat in CEREAL-YES.

converted into dry matter through a 'radiation use efficiency' (ε).

The fraction of radiation intercepted by a canopy (fi) is calculated according to Beer's law (2). The value of the light extinction coefficient, k, depends on the leaf angle distribution of the crop and effectively determines the response of fi to GAI. Within the model, k has been set to a default value of 0.5 based on findings by other researchers.

$$fi = 1 - e^{-(k \cdot GAI)} \quad (2)$$

Where: k = light extinction coefficient

2.9 *Yield*

Two alternative relationships were used to predict yield. In (3) the number of grains is predicted from total dry matter at anthesis. Yield is then calculated as the product of grain number and kernel weight (4).

$$GN = \beta * W_{anthesis} + GN_0 \quad (3)$$

$$Y = GN * GrainW \quad (4)$$

Where:

Y = Yield (g m^{-2})
GN = Number of grains (kernels m^{-2})
$GrainW$ = grain weight (g $kernel^{-1}$)
GN_0 = minimum number of grains (kernels m^{-2})
β = number of grains per g of dry matter at anthesis (kernels g^{-1})
$W_{anthesis}$ = above ground dry matter at anthesis (g m^{-2})

In the 1998 field season, observations were made in Tierra de Campos on grain numbers (GN) and total dry matter. Estimated dry matter at anthesis in each field involved in this study was plotted against the corresponding observed grain numbers (GN). As a result GN_0 and β were found from linear regression, and set to 2500 and 17.6 respectively. Grain weight increases with time, with a temperature dependence. The basic filling rate was set to 0.56 mg d^{-1}.

The alternative relationship (5) is much simpler and was found to give equally good results for the limited data so far available.

$$Yield = HI * Biomass \quad (5)$$

Where HI is the harvest index and was set to 0.47.

3 SATELLITE IMAGE PROCESSING

Multi-temporal satellite data is used within the Cereal YES system for two purposes: 1) to adjust the crop growth model (through model optimisation) and 2) to estimate crop area. During the initial phase of the system LISS-III data from IRS-1C and XS and Xi data from SPOT 3 and 4 have been used. IRS-1 Wide Field Sensor (WiFS) which acquires data at 188-metre resolution in two spectral bands, visible red and near infrared and has a swath width of 804km enabling revisits every five day, has also been used for monitoring the crops' evolution. This leads to the possibility of a two-level Vegetation Index scheme, with the 'big picture' provided on a frequent basis from the WiFS data, complemented with the less frequent but more localised information from LISS-III.

3.1 *Crop classification and area estimation.*

Following quality checking the data is pre-processed to incorporate radiometric correction and geometric transformation to a standard map projection. The next stage is to produce a multi-spectral and multi-temporal classification of the data to provide a land-cover map of the region, identifying the locations of the crops of interest. Supervised classification of the satellite images using ground truth information has been used to map crop areas. GIS analysis then allows the area of each crop covered by the imagery to be calculated for each municipality. This area estimate can then be combined with the yield estimate from the growth model (in Tonnes per hectare) to produce a yield prediction (in Tonnes) for each municipality. These figures can then be aggregated on a provincial or district basis.

3.2 *Adjustment of crop growth model.*

The crop growth model has been designed to be insensitive to a range of factors that may cause crop yields to vary in time and space (e.g. crop establishment, tillering and growth reducing factors like pests and diseases). The inclusion of such spatially variable factors in regional crop modeling exercises would require very extensive data sets as initial input and require a large number of model runs. The incorporation of remote sensing data, however, makes it possible to take the cumulative effect of these factors into account through their influence on the physical and biophysical (and hence reflectance) properties of the canopy. This variability can be quantified through the use of vegetation indices.

The assumption underpinning the use of vegetation indices to adjust the model is that a relationship exists between the vegetation index and GAI. This relationship has been defined as an empirical mathematical function based upon a detailed analysis of field and satellite data.

Optimisation is the term used to describe the

adjustment process whereby a limited number of model parameters are systematically varied to find the best fit between values calculated by a model and values derived from an external data source (in this case a vegetation index). Due to the non-linear nature of the crop growth model, simple analytical inversions (i.e. regressions) are not applicable and so an iterative numerical approach must be employed. In this process, an error function based upon a weighted sum of the square errors is minimised using the Marquardt method (see 5).

$$\chi 2 = \Sigma i \text{ (model - data)}^2/\text{error}^2 \qquad (6)$$

The number of model parameters that may be varied in this optimisation process is governed by the number of external measurements available. The latter is governed by the number of satellite image acquisitions. Note that in an ideal scenario, there should be more external observations than there are parameters to be adjusted.

The most suitable parameters to optimise are those which control the GAI curve, namely GAD, μ, and σ. The parameters obtained following optimisation are updated within the model and an adjusted yield figure calculated.

3.3 *Vigour partitioning of satellite images*

Within the land area covered by a satellite image, the vegetation index varies, between fields and within fields, due to differences in crop establishment and growth. In order to capture the spatial variability in crop development (or vigour) detected by the satellite imagery, the vegetation index data for each crop is partitioned into 5 categories of vigour, following a methodology developed by Logica. Figure 5 illustrates the typical distribution of VI values that is observed for all of the crops. In this case the distribution for non-irrigated spring barley in a SPOT image of 12th June 1998 is presented. The VI data can be seen in this figure to conform well to a Normal distribution.

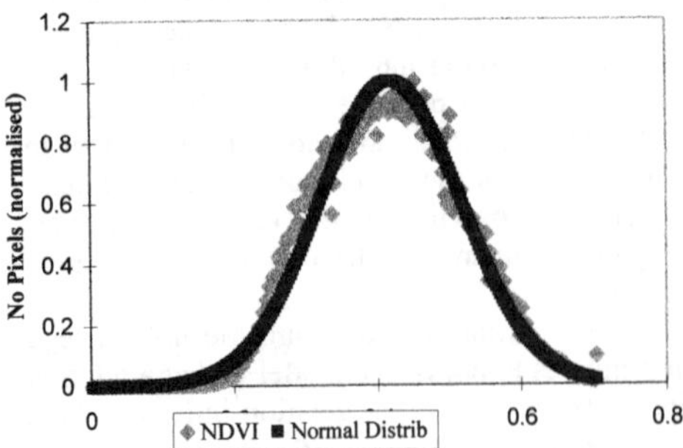

Figure 5 Distribution of VI values for non-irrigated Spring Barley in May 1998 satellite image. Comparison of distribution of VI values with a Normal Distribution

Once the distribution of VI values has been established, the data is partitioned into the 5 categories on a semi-statistical basis (see Figure 6). This process has been carried out for each crop and for each satellite image as it became available. The VI value at the mid-point of each partition is used to optimise the crop growth model and produce a yield estimate (in Tonnes/hectare) for that VI category.

The upper and lower limits of each category form the basis for a VI distribution map. Each pixel of the satellite image is assigned to a VI category 1-5. Subsequently the area estimate for a certain category is based on the number of pixels assigned to that category. The yield estimate and area estimate of each category can then be combined to give a productivity estimate (in Tonnes) for that category. Summing the productivity estimates for all categories gives an overall productivity estimate.

Figure 7 shows the temporal evolution of the 5 vigour categories for non-irrigated spring barley during the 1997/8 growing season.

4 YIELD MODEL

The crop growth model outputs a yield estimate (in tonnes per hectare) for a crop for a particular set of

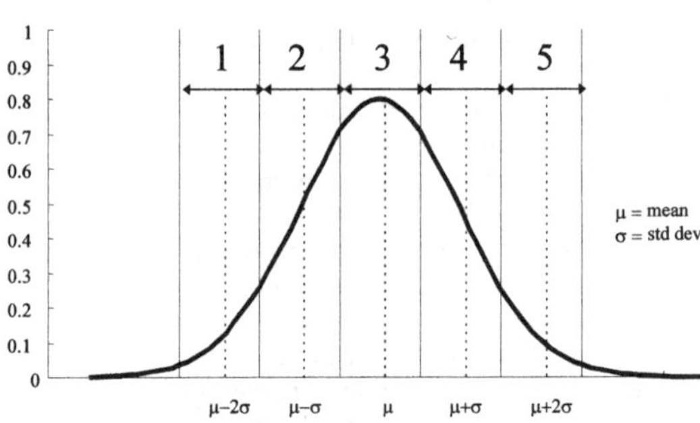

Figure 6 Overview of VI distribution partitioning strategy.

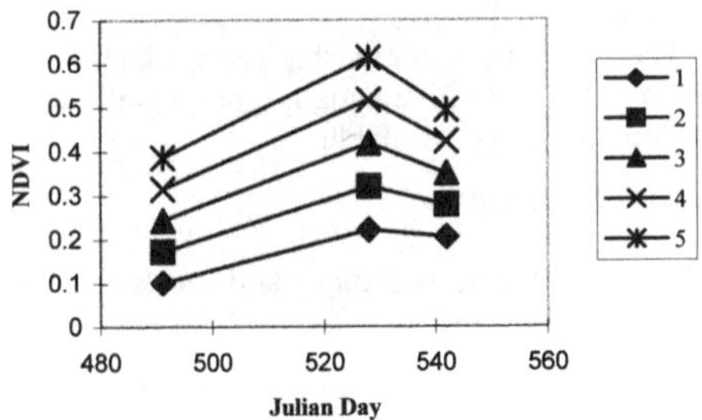

Figure 7 Temporal profile of VI vigour partitions. (Julian Day 490 is equivalent to 5 May 1998.)

input model parameters and the most up-to-date meteorological data available. The input parameters may be the default values contained with model, or those produced by an optimisation process. During the optimisation process, 5 crop-specific yield estimates are produced, one for each of the VI vigour categories. A total yield figure for each municipality can then be calculated by summing the product of the yield figure for each VI category with the area of the crop within each region that falls into that category (see 6). The latter is calculated using a GIS functionality:

$$Y_{ki} = \Sigma a_{ki} y_i \qquad (7)$$

Where: Y = yield (T)
k = region
a = crop area (ha)
y = crop yield (T/ha)
i = vigour class(low/medium/high)

5 PRELIMINARY RESULTS

The system has been tested against the field data collected in 11 fields in Tierra de Campos in 1998. The preliminary results presented here are graphical presentations of the predictions made by the model versus the observations made on the ground.

The model was run for each field and output was compared to the observed values. Before running the model for a particular field the three leaf area parameters (σ, μ ,and GAD) were adjusted in order to obtain an optimum fit of the leaf area curve against the observed green leaf areas. This adjustment imitates that made when using satellite image derived vegetation indices to adjust the model. Thus for each field three points (late April, late May and late June) were available for this fitting procedure. The figure below compares the predictions, after adjusting the model, against the field observations. Once the leaf area module had been fit to the observed values of Green Leaf Area the model was run without further adjustments. The four outputs from the model; Biomass, Yield, grain number and 1000 grain weight, were compared against the field observations. Two of these comparisons are shown in the following figures.

These graphs show that the order of magnitude of the predictions is good. The model also gave reasonable estimates for average predictions but failed to explain much of the variation between fields. This may be the result of the high variation in rainfall across the region. The model predictions are based on average rainfall figures from 7 weather stations in the region for which the cumulative precipitation from 1 January 1998 to 31 July 1998 varied from 600 and 2200 mm.

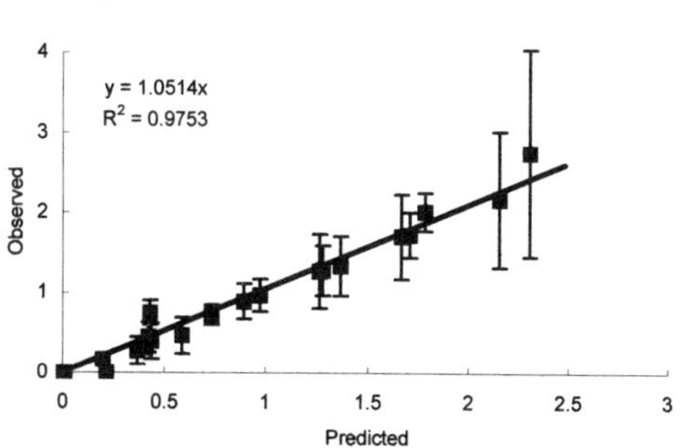

Figure 8 Green Leaf Area, predicted versus observed

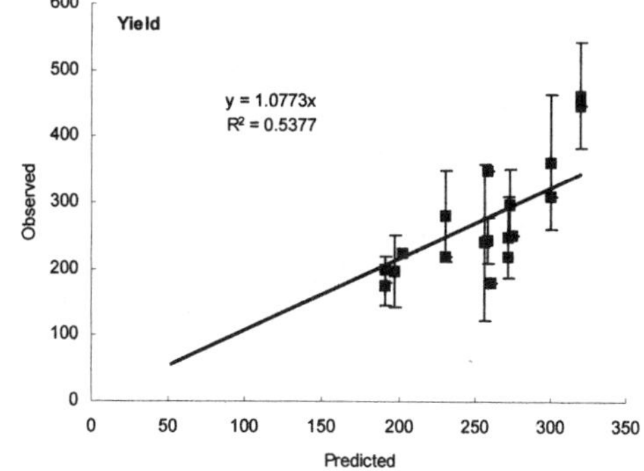

Figure 9 Yield, predicted versus observed

6 DISCUSSION

The current version of the system has demonstrated the feasibility of estimating yields based on a crop growth model and satellite data. Furthermore the scale and resolution of the images have been shown to be adequate for agricultural district level area and yield estimates. However it has not yet been possible to operate the model in "real-time" to produce yield estimates prior to harvest. A critical factor to be considered in 1999 is the logistics of data collection especially the meteorological data collection and the image acquisition.

The crop growth model seems to be appropriate to the conditions of the region and to cereal production in similar conditions. A series of sensitivity analyses, together with the validations that have been carried out so far, indicate that the model is sensible, sensitive and robust. Nevertheless the model needs to be evaluated against considerably more real yield data and information on the evolution of crops through the season before any objective assessment of the model's value can be made.

The structure and implementation of the model is such that it could be easily transferred to another region and possibly to different crops, thus enhancing its value. Indeed the model is currently tested against wheat data collected at six locations in the UK in 1993 and 1994.

7 CONCLUSION

So far the main obstacle to the operational viability of the system is the difficulty of obtaining quickly and with a high level of assurance the required input data: from satellites and from weather stations. Data from both the IRS-LISS (1997) and SPOT-XS/Xi (1998) sensors have been used and two IRS-WiFS scenes have also been acquired. By using different satellites and the combination of medium and high resolution images the risk of non-acquisition of images during critical time windows is greatly reduced.

An analysis of the 1998 meteorological data suggests that average data from several weather stations in the area is required to be sufficiently representative of the whole of the area. Where there are important differences in some meteorological parameter - for instance if part of the area receives more rainfall than others at a critical time of the crop development - then the effects of these differences are seen in the satellite images.

The methodology for optimising the agro-meteorological model's yield estimates using the observed vegetation indices in the images seems to be sound and to produce good results. However this will require further testing and analysis.

Acknowledgements
The work reported in this paper was carried out within the Cereal YES Project, a shared cost action financed by the European Commission through the Environment and Climate Programme of the Fourth Framework Programme for Research and Technological Development.

REFERENCES

Azam-Ali, S.N., Crout, N.M.J. and Bradley, R.G., 1994. Perspectives in modelling resource capture by crops. In Monteith J. L., Scott R.K., & Unsworth M.H.: (Eds.). *Proceedings of 52nd Easter school in agricultural science*, University of Nottingham, UK. pp125-148. Nottingham: Nottingham University Press.

Bouman, B.A.M. 1995 Crop Modelling And Remote Sensing For Yield Prediction. *Netherlands Journal of Agricultural Science,* Vol. 43, pp. 143-161.

Bradley, R.G. and Crout, N.M.J. 1993 The PARCH model, *PARCH User Guide*, University of Nottingham, UK. 122 pp..

Gallagher J.N. and Biscoe P.V 1978, Radiation Absorption, Growth and Yield of Cereals, *Journal of Agricultural Science,* Vol. 91, pp. 47-60,.

Gillet,. A.G. 1997 Modelling the response of winter wheat to different environments: A parsimonious approach. *PhD. thesis* University of Nottingham. pp. 219,.

Gillett, A.G., Crout N.M.J., Stokes D.T., Sylvester R. - Bradley and Scott R.K.. 1998. A simple model for green area index of winter wheat under UK conditions. *Submitted to Journal Agricultural Science*, Cambridge.

Goudriaan, J. and van Laar H.H. 1994. *Modelling crop growth processes: Textbook with exercises.* Kluwer academic publishers. Dordrecht, Boston., London. 238 pp

Jaggard K.W. and Clark J.A, 1990 Remote Sensing to Predict Yield of Sugar Beet in England In: M.D. Steven and J.A. Clark (eds.), *Applications of Remote Sensing in Agriculture,* Butterworths London, pp. 201-207..

Kiniry J.R., Major D.J., Izauralde R.C., Williams J.R., Gassman P.W., Morrison M., Bergentine R. and Zentner R.P, 1995 Epic Model Parameters for Cereal, Oilseed, and Forage Crops in the Northern Great-Plains Region, *Canadian Journal of Plant Science*, Vol. 75, pp. 679-688.

Lövenstein, H., Lantinga, E.A. Rabbinge, R. and Keulen, H.van 1992. Principles of theoretical production ecology. *Lecture Syllabus F300-001*, Wageningen Agricultural University, pp. 117.

Monteith J, 1977. Climate and the Efficiency of Crop Production in Britain, *Philosophical Transactions of the Royal Society*, London, Vol. B 281, pp. 277-294.

Monteith J.L., 1981 Does Light Limit Production? In: C.B. Johnson (ed.), *Physiological Processes Limiting Plant Productivity*, Butterworths London, pp. 23-85.

Ritchie, J.T. 1972. Model for predicting evaporation from a row crop with incomplete cover. *Water Resources Management* 8 (5), 1204-1213.

Supit I., Hooijer A.A. and van Diepen C.A., 1994 *System Description of the WOFOST 6.0 Crop Simulation Model Implemented in CGMS*, European Commission, Directorate General XIII, L-2920 Luxemburg. pp. 145,.

van Diepen C.A., Wolf J., van Keulen H. and Rappoldt C., 1989 WOFOST a Simulation Model Of Crop Production, *Soil Use and Management*, Vol. 5, pp. 16-24..

Werker A.R. and Jaggard K.W, 1997 Modelling Asymmetrical Growth Curves That Rise and Then Fall: Applications to Foliage Dynamics of Sugarbeet (Beta vulgaris L.), *Annals of Botany.* Vol. 79, pp. 657-665.

Remote Sensing in the 21st Century: Economic and Environmental Applications, Casanova (ed.)
© 2000 Balkema, Rotterdam, ISBN 90 5809 096 5

Integration of satellite data and administrative information in crop yield monitoring

Risto Kuittinen & Leena Matikainen – *Finnish Geodetic Institute, Masala, Finland*
Ville Keskisarja – *Finnish Environment Institute, Helsinki, Finland*
Eija Parmes – *VTT Automation, Espoo, Finland*
Vanda Perdigao – *Joint Research Centre, Space Applications Institute, Regional and Agricultural Information Systems Unit, Ispra, Italy*

ABSTRACT: The possibility to use geographic information, satellite and administrative data in crop yield modelling and production estimation was studied in Finland in 1994-1997. The research and development work was based on crop species interpretation using ERS- and SPOT-data, the use of the Crop Growth Monitoring System database structures in which a Finnish crop growth model CropWatN and satellite derived leaf area indices were integrated in modelling. The research showed that in crop species interpretation an overall accuracy of 83% can be achieved when SPOT and ERS images and administrative data are used in interpretation. It also showed that modelling methods with additional GIS and satellite data can be applied in Finland for crop production estimation. The results of yield estimation were similar or even slightly better than the forecasts made by subjective basis (traditional method in Finland; test farms and agents) in two out of three tested rural advisory centres.

1 INTRODUCTION

During the past two decades the development of geographical information systems and satellite image processing has offered improved opportunities for crop yield modelling and crop production prediction in large areas. In Europe the MARS-project of the JRC/Space Applications Institute produces area and yield estimates for the European Union. The main goal of this work is to produce estimates of crop production throughout the growing season (Vossen & Rijks 1995). The annual early crop production forecasts in Finland are made by statistical agents of the Information Centre (IC) of the Ministry of Agriculture and Forestry and the estimates are based on their experiences of local conditions and personal opinion about the growing season. Under ordinary conditions these estimates are quite reliable, but if the growing season is exceptional the subjective estimates of the statistical agents appear to be uncertain. To obtain a faster and more objective method for the estimation of Finnish crop production a research project was carried out in 1994-1997. The Finnish Geodetic Institute, the Information Centre, VTT Automation, some other Finnish institutes and the JRC in Ispra Italy participated to the project. A method was also developed for crop species interpretation, and this method was tested by using satellite image data from 1996 and MARS Rapid Estimates data from 1995 and 1996.

The developed method and the results are presented in the following chapters.

2 METHOD

2.1 General

Geographic information and satellite data

The geographic information used in the crop production estimation consists of the following data:

1) Land use map which is made using topographic maps, results of national forest inventory and satellite images. Updated agricultural area data are obtained from this database and it is used in leaf area index estimation from satellite data and crop production estimation.
2) Soil data in agricultural areas which is obtained from different types of soil maps is used in yield modelling.
3) Meteorological data which consists of information about precipitation, temperature and incoming solar radiation, air humidity and wind speed are used in yield modelling.
4) Location and area of crop species which can be interpreted from satellite data (SPOT and ERS) using a rule based system. This information can be used in leaf area index and crop production estimation.
5) Crop statistics of the last 15 years which consists of yield information and crop calendar.

6) Leaf area index estimates which can be interpreted from satellite images (SPOT, NOAA).

Administrative information

Administrative information is used in interpretation of crop species from satellite images and in crop production estimation. It consists of the following data:

1) Areas of the rural advisory centres (RAC) in the country for crop production estimation
2) Areas of different crops which are based on the information collected from farmers' dossiers. Data cover the whole country and can be used for rule-based interpretation (previous year) and crop production estimation (ongoing year).
3) Land parcel identification map which covers the whole country. Data can be used as input in rule-based interpretation of satellite data.

Yield Modelling System

The system was developed on the basis of the Crop Growth Monitoring System (CGMS) of the MARS project (JRC/SAI/RAIS) by integrating a new growth model in the CGMS and adding new sources of information in it. The developed system (CGMSF) is drafted below.

2.2 Method for crop species interpretation

A method for crop species interpretation was developed at the Finnish Geodetic Institute (Matikainen et al. 1998). The method is parcel-based, which means that parcels obtained from maps or by digitisation are interpreted instead of single pixels. The main source of data in the interpretation is SAR images, which can be obtained under all weather conditions. Information on crop species grown in the parcels during the previous year and a SPOT image from the spring or previous autumn are also used. Crop species information is obtained from administrative databases, and together with knowledge of crop rotation it gives valuable information on probable crop species. The SPOT image can be used to distinguish grasslands, spring crops and winter crops from each other. The interpretation can be repeated several times during the growing season as new satellite images become available. The accuracy of the interpretation increases as new images are added, but some information is already obtained early in the growing season. A rule-based approach is used to combine information from various data sources in the interpretation. Field survey data are needed for training area selection and rule development. The method is presented in Figure 1.

2.3 LAI estimation from satellite data

Satellite images from 15 June - 30 June are the most optimal for LAI estimation from satellite images in Finnish areas. The pre-programming option of the SPOT satellite increases the possibility for cloud free imagery. The satellite images are rectified to the Finnish National Co-ordinate System, calibrated to reflectances and corrected for atmospheric effects. The atmospheric parameters are calculated by the 6S code (Vermote et al. 1994). LAI can be estimated from satellite images by the GEMI vegetation index (Global Environment Monitoring Index, Pinty & Verstraete 1992). GEMI values are calculated for each pixel of the satellite image from the red and near infrared reflectance values:

$$GEMIn = \eta(1-0.25\eta) - \frac{\rho_R - 0.125}{1-\rho_R} \qquad (1)$$

where:

$$\eta = \frac{2(\rho_{NIR}^2 - \rho_R^2) + 1.5\rho_{NIR} + 0.5\rho_R}{\rho_{NIR} + \rho_R + 0.5} \qquad (2)$$

and

$GEMIn$ is the global environment monitoring index value at pixel level
ρ_R is the reflectance in red band at pixel level
ρ_{NIR} is the reflectance in near infrared band at pixel level

The LAI estimates are calculated from the GEMI values for each pixel by crop specific linear equations, which are valid for LAI values less than 3.0 :

$$LAIn = a + b \cdot GEMIn \qquad (3)$$

where:

$LAIn$ is the leaf are index estimated from satellite data at pixel level
a is the coefficient of the linear model
b is the coefficient of the linear model

If field measurements are available, the case specific coefficients of the model are estimated by linear regression. Before the modelling, the field measurements are interpolated to the date of the satellite image by crop specific reference profiles from LAI profile database. If field measurements are not available, the crop specific coefficients presented in 'Results' can be used. The pixel level LAI values are integrated to the 10×10 km^2 grid cells for input to CGMSF. LAI estimates from NOAA data can be

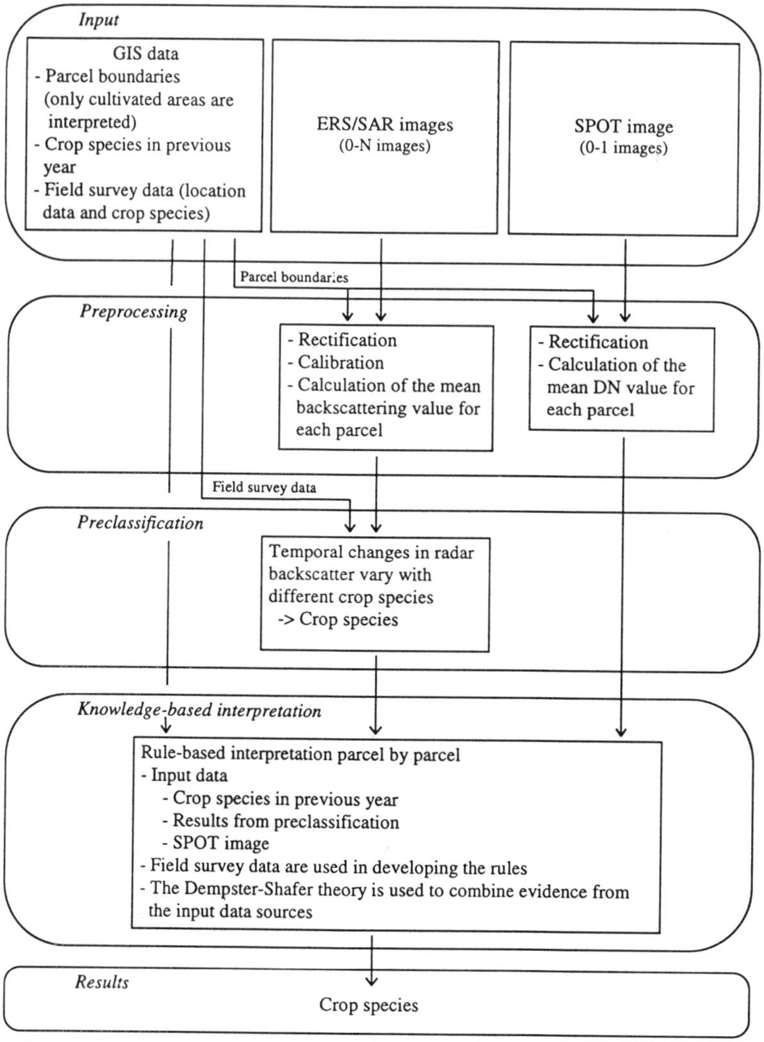

Input

GIS data
- Parcel boundaries (only cultivated areas are interpreted)
- Crop species in previous year
- Field survey data (location data and crop species)

ERS/SAR images
(0-N images)

SPOT image
(0-1 images)

Parcel boundaries

Preprocessing

- Rectification
- Calibration
- Calculation of the mean backscattering value for each parcel

- Rectification
- Calculation of the mean DN value for each parcel

Field survey data

Preclassification

Temporal changes in radar backscatter vary with different crop species
-> Crop species

Knowledge-based interpretation

Rule-based interpretation parcel by parcel
- Input data
  - Crop species in previous year
  - Results from preclassification
  - SPOT image
- Field survey data are used in developing the rules
- The Dempster-Shafer theory is used to combine evidence from the input data sources

Results

Crop species

Figure 1. Method for crop species interpretation.

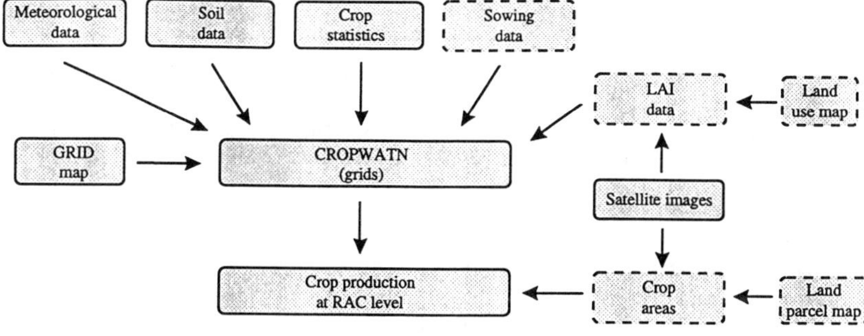

Figure 2. System description of the crop growth modelling system CGMSF.

estimated by using NDVI and linear regression equations but the accuracy is only half of that of SPOT data.

2.4 The Finnish version of the Crop Growth Monitoring System

The kernel of CGMSF is the CropWatN crop growth model (Karvonen & Varis 1992). Additional input information included in CGMSF but not in CGMS are presented by broken lines in Figure 2. The Finnish system differs from the CGMS of the JRC/Space Applications Institute as follows:

- The size of the grid used is 10 × 10 km^2. Only selected grid cells within a region are used in crop production estimation. The selection of grids is done using the land use map.
- Meteorological data are interpolated for grids using a procedure developed at the Finnish Meteorological Institute. This method accounts for the local variations in the weather caused mainly by lakes and elevation differences.
- The CropWatN model is used in estimating yield at the grid level. The soil water submodel of this model has been developed by Helsinki University of Technology/Laboratory of Water Resources for Finnish conditions.
- Leaf area index can be estimated from satellite data. In the process of forecasting it is enable to calculate crop growth in three different ways: (1) conventional method (no LAI values from satellite images), (2) the model calculated value of LAI is replaced with LAI values determined from satellite data, and (3) the combination of these.
- Crop production can be estimated for areas of 20,000 km^2 or larger, which allows the estimates at the rural advisory centre level.
- Sowing dates for the model are obtained from reports from the agents of the Information Centre.

3 RESULTS

3.1 Crop species interpretation

The crop species interpretation method was tested in a study area situated near Seinäjoki in western Finland. A SPOT image from spring 1996, four ERS-2 SAR images from summer 1996 and information on crop species grown in the area in 1995 and 1996 were used. The crop species information was obtained from MARS Rapid Estimates data. Crop species from 1995 were used as one data source in the rule-based interpretation and crop species from 1996 as reference data. The best overall accuracy, 83%, was obtained when all the data sources were used. When using the SAR images alone, the accuracy varied from 33% to 72%, depending on the number of images. The crop species interpreted and used in the calculation of accuracy estimates included grassland, rye, oats, barley and turnip rape. These crops cover about 95% of the agricultural land in Finland.

3.2 Leaf area index estimation from satellite images

Eleven SPOT satellite images from June and July 1994-1996 were used to study the relationship between satellite image derived vegetation related indices and the leaf area index (LAI). The analysis covered LAI observations from spring wheat, oats, barley, turnip rape, potato, and sugar beat from Kirkkonummi, Jokioinen and Lapua test sites. The incidence angle of the SPOT satellite images included in the analysis varied from -21.6 to 24.5 degrees.

Six different vegetation indices from satellite images were included in the tests: the Normalised Difference Vegetation Index (NDVI), the Simple Ratio Vegetation Index (SRVI), the Transformed Vegetation Index (TVI), the Global Environmental Vegetation Index (GEMI), the red reflectance (RED) and the near infrared reflectance (NIR).

The best predictors were the GEMI index and the near infrared reflectance from SPOT image, with atmospheric corrections applied. Figure 3 shows their relation to the field measured LAI values.

The accuracy of LAI estimation from satellite images increased with crop specific modelling. Table 1 shows the LAI models for different crops and the coefficients when the crop type is not known (last line 'All').

The GEMI index was chosen as the best predictor variable for LAI estimation from satellite images, although the near infrared reflectance gives almost as good results. With the GEMI index, the coefficients of the model are quite consistent between spring wheat, oats and barley. Turnip rape, potato and sugar beet have different models, but because of the relatively small number of LAI observations on these crop species, the reason for the different models is not clear.

3.3 Crop yield estimation

The research was carried out in three RAC's which cover about 20 % of the arable land in Finland. Both SPOT and NOAA satellite derived leaf area index can be used to calculate yields and yield estimates. Table 2 shows the influence of these LAI inputs on the simulated yields. For all crops the satellite LAI input has heightened the LAI curves calculated in modelling.

Table 3 shows that the effect is levelled down when the coefficients from regression analysis of previous years are used to compute the yield estimates. The average effect is anymore only 2% whereas for the simulated yields (Table 2) the

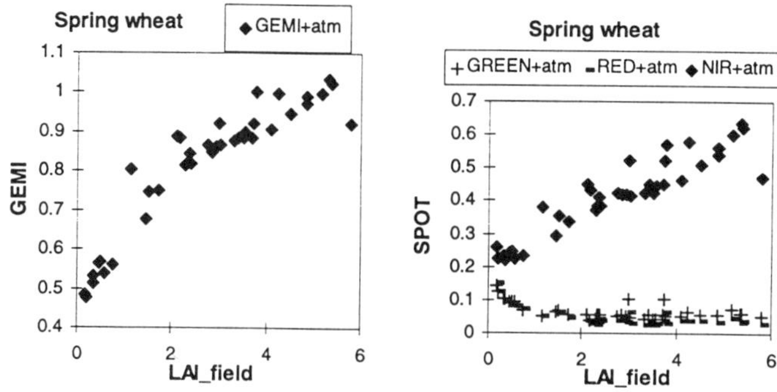

Figure 3. The relation between field measured LAI and GEMI (left) and between LAI and SPOT image reflectances in the three wavelength bands (right). Spring wheat observations from Kirkkonummi, Jokioinen and Lapua, 1994-1996.

Table 1. Coefficients of the crop specific linear models for LAI estimation from GEMI.

Crop	LAI(< 3.0) = a+b*GEMI a	b	Number of obs.	Mean LAI	Std. Dev. LAI	R^2	RMSE
Spring wheat	-3.1239	6.6049	14	1.30	1.01	0.96	0.21
Oats	-2.6706	5.9062	35	1.36	1.08	0.84	0.44
Barley	-3.0622	6.4075	47	1.36	1.04	0.85	0.42
Turnip rape	(-0.4300	3.5740)	8	1.99	0.95	0.54	0.74
Potato	(1.1703	0.5907)	3	1.62	0.22	0.13	0.34
Sugar beet	(0.1042	1.4030)	5	1.17	0.24	0.57	0.20
All	-2.4132	5.5183	113	1.40	1.01	0.75	0.51

Table 2. Comparison of the simulated yields and influence of LAI input in Etelä-Pohjanmaa (1997).

Crop	Model only kg/ha	SPOT LAI input kg/ha	%	NOAA LAI input kg/ha	%
Spring wheat	4531	4754	+5	5051	+11
Barley	3959	4192	+6	4440	+12
Oats	4952	4992	+1	5366	+8

average effect was 7%. Presumable, if a sufficient number of years (8) with satellite derived LAI data were available for regression analysis, the effect of LAI input on yield estimates would be between these values. Some data are missing from the table because SPOT images were available only for Etelä-Pohjanmaa.

4 CONCLUSIONS

Geographic and administrative information

The information listed in chapter 2 except satellite data are available in whole Finland in a uniform co-ordinate system. Satellite images can be rectified in the same co-ordinate system. However to establish a countrywide database is relatively expensive because the data have to be bought for the owners. Administrative data can be used by IC for statistical purposes. The crop areas from administrative data are available only in late July and these can be used for the late crop production estimates.

Satellite images for leaf area index estimation and crop species interpretation

The spatial resolution of NOAA/AVHRR is problematic, especially in those cases in which the pixel includes land use classes other than

Table 3. Comparison of the yield estimates and percentage deviation from the IC estimate (1997).

Crop / Rural Advisory Centre	IC Estimate	CGMS model		CGMSF model		CGMSF SPOT input		CGMSF NOAA input	
	Yield	Yield	%	Yield	%	Yield	%	Yield	%
Spring wheat									
Nylands Svenska	3590	4030	12	3650	2	-		3781	5
Häme	3450	4017	16	3446	0	-		3459	0
Etelä-Pohjanmaa	3160	4493	42	4611	46	4609	46	4605	46
Barley									
Nylands Svenska	4210	3629	-14	3401	-19	-		3503	-17
Häme	3560	3171	-11	3590	1	-		3477	-2
Etelä-Pohjanmaa	3350	4216	26	3955	18	3984	19	4022	20
Oats									
Nylands Svenska	3540	3730	5	3291	-7	-		3198	-10
Häme	3600	3404	-5	3598	0	-		3433	-5
Etelä-Pohjanmaa	3400	4131		3746	10	3752	10	3811	12
Turnip rape									
Nylands Svenska	1750	2021	15	1980	13	-		1986	14
Häme	1770	1755	-1	1643	-7	-		1550	-12
Etelä-Pohjanmaa	-	-		-		-		-	

agricultural land. The error of LAI estimates is in NOAA data 1.0 LAI units and in SPOT data 0.4 units at parcel level but smaller at grid level. The price of SPOT images can be too high for operational purposes. The accuracy of crop species interpretation using SPOT and ERS data is about 80% which is not reasonable for statistical purposes. Interpretation methods should be developed and tested in whole Finland. Cloud cover limits the use of optical satellite data.

CGMSF for yield estimation

The present accuracy of CGMSF compared with the accuracy of the existing yield forecast method in Finland seems to be sufficient. More investigations and especially a sensitive analysis of the factors affecting the accuracy are needed if improvement of reliability and accuracy of CGMSF are desired. For large scale yield estimation in Finland more data for calibration (other RAC's) should be made available, especially crop growth data from various locations over a number of years.

Operational use of CGMS and CGMSF in Finland

The CGMS and the CGMSF can be applied in Finland for crop production estimation, because all necessary information and input data are available. A longer period of five additional years is needed for accurate integration of the methods in IC. The present method used at IC for annual crop production prediction is cheap, and the costs of CGMSF will exceed these costs, however, the benefits of the system compared with the present method are greater. These benefits are mainly indirect, because the CGMSF provide more objective and accurate crop production estimates earlier in the season. The most relevant difference of the CGMSF compared with the present method used at IC is its objectivity and capacity for early results. This provides a good basis for operational yield production estimation when the crop areal data from the Integrated Administration and Control System are combined with yield estimates. The CGMSF also provides results that are comparable with the MARS project estimates and the forecasts made for the EUROSTAT.

ACKNOWLEDGEMENTS

The authors thank the Ministry of Agriculture and Forestry in Finland as well as the JRC in Ispra, Italy for financing the research project.

REFERENCES

Karvonen T. & Varis E. 1992. Mathematical models in crop production. University of Helsinki.

Kuittinen, R., Ahokas E., Granqvist M., Ikäheimo E., Heikinheimo M., Venäläinen A., Jänne S.,

Keskisarja V., Parmes E., Perdigao V., Goot van der E. 1998. An early crop yield estimation method for Finnish conditions: The crop growth monitoring system of the Joint Research Centre with and without remotely sensed and other additional input data. *Reports of the Finnish Geodetic Institute* 98:2. Kirkkonummi.

Matikainen L., Karjalainen M., Kuittinen R. 1998. SAR images and ancillary data in crop species interpretation. *Reports of the Finnish Geodetic Institute* 98:1. Kirkkonummi.

Pinty, B. & Verstraete M.M. 1992. GEMI: A non-linear index to monitor global vegetation from satellite. *Vegetatio* 101, 1992. pp. 10-15.

Vermote, E., Tanre,D., Deuze, J., Herman, M., Morcrette, J. 1994. Second Simulation of the Satellite Signal in the Solar Spectrum. 6S User guide, Version 0, April 18, 1994.

Vossen P. & Rijks D.1995. Early crop yield assessment of the EU countries: The system implemented by the Joint Research Centre.

Remote Sensing in the 21st Century: Economic and Environmental Applications, Casanova (ed.)
© 2000 Balkema, Rotterdam, ISBN 90 5809 096 5

Operational use of NOAA data for crop condition assessment in Poland

Z. Bochenek
Remote Sensing and Spatial Information Centre, Institute of Geodesy and Cartography, Warsaw, Poland

ABSTRACT: This paper presents stages of construction of a Crop Condition Assessment System for Poland within joint project with Canada Centre for Remote Sensing and first results of its use for the whole country. The system is based on 1 KM NOAA AVHRR data, which are collected daily to produce 10-day composite images. The customized software has been prepared for comprehensive processing of raw NOAA images, i.e. for radiometric , geometric and atmospheric corrections, as well as for cloud thresholding and image compositing. At the same time specialized procedures based on ArcView software have been prepared to produce various outputs from the System, e.g. comparative images, maps, graphs and tables. Historical database, consisting of 1 KM Global Land Data collected within IGBP Programme, was constructed for the Poland's area. In the course of 1997 – 1998 growing seasons two sets of AVHRR NDVI composites were prepared and the reports were produced for the particular administrative units in Poland. NDVI indices were further integrated with agricultural information obtained through ground surveys. As a result of regression analyses some relationships between satellite based indices and agricultural parameters were derived. The outputs from the Polish Crop Condition Assessment System were operationally delivered to the Central Statistical Office, to be further analyzed by decision makers and statistical specialists.

1 INTRODUCTION

Reliable and rapid information about crop development is an indispensable element of proper decisions concerning agricultural economy. Therefore, some countries make efforts in order to find optimum methods for monitoring vegetation growth, which would allow for reliable yield assessment. These methods are based on using satellite images with high temporal resolution, which enable to monitor effectively dynamics of changes of vegetation cover. In late eighties remote sensing based Crop Condition Assessment System was created in Canada. It is based on using 1 KM NOAA AVHRR satellite images, which are collected daily for agricultural provinces of Western Canada. The raw images are next thoroughly processed in order to produce NDVI composites, to be further compared with historical database. Comparative images, maps, graphs and tables are the final outputs from the System; they are delivered to the Statistics Canada and Canadian Ministry of Agriculture for final analysis and decision making (Bullock 1992; Hochheim et al. 1993; Reichert et al. 1991)

In 1996 the Polish Remote Sensing and Spatial Information Centre started the joint project with Canada Centre for Remote Sensing and Canadian Intermap company. The main aim of this project was to implement remote sensing based Crop Condition Assessment System in Poland, using Canadian experience. Due to different types of agriculture in both countries several research problems, related to complicated spatial and crop-type structure, as well as different plant phenology in Poland, had to be solved, in order to prepare the system adequate for conditions of Polish agriculture. Moreover, numerous technical tasks, comprising preparation of the customized software for processing of NOAA images of Poland, have to be performed, to make system fully operational. As a result of joint, research and implementation works the system for crop condition assessment in Poland was created and operationally used in the course of 1997 and 1998 growing seasons. The subsequent stages of building the system, as well as first results of its application for monitoring crop growth conditions and for yield forecasting will be presented in the next chapters of this paper.

2 METHODS AND MATERIALS

Construction of the Crop Condition Assessment System for Poland on the basis of NOAA AVHRR satellite data consisted of the following stages:

- organization of regular acquisition of NOAA AVHRR images throughout the whole vegetation period
- preparation of the specialized software used for processing of NOAA AVHRR images
- collection and processing of archival NOAA satellite images for Poland's area
- acquisition and processing of NOAA AVHRR images for 1997 and 1998 growing seasons
- creation of GIS database for making spatial analyses
- preparation of software for producing final outputs from the Crop Condition Assessment System.

In order to ensure continuous acquisition and effective processing of satellite images the receiving station was purchased and installed at the Institute of Geodesy and Cartography. It enables regular capture of NOAA AVHRR images in HRPT format and their conversion to Level-1b format.

Specialized software for precise geometric and radiometric correction of NOAA AVHRR images, as well as for calculating normalized difference vegetation index – NDVI, was prepared at the next stage of the works. The programs were customized within PCI software environment.

The System assumes assessment of crop conditions at 10-day's intervals, so the specialized program for making decade, maximum-value composites was prepared. In order to avoid cloud impact on NDVI values in further steps of analysis the procedures of cloud thresholding were created.

Historical NOAA database is an indispensable element for proper functioning of Crop Condition Assessment System. So, the set of archival NOAA AVHRR images was created at the next stage of the works. 1 KM Global Land Data, collected within International Geosphere Biosphere Programme, covering 1992 – 1995 period, were mainly used for this work. These images were processed to adjust them to cartographic projection applied for georeferencing current NOAA data for Poland. In addition, some raw NOAA images for 1996 growing season were acquired from the Institute of Meteorology and Water Management and processed to format accepted by the Polish Crop Condition Assessment System. All historical NOAA data were next utilized for creating so-called normal NDVI images, which represented average values of vegetation index, calculated on the basis of five year's period.

In order to be able to make spatial analyses within Crop Condition Assessment System, GIS database had to be created. This database, formed within ArcView environment, comprises the following information layers:

- map of administrative division of Poland
- land use / land cover map
- current processed NOAA NDVI and Channel 1 images
- normal processed NOAA NDVI images
- statistical agricultural data (yield, production, valorization of arable land, crop structure)
- agrometeorological data

Information contained in the GIS database was used for making regression analyses and for determining relationships between remote sensing derived indices and parameters characterizing agricultural production (Walker, 1998, Wood, 1993, Yang et al., 1997).

Production of final outputs was the final stage of construction of Crop Condition Assessment System. In order to accomplish this task, specialized module for making spatial analyses has been prepared within ArcView environment. This module enables to create four image products, which inform about NDVI values for current year's decade as compared to these values for normal and previous year, previous decade and highest value in normal year. Two map products are also delivered, which present NDVI values aggregated within administrative units as compared to these values for normal and previous year. In addition, comparative graphs, representing current NDVI curves against those of normal and previous year, can be produced for each administrative unit.

The above presented Crop Condition Assessment System was applied for monitoring crop growth conditions in Poland during 1997 and 1998 vegetation periods. As a result of its utilization comparative maps, graphs and images were produced for each decade covering growing season in Poland, starting from April 1st till September 30. Graphs, which represented changes of NDVI values throughout vegetation period for all main administrative units, were also prepared. Administrative units were grouped into regions, in order to study regional differences in NDVI levels, related to physical – geographical conditions. All above mentioned materials were used for making analysis of crop development in Poland, based on study of the processed NOAA AVHRR images.

3. RESULTS OF ANALYSIS OF CROP GROWTH CONDITIONS IN 1998

The following conclusions were drawn on the basis of analysis of NDVI curves, maps and images produced for 1998 growing season within Crop Condition Assessment System:

1. At the beginning of growing season (April - mid-May) crop conditions in almost whole country were favourable, with exception of some districts in northeastern and southwestern Poland, which

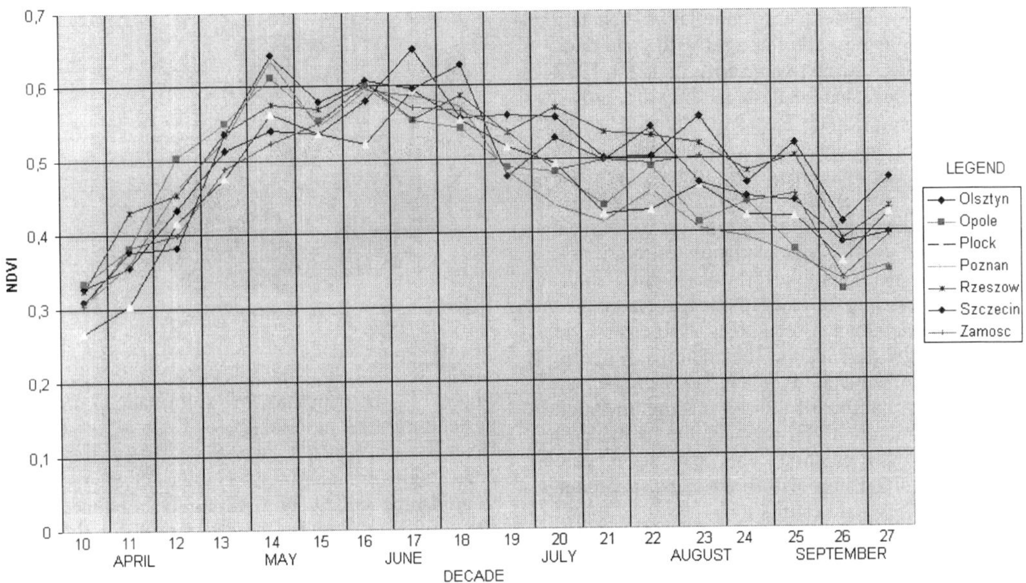

Figure 1. Changes of NDVI values throughout growing season for seven representative districts in Poland

revealed some stress conditions in second decade of April.

2. In all regions of the country (excluding easternmost areas of eastern Poland) NDVI reached its maximum values in second decade of May. Hence, full crop development appeared two weeks earlier, than in 1997 growing season.

3. At third decade of May stress conditions, evidenced by drop of NDVI values, were observed in most of Poland's territory, while in June vegetation index returned to its maximum values.

4. Substantial decrease of NDVI values appeared in the whole Poland at first decade of July, due to maturity phase of cereals.

5. At second part of growing season (mid-July - September) some fluctuations of NDVI values were observed, due to phenological phases of roots crops and agricultural practices (moving, harvesting). Significant decrease of NDVI values appeared for all districts in second decade of September.

Changes of NDVI values for seven districts (voivodships) representative for the main geographic regions in Poland were presented in figure 1.

Simultaneously, the spatial / temporal analysis was done for 1998 growing season as compared to situation in normal year. The following conclusions can be drawn from this analysis:

1. At first part of growing season (April – May) in almost whole Poland crop growth conditions were better (or much better) than in corresponding decade of normal year.
2. In June level of crop development was generally similar or better comparing to normal year; better conditions were observed in northwestern and southern Poland.
3. At first decade of July crop status was in general close to that of normal year, with a few voivodships in southern and northwestern Poland revealing comparatively worse situation.
4. In August almost the whole Poland, except of its northeastern part, was characterized by better crop situation than in normal year.

4. ANALYSIS OF RELATIONSHIPS BETWEEN NDVI VALUES AND AGRICULTURAL PARAMETERS

4.1. Relation between NDVI values versus crop yield and species structure

Numerous correlation analyses were conducted at final stage of the works, in order to determine relationships between remote sensing based vegetation indices and agricultural parameters characterizing crop growth conditions. Large database, which comprises NDVI values covering 1992 – 1997 period and yield data delivered by the Central Statistical Office (CSO), was constructed in order to accomplish this task. This database also includes other relevant agricultural data, like crop structure and agrometeorological parameters.

At first phase regression analyses were conducted for all decades of growing season, covering development of cereals. Various approaches were used in these analyses, e.g. cumulative vegetation index was applied for correlating it with yield data. Detailed analysis, which was carried out for 1992 – 1997 period revealed, that depending on growing season optimum decade for yield inferring, as well as strength of relation varied. Generally, the best relationship occurred in May, when correlation coefficients were around 0.7. Use of cumulative index calculated from first part of growing season (April – May) usually enhanced yield estimation.

At second phase some efforts were undertaken in order to assess impact of crop species structure on strength of relations between NDVI values and yield data. First, 24 voivodships characterized by high percentage of cereals (over 60 %) were selected from the total number of 49 voivodships in Poland. Next, correlation analysis was conducted, based on the selected representation. This analysis revealed, that use of voivodships with high cereal percentage greatly enhances yield estimation.

On the basis of cumulative NDVI data for 1997 growing season regression equation was derived for assessing cereal yield values. Next, map of distribution of yield estimates based on remote sensing derived indices was produced and compared with the map prepared on the basis of conventional yield data. Generally, there was quite high consistency between both types of estimation, especially for highly productive voivodships in northern and western Poland. The mean relative error of 7.5 % was achieved from this comparison, which is recognized as acceptable by statistical specialists. Chart illustrating comparison of yield estimation with the use of conventional method (yield CSO) and remote sensing based method (yield NDVI) was presented in figure 2.

4.2. Relation between NDVI values versus crop yield and valorization of agricultural land

It was decided to check in this part of the work, if stratification of agricultural land into categories of cropland quality will enhance relationship between NDVI values and yield data. In order to do that, the whole territory of the country was divided into four categories, characterized by different quality of agricultural land. Value of cropland valorization index was the basis for stratification. This index is based on soil parameters, relief and agrometeorological conditions. As a result of stratification voivodships characterized by high valorization index were included in first category, voivodships with medium indices to second and third category, while those with the lowest index to fourth category, respectively. Spatial distribution of this index on the territory of Poland is presented in figure 3.

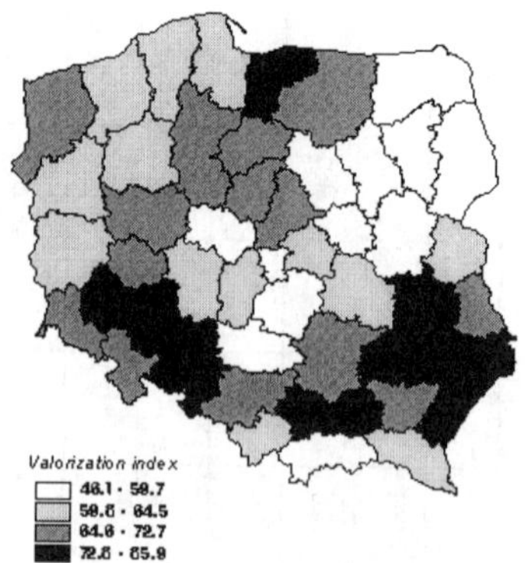

Figure 3. Distribution of valorization index of agricultural land in Poland

Next, regression analysis was done for each group of voivodships separately. This analysis revealed, that the significant relation appeared only for the group of voivodships characterized by high valorization index. The regression equation was derived on the basis of this group and used for yield estimation. The results of this estimation were compared with official yield data collected by the Central Statistical Office. The mean relative error of 8.4 % was achieved from this comparison. So, the accuracy of yield assessment is very similar to that obtained, while using voivodships with high cereal percentage. It means, that application of stratification of cropland for setting-up regression equation does not make important contribution to precision of final yield estimation.

4.3. Relation between NDVI values and agrometeorological parameters

The aim of final part of the work was to study relationships between NDVI values and the most important agrometeorological parameters, which influence crop development: rainfall and air temperature. In order to make this analysis the database, comprising rainfall values in all decades of growing season, as well as mean and maximum temperature values, was created for 1992 –1998 period. At first stage relation between NDVI and rainfall values were studied. As a result of these studies it was found, that for 1992 growing season, characterized by great variability of precipitation within vegetation period, there is significant relation between sum of rainfall values, which appeared in May and NDVI values for first decade of June. Similar correlation was obtained, while summing up rainfall values from first two months of growing season (April – May), whereas the correlations for individual decades were lower. So it can be concluded, that cumulating of rainfall within month's (or longer) period significantly affects NDVI values.

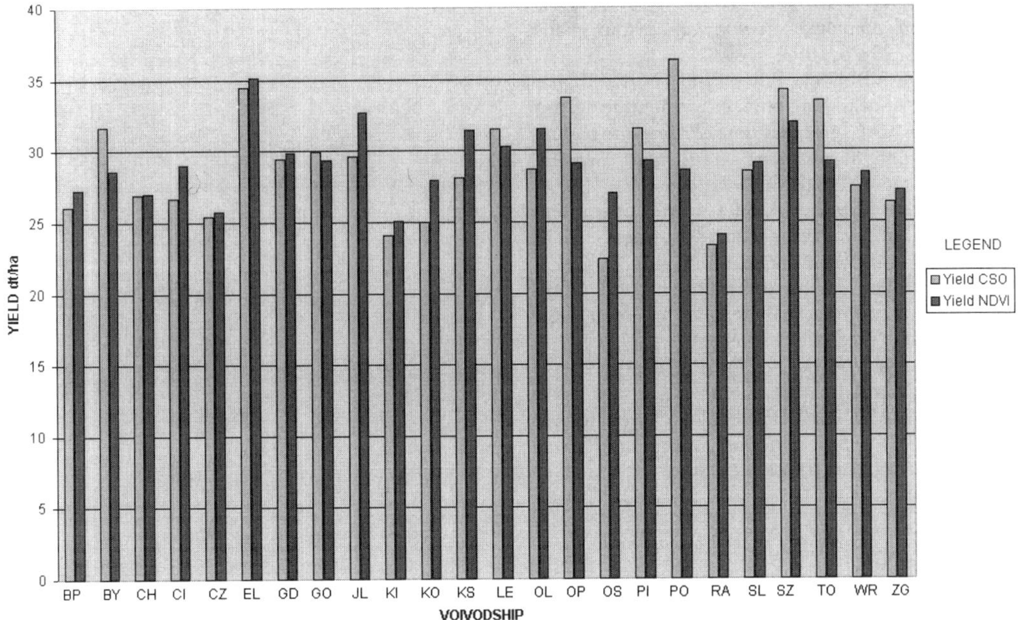

Figure 2. Comparison of yield estimation with the use of conventional and remote sensing based method

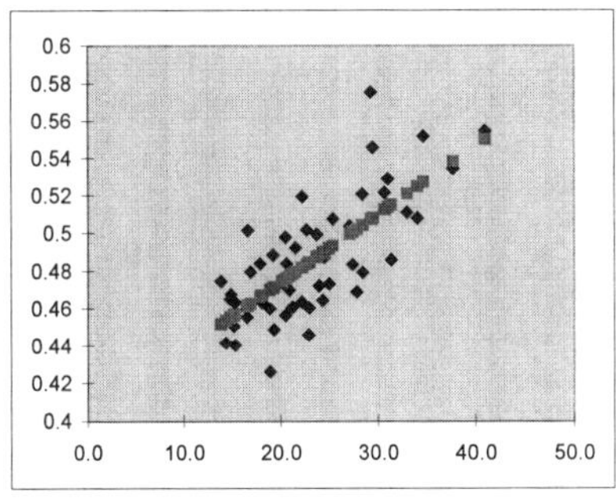

Figure 4. Relation between NDVI values and rainfall in May 1992

Exemplary relationship between rainfall and NDVI values is presented in figure 4.
Similar analysis was done for studying relations between NDVI and air temperature values. It was found for all examined years, that there is significant relation between sum of May temperatures and vegetation index – NDVI, cumulated through first part of growing season (April – May). Strength of relation depends on synergistic impact of both agrometeorological parameters – rainfall and air temperature.

5. CONCLUSIONS

The work presented in this paper resulted in creating the Crop Condition Assessment System in Poland, based on use of NOAA AVHRR images. The specialized software for processing NOAA satellite data has been prepared and historical database was constructed, as well as current NOAA images were collected, in order to ensure proper implementation of the system. The regression analyses conducted between remote sensing based vegetation indices and various agricultural parameters revealed impact of crop species structure and changes of agrometeorological data on NDVI values. They also demonstrated possibilities of inferring yield of cereal crops from normalized vegetation index, cumulated from early stages of crop development. The system for monitoring crop growth conditions became operational, thus enabling delivery of near real time information to the decision makers responsible for agricultural production in Poland.

REFERENCES

Bullock P.R. 1992. Operational estimates of Western Canada grain production using NOAA AVHRR LAC data. *Canadian Journal of Remote Sensing.* Vol. 18, No. 1.

Hochhheim K.P., Bullock P.R. 1993. Operational estimates of Western Canada spring wheat using

NOAA AVHRR LAC data. *Proceedings of the 12th W.T. Pecora Memorial Symposium.* Sioux Falls, USA, August 23-26, 1993.
Reichert G.C., Bullock P.R., Prout N.A., Walker G.K. 1991. A near real time crop monitoring tool using NOAA LAC AVHRR data. *Proceedings of the 14th Canadian Symposium on Remote Sensing.* Calgary, Canada, May 6-10, 1991.
Walker G.K. 1988. Model for operational forecasting of Western Canada wheat yield. *Agricultural and Forest Meteorology.* 44, 339-351.
Wood D. 1993. The use of multitemporal NDVI measurements from AVHRR data for crop yield estimation and prediction. *International Journal of Remote Sensing.* Vol. 14, No. 2, 190-210.
Yang w., Yang L., Merchant J.W. 1997. An assessment of AVHRR/NDVI ecological relations in Nebraska, U.S.A. *International Journal of Remote Sensing.* Vol. 18, No. 10, 2161-2180.

7 Hazards

Remote Sensing in the 21st Century: Economic and Environmental Applications, Casanova (ed.)
© 2000 Balkema, Rotterdam, ISBN 90 5809 096 5

Monitoring the distribution of shallow landslide-prone areas using remote sensing, DTM and GIS: A case study from the tropical Andes of Venezuela

H.Gómez, R.P.Bradshaw & P.M.Mather
School of Geography, University of Nottingham, UK

ABSTRACT: This paper describes progress in the development of a landslide susceptibility model which is based on slope stability equations. Using a test area in the Venezuelan Andes, we show how model inputs can be derived from suitably pre-processed Landsat TM data and maps of the area. Pre-processing steps include geometric and terrain corrections. A combination of unsupervised and supervised classification algorithms is applied to the corrected TM data to derive a land cover map. The paper concludes with a summary of future research activity.

1 INTRODUCTION

Amongst environmental hazards, landslides are considered to be some of the most catastrophic and morpho-dynamic processes that affect mountainous terrain (Smith 1992; Grupta & Joshi 1990; Cooke & Doornkamp 1997). Several landslide definitions exist; the one used in this research is based on Cruden (1991) and quoted in Turner & Schuster (1996: 4) who defines these phenomena as "the movement of a mass of rock, debris or earth down a slope". Various other terms are used to denote landslides, including mass wasting and landslips.

The disasters associated with landslide occurrence are manifold. Amongst the most important effects are loss of human life and severe negative impacts on natural resources. Landslides also cause damage to man-made features such as dams, aqueduct systems, roads, buildings and so on.

The Andes are prone to landslide occurrences. Indeed, a large number of Andean countries have been seriously affected by these phenomena (Ericksen et al. 1989). The severity of the damage caused by landslides can be massive, including the destruction of whole towns in both Peru and Colombia. For instance, in Peru between 1962 and 1983, landslides killed an estimated 29,634 people. In Colombia in 1985 alone some 22,000 people died from landslide events (Cooke & Doornkamp 1997).

Although of less magnitude, landslides have also affected the Venezuelan Andes. In 1994, an intensive precipitation with 120 mm of rainfall in six hours in the State of Táchira triggered several landslides that caused not only ecological impact but also severe damage to the regional aqueduct system, which supplies water to a population of some 900,000 inhabitants. The Venezuelan government had to invest more than 600 million dollars to repair the damage. A similar situation is repeated every year, mainly during the rainy season, when landslides not only destroy the water-supply system, communications, bridges and houses but can also kill people. Despite being a recognised calamity, landslides remain a little studied phenomena in Venezuela. Moreover, no official records are kept of lives lost nor the value of property damaged due to landslide impacts.

There are a large number of methods for analysing landslides. One of the traditional methods has been to take into account slope stability, which may be defined by a factor of safety *f*. This method has been successfully applied using on-site measurements or through analysing samples of rocks and soils in the laboratory. However, this type of analysis is prone to fail because of external parameters that can also influence slope stability (Cooke & Doornkamp 1997). The aim of this paper is to present the early stages for developing a model for predicting extensive landslide-prone areas applying remote sensing techniques, DTM and GIS and using as a frame of reference the slope stability equations.

2 STUDY AREA

The study area is located in the western part of Venezuela, approximately 750 km from Caracas (Figure 1). It covers the river Jabonosa basin in the

State of Táchira, with a surface area of about 24,418 hectares and a perimeter of 64.7km.

The area's topography is mainly mountainous and rugged, dissected by many rivers and streams. Elevation ranges from 550m to 3,500m above mean sea level. The main water-course, the river Jabonosa, has a length of 18.5 km.

Most of the rocks exposed in this valley are of Mesozoic age including sandstones, limestones and clays. A thick, discontinuous mantle of detritus covers the slopes. Most of these superficial deposits and landforms are the result of intensive erosive processes that affected the Venezuelan Andes during the Jurassic period (MARNR 1990). The Jabonosa basin is structurally controlled by a group of active geological faults, with the Boconó fault one of the most conspicuous.

Under these conditions, the surface material has been seriously affected. Thus, different types of mass movements including mudslides, landslides and rock failures are active on the side-slopes of the valley (Gómez & Guerra 1996).

3 THE MODEL TO PREDICT LANDSLIDE-PRONE AREAS.

According to Abramson et al. (1996) the most important factors required in order to analyse the stability of a slope include: 1) the shear strength of the soils; 2) slope geometry; 3) pore-water pressure or seepage forces and 4) loading and environmental conditions. These factors can be evaluated using formulae such as the safety factor and the Mohr-Coulomb equations. The safety factor can be defined either as the ratio of the shear strength divided by the shear stress or the sum of forces resisting slope failure divided by the sum of disturbing forces (Cooke & Doornkamp 1997). The following safety factor formula has been used for infinite slope analysis. It will be used as the starting point in the development of this model:

$$f = \tan \phi / \tan \beta \qquad (1)$$

where f = safety factor; ϕ = internal friction angle of rocks and soils; and β = slope angle.

If $f > 1.0$ stability is likely, but if $f < 1.0$ instability exists and the slope is considered to be prone to failure.

An extra variable that has been added to the safety factor equation is vegetation. Indeed, Wu et al. (1979) pointed out the important role played by vegetation within slope stability analysis. Thus, parameters such as cohesion, the internal friction angle, the weight of the soil, and pore-water pressure, all tend to be substantially modified by the presence of vegetation. For instance, vegetation can both enhance effective soil cohesion due to root matrix reinforcement, and also enhance effective soil cohesion due to soil suction or negative water pressure through evapotranspiration and interception (Coppin & Richards 1990; Greenway 1987). The root network provides mechanical reinforcement to the soil. Indeed, apparent cohesion of the soil provided by root systems frequently falls in the range of 1.0 to 17.5 C_R(kN/m^2) (Coppin & Richards 1990).

Gray (1970) analysed the effects of vegetation on shear strength and shear stress behaviour. It was found that vegetation has an influence through the mechanical reinforcement of the root system, slope surcharge, wind leverage and root wedging. This finding was later analysed by Brown et al. (1975) who also established that the cutting and removal of trees, with the consequent reduction in overburden and wind loading to zero, can significantly increase slope stability. Wu et al. (1979) observed that the beneficial effects of a root system may persist for some time after a tree has been cut down. However, this beneficial effect only lasts for a few years due to the gradual decay of the root networks, with approximately 50 per cent of the root strength being lost within five years (Selby 1993).

From the discussion above it can be seen that the presence of vegetation, mainly through the root system, results in an overall increase in the strength and compaction of a soil (Coppin & Richards 1990). Indeed, tree effects on hillslopes with shallow soils are thought to increase soil shear strength by more than 60 per cent (Selby 1993).

It is clear that vegetation plays an important role in maintaining slope stability. This has been shown by analysis of details such as root tensile strength, root depth, root diameter, and so on, which have been carried out in both the field and laboratory (Coppin & Richards 1990).

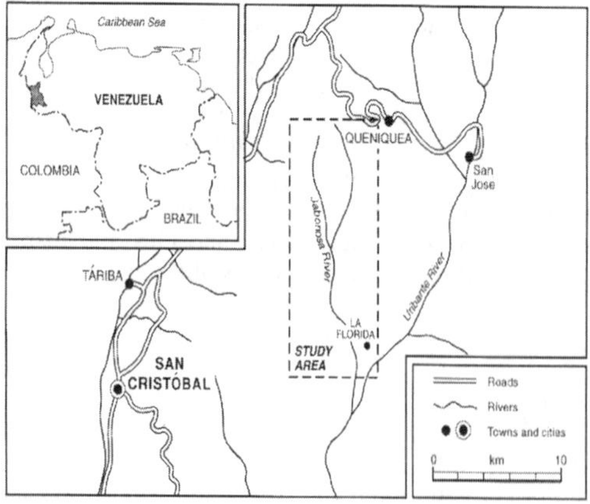

Figure 1. Study area

In addition to the root parameters used in the safety factor formulae, other vegetation elements have been introduced into the equations used for analysing slope stability. Amongst such elements it is possible to mention the transverse width of the crown of the tree, the surcharge due to vegetation and the reduction in pore-water pressure and the enhancement of effective soil cohesion due to evapotranspiration (Greenwood 1983).

In general, it can be seen that the effects of vegetation on slope stability have been analysed by taking into account individual features of trees. Conversely, in this study, vegetation will be considered as a single and integrated protective mantle of the soil, rather than as a series of individual elements. Vegetation will also be considered in relation to the surface area occupied (density).

Previous findings show that many external parameters that control the stability of slopes, including vegetation cover, lineaments, slope angle and soil moisture, and which in turn influence the angle of friction, pore-water pressure, soil cohesion, can all be analysed using remote sensing, DTM, and GIS.

Taking into account all of these previous studies, and since many researchers have also found that deforestation produces adverse effects on slope stability (Bishop & Steven 1964; Brown et al. 1975; Gray 1970; Greenway 1987), vegetation cover is considered in this model as one of the principal variables that influence shear strength forces. In addition to vegetation cover, other variables are included within the model, such as slope angle, lineaments and soil moisture.

The effect of gravity on the movement of material on slopes is a function of the slope angle, which has an important influence on shear stress, contributing to mass movements on hillslopes. Simply put, the steeper the slope the more liable it is to be unstable (Cooke & Doornkamp 1997). For this reason a slope variable has been considered in most slope stability equations as a contributing instability factor.

The water content within a soil has an important impact on slope stability. For instance, Barton & Choubey (1977) found that the internal friction angle decreases in most materials when water enters the surface horizon of the soil. Also, in a fully saturated soil, the pore-water pressures causes a loss in soil cohesion. Furthermore, a rise in pore-water pressure reduces soil strength because, although confined water can withstand a compressive stress, it cannot withstand a shear stress (Selby 1993).

Lineaments such as geological faults, joints and other discontinuities can produce zones of weakness, ones of altered permeability. Furthermore, lineaments can be correlated with the landslide triggering mechanism of earthquakes. Greenbaum et al. (1995) found an increased probability of landslides originating in the vicinity of a lineament.

On the basis of the above considerations it was possible to build the following model using the engineering safety factor equation f:

$$f = \text{resisting forces} / \text{disturbing forces} \Rightarrow \text{vegetation cover}/\Sigma\beta + L + SM \quad (2)$$

Since different types of vegetation appear to give different levels of protection to the soil against instability, equation (2) may be modified to find f:

$$f=\Sigma[(T*Si)+(S*Si)+(G*Si)]-(BS*Si)/\Sigma\beta+L+SM \quad (3)$$

where T = tree, S = shrub, G = grass, B = bare soil, Si = surface in per cent class (i = 0-100), β = slope angle, L = lineament, SM = soil moisture.

Both trees and shrubs, with tensile strengths in the range of 5 to 68 MN/m^2 and root depths up to 3 m deep, are assumed to offer more soil protection on steep slopes than grass. The latter has a tensile strength in the range between 0.0 and 18.5 MN/m^2 and root depths up to 40 cm deep (Coppin & Richards 1990).

As a first step the following provisional weighting values were assigned: T = 4, S = 3, G = 2, and BS = 1.

The following flow-chart shows the different stages of the model (Figure 2):

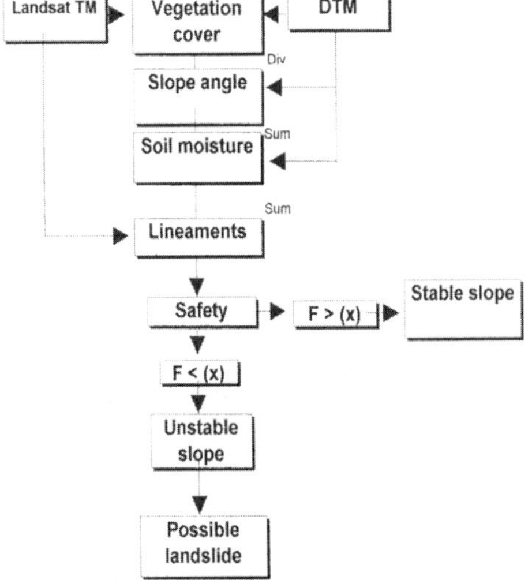

Figure 2. The model flow chart.

4 THE DIGITAL TERRAIN MODEL (DTM) FOR THE JABONOSA BASIN

A DTM was developed for this study. The DTM was based on x,y,z point format data obtained from digitising both the contour lines and the drainage network from a hardcopy 1:100,000 scale map. This map provides contour information at 40 metre intervals.

The digitising of the topographic survey map information was undertaken using the TOSCA package and used a manual, line-following method. A total of 10,175 points and 572 arcs were selected. The binary data obtained were converted to an ASCII file and transferred into the PANACEA package. The DTM was created within the PANACEA package using a grid-based model. The dimensions of the final grid were 315 rows by 185 columns and the cell size 65.3 by 64.9 metres.

Derived DTM data were used to estimate the slope angle and slope aspect. The digitised drainage network was used to refine the DTM by ensuring that valley bottoms were correctly positioned to coincide with the location of streams. Both the DTM and satellite images were draped over the model to determine, qualitatively, the accuracy of the DTM. The DTM was transferred into ArcInfo in order to export it into ERDAS. In ERDAS the DTM was used to create slope and aspect images. Also, the DTM was utilised in the application of the Minnaert model to topographically normalise the Landsat TM image (Section 6.1).

5 THE LANDSAT TM SATELLITE IMAGE

A Landsat Thematic Mapper image taken on September 2nd 1992 was used in this study. The coordinate centre was 72°20′39″W and 07°14′05″N and the sun elevation and azimuth were 52°and 87° respectively.

The image was georeferenced to a UTM coordinate system. The rectification undertaken here consisted of three steps, which included; a) location of ground control points (GCPs), b) image transformation techniques, and c) resampling of the image.

The number of GCPs selected here were close to the range suggested by Mather (1996). Sixteen points were selected and were widely spread throughout the study area. Although different band combinations were tried it was found that bands 3, 2, and 1 were the most useful for selecting the GCPs.

Using the ERDAS IMAGINE program, a transformation matrix was computed taking into account the GCPs obtained above. Coefficients derived from this transformation were used in polynomial equations to convert the coordinates. The total root mean square error (RMS error) was 0.0885 for the sixteen selected points. The X RMS error and the Y RMS error were 0.0536 and 0.0703 respectively.

The nearest neighbour method provides efficient procedures with the ability to produce output images with real pixel values. It has the ability to discriminate between different types of vegetation and to determinate edges associated with geological lineaments (Jensen 1986). For these reasons this method was used for resampling the TM image. From the full image an area of 401 columns and 681 rows was selected. This image area fits within the study area, in which the river Jabonosa basin is located.

6 LAND COVER CLASSIFICATION

This classification was carried out in order to elaborate a land cover image of the Jabonosa basin as input data for the model. To undertake this classification the following two steps were carried out:

6.1 *Correcting for topographic effects*

Since this is a mountainous area, it was necessary to correct for topographic effects within the scene. The Non-Lambertian assumption and Minnaert model were utilised in order to address this problem. The following formula was used to derive the Minnaert constant (Smith et al. 1980 and Colby 1991):

$$L(\lambda,e) = L n(\lambda) \cos^{k(\lambda)-1} i \cos e^{k(\lambda)-1} \quad (4)$$

where L = radiance, λ = wavelength, e = exitance angle = slope, Ln = radiance angle, k = Minnaert constant. Hence, the Minnaert constant (k) was obtained by linearizing (4):

$$L \cos e = Ln \cos^{k} i \cos^{k} e \quad (5)$$

and then obtaining the regression value for k. That is,

$$\log (L \cos e) = \log Ln + k \log (\cos i \cos e) \quad (6)$$

where y = log (L cos e), b log (Ln), and x = log (cos i cos e).

The linear form is y = a +bx.
The k value was calculated using a simple linear regression model:

$$b = n\sum xy - \sum x \sum y / n\sum x^2 - (\sum x)^2 \quad (7)$$

where n = the number of data points.

To estimate the k value and apply this formula, the following models were developed using the spatial modeller in the ERDAS package (Figs 3 and 4).

The k values obtained from the model can be seen in the following table (Table1):

Table 1. Minnaert model: k values

Bands	Values
1	0.505
2	0.504
3	0.493
4	0.564
5	0.549
6	0.522

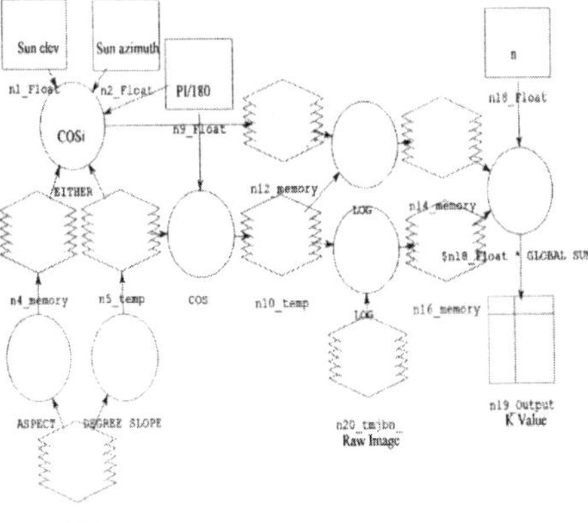

Figure 3. Minnaert model.

Using the data estimated from the model, a new model was developed in order to normalise the image (Figure 4). That is, using the Non-Lambertian assumption:

$$Ln = (\mathrm{L} \cos e) / (\cos^k i \ \cos^k e) \tag{8}$$

Cosine i can be defined by

$$\cos i = \cos(90\text{-}\theta_s) \cos\theta_n + \sin(90\text{-}\theta_s) \sin\theta_n \cos(\phi_s - \phi_n) \tag{9}$$

where θ_s = solar elevation, θ_n = surface normal zenith angle or slope of the terrain surface, ϕ_s = solar azimuth angle, and ϕ_n = surface aspect of the slope angle.

The following input data were used to carry out the equations:

- Landsat image bands 1,2,3,4,5, and 7.
- DTM.
- Sun elevation
- Solar azimuth

Partial correction for topographic effects was carried out, which greatly improved the image quality. The main problems encountered during the application of this correction were in those areas where the slope angles are greater than 65° and slope aspects greater

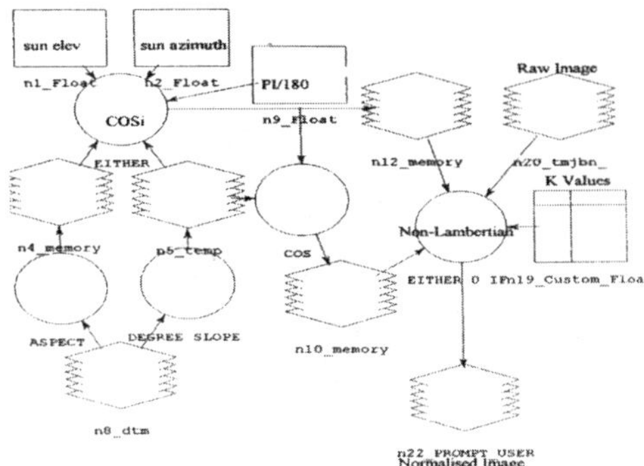

Figure 4. The Non-Lambertian Model

than 245°. In these areas the pixel value was 0, which appears to be a problem with the software. As a result a slight modification was introduced into the original model in order to include only those slopes and aspects whose values were less than the angles noted above. The total area affected was less than 20 pixels.

6.2 Performing the classification

The classification used in this study was a combination of both unsupervised and supervised methods. The normalised Landsat TM image, including bands 1,2,3,4,5, and 7, was used to perform the classification. Using an unsupervised classification 50 clusters were automatically selected by the ISODATA algorithm (Tou & Gonzalez 1974) with a maximum of 10 iterations for re-clustering the data.

Using signature files and field experience, these 50 spectral classes were merged into the four major classes required for the model, i.e. trees, shrubs, grasses and bare soils. Other signatures (clouds and river) were also discriminated from the image but were without significance for the model.

The classes were evaluated using feature space objects and signature separability within the ERDAS program. This evaluation was carried out in order to rule out those bands that were not useful for the results of this classification. The transformed divergence was also computed. Applying this technique bands 1, 2, and 5 were suggested as the best band combination to perform the classification. The separability value obtained (2,000) is located in the upper bound of the transformed divergence. It means that in these three bands the selected signatures are totally separable.

The supervised classification was undertaken using the maximum likelihood procedure, and a classified image containing six classes was obtained (Figure 5, - note: river omitted).

Figure 5. The Jabonosa basin. Land cover classification using Landsat TM data. (colour plate, see p. 602).

This classification was also evaluated. In this case, accuracy assessment was used and 256 points were randomly selected. With the 50 unsupervised classes and field knowledge, the random points were contrasted, keeping on the screen both images and the 'inquire' cursor to locate the coordinate system for each point. The classes selected in this way were introduced as input in the accuracy assessment table. Error matrix, accuracy total and the Kappa statistic were estimated. The overall accuracy was 89.50% and the overall Kappa was 86%, which is acceptable.

7 LINEAMENT INTERPRETATION

The lineament analysis was performed using the Landsat TM image. This interpretation was carried out utilising both the ERDAS and IDRISI packages. Since band 3 is useful for determining geological boundary delineation (ERDAS 1994) it was used in this analysis.

A low-pass filter was used in order to block the high spatial frequency detail in this band (Jensen 1986). The low-frequency kernel averaged the values of the pixels. The resulting image was smoother. The size of the convolution kernel utilised was 3x3 (Table 2):

Table 2. Low-frequency kernel

1	1	1
1	1	1
1	1	1

Afterwards, a similar edge enhancement technique and weighted line with a mask size 5x5 was used. The following kernel definition was applied (Table 3):

Table 3. Edge enhancement. Kernel 5x5

N-S	N-W	W-E	SW-NE
-1-1 4 -1-1	4-1 -1-1-3	-1-1 -3-1-1	-3-2 -1-1 4
-2-2 8 -2-2	-1 8 -2-3-2	-1-2 -3-2-1	-2-3 -2 8-1
-3-3 12 -3-3	-1-2 12-2-1	4 8 12 8 4	-1-2 12-2-1
-2-2 8 -2-2	-2-3 -2 8-1	-1-2 -3-2-1	-1 8 -2-3-2
-1-1 4 -1-1	-3-2 -1-1 4	-1-2 -3-2-1	4-1 -1-2-3

In this manner the different directional components were generated.

The new image was exported into the IDRISI package. To perform the visual analysis and lineament extraction, a black and white colour palette was used. Taking into account the image histogram a reclassification (scaling) was carried out. Using a user-defined reclassification the value 0 was assigned to the range 0-175 and the value 1 was assigned to the range 175-250. In this way, the directional components created earlier were improved for visual interpretation.

To analyse this new image it was necessary to magnify the image using a zoom display. The image was expanded by a factor of 3. From this zoomed image the lineaments could be easily identified within the scene. Using on-screen digitising techniques, the most important lineaments were selected in each direction. A vector file containing the different directional lineaments was generated for each image.

Finally, the four vector layers were merged. Thus, a single vector file was created, containing the entire lineaments for the different directions (Figure 6). This vector file was transferred into ERDAS.

8 DISCUSSION

These early stages of this project are an attempt to investigate the benefits of applying various techniques using remote sensing and DTMS, in order to analyse landslides-prone areas. Several results are given which demonstrate the progress in the elaboration of this model. The utilization of

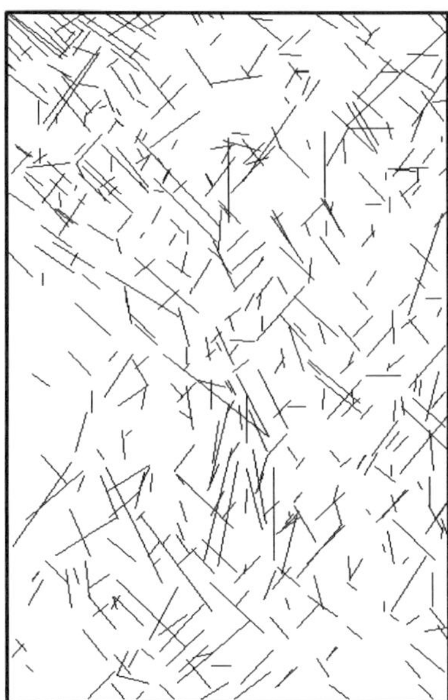

Figure 6. Lineaments within the Jabonosa basin identified from Landsat TM image.

different methodologies, data sources and techniques, as shown here, is not a simple task, and great care is needed if mistakes to be avoided. The next steps in this investigation will be to estimate the density of the lineaments and to calculate soil moisture zones using DTM and hydrological models. Afterwards using GIS, it is expected to run the full model. The last stage in this study will be a validation *in.situ* of the proposed model.

9 CONCLUSIONS

The aim of this paper has been to describe the development of a model that can be used for predicting susceptibility to landslides in areas of mountainous terrain. It is expected that such a model could be used as a basis for guiding the precise location of sampling points and testing in specific geo-technical site studies.

One of the main benefits of this approach is that it has the potential for minimising the need for exhaustive fieldwork and time-consuming analysis of samples in the laboratory. A second, but no less important, benefit of this strategy is that this model can be applied in areas of scarce information where detailed geological and other related information is lacking. Indeed, the early results obtained during the development of this model show how a large volume of data can be gained using techniques such as remote sensing and DTMS

10 ACKNOWLEDGEMENT

We would like to acknowledge the Venezuela Petroleum Company (PDVSA) for providing the Landsat TM image that has been used in this study.

REFERENCES

Abramson, L. *et al.* 1996. *Slope stability and stabilization methods.*, New York, USA: John Wiley & Sons, Inc.

Barton, N. & V. Choubey 1977. The shear strength of rock joints in theory and practice. *Rock Mechanics.* 10 : 1-54.

Bishop, D. & M. Stevens 1964. *Landslides on logged areas in south-east Alaska.* US Department of Agriculture, Forest Service Research Paper NOR- I.

Brown, C. & M. Sheu 1975. Effects of deforestation on slopes. *J of the Geotechnical Eng. Division.* 101 (GT2): 147-165.

Colby, J., 1991. Topographic normalization iii rugged terrain. *Photogrammetric Engineering & Remote Sensing.* 57(6): 531-537.

Cooke, R. U. & J. C. Doomkamp 1997. *Geomorphology in environmental management. A new introduction.* Oxford University Press.

Coppin, N. & I. Richards 1990. *Use of vegetation in civil engineering.* London: Butterworths.

ERDAS 1994. *ERDAS Imagine Version 8.1. Field Guide.* Atlanta: ERDAS Inc.

Ericksen, G. *et al.* 1989. Landslide hazards in the central and southern Andes. In E. Brabb & B. Harrod (eds.), *Landslides: Extent and economic significance - Proc. 28th Int. Geol. Congress - Symposium on landslides.* Washington D.C. 17 .luly 1989: 111-117. Rotterdam: Balkema.

Jensen, J. H. 1986. *Introductory digital image procssing. A remote sensing perspective.* Englewood, NJ: Prentice-Hall.

Gómez, H & F. Guerra 1996. Morfodinámica andina: microcuenca de la Qda. La Jabonosa (Cuenca alta del Pereño-Uribante). Edo. Táchira, Venezuela. *Memorias V Congreso Internacional de Ciencias de La Tierra.* Santiago de Chile: Sociedad Geográfica de Chile & Instituto Geográfico Militar de Chile.

Gray, D. 1970. Effects of forest clear-cutting on the stability of natural slopes. *Bull. Association of Eng. Geol.* 3 (1-2): 45-66.

Greenbaum, D. *et al.* 1995. *Rapid methods of landslide hazard mapping.- Papua New Guinea case study.* Technical Report WC/95/27. Nottingham: British Geological Survey.

Greenway, D. 1987. Vegetation and slope stability. In M. G. Anderson and K. Richards (eds.), *Slope Stability.* Chichester: John Wiley & Sons.

Greenwood, J. 1983. A simple approach to slope stability. *Ground Engineering* 16 (4): 45-48.

Gupta, R. & B. Joshi 1990. Landslide hazard zoning using the GIS approach - A case study from the Ramganga catchment, Himalayas. *Engineering Geology.* 28 :1 19-131.

Mather, P. M. 1996. *Computer processing of remotely-sensed images. An introduction.* Chichester: John Wiley & Sons.

Ministerio del Ambiente y de los Recursos Naturales Renovables (MARNR) 1986. *Atlas del Estado Táchira*. San Cristóbal. Venezuela.

Selby, M. 1993. *Hillslope materials. and processes*. Oxford: Oxford University Press.

Smith, J., *et al*. 1980. The Lambertian assumption and Landsat data. *Photogrammetric Engineering & Remote Sensing*. 46(9): 1183-1189

Smith, K. 1992. *Environmental hazards. Assessing risk and reducing disaster*. London: Routledge.

Tou J. & R. Gonzalez 1974. *Pattern recognition principles*. Reading, MA: Addison-Wesley.

Turner, A. & G. Jayaprakasah 1996. Introduction. In A. Tumer and R. Schuster (eds.), *Landslides Investigation and Mitigation*. Washington, D.C.: National Academy Press.

Wu, T. *et al*. 1979. Strength of the tree roots and landslides on Prince of Wales Island, Alaska. *Canad. Geotech. J*. Vol 16.

Remote Sensing in the 21st Century: Economic and Environmental Applications, Casanova (ed.)
 ISBN 90 5809 096 5

Towards an operational monitoring of drought conditions in Spain using satellite images

F.González-Alonso, A.Vázquez & J.M.Cuevas
INIA, Ministry of Agriculture, Madrid, Spain

A.Calle & J.L.Casanova
Remote Sensing Laboratory, University of Valladolid, Spain

ABSTRACT: Spain is a country which is particularly affected by drought given that during the period from 1880 to 1980 over 50% of years were dry or very dry. During the 1990s the period from 1992 to 1995 was one of the driest in the century. In 1992 losses caused by drought in cereal crops were put at 50 655 million pesetas (47% of the total production value). In 1995 losses in crops were estimated at 27 971 million pesetas (30% of the total production value). The identification of areas affected by drought using information supplied by the AVHRR sensor on the NOAA satellites is viable if a sufficiently large data base is available and could prove useful for those in charge of monitoring this type of phenomena.

1 INTRODUCTION

Definitions of drought vary although in general drought can be considered to be present when there is a significant reduction in precipitation which leads to an important drop in plant growth and development, both in the case of crops or vegetation.Spain is a country which is particularly affected by drought given that during the period from 1880 to 1980 over 50% of years were dry or very dry.

During the 1980s seven years were classed as dry or very dry and in the 1990s five years have so far been given the same classification.

The 1994-1995 hydrologic year has been the worst since the beginning of the decade, as in the southern half of the country precipitation was below 50% of normal values and in the northern half of the country ranged between 75% and 50%, with the exception of isolated areas of Galicia, The Basque Country and Catalonia.

In July 1995 drinking water restrictions affected 12 million people in Andalusia.

According to agrarian trade union organisations during the period from 1992 to 1995 losses caused by drought in the whole of the agrarian sector amounted to over one and a half billion pesetas. According to the national agrarian insurance company losses caused by drought in cereal crops in 1992 and 1995 reached 50 655 and 27 971 million pesetas respectively, meaning a loss in total crop production for those years of 47% and 30 % respectively.

Various authors have tested the usefulness of vegetation indices derived from NOAA-AVHRR images in the identification and monitoring of areas affected by drought at a regional scale with fairly satisfactory results (*Tucker and Choudhury, 1987; Gutman, 1990; Teng, 1990, Liu et al., 1994*).

The goal of this work is to develop a methodology and software to identify and carry out operationally monitoring of areas affected by drought in Spain using the calculation of the Normalised Difference Vegetation Index (NDVI) derived from NOAA-AVHRR images, in such a way that this information may be used by decision makers in agrarian, forest and environmental policy.

2 METHODOLOGY

Since 1993 NOAA-AVHRR images in HRPT format have been received by the aerial installed at The University of Valladolid Remote Sensing Laboratory (LATUV).

The methodology which has been developed to identify and carry out monitoring of areas affected by drought in Spain involves the following stages:

a) Radiometric and atmospheric correction of channels 1 and 2 of the AVHRR sensor.
b) Daily calculation of the Normalised Difference Vegetation Index (NDVI) using channels 1 and 2 of the AVHRR sensor.
c) Geometric Correction of the corresponding images to the daily NDVI.
d) Calculation of the ten day composite corresponding to the maximum NDVI value in each

pixel in the ten day period considered. This image is called MNDVI.

e) Generation of a multitemporal MNDVI image file from 1993 to the present day for each ten day period from February to October.
f) Calculation of the mean MNDVI value in the 1993-1999 period for each ten day period from February to October, excluding the year under study. This image is called MEANMNDVI.
g) Calculation for each ten day period of the year under study of the ratio between the MNDVI value of the year in question and the MEANMNDVI mean value for the period 93-99 (excluding the year under study).

Expression of this ratio in percentage terms and grouping in different types assigning them a specific colour scale (yellows for values below 100% and greens for values above).

3 RESULTS

In order to evaluate the methodology developed 1995 and 1998 have been taken as examples. 1995 was the driest year during the 1993-1997 period and 1998 was one of the hydrologic years with precipitation above normal in Extremadura, Andalusia and most of Castile and Leon, with the exception of the north. In the Cantabrian cornice, Catalonia and Levante, accumulated precipitation was below normal.

Figure 1 shows the mean value of the MNDVI anomalies during the period from February to July 1995. The prevalence of yellow shades in the southern half of Spain, except in the Cantabrian cornice can clearly be seen which is in perfect agreement with the pattern shown by the negative anomalies that occurred in the precipitation in that year, which in turn were the cause of the considerable losses in agriculture and in the lack of drinking water in Andalusian cities.

Figure 2 shows the mean values of the MNDVI during the period from February to July 1998. The prevalence of green shades over practically the whole of Spain can clearly be seen, especially in Extremadura, Castilla la Mancha and Andalusia, in agreement with the abundant precipitation recorded in the whole of that year. However, as can be seen in Figure 3, in the centre of Catalonia, there was a significant negative anomaly MNDVI during the third ten day period of June 1998 and in fact in this area an important fire broke out between July 18 and 21 which destroyed 27 000 hectares in the area of Solsonés, as can clearly be seen in Figure 4 in intense yellow.

Figure 1 and 2: Mean value of the MNDVI anomalies during de period from February to July. Figure 1 is 1995 year and figure 2 is 1998 year. (colour plate, see p. 602).

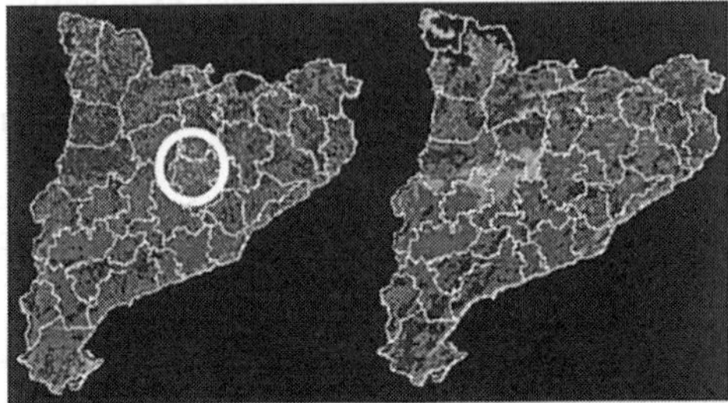

Figure 3 and 4: Catalonia images before forest fire, in July 1998, and after forest fire.

4 CONCLUSIONS

The results obtained in this work reflect the feasibility of carrying out the identification and monitoring of areas affected by drought using the information supplied by the AVHRR sensor on the NOAA series satellites.

The operational application of this methodology in the future will allow the rapid identification of those areas which might be affected by drought and quick decision making concerning palliative measures to reduce as much as possible the harmful effects on crops and forests.

Obtaining long temporal series of NDVI values will allow more and more representative mean values to be calculated enabling greater accuracy and robustness in the determination of the type of anomalies which might be occurring in the development of the vegetation in a given year.

5 ACKNOWLEDGEMENTS

This work as been carried out within the framework of the development of the SC96-096 research project, financed by the Sectorial Research and Development Program of the Spanish Ministry of Agriculture, Fisheries and Food.

6 REFERENCES

- Gutman, G.,1990. Towards monitoring droughts from space. Journal of Climate, 2: 282-295.

-Liu, W.T., Massambani, O., and Nobre, C.A., 1994. Satellite recorded vegetation response to drought in Brazil. International Journal of Climatology, 14: 343-354.

-Teng, W.L., 1990. AVHRR monitoring of U.S. crops during the 1988 drought. Photogrammetric Engineering and Remote Sensing, 56: 1143-1146.

-Tucker, C.J. and Choudhury, B.J., 1987. Satellite remote sensing of drought conditions. Remote Sensing of Environment, 23: 243-251.

Remote Sensing in the 21st Century: Economic and Environmental Applications, Casanova (ed.)
© 2000 Balkema, Rotterdam, ISBN 90 5809 096 5

Land degradation assessment in a semi-arid environment through the use of spectral indices and spectral unmixing

D. Haboudane, F. Bonn & A. Royer
CARTEL, Université de Sherbrooke, Qué., Canada

S. Sommer & W. Mehl
EGEO, Space Applications Institute, JRC, Ispra, Italy

ABSTRACT: The Guadalentin basin, located in the SE of Spain, has a semiarid climate and presents the typical characteristics of mediterranean landscapes vulnerable to land degradation processes and desertification risks. In such an environment, when the vegetation cover is low, the signal received by satellites is dominated by the spectral properties of soils. Changes in these properties can be interpreted in terms of varying soil surface conditions. These optical changes underline the major modifications affecting soil surface under land degradation processes. This research uses remote sensing techniques to characterize land degradation based on two approaches: the spectral mixture analysis and a set of indices describing the spectrum shape. Specific objectives consist in evaluating the potential of extending the color-derived indices to the short-wave infrared region, and in identifying landscape units according to their sensitivity to land degradation.

1 INTRODUCTION

Water erosion is a major source of soil degradation in semi-arid countries. Combined with the destruction of vegetation cover, it contributes to the increase of land degradation and desertification risks. The definition and identification of areas sensitive to these phenomena require the use of major descriptive parameters of climate, land and vegetation. Previous research activities developed methodologies based on analyzing correlations between agents of climate, physiography, lithology, vegetation and land use. While few of these parameters can be measured directly, previous studies have shown that remote sensing gains increasing importance in providing useful surrogate information. However, the extraction of geomorphically and environmentally significant information from this data source remains poorly developed because of the extreme complexity of the relationship between the spectral response and the ecological (environmental) parameters, namely geomorphic characteristics and physiographic conditions (McDermid and Franklin, 1994). In semiarid environments, remotely sensed data are largely influenced by the background soils and rocks. Whereas certain previous research activities developed methodologies to remove the soil contribution to the remotely sensed signal (Huete, 1988; Bannari et al. 1996), others elaborated approaches to enhance information originating from soils (Escadafal et al., 1994; Escadafal & Bacha, 1995; Mougenot & Cailleau, 1995). When the vegetation cover is low, the signal received by satellites is dominated by the spectral properties of soils. The latter vary with soil surface conditions such as texture, composition, moisture, colour and degradation level. They indicate the major changes affecting soils under degradation processes.

2 OBJECTIVES

The aim of this research is to use a remote sensing approach to characterize land degradation based on two approaches: the spectral mixture analysis and a set of indices describing the spectrum shape. Specific objectives consist in evaluating the potential of extending the color-derived indices to the short-wave infrared, identifying landscape units according to their sensitivity to land degradation and desertification processes, and comparing the obtained results to those generated by the spectral mixture analysis.

3 GEOGRAPHICAL SETTING

The Guadalentin basin, our study area, is centered about 20 km to the NW of Lorca, in Murcia and Almeria provinces, SE of Spain. It presents a semiarid climate with precipitation ranging between 260 and 400 mm and a mean temperature of about 17°C, and has altitudes ranging between 260 and 2036 m. These climatic conditions exert a strong influence upon the vegetation cover, which is low and of various types, and control land-use types which are divided into semi-natural dispersed forests, scrubland, abandoned agricultural land and agricultural land with various types of culture like cereals and fruits. Geologically, the basin is located in the eastern sector of the Cordilleras Beticas, which is a part of the western segment of the European Alpine belt (Melia et al., 1993). From the structural point of view, it is divided into two structural domains, the external and internal zones, differentiated by their paleogeographic and tectono-metamorphic evolution. Above these units, there is a filling material of the post-nappe basins developed after the upper Miocene.

The external zone consists of sedimentary units dominated by carbonates and marls. It comprises many lithological units including gypsum, clay, marls, limestones, dolomites, calcarenites and so forth. These outcropping materials cover about 75% of the basin area. The internal zone, which covers the southern part of the study area, consists of schists comprising different lithological units dominated by terrigeneous and carbonate materials, namely: quartzites, conglomerates, phyllites, limestones and dolomites. The Quaternary deposits are made of clays, sands, conglomerates and calcareous crusts of small extension (Melia et al., 1993).

4 MATERIAL AND METHODS

4.1 *Spectral data*

Two types of spectral data were collected for this research: field reflectance spectra of natural materials and satellite optical imagery.

In the framework of the MEDALUS (Mediterranean Desertification and Land Use) project, a field campaign was conducted to collect spectral data needed for modeling and interpretation of remote sensing images. To this purpose, the team of EGEO (Environment and GeoInformation) at the JRC (Joint Research Centre) performed spectroradimetric field measurements in June 1994. The objective was to provide spectral endmembers for the spectral mixture analysis and to construct the basis for a library of reference spectra of surface components such as soils, rocks and vegetation. The reflectances were measured with the GER-IRIS spectroradiometer of the EGEO Unit. Designed for bi-directional reflectance measurements, the instrument measures in 900 wavelength bands between 400 and 2500 nm. It has bandwidths of 1,5 nm in the visible/near infrared (VIS/NIR) and 4,8 nm in the short-wave infrared (SWIR). During the campaign, reflectance spectra were measured both on various types of soils and rocks and on different species of vegetation. They allowed the characterization of different soils conditions (developed, degraded, etc.) and various status of vegetation (green, dry, grey wood, etc.)

Concerning satellite data, they consist of Thematic Mapper (TM) images acquired July 01, 1989 and July 25, 1992, with the following solar illumination conditions: 63° elevation, 115° azimuth and 59° elevation, 118° azimuth, respectively. They have been purchased as 'system processed' products from the Earthnet Program Office (EPO) of the European Space Agency (ESA). Acquired in summer time, the two images are cloud free. As atmospheric conditions were unknown at the times of image acquisition, we used the 'dark-object' approach developed by Kaufman and Sendra (1988) to correct the atmospheric effects on image radiometry.

4.2 *Data processing*

The methodology adopted in this research involved three separate stages: pre-processing steps including geometric and radiometric corrections, spectral mixture analysis of the original image spectra, and use of indices describing the spectrum shape to characterise soil degradation.

A careful analysis of field spectra leads to distinguish various rock types, different vegetation species, and diverse soil conditions. Shape and absorption features analysis allow us to understand and infer soil climax and degradational forms (well-developed, degraded, etc.). When convolved to the band passes of TM, the major characteristics of the spectrum are still preserved. So we selected spectral endmembers (surface components) that will be involved in the spectral mixture analysis. But, the integration of the field measurements into the analysis of optical data requires a conversion of satellite images to physical quantities, such as bi-directional reflectances. Thus, relative radiances (digital numbers) of TM images were converted to reflectance factors by a combined atmospheric and topographic correction (Hill et al., 1995). The images were also rectified to geographic coordinates (UTM) and corrected from the relief displacement effects (ortho-rectification).

Next, the spectral decomposition of the original image was performed by linear spectral mixture modeling (Adams et al., 1989) in order to discriminate ground cover proportions at a subpixel level. It results in an abundance image (fraction image) for each endmember, giving the relative proportion of a surface component at every pixel. Then, erosion status and soil conditions were determined through a combination of automatic classification and cluster analysis of the fraction images. The resulting image was used for comparison with the results obtained by the implementation of spectral indices.

The spectral indices used in this study were derived from the concept of soil color. The latter is considered as an important soil property, because it is an indirect measure of other parameters like soil alteration, organic matter, iron oxides, and texture (Escadafal, 1993). Its changes reflect soil modification occurring through natural and human induced phenomena. Previous studies demonstrated that the Red, Green, Blue color coefficients (R,G,B) are strongly correlated with reflectance values in visible and near infrared bands of Landsat Thematic Mapper (Escadafal, 1993; Mattikalli, 1997). These physically based results are a powerful tool for the interpretation of the relationships between soil color and remote sensing data acquired over non-vegetated or sparsely vegetated regions.

Soil color can be quantitatively defined in terms of three basic variables: intensity (I) which represents the brightness of a color, hue (H) which gives a numerical description of the average wavelength (spectral range) of a color, and saturation (S) which measures the purity of a color (Liu & Moore, 1990). The hue is an important parameter because its values are independent of illumination and are related only to the shapes of spectral signatures (Liu & Moore, 1990). Therefore, it can be used for identification and discrimination of soils and lithological formations (Mattikalli, 1997).

Hence, definition of the indices describing the spectrum shape is derived from the color transformation from the RGB system to the IHS space (Schowengerdt, 1983; Liu & Moore, 1990). In the case of natural surfaces, simplified transformation equations were proposed by Escadafal et al. (1994), giving intensity (Equation 1), coloration index (Equation 2) and shape index (Equation 3) for TM visible bands.

$$I = \frac{1}{\sqrt{3}}(TM1 + TM2 + TM3) \qquad (1)$$

$$IC31 = \frac{TM3 - TM1}{TM3} \qquad (2)$$

$$IF321 = \frac{2TM3 - TM2 - TM1}{TM2 - TM1} \qquad (3)$$

Where TM1,2,3 are expressed in reflectances.

These indices, dedicated to the assessment of land degradation in semiarid environments, use only bands from VIS/NIR region of the solar spectrum. Nevertheless, some pedogenesis indicators, such as clay minerals and carbonates, are well-identified in the short-wave infrared region (Baumgardner et al., 1985). Moreover, the spectral part ranging from 2000 to 2500 nm is well-suitable for the discrimination of rock types. Consequently, it is of importance to evaluate the potential of extending the above indices to the short-wave infrared, and to assess the contribution of this part of the solar spectrum in discriminating different levels of soil degradation.

In remote sensing, soil color designated in its basic constituents I, H and S is usually reserved to the visible domain. But, the corresponding variables intensity (Equation 1), coloration index (Equation 2) and shape index (Equation 3) can be used over the whole range of the solar spectrum. Thus, while intensity corresponds to the average brightness which is related to the presence or absence of organic matter, coloration index represents the slope of the reflectance curve, and the shape index describes the general shape of the reflectance spectrum. The two latter indices can be used to detect other indicators of pedogenesis, such as iron oxides and clay minerals. In fact, these indicators exhibit absorption features at specific wavelengths.

5 RESULTS AND DISCUSSION

Before applying the color-derived indices to image data, we used field spectra with known levels of soil development (degradation). The results obtained are shown in Figure 1 (shape index *versus* intensity) and Figure 2 (shape index *versus* coloration index). In these figures, we denote that in general degraded soils have high values of intensity while the developed soils exhibit low levels of intensity (Figure 1-b). Conversely, coloration and shape indices have high values for degraded soils and low values for developed one (Figure 2-b). As we are interested in evaluating the potential of spectrum shape for soil degradation assessment, our analysis will focus on the shape index which is represented as a function of intensity (Figure 1) and coloration index (Figure 2).

The shape index using only visible bands (Figures 1-a and 2-a) is not able to clearly distinguish different degradational forms of soils. Both degraded and developed soils present similar values of shape index. The latter doesn't reveal any trend in characterizing soil degradation. On the other hand, the integration of short-wave infrared bands (TM5 and TM7) in combination with those of the visible and near infrared (TM1, TM3, TM4) allows a good discrimination between developed, intermediate, and degraded soils (Figures 1-b and 2-b).

Field spectra show that shape index *IF* is negatively correlated with the soil brightness *I* (Figure 1-b) and positively correlated with coloration index *IC* (Figure 2-b). In fact, *I* describes degradational processes in an opposite manner as the *IF*. The cross-plot of *IF* and *I* exhibit a general trend that allows the characterization of soil degradation with a satisfactory discrimination level. Scatter-plots of *IF* versus *IC* (Figure 2-b) show that the values are well-dispersed in the *IF-IC* space. Clouds of values representing different levels of degradation are scattered along an axis slightly parallel to the bisector of the angle formed by the primary axes. Values corresponding to the same degradation level constitute a cloud with principal axis more or less perpendicular to the bisector. Consequently, in the dispersion diagram *IF versus IC*, developed soils are represented by low values (cloud near the diagram origin) and degraded ones by high values (far from the diagram origin) along the bisector.

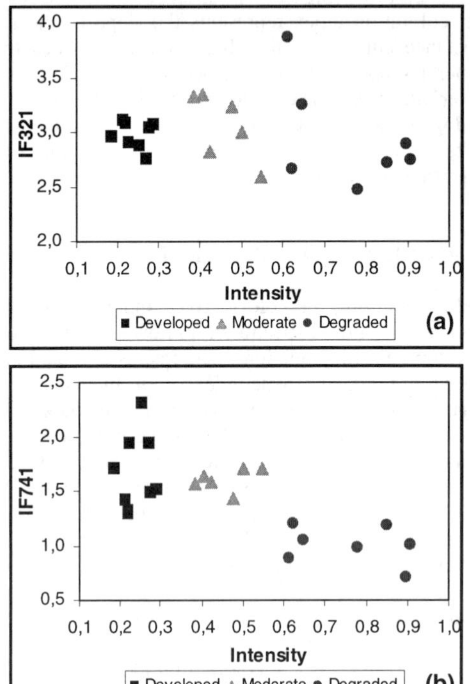

Figure 1. Shape indices combining a) TM bands 3,2, and 1, b) TM bands 7,4, and 1, *versus* Intensity for developed, intermediate and degraded soils. Field data.

Based on the field dataset, the following band triplets (TM1, TM4, TM5) and (TM1, TM4, TM7) generated the best dispersion of the values in the *IF-IC* space. This is in agreement with the findings of the study conducted by Price (1990) on the information content of soil spectra, and with the results of the research undertaken by Frazer and Cheng (1989) to investigate the correlation between TM bands and soils contents of organic matter and iron oxides. Price (1990) showed that broadband spectral measurements corresponding to TM3, TM4, TM5, and TM7 are sufficient to describe the studied soil spectra. Frazer and Cheng (1989) concluded that ratios involving TM1, TM3, TM4, and TM5 provide a good means of mapping soils regarding to their organic carbon and iron oxides contents.

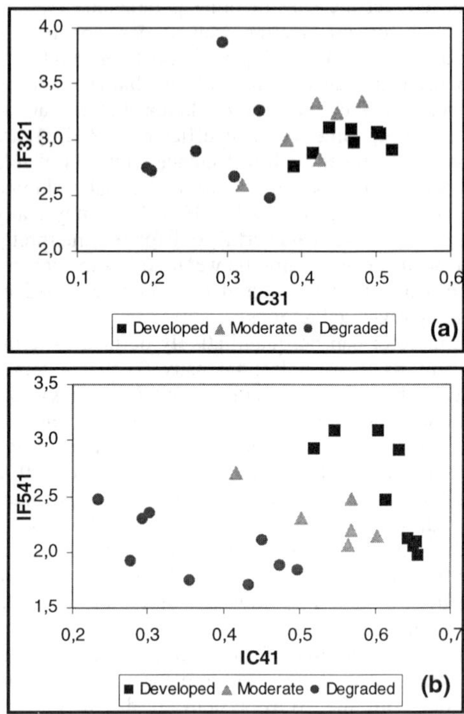

Figure 2. Shape indices combining a) TM bands 3,2, and 1, b) TM bands 5,4, and 1, *versus* Coloration indices for developed, intermediate and degraded soils. Field data.

Afterwards, the color-derived indices were applied to Landsat TM data to map soil surface conditions in the Guadalentin basin. We first extracted image spectral signatures representing three soil degradation classes similar to those defined with field measurements. Identification of these classes is based both on comparing image signatures to those measured in field, supported by our knowledge of the study area, and on the spectral and contextual evidences observed in TM images (agricultural zones, bare soils, vegetation, brightness indices, etc.). Thus, about 320 samples representing the three categories of soil were involved to characterize the degradation status over the study area. Obtained results are plotted in the Figures 3 and 4.

Scatterplots of shape index *IF* versus intensity *I* (Figure 3) and coloration index *IC* (Figure 4) are effective at confirming the trends established with ground measurements. While discrimination of degradation levels is better performed with indices integrating both VIS/NIR and SWIR bands (Figures 3-b and 4-b), indices using only VIS/NIR bands have a limited potential to distinguish soil degradation variations (Figures 3-a and 4-a). Therefore, indices including SWIR bands give a good description of the general shape of soil spectra the main features of which (slope, absorption bands, reflectance amplitude) are correlated with the presence or absence of pedogenetic processes. This is due to the combined effects of clay minerals, iron oxides, and organic matter.

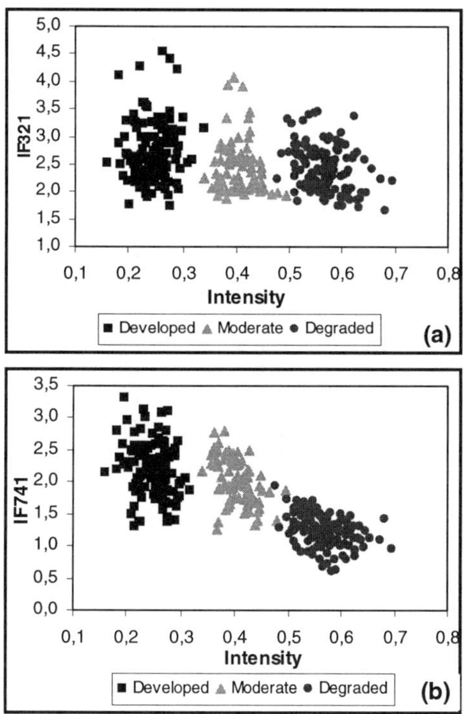

Figure 3. Shape indices combining a) TM bands 3,2, and 1, b) TM bands 7,4, and 1, *versus* Intensity for developed, intermediate and degraded soils. Image data.

However, the use of TM sensor to detect indicators of pedogenesis and, therefore, to predict stages of soil development faces the constraint imposed by the spectral resolution of TM bands. In fact, even if important characteristics of spectral continuum are preserved when convolved to band-passes of TM, the narrow absorption features disappear. TM bands are too broad to precisely detect absorption features of iron oxides and clay minerals. For that reason, it seems reasonable to infer that the color-derived indices describe general trends with satisfactory levels of discrimination between different conditions of land degradation. The nest generation of sensors designed to monitor the environment, like MODIS and MERIS will have several narrow bands with a better spectral resolution. They will provide a better characterization of soil spectra shape through a judicious selection of spectral bands that coincide with the absorption bands of pedogenesis indicators.

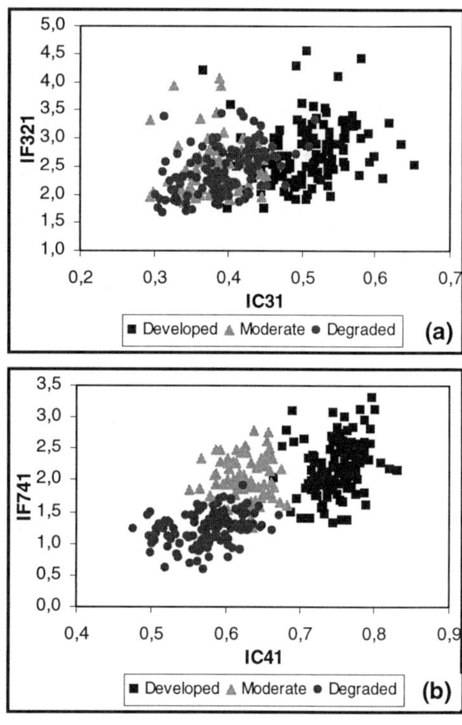

Figure 4. Shape indices combining a) TM bands 3,2, and 1, b) TM bands 7,4, and 1, *versus* Coloration indices for developed, intermediate and degraded soils. Image data.

In conclusion, the above results have shown that the indices describing soil spectra shape exhibit wide variations with degradation levels. They can be considered as indicators illustrating and measuring the changes affecting land surface under degradational processes. Their application to satellite images will contribute to map the environment status in general, and the soil conditions (development, deterioration) in particular. To this purpose, the Figure 5 presents the color composite obtained by encoding shape index *IF* in red, coloration index *IC* in green, and intensity *I* in blue. So, when used together these three variables enable the major vegetation types and different soil conditions to be discriminated by distinctive colors.

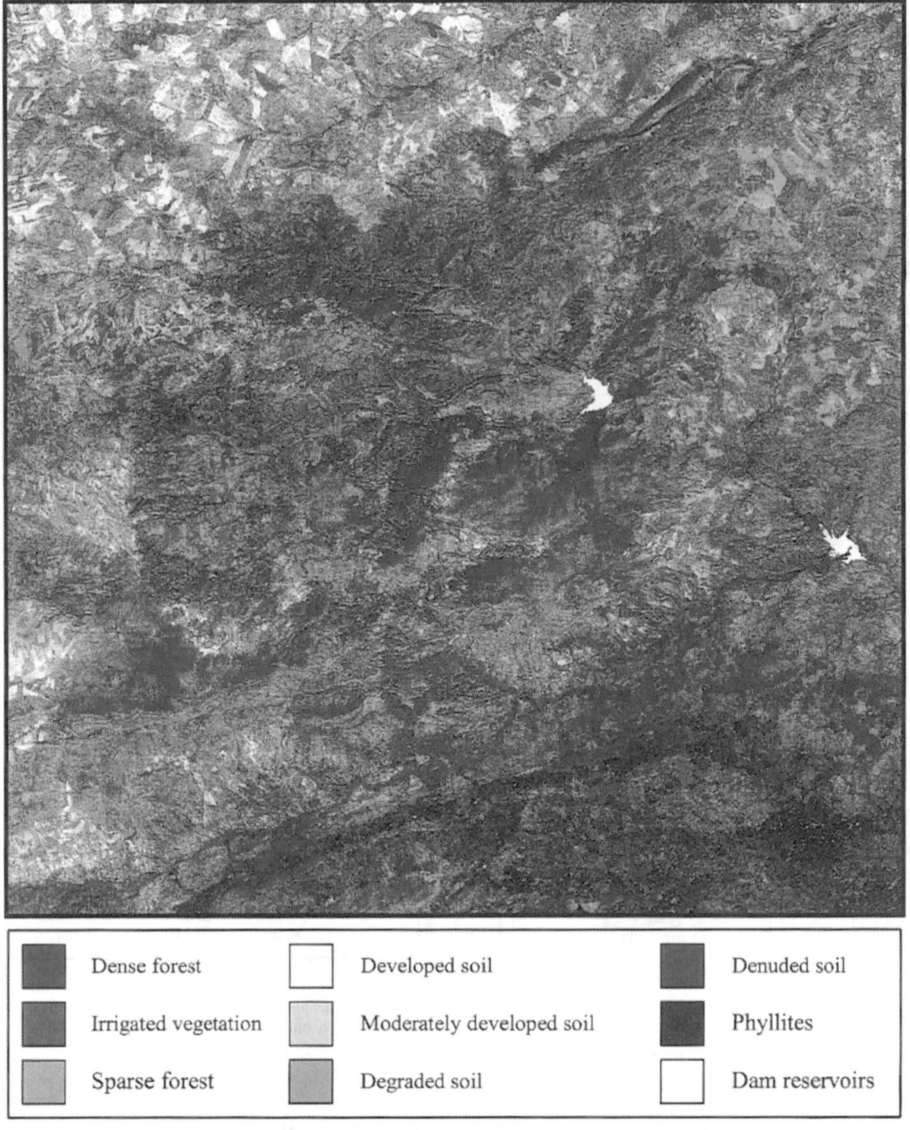

Figure 5. Color composite obtained by encoding shape index *IF* in red, coloration index *IC* in green, and intensity *I* in blue, for TM bands 7, 4 and 1. Image of 45 km by 45 km. (colour plate, see p. 603).

Concerning the vegetation cover, three categories were identified: dense coniferous forest represented in green, sparse forests indicated in dark green, and irrigated vegetation outlined by the cyan color. Regarding soil conditions, four degradation levels were distinguished: well developed, moderately developed, degraded, and denuded soils. In the image (Figure 5), they respectively correspond to the following colors: yellow, orange, magenta, and blue.

In the Paleozoic part of the study area (southern portion of the image), the dominating lithological formations (phyllites and schists) were clearly differentiated (Figure 5). In the phyllites sector, represented by dark colors, soils are rare and vegetation is very sparse so that only outcropping materials were identified. Inversely, in the schists sector, the vegetation cover appears dense and soils are identified as moderately developed. This can be explained by the fact that soils are usually developed inside the schist fissures and generally covered by rock fragments. The latter protect soils against water erosion, improve rain infiltration, and limit evaporation rate

(Geeson and Thornes, 1996). The vegetation abundance is due to the fissures which favor soil development in depth, and water accumulation in subsurface.

Concerning the spectral mixture analysis, the linear mixture model applied in Guadalentin allowed to use up to six surface components to characterise the study area. It used spectral characteristics of ground cover classes as known from field measurements. It yielded abundance estimates of soils, different parent materials (marls, carbonates, gypsum, schists, etc.), and vegetation. These fractions were rescaled to emphasise the relative abundance of developed soils and parent rock material, allowing image interpretation in terms of substrate components only.

An image of soil degradation levels was then obtained through a combination of automatic classification and cluster analysis. These approach generated results similar to those obtained by the color-derived indices. Comparisons with available geologic maps and ground truth observations confirmed that both methods (spectral indices and unmixing) lead to soil conditions corresponding fairly well to the spatial patterns of soils in the Guadalentin.

6 CONCLUSION

This research has shown that regional patterns of soil and vegetation conditions can be reliably mapped by using color-derived indices. The approach, which is based on the characterization of spectra shapes, provides results similar to those of the spectral mixture analysis. It reveals striking features: (i) the soil surface optical properties can be used as indicators of the degradation status in semiarid environments, and (ii) the indices integrating short-wave infrared bands are effective in detecting pedogenetic indicators and degradational features in sparsely vegetated regions.

ACKNOWLEDGEMENTS

The authors wish to thank the National Sciences and Engineering Research Council of Canada (NSERC grant A6043), the MEDALUS project, and the Ministères des relations internationales du Québec for their support.

REFERENCES

Adams, J. B., Smith, O. M., and Gillespie, A. R. 1989. Simple models for Complex Natural Surfaces: A Strategy for the Hyperspectral Era of Remote Sensing. Proceedings of IGARSS'89 Symposium, 10-14 July 1989, Vancouver, Canada,: 16-21.

Bannari, A., Huete, H. R., Morin, D., et Zagolski, F. 1996. Effets de la couleur et de la brillance sur les indices de végétation. Int. J. Remote Sensing, 17(10): 1885-1906.

Baumgardner, M. F., Sylva, L. F., Biehl, L. L., and Stoner, E. R. 1985. Reflectance properties of soil. Advances in Agronomy, 38: 1-44.

Demangeot, J. 1995. Les milieux naturels du globe. Édition MASSON Géographie, 5e édition, 313 p.

Escadafal, R. 1993. Remote sensing of soil color : principles and applications. Remote Sensing Reviews, 7: 261-279.

Escadafal, R., Belghith, A., et Ben Moussa, H. 1994. Indices spectraux pour la télédétection de la dégradation des milieux naturels en Tunisie aride. Sixième Symposium International "Mesures physiques et Signature spectrale en Télédétection", 17-21 janvier 1994, Val d'Isère, France: 253-259

Escadafal, R., and Bacha, S. 1995. Strategy for the dynamic study of desertification. Proceedings of the ISSS International Symposium (working groups RS and MD), Ouagadougou, Burkina Faso, 6-10 February 1995: 19-34.

Frazier, B.E., and Cheng, Y. 1989. Remote sensing of soils in the Eastern Palouse region with Landsat Thematic Mapper. Remote Sensing of Environment, 28: 317-325.

Geeson, N. A., and Thornes, J. B. 1996. MEDALUS II: Mediterranean Desertification and Land Use. Executive Summary Phase II. European Commission, DG XII, Environment and Climate Programme.

Hill, J., Mehl, W., and Radellof, V. 1995. Improved forest mapping by combining corrections of atmospheric and topographic effects in Landsat TM imagery. *In : Sensors and environment applications of remote sensing*, Rotterdam, Balkema: 143-151.

Huete, A. R. 1988. A soil-adjusted vegetation index. Remote Sensing of Environment, 25: 295-309.

Kaufman, Y.J. and Sendra, C. 1988. Algorithm for automatic atmospheric corrections to visible and near-IR satellite imagery. International Journal of Remote Sensing, 9(8): 1357-1381.

Liu, J. G.,and Moore, J. M. 1990. Hue image RGB composition. A simple technique to suppress shadow and enhance spectral signature. International Journal of Remote Sensing, 11(8): 1521-1530.

McDermid, G.J. and S.E. Franklin, 1994. Spectral, spatial, and geomorphometric variables for remote sensing of slope processes, Remote Sensing of Environment, 49: 57-71.

Mattikalli, N.M. 1997. Soil color modeling for the visible and near-infrared bands of Landsat sensors using laboratory spectral measurements. Remote Sensing of Environment, 59: 14-28.

Melia, J., Bastida, J., Gandia, S., Gilabert, M.A., Younis, M.T. and Buendia, A.L. 1993. MEDALUS project: Final report, covering the period of January 1991 –December 1992.

Mougenot, B. et Cilleau, D. 1995. Identification par télédétection des sols dégradés d'un domaine sahélien au Niger. Proceedings of the ISSS International Symposium (working groups RS and MD), Ouagadougou, Burkina Faso, 6-10 February 1995: 19-34.

Price, J. C. 1990. On the information content of soil reflectance spectra. Remote Sensing of Environment, 1990, 33: 113-121.

Schowengerdt, R. A. 1983. Techniques for image processing and classification in remote sensing. Academic Press, New York London, 1983: 117-126.

Remote Sensing in the 21st Century: Economic and Environmental Applications, Casanova (ed.)
© 2000 Balkema, Rotterdam, ISBN 90 5809 096 5

A study on deposit and erosion phenomena along the southern coast of Sicily using satellite remote sensing

Alberto L.Geraci, Francesco F. Landolina & Alessandro O.Trovato
Remote Sensing Group, Faculty of Engineering, University of Catania, Italy

ABSTRACT: Comparing images in the visible bands with maps from *Istituto Geografico Militare Italiano*, it was possible to localise many changes in the shape of the coastline of southern Sicily, Italy, from Porto Empedocle to Capo Passero, on the Italian side of the Malta Channel due to erosion and deposits. It was possible, also, to understand the effects of shore protection structures on coastline changes and on the ports development from a large scale point-of-view.
Since behind the study area, along the coast, the land is very sensitive to rain erosion, significant images were selected, among many others, showing the effects of meteorological events on the sea. In particular, images showing the effects of rivers silt flooding along the coast were processed, in order to obtain information on the changes of the distribution in the above mentioned parameters in the sea water. From the analysis of these images, much useful information on the coastal areas sensitive to deposit and erosion was acquired.

1 INTRODUCTION

Most of the economic activity in Sicily is related to the sea or in any case to the coastal zone.

Erosion and deposit phenomena along the coastline are extremely important. In particular, the deposit of material due to terrain properties and river characteristics has many significant aspects. The case study presented involves the coastline from Porto Empedocle to Capo Passero, and, in particular, data related to a day of significant rain intensity were used to highlight the current behaviour.

The analyses of biological, chemical and physical parameters distribution led us to obtain information on water mass displacements, rich in biomass. In particular, the following standard parameters for water quality were analysed: chlorophyll *a*, suspended solids, turbidity, temperature, and Secchi depth.

Chlorophyll *a* give us useful information on biomass property, the analyses of turbidity and Secchi depth allow us to obtain information on the range of sediment spreading.

In the last 20 years, the increase of human pressure has brought about an intense use of coastal areas, so that it has became necessary to install groins and breakwaters to protect the coastline from moving backward and forward because of current erosion and sediment deposits. Although in the last 20 years many coastal protections have been put in place, in some places the coastline has been substantially affected.

The aim of the present study is to realise a low cost method to forecast the interaction between marine currents and coastal protection highlighting which areas could be at risk of undermining or sediment buildup. A comparison between physical, biological, and dynamics parameters in standard meteorological conditions and after an intense rain event, allowed us to highlight the dynamic interaction between the currents and the sediments borne mainly in terms of dimensions.

The general approach involved near-simultaneous acquisition of water quality samples from boats and the study of LANDSAT Thematic Mapper (TM) overpass data at the location of a sample site on the LANDSAT TM scene. It also involved the development and verification of regression models relating concentrations of selected water quality variables to the spectral data; the application of models to the entire study area; and the production of colour-coded maps, each depicting the distribution of selected water quality parameters.

2 BASIN HYDRO-GEOLOGICAL CHARACTERISTICS

The coastline which interests us is included in both the "Platani" and the "Agrigentino" basin. Both basins have low permeability. The "Platani basin shows a sea front of 4 km and has several geological characteristics.

The surface break up of Flysch's sandstone and chalky sandstone series produces local changes of physical and biological characteristics in terms of both salt and sediment content of the water.

Since the rivers involved in the present study are very small, the difference between the standard flow rate and the same flow rate in the case of an intense rain event is very high. This produces very significant erosion and dragging on the riverbanks, which is also related to the geological properties of the terrain.

3 MATERIALS AND METHODS

To obtain the images we carried out the following steps:

- Acquisition of water quality data from boats, near-simultaneous with the LANDSAT-TM overpass.
- Extraction of digital numbers from TM bands.
- Development and verification of regression models, relating values of selected water quality parameters to the spectral data.
- Application of models to the entire study area.
- Production of colour coded maps, each depicting the distribution of selected water quality parameters.

The digital data of the study area were collected on 5 October 1997. A quarter scene of approximately 90 x 90 km (orbit 189 frame 34) was acquired from Nuova Telespazio. The crews on the boats collected 60 samples nearly simultaneously with the satellite overpass.

The sea sampling was carried-out by trained and experienced personnel in order to remain in a range of 2 hours across satellite overpass using no more than 2 crews of 2 persons each maintaining, as much as possible, a definite standard in sampling procedure. Instruments were used to measure *in situ* the selected parameters, in particular a Turner Design 10-AU field Fluorometer to measure chlorophyll *a* was used and a multisonde YSI 6820 to measure temperature salinity and turbidity. The standard method. for measuring Secchi depth was used of course.

All the statistical analyses were done using the Statgraphics. The parameters studied were treated as dependent variables; whereas the DNS collected were independent variables in the regression analyses. In a second stage, correlation and regression statistical procedures, between dependent variables and independent variables, were applied.

For example, correlation coefficients indicated that chlorophyll *a* and Secchi depth can be highly c6rrelated to the radiance values in bands 1 and 2. Plots of each water quality variable versus each TM band were also produced. Examination of these plots and preliminary models indicated that a natural logarithm transformation would be appropriate for all variables, except for the thermal data. Each statistical sample has been divided into two sub samples, to be used for model development and verification respectively.

The final selection of the model was based on R^2, probability of a greater F-value, and significance of the T-test for regression parameters. A significance level of 0.05 was used for all models and regression parameters.

Results included statistical summaries and colour-coded imagery, showing the distribution of the selected water parameters. The models obtained for each study area and for the various water parameters were as follows:

$$SSLD = 5.81Ln(B2) - 14.038$$
$$R^2 = 0,65$$

$$Ln(SECCHI) = - 11.64 - 1.39Ln(B2) + 1.36Ln(B1)$$
$$R^2 = 0.72$$

$$TEMP = 0.23B6 + 2.57$$
$$R^2 = 0.54$$

$$CHLA = -2.49 + 0.019Ln(B1) + 0.89Ln(B2)$$
$$R^2 = 0,84$$

$$TURB = 2.9Ln(B2) - 7$$
$$R^2 = 0.71$$

Figures 1 to 5 show the images generated using the model predictions as above.

During the image processing the same filter has been applied to mitigate the striping for all parameters excect temperature; in fact for this we applied an unlike and stronger filter.

The results obtained in this special condition are

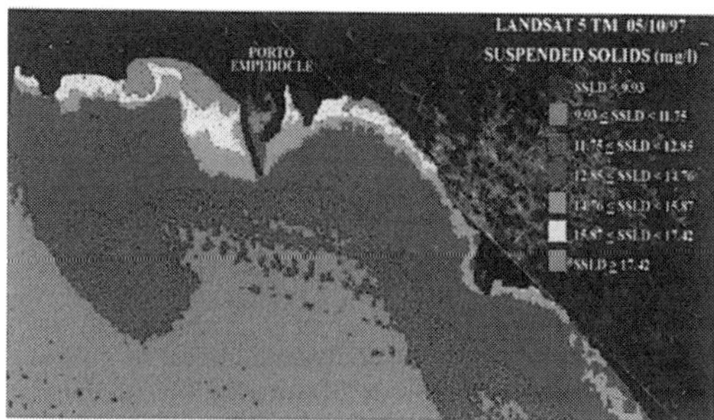

Fig 1. Image generated for marine currents, as measured by suspended solids, in front of Porto Empedocle. (colour plate, see p. 604).

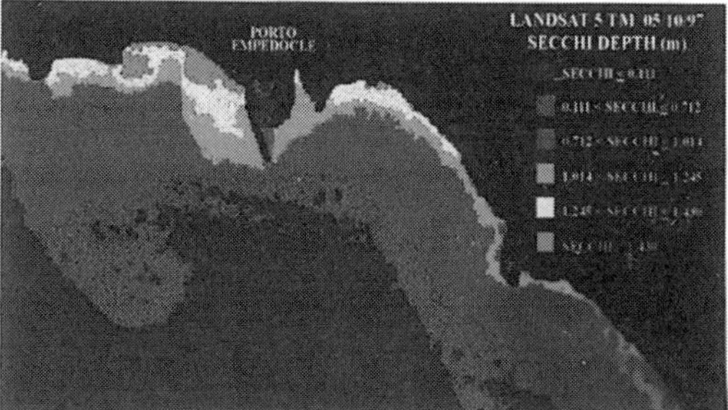

Fig. 2. Image generated for water transparency, as measured by Secchi disk, in front of Porto Empedocle. (colour plate, see p. 604).

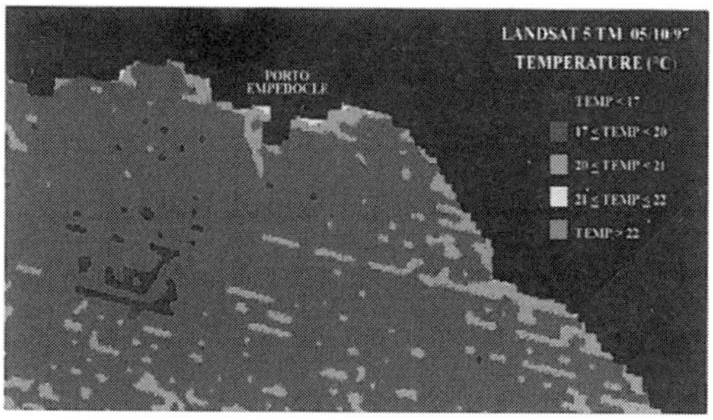

Fig.3. Image generated for surface water temperatures, in front of Porto Empedocle. (colour plate, see p. 604).

compared with standard values measured during normal meteorological conditions. We have obtained the most significant results for chlorophyll *a*, temperature, and turbidity.

For chlorophyll *a* we found an increase about of

Fig. 4. Image generated for chlorophyll *a* content of surface waters in front of Porto Empedocle. (colour plate, see p. 604).

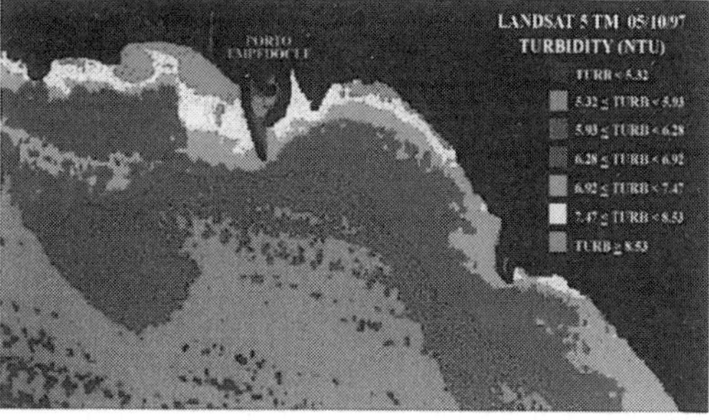

Fig. 5. Image generated for water transparency, as measure by turbidity, in front of Porto Empedocle. (colour plate, see p. 604).

0.5 µg/l, also the temperature values increase with respect to the normal values, whereas turbidity shows local variations.

4 DISCUSSION

The valuation about plume makes in this particular case of meteorological events can deeping the study between marine current and silt rivers flood or the in general with the products of fluvial erosion.

(In this particular case an evaluation can be made between marine currents and river silt flooding or, in general, with the effect of fluvial erosion.)

This, with a study about typology of solid carry can with little costs about traditional methods of determines the stability of coastal protection.

(This, in conjunction with a study about typology of solids carried can, at little cost, determine the stability of coastal protection installations).

It can to study more better to protect the effects of shore construction in neighbouring areas, and better to catch the influential area of the same neighbouring shore construction with the way marine current and of the suspended solids. All of this could may in the primary phase the effects for long time, avoiding carrying out with usual supporting construction for the unexpected effects about mining areas.

(The method used makes it possible to have a better knowledge of the effects of shore constructions on neighbouring areas by studying marine currents and suspended solids. This primary phase could point to the long-term effects and lead to avoiding certain constructions which can lead to unexpected effects).

A deep examination, could eliminating all the realisation of particular shore structures simply chancing only the present, and then in the local morphology either the erosion current or deposit current.

(An in-depth study could eliminate the realisation of certain shore structures by taking into account local morphology and either the erosion current or the deposit current.)

The sediment charge loads when the current velocity is at its top level, the whole global biomass is always present for a longer time. From the images, some particular areas can be pointed out as being rich in nutrients for several fish species, or as being good areas for the location of fish farms.

5 CONCLUSIONS

Using such LANDSAT TM data may be the start of creating a data bank of information on water quality in these local coastal waters, which may be periodically updated.

This monitoring could allow the checking of coastline changes, and therefore provide information about erosion and deposits, above all in this area, where the particulars hydro-geological characteristics could cause some problems.

REFERENCES

Brockhaus, J., A.L.Geraci, and S. Khorram. (1988), "Multitemporal Resources Complex analysis of Catania Province, Italy, from LANDSAT-TM Data", *Proceedings of the International Geoscience and Remote Sensing Symposium, IEEE Geoscieiice and Remote Sensing Society*, Edinburgh, England, 1, 201-204, 1988.

Geraci, A.L., R.N. Colwell, and S. Khorram. (1980) "Remote Sensing of Water Quality for Estuarine Environment", *Space Sciences Laboratory*, University of California, Berkeley, California, USA, XXI, 23

Geraci, A.L., (1985) "Techniche di Remote Sensing per il Controllo dell'Inquinamento del Mare", Proceedings of the *Ventitreesimo* Congresso Nazionale *della Junior Chamber Italiana su Mediterraneo*, ed. *Economia,* Camarina, Italy, 85-102.

Geraci, A.L. (1992) "The Control of Oil Pollution and Water Quality over Sicily's CoastalWaters Integrating Different Remote Sensing Techniques", *Proceedings of the Morocco* International Space year Symposium *on Evaluation* of *Natural Resources with Remote* Sensing Techniques, Morocco Royal Centre for Remote Sensing, Rabat, Morocco, 357-360..

Geraci, A.L., F. Landolina, L. Pantani, and G. Cecchi.(1993) "LASER and Infrared Techniques for Water Quality Control". *Proceedings of the Oil Spill Conference, American Petroleum Institute, Environmental Protection Agency*, United States Coast Guard, Tampa, Florida, USA, 525-529..

Geraci, A.L., F. Patania, and G. A. Fargione (1995). "The Remote Sensing Approach: A Modern Technique to Control Sea Pollution" *Proceedings of the Marina Conference*, Wessex Institute of Technology, St. Raphaël, France, 127-134..

Geraci, A.L., G.A. Fargione, and F. Landolina.(1996). "Classificazione Tematica del Territorio per lo Studio dell'Erosione tramite Dati LANDSAT". *Proceedings of the Giornata di Studio su POP Sicilia - Cartografia Tematica, Morfologia della Coste, Erosione del Suolo, e Gestione dei Sedimenti*, Ispra Joint Research Centre, Palermo, Italy, 43-50.

8 Oceanography

Remote Sensing in the 21st Century: Economic and Environmental Applications, Casanova (ed.)
© 2000 Balkema, Rotterdam, ISBN 90 5809 096 5

A mean sea level variation in the Mediterranean using Topex-Poseidon, ERS-1 and ERS-2 altimeter observations

M.Catalán-PU, P.Villares, M.Bruno, J.J.Alonso & A.Ruiz-Canavate
Department of Applied Physics, University of Cadiz, Puerto Real, Spain

M.Catalán-M
Real Instituto y Observatorio de la Armada, San Fernando Cadiz, Spain

ABSTRACT: Remote sensing techniques, as the high accuracy altimeter from satellites Topex-Poseidon, ERS-1and ERS-2, provide improved measures of mean sea level variation directly linked, via its satellite tracking stations, to the geodetic International Terrestrial Reference Frame. The aim of present communication is to study the different Mediterranean basins mean sea level tendencies using altimeter data from Topex-Poseidon and ERSs satellites. The study includes the correlation between the sea level variation in the Mediterranean basins and the results of an Empirical Orthogonal Functions analysis.

1 THE MEDITERRANEAN BASINS

The Mediterranean basin is a semi-arid region where, the predominance of dry north-west winds and clear skies conditions, leads to an evaporation that largely exceeds the precipitation and river run off, which transforms the Atlantic water inflows into highly saline and warm waters.

The Strait of Gibraltar is the only significant link between the Mediterranean and the world's oceans. The flows within it affect, fundamentally, the dynamics and the heat and material budget of both, the Mediterranean and the Atlantic. As the evaporative loss is water of zero salinity, the evaporation does not change the salt content in the Mediterranean, although can change the salinity if changes the amount of water in the basin. So, the excess of evaporation over precipitation and runoff over the basin, as a whole, should be balanced by a small net of Atlantic fresh water inflow through the Strait of Gibraltar.

However since both, Atlantic waters inflow and the Mediterranean outflow, have substantial salinity differences a much larger volume of Atlantic water masses, about 4%, must be exchange to maintain the balance of salt. In conclusion the water outflowing through the Strait of Gibraltar is a mixture of Levantine Intermediate water, formed by deep convection during the winter in the eastern Mediterranean which flows westwards as an intermediate layer, and the Western Mediterranean Deep Water, which originates in winter in the northern part of the western basin, under the dynamic and cooling action of the dry and cold 'Mistral' cyclone wind.

2 THE MEDITERRANEAN SUB-BASINS

After leaving the Alboran sea the Atlantic water masses flow into the western and eastern Mediterranean basins, becoming saltier and denser under the influence of intense air-sea interactions, forming the levantine intermediate water, during the winter convection in the Levantine basin. These water masses recirculates the Mediterranean at an intermediate depth and eventually exit towards the Atlantic via the Strait of Gibraltar.

The Mediterranean Deep Water (MDW) formation occurs, during the winter convection, in the Gulf of Lion and circulates through the channel of Ibiza towards the Atlantic.

All Mediterranean basins are connected by straits that allow the water mass transportation to the adjacent one. We could consider that the Mediterranean circulation is a combination of mesoscale and seasonal scales. In the west, the Atlantic water masses, after flowing eastward along the Spanish coasts veers south from Almería to Oran crossing through the Alboran sea, to form the Algerian current with an intense mesoscale activity of eddies and

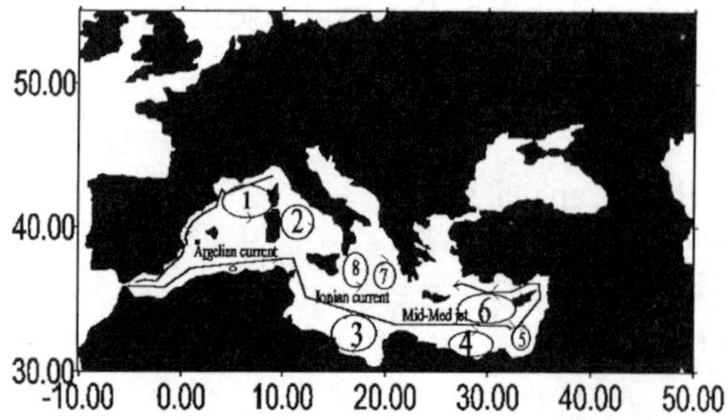

Figure 1: (1) Lion gyro, (2) Tyrrhenian gyro, (3) Ionian current eddies, (4) Mersa-Matruh gyro, (5) Shikmona gy ro, (6) Rhodes gyro, (7) Peloponesus gyro, (8) Western Ionian gyro.

meanders. Other components of the western Mediterranean system are the cyclonic gyros in the Tyrrenian sea and in the Balearic and Liguro-Provenzal basins.

The Atlantic waters that pass east through the straits south of Sicily enter the Ionian sea and forms the Mid-Mediterranean jet. Several sub-basin gyres can be identified on both sides of the jet as the cyclonic Rhodes and Peloponesus gyros at north and the Mesa-Matruh and Shikmona gyros at south. The seasonal variability of the general circulation described, could be related with wind stress, heat flux, rivers outflows and changes in the Atlantic water masses through the Gibraltar Strait.

For the purpose of this study we have divided the Mediterranean in the following basins:

1- The western basin between the Strait of Gibraltar and the Strait of Sicily. 2- The central and eastern basin including Aegean and excluding Adriatic.3- The Adriatic basin 4- The Black Sea.

3 THE ALTIMETER DATA.

Satellite altimeter measurements should provide improved measurements of sea level changes because their direct tie to the geocentre allow to compare, the absolute sea level, with the precise reference frame realised through the Satellite tracking Stations, whose origin also coincides with the Earth centre of masses. In these conditions we can use the altimeter time series to analyse the mean sea level variation in the different Mediterranean basins, to conclude their time and spatial variations in relation with its physical characteristics and possible interactions. Five years of Topex-Poseidon altimeter data and, when available from ERS 1/2 satellites, have been analysed in this study to compute the mean sea level tendencies variation in the Mediterranean basins between October 1992 and October 1997. Figure 2.

The data and corrections applied to the altimeter range measurements were:

1- Orbit: T/P NASA JGM3 orbits. ERS 1/2 from Corrected Sea Surface Height (CORSSH) AVISO Products. 2- Dry Troposphere and Inverse Barometer: T/P and ERS 1/2. From the European Centre for Medium Range Weather Forecasting (ECMWF). 3- Wet Troposphere: T/P from Topex Microwave Radiometer (TMR). ERS-1/2 from ATSR radiometer. 4- Ionosphere: TOPEX from dual frequency altimeter ranges. Poseidon from DORIS beacon net data. ERS 1/2 from Bent empirical model. 5- Sea State Bias: Topex-Poseidon and ERS for semi-empirical BM4 formula. 6- Ocean tides: CSR3.0 for Topex-Poseidon and ERS 1/2. 7- Loading effect, solid earth and pole tides: All corrections were applied as recommended by AVISO in GDR serie C and CORSSH products. 8- Instrumental bias and drift: Applied. 9- The height of the HMSS-OSU95a mean sea reference surface has been removed from SSH data at each location.

4 MEAN SEA LEVEL VARIATION IN THE MEDITERRANEAN

To study the different basins sea level time series variation we have averaged the satellite altimeter data, including its corrections, over a window of 1°X1° around each satellite cross point.

An analysis of Topex-Poseidon sea surface heights on the cross points over the area shows that all points follow, in the Black Sea, a similar behaviour reaching yearly maxims of 10 cm and minima of -10cm, in coincidence with June and November, respectively.

All records usually presents sharp peak at the beginning of 1993 and 1997. Maximum recorded appears lightly higher south-east of Crimea area, reaching 20 cm between Topex repeating cycles 60 to 65 and 90 to 100.

Minima are higher at the east part of parallel 43°N, with extreme values that reaches -28 cm at cycle 28 and sudden changes in the sea level, alternating rises and drops, between cycles 112 and 140.

Sea level tendencies show strong variations going from 20mm/yr, in the east part of the basin, to 3 mm/yr toward the west, close to Marmora sea entrances, probably related with Bosforus water masses interchange.

As can be seen in Figure 2 most of the Mediterranean sea level rises originate in the Levantine basin, with a well-defined maximum of about 2.5 cm/yr, located south-east of Crete and in coincidence with the Rhodes gyro zone and Mersa-Matruh gyro. Northwest

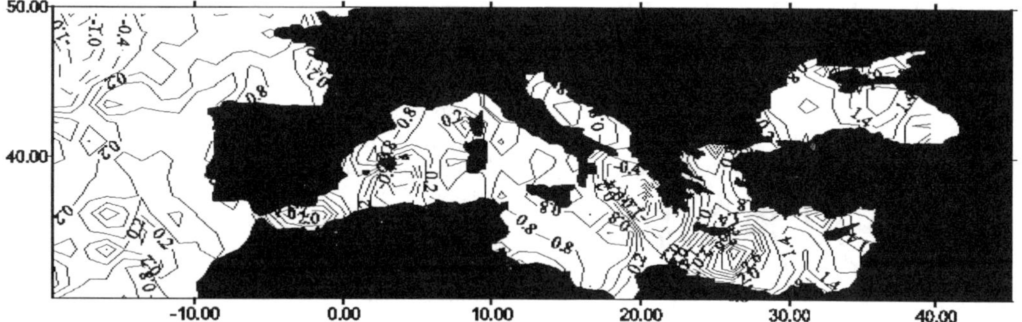

Figure 2: Mean sea level tendencies in the Mediterranean basins, computed from Topex-Poseidon and ERSs satellites observations between October 1992 and October 1997.

of Crete, towards the Adriatic entrances and covering the area from the Western Ionian and Peloponesous gyros, the sea level tendency are falling down between -0.3 and -0.8 cm/yr. Figure 3.

Another fact that could be pointed out is that Black Sea sea level variation has a phase variation in relation with the other Mediterranean basins, as is shown in figure 4 in a direct comparison with Aegean. This singular behaviour could be explained because the isolated form of the Black Sea.

The West Mediterranean and Tyrrhenian basins extend west of Sicilian Strait to Alboran sea including, at south, the unstable Algerian current whose instabilities generate an intense mesoescale activity, in the form of anticyclone eddies, along the Africa coasts.

Other big gyro present in the zone is the cyclonic Tyrrhenian gyro, close to the Italian coasts north of Sicilian and east of Sardinia. North of Balearic islands extends the Liguro Provenzal current and the Lion gyro, characterise for a high seasonal variability, and a low sea level variation tendencies. Alboran sea is dominated by two anticyclone gyros located in the eastern and western parts of the basin.

The western gyros in the eastern side of the Strait of Gibraltar, show the direct influence of the Mediterranean and Atlantic water masses. The eastern gyro is closely related with the dynamic of the Algerian current, which connects the Alboran sea to the rest of the South West Mediterranean . Vazquez et al (1996). An Empirical Orthogonal Function analysis (EOF) of ERS-1 and ERS-2 satellites dynamic signal time series in the basin suggest, in its first mode, an spatial structure with heights at 4,5 W, close to the coasts of Spain and Africa, and 2° W in the coasts of Africa, showing that most of the energy of the mode is associated with the formation and variability of the Alboran anticyclone gyros.

Second EOF mode presents two extended highs one at west and another in a band crossing from Africa to Spain at meridian 2°W.

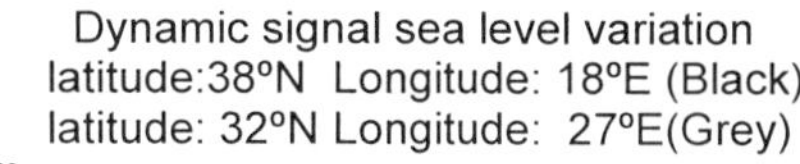

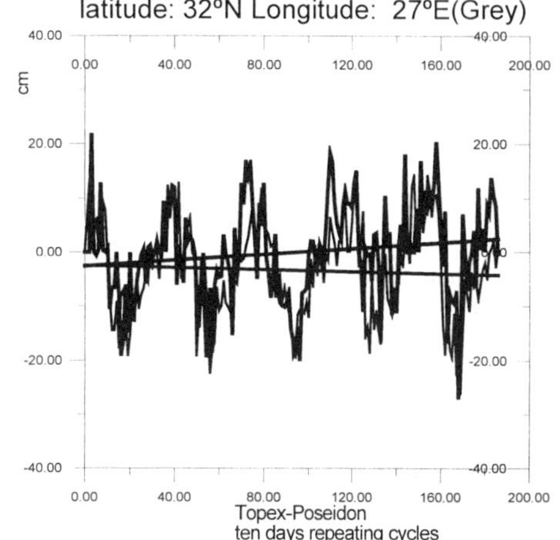

Figure 3: Dynamic signal time series against Topex-Poseidon ten days repeating cycles. Two points were selected in the Aegean basin. One at latitude: 32° N Longitude: 27E ,South of Crete over the Mid-Mediterranean jet. The second at latitude 38° N and Longitude 18°E, east of Greece, close to the Western Ionian gyro. The sea level tendency at the first point, averaged between October 1992 (cycle 1) and October 1997 (cycle 186), records a rise that overpass one cm/yr. At the second selected point the sea level decreases at -0.4 cm/yr. The trends recorded agree with the values predicted in Figure 2.

Most of the energy explained for this mode, up to 45%, is related with the formation and variability of the eastern Alboran gyro.

Third EOF mode presents a low in the coasts of Africa, in the zone of the eastern Alboran gyro, and a high towards the coasts of Spain at latitude 37°,5N. The percent of energy explained by the third mode

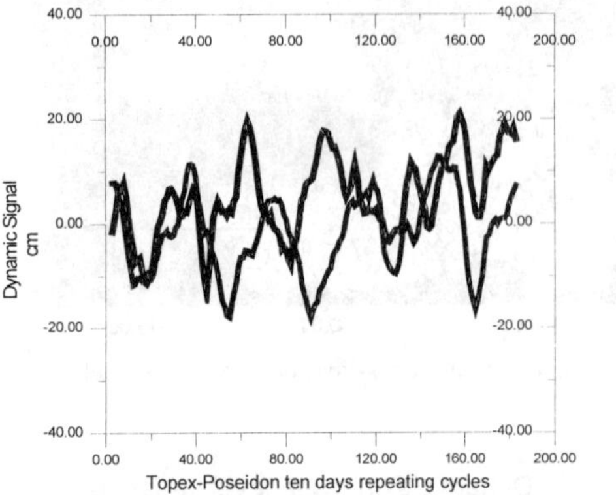

Figure 4: Figure presents the dynamic signal sea level time series, at two points selected in the Black Sea and Libyan Gulf. It can be observed the time difference in phase between the two basins responses. The different behaviour could be explained in the independent behaviour of the Black Sea, as an isolated close basin, with strong influence of big rivers inflows.

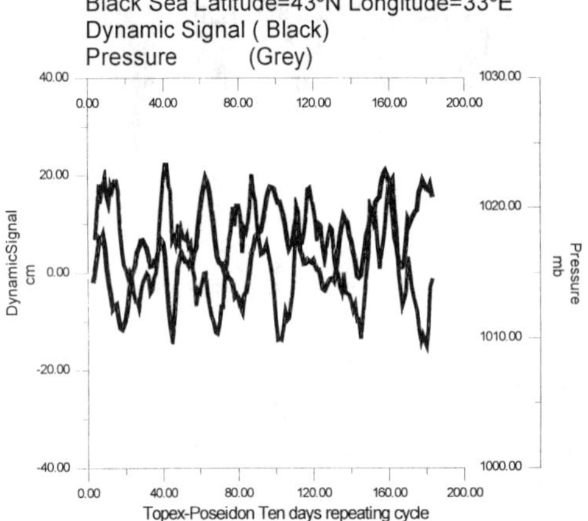

Figure 5: Figure shows the correlation between the dynamic signal time series, at a selected point in the Black Sea, and the pressure variation over the area.

reaches between 20 and 30% of its variability with a lower percent , about 10%, in the western gyro.

5 THE INVERSE BAROMETER CORRECTION.

As is well known the inverse barometer (IB)correction is designed to remove sea level variations from the altimeter data, related with the variations atmospheric pressure and modellice according with an static model. Figure 5. The correction applied is approximately 10 mm of sea level height for every mbar change in the atmospheric pressure.

Significant departures of this rule have been observed in altimeter and tide gauges at different geographic places, suggesting the necessity of developing most complex dynamic and regional models that allows, not only a dynamic model response, sed also allow for the effect of the regional geophysical structures.

The validity of this correction has also been questioned, because of the absence of direct pressure measurements across the oceans, which forces to derive it from atmospheric circulation models and continental surface meteorological measurements, which could be available.

The pressure values provided for Topex-Poseidon by AVISO/TP/GDRs are derived, each 6 hour interval, from received local measurements, at the European Centre Weather Forecast (ECWF) , or computed indirectly from the dry troposphere correction .

6 CORRELATION BETWEEN BASINS

To study the interactions between the different Mediterranean basins, and the effect of the inverse barometer model (IB), to correct for the atmospheric pressure forcing, we have analysed the correlation's between the different Mediterranean basins, computing the sea level variation at the Topex-Poseidon orbit cross points, with and without inverse barometer correction (IB).

In the Black Sea, figure 6 left, the study shows that, before applying to the dynamic signal the inverse barometer correction, the two time series are in antiphase. After correcting the dynamic signal with the IB model the correlation decreases and becomes positive, suggesting that the serie was overcompensated for the effect of pressure.

A study of the dynamic signal time series in a point located in the north of the basin (47º.6 N, 32º.3E), with respect to all the other Topex cross points in the Black Sea shows that the correlation's were maintained always between 0.8 and 0.6, showing there is a uniform and high relationship in the dynamic behaviour of the whole basin area.

Figure 6 left, also presents the results of a correlation analysis between the Black Sea dynamic signal time series, and the same variable in all the others Mediterranean basins showing that the Black Sea behaves as a close and independent area.

Figure 6 right presents the results of a similar study

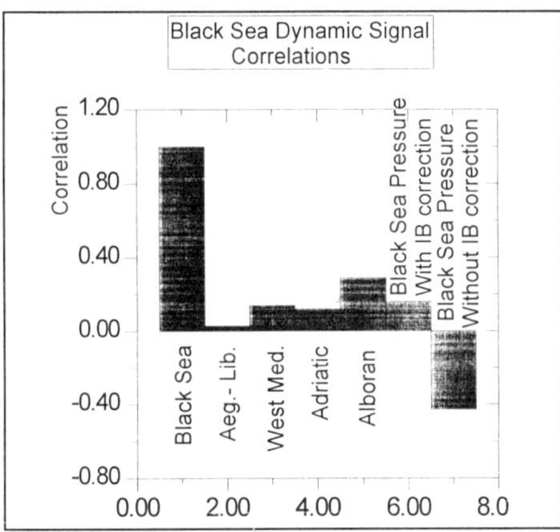

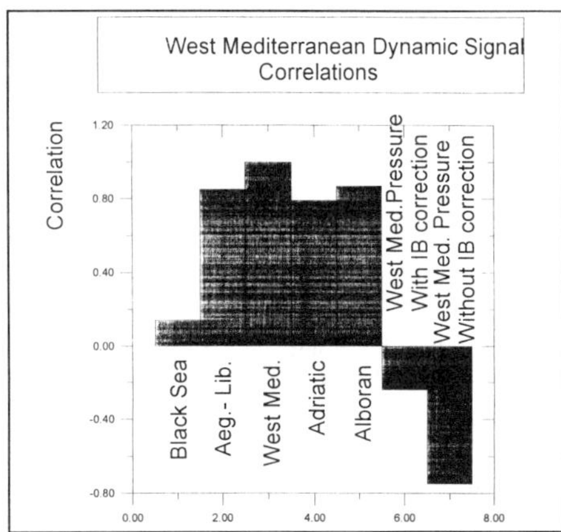

Figure 6 (Left) Correlation of Black Sea dynamic signal time series and the same variable series in the other Mediterranean basins. Figure shows also the correlation of pressure with the Black Sea dynamic signal time series, with and without inverse barometer correction. The analyses suggest that the inverse barometer be overcompensated in the Black Sea, as the correlation decreases but change its sign becoming positive. Figure at the right shows the same analysis for the West Mediterranean basin, including the correlation with pressure before and after applying the inverse barometer correction. The analysis shows that, after applying the correction the correlation decreases, without changing the sign, suggesting that the model undercompensate the correction in the West Mediterranean basin.

that correlates the West Mediterranean behaviour with all the other basins. The studies confirm two facts:

1- Exception made of the Black Sea, all Mediterranean basins behave as highly correlated.

2- The IB model decreases the pressure influence in the dynamic time series.

7 CONCLUSIONS

From the studies made it could be concluded the following :

1- Black Sea behaves as an independent basin, low correlated with the rest of the Mediterranean.

2-In Black Sea the altimeter series, corrected with the Inverse Barometer, is overcompensated, questioning the validity of this static model, mainly for high frequencies.

3- There is an important difference in the time phases, which can reach around five months, between the time dynamic signal series recorded in the Black Sea, with respect all the other Mediterranean basins.

4- All Mediterranean basins mean sea level time series, maintains a high correlation between.

5- A very energetic gyro, south east of Crete, appears stationary and well defined in the Topex-Poseidon data. Its amplitude presents annual variations that undergo a maximum of 15 cm at the beginning of the autumn, and weakens to a minimum in spring, over a diameter of more than 150 km. This strong structure, called the Iapetra anticyclone, is described in the literature as a permanent feature that intensifies in late summer, probably due to the strong interactions of the northerly winds and the Crete orography.

6- South Balearic Islands another strong annual structure overpassing the amplitude of ten centimetres, develops during late summers under the influence of winds forcing and thermohaline flux.

5- The Mediterranean mean sea level variation analysis with altimetric data between October 1992 and October 1997, shows important rise tendencies South East of Crete (> 2cm/yr), almost balanced with the extended area of mean sea level drop trends (>-0.4 cm/yr) at the Adriatic entrances and South of Balearic Islands.

ACKNOWLEDGMENTS

Authors thanks to the French Agencies AVISO and IFREMER for the CD roms received with the altimeter satellite data.
Authors also thanks to the European Space Agency, the information received to collaborate in the satellites altimeter calibration.
Authors also thanks the EU research Programs SELFII and CANIGO its support in the different phases of the research.

REFERENCES

AVISO USER HANDBOOK , AVI-NT-02-101-CN, Ed-3. CNES, Toulouse. France.

Bryden HL,Stommel HM. 1982 Origin of the Mediterranean outflow *J. Mar.Res.* , 40,55-71

Chelton DB, Schals MG, Witter D, Richman JG , 1990 Geosat altimeter observations of the surface circulation of the southern ocean *J. Geophys. Res.* , Vol 95, C10, 17877-17903, Oct 15

Fu LL., Pihos G. 1995 Determining the response of sea level to atmospheric pressure forcing using Topex-Poseidon data *J. Geophys. Res.* ,99, 24633-24642

Nerem RS 1995 Measuring global mean sea level variations using Topex-Poseidon altimeter data *J. Geophys. Res.* Vol 100, n° 113, 25135-25151

Vazquez J., Font J.,Benjamin JJ, 1996 Observations on the circulation in the Alboran sea using ERS 1 altimetry and sea surface temperature data *J . Geophys. Ocean* , Vol 26, pp 1426-1439

Remote Sensing in the 21st Century: Economic and Environmental Applications, Casanova (ed.)

 ISBN 90 5809 096 5

Red tides GIS monitoring the Galician Coast

P. Conde Pardo, M. M. Sacau Cuadrado, J. Martín Herrero & J. M. Torres Palenzuela
Department of Applied Physics, Science Faculty, University of Vigo, Spain

ABSTRACT: A Geographic Information System has been developed for the Galician Coast, aiming to study the environmental parameters which govern the upwelling and downwelling episodes. Daily sequences of thermal images from NOAA-AVHRR, as well as geostrophic winds calculated from pressure maps and coastal bathymetry, have been used in the prediction of toxicity blooms (Red Tides) inside the *Rias*.

1 INTRODUCTION

The proliferation of toxic microscopic algae *Gymnodinium Catenatum* in the *Rias Bajas* has been the subject of several studies (Figueiras & Pazos 1991; Anderson et al. 1989). These estuaries are located in the Northwest of the Iberian peninsula. Mussel production is one of its most important economic resources (nowadays Galicia is the first European producer). The high productivity of the Galician coasts is a consequence of its western oceanic bound together with the seasonal northerly winds, generating strong upwelling of rich nutrient cold waters in the period from May to September (Otto 1975; Tenore & Gonzalez 1975). In opposition to these favourable conditions, during a period in late September, winds change to a southward direction resulting in a downwelling phenomenon which makes the water renewal difficult inside the *Rias*. The result is the appearing of toxic blooms known as Red Tides. This occurs during these months when a cessation or relaxation of upwelling is common (Price et al. 1991).

Harmful Algal Blooms (HABs) result in mass ecosystem dysfunction, risks to public health, and enormous economic losses. Obviously, this is a very important reason for the development of a system that integrates all the known causes of HABs, as well as the volume of information given by satellite imagery, geostrophic winds and upwelling indices in order to get a reliable prediction method.

2 MAIN PARAMETERS IN THE STUDY

2.1 *Pressure Gradients and Associated Winds*

The horizontal pressure gradients which result from the uneven distribution of mass within the fluids are the main driving force of the motions in the atmosphere and ocean. The pressure may suffer variations along an horizontal surface due to changes on the height or the density of the fluid above it. We talk of an *isobaric* surface as that which shows the same pressure at every point, and any line on this surface is an *isobar*.

The pattern of air masses over an ocean is not uniform and the mean pressure follows a seasonal cycle which forces another one on the sea surface. This cycle is partly generated by seasonal changes on the mean pressure over the oceans.

The yearly evolution of the area of study is closely related to the diverse climatology of the northeastern Atlantic. In the summer season, the low-pressure cell centered in Greenland loses intensity while the high pressure from The Azores moves to the central Atlantic, resulting in a pressure gradient which forces air flowing southward along the occidental coast of Iberia. In winter the high-pressure cell appears at the northwestern African coast and low pressures are located off the southeastern coast of Greenland. The relative location between these pressure systems leads to a pressure

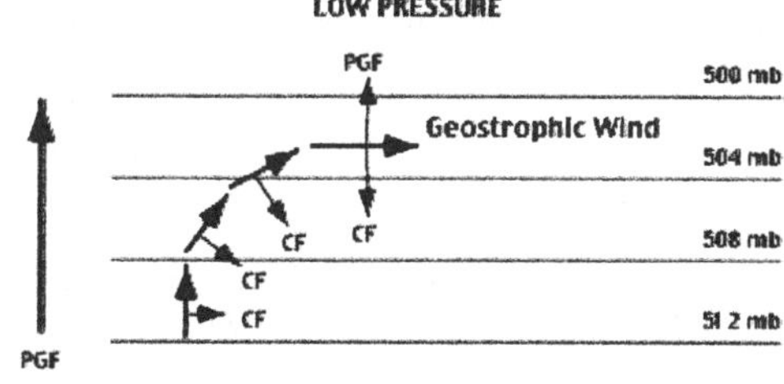

Figure 1. Geostrophic balance

gradient which results in an onshore wind with a strong northward component.

The winds initiated by the Pressure Gradient Force (PGF) will be modified by secondary forces being the Coriolis Force (CF) the most important one. The PGF is always perpendicular to the isobars and is directed towards lower pressures (Fig. 1). The CF always acts at right angles to the wind, clockwise in the northern hemisphere, anti-clockwise in the southern hemisphere.

Away from the ground, when both forces are equal strengths and opposite directions, geostrophic balance is achieved being the wind speed proportional to gradient.

Near the surface, we must also consider the frictional force. When friction is added to the force balance, the wind is not considered to be in geostrophic balance anymore and since friction slows the air motion, it also causes the CF to get weaker. The PGF becomes, therefore, greater than the CF. From the condition of forces equilibrium, it results in a deviation of wind direction with respect to the isobars.

2.2 *Sea Surface Temperature(SST)*

The Sea Surface Temperature (SST) is an important factor on physical processes dealing with the surface energetic balance, the heating exchange between atmosphere and ocean and the atmospheric and oceanic circulation.

SST measures obtained from satellite data rested on the information captured by onboard sensors in the infrared and microwave bands where it is possible to detect and quantify the thermal emission from the surface.

In order to minimize the atmospheric effects, measures must be taken over wavelengths with low atmospheric attenuation.

In general, the oceans tend to maintain their average temperatures with notably less variations than the atmosphere to which they are coupled.

During the day, land will heat up more rapidly than the adjacent body of water, causing unequal heating. The air above the land will be warmed causing air to rise, creating an area of low pressure. Air from over the ocean will rush in to replace the rising air. The air deficit over the ocean is replaced by higher elevation air. Eventually, this sets up an atmospheric circulation pattern. The opposite can happen at night when the land cools faster than the ocean, creating unequal heating and again a circulation pattern opposite of the previous one.

The obvious relation between SST, pressure differences and wind activity makes it a good reason to integrate all these parameters into a GIS study.

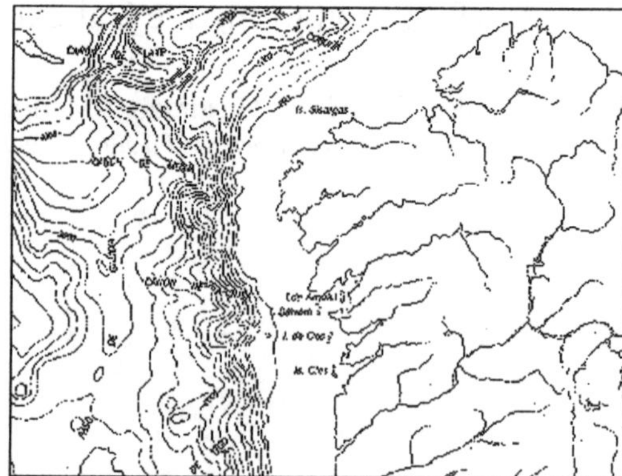

Figure 2. Coastal bathymetry of Galicia

2.3 *Bathymetry*

Coastal dynamic processes are closely related to capes and submarine ridges as well as temporal variations of the wind stress. This close relation becomes a good reason for further studying the coastal structures in order to get a useful tool on prediction studies.

Along the Galician coast we found a multitude of large capes. The two major sites are Cape Finisterre and Cape Ortegal (shown in fig. 2), where clear upwelling structures has been observed.

The effect of wind stress variation caused by coastal morphology may lead to strong offshore flow (Jury 1985). The soft winds increase their speed around the major capes enhancing the upwelling conditions. In the northern part of Galicia, the steep-sided *Rias* deflect the wind to the southwest (McClain et al 1981). However, to the south of the region, where the coast is not so rugged, this phenomenon is of less significance.

3 COASTAL DYNAMIC PROCESSES ASSOCIATED

3.1 *Geostrophic Winds*

As it's been said before, the geostrophic wind results from the balance between Pressure Gradient Forces (PGF) and the Coriolis Force (CF).

In the northern hemisphere, geostrophic wind blows parallel to the isobars, leaving the area of high pressure on its right (Fig. 3). In the southern hemisphere the high pressure area is on its left.

Wind speed, V_g, can be obtained equating the magnitudes of PGF and CF.

$$\frac{m}{\rho}\frac{\Delta p}{\Delta n} = 2\, m V_g \Omega \sin\varphi \qquad (1)$$

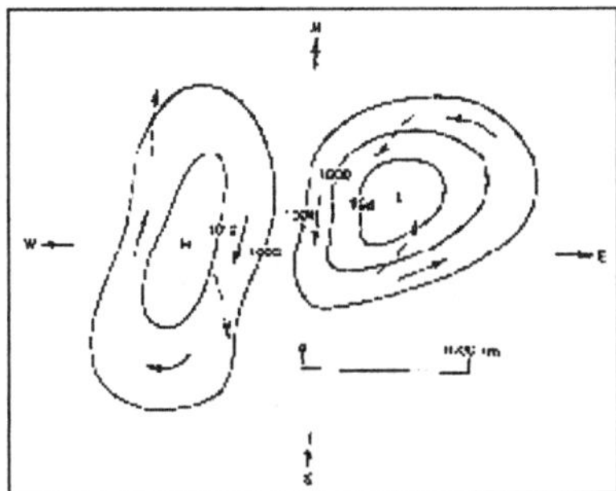

Figure 3. Geostrophic winds directions

i.e.

$$V_g = \frac{1}{\rho\, 2\,\Omega \sin\varphi} \frac{\Delta p}{\Delta n} \tag{2}$$

where m = air mass; ρ = air density (typical value on surface pressure 1.2 kg m^{-3}); Ω = Earth angular velocity; φ = latitude; and Δp = pressure difference between two points separated by Δn.

Vector geostrophic wind velocity can be expressed with two components:

- East-west component

$$u_g = -\frac{1}{\rho f} \frac{\Delta p}{\Delta y} \tag{3}$$

where Δy = north-south distance; and f = Coriolis parameter = $2\,\Omega \sin\varphi$.

- North-south component

$$v_g = \frac{1}{\rho f} \frac{\Delta p}{\Delta x} \tag{4}$$

where Δx = east-west distance; and f = Coriolis parameter.

So we can characterize the geostrophic wind as:

- Directly proportional to the PGF module (the nearer the isobars are, the stronger the wind results).
- Directly proportional to the specific volume, i.e. inversely proportional to air density.
- Inversely proportional to the Coriolis parameter, i.e. to the latitude.

3.2 *Upwelling*

Upwelling is the most important physical event on the contribution of basic nutrients for developing of marine life. The parameter that gives the amount of upwelling water is Ekman´s index obtained from geostrophic wind in the area of study.

Upwelling indices allow an estimation of offshore near-surface transport, from the only knowledge of local wind stress,τ (Bakun 1973). Ekman´s index gives the water flux by unit of shore length as:

$$T = \frac{\tau}{\rho f} \tag{5}$$

where τ = surface wind stress; ρ = water density ; and f = Coriolis parameter.

According to Ekman (1905), wind acts over the sea surface in the same direction in which it blows. Due to Earth rotation, the water layers are acted upon by the Coriolis force which follows the normal direction of the movement and causes the water currents to deviate 45° *cum sole* from the wind direction at the surface (Fig. 4), and the angle of deviation to increase with increasing depth. The current vectors form a spiral pattern known as Ekman spiral.

In these areas, the increase in nutrients is due to the upwelling of waters from depths of 50 to 200 metres to the surface. The upwelled water is cooler and saltier than the original surface water, and typically has much greater concentration of nutrients such as nitrates, phosphates and silicates that are the key to sustaining biological production. It is for this reason that marine ecosystems in the ocean's eastern boundary currents are highly productive, and capable of maintaining large standing crops of plankton and massive fish stocks.

The cooler upwelled waters (two or three degrees lower than the original surface water) appear as a characteristic band of low temperatures close to the coast. These coastal events can be detected by satellite sensors using thermal infrared bands.

3.3 *Red Tides*

Under some circunstances (low rate of water renewal in the stuaries), certain phytoplancktonic species containing reddish pigments (e.g. *Gimnodinium catenatum* , *Dinophysis tripos* , *Dinophysis rotumdatum* , *Dinophysis acuminata*, *Dinophysis acuta*, etc) experiment a proliferation known as Red Tide. It is preferred to use the term HAB, to refer to bloom phenomenon that contains toxins or that causes negative impacts.

Remote sensing is a tool for studying the distribution of HAB over large spatial and short time scales (Keafer and Anderson, 1993). Remotely-

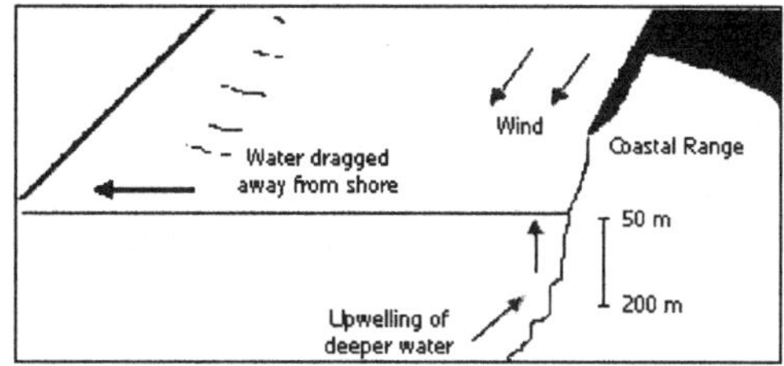

Figure 4. Schematic sample of the upwelling phenomenon.

sensed sea surface temperatures (SST) have been used to follow the movement of fronts, water masses, and other physical features where HAB species accumulate.

A few years ago, Remote Sensing images were expensive and difficult to obtain. The National Oceanic and Atmospheric Administration (NOAA) Coastwatch Program has brought this technology on a near real-time basis to the eventual user (Keafer and Anderson, 1993), so the prospects are bright for an expansion in the application of Remote Sensing to HAB issues.

Nowadays, it's highly known that the increase in intensity, frequency, life length and geographical extension for these toxical episodes (Smayda 1990).

Some of the factors considered to be the main reasons for this increase are: Strong meteorological changes, pollution or other human activities (Anderson 1989), the Greenhouse effect (Fraga et al.1988), and the transport of toxic species in ship ballast water (Fraga et al. 1988).

4 GIS TOOLS

4.1 *Introduction*

GIS (Geographic Information System) is defined as the sequence of capture, storage, analysis and data display processes with a clear spatial reference in order to get a useful tool applicable to the problem-solving effort.

Remote Sensing allows a continuity of new images from the Earth surface, adopted by GIS as a means of data updating.

The stored information, such as statistical maps or climatic data, is georeferenced to make easy the combination between all the data available.

4.2 *Pressure maps*

We obtain the geostrophic wind from isobaric maps.

These maps must be georeferenced. Due to the Earth's curvature, points either side of the centre of the image become distorted with a corresponding loss of detail. Georeferencing is a geometrical correction using earth control points.

After georeferencing we assign the corresponding pressure value to each isobar in the bidimensional image. These assignations are inputs to an interpolation algorithm which outputs a data file with each pressure value at each image pixel.

We represent the pressure maps in 3D images (Fig. 5), where anticyclones correspond to peaks and storms to valleys.

From the interpolated data we can obtain geostrophic winds for a given image and then calculate the average of the geostrophic winds for a

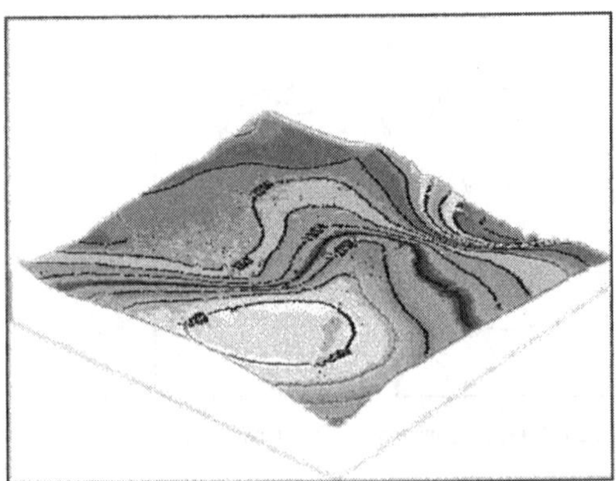

Figure 5. 3D isobar image

known period of time. This let us know wind direction and speed in any point in the map.

We make a vectorial representation of the winds for each point in the map where each vector has the direction of the geostrophic wind for that point, and which has a module proportional to wind speed.

Pressure maps, wind information and vectorial representation are the first inputs to our GIS.

4.3 *NOAA-AVHRR SST images*

The AVHRR (Advanced Very High Resolution Radiometer) is one of the instruments onboard the NOAA satellites operated by the National Oceanic and Atmospheric Administration.

The AVHRR is a 5-channel instrument. Channels 1 and 2 are used for identification of clouds and land/water boundaries, vegetation monitoring and snow and ice coverage. The thermal infrared bands (channels 3, 4 –fig. 6- & 5) are used for estimation

Figure 6. Image from NOAA-14 AVHRR channel 4

of cloud top temperatures and for estimation of sea surface temperatures.

We have developed specific software for obtaining SST imagery. The processing steps are:

- *Conversion from digital counts to radiances.*

The radiance for each pixel is calculated per channel using the formula

$$L = S_i C + I_i \quad (6)$$

where L = radiance (mW m^{-2} sr^{-1} cm); C = input data value (ranging from 0 to 1023); S_i and I_i are the derived mean calibration values.

- *Conversion from radiances to brightness temperatures.*

Radiance is converted into brightness temperature using the inverse of Planck's radiation equation

$$T = \frac{c_2 \eta}{\ln\left(1 + \frac{c_1 \eta^3}{L}\right)} \quad (7)$$

where T = brightness temperature (°K); L = radiance from equation (6); η = wavenumber for the AVHRR channel (cm^{-1}); $c_1 = 1.1910659 \times 10^{-5}$ (mW m^{-2} sr^{-1} cm^{-4}); and $c_2 = 1.438833$ (cm °K).

- *Conversion from brightness temperatures to sea surface temperatures.*

The existing operational algorithms for providing SST from AVHRR data are based on infrared multichannel methods, employing "empirical" formulas (McClain et al.) whose coefficients, a_0, a_1 and a_2, are obtained by linear regression between pairs of simultaneously measured "in situ" temperatures and satellite radiances.

In the absence of clouds and for observations at nadir, sea surface temperature may be written as:

$$T_s = a_0 T_i + a_1 (T_i - T_j) + a_2 \quad (8)$$

where T_i and T_j are brightness temperatures determined from the radiance values in two different infrared window channels i and j.

In general there are three classes of SST algorithms. The "split-window" algorithm uses channels 4 & 5 brightness temperatures (T_{11} and T_{12}). The "dual-window" algorithm uses channels 3 & 4 brightness temperatures ($T_{3.7}$ and T_{11}). Finally, the "triple-window" algorithm uses channels 3, 4 & 5 brightness temperatures ($T_{3.7}$, T_{11} and T_{12}).

In addition, there are correction terms that must be applied to adjust the measurements made off nadir taking into account the sensor zenith angle.

These algorithms provide SST data with a resolution of 0.1°C.

Sequences of temporal consecutive SST images were compared to detect warming of water masses offshore as well as their approach to the coast.

SST imagery are other inputs to our GIS.

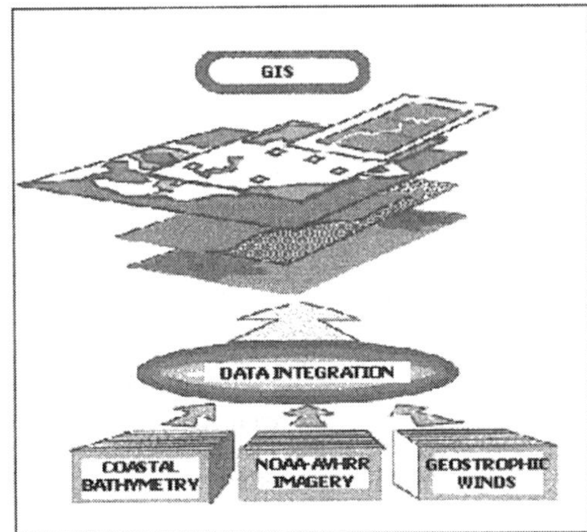

Figure 7. GIS

4.4 *Data Integration*

The data and results obtained in points 4.2 and 4.3 can be studied through a GIS (fig. 7) developed with these different information layers.

The relation between analysis from different sources (winds, upwelling indices, SST, bathymetry) lets us study a bloom of *Gymnodinium Catenatum* from its appearance through its development and finally to its extinction.

5 RESULTS

Comparative analysis in summer 93 and summer 98 shows that:

- In later July, favourable winds cause cool water to rise to the coastal sea surface.
- North winds, favourable to upwelling processes ocurred in August. However, at the beginning of September, the wind changes from north to south. This results in a downwelling or an upwelling relaxation.
- In early September 93, a toxic bloom was detected in the mouth of *Ria de Vigo* as a result of a downwelling period. The SST analysis showed warmer waters close to the coast.
- Both 93 and 98 studies have shown a band of upwelled water consisting of many narrow fingers closely related to the coastal bathymetry.
- The same growth trend has been observed over a similar time period in both years.
- Fingers of cool water found near major capes and ridges can appear as a result of the interaction between the coastal irregularities and the upwelling system.

– Upwelling indices are closely related to wind stress.

6 CONCLUSIONS

1. The geostrophic winds and upwelling indices obtained were very useful for establishing relationships between shelf processes and those that occur inside the *Rias*, which are closely related to Red Tide blooms.
2. The appearance of dinoflagellates is related to the presence of earlier south, southwestern winds following the periods of oceanic upwelling.
3. The appearance of toxicity was observed in all cases to be located at the mouth of the *Rias*. The most important outbreaks coincided with the fading of northerly winds and the presence of upwelling, together with the entrance of warmer waters from the south, which always appeared first in the mouth of *Ria de Vigo* before spreading to the other *Rias*.
4. From the knowledge of geostrophic winds, we were able to predict the formation of oceanic processes favourable to the appearance or disappearance of such blooms.

REFERENCES

Anderson, D.M. et al. 1989. Toxic algal blooms and red tides: a global perspective, in *Red Tides: Biology Environmental Science and Toxicology*, pp. 11-16. New York : T.Okaichi, D.M. Anderson & T. Nemoto

Bakun, A. 1973. *Coastal upwelling indices, west coast of North America 1946-71*. NOAA Tech. Rep. 103 pp. NMFS SSFR-671. U.S. Department of Commerce.

Barton. E.D. 1993. Near Surface Dynamics of Coastal Upwelling. *ICES Statutory Meeting*. C.M.

Blanton, J.O. et al. 1984. Coastal Upwelling of the Rias Bajas, Galicia, NW Spain. In *Hidrography Studies. Reun. Cons. Int. Explor. Mer*, 183, pp. 79-90

Brown, J. 1993. *Ocean Circulation*. The Open University. Oxford-New York-Tokyo: Pergamon Press

Ehlers, M. et al. 1989. Integration of Remote Sensing with Geographic Information System. A necessary evolution. *Photogrammetric Engineering and Remote Sensing*, 55, pp.1619-1627

Figueiras, F.G. & Y. Pazos 1991. *Hidrography and phytoplankton of the Ria de Vigo before and during a Red Tide of Gymnodinium Catenatum*. Graham, J. Plankton Res., 13, pp. 589-608

Fraga, S. et al. 1988. Influence of upwelling relaxation on dinoflagellates and shellfish toxicity in Ria de Vigo, Spain. *9*, 27, pp. 349-361.

Harvey, J.G. 19.. *Atmosphere and Ocean: Our Fluid Environments*. Open University Set Book. Artemis

Haynes, R., E.D. Barton & I. Pilling 1993. Development, Persistence, and Variability of Upwelling Filaments off the Atlantic coast of the Iberian Peninsula. *Journal of Geophysical Research*, vol. 98, NO. C12, pp. 22,681-22,692

Jury, M. 1985. Wind shear and upwelling along the SW tip of Africa. In *International Symposium on Upwelling of West Africa*: 149-159. Inst. Inv. Pesq. Barcelona

Keafer, B.A. & D.M. Anderson. 1993. Use of remotely-sensed sea surface temperatures in studies of Alexandrium tamarense bllom dynamics, in Toxic Phytoplankton in the Sea, pp. 763-768. Proc. 5th Int'l Conf. Amsterdam: T.M. Smayda & Y. Shimizu

McClain, E.P. et al 1985. *Comparative Performance of AVHRR- Based Multichannel Sea Surface Temperatures*. J Geophys. Res., 90, pp. 11,587-11,601

Medina, M. 1976. *Meteorologia Básica Sinóptica*. Madrid: Paraninfo

Otto, L. 1975. *Oceanography of the Ria de Arosa (NW Spain)*. Meded. Verh. K. Ned. Met. Inst., 96, pp. 145-148.

Pickard, G.L. & W.J. Emery 1990. *Descriptive Physical Oceanography – An Introduction*. Butterworth / Heinemann.

Price, D.W. et al. 1991. Hansgen, California's paralytic shellfish poisoning prevention program. *Journal of Shellfish Research*, 10, 119-145.

Smayda, T. 1990. Novel and nuisance phytoplankton blooms in the sea: Evidence for a global epidemic, in *Toxic Marine Phytoplankton*. New York: E. Graneli, B. Sundstrom, L. Edler & D.M. Anderson

Stewart, R.H. 1985. *Methods of Satellite Oceanography*. Berkley: University of California Press.

Tenore, K.R. & N. Gonzalez 1975. Food chain patterns in the Ria de Arosa, Spain: An area of intense mussel aquaculture, in *10th European Symposium of on Marine Biology*, vol. 2, pp. 601-619. Wetteren: Universa Press.

Torres, J.M. et al. 1993. *Follow-up and prediction of upwelling in the Iberian Peninsula West Coast using Meteorologic Parameters, Ekman Flux and Satellite Remote Sensing*. In 81st Statutory Meeting of the International Council for the Exploration of the Sea (ICES). Dublin, 23 September-1 October 1993.

Remote Sensing in the 21st Century: Economic and Environmental Applications, Casanova (ed.)

Location of tuna-fish banks in the Mediterranean Sea using NOAA images

A. Romo, J. L. Casanova & A. Calle
Departamento Física Aplicada I, Facultad de Ciencias, Universidad de Valladolid, Spain

ABSTRACT: This works proposes an operative method for the location of tuna fish banks using NOAA-AVHRR and TOVS images. The area to which the method is applied is the Mediterranean Sea. The importance of the Ocean–Atmosphere exchange, the basis of the method, is clearly inferred and it is shown how the areas to which tuna fish move can be located to within a probability above 80% by using the SST (Sea Surface Temperature) and meteorological parameters derived from the TOVS sonde.

1 INTRODUCTION

Given the high spatial and temporal variability of the phenomenon that occur on the sea together with the difficulty involved in their description and analysis by means of in situ measurements, remote sensing is an essential tool in oceanography. Likewise, remote sensing has meant an advance in the exploitation of fishing resources, providing the answer to the demand for information on the part of fishing fleets. Fishing is of enormous economic importance within the EU and especially in Spain, which has one of the largest fleets in the world. Red tuna fishing in the Mediterranean has undergone a notable increase due to the fact that this species is highly appreciated on the Japanese market, its price having risen ten-fold in recent years. Between the months of April and August the tuna (*Thunus thynnus*) banks pass through the Straits of Gibraltar (Spain) and after having spent the winter in the Arctic Circle and on the coast of Norway, undertake a long journey towards the Mediterranean Sea, where the high temperatures, high salinity and marine currents facilitate the movement of sperm during the egg fertilisation process therefore favouring reproduction. The basins in the Mediterranean are favourable to the catch of the red tuna when they emigrate to spawn.

Given the economic importance, various methods of remote sensing are used to detect their location. One of these is based on the interpretation of the SST (Sea Surface Temperature), for the location of pelagic banks, situated on the warm side of the thermal front (Uda, 1973; Laurs et al, 1984; Fiedler&Bernard, 1987; Petit, 1991; Stretta, 1991; Ramos, 1992; Santiago et al, 1993; Ramos et al, 1996; Romo et al, 1998;).

In the case of the Mediterranean, where the greatest fishing efforts are linked to fronts caused by the invasions of warm and dry continental type air near to the coasts and to those caused by strong exchanges between ocean and sea, this methodology is insufficient, as the high evaporation rates in its interior give rise to the formation of clouds that obscure the surface.

This work proposes an operational methodology that serves to reveal which areas of the sea surface are most apt for catching red tuna using a series of meteorological parameters and the interpretation of sea surface temperature. These data are obtained exclusively from NOAA-AVHRR and TOVS images making the method more economical in comparison with other airborne sensors.

2 INTERPRETATION OF THE SST: THERMAL STRUCTURES FAVOURING TUNA

During the 1996 and 1997 fishing campaign the LATUV co-operated with the team of the Illa de Buda fishing vessel located in Cambrils, to guide it towards those areas of the western Mediterranean where catches are most likely and we receive information on their catches made. The SST was obtained using the split-window method. A comparison of the SST images with the catches made revealed the thermal fronts in which catches normally occurred. It was also observed that 100% of the catches were made within an SST range of between 18 and 23°C.

An analysis of the catches showed that 75% took place in the areas indicated. Research into the areas in which there were and were not catches revealed the following:

1.- There were certain surface thermal structures without cloud cover in which catches took place. Three types could be distinguished:

1.1- SHALLOW FRONT**.** This is a type of front controlled and formed by the strong daily warming and the energy of the tide. In the Mediterranean this type of unusually high warming is the cause of strong irradiation with the presence of light winds. It is also quite normal to find this type of warm pockets, characterised by magnificent stability and variable size. It has been observed that on the edge of the continental shelf, where it is characterised by its bathymetric discontinuity, the clash caused by the water from the shelf itself causes a break in the stratification of the high temperature and weak density mass which in turn leads to sea swells that mix the water downwards, with the resulting upsurge of nitrate and salt from the bottom. The front defined by these conditions favours the presence of tuna.

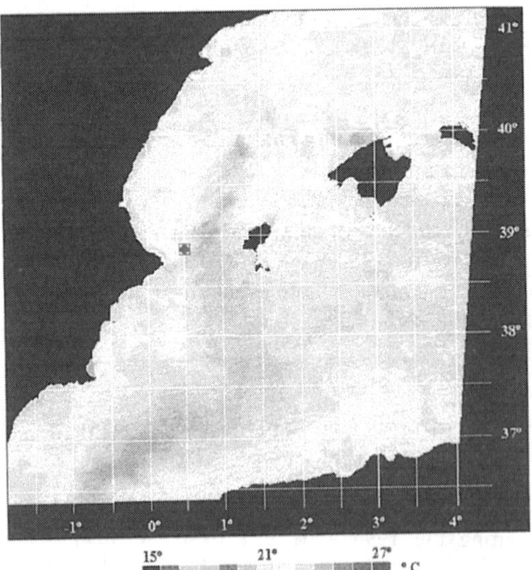

Figure 1. SST 20 June 1997. The grey spot shows where red tuna catches were made.

This can be seen in figure 1, the SST image corresponding to 20 June 1997, where our pilot boat made a catch of red tuna. Its duration is subject to the daily thermocline.

1.2- SHELF BREAK FRONT. A front created once again on the edge of the continental shelf. It is created due to differences between waters on the continental shelf and the abyssal plain. The difference in slope between both leads to a strong gradient between both types of water, creating a circulation from the abyssal plain towards the coast and vice-versa. The front defined is considered by fishing fleets working in the Mediterranean to be one of the most productive. On occasions the rotational effect of the earth makes the waters of the plain, which are heading for the coast, turn and not reach it. The fishing nucleus can be seen in the eddies formed by the clash of these kinds of waters. This type of front is very common on the coast of the African continent.

1.3- FRONTS FORMED BY THE CLASH OF ATLANTIC WATERS AND MEDITERRANEAN WATERS**.** Between the months in which tuna emigrate, it is common to see a fishing front which adopts the shape of a mushroom in May starting from the African coast and reaching the Spanish coast, formed by an anticyclonic movement on its eastern part and cyclonic on its western part, the latter reaching the Mar Menor. From this month until October, this front moves in parallel fashion until it reaches meridian 2°E. (López García et al 1991).

2.- 63.63% of the catches were made in an area corresponding to a cloud edge. It was observed that in many points at which there was no cloud, normally there had been a cloud or cloud edge on previous days. In contrast to other authors who feel that the break in light in the oceanic column leads to an accumulation of tuna, we believe the cause is purely mechanical and thermodynamic as cloud formation would have been caused by the cooling of the sea surface, leading to the formation of upwelling on subsequent days.

3.- On 72.72% of the days on which there were no catches, the areas were cloudy as before. However, on these days, the places of non-catches were directly below the cloud, away from the edge, in contrast to what happened at the points at which there were catches.

3 OBTAINING SST IN CLOUDED AREAS

Evidently in areas which are covered by clouds it is not possible to know the SST. Therefor limit the location of the fishing front. In order to overcome this

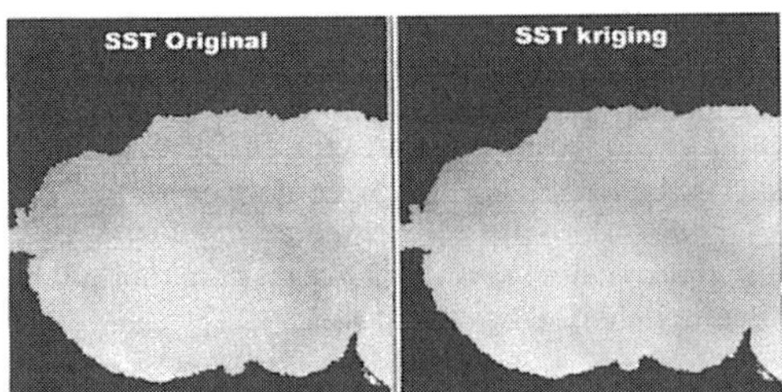

Figure2. SST original and SST Kriging. (colour plate, see p. 605).

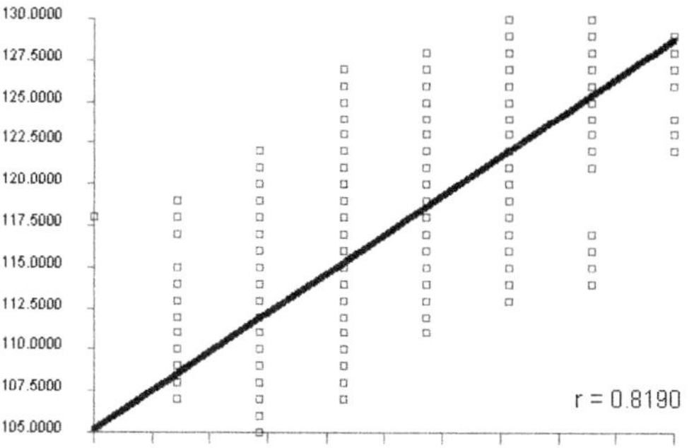

Figure 3. Lineal regression between the real image and that estimated by Kriging. A correlation value of r=0.8190 is obtained. (colour plate, see p.605).

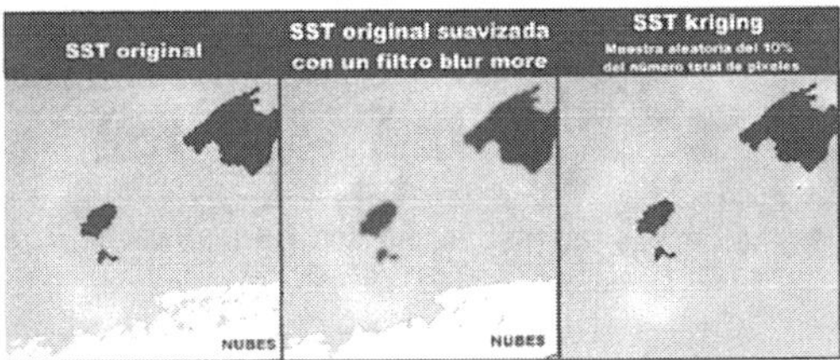

Figure 4. Original SST original, SST smoothed with a Blur filter and SST estimated by Kriging.

a Kriging method has been developed to obtain the temperature of covered areas using interpolation.

An alternative has been sought to resolve the problem of cloud cover, when this is partial. To this end, the validity of geostatistical methods has been seen. The marine surface is divided into 134x134 mile grids, and through a random function 10% of the data not contaminated by clouds are chosen at random, by which a new grid with identical SST dimensions is generated, an SST by the geostatistical Kriging method.

In figure 2, we show an example of the original SST and that generated by Kriging, using SURFER (© Golden Software Inc).

For the example shown, a calculation of the correlation coefficient between the original and the estimated image has been made. The result can be considered good and therefore valid for our aim. (Figure 3).

Figure 4 shows the resulting image, for another image where we have the real case of the existence of clouds. It can be observed how the resulting Kriging image is practically the same as the original if we apply a low pass Blur filter.

4 ATMOSPHERIC INDEX: LOCATION OF FISH USING GEOPHYSICAL PARAMETERS

The Mediterranean is located between the transition of two different climatic domains. In summer it is subject to the influence of high subtropical pressures in their maximum displacement northwards, and in winter to the depressions from the polar fronts that cross it from north-west to south-east. (Jansa, 1986) Cyclogenetic activity is more intense in winter, as the air masses that arrive in the Mediterranean from the north have a lower temperature than that of the sea surface which, together with the injection of humidity they receive, increases their instability. On the other hand, in summer due to the warming they undergo on their continental journey, the air masses that enter the basin are warmer than the sea. At this time of year, the contrast of masses is virtually zero as a result of which there is no development of instability and little cyclogenetic activity (Clavero, 1977). However, the invasion by these warm and dry air masses leads to strong upwelling in the whole of the southern Mediterranean and together with the strong warming effect of the sun, leads to higher evaporation rates, a key element in the hydric deficit suffered by the Mediterranean and at the same time a mechanism that explains the water exchange with other seas. SST calculation shows that, these high evaporation rates are a problem, since under certain meteorological conditions they give rise to the formation of overcast conditions which do not allow the signal from the sea surface to be detected clearly, as was seen in the cloudiness percentages previously presented.

If we consider the heat balance in a marine column, there is a vertical uptake of enthalpy from the bottom to the surface, a part of which is used to warm the air (sensible heat) and another to evaporate marine water (latent heat). Therefore, where there are strong gradient from the sum of these two heats, they will be linked to strong horizontal SST gradients $\frac{\delta(SST)}{\delta z}$, provided there are the necessary relative humidity gradients on the surface.

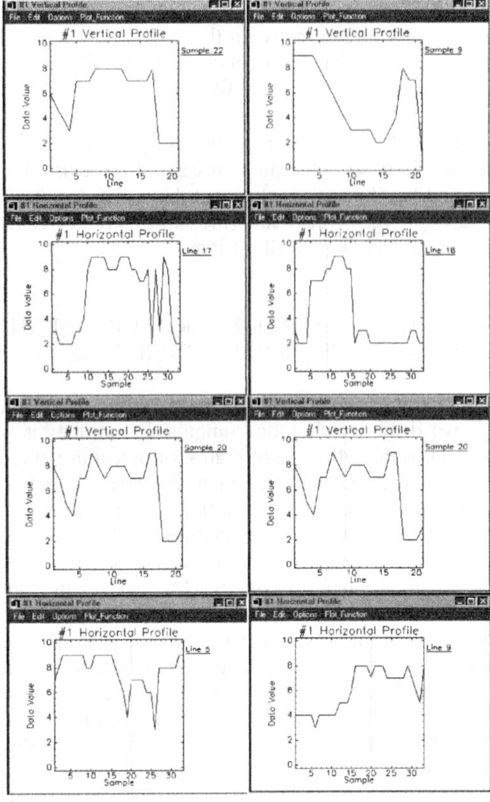

Figure 5. Variability of relative humidity in pixels where thermal fronts can be observed in the SST of the same day and at the same time. (colour plate, see p. 605).

Taking into account the results of the fishing campaigns, a detailed analysis has been carried out of the relation between relative humidity and catches in order to obtain a parameter which would serve to indicate the likely presence or not of tuna.

In order to do this an analysis was performed of the humidity data generated by the ARPPEGE prediction system for the different days of the 1994 fishing campaign, comparing these in turn with the fronts located by the SST at the same time. The result is shown in figure 5, in which an analysis is made of the variation of the humidity for all the pixels that comprise the line and the column, which have their cut-off point on a pixel where there is a thermal front. It can be observed that the pixel where the front is located (marked with a blue line) is in an area of strong humidity gradients in which there is a change from pixels with a high humidity content to neighbouring ones with a low content.

These observations demonstrate that ocean-atmosphere exchange is directly linked to the presence of tuna.

Investigating the meteorological variables supplied by the TOVS sonde from the NOAA satellites and especially the variables linked to humidity and cooling of the surface an ATMOSPHERIC INDEX OF TUNA PRESENCE has been defined as the difference between the air temperature at the dew point, the so-called dew point depression, multiplied by the module of the wind speed.

$$AI = (T_a - T_d) \cdot Vel \quad ;$$

where T_a is the air temperature in C °, T_d is the dew point temperature in C° and Vel is the wind speed in m/s. It can be said that this index is proportional to evaporation.

Representing this INDEX in terms of the dew point temperature at those points in which the fishing boat had been able or not to catch tuna during the 1996 and 1997 tuna fish campaigns, Figure 6 is obtained.

It can be seen that 60% of the points belong to catches made (in red), as opposed to 40% where there were no catches (in blue). Although the point cloud appears random, by observing the area marked out by two straight lines indicated in Figure 7, the total percentage of catches in the area outlined can

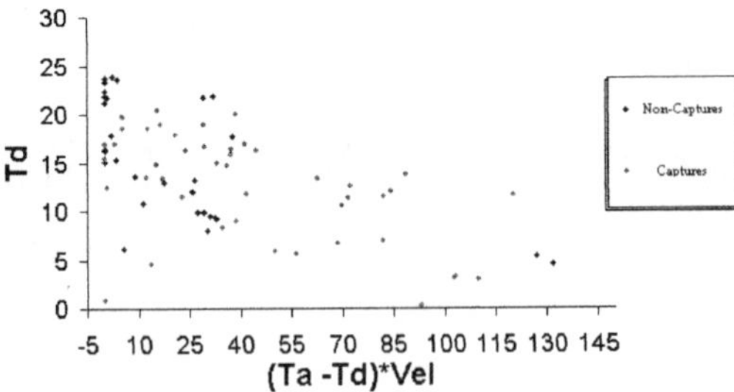

Figure 6. Graph for the prediction of tuna in the Mediterranean. Data from the 1996 and 1997 fishing campaigns in our pilot boat. (colour plate, see p. 605).

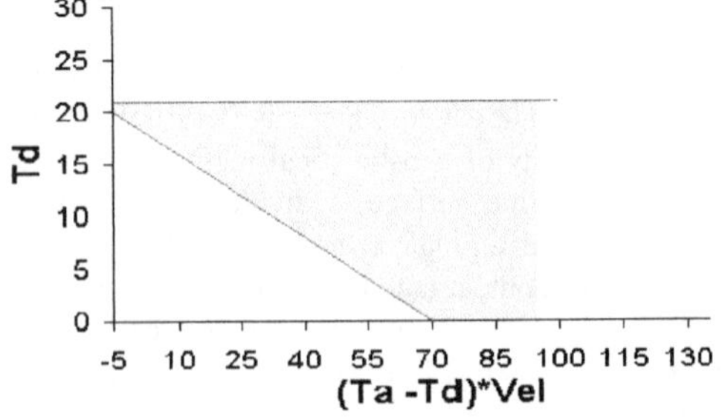

Figure 7. Representation of the most significant areas defining meteorological conditions favouring the presence of tuna. (colour plate, see p. 605).

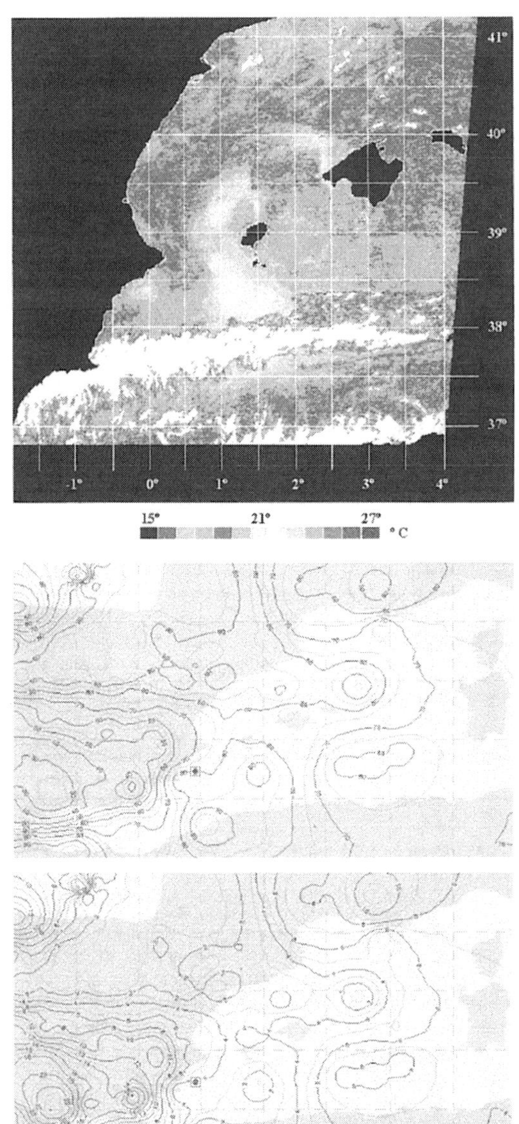

Figure 8. Situation corresponding to 22 May 1997. Upper.On this day red tuna catches were made at the points marked with a grey cross. Middle. Dew depression. Left overlay. Western Mediterranean SST. Lower. Relative humidity at surface level. (colour plate, see p. 606).

be seen to be 84%, whereas for outside the total for non-catches is 75%.

5 OPERATIVE PROCEDURE

Every day the LATUV receives the NOAA image corresponding to the mid-day pass and processes the

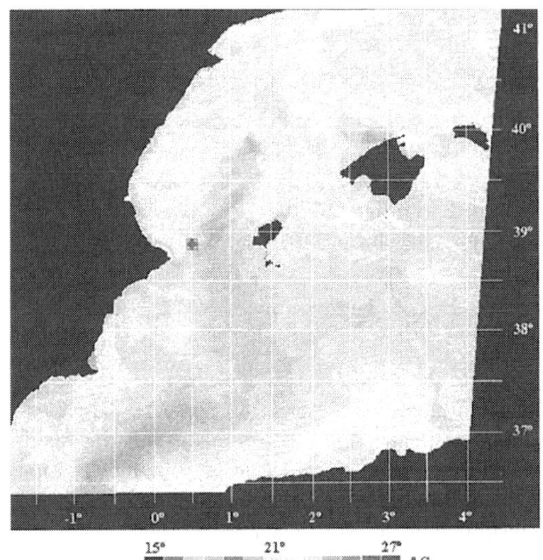

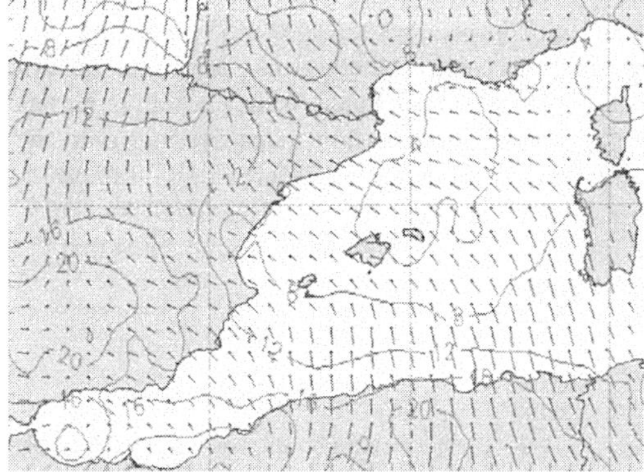

Figure 9. SST and superimposing dew depression on wind fields, 20 June 1997. (colour plate, see p. 606).

information extracting the TIP-DATA (which contain the TOVS sonde data) as well as the AVHRR sensor data. The former are converted into geophysical parameters through the ITPP and the latter into SST images. To do this the sensor data are previously calibrated, and geo-referenced using software created at our own laboratory.

The following maps are then generated:

1. 1 Km^2 resolution SST images, which are rounded to 1°C.

2. Dew depression maps: these show us the type of air mass (humid or dry) that is present in the study areas.

3. Wind field maps. These are overlaid onto the depression maps to see if there are possible invasions of dry and warm air that generate cooling in the sea water with the subsequent formation of thermal fronts near to the coasts.

4. Humidity maps. These allow to see the

areas where there are strong humidity gradients which, from what was previously mentioned, help locate possible thermal fronts on the sea surface.

These maps are shown for a few days in figure 8 and figure 9, where our pilot boat made and did not make tuna catches. The fusion of the previous maps generates an ATMOSPHERIC INDEX map which, combined with the SST maps, helps deduce the areas where tuna catches are most likely to occur.

6 CONCLUSIONS

Based on the conditions of the sea surface and on the atmospheric conditions that characterise the existence of fishing in a determined area, this work presents an operative methodology for the location of areas favourable to the capture of tuna using data transmitted by the NOAA Satellites.

By means of the SST, information is taken based on the thermal fronts observed. For those areas where cloudiness does not allow SST calculation a Kriging interpolation method has been applied. An analysis has also been made of the humidity maps as well as the dew depression maps overlaid onto the wind fields, thus extracting information on those areas where the ATMOSPHERIC INDEX- DEW POINT TEMPERATURE relation indicates that the probability of catches is high.

REFERENCES

Au, D., 1986, *Skipjack* population dynamics is it qualitatively different from that of other tropical tunas? Proceedings of the ICCAT international Conference on Skipjack Year Programme, edited by P. Symons, P. Miyake and G. Sakagawa (Madrid: International Commission for the Conservation of Atlantic Tunas), 388pp

Bakun, A. and Parrish, R., 1982, Turbulence, transport and pelagic fish in the California and Peru Current Systems. CALCOFI Reports, 23.

Barkley, R A., Neill, W. H and Gooding, R. M, 1978, *Skipjack* tuna *Karsumanus pelamis*, habitat based on temperature and oxygen requirements Fishry B 11 t 7'

Bethoux, J.P. (1977): Contribution à l'étude thermique de la mer Méditerranée Rapport n° 20. Laboratoire de Physique et Chemie Marines, Université Piette et MarieCurie, Patis,199pp.

Casanova, J. L. Romo, A. and Calle, A. 1998. Locating fish banks in the Mediterranean using NOAA images. International Symposium. Satellite-Based Observation: A tool for study of Mediterranean Basin. CNES. Tunis

Clavero, P L. (1977): Los clímas de la región valenciana, Tesis doctoral, Facultad de Geografía e Historia, Universitat de Barcelona, 548 ff

Evans. R H., Maclain. D R and Bauar, R. A., 1981, Atlantic skipjack tuna: Influence of the environment on their vulnerability to surface gear. Collection of Scientific Documents (ICCAT), 9, 264-274.

Font, J., J. Tintore y P. La Violette (1988a): Localización de frentes oceanicos por teledetección infrarroja. El caso del Mar de Alboran, II Reunlón Naclonal del Grupo de Trabajo en Te/edetección Valencia,17- 18 diciembre 1987,pp 327-336, (Editado por C.D.T.I)

Font, J., J. Salat y J. Tintoré (1988b): Permanent features of the circulation in the Catalan sea, Oceanologica Acta S-9,51 -57.

Font, J., Salat, J. and Julia, J. (1989): Marine circulation along the Ebro continental margin, en: C.H. Netson y A. Maldonado (Eds.) Marine Geology of the Ebro Continental Margin NW Mediterranean Sea Marine Geology 95 (in press).

Iccat. 1986 Proceeding of the International Commission for the Commission of Atlantic tunas (ICCAT) Conference on the International Skipjack Year Program , edited by P Symon P. Miyake and G Sakagawa (Madrid: International Commission for the Conservation of Atlantic Tunas) 388pp

Jansa Guardiola, J M (1966): Meteorología del Meditenáneo occidental. Publicaciones del Servicio Meteorológico Nacional, serie A, 43, 11, 35 pp. Madrid.

Kelly, A. K. 1989. An inverse model for near-surface velocity from infrared images. J. Phys. Oceanograp. 19: 1845-1864

La Violette, P. E. 1984. The advection of submesoscale thermal features in the Alboran Sea Gyre. J. Phys. Oceanograp. 14: 450-505

Laurs M. Fiedler, P C and Montgomery, D. R.. 1984. Albacore tuna catch distribution relative to environmental feature observed from satellite. Deep Sea Research, 31. 1085-1099

Lopez Garcia, M.J. (1985): Estúdio de la temperatura superficial del mar mediante satélites Aplicacián en el Golfo de Valencia Tesis de Licenciatura, Universitat de Valencia, 166 ff.

Lopez Garcia, M J. y Caselles, V. (1986): Aplicaci6n de la teledetecci6n al estudio de la temperatura superficial del Mar Balear y Golfo de Valencia, Cuadernos de Geagrafía 3g, 75-92.Use of Thematic Mapper data to assess water quality in Albufera Lagoon of Valencia (Spain), Advances in digital Image processing Proc. of the Thirteen Annual Conference of The Remote Sensing Society, Notligham,7-11 September 1987, pp. 510-519

Loup J. (1974): Les eatt terrestes Hydrologie continentale (Paris,Masson et Cie),174pp.

Ramos, A, Sangra, P, Hernandez-guerra and Canton, M, 1991, Large and small scale relationships between skipjack tuna (*Katsuwonus pelamis Linnaeus, 1758*) and oceanographic features observed from satellite imagery in the Canary Islands area ICES 79th Statuary Meeting. C M.91/L:78 Session Q (La Rochelle ICES), 17pp

Ramos, A, 1992, Bioecología del listado (*Karsuwanus pelamis Linnaeus' 1758*) en el área de Canarias. Modelo de gestión y explolación mediante el uso de la teledetección. Pb D Thesis Universidad de Las Palmas de Gran Canarias, Spain, 198pp

Romo, A. Casanova, J. L. and Calle, A. 1998. International Conference on Satellite, Oceanography and Society. Lisboa (Portugal).

Romo, A. Casanova, J. L. and Calle, A. 1999. Relacion entre las condiciones meteorologicas y la captura del atun rojo (*thunus thynnus*), a partir de medidas de los sensores TOVS y AVHRR de los satelites NOAA. Reunión de la Asociación Española de Teledetección. Albacete (Spain).

Sund, P., Blackburn, M and Williams, F, 1988, Tunas and their environment in the Pacific Ocean: a review oceanography and Marine Biology Annual Reviews, 19, 443-512

Tintore, J, LA Violette, P E, Blade, I and Cruzado, A, 1988, A study of an intense density front in the eastern Alboran Sea: the Almeria-Oran front. Journal of Physical, Oceanography, 18, 1384-1397

Uda, M, 1973, Pulsative fluctuation of oceanic fronts in association with the tuna fishing grounds and fisheries. Journal of the Faculty of Marine Science and Technology. Tokio University, 7, 245-265.

Remote Sensing in the 21st Century: Economic and Environmental Applications, Casanova (ed.)
© 2000 Balkema, Rotterdam, ISBN 90 5809 096 5

Developing and testing of an atmospheric correction model for water leaving radiance retrieval in the area of the Gulf of Cádiz based on experimental measurements

V.E.Cachorro, R.Vergaz, A.Berjón & A.M.De Frutos
GOA-UVA, Grupo de Optica Atmosférica-Universidad de Valladolid & Departamento de Optica y Física Aplicada I, Facultad de Ciencias, Valladolid & Departamento de Física Aplicada I, E.T.S.I. Agrarias de Palencia, Spain

J.Morales
CICEM, Centro de Investigación y Cultivos de Especies Marinas 'Agua del Pino', Consejeria de Agricultura y Pesca, Junta de Andalucia, Lepe, Spain

ABSTRACT: Correction of atmospheric effect on the satellite ocean color images is an important prerequisite task in order to retrieve accurate information of water constituents from remote sensing. Atmospheric signal occupies about 90% or more of the total satellite sensor radiance and is highly variable, both temporally and spatially. Here we show an atmospheric correction model for water leaving radiance retrieval based on accurate aerosol characteristics derived from in situ spectroradiometric measurements. Optical aerosol parameters are firstly determined and their behavior in the atmospheric correction algorithm are assessed and compared with current assumptions under different atmospheric conditions. The methodology here applied also allows the calibration of radiometric satellite sensors. An operational algorithm has been developed and applied to SeaWiFS color sensor images in the area of the Gulf of Cádiz.

1 INTRODUCTION

Remote sensing of ocean color from space provides the means to monitor sea water constituents, like pigment concentration, total suspended material, colored dissolved organic material (CDOM), etc. (McClain et al., 1980).

To derive marine spectral reflectances from top of the atmosphere (TOA) radiances in the visible and near infrared is not easy because of the interference of atmospheric effects, namely, scattering and absorption. Typically, 90% of the TOA signal is composed of photons that not interacted with the water body and, thus, do not contain information on water composition. Contrarily it has a rich information on atmospheric constituents, which can be determined by these remote sensing techniques (Tanré et al., 1998) and can be used at the same time in a self consistent method to correct the atmospheric effect.

As it was already mentioned (see Cachorro et al.,1999b, this issue), atmospheric correction algorithms for ocean color applications attempt to retrieve the water leaving radiance from TOA measured radiance. The most complex problem is the correction of aerosol effect, mainly due to their spectral behavior and their spatial and temporal variability. The atmospheric correction effect can be accomplished by in situ atmospheric optical measurements, from a ship or any platform, but this is difficult. However, coastal sea regional areas allow this kind of measurements.

Another approach is based on the same atmospheric information carried out by the satellite signal in order to correct it. Obviously, this is the most current approach in ocean remote sensing applications due to the difficulty of in situ measurements over the ocean and the great spatial coverage. Although the efforts during last years, validation of atmospheric correction algorithms is a current task in ocean color community (Isaacman, 1999, SIMBIOS Project).

On the other hand atmospheric correction algorithm developed to open ocean areas (Case water 1) fails on coastal zones (Case water 2) and specific algorithm need to be developed under different assumptions or based on experimental in situ measurements.

We are developing an extended project with the objective of performing an operational measurements data-base for the calibration validation of atmospheric correction algorithm for satellite sensor color application in the area of the Gulf of Cádiz (Spain). Application to satellite ocean color sensors, such us the SeaWiFS and MOS is being performed.

Operational algorithms differ depending on the characteristics of the satellite sensors. The developing of an operational atmospheric correction algorithm for this area to the SeaWiFS (Hooker et al., 1991) sensor is presented in this work. Spectral radiation measurements in the visible and near-infrared range are being carried out in order to determine the main optical atmospheric parameters needed for the atmospheric correction.

Second step of this research is relative to developing algorithms which link reflectance ratios of the water leaving radiances with pigment concentration or other bio-optical water constituents. This work can be accomplished as a separate task, once the image water leaving radiances for each channel has been retrieved.

2 ATMOSPHERIC CORRECTION MODEL

2.1 *Physical Background*

The radiance measured by an ocean color sensor at the top of the atmosphere (TOA) at a wavelength λ, $L_t(\lambda)$ is composed of the following contributions: $L_{path}(\lambda)$, the radiance generated along the optical path by scattering in the atmosphere. Gordon and Wang (1994) also consider in this term the specular reflection of skylight from the sea surface; $L_g(\lambda)$ is the contribution arising from specular reflection of direct sunlight from the sea surface (sun glitter); $L_{wc}(\lambda)$ is the contribution arising from direct and diffuse downward irradiances reflecting from individual whitecaps on the sea surface, and finally $L_w(\lambda)$ the water leaving radiance, the quantity to be retrieved, which is the contribution of scattering by water body and carries the information of water constituents.

$$L_t(\theta_o, \phi_o; \theta_v, \phi_v; \lambda) = L_{path} + T_d L_g + TL_{wc} + TL_w \quad (1.1)$$

where, L_t depends on the two angles of solar position (θ_o , ϕ_o; vector from the sea point (pixel) to the sun) and that of sensor observation (θ_v , ϕ_v; vector from the pixel to the sensor).

We remove the dependence on wavelength and angles for the sake of brevity. T_d is the direct transmittance of the atmosphere and T the total (direct plus diffuse transmittances). Both transmittances contain the scattering and the absorption effects.

For convenience we will work with reflectances ρ instead of radiances, that is the radiance normalized by the solar spectral irradiance outside of the atmosphere, E_o

$$\rho = \pi L / \mu_o E_o \quad (1.2)$$

where μ_o is the cosine of the solar zenith angle θ_o. Therefore equation (1) is reduced to

$$\rho_t = \rho_m + \rho_a + T_d \rho_g + T\rho_{wc} + T\rho_w \quad (1.3)$$

where ρ_{path} have been decomposed in molecular (m) and aerosols (a) contributions. Other effects as coupling term between aerosol and molecules interaction (multiple scattering, etc.) are neglected.

The contribution of ρ_m can be easily computed by theoretical considerations, by means of the Rayleigh optical thickness and sea level pressure.

The term ρ_g is not considered because, these image pixels are removed by a sun-glitter mask. ρ_{wc} is estimated according the formulation of Koepke (1984), given by

$$\rho_{wc} = \text{cts } W^{3.52}$$

where W is the wind speed in m/sec., measured 10 m. above sea surface. In this formulation ρ_{wc} is independent on wavelengths, however Fruin et al. (1996) have reported measurements where whitecaps reflect considerably less in the NIR that in the visible wavelengths. Current status shows that ρ_{wc} has the potential of producing errors of a magnitude similar to the magnitude of the acceptable error in ρ_w. However, under low wind speed (W< 10 m/sec.), this term can be neglected.

The ρ_a is computed according simple scattering approximation according Schwindling & Deschamps, (1998)

$$\rho_{as} = w_o \tau_a P(\varphi) / \mu_o \mu_v \quad (1.4)$$

where $\mu_o = \cos(\theta_o)$ and $\mu_v = \cos(\theta_v)$, respectively. The aerosol parameters are: w_o is the aerosol simple scattering albedo, τ_a is the aerosol optical depth and $P(\varphi)$ is the aerosol simple scattering function, which depends on the scattering angle φ, given by

$$\cos(\varphi) = -\cos\theta_o \cos\theta_v - \sin\theta_o \sin\theta_v \cos(\phi_v - \phi_o) \quad (1.5)$$

The expression (1.4) overestimates the simple-scattering effect and may be a good estimation of the multiple-scattering effect (Tanré et al., 1990). The evaluation of ρ_{as} can be carried out, well by the knowledge of the earlier aerosol parameters or by assuming given approaches in order to determine the aerosol type and amount. Here we have followed the first approach. The second approach is being discussed in detail in Cachorro et al., (1999b, this issue).

The limitations and uncertainties of the algorithms are obviously involved about these approaches. In the case of SeaWiFS sensor the radiance at channels 765 nm and 865 nm may be non zero in coastal zones, as it is assumed in open ocean areas. Also the wavelength dependence of real aerosols not always follows a power (or exponential) law in the extrapolated values of ε or α to other wavelengths (1999b, this issue) as it is assumed in SeaWiFS current algorithm (Gordon & Wang, 1994). Furthermore for high aerosol optical depth ($\tau a \geq 0.4$) multiple scattering contributions must also be accounted for and the very interaction terms (aerosol-molecules, etc). Another remainder problem seems to be related with the absorption of atmospheric aerosols (Gordon & Wang, 1994; Gordon, 1996), but this point needs further research.

As we mentioned earlier, in situ measurements are difficult and expensive for extended areas, but in the case of regional areas the problem is only at what spatial extent the aerosol parameters can be applied.

According our aerosol measurements and methods developed in earlier works (Cachorro et al., 1993; 1997; 1998a, b; Durán, 1997) we have determined all aerosol parameters needed for the atmospheric correction in the image processing.

2.2 *Normalized water-leaving radiances*

Gordon and Clark (1981) define the normalized water-leaving radiance or its equivalent normalized reflectance $[\rho_w]_N$ according to

$$\rho_w = [\rho_w]_N \cos\theta_o \{\exp(\tau_R/2 + \tau_{o3})\mu_o\} \qquad (1.6)$$

where τ_R and τ_{o3} are the scattering Rayleigh and absorption ozone optical depth respectively. The normalized water-leaving radiance is approximately the radiance that would exit the ocean in the absence of atmosphere with the sun at the zenith. This definition was motivated by the desire to remove, as much as possible, the effects of the atmosphere and the solar zenith angle from ρ_w. However Morel and Gentili (1991, 1993) have shown that a residual dependence on θ_o remains in the normalized value.

On the other hand the approach represented by the exponential function as the transmittance of the atmosphere is not very appropriate if we take into account the expression (1.3) and considers the correct value of the total atmospheric transmittance T (Gordon, 1997; Wang 1998, Cachorro et al., 1998c).

In spite of all these remaining problems, the earlier and current algorithms of ocean color sensors (CZCS, OCTS, SeaWiFS, MODIS, etc) for retrieving total pigments, chlorophyll, etc., are developed by relating the nadir viewing water-leaving radiance to the quantity in question. Up to now the approach that $[\rho_w]_N$ is independent of viewing angle has been usually considered in the analysis of ocean color imagery, however the studies of Morel and Gentili (1993, 1996) about the bidirectional characteristics of the in water upwelling radiance an hence on the water-leaving radiance (Voss 1989; Morel et al., 1995) clearly concluded that this subject is an open topic in color imagery research.

Therefore, we will evaluate the normalized water-leaving radiance in order to compare our results with those given by the current SeaWiFS products, and with other works.

3 SEAWIFS SENSOR SPECTRAL BAND CHARACTERISTICS

SeaWiFS ocean color sensor are composed of eight channels centered at: 412 nm (yellow sustance), 443 (Chlorophyll absorption), 440 nm (Pigment absorption (Case 2), 510 (Chlorophyll absorption) 555 nm (Pigments, optical properties, sediments), 670 nm (Atmospheric correction for CZCS heritage), 765 nm (Atmospheric correction, aerosol radiance), 865 nm (Atmospheric correction, aerosol radiance) wavelengths. Parenthesis indicates the main features of each band. Band 1-6 have 20 nm bandwidth and 7 and 8 have 40 nm. The selection of bands is based on considerations about the absorption characteristics of some common in-water optical constituents, extraterrestrial irradiance and spectral transmittance of water vapor, ozone and oxygen transmittance.

Figure 1 shows the spectral features of the eight transmittance filters together with the total optical depth with and without the Rayleigh scattering correction for a typical irradiance spectrum measured by our spectroradiometer. We can observe the position and width of the bands and how the spectral radiometric response of each filter can be affected by the atmospheric absorption and scattering.

The radiance accuracy is less than 5% for each band with a nadir resolution of 1.1 km for LAC and 4.5 km for GAC products.

4 MEASUREMENTS AND DERIVED ATMOSPHERIC OPTICAL PROPERTIES

4.1 *Optical aerosol measurements*

We refer here only to the optical atmospheric measurements carried out by the “Grupo de Optica Atmosférica GOA-UVA” of the University of Valladolid in collaboration with the group of “Estación de Sondeos Atmosféricos El Arenosillo” belonging to INTA (Instituto Nacional de Técnica Aeroespacial). The station is located in a rural site “El Arenosillo”(Huelva) at sea level (37.1°N; 6.7°W) belonging to INTA. This station is the number 213 of Global Ozone Observing System of the WMO with more than 17 years of ozone Dobson observations.

The optical measurements are carried out with the commercial LI-cor LI1800 portable spectroradiometer that covers the spectral range from 300-1100 nm with a spectral resolution of 6 nm

The area is of special interest due to the influence of maritime aerosols, Africa desert dust event (not very usual) and local industrial environment. The period for the optical aerosol measurements extends from August of 1996 to current date. Lack of measurements is due to calibration and instrument maintenance.

4.2 *Derived Aerosol parameter*

For a given spectrum we determine the spectral aerosol optical depth in order to derive all the aerosol parameters and other atmospheric components

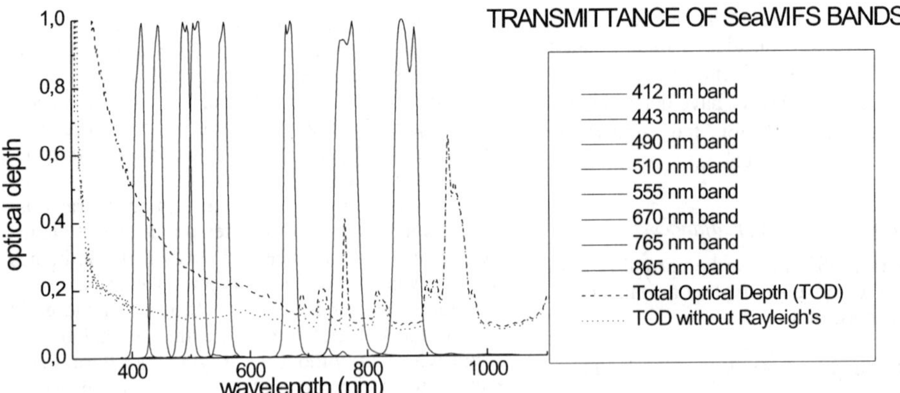

Figure 1. Illustration of SeaWiFS channel transmittance together with the total optical depth (TOD) of the atmosphere (top curve). Bottom curve (dotted) is the TOD corrected of the Rayleigh scattering attenuation 1. Illustration of SeaWiFS channel transmittance

(water vapor, for example) required as input in the atmospheric correction algorithm.

The aerosol optical depth at 550 nm has showed a high variability, ranging from 0.018 to 0.46 with an average value of 0.10 during the 1996-1997 period of measurements (Cachorro et al., 1998b, 1999a). This reveals the low turbidity in this area and permits neglecting the influence of multiple scattering. This low turbidity has been also assessed by the analysis of a great number of images processed by the SEADAS software package in this area.

To process a given image, an irradiance measured spectrum is chosen from our optical measurements data base. We derive by a pseuso-inversion method the particle size distribution function of aerosol n(r). In this process only an assumption must be done relative to the aerosol refractive index, where we have chosen that of maritime aerosol model. From n(r), we can derive the single scattering albedo w_o, the asymmetry parameter g and the simple aerosol phase function P. These parameters are part of our aerosol characterization data base and will be used depending on the required input of the atmospheric algorithm. The two first parameters only depend on wavelength but the phase function also depends on the sun and sensor positions by means of the scattering angle. Therefore this function is the most sensitive to both variables.

Due to the extended range of scattering angles in a given SeaWiFS image this function and hence the atmospheric radiance shown very variable values.

In figure 2 two examples of the derived phase function of our measured aerosols are depicted. Also we have added their geometrical radius r_g and the **Angstrom** turbidity values retrieved from the spectral AOT.

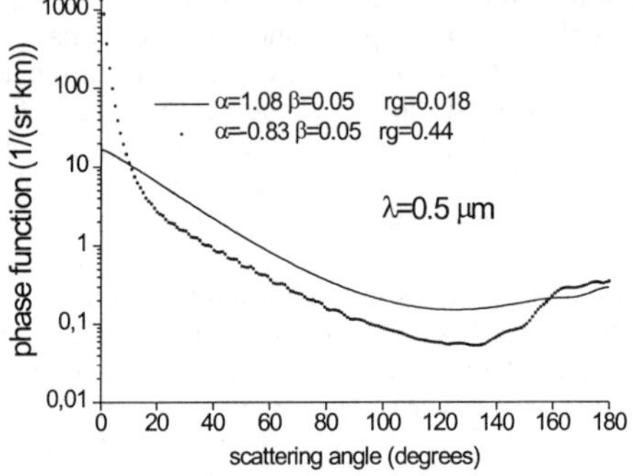

Figure 2. The phase function of aerosol at 0.5 μm as function of the scattering angle for two values of the Angstrom turbidity parameter of our data base. The value of the geometrical radius of the retrieved aerosol size distribution is also depicted.

In figure 3 we show the spectral dependence of the atmospheric aerosol radiance due to the term w_o τ_a $P(\varphi)$ (which gives only the influence of aerosols) for different values of the scattering angle φ, typical of the range of observations in satellite imagery.

We must note the low values of this term due mainly to the low value of the aerosol optical depth (0.04 at 0.5 μm). Figure 3 represents a real case for image processing.

4.3 *Transmitance of the atmosphere*

Scattering and absorption effect are contained in the transmittance functions, direct T_d and total T. T_d is very easy to compute or in our case this is determined directly by our direct spectral irradiance measurements (Cachorro et al., 1985; Cachorro et al. 1998d). The total transmittance T (direct and diffuse transmittances) is evaluated according a two flux radiative transfer code developed by the GOA-UVA group (Durán, 1997; Cachorro et al., 1997; Cachorro et al., 1998c) based on the information given by the same direct irradiance measurements or derived atmospheric parameters.

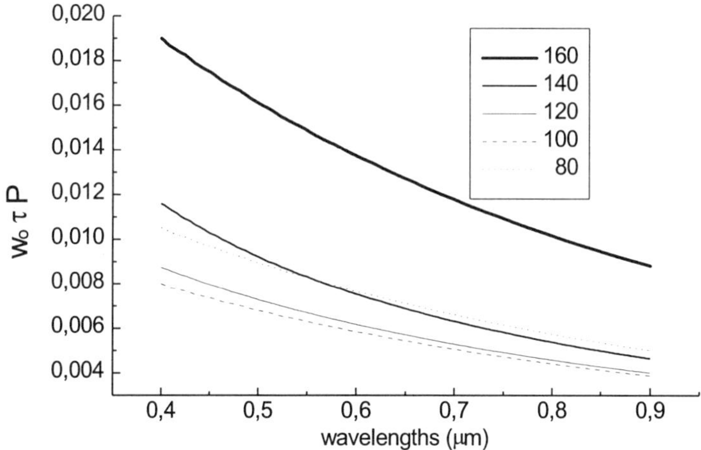

Figure 3. Spectral dependence of the atmospheric aerosol reflectance for different scattering angles (see text).

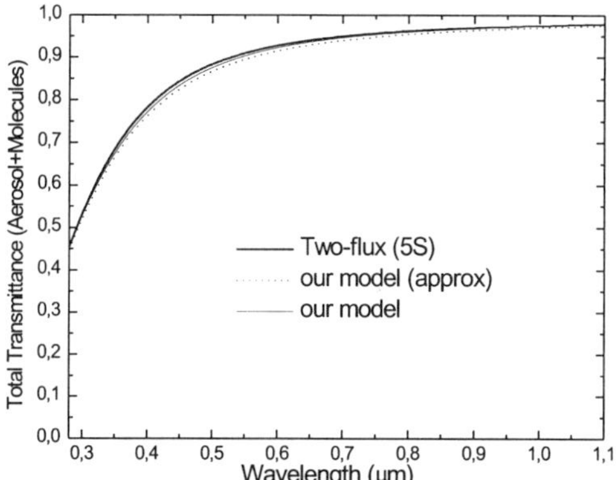

Figure 4. Total transmittance of the atmosphere comparing three different models.

The total transmittance generated by this radiative transfer code has been checked by means of a comparative evaluation with different transmittances formulation given by other authors (Durán, 1997; Cachorro et al. 1998c; Wang 1999; See figure 4). The evaluation of this transmittance is very important since it multiplies to the water leaving radiance. Errors in this function affect strongly in the accurate retrieval of it.

In figure 4 we show the total transmittance of a mixed atmosphere of aerosols and molecules for a SZA of 30 degrees where three different models are compared. The first model is the 5S (solid line), the second model is our model (thin line) and a third approach is the earlier one but considering the product of Rayleigh and aerosol transmittances (dotted line).

To assess the performances of our atmospheric correction algorithm, different simulations was carried out, where the dependence of the different parameters has been analysed. (Cachorro et al., 1997; Cachorro et al., 1998d). This algorithm, called 'sea simulation algorithm (SEASIM)', simulates TOP reflectances assuming given water leaving reflectance values or other surface reflectances, allowing the ana lysis of the features of every relevant parameters in the atmospheric correction.

5 OPERATIONAL ALGORITHM

Besides the SEASIM code, two current operational algorithms are now available, a time-consuming algorithm and a fast algorithm. The first computes all correction terms at step of 1 nm and integrated these values to SeaWiFS band transmittances at the last step. The fast algorithm evaluated at the nominal wavelengths of SeaWiFS bands with a weigh parameter deduced from the above algorithm and uses previously generated look-up tables (mainly for the aerosol phase function) of aerosol parameters in order to avoid computer-consuming time. On the other hand, our images are not very great (usually 200x400 pixels). Note that the application on a given AOT has a limited spatial range.

As first step the SeaWiFS image of Level-1A LAC is processed by SEADAS package to get the Level-1A calibrated radiances. In this process we also obtain the angles of the sun and sensor respectively. These files are our base files to apply the atmospheric correction algorithm, once the required area of study has been cut away of the provided total image.

An irradiance measured spectrum is chosen from our optical measurements data base which corresponds timely to the image that must be processed in our area of study. This file is (or has been previously) processed to evaluate the complete set of atmospheric parameters required as input in the atmospheric correction algorithm.

The application of the atmospheric correction algorithm generated the following product files or images: Atmospheric radiance and reflectance at 765 and 865 nm bands. The radiances and reflectances at the firsts six SeaWiFS bands, according above definitions plus the normalized radiances and reflectances according expression (1.6).

As we have discussed above, the normalized values are obtained in order to be compared with those given by the SEADAS processing package level 2 products. We also evaluate the ε(765/865) coefficient given by the ratio of the corresponding atmospheric radiances for this comparison process.

Furthermore, in this process we have evaluated an average pixel aerosol optical depth over the processed SEADAS image around our site station in order to be compared with our measured AOT data (Wang et al., 1999). However, this work has a true meaning in the case of a great number of processed images. This is a task to be done, as part of a long-term process of comparison between our experimental data and the image processed by SEADAS algorithm.

This task is not only important for ocean color imagery applications, also for aerosol studies by satellite remote sensing techniques in order to developing and improving the very atmospheric algorithm by extracting aerosol information of the image itself (Land, 1999).

6 CONCLUSIONS

The development of an atmospheric correction algorithm, specific for coastal regional applications, as that of Gulf of Cádiz (Huelva, Spain) has been presented in this work. The advantage of the algorithm is that it is based on experimental in situ measurements for aerosol parameter determination, therefore not assumption about aerosol models and image information is required.

The atmospheric model, as a whole, can be applied to other areas of interest but the operational algorithm requires previous processed information about the aerosol parameters. The preliminary applications of our algorithm to SeaWiFS images gives very promising results but further research is necessary for accurate assessment, like comparison with in situ sea water reflectance measurements.

A long-term task of comparison of our results with those given by the processed images by SEADAS is being carried out. First results shows great time variability, which need a classification based on atmospheric conditions, whether patterns and climatological information.

7 ACKNOWLEDMENT

This work is supported by the CICYT ESP97-1771-CO2-O2. We thank to "Junta de Castilla y León" for supporting the grant of R. Vergaz.

8 REFERENCES

Cachorro, V.E., A.M. De Frutos & M.J. González 1993. Analysis of the relationships between Junge size distribution and Ångström (turbidity parameters from spectral measurements of atmospheric aerosol extinction Atmos. Environ. 27A,1585-1591.

Cachorro, V.E., P. Durán, R. Vergaz, A.M. De Frutos & S. Hernandez 1997. A study of atmospheric effect on satellite signal based on ground solar radiation measurements of high spectral resolution. In G. Guyot & T. Phulpin (ed), In Physical Measurements and Signatures in Remote Sensing: Vol. 1: 187-192,. Rotterdam: Balkema.

Cachorro, V.E., P. Durán, R. Vergaz, and A.M, De Frutos, 1998a. Vertical radiative properties of atmospheric aerosols in a representative continental area of the north-center of Spain during 1995. Satellite remote Sensing of Clouds and the Atmosphere II. European Symposium on Remote Sensing, SPIE vol , 3495.

Cachorro, V.E., R. Vergaz, J.M. Vilaplana, A.M. De Frutos, R. Margarida & B. De la Morena B. 1998b. Physical and radiative properties of atmospheric aerosols derived from spectral solar extinction measurements in the Golf of Cádiz (Huelva, Spain). In Proc. of the 25th Annual European Meeting on Atmospheric Studies by Optical Methods. Granada. Spain.

Cachorro, V.E., R. Vergaz, A.M. De Frutos, M.J. Martin & R. Margarida 1998c. Modelling solar spectral irradiances measurements by a single layer two flux radiative transfer model. In Proc. of the 25th Annual European Meeting on Atmospheric Studies by Optical Methods. Granada. Spain.

Cachorro, V.E., R. Vergaz, J.M. Vilaplana, A.M. De Frutos, R. Margarida & B. De la Morena B. 1999a. Turbidity parameters and radiative properties of atmospheric aerosols derived from spectral solar extinction measurements in the Golf of Cádiz (Huelva, Spain). *In ALPS99: The Contribution of POLDER and New Generation Spaceborne Sensors to Global Change Studies.* WK1-p-05. Meribel France.

Cachorro, V.E., R. Vergaz & A.M. De Frutos 1999b. Evaluation and behavior of the aerosol (Ångström coefficient from spectral solar irradiance measurements in the Gulf of Cádiz (Huelva, Spain) to assess ocean color atmospheric correction. Earsel 19th Symposium. Valladolid, Spain.

Durán, P. 1997. Medidas espectroradiometricas para la determinación de componentes atmosféricos (ozono, vapor de agua y aerosoles) y modelización del intercambio radiativo en la atmósfera, Ph.D. Thesis Dissertation. E.T.S.I. Agrarias de Palencia. University of Valladolid, Spain.

Fruin R., P-Y. Schwindling, M & Deschamps, P-Y. 1996. Spectral reflectance of sea foam in the visible and the near-infrared: In situ measurments and the remote sensing implications. J. Geophys. Res. 101: 14361-14371

Gordon, H.R., & M. Wang 1994. Retrieval of water-leaving radiance and aerosol optical thickness over the oceans with SeaWiFS: a preliminary algorithm. *Appl. Opt.* 33: 442-453.

Gordon, H.R 1996. MODIS normalized water-leaving radiance algorithm theoretical basis *document (Mod 18, version 3). Departament of Physis.* Univ.of Miami. USA.

Gordon, H.R. 1997. Atmospheric correction of ocean color imagery in the Earth Observing System era. *J. Geophys. Res.* 102: 17081-17106.

Hooker, S.B., W. E. Esaias, G.C. Feldman, W.W. Gregg & C.R. McClain 1992. An overview of the SeaWiFS and ocean color. SeaWiFS Technical Reports. Vol. 1.

Isaacman, A. 1999.An investigation of time variability in water-leaving radiance retrieved from ocean color measurments. *In ALPS99: The Contribution of POLDER and New Generation Spaceborne Sensors to Global Change Studies. Ocean Color* p-13. Meribel France.

Koepke, P. 1984. Effective reflectance of oceanic whitecaps. *Appl. Opt.* 13: 1816-1824.

Land , P 1999. Atmospheric correction of ocean colour an aerosol model independent approach. *In ALPS99: The Contribution of POLDER and New Generation Spaceborne Sensors to Global Change Studies. Ocean Color* Meribel France. (communication).

McClain, C.R., M.L. Cleave, G.N. Feldman, W.W. Gregg & S.B. Hooker 1998. Science quality SeaWiFS data for global biosphere research. *Sea Technology*, September.

Morel, A. & B. Gentili 1991. Diffuse reflectance of oceanic waters: Its dependence on sun angle as influenced by the molecular scattering contribution. *Appl. Opt.* 30:4427-4438.

Morel, A. & B. Gentili 1993. Diffuse reflectance of oceanic waters, II, Bidirectional aspects. *Appl. Opt.* 30: 6864-6879.

Morel, A. K.J. Voss & B. Gentili 1995. Bidirectional reflectance of oceanic waters: A comparison of modeled and measured upward radiance fields. *J. Geophys. Res.* 100: 13143-13150.

Schwindling, M & Deschamps, P-Y. 1998. Verification of aerosol models for satellite ocean color remote sensing. *J. Geophys. Res.* 103: 24919-24935..

Tanré, C., C. Deroo, P. Duhaut, M. Herman, J.-J. Morcrette, J. Perbos & P.-Y. Deschamps 1990. Description of a computer code to simulate the signal in the solar spectrum: The 5S code, *Int. J. Remote Sens.* 11, 659-668.

Tanré, D. Y.J. Kaufman, M. Herman & S. Mattoo 1997. Remote sensing of aerosol properties over oceans using the MODIS/EOS spectral radiances. *J. Geophys. Res.* 102, 16971-16988.

Voss, K.J. 1989. Electro-optic camera system for the measurement of the underwater radiance distribution. *Opt. Eng.* 28: 241-247.

Wang., M. 1999. Validation study of the SeaWiFS oxygen A-band absorption correction: comparing the retrieved cloud optical thicknesses from SEAWIFS measurements. *Appl. Opt.* 38: 937-944.

Wang, M., S. Bailey, C. R. MacClain, C. Pietras & T. Riley 1999. Remote sensing of the aerosol optical thickness from SeaWiFS in comparison with the in situ measurements. *In ALPS99: The Contribution of POLDER and New Generation Spaceborne Sensors to Global Change Studies.* WK1-O-08. Meribel France.

Remote Sensing in the 21st Century: Economic and Environmental Applications, Casanova (ed.)

ISBN 90 5809 096 5

The use of Topex-Poseidon, ERS-1 and ERS-2 altimeters to detect meddies

P.Villares, M.Bruno, J.J.Alonso, A.Ruiz-Cañavate & M.Catalán-PU
Department of Applied Physics, University of Cadiz, Puerto Real, Spain

M.Catalán-M
Real Instituto y Observatorio de la Armada, San Fernando Cadiz, Spain

ABSTRACT:The Mediterranean water outflow, not only alters the thermohaline field of the Atlantic, but also plays an important role as a carrier of dissolved matter and living organisms from the Mediterranean. Such action is reinforced by the presence of submesoscale eddies (MEDDIES) which detach from the main streams. The use of remote sensing techniques in the study of MEDDIES, their detection and tracking on the sea surface, could be considered as a matter of relevant interest to improve our knowledge about the mechanisms that explain their origin and evolution. This communication presents the results that we have obtained in the detection of MEDDIES off the portuguese coast, using both, the altimeter of Topex-Poseidon and ERS satellites

1 INTRODUCTION.

The water mass exchange through the Gibraltar Strait can be characterized by a two-layer model. Atlantic water flows into the Mediterranean in the upper layer, while a flow of colder and saltier water exits through the Strait in the lower layer. (Candela et al. 1989; Lacombe and Ritchez, 1982). Although qualitatively this simple model is useful, it does not show important features related with velocities and other physical properties across the interface.

In the initial stages after leaving the Strait, the Mediterranean water outflow descends close to the bottom of the westwards slopping topography, towards the Gulf of Cadiz, being deflected to the right as a consequence of Coriolis force and mixed with the surrounding Atlantic water. Consequently, the termohaline characteristics of the Mediterranean outflow are gradually modified. The salinity decreases and becomes less and less dense reaching, at a certain stage, the density values corresponding to mid-depth layers.

In the Gulf of Cadiz, some studies (Madelain 1970; Zenk 1975; Ambar et al. 1976) describe the Mediterranean outflow as a gravity current along the Iberian continental slope. Further west the flow splits in two branches. The denser branch goes westward off the continental slope and the less dense one follows a way several tens of kilometers onshore from the former.

Although the branching is observed at about half distance between the Strait and Cape St. Vincent, eventually both branches meet again as a consequence of the isobaths convergence occurring at about Portimao Canyon (60 km Southeast of Cape St. Vincent). Several meddies reported in the literature show that double vertical structure associated to each branch.(Richardson et al. 1989; Zenk et al.,1991; Ambar and Howe 1979; Ochoa and Bray, 1991).

In general, we can consider that the main water masses present in the Gulf of Cadiz under the seasonal thermocline, that is located between 100 and 200 m depth, can be classified as:

1- Eastern North Atlantic Central Water (ENACW), which extends down to 400-500 m depths, showing a decrease with depth in both, temperature and salinity.

2- Mediterranean water (MW), which shows a higher temperature and salinity, liying inmediatly under the ENACW and extending down to depths between 1200-1300 m.

3-North Atlantic Deep Water (NADW), located below the MW, shows a slight decrease with depth in salinity and temperature.

2 THE MEDITERRANEAN EDDIES IN THE ATLANTIC. ORIGIN AND FATE.

Most of the MW outflow turns norhtward through a gateway east of the Gorringe Banks (Zenk and Armi, 1990). This arrival of Mediterranean Water in the Tagus basin causes a high turbulence and also feeds a reservoir of Mediterranean Water located east of the longitude of 13°W and between the latitudes of 37°N and 40°N. Part of this Mediterranean Water proceeds from northwest away from the shelf break. After it goes to north of 42°N and although the Mediterranean water is flowing shallower than 1500m, a part of that water mass seems to be guided westward following the isobaths between 3000 and 4000 m until it feeds the narrow Mediterranean water branch wich is observed west of Galicia Bank. The skirting of Galician continental slopes causes a new widening and salinity decreasing of the Mediterranean water flow, that finally becomes hardly distinguible. A westward exit from the reservoir into the Atlantic occurs mainly between 37°N and 39°N, but other observed paths can go to southward through a southern exit west of Gorringe Bank.

In the late seventies many studies have been carried out about the origin and fate of MW lenses characterized by high anomalous values in temperature and salinity (MEDDIES), these studies have adquired a particular interest in the region situated off the western Portuguese coast in the Iberian basin.

These eddies shows an anticyclonic gyre and their typical lengths and time scales are of 50 km in diameter, 600-800 m in the vertical dimension and rotation period between 6 and 8 days (Ambar *et al.*, 1991). The life time span of an individual Meddy in the NE Atlantic and the mechanisms that explain its decay were investigated by tracking of it during several years with SOFAR floats. From these studies it have been found that the meddy mantains its shape and angular rotation during at least two years, even after the most of the original salinity and heat are diffused out of the meddy (Armi and Zenk, 1984).

Some observational evidence of a signal of the Meddies in the ocean surface was recently reported by several authors , either by comparison of the sea surface height anomaly field, combining GEOSAT altimetry with 'in situ' hydrographic measurements and satellite radiometer images, or by combining the trajectories of surface drifters and Rafos floats (Tokos *et al.*,1994).

Kase (1989), employed a quasi geostrophic model to show that a northward flow paralell to the Iberian continental slope is a possible source of Meddies, being the generation mechanism their baroclinic instabilities. Once formed these anticyclonic vortices, they follows a southward motion which is associated to its internal dynamics.

The generation mechanism of Meddies is a subject of active investigation, some authors suggest baroclinic and barotropic instabilities while other propose the topographic rectifications of the flow imposed by the submarine canyons.

3 THE DETECTION OF MEDDIES USING SATELLITE ALTIMETRY. COMPUTATIONAL PROCEDURE AND RESULTS.

A satellite altimeter is a radar that measures precisely the range from the radar antenna to the ocean surface.The altimeter transmits a short pulse of microwave radiation with known power toward the sea surface at the satellite nadir. This pulse interacts with the sea surface, and part of the incident radiation reflects back to the satellite. In this way, knowing the travel time of the pulse it is possible calculate the distance from the altimeter to the sea surface. The altimeter of the satellites launched in the previous missions, Skylab, Geos-3, Seasat and Geosat, were poorly used because of its small orbit accuracy. Presently the ERS-2 (35 days time resolution, 100 km track distance at 35°latitude) and Topex -Poseidon (10 days spatial resolution, 270 km track distance at same latitude) missions can measure the sea level topography with the spatial and time resolutions shown in the figure 1. The precissions limits are in the order of 7 cm for unaccuracies in the orbit and significant wave height that affects the shape of the radar reflexion pulse. Besides we have to take into account the systematic and random instrumental errors and the mismodelling in the effects of the different tides (Oceanic, solid earth, pole,ocean loading) and the sea level static response model to atmospheric pressure (inverse barometer).

Satellite altimetry measurements should, in principle, provide improved information for global or local sea level changes because of their global coverage and direct tie of its measurements to the geocenter. Also,

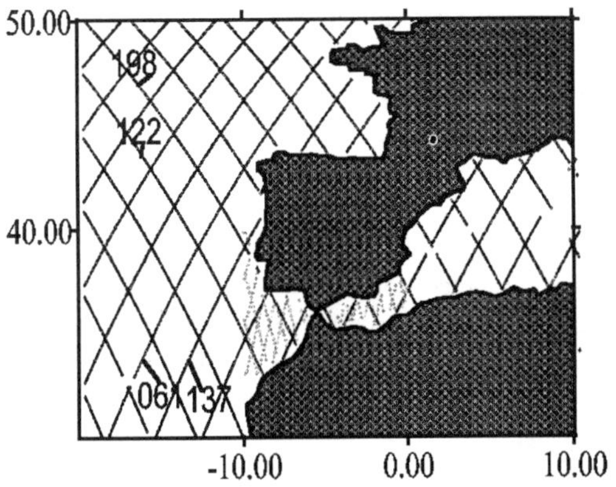

Figure 1: Topex-Poseidon (solid line) and ERS (dotted lines) altimeter satellites tracks in the Atlantic Iberian area. ERS spatial and time resolutions, at a latitude of 38°, are about 100 Km and 35 days. Topex-Poseidon resolutions, at same latitude, are about 300 Km and 10 days. Some Topex-Poseidon orbits are identified by their track number.

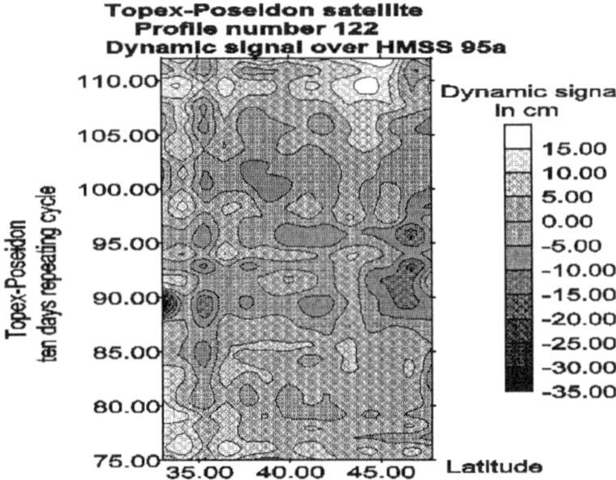

Figure 2: Dynamic signal histogram contour between cycles 076 and 113 (October 1994 October 1995). Topex-Poseidon orbit number 122. Reference surfaces HMSS95a. It can be observed between cycles 94 to 97 the pass of depressions overpassing -10 cm between latitudes 39° and 42°N, Between cycles 98 to 103 there are also depressions crossing the ground path between latitudes 37° and 39° N. It can also be observed a large depression between cycles 87 and 100 covering latitudes 45° to 47°N.

satellite altimeters can provide a measurement of the absolute sea level relative to a precise reference frame when satellite-tracking stations, whose origin coincides with the earth centre of masses, are used.

To detect possible Meddies we have analysed the quantitative values provided by AVISO for Topex-Poseidon, GDR-M version C, and the CORSSH also provided by AVISO for satellites ERS-1/2 (35 days repeating cycle phases). The records of our analysis span from October 1992 to October 1997. They are obtained after application of the following corrections: Topex Microwave Radiometer (TMR) for the wet troposphere correction, European Centre for Medium Range Weather Forecast(ECMWF) for the troposphere and inverse barometer effect, bi-frequency ionospheric correction for Topex and DORIS for Poseidon.

For the ocean tides, the University of Texas model CSR3.0 have been applied and it is also corrected for solid earth and pole tide, ocean loading, electromagnetic bias (Topex-Poseidon BM4 formula from Gaspar et al,1994), and including the bias and instrumental drift according with the models sugested by the Agency AVISO, AVISO 1992-1996.

As the Meddies dynamic signature has a magnitude order of one decimeter, we must to avoid any spatial interpolation that could filter out its signal, it is also recommended consider carefully any possible error in the model suggested by the agencies to correct the local altimeter data.

So, it would be necessary to introduce a different treatment that could take into account above facts.After the above corrections were applied the methodology used to detect possible meddies over the studied area follows the steps:

- Performation of the histograms for all satellite profiles crossing the area of interest. The histograms show the sea level time series, as directly measured by the satellite, detecting the sea level signature evolution when a Meddie eventually crosses the satellite track (see Figure 2).
- Compute the average sea level over the HMSS-95a reference surface around each satellite cross point for each cycle of the time series. Average was applied over two windows of different size. The larger window covers an area of 3° X 3°, and the smaller 0.5° X 0.5°, both centred in each selected cross point. The larger window filter out any local sea level variation, representing only the regional features. The smaller window, with the size of a Meddie, show the local sea level variation which can be related to the dynamical signature of a possible Meddie (see Figure 3).
- Compute the dynamic topography maps, for Topex and ERS repeating cycles, using together both

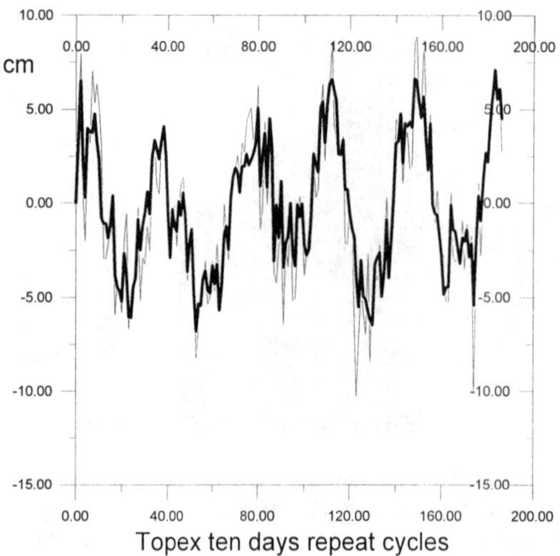

Figure 3: Topex-Poseidon time series of the average dynamic signal computed over a large window of 4° X 4°, black, and over a second window covering a smaller surface ocean area of 1° X1°, in grey. Both windows were centred around the satellite cross point over the studied area. Figure present the two windows analyses around the cross point of orbits 122 and 239 at latitude 41° N and longitude 346°W. It can be observed a possible anticyclonic eddies crossing the area in cycles 95-99 and 119-123.

satellites observation data, taken also as reference surface the HMSS-95a. Figure 4. It would be convenient to point out that the computed surfaces were interpolated not only between spatial tracks but also between time epochs.For Topex the time interpolation could have a time resolution of ten days, reaching eventually 35 days for ERS satellites.

Besides the altimeter data we have also received information from the Portuguese group of IOCEANO, about RAFOS drifters released off the Portuguese coast, providing near real time velocity information, calculated from the buoy position.

Drifters, when available, provide an excellent constrain to compute the components of the velocity field induced for any possible Meddie detected from space in the area.

We have found that the assimilation of both, altimetry and RAFOS, can be an useful tool to detect the presence and evolution of sea surface anomalies that could be identified as possible Meddies proceeding through the following steps:

1- If RAFOS information is available the location of a possible Meddie could be known.

2- With the previously exposed information it is possible to investigate in the Topex-Poseidon and ERS dynamic topography maps, figure 4, and the sea surface dynamic signal satellite tracks histogram, figure 2, looking for an anticyclonic dynamic signature.

3- With the two windows cross satellite tracks analysis, selected around the area geographically close to the presence of a possible Meddie, we can study the local dynamic signal time evolution in the small window, against the regional behaviour represented in the larger one.

The differences of the two histograms could show (as it was seen in Figure 3) the local influence of the dynamic signature of a Meddie.

4- As Meddies are masses of Mediterranean water, with different temperature and salinity whit respect to the surrounding Atlantic masses, additional information could be considered analysing SAR (Synthetic Aperture Radar) images of the ERS satellites.

These images could detect the contrast of the water masses with different electrical properties, which affects the reflectivity of the radar pulses. NOAA and ERS thermal radiometer images, with 1 Km spatial resolution, could allow the detection of Meddies for a limited period before heat loss at the atmosphere makes them indistinct of the surface Atlantic water temperatures.

4 CONCLUSIONS

Although the present study basically suggest that the join use of RAFOS and satellite information can be used to detect the presence and evolution of Meddies off the Portuguese Atlantic coasts, some facts must be considered carefully:

1-The dynamic signatures of a Meddie are in the decimetre level and extend, at most, over 100 Km in diameter.

2-Altimetric observations are local data that follows exclusively the satellite tracks. These path tracks can be spaced, for Topex - Poseidon, up to 270 Km in latitudes of 37° and for ERS satellites about 100 Km.

3-The data from altimeter, although corrected with semi-empirical models, includes instrumental delays and orbit unaccuracies, which locally could reach and overpass the subdecimetric level.

4- Figure 2 presents the profile histograms as directly

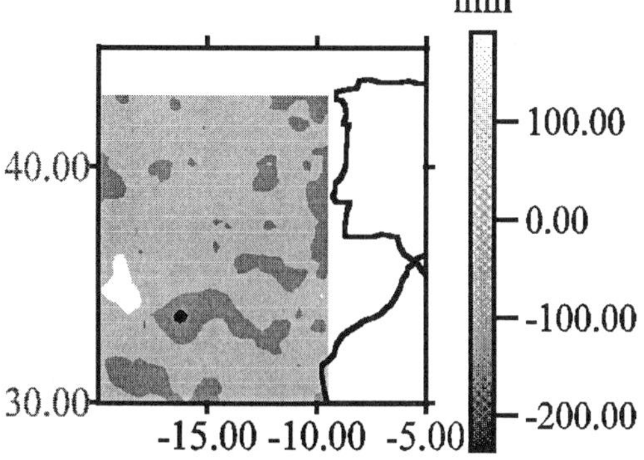

Figure 4: Sea surface dynamic topography contour map, over the area of study, computed with the 10 days repeating cycle number 098 of satellite Topex-Poseidon. The contour map also incorporates the altimeter observation data of the 35 days cycle 1 of ERS-2, between the CNES days 16568 to 16578 (23 May to 2 May 1995).

computed from the altimeter satellite observations, then, the time sequence of histograms can be show some sea surface structures which can be assocciated with medies. Figure 3 shows averages of the observations, directly made, in the area of the selected windows, around the cross-points of the satellite tracks. The information is statistically treated and the two windows comparison, although filtering out models, instrumental and orbit errors which equally affects each window data, mantains the different local sea surface information of each window. This different dynamic signal behaviour could be related, for the smaller window, with the rugosity associated with Meddies.

5- Dynamic topography maps, figure 4, are computed from data observed along each satellite repeating cycle, which are later interpolated between satellite tracks. Nevertheless and although these interpolations could filter out the Meddies time and spatial dynamic rugosity, it could mantain the Meddie signature in geographically places close to the satellite cross points.

ACKNOWLEDGEMENTS

The authors would like to thank the European Space Agency, for its support during the development of the research programm for validation of the ERS satellite altimeter data.

Thanks also to the French Agencies AVISO and IFREMER for the information received with the CD-rom altimeter data.

Thanks also to the EU Research programm CANIGO, for its support.

REFERENCES

Ambar I., M. R. Howe and M .I. Abdullah, 1976 A physical and chemical description of the Mediterranean outflow in the Gulf of Cadiz. *Deutsche Hydrographische Zeitschrift,* 29, 58-68

Ambar I.,M.R.Howe, 1979 Observations of the Mediterr anean outflow.I:Mixing in the Mediterranean Outflow. *Deep Sea Research,* 26 A,535-554

Ambar I, Fiuza A.,Costa P.;1991 Time series of currents off Algarbe: Evidence of the formation of Meddies? Proceedings "*International Workshop on Outflows and Inflows in the Atlantic and their role in the Eastern boundary current system*", Lisboa, 1991.

Armi L.,and W Zenk, 1984 Large lenses of highly saline Mediterranean water. *J. Phys.Oceanogr.*, 14,1560-1576,1984

AVISO *Handbook Merged Topex-Poseidon products.* AVI-NT-02-101-CN ed 2.1,1992

AVISO *Handbook Merged Topex-Poseidon products.* AVI-NT-02-102-CN,1996

Candela J., Winant C., Bryden H. 1989 Meteorologically forced subinertial flow through the Stait of Gibraltar *J. Phys.Oceanog.*, 94, 12667-12674.

Fiuza A., Ambar I., Dias J., 1990 Variability on slope currents off Southern Portugal and the formation of Meddies *XV General Assembly of the European Geophysical Society*.Copenhaguen.

Gaspar P.,Ogor F.,Le Traon P.Y. and Zanife O 1994 Estimating the sea state bias of Topex-poseidon altimeters from cross over differences' *J. Geophys. Res.*, 99, 24981-24994.

Kase R. H.,A. Beckmann,H.H: Hinrichsen 1989 Observa tional evidence of salt lens formation in the Iberian basin. *Journal of Geophysical Research,* 94 4905-4912

Lacombe H., Richer C. 1982 The regime of the Gibraltar Strait Hydrodinamic of semiclosed seas ed. *J.*

Madelaine F.,1970 Influence de la topographie du fond sur l'eculement mediterraneen entre le Detroit de Gibraltar et le Cap St. Vincent.*Cahiers Oceanographiques* 22, 43-61

Ochoa J. and N.A.Bray 1991 Water mass exchange in the Gulf of Cadiz *Deep Sea Research* 38 Suppl.1 S465-S503

Richardson P.L.,Walsh L.,Armi M., Schoreder M.,Price F. 1989 Tracking three Meddies with SOFAR floats' *Journal of Physical Oceanography,* 19,371-383

Tokos K., Hinrichsen H., Zenk W 1994 Merging and migation of two meddies *J. Geoph. Res.* 24 2129-2141

Zenk W. 1975 On the Mediterranean outflow west of Gibraltar. *Meteor Forsch-Ergebnisse,* A 16 23-34

Zenk W.,Klein B., Scroder M 1991 Cape Verde frontal Zone *Deep Sea Research* 38,

Remote Sensing in the 21st Century: Economic and Environmental Applications, Casanova (ed.)
 ISBN 90 5809 096 5

Environmental study in the estuary of river Miño

I. Bravo Villar, J. Martín Herrero, J. M. Torres Palenzuela & L. Fernandes Fernandes
Department of Applied Physics, University of Vigo, Spain

ABSTRACT: A remote sensing study was carried out to determine the estuarine circulation, the sandbank patterns, the River Miño plume and classification of coastal lands and shorelines using TM-Landsat images, ERS-1 SAR image and a 3D projection of submarine relief. The circulation patterns in the area of study were deduced from the bathymetric study, the sediment patterns, the thermal plumes and the radar image. A classification using neural nets and a Landsat image for the Miño area has been carried out.

1 INTRODUCTION

1.1 *Coastal zone*

The coastal zone is a highly complex area containing a varied continuum of many different spatial, temporal and spectral scales: spatial variability comes from the varied landscape formed by a coast and its often intricate relief; temporal variations come from the periodic tides during the day together with the periodic seasonal changes over the year and with the long term evolution of the shoreline; spectral variability is due to the great variety of materials and structures that are found on the coast.

Nowadays, a great number of remote sensing platforms, both airborne and satellite, fly above us tracking the ground beneath. All of the expensive instruments and resources used would be in vain if there were no meaningful development of adequate applications to extract as much information as can possibly be obtained. Therefore, a number of techniques have and are being developed to apply the satellite and aircraft data to all of the possible fields of research. A lot of work has already been done on land cover classification from remote sensing images, which varies from visual inspection to automatic methods taking advance of modern algorithmic techniques, using highly efficient computers. Among these, neural nets outshine by outperforming more traditional classifiers, such as the Bayesian Rule. Neural nets are no more a novelty in the field of image classification and pattern recognition thanks to the work of various researchers over the past years.

Classification and interpretation would be improved substantially by integrating different data sets from a number of sensors, wherein each one would contribute its own advantages. Both the spectral signature information on solar energy reflectivity like that obtained from the Landsat Thematic Mapper or by using aircraft, to cope with of a Digital Elevation Model, permitting a correction of the reflectivity measured by the passive sensors. The DEM will provide us with an enhanced tool to present the monitoring results.

The Miño Estuary is located on the NW Atlantic coast of the Iberian Peninsula at 41°52'N and 8°52'W. It is the natural boundary between Portugal and the Spanish region of Galicia. Remote sensing has been used to determine the bathymetry and to study the estuarine circulation and its sedimentation patterns. Thematic Mapper (TM) images from the Landsat satellite have been used to study the sediment patterns for the years 1991 and 1995, which show the changes resulting from the dredged channel. The studies of the thermal band from the TM-Landsat image show that the structures of the plume are correlated with the bathymetry of the Estuary. The bathymetry of the Miño River was obtained from a navigation map of the area and its 3D projection was produced for this study. A SAR image from the ERS-1 satellite was used to study the circulation in the mouth of the Estuary. Shoreline land classification maps have been obtained using neural nets and a Landsat image.

1.2. *Classification.*

The application of automatic image classification techniques to coastal lands and shorelines is still a novelty due to the specific characteristics of these border areas. The coastal zone is a highly complex area containing a varied continuum of many different spatial, temporal and spectral scales: spatial variability comes from the varied landscape formed by a coast and its often intricate relief; temporal variations come from the periodic tides during the day together with the periodic seasonal changes over the year and with the long term evolution of the shoreline; spectral variability is due to the great variety of materials and structures that are found on the coast. Thus, a high proportion of the discretely sampled pixels within an image contain mixed spectral signatures which are not easily placed into fixed thematic classes using conventional classification techniques. Fuzzy techniques have to be developed for the analysis of such areas from a more realistic view point which would produce more accurate results, with the advantage of permitting a good interaction with Parallel Distributed Models as is the case with neural nets.

Even though these tools are useful and necessary for coastal land monitoring, they are not enough from the applicant's view point. The possibility of developing appropriate object recognition techniques that permit tracking sand bank displacements and changes in the shape of the shoreline shall be studied, for an accurate study of the evolution of such a changing environment like the coast. Here, again, neural nets will play a major role as one of the few possible solutions to permit an automatic monitoring over time of such changes in size, shape and position.

The objective therefore is to develop a system which permits us to obtain information relative to land uses, their type, extension and evolution, human pressure and changes over time in a coastal area for which adequate remote sensing data is available. This is especially important considering, on the one side the intrinsic difficulties of such studies in these zones, and, on the other side, the primary importance of coasts as human settlements, sources of richness and that of extremely fragile ecosystems.

2 METHODS

2.1 *Study of sand banks and river circulation.*

The TM1, TM2, TM3 and TM6 bands from the Landsat satellite have been used in this study. These have the spectral ranges: TM1 (0.45-0.52μm), TM2 (0.52-0.60μm), TM3 (0.63-0.69μm) and TM6 (10.4-12.5μm). It has been assumed that the radiance reaching the satellite sensor for a given satellite band is:

$$Lt(\lambda) = La(\lambda) + Lw(\lambda)$$

where La is the radiance from atmospheric effects and Lw is the radiance from the surface. An atmospheric correction was made using histogram correction for each band by the minimum value method (Campbell 1987). Principal Component analysis was carried out using the first three bands from TM Landsat. The land cover was masked in the final image. A land cover mask and an increase in contrast were made to the thermal image in order to study the riverine plume. Land cover was masked in the radar image and all images were geo-referenced.

The Bathymetry has been obtained from isobath maps and the intermediate points were obtained using the minimum curvature method. The 3D projection was produced by the ray tracing method applied to the surface which was generated by B-splines in specially developed graphical software (Foley et al. 1996). The submarine relief scale is exaggerated with regard to land cover scale.

2.2 *Classification.*

We used a two layer perceptron with 18 hidden nodes and 12 output nodes corresponding to the 12 categories of interest, listed in Table 1. For the coding of the input vectors (8th dimensional), we adopted the method called coarse coding (Bischof et al. 1992), using the same 13 nodes per input band, though we did not observe any significant variation in the performance of the net due to alterations in this parameter. For the learning process, we selected 3650 pixels for the training set and 1130 for the validation set. The rest of the labelled pixels (up to 14,000, after removing the redundancy) were used as test set, as an indicator of the net's final performance.

The well known Backprop (Generalised Delta Rule) learning algorithm with momentum (Bertsekas 1995) was selected as the one which offered the best results in speed and quality of convergence (although it was seasoned with a pair of additional tools to achieve a better general performance) after a comparison between the performance of three of them: Backpropagation (Rumelhart 1986), Quickprop (Fahlman 1988) and RProp (Riedmiller 1993) on this precise subject. Firstly, there exists the possibility of changing the learning rate and the momentum factor during training in order to adapt the dynamic behaviour of the system to the local characteristics of the error surface. This requires some skill from the operator, which is obtained after a repeated series of training, but ends up with very fine learning curves, that are smooth and without oscillations. Secondly, what we have called "electroshock", can be summarised as a very simple but highly efficient (with improvements in classification accuracy up to 1%) operator tool, which introduces some randomness in the behaviour of the net, to cope with the problem of a local minima in the error surface. It consists just of the ability to do a random search at the final stages of the training, reducing the long training times derived from the habitual repeated entire training cycles from different initial points (random initial weights) in search of good minima. In as far as output of the classifier, we not only store the highest output and its level for each pixel, but also the next one in level together with its value, in order to do the subsequent smoothing of the classification, using these values as "confidence" levels.

3 RESULTS

3.1 *Sediment study*

The Figure 1 shows the River Miño Estuary and the sedimentation patterns for 1991(a) and 1995(b). In both these images the histogram was inverted hence the sedimentary banks are shown in a dark tone. The dredged channel in the second image can be detected by a comparison between points A and B in both images: these correspond respectively to A Guarda (Spain) and Caminha (Portugal).

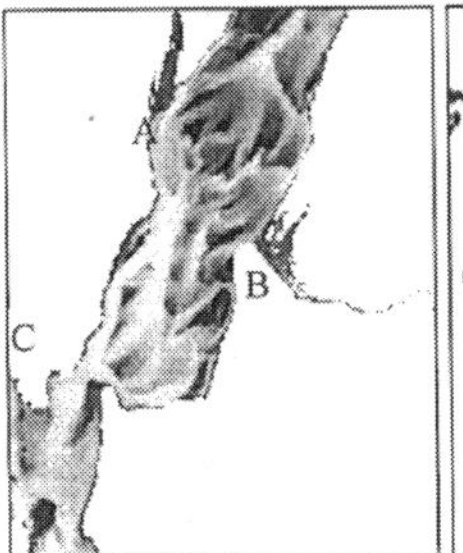

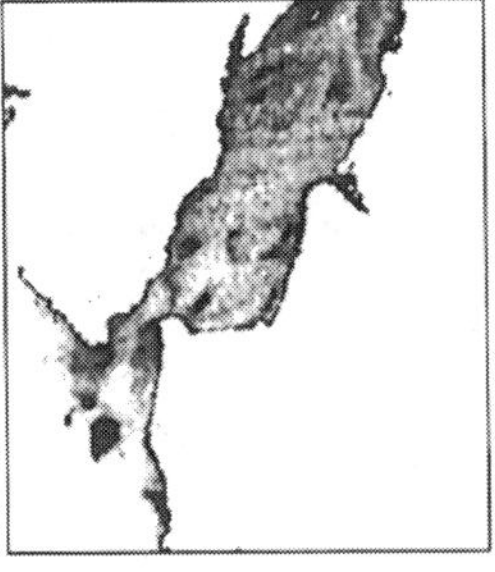

Figure 1. The results of the Principal Component analysis can be seen in two Landsat images using TM1, TM2 and TM3 bands corresponding to (a) 1991 and (b) 1995 where a land cover mask and an inverted histogram has been made. Thus the sediment banks are shown in dark tones and a dredged channel can be seen in the second image between the points A and B.

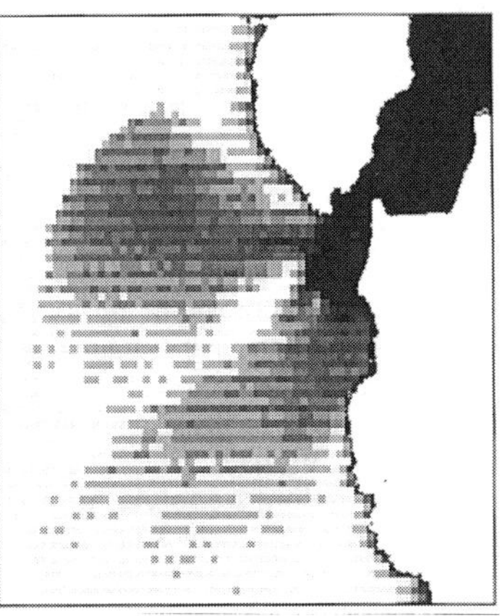

Figure 2.Thermal image from the TM6 Landsat satellite, which shows two river plumes, one towards the west and the other to the south-west. The image has been masked and the histogram inverted so the dark tone corresponds to the river water.

We can also see how the sandbanks (points D and C) have increased at the river mouth due to was necessary to dredge the riverbed at regular intervals to keep a channel free because a new ferry crossing was inaugurated in 1994 across the river as a result of tourist interest in the area.

Obstruction of the river's normal flow results in the production of several plumes at the mouth of the river, as will be seen in the second part of this study.

3.2 *Study of the River Miño Plume*

The thermal plume of the River Miño can be seen in the Figure2 where the sedimentary processes have produced two plumes, one towards the west and the other to the south-west. The thermal image (Fig. 3) from TM6 Landsat has been masked by land cover and an increase in the contrast was applied to the histogram. The image was then inverted such that the warm water from the river is shown in dark tones.

Finally, the SAR image from the ERS-1 satellite gives us information about the circulation patterns (Fig. 3). In the radar image we can see the main channels referred to above as well as the islands at the river mouth. A dark area can also be seen opposite the river mouth. The bathymetric study and the sedimentary patterns lead us to believe that there is a clockwise circulation vortex at the river mouth.

3.3 *The Neural Net*

The final results are shown in Table II, where one can see the percentage of correct classification together with the percentage of confusion between classes. We can make several interesting remarks from a detailed inspection of the data.

Firstly, to emphasize the coherence of confusions; that is, if we discard the percentages that are less than 0.5% as non-significant, one can view how all the confusions in the classification agree with the aforementioned in this paper. Thus, the confusion amongst forest classes mainly takes place between each other: eucalyptus has a 7.4% of confusion to pinewood and a 1.6% to deciduous wood; this a 2.7% to pinewood and a 2% to agricultural lands; pinewood a 2.3% to deciduous wood and a 1.81% to eucalyptus. Agricultural lands are incorrectly classified as deciduous wood in a 2.48% of the cases and as pinewood in a 0.92%. The expected confusion between urban and industrial classes has been mainly internal, because of the imprecise separation, between dense urban fabric and industrial area, about a 9% in both senses; dense urban and disperse urban confuse with each other in a 5% and a 3.83%. The border roll of disperse urban makes the rest of the confusions go to scrub, a 5.3%, and to rocky ground, a 2.4%.

The remaining confusions in industrial area go to rocky ground, 3.57%, and to disperse urban, 1.36%, and in dense urban, only to rocky ground, a 2.96%. In sandy ground, a small 0.8% goes wrongly to rocky ground, probably because of the real presence of rocks in areas selected and marked as sandy; the same in the class scrub (brushwood), where a 3.1% goes to rocky ground: the areas of scrub in Galicia bear a high degree of erosion, causing the ground under and between thickets to be covered by stones; the rest of the confusion in scrub is with disperse urban, a 1.31%, probably because the mixture between rocks and vegetation in some cases resembles that of small buildings and vegetation which characterises disperse urban pixels.

In the category of intertidal areas the source of error seems to be different from that of the rest of the categories, in which, as mentioned, the principal source of confusion would be the mixture of the terrain classes and their imprecise definition. This category, however, has a precise definition and it generally appears clearly separated on the terrain: in the middle of a marsh that gets covered by the waters at high tide it is difficult to find trees (there are no mangroves in Galicia) or houses. But there is a relatively high percent of confusion with forest classes, such as pinewood, a 3.35%, or eucalyptus, 1%. So, we suspect as possible cause of error as a poor capacity of the TM bands to precisely characterise this type of land cover, especially in areas covered by thick layers of algae or where drainage is poor. This consideration must be kept in his

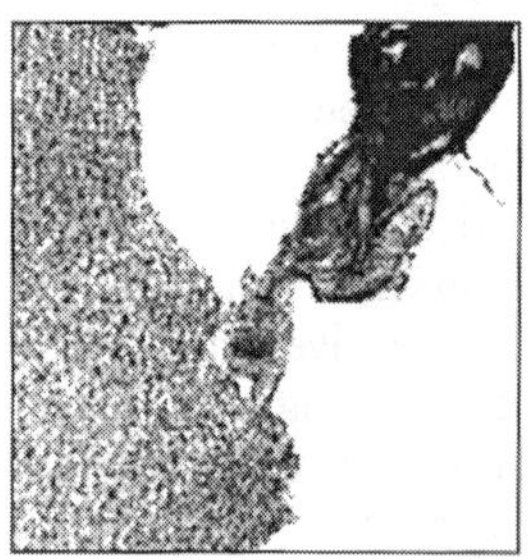

Figure 3. SAR image from the ERS-1 satellite gives us information about the circulation patterns.

Table 1. Classes of land cover.

Class	Characteristics	N. of selected píxels
Dense urban area	Urban centres, areas with high density of buildings.	1049
Disperse urban area	Rural zones and disperse, low density village settlement. Farm houses in agricultural or forest lands.	779
Industrial area	Industrial facilities, big factories.	588
Agricultural lands	Agricultural lands without distinction of crops: yards, vineyards, cornfields, pastures.	1085
Brushwood	Scrub areas: gorse, heather, fern...	3668
Deciduous wood	Broad-leaf forest species different from eucalyptus: chestnut tree, oak tree, beech tree...	351
Eucalyptus	Forest areas with prevalence of eucalyptus.	1801
Pinewood	Forest areas with prevalence of pines.	674
Bare soil	Rocky ground, stone quarries, highly eroded mountain areas.	486
Sandy ground	Beaches. Sandy areas without vegetation or with scarce vegetable covering.	206
Intertidal zone	Marshes, quagmires, slimy areas unveiled by water. Often with big areas covered by alga.	1404
Water	Water courses and bodies: sea, dams, rivers, flooded lands.	4054

Table 2: Accuracy and confusion percentages in each class.

Class	Correctly classified	Confusion	Classified as
Agricultural lands	95.66 %	2.48 %	Deciduous wood
		0.92 %	Pinewood
		0.46 %	Brushwood
Water	99.24 %	0.76 %	Bare soil
		0.08 %	Decid. wood
		0.06 %	Eucalyptus
		0.06 %	Pinewood
Sandy-ground	99.24 %	0.76 %	Bare soil
Eucalyptus	90.43 %	7.42 %	Pinewood
		1.68 %	Decid. wood
		0.32 %	Bare soil
		0.16 %	Interti. Zone
Dcid. wood	94.35 %	2.66 %	Pinewood
Decid.wood	94.35 %	1.99 %	Agricu.lands
Class	Correctly classified	Confusion	Classified as
		0.33 %	Bare soil
		0.33 %	Eucalyp-tus
		0.33 %	Brush-wood
Industrial area	85.37 %	9.01 %	Dense ur-ban area
		3.57 %	Bare soil
		1.36 %	Disperse urban area
		0.34 %	Sandy ground
		0.34 %	Intertidal zone
Intertidal zone	94.23 %	3.35 %	Pinewood
		0.99 %	Eucalyp-tus
		0.42 %	Dense ur-ban area
		0.42 %	Brush-wood
		0.21 %	Zona In-dustrial area
		0.14 %	Disperse urban area
		0.14 %	Bare soil
		0.07 %	Water
Brushwood	95.07 %	3.10 %	Bare soil
		1.31 %	Disperse urban area
		0.30 %	Pinewood
		0.13 %	Deciduous wood
		0.11 %	Agricul-tural lands
		0.03 %	Intertidal zone
Pinewood	94.35 %	2.27 %	Deciduous wood
		1.81 %	Eucalyp-tus
		0.45 %	Bare soil
		0.45 %	Brush-wood
		0.45 %	Disperse urban area
		0.23 %	Intertidal zone
Bare soil	94.65 %	2.26 %	Disperse urban area
		1.44 %	Pinewood
		0.41 %	Dense ur-ban area
		0.41 %	Sandy ground
		0.21 %	Eucalyp-tus
		0.21 %	Intertidal zone
		0.21 %	Brush-wood
		0.21 %	Deciduous wood
Dense urban area	82.08 %	9.53 %	Industrial area
		5.05 %	Dis. urban
Class	Correctly classified	Confusion	Classified as
		2.96 %	Bare soil
		0.38 %	Interti. zone
Disperse urban area	86.43 %	5.30 %	Brushwood
		3.83 %	Dense urban area
		2.36 %	Bare soil

relative value, because, nevertheless, we have obtained a 94.23% of accuracy which should be enough for most of the applications.

We can make an analysis by grouping classes, to obtain a global accuracy of 92.88% in forest classes, a 94.54% in vegetation type classes (including forest ones), a 95.25% in "natural" classes (all but urban and industrial areas), and a 84.63% in urban and industrial classes. Another possible viewpoint is by computing the groups as superclasses, considering these wrongly classified pixels as belonging to other correctly classified classes of the same group. In this way, the accuracy in the superclass "forest" is 98.43%, in the bigger superclass "vegetation", 98.63%, and in the even bigger superclass "natural", 99.42%; the accuracy in the superclass "urban and industrial" is 97.35%. These are very good results, bearing in mind that the net was trained to distinguishing the twelve classes: it is reasonable to think that training the net to separate only some of these superclasses would permit a still better performance.

4 CONCLUSIONS

The images from the TM Landsat sensor were useful for studying the sedimentary patterns since they showed the changes due to the dredging of the ferry channel. The thermal image from the TM6 Landsat gives us information about the circulation patterns in the river mouth with two plumes that correspond with the bathymetry at the site.

The radar image showed a clockwise vortex circulation in the mouth of the estuary, which corresponds with the thermal image and the river plumes as well as the bathymetry relief.

From the above results of classification, we can deduce a very good ability of the tandem net-sensor to separate between vegetation types and between what we could call "natural" soils from "artificial" soils, and a slightly worse performance in distinguishing between different kinds of "artificial" soils.

REFERENCES

Bertsekas, D.P. 1995. *Non-Linear Programming.* Belmont, Massachussets: Athena Scientific.

Bischof, H., W. Scheneider & A. J. Pinz 1992. *Multispectral Classification of Landsat Images Using Neural Networks.* IEEE Transactions on Geosciences and Remote Sensing, Vol. 30, 3:482-490.

Campbell, J. B., 1987. *Introduction to Remote Sensing.*, New York: The Guilford Press.

Cracknell, A. P. et al. 1982. *Remote Sensing in Scotland Using Data Received from Satellites. A Study of the Tay Estuary Region Using Landsat Multiespectral Scanning Imagery.* International Journal of Remote Sensing, 3:113-137.

Fahlman, S.E. 1988. *An Empirical Study of Learning Speed in Backpropagation Networks.* Technical Report CMU-CS-88-162, School of Computer Science, Carnegie Mellon University, Pittsburgh.

Foley, V.D., Feiner & Hughes 1996. *Computer graphics: principles and practice.* Addison-Wesley.

Jensen, J.R. 1986. *Introductory Digital Image Processing: A Remote Sensing Perspective.* Englewood Cliffs, New Jersey: Prentice-Hall.

Lodge, D. W. S., 1983. *Surface Expression Of Bathymetry On SEASAT Synthetic Aperture Radar Images.* International Journal of Remote Sensing. 4: 639-653.

Kennie, T.J.M. & M. C. Matthews 1985. *Remote Sensing in Civil Engineering.* New York: John Wiley.

Lillesand, T.M. & R. W. Kiefer 1994. *Remote Sensing and Image Interpretation.* New York: John Wiley

Riedmiller, M. & H. Braun 1993. *A Direct Adaptative Method for Faster Backpropagation Learning: The RPROP Algorithm.* IEEE Procs. of the International Conference on Neural Networks, San Francisco:586-591.

Ritchie, J. C., C.M. Cooper, C. & F.R. Schiebe 1990. *The relationship of MSS and TM digital data with suspended sediments, chlorophyll, and temperature in Moon Lake, Mississippi.* Remote Sens. Environ., 33: 137-148.

Rumelthart, D.E., G. E. Hinton & R. J. William 1986. *Learning Internal Representations by Error Propagation.* Parallel Distributed Processing: Exploration in the Microstructure of Cognition, Vol. 1, 8: 318-362. Cambridge: M.I.T. Press.

Shuchman, R.A., D.R. Lyzenga & G.A. Meadows 1985. *Synthetic Aperture Radar Imaging Of Ocean - Bottom Topography Via Tidal - Current Interactions: Theory And Observations.* International Journal Of Remote Sensing, 6 : 1179-1200.

Wagle, B.G., 1985. *Cover, distribution and dispersion of turbid waters on the western continental shelf of India. An interpretation of LANDSAT images.* International Journal of Remote Sensing, 6: 1681-1684.

Remote Sensing in the 21st Century: Economic and Environmental Applications, Casanova (ed.)
 ISBN 90 5809 096 5

Applications of SAR images in the western Mediterranean Sea

J.J. Martinez Benjamin, J.M. Redondo, J. Jorge & A. Platonov
Departamento de Física Aplicada, Universidad Politecnica de Catalunya, Barcelona, Spain

ABSTRACT: The Synthetic Aperture Radar (SAR) has made significant contributions to open-sea and coastal oceanography. Extremely valuable data has been obtained on surface and internal waves, fronts, eddies, variations in bottom topography and bathimetry, small-scale and mesoscale currents, surface wind speed and a variety of other processes that give rise to subtle variations in surface roughness detectable by SAR. We present here results stemming from the European Union MAST projects CANIGO and CLEAN SEAS that reveal significant trends in ocean surface dynamics and in oil pollution, which have an important incidence in their economic and environmental applications. Both images from RADARSAT (CSA) and ERS-2 (ESA) have been used.

1 INTRODUCTION

The Synthetic Aperture Radar is a valuable instrument in to Space Oceanography. The polarization , frequency and incidence angles must be considered in assesing the specific oceanographic value of SAR data. Observations from ERS satellites of the European Space Agency (ESA) and RADARSAT of the Canadian Space Agency (CSA), allow detailed analysis of oceanic mesoscale and microscale processes.

CANIGO (Canary Islands Azores Gibraltar Observations) is the North East Atlantic MASTIII regional sea project of the European Union (Contract MAS3-CT960060), for investigating the marine system in the Azores-Canary-Gibraltar region and its links with the Mediterranean Sea, through comprehensive interdisciplinary basin scale studies.

CLEAN SEAS (ENV4-CT96-0334) is a collaborative project supported by eight research organisations drawn from 6 european countries under the Environment and Climate topic of the EU Fourth Framework Programme. The project aims to assess the efectiveness of the current Earth observation system with reference to the practical application of marine pollution monitoring.

Satellite data derived from SAR, ocean colour and imaging radiometer instruments are being analysed in the western Mediterranean Sea. Some previous work can be seen in Martinez-Benjamin et al. (1999a,1999b) and Redondo et al. (1998,1999).

2 SAR IMAGES ANALYSIS

SAR observations from ERS-2 and RADARSAT satellites allow detailed analysis of oceanic mesoscale and microscale processes. Due to their different polarization (ERS-2: VV, RADARSAT:HH) they can give complementary information.

The dominant synoptic eddies, fronts and currents that affect the western Mediterranean Sea may be analysed with much further detail by using SAR images such as this presented in figure 1 where is possible to see a front and atmospheric features. The image of the NE Spain is from ERS-2 taken on 15 February 1998 at 10:34 UTC.

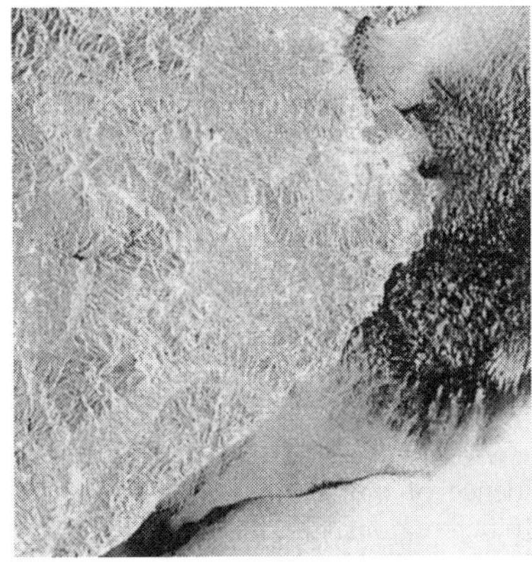

Figure 1: ERS-2 image taken on 15 February, 1998.

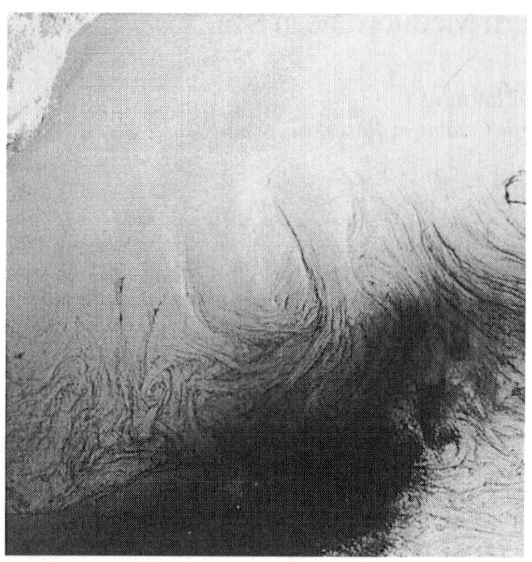

Figure 2: ERS-2 image taken on the 9th August 1998.

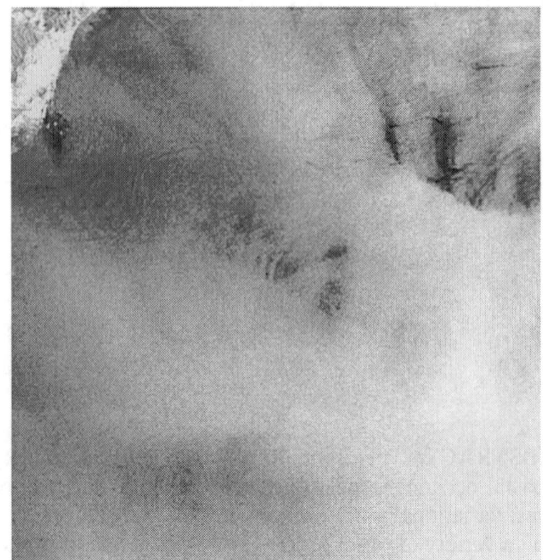

Figure 3: ERS-2 image taken on the 31th May 1998.

From a detailed image analysis of the SAR images showing significant features that reveal aspects of the topography of the sea surface, calculating both the spatial distribution of eddy integral lenghtscales and the spectrum of the SAR intensity crosscorrelations, akin to a turbulent energy spectrum, it has been possible to estimate local seasonal eddy diffusivity values for the ocean surface.

An statistical analysis of the SAR observations performed within the CLEAN SEAS in the NW Mediterranean Sea for the coastal area between Cape Creus and the Ebro delta made by Gade et al. (1999) made an extensive study of the marine pollution of this area. He showed that the statistical analysis of the ERS SAR images reveals that the most frequent pollution occurs along the main ship traffic routes and the SAR images acquired during descending, morning, and ascending, evening, satellite passes show different percentages of oil pollution, which can be explained by the fact that most of the pollution occurs during night time.

Figure 2 show an ERS-2 image taken on the 9th August, 1998, at 10:35 UTC where is possible to see a mixture of slicks and spills.

Figure 3 shows the same frame with Barcelona in the upper left of the image. It was taken on the 31th of May, 1998, at the 10:35 UTC and it is very clear a pollution event off Barcelona harbour. It has been showed that the shiping routes may be easily detected in spite of the low frequency of observations, on average one SAR image every 35 days. There is clearly an incidence of the ship routes. Comparing this area with other regions like the Gulf of Lions, the North Sea and the Baltic there is a higher proportion of oil spill detection in this area.

This method is envisaged to be extended in the future and there is a strong European Union support for the interaction between the scientific community and the Harbour and Environmental protection agencies.

This region is part of a more general study made within the CLEAN SEAS Project. A more complete schema of the geographic distribution of frames in the NW Mediterranean is given in figure 4.

RADARSAT, from the Canadian Space Agency uses a single frequency C-band, HH polarisation and has access to a variety of beam selections that can image swaths from 50 to 500 km with resolutions from 8 to 100 m., respectively. Incidence angles

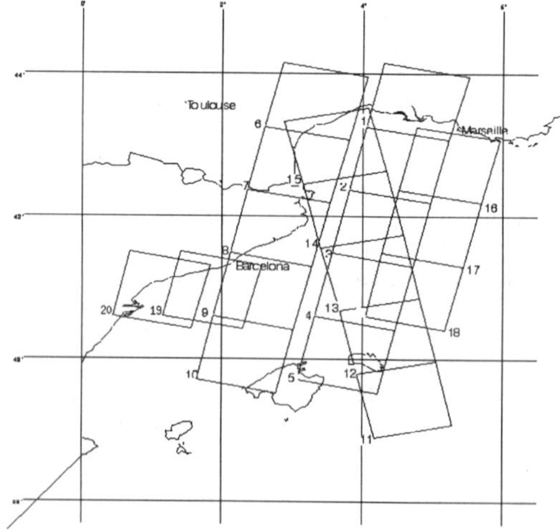

Figure 4: Frames of the Gulf of Lions area (CLEAN SEAS)

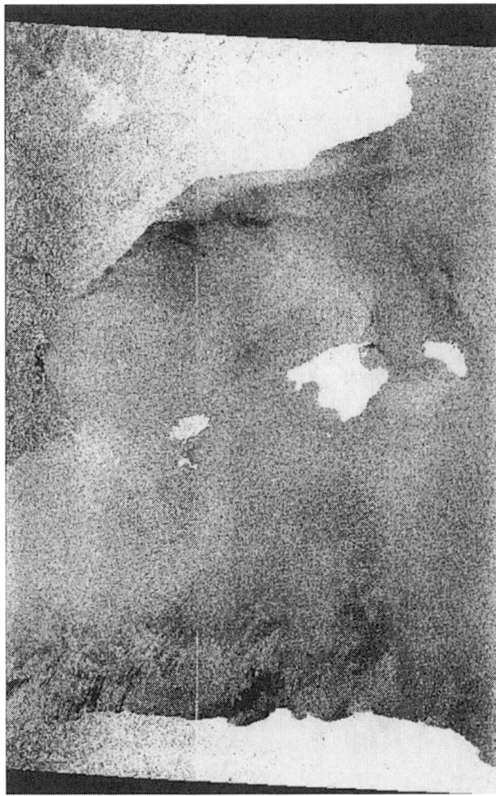

Figure 5: NW Mediterranean Sea @Canadian Space Agency

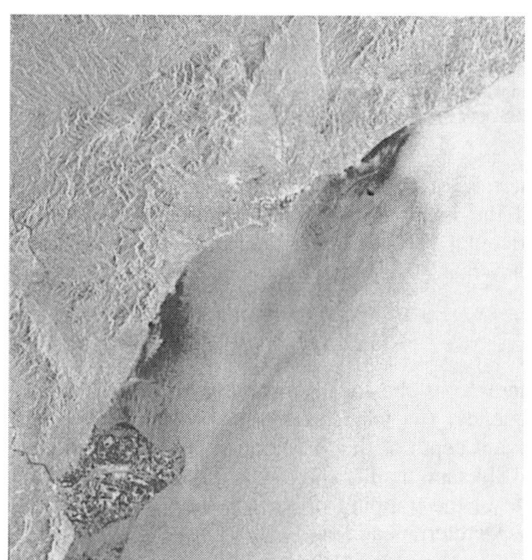

Figure 6:ERS-2 image taken on the 3rd of June 1998

range from less than 20 degrees to more than 50 degrees. The wide ScanSAR swath gives RADARSAT a major advantage in ocean applications as mapping

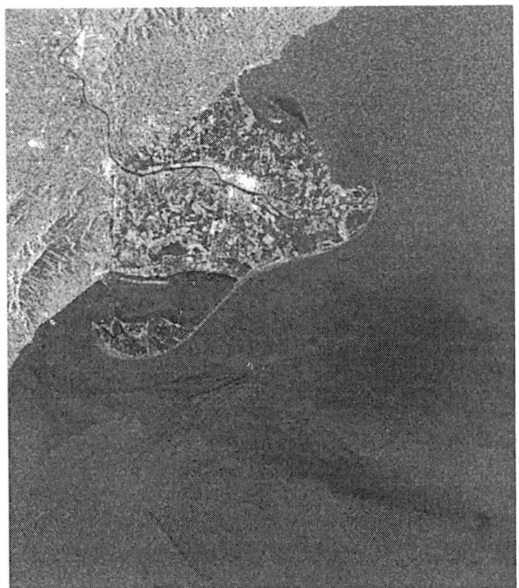

Figure 7: RADARSAT view of Ebro delta @Canadian Space Agency

mesoscale features and for rapid and more frequent coverage.

A RADARSAT Wide ScanSAR image taken on 29th May 1998 is shown in figure 5, giving a global synotic view of a large part of the western Mediterranean Sea.

Finally figures 6 and 7 show the region near the Ebro delta as seen by the ERS-2 and RADARSAT.

3 OCEANOGRAPHIC ANALYSIS

We present in figures 8 - 11 some results of an statistical analysis of the SAR observations performed within the CLEAN SEAS in the NW Mediterranean Sea for the coastal area between Cape Creus and the Ebro delta.

In figure 8 we show the total number of detected oil spills measured on a 5' x 5' grid. The different frames near Barcelona like those are depicted in figure 4 (frames number 8, 9, 10, 19 and 20) were used during the period 1997 - 1998. Not all frames had the same availability and a selection of 81 SAR images were used from the more than 700 images.

To compensate for the different sampling of the frames a normalized plot is shown, where the numver of detected spills is divided by the number of observations, giving an estimate of the probability of an oil spill detection in a certain area as appears in figure 9.

In figure 10 we show the regions where vortical slicks are detected, and is thought that the coastal bathimetry plays an important role in their generation mechanism.

In figure 11 we show the location of the detected vortices during the period 1997 - 1998, their distribution of angles and directions, shapes and size.

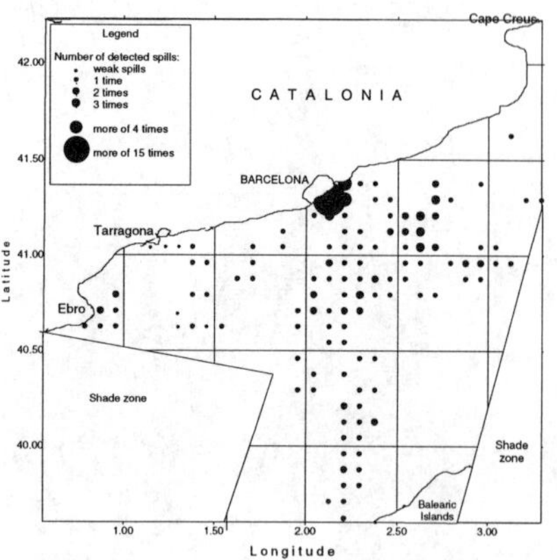

Figure 8: total number of detected oil spills during the period 1997 - 1998 near Barcelona.

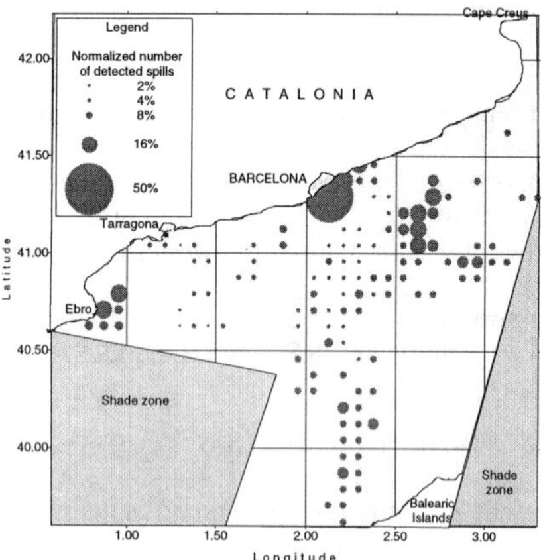

Figure 9: normalized number of detected oil spills during the period 1997 - 1998 near Barcelona

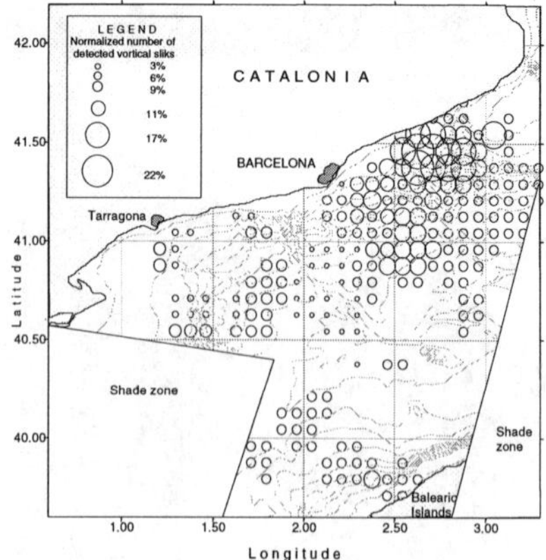

Figure 10: normalized number of detected vortical sliks during the period 1997 - 1998 near Barcelona

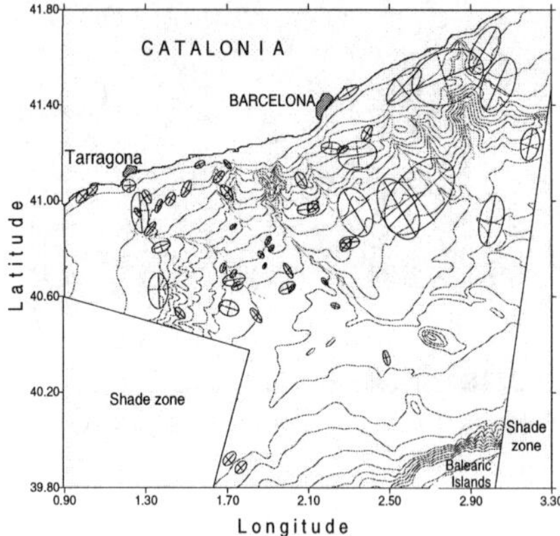

Figure 11: location of the vortices, their distribution of angles,shapes and size during the period 1997 - 1998 near Barcelona.

4 CONCLUSIONS

Analysis of SAR images from RADARSAT and ERS-2 has been made in the framework of the CLEAN SEAS and CANIGO European Union Projects. The detection of several kinds of oceanographic features in the ocean surface has been made. The monitoring of pollution events caused by ships or coastal emisions that release oil spills provides an effective way of assesing the environmental impact of human activities. The detection of other types of features such as the relevant lengthscales associated to natural surface slicks detected by SAR also gives information on mesoscale eddy diffusivities, Redondo and Platonov(1999), showing the important role of coastal bathimetry in eddy generation and the break up mechanism that takes place for horizontal vortices larger than the Rossby deformation radius

$$R_D = \frac{N}{f}.h$$

Where N is the local buoyancy, or Brunt-Vaisalla frequency, f is the Coriolis parameter and h is the relevant depth of the thermocline. The detected vortical sliks are mainly shallow and seem to be responsable of the trapping of surface buoy tracers in the NW Mediterranean Sea.

These observations may be used to prevent pollution as well as to predict the dispersion of pollutant sources using an integrated system where satellite data is used to feed initial conditions to numerical models, this capability is very important on economic and environmental applications.

ACKNOWLEDGMENTS

We would like to acknowledge the work of Dr. Martin Gade from Hamburg University and the financial support from both the CLEAN SEAS (ENV4-CT96-0334) and CANIGO (MAS3-CT96-0060) European Union Projects. Special thanks to the Canadian Space Agency for the RADARSAT images provided under Project ADRO#321.

REFERENCES

Gade, M., Ufermann, S., Rud, O., Ishii, M. & Jolly, G.W., 1999. Using ERS-2 SAR images for routine observation of marine pollution in European coastal waters.*Operational Remote Sensing for Sustainable Development, Nieuwenhuis G.J.A., Vaughan R.A. & Molenaar M. (eds) © 1999 Balkema, Rotterdam,* 157-163.

Martinez Benjamin, J.J., Redondo, J.M., Van der Graaf, G., Font, J. & Chic, O. 1999a. Use of RADARSAT images to characterize oceanic features in the western Mediterranean Sea. *Symposium, CSA, Radarsat International, NASA, Montreal, Canada, in press.*

Martinez Benjamin, J.J., Redondo, J.M., Van der Graaf & Platonov, A. 1999b. Monitoring of the NW Mediterranean Sea by SAR observation. *Geophysical Research Abstracts, Volume 1, Number 2.* 413.

Redondo, J.M.& Platonov, A. 1999. Aplicación de las imágenes SAR en la monitorización ambiental de las aguas y de la polución del Mar Mediterráneo cerca de Barcelona. *Ingenieria del Agua.*Univ.Politacnica de Valencia.

Redondo, J.M., Sanchez, M., Martinez Benjamin, J.J. & Jolly, G. 1998. Spectral study of the ocean surface with SAR.*EUROPTO Conference on Remote Sensing of the Ocean and Sea Ice IV, Barcelona, Spain, Volume 3496,* 217-222.

Redondo, J.M., Platonov, A. & Martinez Benjamin, J.J. 1999. Turbulent tracer diffusion in the NW Mediterranean. *Report to the European Union _ Clean Seas (*http://bachelor.upc.es/jose/redondo.html*).*

Remote Sensing in the 21st Century: Economic and Environmental Applications, Casanova (ed.)
© 2000 Balkema, Rotterdam, ISBN 90 5809 096 5

Pressure ridges extraction using Radarsat fine beam mode

E. Hudier
Université du Québec, Rimouski, Qué., Canada

ABSTRACT: We present data recorded before and during the spring melt onset. This time period was chosen due to the fast evolution of snow and ice characteristics and, therefore, of the contrast between ice plate and ridges. With the development of a wet film on the snow and ice surface, reflection on some ice blocks enhances the ridge return while due to the increased forward scattering, the signal intensity from flat ice areas is reduced. As the melt goes on, this produces an image on which the ridge-plate contrast allows the extraction of a larger proportion of the ridge network. This data was then used to position ridges on an image recorded before the melt and compute ridge statistics. We show that ridges are not only characterized by bright linear structures. The bright structure is paralleled by a black structure in which the return is lower than the background. This explains why standard resolution images do not give good results in ridge detection. Pixels from both structures are integrated, producing an image with a reduced ridge-plate contrast.

1 INTRODUCTION

Snow-covered sea ice is an important component of high-latitude environments. The solid frontier it forms between the ocean and the atmosphere not only reduces, but controls the intensity of vertical energy fluxes. In mapping sea-ice characteristics for ship navigation and climate modeling, one of the key tasks is to estimate ridging intensity. According to Lytle et Ackley (1991) ridges and rubble fields may account for up to 30% of the ice surface and up to 50% of the total ice mass (Similä, 1991 ; Wadhams, 1984 ; Williams et al., 1975) with densities up to 20 ridges/kilometer (Gow and Tucker, 1990, Wadhams, 1992).

In recent years, the need for extensive data collection, especially in polar regions have underlined the necessity to develop satellite and airborne remote sensing technologies and automated algorithms for image interpretation. With the advent of microwave sensors (ERS, JERS and RADARSAT), it is now possible to acquire data on a regular basis independently of cloud cover.

Pressure ridges are characterized as long, linear features composed of tilted blocks of broken ice. Locally, they induce accumulations of snow which thickness and shape may vary with time depending on wind force and direction. In the literature, first-year ridges are described as bright structures, which strong back-scattering is associated with tilted surfaces. It has been observed that, as a consequence of weathering, multiyear pressure ridges may not be detectable if observed along track. The fact that they are most easily detected if observed broadside suggests that the reflection on flat ice may enhance the intensity of the signal reaching them. Overall, only a few ridges are contrasted enough to allow their extraction.

The microwave signatures of sea ice change dramatically with the season (Onstott, 92). During winter, with the snow layer being almost transparent to SAR frequencies, most changes are related to varying characteristics of the ice sheet and its upper interface. Pressure ridge detection results from an increased surface back-scattering on some individual block faces that are favorably oriented. By early spring, as melt starts and snow wetness increases, the snow layer becomes opaque to microwaves. This, in turn, reduces drastically the number of medium which contribute to the back-scattering. The returned signal becomes then mainly function of the surface roughness and slope. In addition, considering that smooth surfaces, such as wet ice blocks, have an increased normal scattering, it is expected that the non-buried part of the ridge network may have a stronger return than what is usually observed in images recorded before the melt.

In this research, we investigate the evolution of the ridge vs. ice floes back-scattering contrast on first year sea ice through the melt season.

2 METHODS

Data used in this research were obtained from a consolidated, thick (=1.5m) first year sea ice located offshore Kuujjuarapik, Hudson Bay, Canada. They were collected in April and May, 1997 and 1998. In this region, this time period of the year corresponds typically to the transition between late winter and early melt conditions.

The same sampling plans were applied for both 1997 and 1998 field works. A detailed survey of the snow layer and ridge features was performed systematically along a linear path between reference points. These points were positioned +/- 30m using fixed bearings and a GPS. The mapping of the snow layer morphology included: structure orientation, slope, size, lateral extension of the structures relatively to the ridge, and normal vs. satellite incidence angle. Ridge characteristics measurements included: height, width, block size and distribution. These surveys were performed on April 23rd and May 5th 1997 and April 8th, April 28th and May 5th 1998.

Snow temperature and wetness were measured using a thermocouple sensor and a snow fork (dielectric measurement device). These measurements were done for each SAR image on the day and time (+/- 30 minutes) they were recorded. In addition, ridge blocks surface characteristics were documented as a function of face orientations.

Image data discussed in this paper were obtained from 6 Radarsat fine beam mode images (table 1). In its fine beam mode, the SAR looks to the right of the spacecraft at 38.4° average incidence angle (mode F1) and swath an area of 57x50 km. The average local overpass times were 5:30 am (descending) and 4:57 pm (ascending).

Table 1. SAR images acquisition.

Recording date	Orbit
April 23, 1997	Descending
May 10, 1997	Descending
April 4, 1998	Ascending
April 11, 1998	Descending
April 28, 1998	Ascending
May 5, 1998	Descending

In order to reduce non-ice related parameters, both 1997 images (fig. 1a, 1b)were acquired under the same orbit characteristics.

3 RESULTS AND DISCUSSION

3.1 *The 1997 experiment*

Based on recurrent field work we did since 1987, Eastern Hudson Bay transition from winter to spring occurs during the second and third week of April. In this region, spring onset is marked by a net warming of the near-surface atmosphere accompanied by a rapid melting of the snow layer. Typically, by the second week of May, most of the ice sheet snow cover is melted out. Our 1997 campaign was planned to explore the incidence of the snow cover ablation on SAR image outputs.

As anticipated, the acquisition of the two images, on which the following analyses were done, happened after the melt initiation. Considering snow wetness values measured on April 23rd, after Tiuri et al. (1984) and Barber et al. (1994), our first image was recorded as the averaged penetration depth at 5.3 GHz was already below 5 cm. However, overall, the volume of snow was still typical of a winter-like cover. Later, after 17 days with mean high temperatures ranging from -6° to +15°, with an average of +4°, the second image (May 10th) was recorded as almost all the snow layer was melted out. In fact, most of the snow observed on the ice was accumulated along pressure ridges, hardly accounting for one third of ridge heights.

Although it is widely admitted in the literature that ridges form bright linear features, no studies were done to assess the proportion of the ridge network actually detected, let alone the characteristics of these ridges. As part of the field measurements and observations, we realized the mapping of all ridge-type structures on two straight imaginary lines linking 3 accurately positioned reference points A, B and C (fig.1b). In addition, to be positioned using coast-based bearings and GPS coordinates, these points were chosen close or on ice structures which were believed to have a remarkable reading on SAR images. The cross-analysis of those field data and of the May 10 image showed that all, 100 % of the bright linear features were areas of deformed ice. No clear relation was observed between ridge height and signal return. However, despite the limited number of ridges sampled, our observations suggested a relation between pixel density identifying a ridge and the size and density of blocks in a ridge. This will be further discussed in section 3.2.

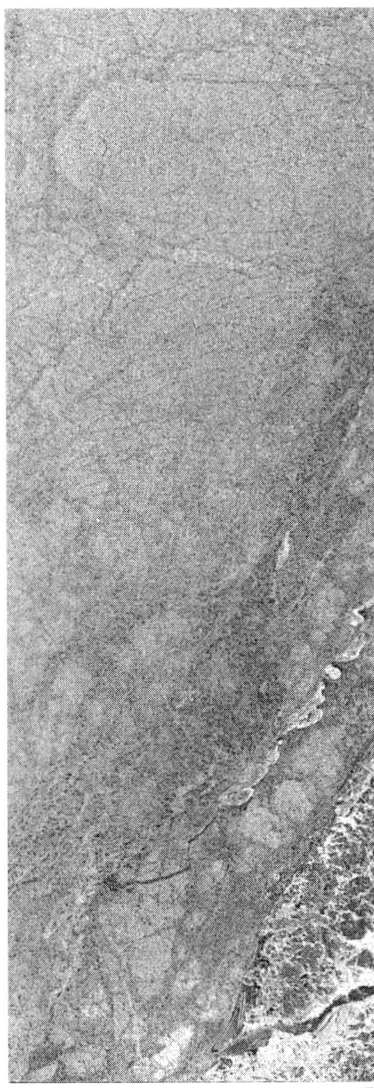

Figure 1a : Image recorded on April, 23rd 1997, Hudson Bay, Canada. The Great Whale river estuary appears at the bottom right corner of the image.

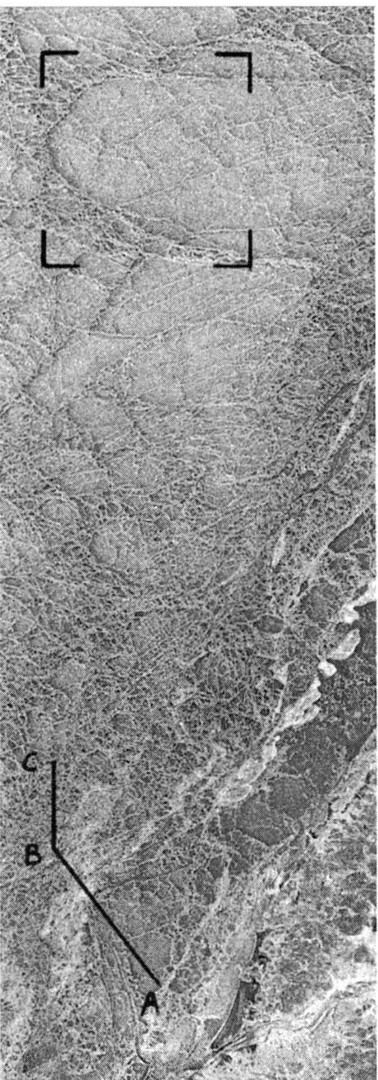

Figure 1b : Image recorded on May, 10th 1997. The letters A, B and C indicate the reference point locations and the square at the top of the image the coordinates of the sub-image.

Unlike what is usually described in the literature, the first image (April 23rd) reveals a network of wide dark linear structures. The second image (May 10th), as anticipated, displays the typical linear ridge network made of individual bright pixels. The cross-analysis of the two images shows that all dark structures from the first image have a corresponding bright structure on the second. Considering that all bright structures observed in regions sampled over the field work were positively associated to deformed ice areas, it may be assumed that the dark bands reveal an indirect, SAR-sensible, influence of ridges on their environment. Also, taking into account that, at the time images were recorded, the snow layer was opaque, it suggests that these bands of reduced back-scattering may be associated to snow cover topographic variations created by wind. After Tucker et al. (1992), this type of features may reach several hundred meters in length downwind of ice blocks. Our field measurements show that such structures developed behind ridges on a 110° Eastward orientation with heights over 1 m. tapering

Table 2. First year sea ice signatures over the early melt season.

σ0 (dB)	April 23rd, 1997	May 10th, 1997
Dark ridges	-17.8	-17.8
Bright ridges	-14.9	-12.9
Ice floes	-15.9	-15.7

away from ridges. Measurements of the slope variability on these structures showed a relative angle between the incident SAR beam and the normal to surfaces ranging from 40° to 90°. After data from Kim et al. (1984), on rough surfaces, an increase of the incident angle of 10° could induce a decrease of the back-scattering of about 3.5 dB. In addition, the dependence of the dark bands' width on ridge orientation with a maximum for North-South features is coherent with the lateral extension of wind generated snow accumulations. Also, the transfer of the bright features from the second image onto the first one, showed that dark bands are located eastward of ridges.

After magnification of selected ridges, it was also observed that both dark and bright structures coexist on both images. Table 2 gives ice floes, dark and bright ridges signatures from a numeric analysis performed on both 1997 images. These statistics were computed on a 3000x3000 pixels sub-image (fig.1) in a region where remarkable points allowed us to identify corresponding structures on both images. Considering that such observations were never made on standard beam mode images, it suggests that the fine beam mode should be preferred for the development of ridge extraction algorithms. In addition to integrating large areas of adjacent ice floes with ridges, standard mode images also integrate their dark and bright components reducing then the ridge-ice floe contrast.

3.2 *Ridge definition and block size; The 1998 experiment*

The 1998 experiment was planned to collect data in order to study the relation between ridge definition and block size and investigate parameters which could explain the reduced back-scattering in the vicinity of pressure ridges. Unlike the 1997 experiment, sampling locations were selected randomly to cover a wide range of ridges based on block characteristics, ridge amplitude and back-scattering intensity. To achieve this, 2 images were acquired early in April to guide our investigation on the ice.

As a first result, none of the four 1998 images displayed the reduced back-scattering patterns observed the previous year. In fact, the two data sets show very different ridge features. On the 1998 images, well defined, circular, pancake-like ice floes form the main structure with little of the typical linear fracture features associated with the formation of ridges. This suggests that the 1998 ice sheet consolidated without much mechanical deformation whereas the 1997 one underwent major compression and shear events. As observed on the ice, this results in a reduction in ridge sizes and therefore a lesser influence of ridges on their immediate environment. The short lateral extension of wind generated snow features may then explain the absence of dark streaks.

A ridge definition index was developed to analyze the relationship between ice characteristics and their response on SAR images. The image is first enhanced using a 7x7 adaptive box filter and resampled into 500x500 sub-images to select the thresholds giving the best results for each of them (Hudier et al., in press). In a second step, on well defined linear, ridge-like structures, a value giving the average distance between each pixel identifying the ridge and its closer neighbor is computed. A non-dimensional ridge definition index (RDI) is then obtained dividing this value by the pixel size.

Figure 2 shows the relationship between ridge characteristics and RDI for 52 ridges. To describe blocks, two dimensions were measured: the block thickness and the perimeter of its larger face. The block diameter is defined as 1/4th of the average perimeter in a ridge. The three graphs plot respectively RDI vs. block thickness, block size and ridge height. These three graphics show a common pattern with a maximum, for the ridge definition index, for a ridge height between 1 and 1.5 m and a block's diameter of about 0.80 m. This is consistent with the fact that block size and ridge height are roughly a function of the thickness of the ice from which the ridge is formed (Tucker et al. 1984). Considering that bright pixels identifying ridges are the result of a close to normal scattering on favourably oriented faces, it may be expected that the smaller the blocks are, the more orientation combinations are offered per unit area covered by a pixel, and then, the higher the probability is that nearby pixels have a back-scattering intensity above the threshold value set for ridges.

Typically, ridge detection is based on the extraction of pixels whose value is above a threshold (σ_{th}) determined from field data and image statistics.

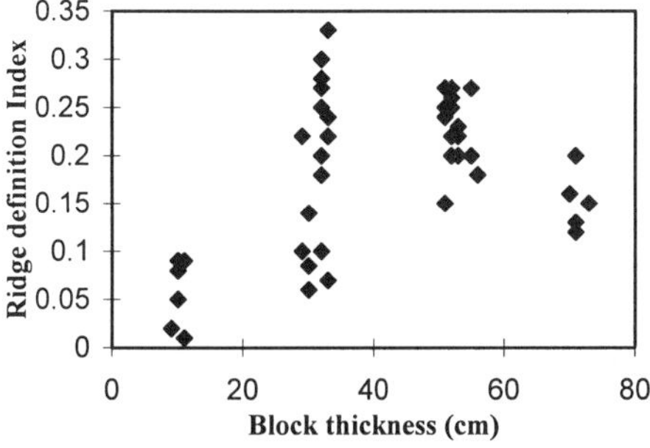

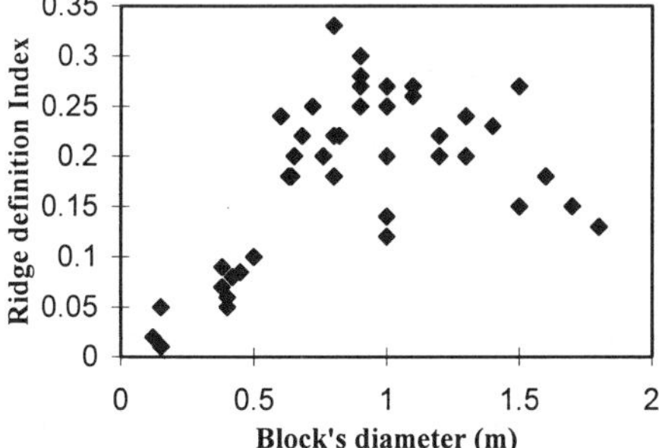

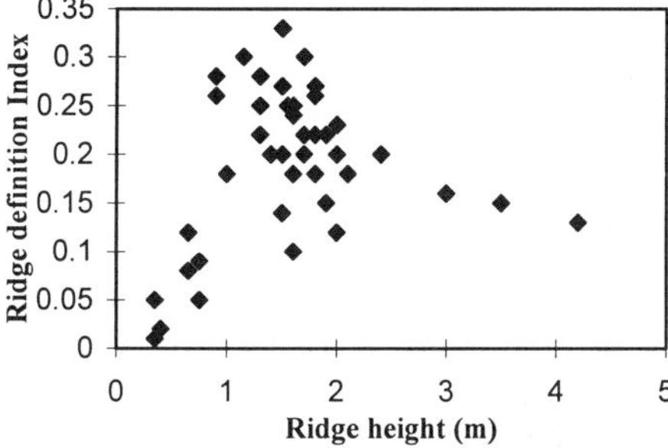

Fig 2 : Quality of ridge definition vs. block size.

Considering that the intensity of these pixels (which will be referred as ridge-pixels) is due to the enhanced return from surface elements which local incidence angles lie closest to 0°, it is expected that their density should be a function of the probability that a blocks' face be normal to the radar beam.

After Hallikainen and Winebrenner (1992), the back-scattering intensity is given by :

$$\sigma^0(\theta) = \frac{4\pi R^2 |E_{sc}|^2}{A|E_i|^2}$$

or :

$$\sigma^0_{sc}(\theta) = C_s |E_{sc}|^2$$

where E_{sc} and E_i are the scattered and incident electric fields; R is the distance from radar to target; and A is the area illuminated by the radar.

The value in decibels being obtained through the relation :

$$\sigma_0\,(\mathrm{dB}) = 10\log(\sigma^0)$$

Then defining E_{bl}, E_{bg} as the electric fields scattered by the ice blocks and the background ice and E_{th} the electric field corresponding to the threshold back-scattering intensity above which a pixel is identified as a ridge-pixel, the relation to extract ridge-pixels becomes:

$$|E_{th}|^2 \leq |E_{bl}|^2 + |E_{bg}|^2$$

After Gohin et al. (1994), and Lewis et al. (1994), σ_0 (in decibels) is well described by a linear function of θ, the angle made between the normal to the ice interface and the radar incident beam:

$$\sigma_0(\theta) = \sigma_0(0°) + C_i\,\theta$$

Which gives :

$$|E_{bl}|^2 = \frac{S_{bl}\cos\theta}{C_s.S_{grc}\cos\theta_i} 10^{\frac{\sigma_0(0°)}{10}+\frac{C_i}{10}\theta}$$

Where, S_{bl} is the block's surface, S_{grc} the ground resolution cell of the SAR beam, θ the incident angle relative to the block's face and θ_i the incident angle of the SAR beam.

Kim et al. (1984) showed that $\sigma_0(0°)$ and the slope C_i are characteristic of the ice roughness. In the case of electromagnetically smooth surfaces the radar return is large at normal incidence and σ_0 decreases rapidly with θ as most of the radar energy scatters forward. In contrast, rough surfaces show a reduced dependence with θ. In this simulation, using data from Kim et al. (1984) for a smooth ice surface (m=0.2), we will set $\sigma_0(0°)$ = 0 dB and C_i = - 0.9 for ice block faces. The background will be supposed flat with a back-scattering σ_{ba} = -15 dB.

As roughness decreases, the near vertical response increases and the middle- and large-angle back-scatter decreases. For smooth ice surfaces σ_0 is dominated by returns from those elements whose local incident angles lie closest to 0°. Our computations show that a single ice block with a 57x57 cm face normal to the radar beam would be recorded as a ridge. In addition, a 250x250 cm block whose incidence angle is superior to 15° would not. Considering all possible orientations of an ice block, we computed the probability that the back-scattering from a single block increase the return value so as

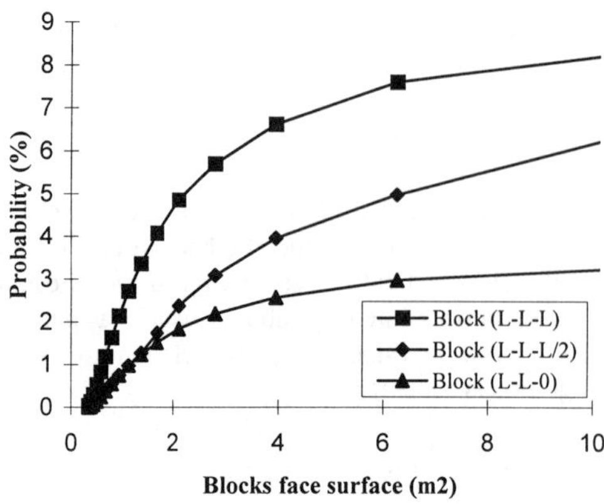

Figure 3 : Probability that a single block increases the back-scattering intensity of a pixel above the threshold value used for ridge extraction

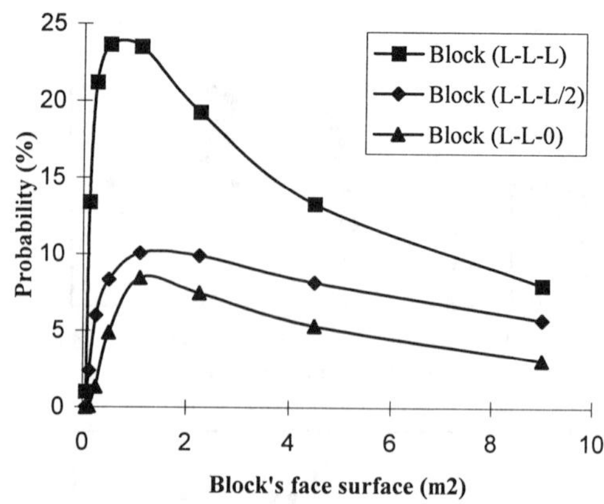

Figure 4 : Probability for a pixel from an area of randomly oriented blocks of identical size to be identified as a ridge-pixel.

the corresponding pixel be identified as a ridge-pixel. In the following simulation blocks are modeled as parallelepipeds (LxLxT) of thickness T. Figure 3 shows results for T=L, T=L/2 and T=0 (negligible thickness) and a threshold set to -13 dB. Assuming homogeneous ridges composed of blocks all the same size, we then computed the probability that the combined return from multiple blocks increased the signal intensity, so as the corresponding pixel be identified as a ridge-pixel (fig. 4).
These results are in good agreement with our field data (fig. 2). The probability that a pixel in a given ridge be identified as a ridge-like pixel may also be considered as a measurement of the density of ridge-pixels in that ridge which is given by the ridge definition Index.

It is also noticeable that the slope of the graph is a function of the angle dependence of the back-scattering (C_i). It follows that an increase of the block's face roughness, as induced by snow coating, would reduce the probabilities to get ridge-pixel readings out of a ridge.

4 CONCLUSION

Shortly after melting starts, the increased wetness in the first centimeters of the snow layer creates a barrier for microwaves which reduces drastically the ice contribution on SAR images. At the initiation of the melt, as the snow cover is still at its thicker and ridges are coated with snow, signal intensity becomes mainly a function of the characteristics of the snow-air interface.

Unlike what is usually reported in the literature, pressure ridges may also be characterized by linear, shaded bands darker than the background. In fact both bright and dark linear structures were observed on all our images, which suggests that standard mode may be inappropriate for the development of ridge extraction algorithms. In addition to integrating large areas of adjacent ice floes in the ridge signal, the standard mode integrates both the bright and dark components of a ridge, reducing, in turn, the floe-ridge contrast.

As melting goes on, and ice blocks slowly emerge from the snow cover, the ridge return gets brighter. Although larger ridges may be the first to display an increased signal, our results show no correlation between height and ridge definition. On the contrary, ridges formed of smaller blocks show a better resolution.

Our simulation confirms that the probability that the return signal from a ridge be identified as a ridge-return decreases with increasing block size. Considering that block size and ridge height are both highly related to the ice thickness when the ridge was formed, it follows that ridge resolution should decrease with ridge height. However, the heterogeneity of the block size in ridges may partly erase the resolution difference with ridge height. In addition, with part of the ridges being covered with snow, smaller ridges may be undetected while larger ones are detected.

Nevertheless, with the development of new satellites, and improved SAR, the relation between ridge block characteristics and ridge definition may offer a way to determine ridge height from remotely sensed sea ice.

Acknowledgments : The author is indebted to P. Larouche for his constant support in all the ice-related research projects I launched. I should also like to thank G. Desmeules for his never failing enthusiasm on the field, and Kendra and Dave Motsinger for their help in reviewing the final draft. This work was made possible by funding from NSERC and the Canadian Ice Center.

5 REFERENCES

Barber D.G., T.N. Papakyriakou and E.F. LeDrew, 1994. On the relationship between energy fluxes, dielectric properties, and microwave scattering over snow covered first-year sea ice during the spring transition period, *J. Geophys. Res.*, C11, 22,401-22,411.

Gohin F. & A. Cavanié, 1994. A first try at identification of sea ice using the three beam scatterometer of ERS-1, *Int. J. Remote Sens.*, 15, 1221-1228.

Gohin F., A. Cavanié & R. Ezraty, 1998. Evolution of the pas sive and active microwave signatures of a large sea ice feature during its 2½-year drift through the Arctic Ocean. *J. Geophys. Res.,* 103, 8177-8189.

Gow, A. J. and W. B. Tucker III, 1990. Sea ice in the polar re gion, in : *Polar Oceanography, Part A*,W.O. Smith, Jr. Ed., Academic Press, San Diago, 47-122.

Hallikainen M. and D.P. Winebrenner, 1992. Chapter 3 :The physical basis for sea ice remote sensing, *in Microwave Remote Sensing of Sea Ice, Edited by F.D. Carsey, Geophys. Monogr. 68, American Geophysical Union*, Washington DC., 29-46.

Hudier E., C. Fuentes and P. Larouche, in press, Extraction des crêtes de pression sur image radarsat en mode fin, Fourth International Airborne Remote Sensing Conference and Exhibition/ 21st Canadian Symposium on Remote Sensing, Ottawa, Ontario, Canada, 21-24 June 1999.

Kim Y.S., R.K. Moore & R.G. Onstott, 1984. Theoretical and experimental study of radar backscatter from sea ice. *Remote Sensing Laboratory, Univ. Kansas Rept.,* RSL TR 331-337.

Lewis E.O., C.E. Livingstone, C. Garrity, and J.R. Rossiter, 1994 . Chapter 2: Properties of snow an ice, in *Remote Sesing of Sea Ice and Icebergs*, Edited by S.Haykin, E.O. Lewis, R.K. Raney and J.R. Rossiter, John Wiley & Sons, Inc., 21-96.

Lytle, V. I. and S. F. Ackley, 1991. Sea ice ridging in the eastern Weddell sea. *J. Geophys. Res., 96*, 18411-18416.

Onstott R.G., 1992. Chapter 5: SAR and Scatterometer signa tures of sea ice, in *Microwave Remote Sensing of Sea Ice*, Edited by F.D. Carsey, Geophys. Monog. 68, American Geophysical Union, Washington, DC., 73-104.

Similä, M., 1991. Statistical description of sea ice ridging in tensity by SAR imagery, *Proceedings of the 11th Annual Int. Geosc. and Remote Sensing Symp.,* vol. 3, Espoo, Finland, June 3-6, 1223-1227.

Tiuri M., A. Sihvola, E. Nyfors and M.T. Hallikainen, 1984. The complex dielectric constant of snow at microwave frequencies, *IEEE Journal of Oceanic Engineering*, OE-9, 377-382.

Tucker W. B., D. S. Sodhi & J. W. Govoni, 1984. Structure of first-year pressure ridge sails in the Alaskan Beaufort Sea, *The Alaskan Beaufort Sea : Ecosystems and Environment,* edited by P.W. Barnes et al., Academic Press, New-York, 115-135.

Tucker W.B., D.K. Perovich, A.J. Gow, W.F. Weeks and M.R. Drinkwater, 1992. Chapter 2 : Physical properties of sea ice relevant to remote sensing, in *Microwave Remote Sensing of Sea Ice*, Edited by F.D. Carsey, Geophys. Monog. 68, American Geophysical Union, Washington, DC., 73-104.

Wadhams, P. 1984. Sea ice morphology and its measurements, *Arctic Technology and Policy*, edited by I. Dyer and C. Chryssostomidis, Hemisphere publishing corporation, Washington DC. 179-195.

Wadhams, P. 1992. Sea ice thickness distribution in the Greenland sea and Eurasian Basin May 1987. J. Geophys. Res. C4, 5331-5348.

Williams, E., C. W. M. Swithinbank & G. deQ Robin, 1975. A submarine study of Arctic pack ice, J. of Glaciology, 15, 349-362.

9 Geology

Remote Sensing in the 21st Century: Economic and Environmental Applications, Casanova (ed.)
© 2000 Balkema, Rotterdam, ISBN 90 5809 096 5

A GIS for the Natural Resources Authority, Amman, Jordan

H.J. Baxendale
British Geological Survey, Kingsley Dunham Centre, Keyworth, Nottingham, UK
L.A. Abu Saad
Natural Resources Authority, Amman, Jordan

ABSTRACT: In recent years Geographical Information Systems (GIS) have become an integral part of many UK businesses, Local Authorities and research organisations and are increasingly popular as study topics for undergraduates and postgraduates. As concern grows for the environment and Earth's resources, work involving GIS is now growing in developing countries. The British Geological Survey (BGS) has co-operated with the Natural Resources Authority (NRA) in Amman, Jordan since 1984 in a variety of geoscientific and training projects. With funding from the Department for International Development (DfID), the BGS and the NRA set up a project, in April 1997, with a view to developing an Environmental Geology GIS for the greater Aqaba area in south Jordan, which could eventually be extended to other parts of the country. A range of issues had to be considered: the geological setting; what geological factors are important compared to those considered in a UK based project?; what are the environmental issues involved?; what level of training would be required to enable the user to become self-sufficient? This paper describes the various stages of the project from initial training of participating geoscientists through to the development and subsequent use of the system with particular emphasis on the level of training needed to enable the user to work unaided at the close of the project. The possible direction of future work is also discussed.

1 INTRODUCTION

The NRA is a governmental organisation involved in the collection, management and interpretation of geoscientific information in Jordan. Currently funded by the DfID, the BGS has, over the years, developed close ties with the NRA. A project was set up in 1984 to assist and train NRA staff in carrying out geological mapping at 1:50,000 scale to produce high quality printed maps. This initially involved a residential team of BGS geologists and up to 20 geologists from the NRA. Training was also given in earth science, mineral resource evaluation and information technology by the residential team and visiting BGS specialists. In addition to this, 20 NRA staff have received either long-term (MSc) or short-term training in the UK in subjects such as hydrogeology, mineral exploration, sedimentology, remote sensing and engineering geology. The DfID has subsequently supported a 3 year professional linkage between the NRA and the BGS. This involved short term visits by BGS experts who assisted the NRA in maintaining the production of the 1:50,000 scale map series and advised on the analysis of and exploration for industrial minerals. Further training courses included environmental geology, hydrogeology and seismic interpretation.

Following on from this initial co-operation, a project was set up with the aim of incorporating geoscientific data held at the NRA in one standalone desktop GIS, and supporting MS Access™ databases, with scope for expansion where necessary. The purpose of this GIS was primarily to provide planning authorities and developers with up-to-date information on the natural environment to help them make informed decisions on the development of any region and its natural resources on the basis of accurate scientific data. Examples of the ways in which this information could be used include guidance on the most suitable land for construction, optimum locations for quarries and the establishment of protection zones around water wells. Future uses might include the identification and analysis of hazards such a landslides.

The threat of landslides is a major concern in Jordan. They are often induced by man as a result of road cuttings, excavated at or near the foot of slopes, triggering movement. Attempts to counter this movement have involved the construction of gabions (walls constructed of wire netting filled with rocks) and drainage ditches excavated in order to carry away excess water.

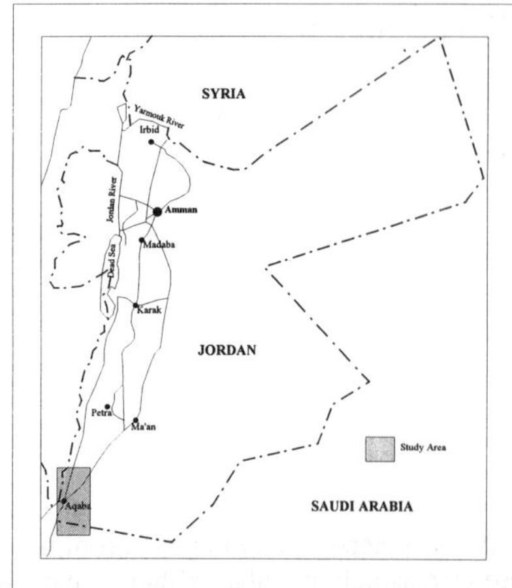

Figure 1 - Study area

The chosen study area covers two 1:50,000 scale geological map sheets depicting the Greater Aqaba region (see figure 1).

The geology of this area comprises basement igneous rocks and sedimentary rocks with some Quaternary deposits; mainly alluvium and Pleistocene gravels. This area experiences environmental pressures because the coastal development zone includes tourist sites as well as industrial and port complexes in close proximity to each other.

This paper describes the training and development involved in implementing the GIS. The development of the supporting MS Access™ databases will not be covered here.

The agreed objectives of the project were:

i) identify available datasets,
ii) give initial training in GIS and Remote Sensing to two NRA geoscientists,
iii) develop a desktop GIS for the Aqaba region,
iv) implement the GIS at the NRA and give further training where necessary.

2 IDENTIFICATION OF SUITABLE GIS SOFTWARE

A number of criteria were considered when choosing a suitable GIS software package for this project, but the most important consideration was that the GIS be easy to use. The time available for training during the project was limited therefore the GIS software chosen needed to be intuitive and easy to learn so that the eventual users of the GIS could master the basics of the system quickly and become operational within a short space of time. The GIS would need to be compatible with a range of other systems as the NRA anticipated receiving digital data in a variety of GIS and database formats from external sources. The system would be used for demonstration of NRA data and digital capabilities and therefore a high quality screen display was necessary. The GIS would need to allow a range of functionality enabling analysis of the geoscientific datasets. The output of hard copy maps was also considered a high priority. In summary the GIS should:

i) be easy to use with a short, steep learning curve,
ii) allow compatibility with many other data formats,
iii) provide attractive and clear display of geological data,
iv) provide a good range of functionality,
v) produce good quality hard copy output to a range of colour printers at a variety of sizes and scales.

MapInfo™ was chosen as the most suitable GIS for this project.

3 IDENTIFICATION OF AVAILABLE DATASETS

The primary dataset to be included in the GIS was the geological map. The NRA holds most geological information as paper maps. Geological mapping is undertaken by field geologists mainly at a scale of 1:50,000. The geological maps comprise bedrock and superficial (or drift) deposits mapped at surface. These geological maps hold lithological and lithostratigraphical information in the key. A further four detailed geological maps for the coastal strip, at 1:10,000 scale, were also made available. Fault lines and dykes (minor igneous intrusions), also captured during geological field mapping, were to be included as separate data layers. Other paper maps available included an Aqaba town planning map and an Aqaba region planning map supplied by the Aqaba Region Authority.

MS Access™ databases, developed during an earlier part of this project, containing hydrogeological, minerals and engineering geology data were also made available for input to the GIS.

Other datasets included 1:50,000 scale topographic bases, used as backdrop information, and two Landsat Thematic Mapper (TM) scenes acquired by the NRA and BGS. The Landsat TM scenes were merged and various false colour composites produced to discriminate the different lithologies resulting in a TM 754 decorrelation stretched false colour composite (pixel size 30m) for input to the GIS (Dennis, 1997).

4 TRAINING

Two geoscientists from the NRA visited the BGS over a 6 week period. During this time they received intensive training courses. One geoscientist received training in Remote Sensing techniques (Dennis, 1997) while the other received training in GIS development.

It was felt that training should initially concentrate on digitising techniques; Microstation 95™ was used for this due to the experience available in the BGS. This training was carried out by staff in the BGS Cartographic Services.

4.1 *GIS Training - phase one*

GIS training involved an introduction to the principles of GIS followed by an introduction to the basic functionality of the MapInfo™ GIS software package, including data input, editing, manipulation and hard copy map output. Further training involved the use of project data that would be used for development of the GIS itself. This included:

i) data input
ii) data editing and manipulation of spatial and non-spatial data
iii) hard copy map output

4.1.1 data input

The NRA is likely to receive datasets in a variety of formats from different sources therefore training covered a number of different methods of data input. Methods covered included transfer from Intergraph MGE™ files to MapInfo™ format with attached attribute information (Baxendale, 1997), transfer from Microstation 95™ to MapInfo™ format via DXF format, connection to existing external databases such as MS Access™, input from MS Excel™ and ASCII data files in MapInfo™ Interchange Format (MIF) and 'heads up' digitising within the MapInfo™ GIS software.

Use of the project data involved the following work flow. A 1:50,000 scale geological map sheet provided by the NRA was digitised within Microstation 95™ (training provided by BGS Cartographic Services) and then transferred to MapInfo™ format, via DXF format, as digitised polygons containing no attached geological attribute information. Fault lines were also input to MapInfo™ from Microstation 95™ via DXF format. The trainee was then shown how to colour code the geological polygons and assign lithostratigraphic codes and names with reference to an existing paper map. Training was then given in how to use these lithostratigraphic codes to update the rest of the geological information using the **SQL Select...** facility and the **Update Column...** option in MapInfo™. See figures 2 and 3.

Trial areas of the same data were also used to demonstrate how the Intergraph MGE™ to MapInfo™ translator (Baxendale, 1997) could transfer spatial data with attached geological attribute information. In addition to this, parts of the geological map were

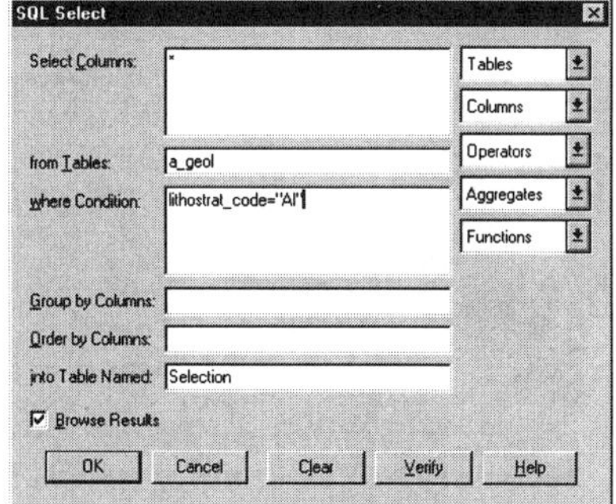

Figure 2 – SQL Select... facility

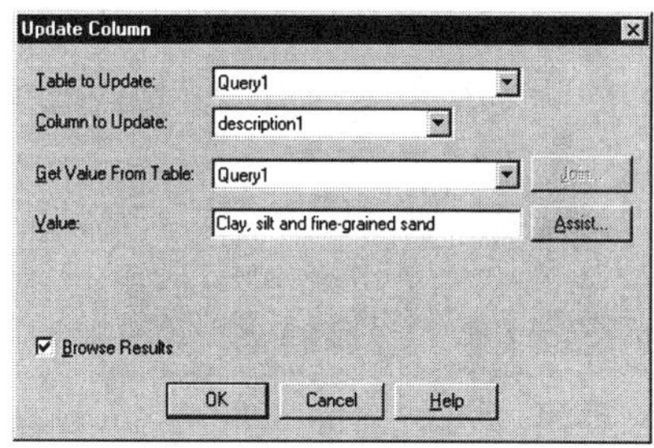

Figure 3 – Update Column... facility

scanned, saved as TIF raster images and used to give instruction in how to register a scanned image within MapInfo™. These images were then used to demonstrate how to digitise the geological polygons from the scanned image using 'heads up' digitising.

Topographic information for the study area was scanned from two 1:50,000 scale paper maps which were saved as TIF raster images and registered within MapInfo™. Training was given in registration of the images within MapInfo™ (Baxendale, 1997). A Landsat Thematic Mapper image of the study area was also included as an extra data layer and training was given in its registration. Both raster datasets were registered using the Palestinian National Grid for Jordan.

4.1.2 data editing and manipulation

Training was given in the following subjects: editing existing data; querying the data using SQL; creating zones around map objects; editing objects using 'cookie cutting' operations.

i) editing existing data

The trainee was shown how to edit the colour and pattern of map objects using those available within MapInfo™ together with techniques for updating the geological attribute information of the objects using the **Update Column...** facility. Data input from MS Excel™ spreadsheets and MS Access™ database ta-

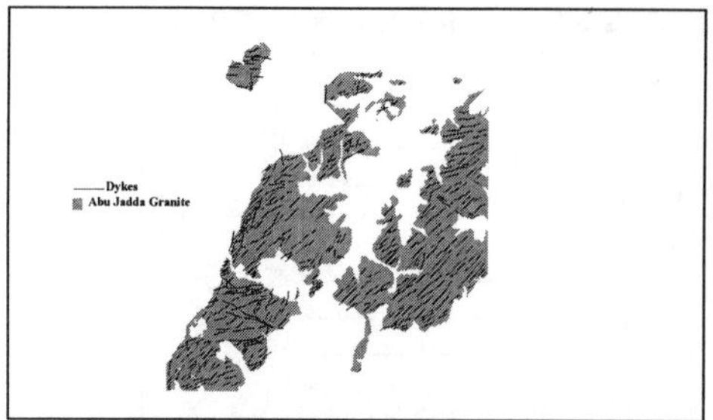

Figure 4 – SQL query – Find areas of Abu Jadda Granite that are intersected by dykes

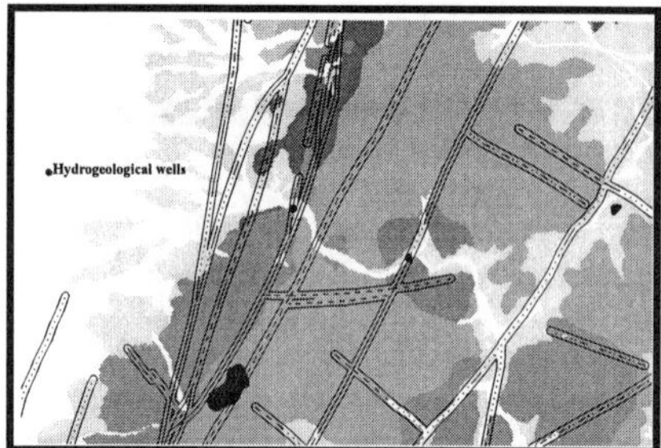

Figure 5 – Buffer zones – Find wells that lie within 100 metres of faults

bles, developed during an earlier part of the project, was also covered.

ii) querying the data using SQL

After an initial introduction to the general application of SQL, the trainee was shown how to construct SQL queries specific to geoscience applications. This training included the query of spatial (graphic) data and non-spatial (attribute) data. An example of this type of query might be **'Find areas of Abu Jadda Granite that are intersected by dykes'**, see figure 4.

It was stressed that finding the most appropriate SQL query to solve a particular problem was important as was checking the result of the query to ensure its validity.

iii) creating zones around map objects

The user was then introduced to the concept of 'buffer' zones and the relevance of these to geoscience applications. The creation and subsequent use of these buffer zones, in conjunction with SQL query facilities, demonstrated the increased applicability of the SQL functionality. An example of this might be **'Find wells that lie within 100 metres of faults'**, see figure 5.

iv) editing objects using 'cookie cutting' operations

The trainee was shown how to perform the different object editing techniques available within MapInfo™. He was then shown how to use these object editing techniques in conjunction with the SQL

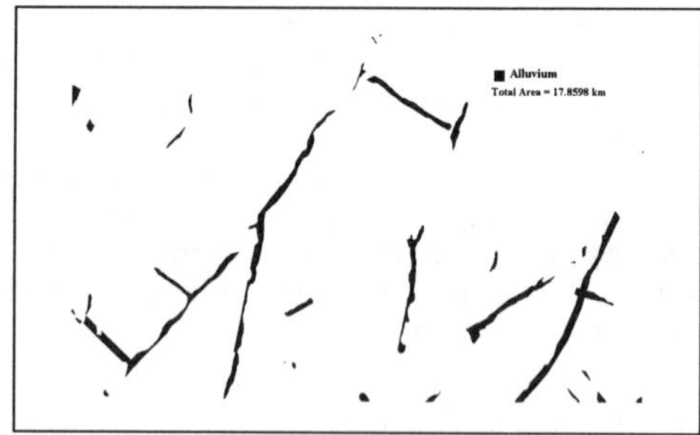

Figure 6 – 'cookie cutting' – Find the total area of alluvium within 100 metres of faults

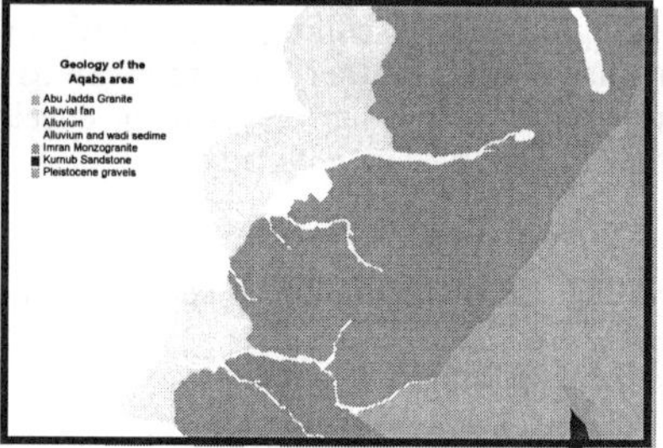

Figure 7 – Example of hardcopy map output

query facility and the buffer zones creation facility, hence increasing the functionality available to the user for geoscience applications. An example of this type of operation was **'Find the total area of alluvium within 100 metres of faults'**, see figure 6. It was again stressed that this functionality must only be used in appropriate situations and that any results should be checked for their validity.

4.1.3 hard copy map output

An essential part of any GIS project is the ability to output good quality, clear maps. Initially, training was given in the creation of thematic maps with legends using information held in the attribute tables such as geological age, aquifer class or mineral potential. Labelling of geological polygons using information held in the attribute tables was also covered. The auto labelling facility within MapInfo™ and the manual addition of labels was used for this purpose.

The trainee was then shown how to create and design a map for output to various printers and plotters. This involved creating a layout showing the map and legend positioned on the desired page size (page size dependant on printer/plotter chosen) and the addition of a title, north arrow etc. to the map face, see figure 7.

5 DEVELOPMENT OF THE GIS

Not all the available data were input to the GIS during the initial training exercises so continued development of the GIS was necessary to ensure a complete and functioning system. This development was performed by BGS staff in the UK. Datasets input to the GIS included two 1:50,000 scale geological map sheets including bedrock geology and superficial (drift) deposits, fault lines and dykes (using the same method used for input of the fault lines). Also included with the 1:50,000 scale map data were an Aqaba town planning map and an Aqaba region planning map supplied by the Aqaba Region Authority. In addition to this geological information at 1:50,000 scale, were four 1:10,000 scale geological maps covering the coastal area around Aqaba.

Three MS Access™ databases, a minerals database (Collins, 1997), a hydrogeological database (Hargreaves, 1997) and an engineering geology database (Hallam, 1997), set up during an earlier phase of the project, were input to the GIS also. Point map objects were created from these databases using coordinate information held in the database tables registered to the Palestinian National Grid for Jordan. A decision was made to keep these data in MS Access™ as the master dataset and add subsets of the up-to-date information to the GIS when required. This would ensure that duplication of these data was kept to a minimum.

Two topographic maps and a Landsat TM image were also input to the GIS to provide background information for the study area.

An extra layer of point objects within the GIS were linked to photographs taken of the geology at each location (including views of landslides). Clicking these points displayed the photograph at that position. The MULTIMED.MBX file, downloaded from the MapInfo™ World Wide Web site and created by Kevin G. Mahoney, was used to implement this.

6 FURTHER TRAINING AT THE NRA

During a subsequent visit to the NRA, the GIS was installed as a functioning system on the project computer. Further training was then given where required.

6.1 *One to one intensive training*

This was given to ensure that the geoscientist involved was able to continue to develop the GIS unaided. Subjects covered in Part 4 were repeated and more intensive training given in areas where the trainee was still unsure. The understanding of these methods was then tested by updating existing information and inputting of new information to the GIS.

6.2 *Tutorials*

In order for the GIS to become a more widely used tool (and to include other geoscientific datasets), other staff at the NRA needed to be aware of its presence and capabilities. Tutorials demonstrating the functioning GIS were given to NRA staff within the Geology Directorate and from other Directorates not directly associated with the project. As a result, detailed discussions concerning the GIS and its datasets were held within the NRA. Staff from the Cartography Group were interested in data conversions between their cartographic systems and the GIS and the cartographic capabilities of the GIS, and members of the Remote Sensing group were interested in data conversion between the GIS and the remote sensing systems.

7 CONCLUSIONS

As a result of long standing co-operation between the BGS and the NRA, and with funding from the DfID, a GIS training programme enabling geoscientists at the NRA to become proficient in the use of the MapInfo™ GIS software has been successfully concluded. The programme allowed the transfer of skills from BGS to NRA staff and raised awareness of the capabilities of such systems amongst other NRA staff. The continued development of a GIS for the Greater Aqaba region can now be contemplated by geoscientists at the NRA, for instance, in areas such as hazard mapping and analysis.

The Aqaba Region Authority has received a copy of the GIS and supporting databases for use in planning and development of the Greater Aqaba region and the methodologies and techniques developed may now be applied to other environmentally sensitive areas in Jordan, including the capital Amman.

ACKNOWLEDGEMENT

This paper is published with the permission of the Director of the BGS and is NERC copyright. Funding for the project was provided by the Department for International Development (DfID).

Thanks to Dr John H. Powell CGeol and Mr Kevin J. Northmore CGeol for their advice and guidance throughout.

REFERENCES

Baxendale, H.J., 1997. A Geographic Information System to facilitate geoscience data integration and availability. In: AGI 97 Conference Proceedings, Birmingham, 1997.

Collins, G.L., 1997. Report on the implementation of a Mineral Information Database for the Natural Resources Authority, Amman, Jordan. British Geological Survey Technical Report WF/97/8C.

Dennis, A.M., 1997. pers. com.

Hallam, J., 1997. pers. com.

Hargreaves, R.L., 1997. Natural Resources Authority, Hashemite Kingdom of Jordan. Hydrogeological Database.

Remote Sensing in the 21st Century: Economic and Environmental Applications, Casanova (ed.)
© 2000 Balkema, Rotterdam, ISBN 90 5809 096 5

Developing a technical computing facility at the Geological Commission in Guyana

A. P. Marchant & J. R. Gibson
British Geological Survey, Keyworth, Nottingham, UK

ABSTRACT: The purpose of this paper is to document the development of an assistance program to introduce technical computing to the geological commission in Guyana, South America. As well as describing the achievements of the project, the paper highlights the approaches used to undertake training and software development in an organisation with very little previous IT experience. It also discusses the problems encountered during the course of the project and the lessons that have been learned.

1 INTRODUCTION

This paper documents the development of a technical computing facility at the geological commission in Guyana, South America. The project, funded by the Commonwealth Fund for Technical Cooperation (CFTC), involved training, and the development of geological database and GIS products. The project started in February 1997, and has consisted of several visits to Guyana by staff of the British Geological Survey (BGS). The project was completed in the summer of 1998.

The Guyana Geology and Mines Commission (GGMC) is Guyana's national body for the collection of geological information and for the regulation of mining activities.

This paper begins with a background to the situation in Guyana, a fairly typical developing nation, which is currently burdened with a large external debt. The goals of the project are discussed, followed by sections on how these goals were achieved by means of teaching and product development. Lessons learned from the successes and failures of the project are discussed and finally, the conclusions sum up the results of the project activities.

2 BACKGROUND

Guyana is situated on the northern seaboard of South America. it is bounded by Venezuela to the north and northwest, Brazil to the south and southwest, and Suriname to the east. The population of Guyana is roughly 0.5 million (with about 250,000 in the capital Georgetown), whilst the land area is 215 000 km^2 or slightly less than that of the UK.

Guyana's economy is mainly based on exporting bauxite, sugar, rice, and, more recently, gold and forestry products. During the 1970's and 1980's Guyana suffered from severe economic problems, with a falling GDP and a rapid increase in foreign debt. Extensive outward migration occurred, causing a loss in skills. As a result, education and health indicators fell to amongst the lowest in the region. Since the late 1980's the government has embarked on a recovery program supported by the IDA and the IMF. Nevertheless, severe problems remain in terms of developing the economic infrastructure and in raising the standard of living

The Guyana Geology and Mines Commission (GGMC) is Guyana's national body for the collection of geological information and for the regulation of mining activities. GGMC was created in 1979, from the Department of Geological Surveys and Mines. During the economic decline of the 1970's and 1980's the organisation suffered from low investment which resulted in the end of geological mapping.

The major part of GGMC's work now consists of the regulation of the mining sector, including the issuing and management of licences for mining. In addition, GGMC maintains the national archives of geological information, and organises periodic field survey expeditions to gather new information.

The major data holdings of GGMC are the collection of mining license records (divided into small, medium and large-scale claims), the map collection, and the library document holdings. Prior to this project, all of these data collections were indexed by manual methods only. The use of IT within the organisation at this time was very limited, with only the secretaries of senior managers having access to PC's.

3 GOALS OF THE PROJECT

The problems faced by GGMC are typical of those in geological surveys and other government bodies in developing countries, for example bureaucracy and poor data management.

During initial visits to GGMC, it became apparent that activities were hampered by the fact that administrative tasks tended to be slow and cumbersome. Due to poor storage, indexing and integration of data, it was often difficult to find accurate and comprehensive information about mining claim areas, new mineral prospects or mine records. In addition, points of access to GGMC information, for the mining industry and for the general public were unclear; there was often no obvious single source of contact for external enquiries.

GGMC deals with a continuing influx of materials in terms of mining claims, and technical reports from mineral exploration, etc. Due to the rapid expansion of mining in the last few years in Guyana, GGMC is hard pressed to cope effectively with all of this extra information. In some cases, materials are stored in cramped presses, indexes are not able to be kept up-to-date, or valuable materials are not publicly accessible.

These problems are largely due to GGMC's difficulties in acquiring and retaining a sufficient body of trained staff, since wages for government bodies cannot compete with those in the private sector. However, it was thought that some of the problems might be ameliorated by use of computers in key areas to improve data storage and distribution.

After these initial visits to GGMC, it became apparent that the major goals of IT assistance should be:

1. To assist in cataloguing existing data holdings.
2. To improve accessibility of information to the mining sector.

In discussions with GGMC management, it became clear that a large component of staff training would be required if the results of the project were to be sustainable. Thus the success of the project would depend on the ability to balance limited time and resources between training, equipment installation and maintenance and product development.

The broad goals of the project were as follows:

- Improve the accessibility of data.
- Improve administration procedures.
- Improve revenue collection.
- Improve security and preservation of data.
- Improve integration and understanding of data.
- Improve customer service and advice to clients.
- Raise the profile of GGMC and advertise the services it provides.

In order to achieve these goals the following project activities were planned:

- Provide and install equipment.
- Train staff in use of computers, database & GIS.
- Create computerised index of key data holdings.
- Create integrated digital geology data set for one medium-sized area of Guyana.

The project plan envisaged two phases to the project. Phase One would involve the development of a database of selected holdings, with a graphical 'front-end' based on GIS. Phase Two would involve the creation and integration of digital mapping products for one area of geological and mining interest in Guyana.

It was also planned to link project developments as closely as possible to other IT developments within Government institutions. This included the adoption of common software standards, e.g. for GIS.

4 TEACHING ACTIVITIES

At the start of the project, six members of staff from a range of departments were chosen to receive IT training. In this way it was hoped that IT skills would be disseminated to as wide a range of GGMC staff as possible. The competition to be selected was strong between the staff of GGMC, indicating an enthusiasm for acquiring new skills, which had been suppressed by the inability of GGMC to pay for training.

When planning the training courses, the following factors had to be considered:

- The 6 project staff had little or no IT experience at the start of the project, and a relatively high level of competence would be required to sustain the project.
- The time available was limited (i.e. seven two week trips).
- Resources were severely limited (only one PC was available at the start of the project, with another purchased shortly afterwards).

With this in mind training was carried out using the techniques described below:

- To use local computer schools where possible to perform non-specialised IT training e.g. Windows 95. The advantages of using external courses are that the staff have access to a greater range of IT resources, whilst BGS staff are left free to carry out development tasks.
- Formal two or three day BGS run training courses with all six members of staff present, for specialised IT training e.g. GIS.
- Tutorials with one or two members of staff present, for skills with a high practical content e.g. digitising.
- Work assignments to be carried out between visits, to help to reinforce ideas learnt during the training.

Listed below are the courses attended by the project staff (external courses, BGS run courses and tutorials). In the cases of external and BGS run courses the staff were presented with certificates in order to record their achievement in completing the course. All courses were taken as much as possible from standard texts, and full documentation was left with the project staff.

- Windows 95: run by local computer school.
- Text scanning & processing: BGS course.
- Spreadsheets & data entry (Microsoft Excel): BGS course.
- Databases (Microsoft Access): BGS course.
- GIS (ArcView): BGS course.
- Digital mapping & digitising: BGS tutorials.
- Use of Internet: BGS tutorials.
- Grid interpolation & surface modelling: BGS course.
- Web page development & maintenance: BGS tutorials.

5 SOFTWARE PRODUCTS

This section will describe the products developed during the two phases of the project, and how they fit into the working practices of GGMC. Annex A lists the equipment and software packages used during the project.

5.1 *Phase One products*

5.1.1 *Database of selected data holdings in GGMC*

A database of GGMC map and document holdings was implemented in Microsoft Access with an easy-to-use menu-driven user interface. The database design was validated through discussions with specialists in map library cataloguing and relational databases at BGS. The contents of the database were checked with the officer in charge of Cartographic Cataloguing at GGMC, and the head GGMC librarian.

The database contains indexes to the following information:

- Geologists' field report maps
- Geological Atlas (1:200 000)
- Geophysical maps (Aeromagnetic maps from the 1960's and 70's)
- Aerial photography holdings
- Company report maps (when released)

The user can search the database by specifying information such as author, subject or title. Each search results in the production of a report containing details of the relevant data holdings, including their physical location within GGMC.

Data entry consisted of typing in records from existing catalogue sources and from the results of searches of existing map collections. The data entry was performed jointly by GGMC project staff, a private data entry bureau in Georgetown, and BGS staff. The database template was designed to allow later entry of other map holdings, when required.

5.1.2 *National Graphic Index*

The Guyana National Graphic Index provides a rapid, easy-to-use, method of obtaining descriptions of GGMC's spatially indexed data holdings for any selected area of Guyana. The system is PC-based and uses a customised version of the ArcView GIS.

A complete list of the data sets included in the Graphic Index is given below:

- Map information taken from the database of selected data holdings in GGMC, indexed by half degree square.
- Flight Lines and photograph locations taken from the most complete aerial survey of Guyana (conducted by Terra Survey).
- Possible mineral exploration sites identified in GGMC 'Bulletin' 38 (A guide to Mineral Exploration in Guyana).
- UN Aeromagnetic and Electromagnetic survey sheets.
- Terra Survey Aeromagnetic sheets at 1:50,000 and 1:200,000.
- Extents of SLAR (Radar) and Satellite images.

Each data set (other than the flight lines) consists

of digitised map extents, and basic information about each map sheet such as title, scale, date etc.

The graphic index system allows the user to search these data sets using any of the following methods:

- Searching within a particular degree square or half degree square.
- Searching within a shape defined by the user.
- Searching a particular latitude/longitude co-ordinate.

The output from the system comes in the format of a printed report. This report lists the items found in the search, including information on where the relevant maps and images are physically located within GGMC.

5.1.3 *GGMC 'Bulletin 38'*

GGMC 'Bulletin 38', 'A Guide to Mineral Exploration in Guyana', was scanned using a small desktop scanner. It was then converted to digital text format for reformatting and reprinting in Microsoft Word. This document is the most popular publication requested from the GGMC library, but existing copies had been in very poor condition and missing most of the maps and diagrams. The digital document was also converted into PDF format for easy portability and reproduction on CD.

5.1.4 *How Phase One products fit into the organisation*

The results of work carried out in Phase One are products which allow the user to search GGMC's key data holdings both by specifying information such as author, subject and title, and using spatial parameters to specify a particular area in Guyana. The user is presented with an automatically generated report showing the textual results of these searches.

These products have therefore met the major goals of the project, to improve the cataloguing of existing data holdings, and to speed up the accessibility of information to both GGMC staff and to external mining consultants.

The intended users of these products are any staff and visitors to GGMC who wish to locate maps or reports, and not just the project staff. The typical user will therefore have little or no IT experience, and will rely on the product having a simple and intuitive user interface. Thus for all software produced, ease of use was the most important factor driving the design and development.

5.2 *Phase Two products*

5.2.1 *Integrated digital data set*

Phase Two resulted in a set of digital geological maps compiled from existing records at GGMC for one medium sized area of Guyana. Each map was digitised using a customised version of MapInfo, and printed using a large-scale plotter left from a previous attempt to introduce IT to the organisation.

The last published geological map series dated from the 1960's. Since this time topographic base mapping has improved and many new mining company maps have become available.

Although updated digital maps were completed, there proved to be great difficulties in producing relatively complex digital maps. This difficulty can be ascribed to the complexity of the task, problems of creating complex linework with desktop GIS and the loss of a cartographic tradition in preceding years.

5.2.2 *The Internet and a GGMC Web site*

During the course of the project an Internet account was set up for project staff at GGMC. Initially staff were trained in the basic use of the Internet, including the use of email. This training was particularly popular, as several members of staff had heard of the Internet, but none had experience in its use. It also meant that communication between project staff and BGS staff could continue between visits, via email.

During a later trip to GGMC, WEB pages for the organisation were created. The content of these pages was agreed during discussions with the Information and Documentation Officer, as well with the senior managers of GGMC. The Sustainable Development Network Programme in Guyana (SDNP), kindly allowed GGMC to place these pages on their server, so that they are accessible to all Internet users. The address of GGMC's home page is given below:

http://www.sdnp.org.gy/ggmc

The GGMC web pages have attracted a lot of attention within the organisation and have resulted in a high profile both to this project, and to GGMC as a whole.

5.2.3 *CD of key data holdings*

A CD was created which contained many of the major software products developed during the project. Included on the CD are:

- The index database of selected data holdings in GGMC.
- The information digitised as part of the national graphic index. These data sets were integrated into Arc Explorer. Arc Explorer is a cut down version of ArcView and provides a very simple and easy-to-use interface to browse the various data sets.
- A digital copy of GGMC 'Bulletin 38', 'A Guide to Mineral Exploration in Guyana'.

All of the software packages required to view this information, such as Arc Explorer and Microsoft Access, are contained on the CD itself.

The aim of producing this CD was to raise the profile of GGMC and to provide a simple way to distribute information about the services provided and data held by GGMC.

5.2.4 *How Phase Two products fit into the organisation*

There are two broad aims of the work carried out under Phase two of the project:

1. To teach techniques of digital mapping to allow staff to create an integrated digital map collection.
2. To raise the profile of GGMC and to advertise its services through connection to the Internet, and by distributing a CD of the work carried out during this project.

These aims were met and provided GGMC with the digital expertise that will continue to enhance their capabilities and assist in the provision of information to their customers. The data they hold and the type of services they can provide are increasingly well known to mining consultants and other important customers.

6 LESSONS LEARNT

One of the most important and challenging tasks of the project was to train the six project staff from an initial level of very little IT experience, up to a point where they become capable of sustaining the project without external assistance. The main problems faced during this training were:

- How to organise the structure of the training sessions i.e. whether to teach all six project staff at once, or to split them into groups of two or three for more informal tutorial sessions.
- The difficulty of knowing at what level to set the training, as from an early stage it became clear that some of the staff were learning more quickly than others.
- Determining how well the project staff have understood the training given. Often the staff would say that they had fully understood a concept, when they were infact still having problems with it.
- Deciding how to make best use of limited resources, i.e. only two PC's between the six project staff and a limited amount of time to perform both training and product development.

As a result of these problems, the following lessons were learned:

- Whenever practical work is involved small tutorial sessions allowed the staff more opportunity to practice and become proficient in that skill. Formal classes with all six staff present were best left to more theoretical work, where practical work was less important than the concept itself.
- Repetition was the key to getting ideas successfully across to all of the staff. One of the most powerful tools was to set work programmes between visits, to allow the staff to put into practice the skills that they had learnt.
- To make the best possible use of the limited time available these work programmes were designed to include tasks relevant to the products that were being developed e.g. to practise digitising, staff were asked to digitise Flight Lines and Photograph Points which were to be included in the National Graphic Index.
- It is important not to make the teaching too advanced. It is vital that the staff gain a thorough understanding of fundamental principals and techniques.
- It is important to make the training focused to the needs of the project. By providing training in packages such as Microsoft Access and Microsoft Excel, which were not widely used by the staff during the project, attention was taken away from the more relevant tasks of GIS and digital cartography. Too broad a range of training results in staff gaining a little understanding in several fields rather than the sound understanding in the more relevant fields that allow the project to become sustainable.
- One solution to the problem of limited training resources was to use outside computer schools for training courses. A reputable computer school in Georgetown was very successfully used in order to teach Windows 95 skills to the staff.

An unfortunate consequence of this training is that it increases the risk of staff leaving the organisation, as the financial rewards in the private sector for skilled IT staff are a lot higher than those in the public sector. In order to sustain the project after staff loss, it is important that IT skills are duplicated amongst the project staff.

When creating IT products, it was noticeable that the most enthusiastically accepted products were not necessarily those with the largest amount of work invested into them, or even those products which potentially were of most use to the organisation. Instead it was the most visible products such as the Internet Web Pages, which took only a few days to develop, that were most popular. It is therefore important to produce a number of these 'quick hit' products in order to create a positive attitude towards IT, and to encourage the staff to work on less motivating tasks such as database work.

When developing products, which involve a significant amount of data entry to be carried out by the project staff (e.g. the production of updated digital maps), it is important not to underestimate the amount of time required to complete these tasks. The staff had little previous experience of such tasks and therefore frequently made errors, which took time to rectify. Tasks that were left for completion between BGS visits often hit trouble with little progress having been made, and therefore had to wait for the next visit of BGS staff before being rectified. The introduction of an Internet connection at GGMC helped prevent this situation since most problems could be solved by email communications between GGMC and BGS.

In terms of software used on the project it is important to use simple, standard Windows software only (i.e. no DOS command-line software). Where possible local standards for software should be used. Two other Government projects in Georgetown use the ArcView GIS package (GTZ and the Lands and Survey Department), so there is a growing level of expertise in this package for the staff to make use of in the future. If there are no local standards for software, the next best solution is to mirror trusted systems already in use in organisations such as BGS.

For this project standard PC desktop GIS packages were adequate (i.e. MapInfo and ArcView). It, generally, would have been inappropriate in this case to use 'industrial strength' software (e.g. workstation Arc/Info), due to the extra cost and complexity of these products. Standard PC desktop GIS packages also have the advantage that they are more likely to get local support, since a number of organisations in Georgetown use ArcView. However a couple of problems were identified when using such packages:

- Software stability. For example ArcView 3.0 crashed a number of times when certain operations were being performed. This lead to the corruption of the files that were currently being updated. To minimise the damage caused by such crashes, it is important that staff are taught to back up work at regular intervals.
- Digitiser support. Both ArcView and MapInfo have digitising capabilities, however both are relatively basic. Simple customisation of both packages provided most of the functionality required for this project, but there were still problems with digitising complex linework.

When digitising, it is important to use very simple map elements, provide automated line-cleaning tools and to train staff initially on simple digitising. This project was hindered severely by technical problems with digitising. In general it was found that "heads-up" digitising (if a suitable scanner is available) performed more reliably than digitising using a tablet and puck.

Some of the major difficulties experienced during the project were concerned with the hardware used. A number of external problems, such as the poor quality of the local power supply and electromagnetic interference caused significant interference to the project. Such problems significantly added to the cost of the project in terms of buying power regulation equipment and performing electrical testing.

There is a problem in terms of sustainability of the project, due to the lack of PC technical experts in GGMC. In the future staff may find problems in rectifying technical failures as they have very little experience of PC trouble shooting. To reduce the risk of this happening it is important to buy robust and recognised PC brands, and to avoid cheaper, less reliable PC equipment. With PC's connected to the Internet there is also the possibility of viruses being downloaded onto the machine. In order to reduce this risk it is important that staff are educated about the problem of viruses and therefore become more careful about what they download.

The production of a CD proved to be one of the major successes of the project as it provided an easy way for GGMC to both demonstrate its IT capabilities and to distribute information on its data holdings to interested organisations.

One of the most significant lessons learned, was the importance of sticking as much as possible to the original plan. Although planning must be flexible in order to meet the changing IT needs of the organisation, it is important that digression does not deflect essential effort from the main tasks. During the course of this project a number of minor tasks caused a dilution of effort e.g. data entry to support another field mapping project, and assistance in the production of the GGMC newsletter. Nevertheless, it is also important to be able to identify extra topics which are worth adding to the project. For example, the introduction of the Internet to GGMC, attracted much interest, and significantly raised the profile of the project.

7 CONCLUSIONS

The primary goal of any project of this type must be sustainability. No matter what activities take place during the course of the project, its success can only be measured in relation to what happens afterwards.

In the case of this project, a number of hardware failures due to the poor quality of the local power supply severely hampered the work of the project staff, after the project had finished. This highlights the importance of having local project staff who have PC trouble shooting skills.

The major successes of this project were the production a CD containing the software products developed during the project, and the creation of a GGMC WEB site. Both have created widespread interest in the activities of this project and of GGMC as a whole.

The training given to the project staff has allowed them to develop the techniques required to make IT an integral part of GGMC's future. It is acknowledged that, in hindsight, some of the training given was not absolutely necessary and may have deflected effort from more relevant topics, however a number of the project staff have developed skills to a high enough standard to sustain the use of GIS and databases in the organisation. It remains to be seen if GGMC can keep these capabilities in the future if trained staff decide to move elsewhere.

The successes of this project would not have been possible without the help of senior managers at GGMC who at an early stage saw the benefits of this work, and provided substantial help and advice.

In general, the key to making such a project sustainable, and therefore successful, is to keep it as simple as possible in terms of the training given and the products developed. The dangers of aiming for too high a level of implementation or of attempting to teach too many new skills are all too easily encountered.

8 ACKNOWLEDGEMENTS

We are pleased to acknowledge support and assistance from the Commonwealth Fund for Technical Cooperation (CFTC) in funding this project. We would like to offer our thanks to the staff of GGMC for their support and enthusiasm, and to Dr Rob Evans and Mike Petterson of BGS for their guidance and helpful advice. Mr Michael Fisher of the GTZ/NRMP project is thanked for his generous support in terms of the loan of a large-scale plotter to the project. Andrew Mancey of SDNP (Guyana) is thanked for agreeing to host GGMC's WEB pages on the SDNP server.

ANNEX A: HARDWARE AND SOFTWARE USED

The following equipment has been installed for use in the project:

- Two PC's running Windows 95.
- Voltage regulators and UPS's.
- A digitiser.
- An A4 printer.
- An A4 scanner.
- A plotter, which remained from an unsuccessful previous IT project.
- A modem to provide access to the Internet.

Where possible, this equipment was purchased in Guyana, so that it would be easier to replace in the case of equipment failure.

Table 1. lists the software packages used during the project:

Table 1. Software packages used

PRODUCT	SOFTWARE
Database of selected data holdings in GGMC	Microsoft Access Microsoft Excel (for data entry)
National Graphic Index	ArcView
GGMC 'Bulletin 38'	Microsoft Word Adobe Acrobat
CD of data holdings	Arc Explorer
Integrated digital data set	MapInfo Surfer
WEB page design	Microsoft FrontPage

Software was selected for the project on the basis of:

1. The software being compatible with other projects in GGMC and Guyana.
2. The software being widely used within BGS.
3. The software being well known and trusted.

Remote Sensing in the 21st Century: Economic and Environmental Applications, Casanova (ed.)
© 2000 Balkema, Rotterdam, ISBN 90 5809 096 5

A remote sensing based approach for the restoration of an open-cast coal mine site in Spain

J.F. Llorens & J.L. Fernández-Turiel
Institute of Earth Sciences, 'J.Almera', CSIC, Solé i Sabaris s/n, Barcelona, Spain

C. Banninger & A. Almer
Joanneum Research Centre, Graz, Austria

ABSTRACT: The application of stepwise linear supervised classification and post-classification automatic logical comparison change detection methods to Landsat TM imagery have allowed to monitor and quantify the restoration process o a large open-cast coal mine in NW Spain. Thus, the multitemporal analysis of 1981, 1991 and 1997 Landsat TM images have shown the progress of the mine exploitation and the advance of the restoration process. A surface of about 1000 Ha has been restored using different vegetation patterns during the considered period of time.

1 INTRODUCTION

Mining exploration is a primary application of remote sensing in mining industry, generally without extension in further steps of the exploitation and restoration (Rathore & Wright, 1993). Surface mining activities in Europe are estimated to occupy an area of between 5-10,000 km^2 and range from large open-cast coal and base metal mines, to much smaller aggregate (rock, gravel, and sand), industrial mineral (potash and clay), and building materials (granite and marble) quarries. Most national governments as well as the EU, have issued directives that require the restoration of mine sites to a semblance of their "pre-mining condition" on the completion of mining activities. The planning and implementation of this task have traditionally been undertaken by ground-based surveys and augmented by aerial photography, which are both labour intensive and time-consuming, and hence costly, especially as regular updating is usually required.

The availability of several satellite-borne remote sensing instruments for Earth observation allows for the collection of environmental and mine-related data for use in the planning and undertaking of mine restoration work on a frequent and cost-effective basis. As the data are acquired digitally, they can be quickly and easily transformed and utilised in various information formats, including thematic, infrastructure, and mine-workings maps, as well as 3-D presentations of the mine site at various points in time and from various perspectives for visualisation purposes.

In this setting, the ASTERISMOS project (funded by the DGXII/CEO of European Commission) is designed to provide a decision support system based on satellite remote sensing data for the mining decision makers in order to protect the landscape from the disturbances created for the open-cast mine works (Fig. 1).

A large open-cast lignite mine in NW Spain that is nearing the end of this economic life has been selected to examine the usefulness of satellite remote sensing data for monitoring and quantitation of mining restoration.

2 METHODOLOGY

2.1 *Study area*

The open-cast coal mine of As Pontes is located in NW Spain (Galicia). The mining area is 13,4 x 7,4 km in size, while the regional area studied is 36,4 x 27,1 km in size. The coordinates of the area centre are 4812850 N and 877200 E (UTM Zone 29).

The mine area is situated on the Eume River basin which has a E-W general flow direction. The river flow is regulated with the Ribeira and Eume dams.

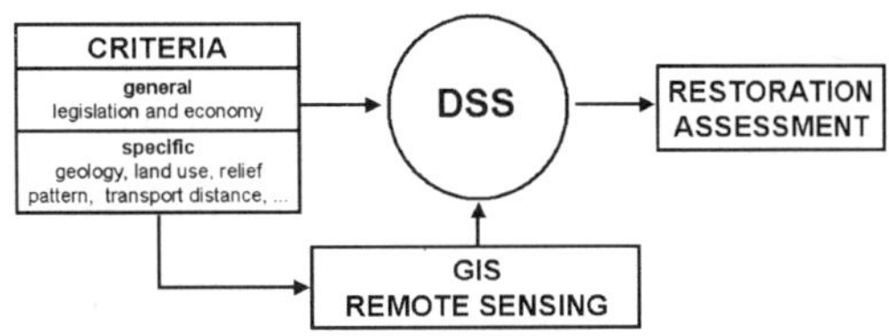

Figure 1. Framework of ASTERISMOS Decision Support System (DSS) for restoration of open-cast mining areas.

The rainfall of this area is 1600 mm per year in average (the maximum in December with 250 mm and the minimum in July with 40 mm), the monthly temperature ranges between 6.8 °C (January) and 17.5 °C (august) in average (FEUGA, 1994)..

The vegetation of this area belongs to the Eurosiberian Region, Cantabrian-Atlantic Province, Asturian-Galician Subprovince, and Galician-Asturian Sector. Oak forest is the climax vegetation, it consists of *Quercus robur*, *Betula pubescens* and *Castanea sativa*. The agriculture, forestry and cattle raising practices have reduced significatively its surface (< 25 %). The main vegetation types occurring at the present time are shrubland (*Ulex europaeus*, *U. Gallii*, *Erica mackaina*, *E. Cinerea*, etc.), crops (corn, wheat, beams, potatoes, and vegetables) and pastureland (Agrostis, Holcus, Cynosorus, Festuca, Trifolium, etc.), coniferous (*Pinus pinaster*, *P. sylvestris*) and eucalyptus (*Eucalyptus globulus*) plantations and relict deciduous climax forests. Forest fires and abandoned crops have increased the shrubland surface.

Geologically, the area basement consist of metasedimentary and igneous materials (Precambrian to Silurian schists, filites, and quartzites, and Hercynian granites). Tectonic events have generated some basins in the basement, which are filled mainly with Tertiary detrital sediments, with coal seams. Furthermore, Quaternary sediments occurs in the bottom of the wider valleys, e.g., the Eume river valley.

Since middle 20th century mining works have been carried out to exploit the lignites of the As Pontes Tertiary basin. A sedimentary threshold separated this basin into two grabens. The reserves of this coal deposit were evaluated in 300 million tons of lignite. The production during the nineties was about 10-12 million tons of coal per year.

The Restoration Plan of the As Pontes mine is aimed to reclaim a surface of about 3000 Ha, including the exploitation area and the dump. The final scenario of this plan combine the existence of a lake with the re-vegetation of the surface affected by the mining works, following the natural patterns of the regional environment. The solution is a compromise among economic, geotechnical and ecological criteria.

2.2 *Data collection*

The remote sensing data used in this work are compiled in the Table 1. The satellite imagery was selected taking into account the mine life, following the standard methodology to assess delivered image quality: cloud coverage –usually high due to the local climatic conditions-, stripping and banding, band-to-band registration and missing lines.

The satellite imagery has been used for classification and change detection purposes, while vegetation maps and aerial photographs has been used to define the training areas and to the classification quality assessment.

Table 1. Remote sensing data used in the study

Data	Pixel size (m) - Scale	Acquisition date
Landsat TM	30 (Bands 1,2,3,4,5,7)	08-09-1987
Landsat TM	30 (Bands 1,2,3,4,5,7)	11-08-1991
Landsat TM	30 (Bands 1,2,3,4,5,7)	05-04-1997
Aerial photography[1]	1:31,000	1956-1957
Aerial photography[1]	1:8,000	09-1986
Aerial photography[1]	1:8,000	03-09-1994
Aerial photography[2]	1:22,000	28-02-1997

[1] Black and white photographs
[2] Colour photographs

2.3 *Data preprocessing*

The data pre-processing include geocoding and radiometric calibration in order to obtain a consistent time series. Image geocoding has been performed using the remote sensing software package GRAZ (RSG) and ERDAS Imagine. Based on the ground control points (GCP's) measured from topographic maps and in the images parametric imaging models were determined and optimised for each of the individual images using at least square parameter adjustment implemented in the software package RSG. Images were georeferenced in UTM coordinates. Resampling was carried out using the nearest neighbour method and a second order polynomial transformation of a relatively large number of well-dispersed ground control points. The average accuracy has been 0.95, 0.82 and 0.78 pixels for 1984, 1991 and 1997 Landsat TM images, respectively. Radiometric calibration was applied to normalise exoatmospheric illumination conditions, using specific coefficients for each date. Midlatitude summer atmospheric model and rural midlatitude summer aerosol model have been used, through ERDAS Imagine ATCOR module, to normalise differences on atmospheric conditions. Constant spatial atmospheric conditions have been assumed. The visibility for 1984, 1991 and 1997 Landsat TM images was 65, 80 and 80 km, respectively.

3 RESULTS

3.1 *Classification*

A classification for the region and another for the mine area have been carried out. The first to monitor the regional vegetation patterns and the second to assess the restoration evolution. The mining area was defined in the most recent image (1997), and its perimeter was used to produce a mask that allowed working separately in the mining area and in its regional setting for the three years considered.

Thirteen classes have been distinguished according to the restoration requirements and the possibilities offered by the Landsat TM spatial resolution (Table 2):

- Unclassified: masked clouds, outside mining area or no satellite data.
- Pasture, grassland and crops: agricultural land cover, natural and anthropogenic pasture, and restored dump with grassland.
- Shrubland: including also sparse forest and grassland mixed, and restored dump with shrubland.
- Deciduous forest: oak, chestnut, birch trees and occasionally poplar, alder and willow trees. Natural, planted and restored dump with pure or mixed deciduous forest.
- Perennial forest: Pine forest and eucalyptus plantations. Pure and mixed with deciduous forest and shrubs. Restored dump with pine forest.
- Unvegetated: built-up industrial and urban area, transportation, roads, barren soil and quarry outside mine area. In the mining area three classes were differentiated: clays, slates and coal covered area according to the lithology. The last one includes coal seams, coal piles, conveyor belts and other transportation ways showing a surface coal cover.
- Water: reservoir and seawater outside mine area. In the mine area clear water and turbid water were differentiated.
- Burnt area: burnt vegetation area. The same spectral signature has been observed for a mixture of shrubland and barren soil (roads, buildings and coal cover soil) near the As Pontes Coal Fired Power Plant.

Two spectral supervised classifiers were tested: Maximum Likelihood (ML) and Stepwise Linear (SL). The ML classifier was inappropriate for the mining area, due to the low extension of perfect end-members to define the training classes. The two classifiers proved similar accuracy at regional scale and the SL classifier was used in subsequent analysis of both regional and mine scales.

A training set raster for the classifications on the outside mine region was established using unchanged areas among 1984 - 1991 - 1997. The burnt area class is defined for each year. In the mine area the training set was defined independently for each year. The sample areas for each of the desired output classes that was used to train the supervised classification were designated with information supplied by aerial photographs of 1984, 1994 and 1997, and vegetation and agriculture maps (Ministerio de Agricultura, 1976a and b, 1977a, b, c and d, 1978a and b). The bands used in the classification have been 1-2-3-4-5-7. In order to increase the classification accuracy, the training areas were optimised using error matrix analysis.

Table 2. Classification classes.

Class	Description
0	Unclassified
1	Pasture, grassland and crop
2	Shrubland
3	Deciduous forest
4	Perennial forest
5	Unvegetated (barren soil, built-up areas)
6	Barren soil: clay [1]
7	Barren soil: coal covered area [1]
8	Barren soil: slate [1]
9	Water (reservoir, sea)
10	Water: mine clear water [1]
11	Water: mine turbid water [1]
12	Burnt area

[1] Classes distinguished only inside the mine area

Table 3. Distribution of land cover and land use classes (Ha) of As Pontes region obtained masking the mine area. Stepwise linear classification of Landsat TM images.

LAND USE	1984	1991	1997
Pasture and crops	28359.8	26870.9	24920.8
Shrubland	11289.1	25765.2	35874.8
Deciduous forest	26397.6	21769.4	16656.4
Perennial forest	20948.2	17337.5	14481.6
Unvegetated	476.4	524.2	546.9
Water	681.3	617.5	697.2
Burnt area	5691.4	959.1	666.1
Total	93843.8	93843.8	93843.8

Table 4. Distribution of land cover and land use classes (Ha) in the As Pontes mine area. Stepwise linear classification of Landsat TM images.

LAND USE	1984	1991	1997
Grassland	359.3	158.7	466.3
Shrubland	837.4	479.1	998.9
Deciduous forest	176.0	104.4	176.6
Perennial forest	143.0	205.8	230.4
Clays	632.4	471.2	526.1
Coal covered area	468.8	1156.8	238.8
Slates	409.5	441.4	380.1
Clear water	0.0	18.8	12.7
Turbid water	9.8	0.0	6.3
Total	3036.1	3036.1	3036.1

Table 5 GIS layers used for As Pontes mining restoration.

Layer	Description
1	Digital Elevation Model (DEM)
2	Slope Map
3	Relief patterns influencing the restoration
4	Drainage
5	Geology
6	Land cover and land use
7	Restoration (Change Detection)

The most significative features showed at regional scale by the classification results (Table 3 and Figure 3) are the increase of the shrubland surface against the surface decrease in all of the other vegetation classes. It is important also the burnt surface decrease, which arisen more than 5,000 Ha in 1984. Inside the mine perimeter (Table 4 and Figure 3), the surface with vegetation obtained in the 1991 classification is lower than that measured in 1984, while in 1997 arise another time as consequence of the restoration works carried out. This fact is verified when the change detection is bearing in mind.

The land cover and land use maps obtained have been implemented as a layer in the GIS that supports the ASTERISMOS Decision Support System (Table 5).

3.2 *Change detection*

Change detection is the process of identifying differences in the state by observing it at different times. Various change detection techniques have been developed for analysing a whole range of alterations in natural and anthropogenic surface features. Depending of the land cover and the objective searched the change detection technique would be selected. For homogeneous landscapes those based on radiometric change between acquisition dates can be applied. These methods include band differencing (Macleod and Congalton, 1998), transformed band -e.g., band ratios and vegetation indices- differencing (Venkataraman et al., 1997), and Principal Components (Richards, 1984; Collins and Woodcock, 1996). All these techniques can be also applied for monitoring vegetation vigour. For heterogeneous scenarios, anthropically dynamic regions or those affected by natural changes in land use (mainly produced by geologic hazards), techniques based on spectral classification of the impute data can be used to detect land cover changes (Jensen et al, 1995; Venkataraman et al, 1997). The later case is applicable to As Pontes mine area.

A restoration map of a open-cast mine is aimed to monitor and quantify the restoration process. In fact, it is a change detection map which indicates the modifications introduced by the advance of the mine exploitation and the restoration carried out in a determined period of time.

The restoration classes are defined bearing in mind the type of the most recent land use class. The restoration classes are divided in two groups, changed and not changed classes (Table 6). The restored classes are those related to planted vegetation on reclaimed area.

In the present case, the change detection in the mine area has been carried out using a post-classification automatic logical comparison (PCALC) method in order to obtain the restoration maps. To discriminate restored and not restored vegetation, a 1956 extrapolated raster was used assuming that the mining area was in pre-exploitation status at this time. The aerial photographs of 1956-1957 show incipient mining works in an area covered mainly by agriculture land and grassland. This extrapolated 'pre-restoration map' contains the restoration class called Not changed original vegetation.

The process of this method consists of to compare the classification obtained in a specific date (Classes of Table 2) with the previous restoration map (Classes of Table 6), using logical operators to obtain a new restoration map. For example, a typical evolution could be to change from the original vegetation (pre-mining status) to barren soil (mine exploitation) and later to changed restored grassland (restoration). Class 5 (unvegetated), class 9 (water reservoir and seawater) and class 12 (burnt area) have not been considered because they do not appear on the mine area.

Table 6. Classes used in the restoration map

	CODE	CLASS	DESCRIPTION
CHANGE	101	CRgrass	Changed restored grassland
	102	CRshrub	Changed restored shrubland
	103	Crdeciduous	Changed restored deciduous forest
	104	Crperennial	Changed restored perennial forest
	105	CbarrenSoil	Changed barren soil
	106	Cwater	Changed water
NO CHANGE	107	NRgrass	Not changed restored grassland
	108	NRshrub	Not changed restored shrubland
	109	Nrdeciduous	Not changed restored deciuous forest
	110	Nrperennial	Not changed restored perennial forest
	111	NbarrenSoil	Not changed barren soil
	112	Nwater	Not changed water
	113	NorVeg	Not changed original vegetation

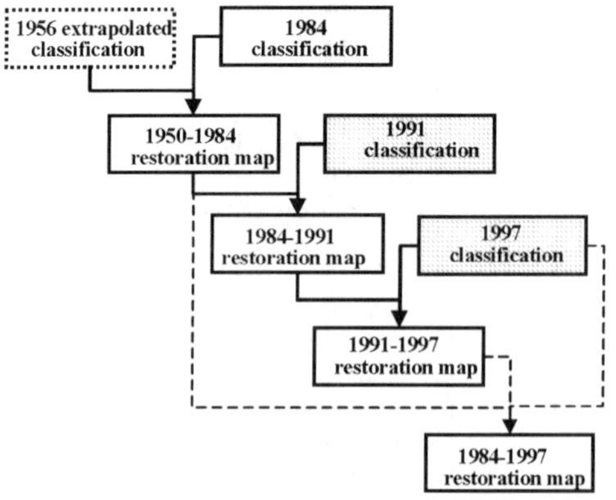

Fig. 2. Sequence of restoration maps generated for the As Pontes open-cast coal mine.

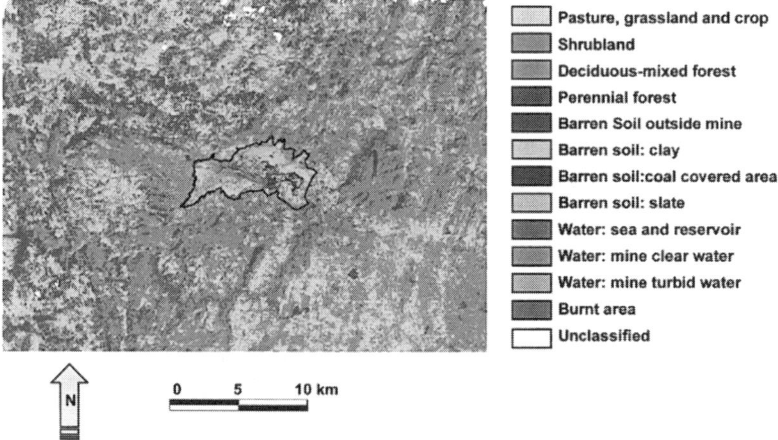

Fig. 3. Stepwise linear supervised classification of As Pontes region. Landsat TM imagery 05-04-1997. Bands 1-2-3-4-5-7. (colour plate, see p. 607).

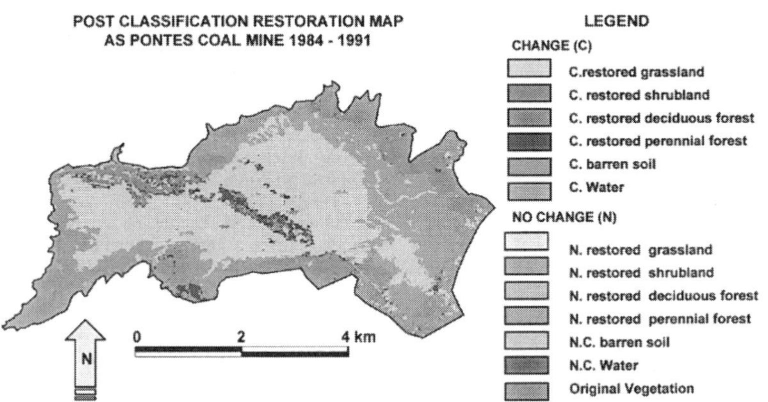

Fig. 4. Restoration Map of As Pontes region for 1984-1991 obtained with Post Classification automatic logical comparison algorithm. Stepwise Linear Classification. Landsat TM imagery 08/09/1984 and 11/08/1991. Bands 1-2-3-4-5-7. (colour plate, see p. 607).

The restoration maps are generated using scripts written in the Spatial Manipulation Language of TNTmips® (Microimages, Inc.). This procedures has allowed to produce the restoration maps for the years 1956-1984, 1984-1991, 1991-1997 and 1984-1997. After obtaining the 1956-1984 restoration map, we achieve the others in the sequence summarised in Figure 2.

These maps are also implemented as layers in the GIS of the ASTERISMOS Decision Support System.

The restoration maps allow to monitor the evolution of the restoration process (Fig. 4) and to quantify it. Thus, the Table 7 shows the restored surface increase produced until 1997, when practically a third part of the reclaimed area has been restored.

Assuming that the planned final lake will fill the mine surface situated under 340 meters after sea level, which represents another third part, and the preserved surface of original vegetation is about 20 %, the part that must be restored to complain with the Restoration Plan is about 450 Ha.

The results obtained on the restored surface monitoring are comparable to those obtained by the mining company using ground – based surveys and aerial photography, and confirm the possibilities offered by satellite imagery for mining restoration. By other hand, the coarse resolution of Landsat TM imagery (30 m) has avoided to analyse the potentiality of satellite-based imagery to monitor the vegetation status, mainly due to the size of the restoration pattern surfaces, in the same way that has been indicated for precision crop management (Moran et al., 1997).

Table 7. Restoration evolution of the As Pontes open-cast mine. Surface expressed in Ha.

Land use	1950-1984	1984-1991	1991-1997
Changed restored grassland	0.0	8.7	236.4
Previous restored grassland	0.0	0.0	3.1
Changed restored Shrubland	0.0	125.9	526.8
Previous restored Shrubland	0.0	0.0	56.3
Changed restored Deciduous forest	0.0	12.3	64.9
Previous restored Deciduous forest	0.0	0.0	30.3
Changed restored Perennial forest	0.0	87.0	103.9
Previous restored Perennial forest	0.0	0.0	2.9
Original Vegetation	1169.3	665.3	581.6
Changed waterbody	9.8	7.5	5.6
Previous waterbody	0.0	0.3	1.2
Changed barren soil	888.3	449.3	86.6
Previous barren soil	0.0	711.2	367.8
Planned lake (< 340 m)	968.75	968.75	968.75
Total mine area	3036.1	3036.1	3036.1
Total restored vegetation	0.0	233.8	1024.6

4 CONCLUSIONS

The analysis of Landsat TM imagery of As Pontes open-cast coal mine reveals:

- Satellite imagery can be applied to open-cast mine management for quantitative monitoring of restoration and for regional vegetation monitoring.
- The use of PCALC change detection algorithms allow to generate easily restoration maps in a regular updating basis.
- The digital format of the products obtained makes very simple their integration in the GIS of the mine.
- The upcoming commercial high resolution satellite-based sensors may be suitable the vegetation status monitoring during the restoration.

5 ACKNOWLEDGEMENTS

This work has been funded by DGXII/CEO of European Commission (ASTERISMOS Project, ENV-CT96-0373). The cooperation and assistance of our colleagues of ASTERISMOS project of CCA, ENDESA, IIS, RA and the University of Thessaloniki are gratefully acknowledged.

6 REFERENCES

FEUGA 1994. *Proyecto Eume: Carcaterización y seguimiento del medio físico, impactos recursos, hídricos y biológicos de la Cuenca del Eume*. 296 pp. Santiago de Compostela. Universidad de Santiago.

Collins J.B. & Woodcock C.E. 1996. An assessment of several linear change detection techniques for mapping forest mortality using multitemporal Landsat TM data. *Remote Sens. Environ. 56*: 66-77.

Jensen J.R., Rutchey K., Marguerite S.K. & Narumalani S. 1995. Inland wetland change detection in the everglades water conservation area 2a using a time series of normalized remotely sensed data. *Photogrammetric Engineering & Remore Sensing 61(2)*: 199-209

McLeod R.D. & Congalton R.G. 1998. A quantitative comparison of change detection algorithms for monitoring eelgrass from remotely sensed data. *Photogrammetric Engineering & Remore Sensing 64(3)*: 207-216.

Ministerio de Agricultura, 1976a. Evaluación de Recursos Agrarios. Mapa de cultivos y aprovechamientos. Escala 1:50.000. Puentedeume (La Coruña), no. 22 (6-4). Madrid, 39 pp. and 1 map.

Ministerio de Agricultura, 1976b. Evaluación de Recursos Agrarios. Mapa de clases agrológicas. Escala 1:50.000. Puentedeume (La Coruña), no. 22 (6-4). Madrid, 48 pp. and 1 map.

Ministerio de Agricultura, 1977a. Evaluación de Recursos Agrarios. Mapa de cultivos y aprovechamientos. Escala 1:50.000. Cedeira (La Coruña), no. 7 (6-3). Madrid, 44 pp. and 1 map.

Ministerio de Agricultura, 1977b. Evaluación de Recursos Agrarios. Mapa de clases agrológicas. Escala 1:50.000. Cedeira (La Coruña), no. 7 (6-3). Madrid, 56 pp. and 1 map.

Ministerio de Agricultura, 1977c. Evaluación de Recursos Agrarios. Mapa de cultivos y aprovechamientos. Escala 1:50.000. Puentes de García Rodríguez (La Coruña), no. 23 (7-4). Madrid, 34 pp. and 1 map.

Ministerio de Agricultura, 1977d. Evaluación de Recursos Agrarios. Mapa de clases agrológicas. Escala 1:50.000. Puentes de García Rodríguez (La Coruña), no. 23 (7-4). Madrid, 42 pp. and 1 map.

Ministerio de Agricultura, 1978a. Evaluación de Recursos Agrarios. Mapa de cultivos y aprovechamientos. Escala 1:50.000. Vivero (Lugo), no. 8 (7-3). Madrid, 44 pp. and 1 map.

Ministerio de Agricultura, 1978b. *Evaluación de Recursos Agrarios. Mapa de clases agrológicas. Escala 1:50.000. Vivero (Lugo), no. 8 (7-3)*. Madrid, 56 pp. and 1 map.

Moran, M.S., Barnes, E.M. & Inoue Y. 1997. Opportunities and limitations of remote sensing for precision crop management. In G. Guyot & T. Phulpin (eds), *Phisical Measurements and Signatures in Remote Sensing*: 629-640. Rotterdam: Balkema.

Rathore, C.S. & Wright, R. 1993. Monitoring environmental impacts of surface coal mining. *Inst. J. Remote Sensing, 14hild 14*(6): 1021-1042.

Richards J.A. 1984. Thematic mapping from multitemporal image data using the principal components transformation. *Remote Sens. Environ. 16*: 35-46.

Remote Sensing in the 21st Century: Economic and Environmental Applications, Casanova (ed.)
© 2000 Balkema, Rotterdam, ISBN 90 5809 096 5

Discrimination of active alluvial sedimentary surfaces in TM images based on laboratory reflectance spectroscopy

E. García-Meléndez & M. Suárez
Departamento de Geología, Facultad de Ciencias, Universidad de Salamanca, Spain

A. Riaza
Instituto Tecnológico Geominero de España, Madrid, Spain

ABSTRACT: The purpose of this work is to discriminate the areas that most probably will be affected by future alluvial sedimentation and erosion in a semiarid quaternary basin located in the southeast of Spain, based on spectral differences shown by Thematic Mapper images. X-ray diffraction analysis shows very similar mineralogical contents, due to the uniformity of the source areas. Carbonate develops during consolidation processes after sedimentation, being the only relevant mineralogical difference. Carbonate contents are responsible for the spectral differences detected both in the laboratory and imagery. Recent areas of sedimentation show flatter spectra and an overall lower reflectance confirmed by the imagery. Different ages of alluvial fans and degrees of sediment consolidation can be mapped using images. Different rock consolidation and geomorphological features help to define the areas most likely suffering future erosion and sedimentation processes in a natural hazard prone area.

1 INTRODUCTION

The study of the actual pattern and characteristics of landforms informs about the natural processes which are active in the present and have been active in the recent past. Observation of the Earth´s surface from space is essential to map the terrain in terms of natural hazards. Detailed surveys of the terrain using satellite images, aerial photographs and field-work help the recognition, inventory and outline of mapping units showing differences in terms of terrain characteristics and active geomorphological and geological processes. This approach using visual interpretation of satellite images has been applied in the same study area (García-Meléndez & Riaza 1997; García-Meléndez et al. 1998).

The purpose of this work is to explain the differences observable by visual interpretation in terms of the spectral response of the Earth´s surface, related to variations in mineralogical composition. Laboratory reflectance spectroscopy analysis has been applied and the spectral response is explained in terms of the mineralogical composition analyzed by X ray diffraction, with reference to absorption features gathered by spectral libraries.

1.1 *Geologic and Geographic setting.*

The study area is located in the northeastern part of the Huércal-Overa basin, which is filled by a variety of Neogene continental and marine sedimentary environments overlayed by Quaternary alluvial sedimentary facies. This basin is located in the Eastern Betic Cordilleras (Southeast Spain) and is affected by faulting along two regional seismic lineations oriented E-W and NE-SW, conditioning the geomorphological pattern. The basin studied is located at the foot of the southern part of the Sierra de Las Estancias. This zone is suffering active denudation in a dry, semiarid climate in an area with active tectonics. The eroded material from the ridges, sediments on alluvial fans in the low areas, with irrigated crops and human settlements. The Almanzora river hydrographic basin hosting the study area is featured by dry water courses only active after infrequent heavy rainstorms causing considerable damage. Non consolidated lithologies and a sparse vegetation cover favour erosion processes.

2 METHOD

Laboratory reflectance spectroscopy involves the measurement and interpretation of the variation with wavelength of the ratio of reflected/incident electromagnetic radiation. This method is based on the physical principles involved in the interaction of electromagnetic radiation and matter.

The morphosedimentary levels under study present a mixture of rock fragments and minerals such as quartzite, shale, micaschists, quartz, among others whose composition is closely associated with the source area. Chemical weathering after sedimentation produces different response depending on the mineralogical composition, climate and local soil drainage. The geomorphology and mineralogy of the surfaces is controlled by the alluvial sedimentation parameters, the source area composition and mineralogical changes related to postsedimentary processes. Representative samples from a total number of 105 were selected for mineralogical analysis.

After sample preparation, X ray diffraction was performed with a Siemens D-500 diffractometer in order to obtain the mineralogical composition through the comparison of results with the ASTM tables. A semiquantification analysis was performed estimating the area of the peaks of the main minerals depicted in the diffractograms.

Spectral measurements were made with a Perkin-Elmer Lambda-9 spectrophotometer provided with an integrating sphere registering reflectance with reference to a standard of Barium sulphate (SO_4Ba); measurements were taken within 1 μm interval in the visible and near infrared range of the electromagnetic spectrum (0.4 – 2.5 μm). Electronic charge transfer, crystal field effects and vibrational effects are the primary physical processes that control scattering, absorption and/or re-emission of electromagnetic radiation on these wavelengths for sedimentary rock-forming minerals (Lang et al. 1990). Electronic processes involving transition metal ions such as Fe^{+3} and Fe^{+2} and vibrational processes such as those involving H_2O, OH^-, CO_3^{-2}, among others, give specific features in reflectance spectra (Hunt 1980), referred as absorption features.

The laboratory spectral analysis helped to identify series of fans mappable through image processing in terms of their different mineralogical contents.

Table 1. Mineral Composition through X-diffraction analysis (%).

Sample	Quartz	Feldspar	Phyllo-silicates	Carbonates
S40	70	5	25	-
S54	62	3	20	15
S60	72	4	21	2
S61	52	10	18	20
S62	58	6	26	7
S63	51	6	21	26
S64	65	3	19	13
S65	62	5	33	-

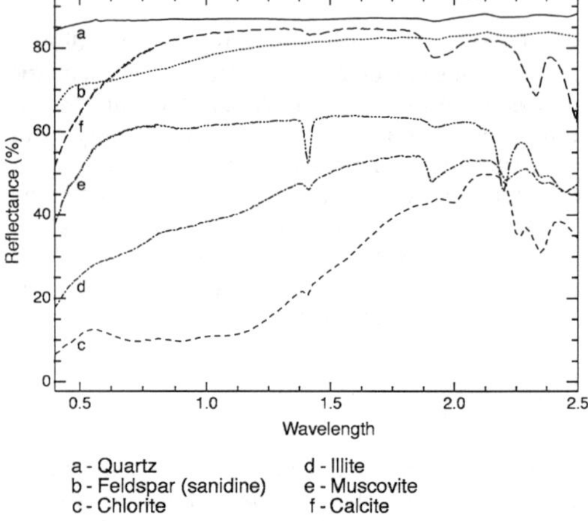

Figure 1. Representative spectra for minerals listed on Table 1. Wavelength in μm. Spectra a, c, e and f, from Clark et al. 1993. Spectra b and d from Grove et al. 1992.

3 RESULTS

All spectral measurements show similar shapes, nevertheless minor but significant variations occur. They can be grouped in two classes: I) spectra with a relative low overall reflectance with weak absorption features, and II) spectra with higher overall reflectance and sharper absorption features.

The mineralogical composition determined by X ray diffraction for the selected representative samples (Table 1) is very similar. Quartz and phyllosilicates (illite and chlorite) with small amounts of feldspars are the main minerals present, being the carbonate content the principal difference.

According to spectral libraries (Grove et al. 1992, Clark et al. 1993), pure quartz is almost devoid of spectral features, milky white or gray quartz show absorption features at 1.4, 1.9 and 2.6 μm due to fluid inclusions (Hunt et al. 1970). Pure feldspars are also spectrally featureless, being the only present spectral features due to small amounts of impurities such as iron and water because of the incipient alteration (Hunt et al. 1973). From X ray diffraction, the main phyllosilicates found are chlorite and illite. Chlorite presents spectral features due to the ferric ion at 0.7 and 0.9 μm, and also hydroxyl absorption features near 2.2, 2.35 and 2.45 μm. The presence of molecular water trapped in the lattice is indicated at 1.9 μm. Illite presents absorption features at 1.4 μm due to OH^-, at 1.9 μm due to molecular water and at 2.2, 2.35 and 2.45 μm due to the OH^- stretch and to the Al-O-H bending mode at 2.2 μm (Hunt & Salisbury 1970; Hunt et al. 1973; Clark et al. 1990). The observed absorptions in carbonates are due to the CO_3^{-2} ion, being the strongest at 2.5-2.55 and at 2,3-2.35 μm (Clark et al. 1990). (Figure 1).

The morphosedimentary units show two distinctive spectral responses according to subtle differences on their mineralogical contents:

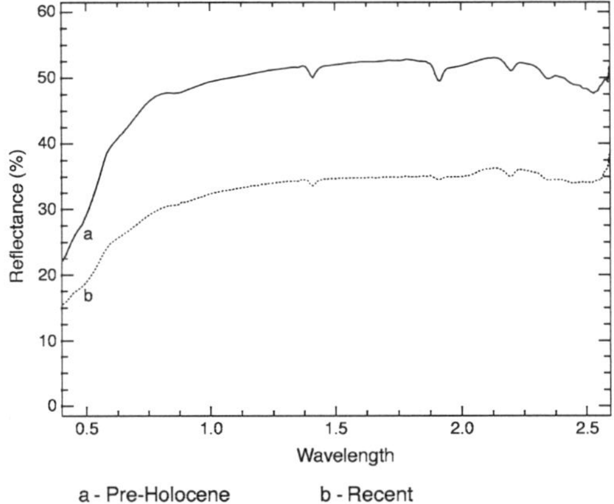

Figure 2. Spectral measurements for a) Recent alluvial sedimentary units; b) old alluvial sedimentary units. Wavelength in μm.

3.1 *Recent alluvial fan spectra (Holocene)*

Spectral measurements from Holocene alluvial fans show an almost flat shape in the wavelength range 1-2.5 μm (Figure 2a). The absorption feature at 1.4 μm is due to the OH^- group present in the minerals, and the almost absent absorption feature at 1.9 μm represents a small amount of molecular water in the mineral structure. The absorption feature at 2.2 μm in the recent alluvial fan spectrum is related to the presence of phyllosilicates with a dioctaedric structure, which together with the feature at 2.33 μm represents a combination of processes associated to the OH^- stretch and with the angle between the cation (Al^{3+}, Mg^{2+}, among others) and the OH^- group. All samples display the 2.2 μm absorption feature, therefore can not be used as a discriminating element through image processing.

3.2 *Old alluvial sedimentary spectra (pre-Holocene alluvial fans)*

The main spectral differences on the older alluvial fans (Figure 2b) compared to the previous sedimentary units (Figure 2a) are displayed on the wavelength range between 1 μm and 2.5 μm. The absorption feature at 1.9 μm is better defined, and there is a new absorption feature located at 2.54 μm which together with the one at 2.3 μm account for the presence of carbonate. The 2.3 μm absorption feature is related in this case, both to the CO_3 ion and the OH^- stretch in phyllosilicates.

The overall reflectance is higher for the old alluvial sedimentary surfaces through all the wavelength range. This is evident in the Thematic Mapper bands, showing lighter gray tones. The X-ray diffraction mineralogical analysis demonstrate that carbonate is responsible for the higher overall reflectance (Table 1, Figure 2).

All spectra show between 0.4 μm and 1 μm a strong decrease of reflectance that may be caused by strong iron absorption features at short wavelengths outside the considered spectral range (Van der Meer 1995).

4 CONCLUSIONS

Landforms created in the study area by alluvial fan sedimentation with similar lithological source areas show mineralogical differences related to their relative ages and consolidation processes. These differences are observed in the TM satellite images in terms of different spectral response which can be mineralogically explained by means of laboratory reflectance spectroscopy and X-ray diffraction analysis, demonstrating the usefulness of such techniques to map natural hazards prone areas. The presence of small carbonate content increases remarkably the overall reflectance of the areas. The precipitation of carbonates is a consolidating post-sedimentary process. Therefore, the areas lacking carbonates indicate a younger sedimentation and a higher possibility of future sedimentary accumulation. Mapping areas where preferred erosion and sedimentation are likely to occur could help the outline of terrain units qualifying future damage.

REFERENCES

Clark, R.N., T.V.V King, M. Klejwa, G.A. Swayze & N. Vergo 1990. High spectral resolution reflectance spectroscopy of minerals. *J. Geophys. Res.* 95:12653-12680.

Clark, R.N., G.A. Swayze, A.J. Gallagher, T.V.V. King, & W.M. Calvin 1993. The U.S. Geological Survey, digital spectral library: version 1: 0.2 to 3.0 microns. U.W. Geological Survey Open File report 93-592.

García-Meléndez, E., I. Molina, M. Ferrer-Julià & J. Aguirre 1998. Multisensor data integration and GIS analysis for natural hazard mapping in a semiarid area (Southeast Spain). *Adv. Space Res.* 21(3): 493-499.

García-Meléndez, E. & A. Riaza 1997. Diferenciación de unidades geomorfológicas con imágenes TM y SPOT en la cuenca de Huércal-Overa y la Sierra de las Estancias (Almería). In J.L. Casanova & J. Sanz Justo (eds), *Teledetección. Usos y aplicaciones*: 331-338. Universidad de Valladolid.

Grove, C.I., S.J. Hook & E.D. Paylor 1992. Laboratory reflectance spectra for 160 minerals 0.4-2.5 micrometers. JPL Publication 92-2. Pasadena, CA: Jet Propulsion Laboratory.

Hunt, G.R. 1980. Electromagnetic radiation: the communication link in remote sensing. In B.S. Siegal & A.R.

Gillespie (eds), *Remote Sensing in Geology*: 5-45. New York: John Wiley.
Hunt, G.R. & J.W. Salisbury 1970. Visible and near-infrared spectra of minerals and rocks, I. Silicate minerals. *Modern Geology*, 2:23-30.
Hunt, G.R., J.W. Salisbury & C.J. Lenhoff 1973. Visible and near-infrared spectra of minerals and rocks, VI. Additional silicates. *Modern Geology*, 3:1-14.
Lang, H.R., M.J. Bartholomew, C.I. Grove & E.D. Paylor 1990. Spectral reflectance characterization (0.4 to 2.5 and 8.0 to 12.0 μm) of Phanerozoic strata, Wind River basin and southern BigHorn basin areas, Wyoming. *J. Sed. Petrol.*, 60(4): 504-524.
Van der Meer, F. 1995. Imaging spectrometry and the Ronda peridotites. Ph. D. Thesis Wageningen.

Remote Sensing in the 21st Century: Economic and Environmental Applications, Casanova (ed.)
© 2000 Balkema, Rotterdam, ISBN 90 5809 096 5

Lithological and geomorphological large-scale mapping using remotely sensed data, GIS and terrain analysis

E.Wolk-Musial & B.Zagajewski
Faculty of Geography and Regional Studies, University of Warsaw, Poland

ABSTRACT: The lithological and geomorphological large-scale mapping is necessary for correct landuse and environmental protection. The Remote Sensing of Environment Department, Faculty of Geography and Regional Studies, University of Warsaw has tackled these problems using remotely sensed data (black and white aerial photographs, satellite imagery – Landsat 5 TM), GIS and terrain analysis. The result was the creation of digital database containing morphometrical, lithological, geomorphological information for Choroszcz regional (Polish cooradinate system 1965). Study area is situated at the north – eastern part of Poland near Białystok. The landscape is dominated by glacial forms.

1 INTRODUCTION

In addition to geological structure, soil cover, climate, water and plant cover, land relief is a major element of the geographic environment. In order to fully explore it, it is necessary to examine all the landforms and elements in both qualitative (how, when and in what conditions individual forms originated) and quantitative terms (what are the dimensions, distribution, frequency and density of such forms, what is the pace of their evolving and maturing). Therefore, large scale geomorphological mapping aims to accurately locate and classify all relief forms taking into account their morphometric and morphographic features, origin and age.

A good geomorphological map cannot be compiled without an up-to-date data base developed using research methods applied in remote sensing and physical geography, with a particular emphasis on geomorphological analysis and Geographical Information Systems (GIS). Especially the latter prove extremely useful in collecting, processing and visualising data in the form of a numerical geomorphological map. All the numerical data are uniformly orientated in the adopted system of coordinates, which makes it possible to correlate them with data plotted in various projections. One of the main advantages of GIS is the possibility to establish connections between data frequently coming from different sources and periods, and to select them according to freely adopted criteria, which sometimes tend to be very complex ones. Besides, digitised information can be constantly updated.

In the geomorphological analysis of Poland, the use of GIS has so far been scarce (Wołk Musiał, E. 1992, 1993, 1994) since it requires considerable financial spending for equipment and software and developing sufficient skills and expertise. At the same time, it has only been a decade since geographical information systems began to mark their presence among the research methods related to the natural environment.

The issues outlined above have been analysed in the area comprised by the Choroszcz map sheet in the 1:50000 scale, covering the plateau areas adjoining to the part of the Narew valley in the vicinity of the Biebrza confluence.

2 RESEARCH ALGORITHM

2.1 *The work was conducted according to the following algorithm:*

The preliminary stage of work comprised defining the researched area, which was the meander of the Narew valley depicted on the Choroszcz map sheet in the 50000 scale (Polish coordinate system 1965). Basic research was conducted in the 1:25 000 scale for the constituent sheets: Szorce, Tykocin, Choroszcz, Długołęka. Preliminary geomorphological survey was conducted using topographic and geological maps as well as relevant literature of the subject, which allowed to specify the terrain dissection and hypsometric relations between landforms of various origin. It was agreed that the scope of infor-

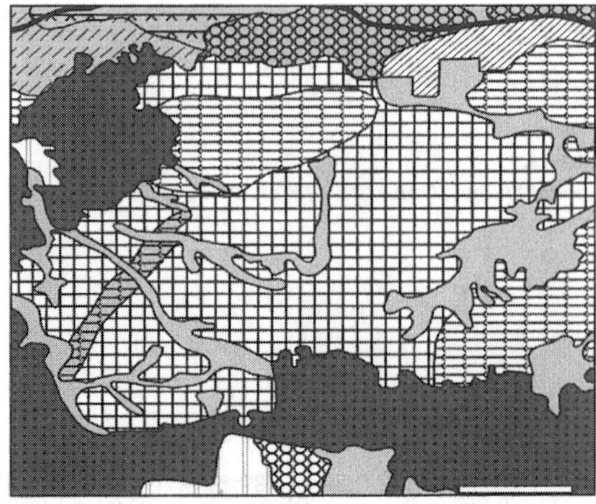

Legend of homogeneous photomorphic areas

Fig. 1. Exemplary part of homogeneous photomorphic areas delimited on the aerial photograph

mation would comprise a catalogue of landforms the origin of which is connected with the activity of the ice-sheet of the Warta stage.

The comprehensiveness of data necessary to correctly identify taxonomic units is extremely important in relief research, while remotely sensed data are not, as yet, used sufficiently frequently. In this research, the satellite image was used (RGB coloured composition) and panchromatic aerial photographs at a scale 1:25 000. They served as the basis for compiling a map of photomorphic units upon the presumption that land sculpture in which the relief often plays a dominant part should be reflected in the photographic image (Fig. 1.). Areas with three phototonal groups (white and light grey, grey and dark grey, and black) were distinguished; in them, the following structures were identified: streaked, spotted, spot, block, striped, wavy, granular and amorphic. It was assumed that the set of forms occurring in a given unit should be genetically coherent.

Subsequently, morphometric classification of relief was made on the basis of maps at a scale 1:25000 for constituent sheets, which was oriented to systematise landforms in glacial landscapes (Raman, K.G. 1965, Richling, A. 1973). The result was a distribution of typological units (Fig. 2.), distinguished on the basis of relief and morphometric criteria of land sculpture, such as: height differences, natural falls and dimensions of the landform base. In order to capture the specific differentiation of the natural environment in the landscapes of great river valley bottoms, an additional category was introduced, covering higher plain strata in valley bottoms and depressions. It embraced individual landforms or groups of landforms which could not be classified as low small- and large-radius mounds due to smaller falls and height differences. However, in the category of plains in valley bottoms and depressions, they constituted landforms which were distinctly different from the surrounding areas. Their specificity was manifested in the differentiation of the depth to ground water which, albeit minute in absolute values with respect to landforms built of loose material led to the evolving of radically different conditions of the natural environment. Very often, against the background of meadows and riverside forests, dry ground forests or forest communities were observed. For this reason, the formal classification criteria were not always applied (such as landform base, natural falls, height differences). Instead, efforts were made to take into account the broadest spectrum of information pointing to differing roles of individual landforms in the landscape. Particu-

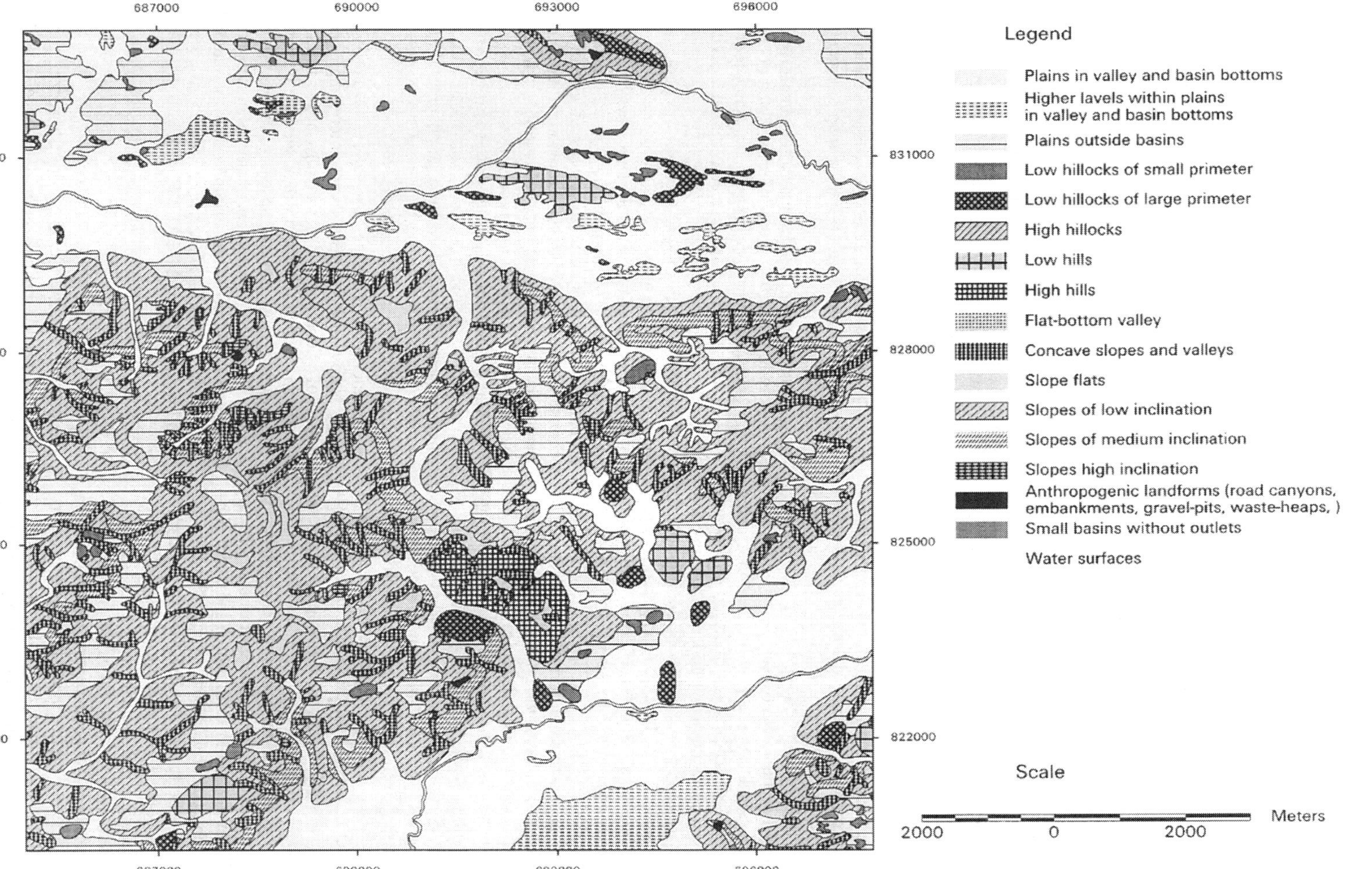

Fig. 2. A fragment of the map of morphometric relief types, Choroszcz sheet.

larly, the analysis of land use in the researched area and the range of built-up areas conducted on a panchromatic aerial photograph proved to be an important indicator. Its differentiating significance could be visible mainly in the zone of large river valleys due to "diluted" forms of relief. As a result, the following relief morphometric types were distinguished:

- plains in valley bottoms and depressions;
- three types of mounds (depending on the base dimensions, slope angle and height);
- two types of hills (depending on their height);
- flat-bottomed and basin-shaped small valleys;
- Piedmonts flattenings and three types of slopes: 1-3, 3-9, > 9;
- small flowless depressions;
- anthropogenic forms.

The basic factor determining the correct identification of the origin of relief is the lithological structure of the researched area. It is often difficult to establish in field research due to the limited number of randomly distributed exposures. Lithological survey was conducted using soil and agricultural maps at a scale 1:25000. The ranges of lithological stand ups were verified using aerial and photographs and field survey research. Four sheets of the lithological map at a scale 1:25000 were compiled (Szorce, Tykocin, Długołęka, Choroszcz); they cover the analysed area and jointly make up the Choroszcz sheet at a scale 1:25000.
The use of geographical information systems (SINUS, Arc/Info) made it possible to conduct lithological examination on three levels (0.5 m, 1.0 m and 1.5 m) for the entire area under research. Fig. 3 is an example of the first information layer (surface layer), covering the south-eastern part of the Choroszcz sheet, where the following materials were distinguished: gravel, loose sand, sand loamy, loamy sand, clay, dust, silt, peat and peat earth. Four sheets of constituent maps with lithological data encoded for three levels were compiled using information technology.

Subsequently, the examination of genetic types of relief was conducted using the information layers developed beforehand in GIS, which included: photomorphic units, morphometric types of relief and lithology. Genetic types of relief (fig. 4) are understood in this context as the result of the activity of morphogenetic factors, such as ice-sheet, postglacial waters, rivers, wind, biogenic and anthropogenic factors. The division of the researched area into genetic types of relief made it possible to identify the origin of individual landforms and to prepare the map key to the detailed geomorphological map at a scale 1:50 000 for the Choroszcz sheet (Wołk Musiał, E. 1994).

The next stage was to prepare a detailed elaboration of the map of genetic types of relief and to distinguish the ranges of individual landforms in accordance with the map key. This process was assisted by an analysis of aerial photographs and general field mapping, which denoted verification of the correct marking of landforms. By combining the morphometric and lithological features and taking into account the analysis of the spatial arrangement of the individual landforms on remotely sensed materials, four sheets of geomorphological maps were compiled in the 1: 25000 scale (Szorce, Długołęka, Choroszcz, Tykocin). Then, maps in the basic scale were subject to numerical processing, which included:

- scanning and vectorisation in the SINUS system;
- creating the typology, coding, merging the constituent sheets into a unified data base (vectoral and screening transformations);
- developing the map key using the ERDAS application. Colours in the map key convey information on morphogenetic factors (glaciers, water, wind, denudation, man), which in the specific period (Middle Poland glaciation, the Warta stage) shaped the area as a result of both destructive and creative activity.

The result of this work is a geomorphological map at a scale 1:50000, the Choroszcz sheet, a part of which is shown in Fig. 5. The relief presented here is the result of area ablation of the ice-sheet, which led to releasing huge masses of glacial and fluvial waters, causing intense erosion activity. The thaw forms are terrain culminations, indicating places where the ablation process was the fastest. Material washed out by glacial waters was deposited in the form of kames, eskers, plains located by ice melt-out areas, which have a large share in the researched area. Also, eolian forms can be recognised (in the western part of the sheet).

The morphology of the vast valleys of the rivers Ślina, Narew and Supraśl is very interesting. They consist of widenings and narrowings in the form of water gaps, which testifies to their lacustrine origin. This phenomenon can be observed in the Narew valley. These lakes were genetically related to marginal basins and melt-outs originated on the internal side of the marginal sequence of the Mława stage (Falkowski, E. 1971; Mojski, J.E. 1972).

3 CONCLUSION

The geomorphological map of the Choroszcz sheet in the 1:50000 scale is a numerical data base and constitutes a separate layer in the GIS system. It can be used in the research of correlation with other

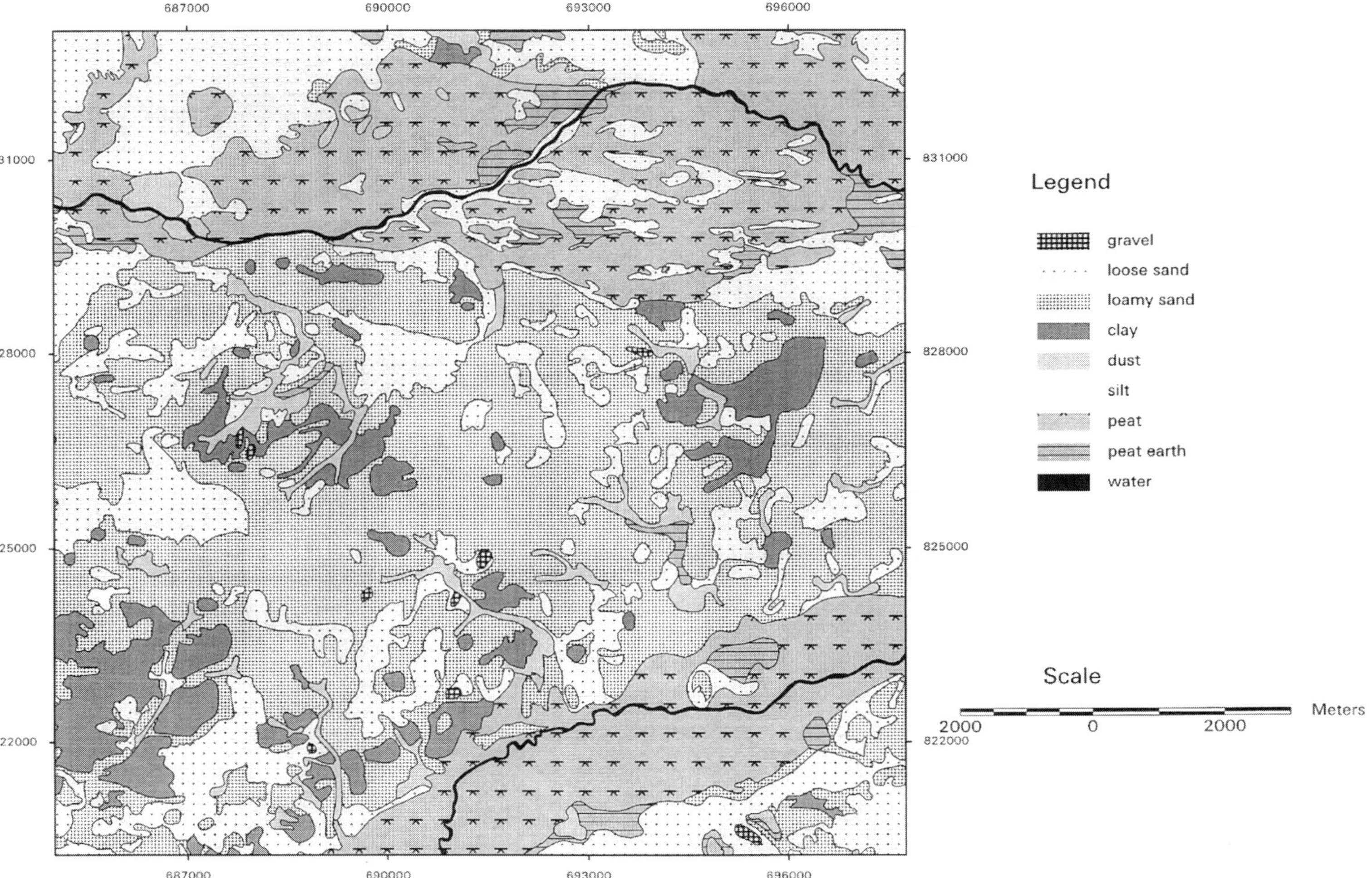

Fig. 3. A fragment of the lithological map (surface layer), Choroszcz sheet.

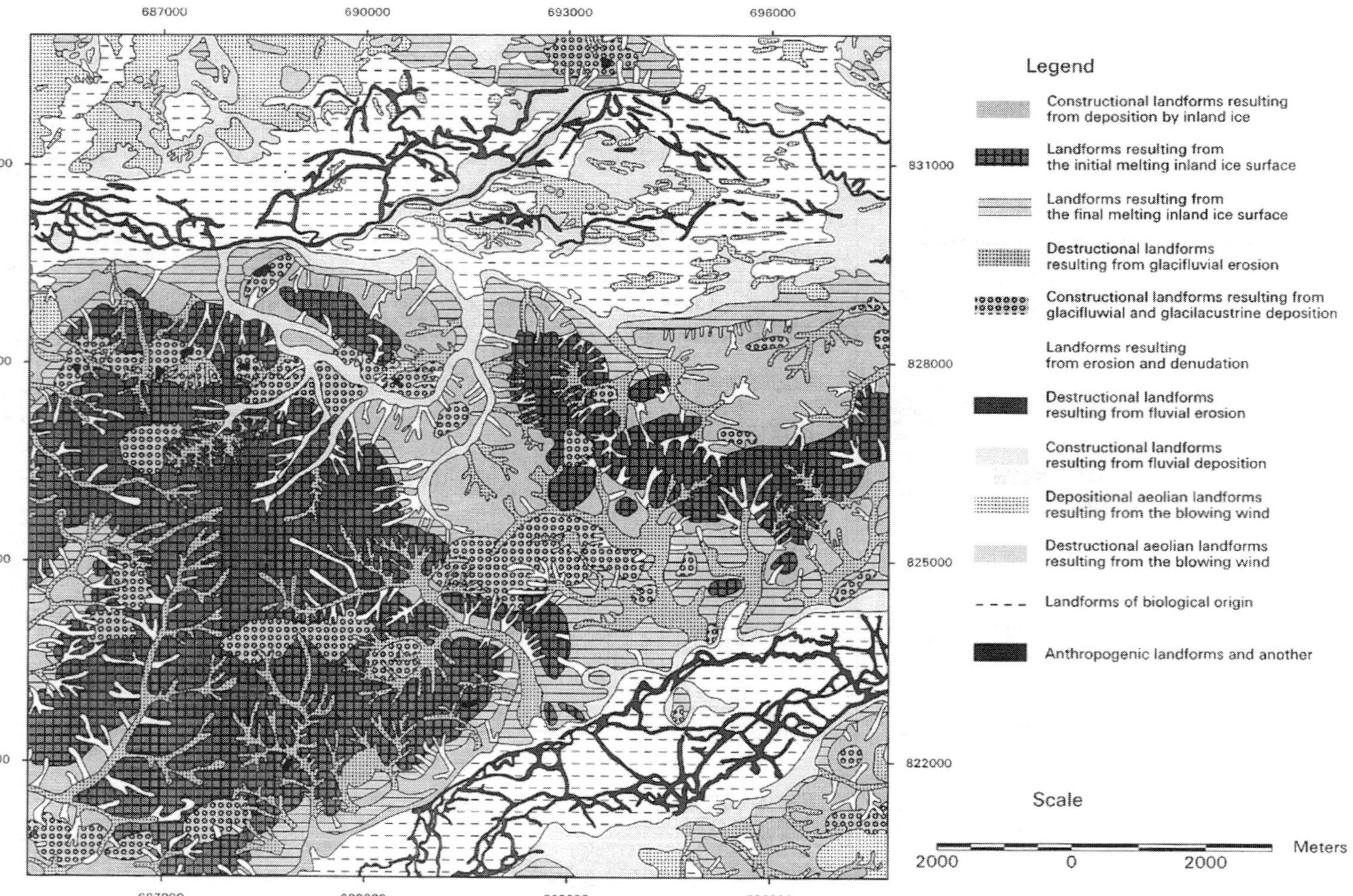

Fig. 4. A fragment of the map of relief genetic types, Choroszcz sheet.

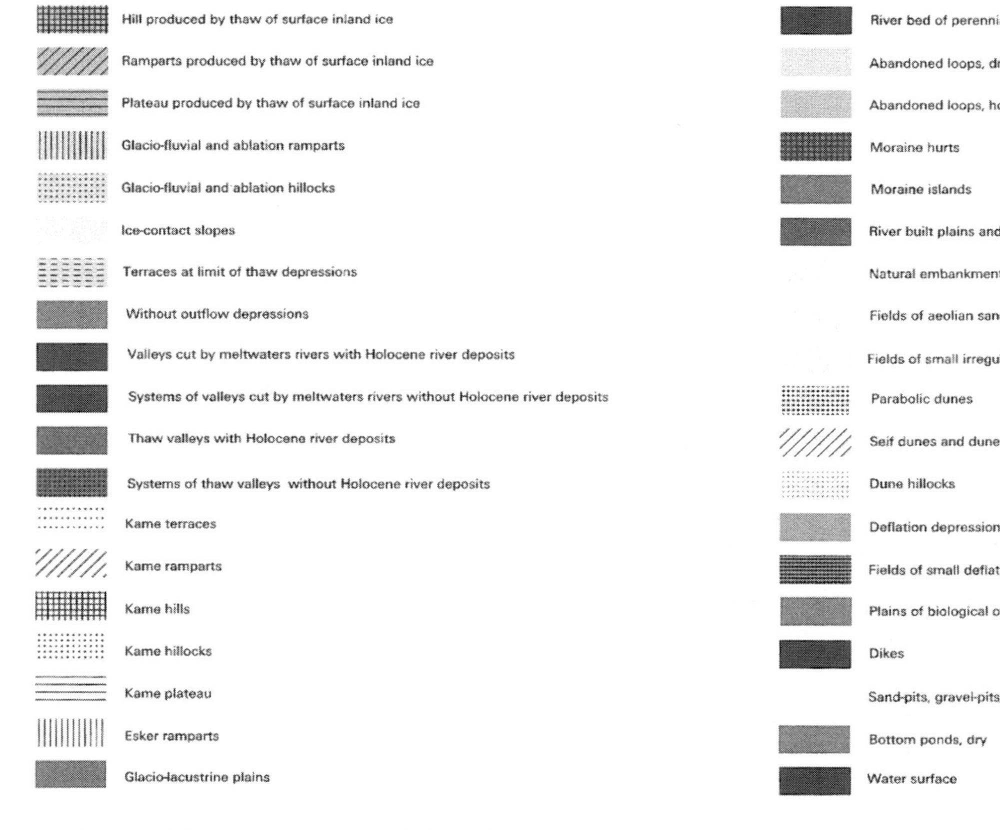

Fig. 5. Legend for accurate geomorphological map, Choroszcz sheet

elements of the natural environment. They include the environmental resistance degree for the people intervention. It is crucial for planning the human activity in environment (Wołk-Musiał E., Zagajewski B., 1999)

4 REFERENCES

Falkowski, E., 1971: "*Historia i prognoza rozwoju układu koryta wybranych odcinków rzek nizinnych Polski*" in: Biuletyn Geologiczny, vol. 12, Wydawnictwa UW

Mojski, J.E., 1972: "*Nizina Podlaska*" in: Geomorfologia Polski, PWN

Raman, K.G., 1965: "*Zagadnienia klasyfikacj i typologii krajobrazów geograficznych jako podstawy dla regionalizacji fizycznogeograficznej*" in: Przegląd Zagranicznej Literatury Geograficznej, vol. 4

Richling, A., 1973: "*O sposobach przedstawiania urzeźbienia w terenach młodoglacjalnych*" in: Przegląd Geograficzny, vol. 45, part 1

Wołk-Musiał, E., 1992: *otointerpretacyjna mapa geomorfologiczna w skali 1:50000, arkusz Choroszcz, opracowana numerycznie* in: Fotinterpretacja w geografii, vol. 22

Wołk-Musiał, E., 1994: "*Systeme d'information spatial SUINUS – instrument de l'analise geomorphologique*" in: Miscellanea Geographica, vol. 1994

Wołk-Musial, E., Zagajewski B., 1999. Geoinformation as en effective tool for environmental protection and landuse optimisation. *Operational Remote Sensing for Sustainable Development.* Rotterdam: Balkema

Remote Sensing in the 21st Century: Economic and Environmental Applications, Casanova (ed.)
© 2000 Balkema, Rotterdam, ISBN 90 5809 096 5

Geomorphological-geological evaluation of the Gazikoy-Saros section of Northern Anatolian Fault zone using remotely sensed data

S. Kaya & O. Muftuoglu
ITÜ Civil Engineering Faculty, Geodesy and Photogrammetry Department, Remote Sensing Division, Istanbul, Turkey

ABSTRACT: Since the early 1970s, parallel to the developments in satellite and computer technology the remote sensing data have been utilized in a variety of subjects, forestry, cartography, geomorphology, geology, hydrology, urbanism, agriculture, etc., for the purpose to study the earth surface. At present, solution of a number of problems have been made easier through the advent of integrating the different data with the remote sensing data by transforming them into the same co-ordinate system, so that they are analyzed together. In this study, Gazikoy-Saros segment of the Northern Anatolian Fault and geomorphological -geological and tectonic structures around the Koru - Isiklar mountain have been analyzed by using the remote sensing data and digital elevation model. For this purpose, topographic maps of 1/25 000 scale and Landsat 5 TM and Spot P satellite data have been used.

1 INTRODUCTION

Splitting vertically the northern section of Anatolia, North Anatolian Fault (NAF) reaches the north of Lake Van after intersecting the East Anatolian Fault near Karliova (Bingol). By forking in its western section the fault system is branched out into some lines which reach the Gulf of Izmit, the Gulf of Saros through the Marmara Sea and which extend to Lake Iznik or to the South of the Marmara through from the Gulf of Gemlik (Barka, 1992, Barka, 1997). Alongside this belt which geomorphologically shows a form of rift valley in most places, stream beds, hills formed by pressed rocks , troughs ponds , hot springs and in some locations prematurely extinct volcanic cones are encountered in Turkey, earthquakes which claimed most lives and caused extensive damage have often occurred on the NAF or its periphery.

In this study, geomorphological assessments in the Gazikoy – Saros segment have been made by using Landsat 5 Thematic Mapper data, Spot Panchromatic data and merge images obtained by merging these two groups of data. These data groups have been compared morphologically with the aerial photograph of 1 / 10 000 scale.

Earlier studies in the area were used as ground information (Altın, 1992, Eldeniz, 1996, Erol 1979, 1983, Erol and Cetin, 1995). All the data used have been transformed into UTM (Universal Transform Mercator) by means of ground control points (Sunar and Kaya, 1997). Furthermore, land use was obtained in a short period of time by classifying the satellite images (Kaya, 1993). By using the Digital Elevation Model and Remote Sensing data, Gazikoy – Saros segment, which was so far shown as a straight line has been re-drawn based on this data.

2 STUDY AREA AND CHARACTERISTICS

Geographical location of the area lies between the 26° 30' – 27° 30' east and 40° 20' – 41° 00' north coordinates. 19 pieces of standard topographic maps 1 / 25 000 scale have been analyzed. Area of study covers approximately 2500 km^2. In the area through which the western section of North Anatolian Fault crosses (Land section between Gazikoy – Gulf of Saros), Koru Mountain, Isıklar Mountain and Doluca Hill exist. There are agricultural lands, forest lands, shrubberies and settlements etc., as well. As satellite images, Landsat 5 TM data with 30 m. resolution and 7 bands and Spot P data with 10 m. resolution and 1 band have been used.

3 DATA ANALYSIS

On the 1/25 000 scale topographic maps, contours have been digitized at intervals of each 20 m. and transferred to computer media. These plots which have been digitized one by one, have been united in

Figure 1. Study area (Landsat 5 TM image, band combination 7, 4, 1). (colour plate, see p. 608)

AutoCAD software and altitude values of each contour have been given. By using Erdas Image 8.3 software, Digital Elevation Model in the form of raster structure has been produced.

Satellite images have been transformed into UTM projection system with the help of ground control points. Due to the fact that the sensors through which remote sensing data are obtained have different characteristics and collect data in different forms for researching the earth, it is necessary to evaluate this data together and compile, within a data group, the images received from different sensors for the purpose of finding out different features of the earth. (Ehlers et al., 1990). In order to enhance the visual interpretability of the satellite images and obtain data with high spatial resolution, Spot P data with 10 m. ground resolution has been merged on the 7,4,1 bands of Landsat 5 TM with 30 m. resolution. The images that has been obtained is a multispectral one, with 10 m. ground resolution spectrally enhanced, featuring the structural characteristics of the area and is easily interpretable. Geomorphological and geological structures are better seen in this image.

Digital Elevation Models produced by means of vector data groups do not contain much information on the surface. Therefore, not much can be said about land surface in interpretation and assessment. They can be interpreted and digital elevation models that can be evaluated by different disciplines could be formed only if they are merged with the data groups defining the information on the surface of earth, having possibility of evaluating information obtained at different times, determining the structural characteristics of the surface and ensuring quick flow of information. At present, use of remote sensing data obtained through satellites and the ability to integrate them with other data shed considerable light on this issue.

4 RESULT

4.1 *Geomorphological Assessment*

Most important source for the geomorphological study of an area are maps featuring the characteristics of the surface. However, searched information may not be available on the maps or these maps may not be up-to date. Therefore, use of satellite images is inevitable.

Geomorphologically, the Northern Anatolian Fault has here and there characteristics of a rift valley. Such valleys cause significant topographic changes and while the block in the middle is rifted, the ones in the sides uplift. This morphological feature is available in the middle section of Gazikoy-Saros segment. These valleys can easily be identified by means of remote sensing data and digital elevation model (Fig.2).

Compared to the aerial photographs, remote sensing data offers considerable advantages. Allow-

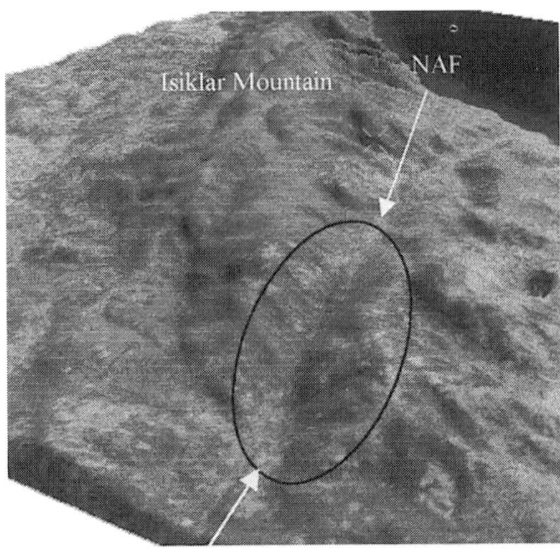

Figure 2. Rift valleys. (colour plate, see p. 609).

Figure 3. Aerial photographs (Scale 1/10 000). (colour plate, see p. 609).

ing to work in large areas, to obtain coloured data indicating land features, to evaluate such data in computerized media and to analyze in the desired area are among the advantages of remote sensing data. However, assessment of satellite data with low resolution can generally be made. When the data obtained as multi-band is merged with high resolution data, they can be better interpreted and better data can be obtained. By using high resolution satellite data, aerial photographs of 1/10000 scale have been approached. This is an important advantage in terms of assessment(Figure 3,4).

Figure 4. Merge image (colour plate, see p. 609).

4.2 *Geological Assessment*

In order to present all the basic geological features of an area, it is of great importance to be able to simultaneously observe all the formations spread over large areas and monitor on the same image tectonic structures of large proportions. In the study, Gazikoy-Saros Segment of Northern Anatolian Fault and its surroundings have been analyzed. Formations affecting the area have been analyzed by means of remote sensing data and digital elevation models. Utilizing the digital elevation models that have been formed and the remote sensing data, geometry and continuity of the fault have been studied. Areas which deform the continuity of the fault along Gazikoy – Saros segment and where the fault is angularly turned were modeled three dimensionally (Fig. 5,6).

The 1912 Murefte-Sarkoy earthquake M=7.4, occurred along the Gazikoy-Saros segment (Macovei,1912, Mihalinovic,1927, Sieberg,1932, Ates &Tabban,1976, Ates,1982, Gundogdu, 1986, Oztin,1987, Ambraseys&Finkel,1987,1991, Ambraseys, 1990, Eyidogan et al.,1991, Tuysuz et al.,1998). Again during this earthquake, the Gazikoy-Saros segment of the northwest of the North Anatolian Fault moved. In recent years, this segment of the North Anatolian Fault has been quiet as far as earthquakes are concerned. This shows that the fault has been locked again. Earthquakes throughout history in the Marmara Region generally occurred on this fault (Ambraseys and Finkel, 1991).

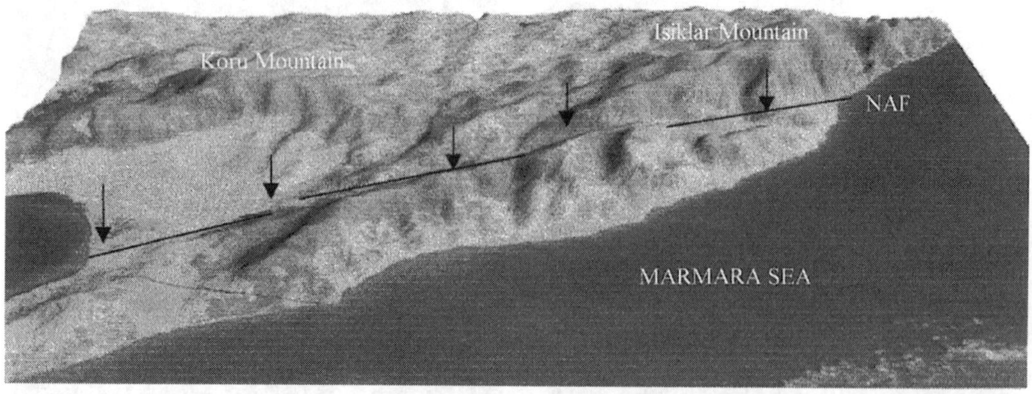

Figure 5. General view of the study area (Digital Elevation Model). (colour plate, see p. 610).

Figure 6. Digital Elevation Model overlaid with merge image. (colour plate, see p. 610).

5 CONCLUSIONS

Remote Sensing data obtained by using satellite technology makes it possible to look at an image of different scale in a large area. For being able to evaluate an area geologically and geomorfologically as a whole, it is necessary to see together all the formations affecting the area and be able to look at the minor formations too on the same data. Therefore, use of satellite images gained extreme –importance today.

It has been show that merge images obtained by merging the Landsat 5 TM and Spot P data can be used in assessing the land morphology and understanding the active fault formations. Also, data at a scale close to 1/10 000 scale aerial photographs has been obtained. Benefits of using satellite images in monitoring the geometry of known active faults on earth have been pointed out. Gazikoy – Saros segment of North Anatolian Fault has been re-drawn based on these images.

This study clearly illustrates that satellite and digital elevation data are very useful tools to analyze many different aspects of geomorphological and geological characteristics of any active tectonic regions.

REFERENCES

Altın, B.N. 1992. Isiklar Dagi ve Cevresinin Neotektonik Donem Jeomorfolojik Tektonik Gelisimi(in Turkish). *Doktora Tezi. IU.Deniz Bilimleri ve Cografya Enstitusu.* Istanbul-Turkey.

Ambraseys, N.N. 1990. Engineering Seismology. *Earthquake Eng. Struct. Dyn.* 17.pp 1-105.

Ambraseys, N.N.&Finkel, C.F. 1987. The Saros-Marmara Earthquake of 9 August 1912. *Earthquake Eng. Struct. Dyn.15*.pp 189-211.

Ambraseys, N.N.&Finkel, C.F. 1991. Long-term Seismicity of Istanbul and the Marmara region. *Eng.Seismol. Earthquake Report.*91/8.Imperial College.Eng.

Ates, R. 1982. Earthquake Activity on the North Anatolian Fault Zone. In:Isikara, A.M. Vogel A.(Eds.).*Progress in Earthquake Prediction.* Vieweg. Braunschweig. pp 95-113.

Ates, R.&Tabban, A. 1976. Preliminary Report for the Sarkoy-Murefte Earthquake of 9th August 1912. *Publ.Earth.Res. Inst.* 24 pp.

Barka A.A. 1997. Neotectonics of the Marmara Region, Active Tectonics of Northwestern Anatolia. *The Marmara Poly-Project.* Hochschulverlag AG an der ETH. Zürih.

Barka, A.A. 1992. The North Anatolian Fault Zone. *Anales Tectonicae.* Special Issue Supplement to Volume 6, pp 164 – 195.

Ehlers, M., Jadkawski, A.M., Howard, R.R.&Brostuen,D. 1990. Application of Spot Data for Regional Growth Analysis and Local Planing. *Photogrammetric Engineering and Remote Sensing.* Vol:56. No:2,pp 175-180.

Eldeniz, S. 1996. Kesan – Malkara – Koru Dagi Dolayinin Jeomorfolojisi (in Turkish). *Doktora Tezi. IU.Deniz Bilimleri ve Cografya Enstitusu.* Istanbul-Turkey.

Erol, O. 1979. Turkiye'de Neojen ve Kuaterner Asinim Donemleri, Bu Donemlerin Asinim Yuzeyleri ile Yasit (Korelan) Tortullara Gore Belirlenmesi. *TJD.* Vol:8.pp 1 – 40. Ankara.

Erol, O. 1983. Turkiye'nin Genc Tektonik ve Jeomorfolojik Gelisimi. *TJD.* Vol:11.pp 1 – 22. Ankara.

Erol, O.&Cetin, O. 1995. Evolution of the Marmara Sea From Late Miocene to Holocene (An interpretation under the light of some new radiometric dating).(in Turkish).*Quaternary Sequence in the Gulf of Izmit.* Istanbul.

Eyidogan, H., Guclu, U., Utku, Z.&Degirmenci, E. 1991. Turkiye Buyuk Depremleri Makro Sismik Rehberi. *ITU Maden Fak. Jeofizik Bol.Yayini.* 198 pp.

Gundogdu, O. 1986. Turkiye Depremlerinin Kaynak Paremetreleri ve Aralarindaki İliskiler. *PhD.Thesis. IU. Department of Geophysics.* 260 pp.

Kaya, S. 1993. The Metropolitan Analysis by Using the Remote Sensing Data Within the Example of Istanbul-Ikitelli. *Master Thesis. ITU Fen Bilimleri Enstitusu.*Istanbul.

Macovei, G. 1912. Sur Le Tremblement de Terre de la Mer de Marmara, le 9 August 1912. *Bull.Sect.Sci.Acad.Roumanie.* Bucharest 1/1.pp 1-10.

Mihalinovic, J. 1927. Les Grandes Catastrophes Seismiques la mer de Marmara. *Inst.Seismol.Univ.*Beograd.350 pp.

Oztin, F. 1987. 9 Agustos 1912 Sarkoy-Murefte Depremi. *Deprem Arastirma Enstitusu Bulteni.* No:56.pp 91-127.

Sieberg, A. 1932. Erdbebengeographie. In:Gutenberg. B.(Ed.).*Handbuch der Geophysik.*Band IV.Borntrager. Berlin.pp 527-1005.

Sunar, F.&Kaya, S. 1997. An Assessment of the Geometric Accuracy of Remotely-Sensed Images. *Int.J.Remote Sensing. Vol.*18.No:14.pp 3069-3074.

Tuysuz, O., Barka, A.A.&Yigitbas, E. 1998. Geology of the Saros Graben and Its Implications for the Evolution of the North Anatolian Fault in the Ganos-Saros Region, Northwestern Turkey. *Tectonophysics.* 293.pp 105-126.

10 Radar and hyperspectral techniques

Remote Sensing in the 21st Century: Economic and Environmental Applications, Casanova (ed.)
© 2000 Balkema, Rotterdam, ISBN 90 5809 096 5

Optical indices as bioindicators of forest condition from hyperspectral CASI data

P.J. Zarco-Tejada
Centre for Research in Earth and Space Science (CRESS), York University, Toronto, Ont., Canada

J.R. Miller
Department of Physics and Astronomy, York University and Centre for Research in Earth and Space Technology, (CRESTech), Toronto, Ont., Canada

G.H. Mohammed, T.L. Noland & P.H. Sampson
Ontario Forest Research Institute (OFRI), Ontario Ministry of Natural Resources, Sault Ste. Marie, Ont., Canada

ABSTRACT: This paper reports on progress made to link physiologically-based indicators to optical indices scaling-up from leaf level to the canopy through *SAIL* and *Kuusk* Canopy Reflectance Models (CR). Hyperspectral CASI data of 2m spatial resolution and 72 channels were collected in 1997 and 1998 deployments over twelve sites of *Acer saccharum* M. in Ontario (Canada). A field sampling campaign was carried out for biochemical analysis of leaf chlorophyll and carotenoid concentrations, and fluorescence along with leaf reflectance and transmittance. Leaf-level relationships obtained between optical indices and biochemical indicators were scaled-up to canopy level through CR models using input model parameters related to the canopy structure and viewing geometry at the time of data acquisition. The result is an algorithm which predicts leaf-level bioindicators from airborne hyperspectral imagery. A modeling study was carried out to determine the influence of CR on the four types of optical indices used in this study.

1 INTRODUCTION

The *Bioindicators of Forest Sustainability Project* (Mohammed *et al.*, 1997; Sampson *et al.*, 1998) aims to develop links between physiologically-based bio-indicators (e.g. pigment concentrations, chlorophyll fluorescence) from field and laboratory data and optical indices from hyperspectral remote sensing data for assessing forest condition. This paper reports on results obtained from 12 sites of *Acer saccharum* M. (sugar maple) in the Algoma Region, Ontario (Canada), where field measurements and hyperspectral *CASI* imagery have been collected in 1997 and 1998 field campaigns. Previous work (Zarco-Tejada *et al.*, 1999) shows that optical indices calculated at three scales progressively more closely represented the observed above-canopy reflectance spectra: from single leaf reflectance data; infinite reflectance data, calculated by using single leaf reflectance and transmittance data in optically-thick simulation formulae; and from canopy reflectance models using single leaf reflectance and transmittance data and nominal canopy architecture data. Input CR parameters defining the canopy structure and the viewing geometry at the time of data acquisition are studied in this paper through the *SAIL* CR model in order to study their influence in calculated optical indices.

2 AIRBORNE AND FIELD DATA COLLECTION AND PROCESSING

Crown cover and LAI measurements were acquired for all the plots using hemispherical photography, a PCA Li-Cor 2000 and a spherical densiometer. A total of 440 single leaf samples were collected at 12 *Acer saccharum* M. sites for biochemical analysis and measurement of leaf chlorophyll, carotenoid concentrations and fluorescence. The ratio of variable to maximum chlorophyll fluorescence (Fv/Fm), a measure of photosynthetic efficiency (Mohammed *et al.*, 1995), was measured in all leaf samples. Single leaf reflectance and transmittance measurements were acquired on all leaf samples using a Li-Cor 1800 Sphere apparatus with an Ocean Optics fibre spectrometer with 0.5 nm spacing and 7.5 nm spectral resolution in the 400-900 nm range. A signal-to-noise study was carried out to choose the optimum passbands for the smoothing and derivative processing to be applied to the single leaf reflectance and transmittance measurements. It was found that reflectance spectra were optimally smoothed with an order-3 Savitzky-Golay algorithm (Savitzky & Golay, 1964) with 25 nm bandwidth; for the calculation of derivative spectra a 13 nm bandwidth was selected using the same Savitzky-Golay polynomial fit algorithm.

The above-canopy data acquisition using the *CASI* sensor was divided into three missions each with a specific sensor mode of operation: the *mapping mission*, with 0.5 m spatial resolution and 5 spectral bands; the *hyperspectral mission*, with 2 m spatial resolution, 72 channels and 7.5 nm spectral resolution; and the *full-spectral hyperspectral mission*, with 288 channels and 2.5 nm spectral resolution. The 12-bit radiometric resolution data collected by *CASI* was processed to *at-sensor* radiance using calibration coefficients derived in the laboratory by CRESTech. Aerosol optical depth data at 550 nm were collected in the study area at the time of data acquisition in order to process image data to *ground-reflectance* using the CAM5S atmospheric correction model (O'Neill *et al.*,1997). Reflectance data were georeferenced using GPS data collected onboard the aircraft. Final registration of the hyperspectral mode imagery was achieved by registration to the mapping mission *CASI* imagery using visual identification of ground-referenced 1 m white targets, which served to accurately identify the location of the sites.

3 SELECTION OF OPTICAL INDICES

Candidate optical indices from reflectance and derivative spectra were identified from literature sources and grouped into 4 categories, based on the spectral region and the type of parameter used:

(a) Visible Ratios:
SRPI (R_{430}/R_{680}); NPQI ($(R_{415}-R_{435})/(R_{415}+R_{435})$); PRI calculated as $(R_{531}-R_{570})/(R_{531}+R_{570})$, $(R_{550}-R_{531})/(R_{550}+R_{531})$ and $(R_{570}-R_{539})/(R_{570}+R_{539})$; NPCI $(R_{680}-R_{430})/(R_{680}+R_{430})$; Carter ($R_{695}/R_{420}$), G ($R_{554}/R_{677}$) and Lichtenthaler ($R_{440}/R_{690}$).

(b) Visible/NIR Ratios:
NDVI $(R_{774}-R_{677})/(R_{774}+R_{677})$; SR ($R_{774}/R_{677}$); Lichtenthaler $(R_{800}-R_{680})/(R_{800}+R_{680})$, and ($R_{440}/R_{740}$); Gitelson & Merzylak ($R_{750}/R_{550}$); and SIPI $(R_{800}-R_{450})/(R_{800}+R_{650})$.

(c) Red Edge Reflectance-Ratio Indices:
Vogelmann (R_{740}/R_{720}), $(R_{734}-R_{747})/(R_{715}+R_{726})$, $(R_{734}-R_{747})/(R_{715}+R_{720})$; Gitelson & Merzylak ($R_{750}/R_{700}$); and Carter ($R_{695}/R_{760}$).

(d) Spectral and Derivative Red Edge Indices:
λ_p, λ_o and σ from red edge *inverted-gaussian* curve fitting, as well as spectral indices calculated from derivative analysis: (D_{715}/D_{705}); DPR1 ($D_{\lambda o}/D_{\lambda o+12}$), DPR2 ($D_{\lambda o}/D_{\lambda o+22}$), DP21 ($D_{\lambda o}/D_{703}$) and DP22 ($D_{\lambda o}/D_{720}$), amongst others.

4 *SAIL* AND *KUUSK* CANOPY REFLECTANCE MODELS: LEAF-LEVEL MEASUREMENTS SCALED-UP TO ABOVE-CANOPY LEVEL

The single leaf reflectance and transmittance data collected from the ground-truth deployment were scaled-up to above-canopy level through *SAIL* (Verhoef, 1984) and *Kuusk* (Kuusk, 1996) canopy reflectance models. Optical indices calculated from the simulated above-canopy reflectance were therefore a function of the canopy structure and viewing geometry. Simulated above-canopy optical indices and the corresponding measurements of bioindicators (pigments and fluorescence) from the 440 leaf samples permitted the derivation of prediction algorithms for site bioindicators to be tested on data acquired over the study sites by the hyperspectral *CASI* sensor.

For the application of the *SAIL* and *Kuusk* canopy reflectance models nominal input parameters derived from the study areas were: LAI=3.5, plagiophile leaf angle distribution function (LADF), soil reflectance data derived from

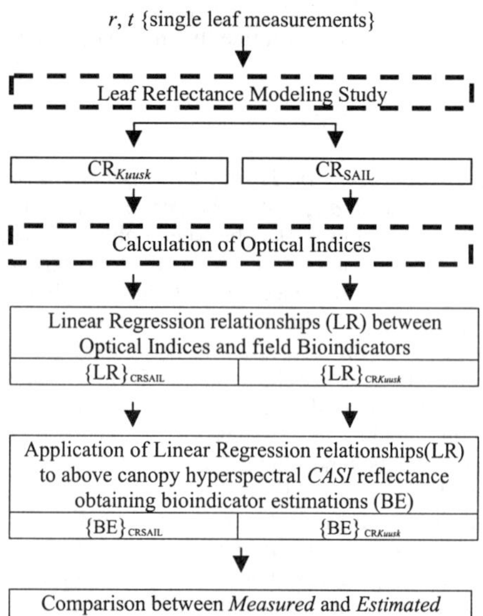

Figure 1. Schematic view of the overall analysis methodology followed in this study. Leaf-level reflectance and transmittance measurements are scaled-up to canopy level through CR models and input parameters related to the canopy structure and viewing geometry. Linear regression Relationships (LR) between optical indices calculated from above-canopy simulated reflectance and ground-truth bioindicators are applied to above-canopy hyperspectral CASI reflectance to obtain bioindicator estimation (BE). Assessment is made comparing ground-truth measured with estimated bioindicators.

CASI imagery and model-estimated skylight irradiance fraction based on conditions during airborne acquisitions. Additional parameters needed in the *Kuusk* model were n=1.4, sl=0.007 & $\theta^*=40°$, and $\varepsilon=0.95$ & $\theta_m=45°$ for the LADF for the assumed plagiophile leaf distribution function.

The schematic view of the simulation study is shown in Figure 1. Single leaf reflectance and transmittance measurements from the field data sampling were used for the simulation of above-canopy reflectance through infinite reflectance and canopy reflectance models *SAIL* and *Kuusk*. Optical indices calculated from above-canopy simulated-reflectance were used to derive linear relationships that are then applied to hyperspectral *CASI* data to obtain bioindicator estimations. Assessment is made comparing *in-field* measured bioindicators with *CASI*-derived estimations.

5 RESULTS AT CANOPY LEVEL FROM HYPERSPECTRAL CASI DATA

CASI data were collected over the study sites within the same period as the field data acquisition. Mean reflectance values per plot were calculated in each *Acer saccharum* M. study site of 20 x 20 m. Data were acquired in the hyperspectral reflectance mode, with 2 m spatial resolution and 72 spectral channels with 7.5 nm spectral resolution. The mean reflectance per plot was calculated selecting the brightest 25% of pixels in the NIR, therefore targeting crowns while minimizing influence of shadows and canopy openings.

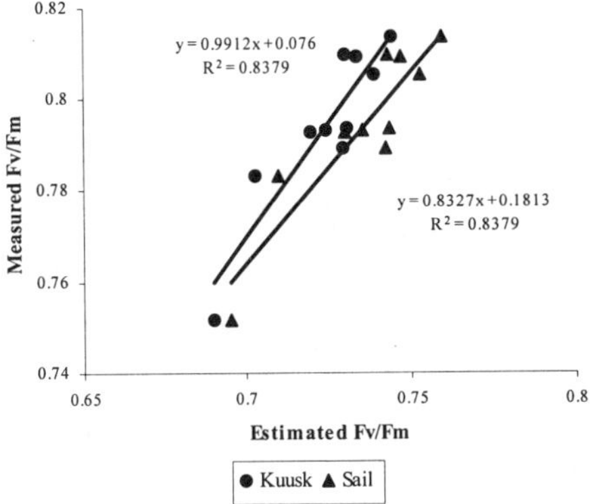

Figure 2. Estimation of Fv/Fm from CASI data using the DP21 ($D_{\lambda o}/D_{703}$) algorithm developed at leaf level through SAIL and Kuusk CR models. Data from *Acer saccharum* M. study sites.

Table 1. Determination coefficients (r>0.4) obtained in chlorophyll-*a*, chlorophyll a&*b*, carotenoids and Fv/Fm estimations applying relationships from *SAIL* and *Kuusk* CR models to *CASI* data collected over *Acer saccharum* M. study sites.

Optical Index	*r* chl-a/ cm²	*r* a&b/ cm²	*r* caro/ cm²	*r* ChFl Fv/Fm
DP21 ($D_{\lambda o}/D_{703}$)	0.64	0.63	0.42	0.91
Vogelmann (R_{740}/R_{720})	0.6	0.6	-	0.9
Vogelmann ($(R_{734}-R_{747})/(R_{715}+R_{726})$)	0.61	0.6	-	0.87
Vogelmann ($(R_{734}-R_{747})/(R_{715}+R_{720})$)	0.61	0.61	-	0.87
Gitelson & Merzylak (R_{750}/R_{700})	0.45	0.44	-	0.84
Carter (R_{695}/R_{760})	-	-	-	0.83
λ_p	0.56	0.56	-	0.82
Vogelmann (D_{715}/D_{705})	0.46	0.45	0.42	0.79
DPR2 ($D_{\lambda o}/D_{\lambda o+22}$),	-	-	-	-0.68
Gitelson & Merzylak (R_{750}/R_{550})	0.45	0.44	-	0.67
DP22 ($D_{\lambda o}/D_{720}$)	-0.55	-0.54	-	-0.63
PRI ($(R_{570}-R_{539})/(R_{570}+R_{539})$)	-	-	-	0.63
DPR1 ($D_{\lambda o}/D_{\lambda o+12}$)	-	-	-	-0.6
PRI ($(R_{531}-R_{570})/(R_{531}+R_{570})$)	-	-	-	0.59
PRI ($(R_{550}-R_{531})/(R_{550}+R_{531})$)	0.59	0.58	-	-
Lichtenthaler (R_{440}/R_{690})	0.46	0.46	-	-
Carter (R_{695}/R_{420})	0.46	0.47	-	-
G (R_{554}/R_{677})	0.46	0.44	-	-

Table 1 shows the determination coefficients between measured and estimated values of chlorophyll-*a*, chlorophyll-*a&b*, carotenoids and leaf fluorescence Fv/Fm derived by applying leaf simulation relationships obtained through *SAIL* and *Kuusk* canopy reflectance models to *CASI* data collected over the 12 *Acer saccharum* M. study sites. As an example, the simulations with the *SAIL* and *Kuusk* CR models when the index DP21 ($D_{\lambda o}/D_{703}$) is used to estimate Fv/Fm is shown in Figure 2. The slope near unity (0.99) indicates the effectiveness of scaling up from leaf-level to canopy level.

6 MODEL PARAMETER SENSITIVITY STUDY

For monitoring of bioindicators by remote sensing changes are expected in θ_s during data acquisition in diurnal and seasonal cycles. LAI variations occur with changes in the canopy conditions, as well as changes in LADF between species. A modeling study was carried out in order to determine the influence of changes in the CR parameters on the calculated optical indices and therefore in the final bioindicator prediction. *SAIL* input parameters were changed to study their sensitivity on the indices derived from the simulated canopy spectra. LAI was stepped from LAI=1 to LAI=8; LADF was changed to simulate 4 different canopy types: planophile ($\theta=0°$, $\varepsilon=0.985$), erectophile ($\theta=90°$, $\varepsilon=0.985$), plagiophile ($\theta=45°$, $\varepsilon=0.95$), and spherical ($\varepsilon=0$). Solar zenith angle θ_s was stepped from $\theta_s=20°$ to $\theta_s=60°$. Two single leaf measurements of reflectance and transmittance collected as part of the experiment were used for the modeling study. The two leaves used in the study were chosen from the 2 study sites that showed highest and lowest Fv/Fm and Chl_a/cm^2 values in the field: Fv/Fm=0.69 and Chl_a/cm^2=17.31 for the high stressed leaf, and Fv/Fm=0.83 and Chl_a/cm^2=32.2 for the low stressed or healthy leaf. The modeling study consisted of the comparison of the predicted bioindicator from optical indices through the *SAIL* CR model with nominal CR parameters relative to the prediction when non-nominal CR parameters are used.

Representative optical indices were selected from the 4 categories used in the modeling study: PRI $(R_{531}-R_{570})/(R_{531}+R_{570})$ and Lichtenthaler (R_{440}/R_{690}) from the Visible Ratios optical indices; Gitelson & Merzylak (R_{750}/R_{550}) and NDVI $(R_{774}-R_{677})/(R_{774}+R_{677})$ from the Visible/NIR ratios; Carter (R_{695}/R_{760}), Vogelmann (R_{740}/R_{720}), Vogelmann $(R_{734}-R_{747})/(R_{715}+R_{726})$ and Gitelson & Merzylak (R_{750}/R_{700}) from the Red Edge Reflectance-Ratio Indices; and DP21 $(D_{\lambda o}/D_{703})$, λ_p and Vogelmann (D_{715}/D_{705}) from the Spectral and Derivative Red Edge Indices. At low LAI (<2) the errors in the predicted bioindicator become large regardless of the type of canopy considered. Figure 3 (top) shows this result using the R_{750}/R_{700} optical index (Gitelson & Merzylak, 1997) for θ_s=30° as a function of canopy type and LAI. For values of LAI higher than 3 the differences between the predicted bioindicator using nominal canopy parameters and the prediction with changed θ_s and LADF canopy parameters is insignificant. Moreover, when the LAI is higher than 3 and the θ_s is a nominal 30° the type of canopy is irrelevant for the bioindicator prediction (Figure 3, top).

For a given plagiophile canopy type (Figure 3 - middle) errors of less than 5% are found in the predicted bioindicator when LAI is higher than 3 and the optical indices used are the Red Edge Reflectance-Ratio Indices and Spectral and Derivative Red Edge Indices (R_{740}/R_{720} optical index, Vogelmann *et al.,* 1993 is shown in the figure). These results indicate that bioindicator prediction in closed canopies with LAI>3 are not affected by LAI variability. This further suggests that LAI variations between the 12 *Acer saccharum* M. study sites (2.85 to 5.17) in a plagiophile canopy have an insignificant effect on the optical indices and on the bioindicator prediction; less than 5% differences are expected relative to predictions made with a nominal LAI=4.

Results also demonstrate the small effect of θ_s especially in red edge spectral and derivative indices (Figure 3, middle and bottom). It shows

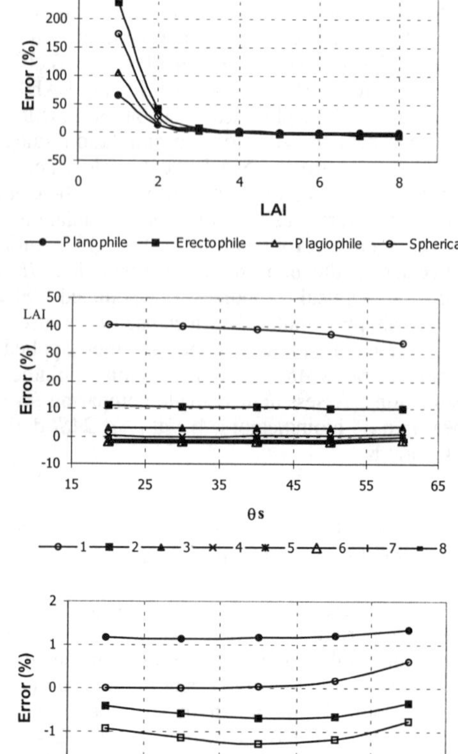

Figure 3. Model parameter sensitivity study shows that low LAI affects bioindicator estimation regardless the type of canopy (top plot, Gitelson & Merzylak index (R_{750}/R_{700}), θs=30°) and θs angle (middle plot, Vogelmann index (R_{740}/R_{720}), plagiophile canopy). When θs=30° and LAI>3 the canopy type is irrelevant (top plot). θs has no effect in the bioindicator estimation as can be seen in the middle plot (LAI ranging from 1 to 8) and bottom plot (DP21 index $(D_{\lambda o}/D_{703})$, LAI=4) for any type of canopy. The % Error shown is the relative error to nominal *SAIL* CR parameters (plagiophile canopy, LAI=4, θs=30°) from the 12 study sites of *Acer saccharum* M. used in the present study, in relation to variable values of LAI, θs and canopy type.

the insignificant variation of the predicted bioindicator when θ_s changes from 20° to 60° where the optical indices used are Vogelmann (R_{740}/R_{720}) and DP21 $(D_{\lambda o}/D_{703})$, respectively. Therefore changes in θ_s from 29° to 41° in the 12 CASI images obtained from the study sites are not expected to affect the bioindicator prediction when θ_s=30° was chosen as nominal input parameter in

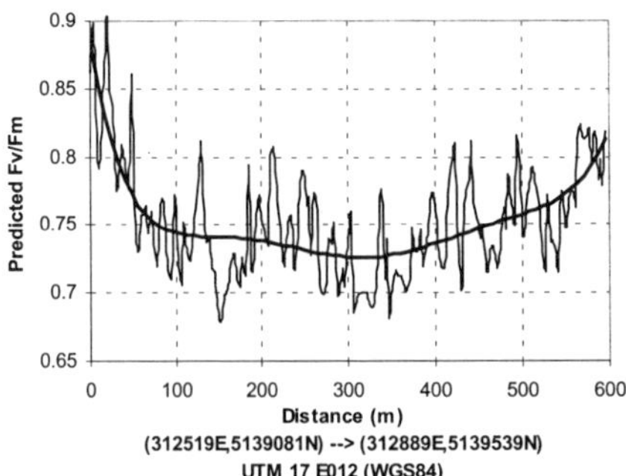

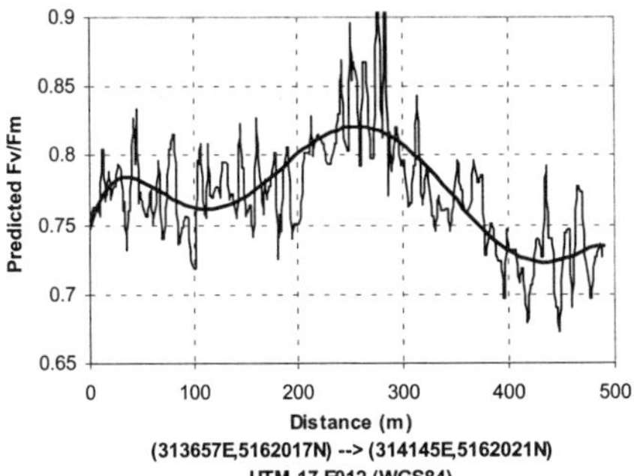

Figure 4. Transect estimation of Fv/Fm in the two study sites showing the lowest (upper plot) and highest (lower plot) values of Fv/Fm. Estimation made from Vogelmann (R_{740}/R_{720}) optical index calculated from *CASI* data scaling-up single-leaf reflectance measurements through *SAIL* CR model.

the CR model when scaling from leaf-level to canopy-level.

This study shows that derivative indices are less sensitive to low LAI values than other optical indices. Vogelmann (D_{715}/D_{705}), λ_p and DP21 ($D_{\lambda o}/D_{703}$) predict the bioindicator with only 15-20% error when LAI=1 relative to the prediction with nominal LAI=4. Non-derivative optical indices show higher errors: for example, Gitelson & Merzylak (R_{750}/R_{550}) 140%, Carter (R_{695}/R_{760}) 150%, Vogelmann (R_{740}/R_{720}) 65%, Gitelson & Merzylak (R_{750}/R_{700}) 200% for LAI=1. This demonstrates that red edge and derivative indices are more suitable for bioindicator prediction and mapping with high spatial resolution hyperspectral remote sensing, where the effects of shadows and canopy openings can be selectively discriminated against in data analysis.

7 VALIDATION OF METHODOLOGY: A TRANSECT DATA COLLECTION

Validation of algorithms in different study sites as well as selective field data acquisition of leaf samples based on the estimations made by optical indices has been planned for the 1999 deployment. An evaluation of the predicative capability of the algorithms and methodology is planned through re-visits of the sites with 1999 field and airborne deployments.

Transect sampling collection across predicted site bioindicator variations has been designed for validation purposes. Two study sites with highest and lowest Fv/Fm have been selected for field data sampling along a 500 m transect. Ground-truth measurements of chlorophyll-*a*, chlorophyll *a&b*, carotenoids and Fv/Fm will be compared to estimations made from airborne *CASI* data through the methodology explained in this paper. Figure 4 shows the estimations of Fv/Fm in the mentioned study sites using the Vogelmann (R_{740}/R_{720}) optical index obtained from *CASI* data through *SAIL* CR model. Both transects from 1998 data show patterns with extreme Fv/Fm values that will allow comparison with the leaf samples.

Additional hyperspectral *CASI* data with 2.5 nm spectral resolution will be collected in the 1999 deployment in all *Acer saccharum* M. study sites. The 2.5 nm fibre spectrometer for leaf-level reflectance and transmittance measurements will allow the study of the spectral resolution influence in the calculation of optical indices. Additional sites of *Pinus strobus* L. and *Pinus banksiana* Lamb. have been included in the deployment in order to study the species dependency of the optical indices and algorithms developed.

8 CONCLUSIONS

This study demonstrates that leaf-level measurements of pigments and fluorescence along with leaf reflectance and transmittance can be used to produce algorithms to estimate these variables through above-canopy spectral reflectance models *SAIL* and *Kuusk*. Canopy structure was shown to play an important role in this link from leaf-level measurements to canopy-level hyperspectral data, affecting optical indices in different ways. A sensitivity study showed that low LAI values are very critical to the accuracy of predicted bioindicator regardless of the considered type of canopy. Differences between the predicted bioindicator using nominal canopy parameters and the prediction with variable θ_s and LADF is insignificant when LAI is higher than 3. Furthermore, the canopy type is irrelevant for the estimations when LAI is higher than 3 and the θ_s is a nominal 30°.

This study demonstrates that the algorithm development methodology presented here using nominal CR parameters from mapping bioindicators at this site is valid: LAI variability from the 12 *Acer saccharum* M. study sites in a plagiophile canopy and θ_s variability in the collected data have an insignificant effect in the optical indices and in the bioindicator prediction.

Finally, this study has shown that derivative indices are less sensitive to low LAI values than other optical indices. This demonstrates that red edge and derivative indices are more suitable for bioindicator prediction and mapping with high spatial hyperspectral remote sensing data.

9 ACKNOWLEDGEMENTS

The authors wish to express our thanks especially to Denzil Irving (OFRI) for invaluable contributions toward the field programme, to Sebastian Martin (*U. of Dundee*, Scotland) and John Harron (CRESTech) for help with the single leaf measurements, and to Paul Shepherd and Lawrence Gray (CRESTech) for contributions toward successful airborne campaigns. Financial support provided by the *Centre for Research in Earth and Space Technology* (CRESTech), the *Ontario Ministry of Natural Resources*, the *Canadian Forestry Service*, and the *Ministry of Environment and Energy* is gratefully acknowledged.

10 REFERENCES

Gitelson, A.A., and M.N. Merzlyak, 1997. Remote estimation of chlorophyll content in higher plant leaves, *International Journal of Remote Sensing*, 18 (12), 2691-2697.

Kuusk, A., 1996. A computer-efficient plant canopy reflectance model. *Computers & Geosciences*, 22(2), pp. 149-163.

Mohammed, G.H., W.D. Binder and S.L. Gillies, 1995. Chlorophyll fluorescence: A review of its practical forestry applications and instrumentation, *Scand. J. For. Res.*, *10*, pp. 383-410.

Mohammed, G.H., P.H. Sampson, S.J. Colombo, T.L. Noland and J.R. Miller, 1997. Bioindicators of forest sustainability: Development of a forest condition rating system for Ontario, Ontario Forest Research Institute, Sault Ste. Marie, ON, Canada.

O'Neill, N.T., F. Zagolski, M. Bergeron, A. Royer, J.R. Miller, and J. Freemantle, 1997. Atmospheric correction validation of casi images acquired over the BOREAS Southern Study Area, *Canadian Journal of Remote Sensing*, 23 (2), 143-162.

Sampson, P.H., G.H. Mohammed, S.J. Colombo, T.L. Noland, J.R. Miller and P.J. Zarco-Tejada, 1998. *Bioindicators of Forest Sustainability Progress Report, Ontario Forest Research Institute*, Sault Ste. Marie, ON, Canada.

Savitzky, A., and M.J.E. Golay, 1964. Smoothing and differentiation of data by simplified least squares procedures, *Analytical Chemistry*, 36 (8), 1627-1639.

Verhoef, W., 1984. Light Scattering by Leaf Layers with Application to Canopy Reflectance Modeling: The *SAIL* Model, *Remote Sensing of the Environment*, 16, pp. 125-141.

Vogelmann, J.E., B.N. Rock and D.M. Moss, 1993. Red edge spectral measurements from sugar maple leaves, *International Journal of Remote Sensing*, Vol. 14 (8), pp. 1563-1575.

Zarco-Tejada, P.J., J.R. Miller, G.H. Mohammed, T.L. Noland, and P.H. Sampson, 1999. Canopy Optical Indices from Infinite Reflectance and Canopy Reflectance Models for Forest Condition Monitoring: Application to Hyperspectral CASI data. *IEEE 1999 International Geoscience and Remote Sensing Symposium*, IGARSS'99, Hamburg (Germany), 28th June - 2nd July 1999.

Remote Sensing in the 21st Century: Economic and Environmental Applications, Casanova (ed.)
 ISBN 90 5809 096 5

Metric relations in INSAR topographic modelling

G. Brandstätter
Department of Photogrammetry and Remote Sensing, Graz University of Technology, Austria
A. I. Sharov
Applied Cosmonautics Faculty, Moscow State University of Geodesy and Cartography, Russia

ABSTRACT: Scalar formulation of metric relations having relevance to satellite topographic interferometry served as a basis for several original concepts, which provided an important aid for significant methodological improvements and simplification to the main operations of the INSAR data processing, including flat terrain phase correction, interferometric phase unwrapping and differential analysis. Topographic heights and slopes of several of the largest ice caps in Franz Josef Land, Russian High Arctic were derived from the repeat-pass INSAR data acquired in September and October 1995 without previous phase unwrapping yet without compromising on accuracy.

1 INTRODUCTION

Satellite radar interferometry (INSAR) is regarded as a very promising technique for precise topographic modelling and monitoring of rapid terrestrial changes in various regions of the globe. Apart from wide geographic and temporal coverage, the INSAR method provides an appropriate ground resolution and high vertical accuracy of spatial modelling, which render it highly suitable for topographic studies of barren landscapes, e.g. in polar regions (Hartl & Thiel 1993).

The generation of a precise digital elevation model (DEM) of some track of land is considered the primary goal of INSAR topographic modelling. Nearly all operational topographic applications of INSAR are accordingly focussed on the direct conversion of an interferogram into DEM, which is generally performed on a pixel-by-pixel basis using sophisticated algorithms for the interferometric phase unwrapping. Current research activities are mostly concentrated on the efficient implementation of the unwrapping procedure (Hellwich, personal communication 1998), and, apart from a few thematic studies, comparatively little attention has been paid to extracting reliable terrestrial information from the interferogram itself.

However, there is a group of special interferometric applications that do not require phase unwrapping, e.g. determination of extreme elevations and measurement of terrain slopes, coherency mapping and differential interferometry. Besides, the relative measurement accuracy in the interferogram with original spatial resolution is generally higher than that in multistage derivatives of the interferometric picture. The structural elements of the ground surface, e.g. highest positions, coastlines etc., primarily detected and measured in the original interferogram could be used for the control of unwrapping errors and may be included into DEM as constraints in order to morphologically optimise the model and improve its accuracy.

Several basic questions concerned with the exploitation of spaceborne interferometric data have yet to be answered. Which terrestrial features can be detected and which topographic parameters can be reliably derived from the interferogram prior to phase unwrapping? How could we simplify the INSAR procedure and obtain rapid access to results, yet without compromising on accuracy? Which basic relationships between the interferometric picture and the three-dimensional surface of the land are of importance in this context?

The present paper generalises the results and experience gained in attempting to answer these questions. Main emphasis has been placed on geometric and radiometric equations relating the quantities observed in spaceborne INSAR products with real topographic unknowns. Theoretical results were verified and illustrated by practical processing of 3 radar image pairs taken on September 3/4 and

October 8/9, 1995 over the Franz Josef Land archipelago (FJL), Russian High Arctic by the synthetic aperture radar (SAR) installed on board the European ERS-1/2 satellites. Several spaceborne photographs acquired over FJL by the KATE-200 film camera from the Russian "Resource-F1" satellite were additionally used in order to verify the results of interferometric modelling. Image processing has been accomplished using the ADOBE PhotoShop 5.0, ENVI 3.0, PHOTOMOD 1.6 and RSG 3.21 software packages.

2 INSAR DATA PROCESSING: BACKGROUND

The INSAR method is based on the generation of an interferogram, i.e. specific "striped picture", by combining two complex radar images, which contain the information on amplitude and phase of radio signals reflected from the Earth's surface. Each fringe in the resultant interferogram corresponds to a certain phase difference between radio signals, which in its turn depends on the elevation of terrain. The path difference δ between two radio signals and the corresponding interferometric phase ψ are related by the simple equation

$$\delta = \frac{\lambda}{4\pi}\Psi \qquad (1)$$

and theoretically can be measured with a relative accuracy of some tenths of a wavelength λ in the centimetre range. Under favourable conditions, e.g. in terrain with a relatively flat topography, the vertical accuracy of spaceborne interferometric determinations may reach up to several meters and even the sub-meter range. It was reported, for example, that a relative elevation accuracy on the order of 2.5 m and absolute accuracy of about 4 m could be achieved over extensive glacier sheets with ERS-1-SAR repeat-pass interferometry (Joughin et al. 1996).

The multistage interferometric processing of complex radar images is often considered a somewhat complicated technique, since it involves up to 7-8 obligatory steps and is exceptionally intolerant of any imperfections in data acquisition and processing. The principal stages of the standard interferometric processing procedure is usually specified as follows (See, for example, Kenyi et al. 1996):

1) preliminary estimation of the relative position of satellites and the interferometric baseline using orbital data;
2) SAR image co-registration and superimposition using automatic correlation procedures, sub-pixel resampling and polynomial interpolation;
3) calculation of a complex interferogram as the product of one complex SAR image with the complex conjugate of a second and determination of phase differences in the image pair;
4) flat terrain phase (FTP) correction, which implies the removal of the phase term related solely to the SAR imaging geometry with respect to the reference surface;
5) representation of intermediate INSAR products, i.e. amplitude image, coherency image and fringe image, using optional phase-noise filtering, interpolation of phase values in unreliable areas, multi-looking and the linear phase coding algorithm;
6) interferometric phase unwrapping, i.e. conversion of the interferometric picture into a continuous (non-striped) topogram* by means of successive integration of phase differences between adjacent fringes;
7) topogram geocoding, separate measurements and model adjustment based on ground control data;
8) DEM generation and quality control.

*) The term *topogram* is suggested here in order to determine a picture, derivative of the fringe image, representing the unwrapped phase, which can be unambiguously related to topographic height on a pixel-by-pixel basis.

The interferometric *baseline*, i.e. the separation of points from which two images are acquired, is one of the most important INSAR parameters, which must be known from the very beginning of data processing. In repeat-pass interferometry, the baseline length is not a constant value over the imaged area because of the non-parallelism of real orbits. Depending on application, the baseline length must be known with a precision of 5-10 cm. However, the uncertainty in the relative position of ERS-1/2 satellites determined on the basis of precise orbit data amount to 30-35 cm at best (Reigber et al. 1996). This makes additional control necessary and the exact local value of the baseline is usually estimated via tie points (stage 7).

The baseline decorrelation effect limits the range of optimal interferometric baselines. According to our practical experience, the length of a baseline lying in the range of 50 to several hundreds of meters is preferable for topographic applications in the High Arctic (Brandstätter & Sharov 1998). Figure 1 shows some typical results of the interferometric processing of complex SAR images acquired by ERS-1/2 satellites on September 3/4 and October 8/9, 1995 with a time interval of 24 hours

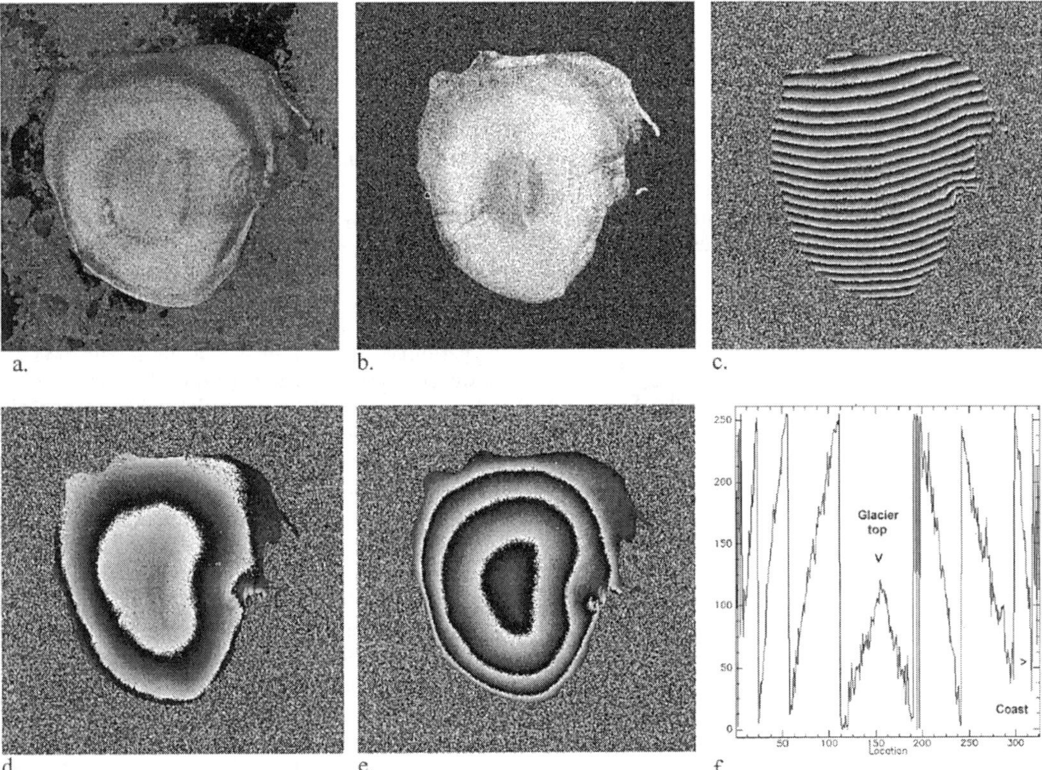

Figure 1. ERS-1/2-INSAR data of Rainer Island, FJL: precision image (a), coherency image (b), primary interferogram (c); fringe images generated from SAR data taken on September 3/4 (d) and October 8/9, 1995 (e); radiometric profile taken across the fringe image (f).

and approximate baselines of 53 m and 130 m, respectively.

Due to the typically high rate of fringes, the direct interpretation of terrestrial features in the primary interferogram is practically impossible (Fig. 1 c). On the other hand, most terrestrial features are represented in the amplitude and coherency image, but the height measurement cannot be performed there (Fig. 1, a, b). Some topographic elements, e.g. points with extreme elevations are better detectable and can be measured in the fringe image after the flat terrain phase correction is performed (Fig. 1 d, e). The analysis of the general shape of an ice cap is also possible in this INSAR product.

Thus, only simultaneous use of all these products can provide the accurate delineation and measurement of glacial surfaces prior to phase unwrapping. The geometric identity of INSAR products ensures the high quality of joint processing and the similarity of metric relations for different combinations. In the next chapters, the basic operations and specific problems involved at stages 2-5 of the INSAR procedure will be treated.

3 COMPLEX CROSS-CORRELATION OF RADAR IMAGES

A cross-correlation between two single-look complex (SLC) radar images is considered the key procedure of INSAR data processing, since it serves as a basis for both the SLC-image matching and the generation of the complex interferogram and coherency image (Raggam et al. 1998). The next consideration is of special interest in this context.

Let SAR emit monochromatic radio signals with a frequency ω

$$s_1(t) = s_2(t) = e^{-i\omega t}, \qquad (2)$$

where $\omega = 2\pi c/\lambda$ and c is the velocity of light. Then two complex radio signals (or images) registered at different instants can be expressed as

$$f_1(t) = a_1(t) + ib_1(t) = u_1 \cdot e^{-i\omega t} \cdot e^{i\frac{4\pi}{\lambda}\rho_1} \quad \text{and}$$

$$f_2(t) = a_2(t) + ib_2(t) = u_2 \cdot e^{-i\omega t} \cdot e^{i\frac{4\pi}{\lambda}\rho_2}, \quad (3)$$

where u_1, u_2 denote the backscattering coefficients of the land surface at different instants of survey, and ρ_1, ρ_2 are the corresponding slant ranges.

Normalised function of cross-correlation between registered radio signals can be written as follows

$$R_{12}(\tau) = \lim_{T\to\infty} \frac{1}{2T} \int_{-T}^{T} f_1(t) f_2^*(t+\tau)dt = \quad (4)$$

$$= u_1 u_2^* e^{i\omega\tau} e^{\frac{4\pi}{\lambda}(\rho_1-\rho_2)} \lim_{T\to\infty} \frac{1}{2T} \int_{-T}^{T} dt = u_1 u_2^* e^{i\omega\tau} e^{\frac{4\pi}{\lambda}(\rho_1-\rho_2)},$$

where * denotes a complex conjugate. If $u_1 = u_2^* = u$, i.e. the backscattering, did not change during the time interval between surveys, then

$$R_{12}(\tau) = |u|^2 e^{i\omega\tau} e^{-\frac{4\pi}{\lambda}\delta} \quad \text{with} \quad (5)$$

$\delta = \rho_2 - \rho_1$ - the path difference for a single point.

When translated into two-dimensional form, the equations (4, 5) represent an interferometric picture with bright fringes corresponding to the maximum value of the complex correlation function and dark fringes corresponding to its minimum (Fig. 3, c, e). The normalised value of complex cross-correlation is often used as an indicator of the coherency, i.e. the capability of two monochromatic signals (or images) to interfere. The complex correlation coefficient is defined as (Lee et al. 1998)

$$r_{12} = \frac{E\{f_1 \cdot f_2^*\}}{\sqrt{E\{|f_1|^2\} \cdot E\{|f_2|^2\}}} = \gamma \cdot e^{i\phi}, \quad (6)$$

where f_1 *and* f_2 denote the pair of SLC images, γ and ϕ are respectively the magnitude (or coherency) and the phase of the complex correlation coefficient. The value of coherency γ between two SLC images determines the fringe contrast in the resultant interferogram; fringe contrast increases with the coherency. The point-wise relation between the visibility of interferometric fringes V and the value of coherency appears as follows (Landsberg 1976)

$$V = \frac{E_{\max} - E_{\min}}{E_{\max} + E_{\min}} = \frac{2\sqrt{I_1 \cdot I_2}}{I_1 + I_2} \cdot \gamma, \quad (7)$$

where E_{max} and E_{min} - the maximal and the minimal brightness of fringes; I_1 and I_2 - the intensity of two complex images at conjugate points.

If $I_1 = I_2$, then $\gamma = V$. The coherency value $\gamma \geq 0.5$ provides $E_{\max} \geq 3E_{\min}$, which ensures acceptable contrast. Spatial and physical changes of the land surface during the time interval between surveys as well as other effects bringing about random variations in the interferometric phase ψ reduce the coherency and all areas undergoing such changes are reproduced by mostly dark-grey values in the coherency image (Fig. 3, b).

In practical work, the procedure of complex correlation is utilised in order to match two radar images constituting the interferometric pair and to accomplish the image co-registration, which has to be performed with sub-pixel accuracy. This task is, however, very problematic due to the speckle (phase) noise, which limits spatial resolution of SAR image data and renders the identification of conjugate points in the interferometric image pair rather troublesome.

Asides from the long computing time the automatic complex correlation of SLC images sometimes brings unsatisfactory results, especially, when image data with extensive areas of low coherency is processed. Frequently, no conjugate points can be identified in such images at all or pseudo-homologous points are detected; the latter leads to false image matching and consequent image misregistration. In both cases, precise image co-registration is impossible, the interferogram generation fails and the overall conclusion of "too low coherency" is drawn all too frequently.

The speckle noise can be suppressed to some degree by filtering or multi-looking. However, this does not help because multi-look processing reduces the spatial resolution and does not improve the coherency. Besides, it is difficult to separately evaluate the impact of phase and amplitude decorrelation on the final result (see Eq. 4), and the tuning of the complex correlator is not easy. An interactive correlation of the single-look amplitude images derived from original SLC image data sometimes provides better results in a complicated case.

4 INSAR GEOMETRIC RELATIONS

For purposes of simplicity the INSAR geometry is treated separately in the range and azimuth direction and the following initial assumptions are made:

- radio "rays" propagate directly along the same path on their way from the radar system to the ground and back;
- side physical effects and signal processing imperfections are negligible;

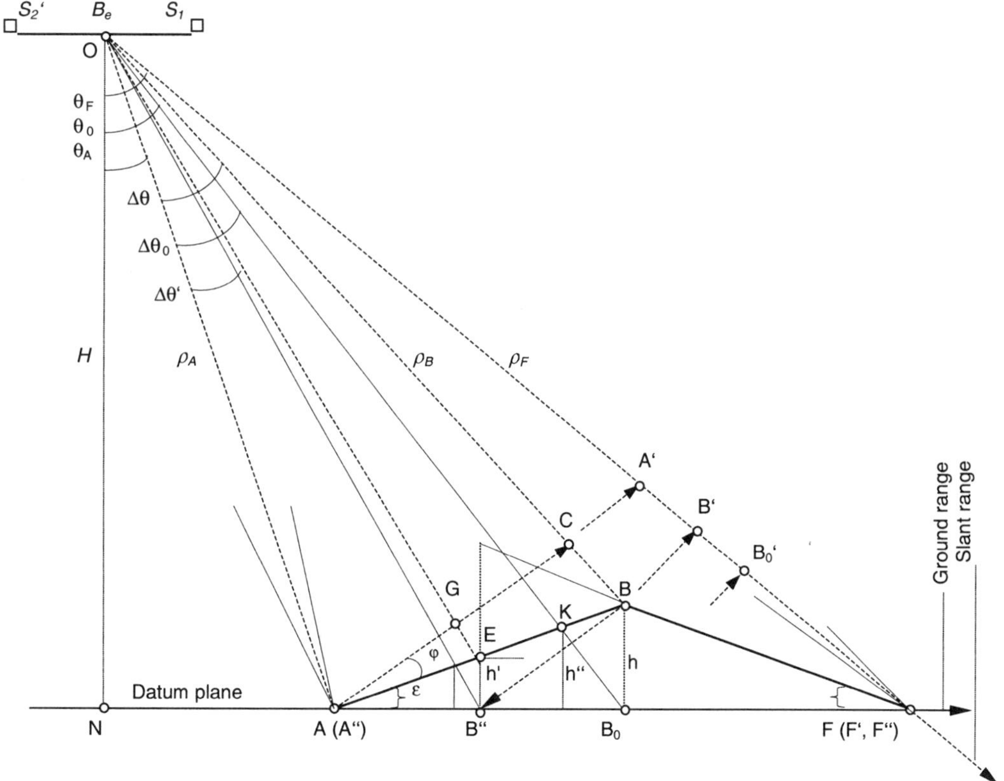

Fig. 2. Range geometry of INSAR survey (not to scale)

- the topographic situation is represented by a single symmetric insular ice cap surrounded by sea;
- terrestrial slope is constant and small, $0 < \varepsilon \leq 5°$;
- maximum elevation of the ice cap is defined as $h = \mathrm{AB}_0 \cdot \sin\varepsilon$; the slope projection $\mathrm{AB}_0 \leq 6$ km.

The latter suppositions conform to the real topographic situation in FJL, where the glacial topography of numerous ice caps and domes is mostly homogeneous and the height increases unvaryingly from sea level to the glacier top (Figure 1, a). The geometric disposition of the INSAR survey in the range direction is shown in Figure 2.

In repeat-pass interferometry radar images are successively acquired from two neighbouring orbital points S_1 and S_2. The real difference in altitude positions of satellites S_1 and S_2 may be taken into account by introducing a virtual baseline B_e as offered in (Hartl and Thiel, 1993)

$$B_e = B_x + B_y \cdot \tan\theta \,, \qquad (8)$$

where B_x and B_y are horizontal and vertical components of a baseline, respectively. The value of the local incidence angle θ is defined as follows

$$\theta \cong 0.5 \cdot (\theta_1 + \theta_2) \,, \qquad (9)$$

where θ_1 and θ_2 are the local incidence angles at the instant of the first and second survey, respectively. Equations (8) and (9) hold for any point of the reference surface (A, B_0, F) and show that the value of baseline B_e varies over the scene depending on the current incidence angle. In Figure 2, points A', B', F' and A'', B'', F'' are respectively the slant- and the ground-range representations of terrestrial points A, B, F.

4.1 *General path difference for two terrestrial points*

The general path difference for the points A and B, both surveyed from two positions S_1 and S_2' can be determined as

$$\delta_{BA} = (\mathrm{S}_2^{'}\mathrm{B} - \mathrm{S}_1\mathrm{B}) - (\mathrm{S}_2^{'}\mathrm{A} - \mathrm{S}_1\mathrm{A}) = \delta_B - \delta_A \cong \mathrm{CB} \,. \qquad (10)$$

The following relations between distances

S'_2B, S_1B, S'_2B, S_1B can be written

$$\left.\begin{aligned}(S'_2B)^2 &= (H-h)^2 + (NA + AB_0 + 0.5\cdot B_e)^2\\ (S_1B)^2 &= (H-h)^2 + (NA + AB_0 - 0.5\cdot B_e)^2\\ (S'_2A)^2 &= H^2 + (NA + 0.5\cdot B_e)^2\\ (S_1A)^2 &= H^2 + (NA - 0.5\cdot B_e)^2\end{aligned}\right\} \quad (11)$$

and

$$\left.\begin{aligned}(S'_2B)^2 - (S_1B)^2 &= (S'_2B - S_1B)(S'_2B + S_1B) = 2B_e(NA + AB_0)\\ (S'_2A)^2 - (S_1A)^2 &= (S'_2A - S_1A)(S'_2A + S_1A) = 2B_e\cdot NA\end{aligned}\right\}.$$

Then

$$S'_2B - S_1B = \frac{2B_e(NA + AB_0)}{S'_2B + S_1B};\; S'_2A - S_1A = \frac{2B_e\cdot NA}{S'_2A + S_1A}. \quad (12)$$

Assuming that $B_e << H$ and $AB_0 << NA$, we obtain

$$S'_2B + S_1B \cong 2\cdot OB = 2\rho_B;\quad S'_2A + S_1A \cong 2\cdot OA = 2\rho_A$$

and

$$\left.\begin{aligned}\delta_B &= \frac{B_e\cdot(NA + AB_0)}{\rho_B} = B_e\cdot\sin(\theta_A + \Delta\theta)\\ \delta_A &= \frac{B_e\cdot NA}{\rho_A} = B_e\cdot\sin\theta_A\end{aligned}\right\}. \quad (13)$$

By using (13), the expression (10) can be rewritten as follow (14)

$$\delta = B_e[\sin(\theta_A + \Delta\theta) - \sin\theta_A] = 2B_e\cdot\sin\frac{\Delta\theta}{2}\cdot\cos(\theta_A + \frac{\Delta\theta}{2})$$

AB_0 is not large. Hence, we can employ a small angle approximation and write

$$\delta \cong B_e\cdot\Delta\theta\cdot\cos\theta_A. \quad (15)$$

The angular measure of the slope AB $\Delta\theta$ can be defined as

$$\Delta\theta = \frac{AC}{\rho_A} = \frac{AC}{H}\cdot\cos\theta_A, \quad (16)$$

with $AC = AB\cdot\cos\varphi$.

In ΔABC, the angle $\varphi = \theta_A - \varepsilon + 0.5\cdot\Delta\theta \cong \theta_A - \varepsilon$, and therefore

$$\Delta\theta \cong \frac{h}{H}\cdot\frac{\cos\theta_A\cdot\cos(\theta_A - \varepsilon)}{\sin\varepsilon}. \quad (17)$$

Finally, the general path difference between points B and A can be calculated from one of the following equivalent formulas

$$\left.\begin{aligned}\delta_{BA} &\cong B_e\cdot\frac{h}{H}\cdot\frac{\cos^2\theta_A\cdot\cos(\theta_A - \varepsilon)}{\sin\varepsilon}\\ \delta_{BA} &\cong B_e\cdot\frac{a}{H}\cdot\cos^2\theta_A\cdot\cos(\theta_A - \varepsilon)\\ \delta_{BA} &\cong B_e\cdot\frac{d}{H}\cdot\frac{\cos^2\theta_A\cdot\cos(\theta_A - \varepsilon)}{\cos\varepsilon}\end{aligned}\right\}, \quad (18)$$

where $a = AB$ and $d = AB_0$.

Practical calculations have shown that for the orbital altitude $H = 785$ km and maximal elevation $h = 600$ m a.s.l., the angle $\Delta\theta$ does not exceed 0.02° and the relative mistake introduced by the small angle approximation will not exceed 0.05%. This means that, in most practical cases, the approximate formula (18) may be used instead of the stringent equation (14).

In accordance with equation (1), the value of interferometric phase Ψ_{BA} corresponding to δ_{BA} can be expressed as

$$\Psi_{BA} \cong \frac{4\pi B_e}{\lambda}\cdot\frac{h}{H}\cdot\frac{\cos^2\theta_A\cdot\cos(\theta_A - \varepsilon)}{\sin\varepsilon}. \quad (19)$$

4.2 *FTP correction*

The procedure of FTP correction is performed in order to remove the phase modulation of the primary interferogram, which results solely from SAR imaging geometry, and to preserve the phase term related to the local topography. In order to diminish the influence of the Earth's curvature and to reduce computing times, only small image fragments, corresponding to separate islands of the FJL archipelago were processed in this work.

The value of the interferometric phase ψ_T related to the local topography (slope) between points A and B is given by

$$\Psi_T = \Psi_{AB} - \Psi_0 = \frac{4\pi}{\lambda}\cdot(\delta_{BA} - \delta_0), \quad (20)$$

where the phase term ψ_0 is called "flat terrain phase". The FTP correction is generally performed for the whole scene (island) and the general path difference corresponding to the "flat terrain phase" is defined as

$$\delta_{FA} = \delta_F - \delta_A = B_e\cdot(\sin\theta_F - \sin\theta_A). \quad (21)$$

Nevertheless, the path difference δ_{FA} can be divided into two components as follows

$$\delta_{FA} = \delta_{B_0A} + \delta_{FB_0} \quad\text{or}\quad \delta_{FA} = \delta_{B''A} + \delta_{FB''}. \quad (22)$$

Point B'' can be directly identified in image data and the second expression in (22) is preferred in our case, i.e.

$$\delta_0 = \delta_{B''A} = \delta_{B''} - \delta_A = B_e\cdot[\sin(\theta_A + \Delta\theta') - \sin\theta_A] \quad (23)$$

and

$$\delta_0 = B_e\cdot\Delta\theta'\cdot\cos\theta_A. \quad (24)$$

The value of angle $\Delta\theta'$ can be determined from the following formula

$$\Delta\theta' \cong \frac{AG}{\rho_A} = \frac{AG}{H}\cdot\cos\theta_A, \text{ with} \qquad (25)$$

$$AG \cong AE\cdot\cos\varphi \cong AE\cdot\cos(\theta_A-\varepsilon) = \frac{h'}{\sin\varepsilon}\cos(\theta_A-\varepsilon).$$

Then $\delta_{B''A} \cong B_e\cdot\frac{h'}{H}\cdot\frac{\cos^2\theta_A\cdot\cos(\theta_A-\varepsilon)}{\sin\varepsilon}$. (26)

Analogously, it can be shown that the path difference between points B_0 and A may be derived from the next formula

$$\delta_{B_0A} \cong B_e\cdot\frac{h''}{H}\cdot\frac{\cos^2\theta_A\cdot\cos(\theta_0-\varepsilon)}{\sin\varepsilon}, \qquad (27)$$

which is very similar to expression (26). In (27), the value of h'' is defined from ΔAKB_0 using the following well-known formula from terrestrial photogrammetry (Horn 1989)

$$h'' = \frac{AB_0\cdot\sin\varepsilon\cdot\cos\theta_0}{\cos(\theta_0-\varepsilon)}. \qquad (28)$$

By subtracting (26) from (18) we obtain

$$\delta_T = \delta_{BA} - \delta_{B''A} \cong B_e\cdot\frac{h-h'}{H}\cdot\frac{\cos^2\theta_A\cdot\cos(\theta_A-\varepsilon)}{\sin\varepsilon}, \qquad (29)$$

where $h - h' = B''B_0\cdot\tan\varepsilon \cong h\cdot\cot\theta_A\cdot\tan\varepsilon$.

Therefore, $\delta_T \cong B_e\cdot\frac{h}{H}\cdot\frac{\cos^2\theta_A\cdot\cos(\theta_A-\varepsilon)}{\tan\theta_A\cdot\cos\varepsilon}$. (30)

Finally, the interferometric phase related to the local topography between points A and B can be derived from

$$\Psi_T = \frac{4\pi}{\lambda}\delta_T \cong \frac{4\pi B_e}{\lambda}\cdot\frac{h}{H}\cdot\frac{\cos^2\theta_A\cdot\cos(\theta_A-\varepsilon)}{\tan\theta_A\cdot\cos\varepsilon}. \qquad (31)$$

Since the incidence angle is always determined with respect to the reference surface (mean sea level), index A will be omitted in the next equations.

4.3 *Height interval. Width of interferometric fringes*

In the fringe image, the interferometric phase of 2π corresponds to one fringe. The height interval e corresponding to one interferometric fringe depends on the terrestrial slope and may be derived from the following formula

$$e = \frac{H\cdot\lambda\cdot tg\theta\cdot\cos\varepsilon}{2B_e\cdot\cos^2\theta\cdot\cos(\theta-\varepsilon)}, \qquad (32)$$

which is obtained by inserting $h = e$ and $\psi_T = 2\pi$ into equation (31) or $h = e$ and $\delta_T = \frac{\lambda}{2}$ into (30).

The width of interferometric fringes in the primary slant-range interferogram W_{s0} and in the fringe image W_{sT} can be determined by inserting $h \cong \frac{c\cdot\sin\varepsilon}{\sin(\theta-\varepsilon)}$, where c = BC, into the equations (19) and (31), respectively, as follows

$$\left.\begin{aligned} W_{s0} &\cong \frac{H\cdot\lambda\cdot\tan(\theta-\varepsilon)}{2B_e\cdot\cos^2\theta} \\ W_{sT} &\cong \frac{H\cdot\lambda\cdot\tan\theta\cdot\tan(\theta-\varepsilon)}{2B_e\cdot\cos^2\theta\cdot\tan\varepsilon} \end{aligned}\right\}. \qquad (33)$$

For ground-range representation the width of fringes in the primary interferogram W_{g0} and in the fringe image W_{gT} is defined by substituting $h = d\cdot\tan\varepsilon$ as follows

$$\left.\begin{aligned} W_{g0} &\cong \frac{H\cdot\lambda\cdot\cos\varepsilon}{2B_e\cdot\cos^2\theta\cdot\cos(\theta-\varepsilon)} \\ W_{gT} &\cong \frac{H\cdot\lambda\cdot\tan\theta\cdot\cos\varepsilon}{2B_e\cdot\tan\varepsilon\cdot\cos^2\theta\cdot\cos(\theta-\varepsilon)} \end{aligned}\right\}. \qquad (34)$$

4.4 *Derivation of terrain slopes. Detection of highest positions*

The fringe width W_{s0} or W_{sT} can be measured directly in the primary interferogram or in the fringe image. Therefore, a derivation of the terrain slope is possible without phase unwrapping. In order to express ε as a function of W_{sT} the equation (33) is transformed as follows

$$\frac{\tan\varepsilon + \tan^2\varepsilon\cdot\tan\theta}{\tan\theta - \tan\varepsilon} = \frac{H\cdot\lambda\cdot\tan\theta}{2B_e\cdot W_{sT}\cdot\cos^2\theta}$$

and
$$\tan^2\varepsilon + (\cot\theta + E)\cdot\tan\varepsilon - E = 0 \qquad (35)$$

with
$$E = \frac{H\cdot\lambda\cdot\tan\theta}{2B_e\cdot W_{sT}\cdot\cos^2\theta}.$$

The solution of quadratic equation (35) is given by

$$\varepsilon_{1,2} = \arctan\left\{-\frac{E + \cot\theta \pm \sqrt{(E+\cot\theta)^2 + 4E}}{2}\right\}. \qquad (36)$$

Positive or negative values of ε imply the direction of slope toward or away from the sensor, respectively.

When the glacier slope ε is known, the height interval can be calculated using (32) and the elevation of an ice cap with respect to the current sea level can be simply determined as follows

$$h \cong k\cdot e, \qquad (37)$$

where k is the fractional number of interferometric fringes enclosed between two points of the fringe image corresponding to the glacier top and the shoreline. Detecting extreme pixel values in the fringe image in the vicinity of the presumed summit allows the highest position on the glacier to be recognised.

The value of the virtual baseline B_e can be selected or changed, e.g. in differential interferometry, so that the height interval e will exceed the maximum elevation in the area. Then $k \leq 1$, which means that the procedure of interferometric phase unwrapping can be avoided, since the ambiguity in phase representation disappears, if a *"single-fringe interferogram"* is generated.

4.5 *INSAR geometric relations in azimuth direction*

In order to describe the INSAR geometry in azimuth direction the following relations may be obtained under assumptions $\varepsilon = 0$, $\varepsilon_a \neq 0$, $h = d_a \cdot \tan \varepsilon_a$

$$\delta_{aT} \cong B_e \cdot \frac{h}{H} \cdot \frac{\cos^3 \theta}{\tan \theta} = B_e \cdot \frac{d_a}{H} \cdot \frac{\cos^3 \theta \cdot \tan \varepsilon_a}{\tan \theta}, \quad (38)$$

$$W_{aT} \cong \frac{H \cdot \lambda \cdot \tan \theta}{2B_e \cdot \cos^3 \theta \cdot \tan \varepsilon_a}, \quad (39)$$

$$\varepsilon_a = \arctan\left(\frac{H \cdot \lambda \cdot \tan \theta}{2B_e \cdot W_{aT} \cdot \cos^3 \theta}\right), \quad (40)$$

$$e_a \cong \frac{H \cdot \lambda \cdot tg\theta}{2B_e \cdot \cos^3 \theta}. \quad (41)$$

5 SOME RESULTS OF PRACTICAL APPLICATIONS AND QUALITY CONTROL

The system of equations (1-41) has been practically applied to the methodological development in performing FTP correction, phase unwrapping and differential interferometry. The development is based on the main principle of differential interferometry, which states that, if properly done, the linear combination of original interferograms results in a new secondary interferogram (Landsberg 1976).

For instance, the FTP correction may be accomplished by differentiating between the primary interferogram and a simulated interferogram with straight fringes being perpendicular to the range direction. The approximate width of interferometric fringes in the simulated interferogram can be derived from equations (22-28).

An iterative approach was used in order to determine the appropriate scale of the simulated interferogram for different slope directions. The width of fringes in the simulated interferogram influences the number of fringes in the resultant fringe image so that a single-fringe interferogram can be generated either for the direct or opposite slope. Some results of the differential approach to INSAR data processing, including the simulated interferogram called a *phase filter,* are shown in Fig. 3 a, b, c. We think that such an approach could provide an original solution for the phase unwrapping problem. The linear phase-coding algorithm should be modernised in order to remove the narrow "traces" of original fringes, which can be seen in the secondary fringe image.

All basic relations (7, 32, 33, 36, 39 - 41) were verified during our practical work by comparing radiometric profiles taken across and along the fringe images of September 3/4 and October 8/9, 1995. All profiles were drawn over the tops of several of the largest ice caps in FJL so that each profile began and ended at the shore. A typical transverse ice cap profile representing the pixel value in the fringe image versus the horizontal location is given in Figure 1, f.

Careful analysis of radiometric profiles has clearly shown the reliability of all equations offered. Moreover, the absolute elevation values of several of the largest glaciers, e.g. Moscow Ice Dome (424.7 m), Ice Cap No. 27 at Mac-Clintock Island (449.6 m), Wüllerstorf Mountains (604.0 m), Peak Parnas (619.5 m) etc., were defined and compared with those given in available topographic maps. However, the insufficient number of INSAR data sets and the lack of reliable ground control did not allow conclusions to be drawn on the absolute accuracy of interferometric measurements in the dynamic glacial environment of FJL.

Original comparative studies using the spaceborne stereoscopic photographs acquired over FJL by the KATE-200 film camera from the Russian "Resource-F1" satellite were recently carried out in order to verify the results of interferometric modelling. Prior to comparison, stereophotographs were oriented, reduced to the normal case, superimposed and represented in anaglyphic colours using dithering functions. The resultant stereo composite was subjected to high-pass filtering, which resulted in the fringe pattern or quasi-interferogram shown in Figure 3, d. The optical fringe pattern was converted into the system of isolines called *isoclinals*, which were overlaid on the fringe image (Figure 3, e, f). The combination showed a good correspondence between the radar interferogram and the optical fringe pattern and thus provided certain evidence on the similarity between the stereophotogrammetric and interferometric method of image analysis.

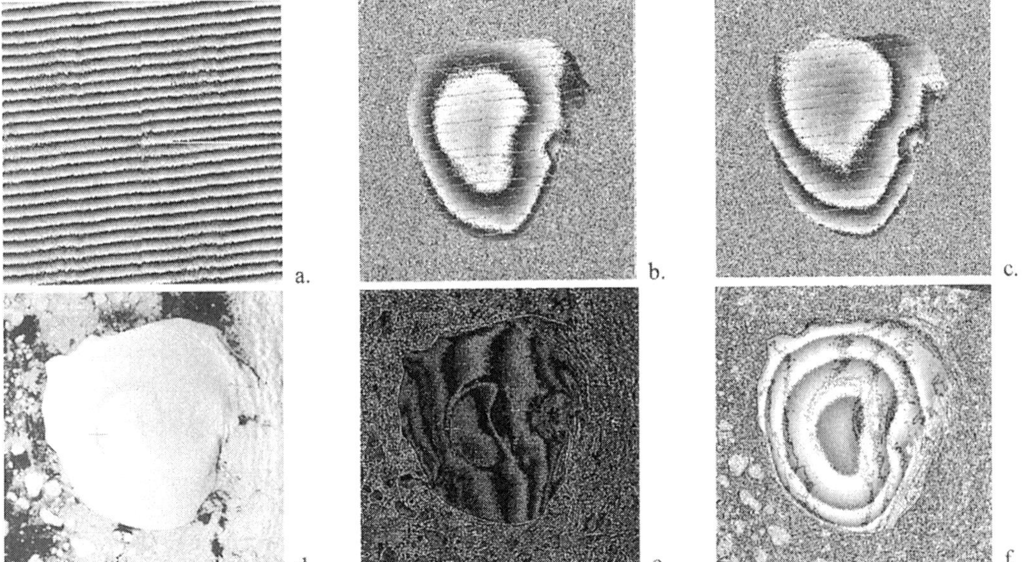

Figure 3. Some results of practical applications: synthetic phase filter (a), fringe images derived from the original interferogram (Figure 1, c) with the phase filter scaled 102% (b) and 109% (c); KATE-200 photograph of Rainer Island; optical fringe pattern (e); fringe image with the isoclinals overlaid (f)

6 CONCLUSIONS

An original set of scalar equations relating general features of interferometric pictures with real topographic unknowns was derived and served as a basis for a simplified theoretical model of INSAR surveying. The knowledge gained from this model, e.g. a new concept of a single-fringe interferogram, combined with experimental results, allow us to propose significant simplifications and methodological improvements to INSAR data processing. Further investigations will be focussed on studying relations between stereogrammetric and interferometric methods as well as the retrieval of reliable topographic information from INSAR data without precise knowledge of satellite orbits.

REFERENCES

Brandstätter, G. & A.I.Sharov 1998. Environmental monitoring in the High Arctic using different types of high-resolution satellite imagery. IAPRS, v.32, Part 7, 201-210.

Hartl, Ph. & K.-H.Thiel 1993. Radar interferometry. Basic concept and applications. ISPRS, v.15, Hamburg, 207-232.

Horn, B. 1989. *Robot vision*. Moscow: Mir (in Russ).

Joughin, I., D.Winebrenner & M.Fahnestock 1996. Measurement of ice-sheet topography using satellite-radar interferometry. *J. of Glac.*, v.42, No.140, 11-21.

Kenyi, L., H.Raggam & M.Shardt 1996. SAR interferometry: experiences with ERS-1/2 SLC data. *Vermessung & Geoinformation,* Heft 2, 157-163.

Landsberg, G. 1976. *Optics*. Moscow: Nauka (in Russ).

Lee, J.S. et al. 1998. A new technique for noise filtering of SAR interferometric phase images. *IEEE Trans. on Geoscience and RS*, v.36, No.5, 1456-1455.

Raggam, H. et al. 1998. *RSG in ERDAS Imagine. Tour guide.* Version 3.22. Joanneum Research, Graz.

Reigber, C. et al. 1996. Impact of precise orbits on SAR interferometry. *Proc.IGARSS'96 Symp.*, 27-31 May, Lincoln.

Remote Sensing in the 21st Century: Economic and Environmental Applications, Casanova (ed.)
 ISBN 90 5809 096 5

Advanced polarimetric SAR data classification for cartographic information extraction

Manfred F. Buchroithner
Institute for Cartography, Dresden University of Technology, Germany

Elke Kraetzschmar
Institute for Cartography, Dresden, Germany (Currently: G.E.O.S. Freiberg Ingenieurgesellschaft mbH, Germany)

Martin Hellmann
German Space Agency (DLR), Institute for High Frequency Technology, Oberpfaffenhofen, Germany

ABSTRACT: Within the last decade several studies using polarimetric SAR data for bio-/geo-physical feature extraction have been reported and have significantly improved the understanding of polarimetric scattering mechanisms. In a comparative view, the approach presented here, which is based on Shane Cloud's Decomposition Theorem, seems to be most promising: the Entropy H and $\propto$ Angle Classification has been extended by using not only the parameters H and $\propto$ but also the first eigenvalue λ_1. A big advantage of this approach is the high correlation between its results and the physical properties of different landsurface materials, which makes this recently adapted method well-suited for both supervised and automated landcover classification. It is also most useful for the derivation of topographic as well as thematic map information, the scale naturally depending on the sensor's spatial resolution.

The well-known DLR Oberpfaffenhofen study site in southern Germany served as a testbed for the exploitation of spaceborne SIR-C/X-SAR data and of L-band data from DLR's airborne experimental ESAR. For validation purposes the classification results have been high-precision geocoded and compared to both ground truth and recent map data, also superimposing the digital national topographic geodata of Germany, ATKIS. The classification proved to be of extraordinary accuracy, as far as can be judged by visual inspection. An exact quantification is underway. The approach presented is being further developed and will - in contrast to the flat test-site used so far - be applied to data flown by a new airborne SAR sensor over mountainous terrain at the northern rim of the Bavarian Alps. First results in this respect look very promising.

1 INTRODUCTION

One of the major problems in information extraction for thematic, but most of all topographic maps from air- or spaceborne SAR data is the fact that the object classes as defined in cartography do not coincide with the information retrievable from active microwave remote sensing data. Thus, in many instances the application purpose of an object cannot be retrieved: a large building can be both an industrial plant, a school or a modern church, a square-shaped water body both a public swimming pool or a fishing pond.

However, as far as bio-geogphysical information is concerend, much can be obtained from SAR data; this even more so, if multiple frequencies and full polarimetry can be used. If such data are applied, much of the information required for – foremost – topographic map-making can be extracted. Within the process to develop tools for a more automated future cartographic information extraction and to extend its limits, the German Space Agency (DLR), Institute for High Frequency Technology, the AeroSensing Radar Systems Company in Oberpfaffenhofen and the Institute for Cartography of the Dresden University of Technology have set up a fruitful cooperation of SAR signal processing experts and SAR application cartographers. Some of the more recent results of this cooperation, which until now have been submitted to a more physics- or geometry-oriented auditorium, are presented here to the remote sensing user community.

Conventional radars operate with a fixed single-polarisation antenna for transmission and reception of radio frequency signals. In this way only a single scattering coefficient is measured for a specific combination of transmitted and received polarisation. Due to this implementation the vector quantity and the additional information about the scattering process contained in the polarisation properties of the signal is lost. To ensure that all the information about the scattering process contained in the polaristion properties of the signal is lost. To ensure that all the information in the scattered wave is retained, the polarisation of the scattered wave must be measured through a vector measurement process. The SIR-C/X-SAR mission provides for the first time spaceborne full polarimetric data. From this full polarimetric data a couple of parameters can be derived and used for classification purposes. First, an algorithm is outlined using the Cloude Decomposition Theorem (Cloude and Pottier 1997). The feature extraction is based on an entropy -α-analysis and a feed forward neural net is used for the classification. In the next step the first eigenvalue λ_1 is used additionally to improve the classification accuracy. As a third possibility a classification based on the 3 eigenvalues is outlined. The data are compared to ground data and an error analysis is given to outline the strengths and constraints of the algorithms.

In order to improve the handling as well as the accuracy of the classification, an Artificial Neural Network and a Fuzzy Rule Base classifier are used to assign the feature vectors to the desired classes.

In the last section the classification algorithms are compared to each other and the advantages and disadvantages are discussed.

2 TEST AREAS AND DATA

It seems there is no need to introduce once again the Oberpfaffenhofen test site, since it has been used for almost two decades for several SAR-related activities. The data used for this test site are both the "classical" spaceborne SIR-C/X-SAR C- and L-band data and the more recent airborne L-band data of DLR's Experimental SAR (ESAR) from April 1997.

In addition to this almost totally flat area, another hilly, or rather mountainous, test site, Edelsberg southwest of Munich has been covered by In-SAR data of AeroSensing's AES-1 sensor. The results from this test site will, however, only be briefly mentioned since here the data exploitation is still on-going. Table 1 shows the parameters of the SAR data used.

3 COHERENT TARGET DECOMPOSITION THEORY

In the case of the SIR-C system as well as the ESAR system the scattering Matrix [S] is measured using two linear polarised antennas where the first index denotes the transmitted polarisation and the second one the received polarisation. In the case of the ESAR and the SIR-C systems the antennas are polarised horizontal H and vertical V. Because of the reciprocity theorem $S_{HV} = S_{VH}$ [1]. Therefore the whole information, contained in the [S]-matrix can be written in a vector (Pauli basis)

$$\underline{k}_i = \frac{1}{\sqrt{2}} \begin{bmatrix} S_{HH} + S_{VV} \\ S_{HH} - S_{VV} \\ 2S_{HV} \end{bmatrix} \quad (1)$$

with this vector the coherency matrix [T] can be formed.

This coherency matrix representation has the advantage of relating underlying physical scattering mechanisms as shown in the following.

By applying an eigenvalue analysis, the coherency matrix can be decomposed into

$$[T] = \lambda_1 \underline{e}_1 \underline{e}_1^{*T} + \lambda_2 \underline{e}_2 \underline{e}_2^{*T} + \lambda_3 \underline{e}_3 \underline{e}_3^{*T} \quad (2)$$

where λ_i are the eigenvalues of [T] and $\underline{e}_i$ the corresponding eigenvectors, which can be written as

$$\underline{e}_i = \begin{bmatrix} \cos\alpha_i \\ \sin\alpha_i \cos\beta_i e^{i\delta_i} \\ \sin\alpha_i \sin\beta_i e^{i\gamma_i} \end{bmatrix} \quad (3)$$

The α angle is a continuous parameter with a range from 0° to 90°. It can be used to represent a wide variety of different scatterers, while β_i is the orientation angle, containing information about the orientation of the scatterer with regard to the incidence line of the radar.

An angle α = 0° can be interpreted as a surface scattering process. As α increases, the surface becomes anisotropic (i.e. HH ≠ VV). In the extreme, α = 45° where we have a dipole for which either HH or VV is zero (which one is zero is determined by β). If α > 45°, then we obtain an anisotropic double bounce scattering mechanism. Note that α is rotation-invariant, i.e. it is decoupled from β. Thus we can identify the scattering mechanism independently of its physical orientation in space. Cloude and Pottier (1997) define a quantity named entropy H. Ranging from 0 to 1, it represents the randomness of the scattering medium from deterministic scattering H = 0 to totally random scattering H = 1. Similar to the entropy H, α is averaged.

Table 1: Important technical parameters of the major spaceborne SAR missions.

Parameters	SEASAT-SAR	SIR-A	SIR-B	ALMAZ-SAR	ERS-SAR	JERS-1 SAR	SIR-C/X-SAR	RADAR-SAT	EOS-SAR	ENVISAT-ASAR
Band and Wavelength	L(23.5)	L(23.5)	L(23.4)	S(10)	C(5.7)	L(23.5)	L(23.9) C(5.7) X(9.6)	C(5.6)	L(23.7) C(5.7) X(3.1)	C(5.3)
Frequency, GHz	1.275	1.278	1.282	3	5.3	1.275	L, 1.25 C, 5.3 X, 9.6	5.3	L, 1.25 C, 5.3 X, 9.6	5.331
Incidence angle (°)	23	50	15-64	30-60	20-35	38	15-50	10-59	15-45	15-50
Polarization	HH	HH	HH	HH	VV	HH	L, C (MP) X (VV)	HH	C, X (HH, VH, phase) L (quad)	HH, VH
Noise eq. (dB)	-24	-32	-28		-24	-20,5	L(-50) C(-40) X(-26)	-23	L(36) C(27) X(22)	
Range res., m	25	40	17-58	15-30	26	18	10-60	10-100	8-250	90-850
Azimuth res., m	25	40	25	15	28	18	30	9-100	20-250	90-850
No. looks	4	6	4	...	6	3	4	1-8	4	8
Swathwidth, km	100	50	10-60	20-45	100	75	15-60	50-500	30-360	100-400
Altitude, km	800	260	225-350	300	785	568	222	798	600-800	800
Launch date	27 Jun 78	12 Nov 81	5 Oct 84	3 Mar 91	17 Jul 91	22 Feb 92	9 Apr 94	4 Nov 95	1998	1999
					4 Apr 95		30 Sep 94			
Mission end	10 Oct 78	14 Nov 81	13 Oct 84	2 years	Oper.	Oper.	each 11 days	5-years	15 years	5-years
Launching country/Agency	USA	USA	USA	USSR	ESA	NASDA, Japan	USA/Italy/ Germany	Canada	NASA, ESA, NASDA	ESA

4 ENTROPY -α CLASSIFICATION

The Entropy H and the α-angle derived from the Cloude Decomposition Theorem provide two features. First of all these two parameters are highly correlated with the physical properties of the scatterer which is a basic demand for unsupervised classification without using a-priori knowledge (e.g. ground data). The second aspect is the fact that the range of both parameters is fixed and therefore not dependent on the data set. These facts allow feature extraction and classification independent from data set and a-priori knowledge. To achieve a more flexible classification a neural network was applied. The network was implemented as a feed forward net with 2 input units, 7 hidden units and 4 output units. As training function a back-propagation with momentum term and flat spot elimination algorithm was used (Rojas 1993). For the training a representative pattern set with 20 000 patterns was used. The classification compares positively to the results of a decision tree classifier (Hellmann 1997). But even using this method only 4 classes (water, settlements, low vegetation and forest) can be detected from spaceborne data sets. This, however, is not satisfying for the purpose of map generation and leads to the extended approach described below.

5 H-α PLUS 1ST EIGENVALUE CLASSIFICATION

Since the H-α parameters of Cloude's Decomposition Theorem are not sufficient for a good interclass resolution if used alone, additional information is needed (Hellmann *et al.* 1997). In this approach the backscatter intensity information contained in the first eigenvalue λ_{L1} is used. With this information a better interclass resolution can be achieved due to the different reflectivity of different scatterers. In the first step the classes are separated by analysis of their backscatter intensity (e.g. buildings show a much higher intensity than all other classes), and in the second step these areas are identified using their scattering properties which can be interpreted by the H-α parameters. This method provides better results but some problems remain.

One problem region is located in the city of Gilching. In this region we observe a high backscatter intensity from fields. The reason for this effect is probably that the fields are freshly ploughed and very wet due to rainfall and show therefore a high backscatter intensity. This ambiguity can be resolved with the H-α parameters. Buildings as well as the fields have a low entropy H. While H-α is very low for the fields (surface scattering), it is quite high (double bounce scattering) for buildings. This fits well to the theory (Cloude and Pottier 1997) and allows a separation for most of these ambiguous regions. Only two fields cannot be classified correctly. This is due to the fact that there are fruit trees on these fields showing double bounce behaviour and raising the α-angle. Other ambiguous regions are those where forest and buildings are mixed like in the vicinity of the Lake Wessling and in the south east of the city of Gilching. In these regions most buildings are hidden under the canopy. Therefore the λ_{L1} parameter is not as high as expected for buildings, and also the entropy is high because of additional scattering processes from the canopy.

Based on the evaluation of λ_{L1} 5 ESAR classes can be separated:

- **very high backscatter:** buildings and corner reflectors.
- **high backscatter:** forest.
- **medium backscatter:** medium vegetation (e.g. bushes/sweet corn) and fresh ploughed fields.
- **low backscatter:** low vegetation (e.g. grass).
- **very low backscatter:** very smooth surfaces (e.g. runway and taxiways of the Oberpfaffenhofen air field, water and harvested fields.

For the spaceborne SIR-C data we find:

- **very high backscatter:** settlements/buildings.
- **high backscatter:** forest.
- **medium backscatter:** low and medium vegetation.
- **low backscatter:** mainly the runway and taxi ways of the Oberpfaffenhofen air field.
- **very low backscatter:** water and very smooth surfaces.

With the additional information from the H-α parameters the classes can be identified, e.g. for the second class high values of H and α lead to the conclusion that this region is a forested area which fits very well to the reality. Thus the ambiguities in the very high backscatter class can be resolved. Also the third class can be split into two classes: low and medium vegetation. For bare soil and the grass on the air field the entropy is lower than for the medium vegetation because only surface scattering processes appear here while in the medium vegetated areas also other scattering processes take place. From the combination of λ_{L1}, H and α, six classes can be detected and identified: buildings, forest, medium vegetation, low vegetation, runway and water.

6 EIGENVALUE CLASSIFICATION

The method described above makes use of the H-α parameters which provide information about the scatterer. For the next method only eigenvalues λ_i are used. If the eigenvalues are not averaged into the H parameter also C-band is useful for several classes (settlement, water and runway). While the C-band data were not always useful for the approach using the H-α parameters, they provide some additional information for this approach. The eigenvalues λ_{Ci} do not provide useful information for the discrimination between different vegetation types due to the limited penetration depth. However, they do provide information for settlements, and surfaces like water or the runway. Based on the intensity information of the eigenvalues and the relation between the eigenvalues of the two bands a classification is possible which also provides information about the scattering processes on ground. From the analysis of the eigenvalues of the L-band λ_{Li} and the C-band λ_{Ci} 6 classes can be separated and interpreted.

- **buildings:**
  Very high values for λ_1, low or medium entropy and high values for the α-angle. This is due to the fact that man-made structures show a high backscatter intensity and only one (or two) dominant scattering process(es) of double bounce behaviour.
- **forest:**
  High values for λ_1, high entropy and high values for α. The trees have a quite good reflectivity, the entropy is high because all kinds of scattering processes take place, e.g. volume scattering in the canopy, double bounce scattering at the trunks and surface scattering at the ground.
- **medium vegetation:**
  Medium values for λ_1, medium entropy and low or medium values for α. In such areas the surface scattering process is dominant but a second one (resulting from the interaction with the vegetation) - even if much lower - is present.
- **low vegetation:**
  Low values for λ_1, low entropy and low values for α. On low vegetation mainly surface scattering processes appear.
- **smooth surfaces (roads, runway taxiways, and harvested fields, water):**
  Very low values for λ_1, low to medium entropy and α values. On smooth surfaces also mainly surface scattering processes appear, but since the signal is reflected away from the sensor, the backscatter intensity is very low and the signal-to-noise ratio (SNR) is degraded, which results in an increase of the entropy and the α values.
- **ploughed fields:**
  Medium values for λ_1, low entropy and low values for α. These fields show also surface behaviour, as expected, but due to the roughness more energy is reflected towards the sensor and the backscatter intensity is higher than for non ploughed fields.

With similar considerations we can obtain the classification results for the SIR-C data.

7 ARTIFICIAL NEURAL NETWORK CLASSIFICATION

Using this approach a couple of network topologies and learning functions were tested. For the classification presented in this paper a simple Feed-Forward neural network (Rojas 1993) with 3 input neurons provided the best results and were therefore implemented. For the learning function we used the Standard Backpropagation Algorithm (Rojas 1993). The ANN was trained and validated using 2 training patterns of 1000 test pixels from the data set (ca. 0.01 % of the data set).

For each pixel a 3-element feature vector [H, α, λ_{L1}] is used as an input vector for the Artificial Neural Network and the Fuzzy Rule Base classifier.

8 FUZZY CLASSIFICATION

The fuzzy classifier uses the Mamdani approach for fuzzy control. A detailed description of this approach can be found in (Kruse et al. 1993). In the first step we fix a set of 'expert rules'.

Every term in these rules such as 'λ is high' is a so-called linguistic variable and is described by a fuzzy set (Zadeh 1965), i.e. by a function μ_A from the universe of discourse X (e.g. X = [0,90] for the α-angle) to the closed unit interval [0,1]. We interpret $\mu_A(\chi)$ as the degree of membership of χ in A.

For entropy and α-angle these fuzzy sets can be

Table 2: Classification rules (from Hellmann et al. 1999).

λ	H	α	Decision
very high	medium		urban
(very) high	very low	medium/high	urban
high	high		forest
medium	high	medium/high	forest
medium	medium	medium/low	veget.*)
medium	(very) low	low	veget*)
(very) low			runway

*) veget. = vegetation

defined data-independent. The fuzzy sets for the first eigenvalue are determined interactively by the user. Six occurring grey levels are picked ranging from very dark grey (low backscatter: smooth surface) to very light grey (high backscatter: buildings). The values of these grey levels $g_1,...,g_6$ determine the shape of the membership functions for λ. We chose in all cases triangular or trapezoid shapes for computational simplicity. Input to the system is a triple of values (λ, H, α) for an image pixel. This triple is first fuzzified, i.e. the values of the corresponding fuzzy sets are evaluated. The degree of the rule antecedents is then computed by taking the maximum of all rules which 'fire' for this class, i.e. all rules having this class consequence. To obtain a crisp classification we have to defuzzify the fuzzy output of the system. This was done by simply assigning the class with the maximum degree to each pixel.

9 RESULTS AND VALIDATION

The main advantage of the H-α analysis over conventional non-polarimetric classification algorithms is the fact that a good correlation between the H-α parameters and the scattering processes provide a good interpretation base without *a-priori* knowledge. Therefore for investigations in areas with no ground information (e.g. the mapping of rain forest) this method can provide valuable information. The use of λ_1 provides a good interclass resolution and thus an improved classification accuracy. On the other hand, the λ_1 parameter is not constrained to a fixed interval and is therefore not easy to handle for data-independent classification algorithms.

Using the H, α and λ_{L1} features, derived from the full polarimetric data, at least 5 different classes can be separated (cf. Hellmann and Kraetzschmar 1998 a, b). These classes represent different types of scattering mechanisms (e.g. double bounce, surface scattering etc.). If the scattering mechanisms of cartographic objects are known, the separated classes can be assigned to cartographic object classes like buildings, forest, etc. The classification approaches presented in this paper are designed to distinguish between 5 different classes, i.e. buildings, forest, low vegetation, rough surfaces (e.g. road-like structures), smooth surfaces (e.g. water). For validation purposes, i.e. to compare the classification results with topographic maps, it is useful to restrict the attention to those 5 classes.

Both the ANN and the Fuzzy Rule Base classification provide homogeneous results in good agreement with reality. Even for small-scale structured classes, i.e. settlements and industrial areas, which cause problems for many classification algorithms, the results were compared to optical data and topographic maps at the scale 1:25 000. Conspicuous is the high percentage of road-like structure classes for the fuzzy approach compared to the ANN approach. For these structures the ANN approach comes closer to reality, but on the other hand, some parts of the taxiways beside the runway disappear, which are clearly visible in the fuzzy approach. Buildings are extracted by detecting the double bounce scattering mechanisms occurring on the illuminated walls of the buildings. Due to this, only the illuminated walls are classified as buildings. In the case of large industrial buildings this means that the position of the building is clearly detectable but the roof of the building may be classified as a surface structure. Application of advanced image processing algorithms may reduce this effect. For both classification approaches problems occur at the edges of the forest. These areas show a clear double-bounce scattering behaviour as well as a low entropy and are therefore sometimes classified as buildings. This problem could be solved by using additional context information. Forest is well detected in all approaches.

To summarise, it can be stated that the classification results achieved are sufficient for the extraction of relevant cartographic information at a scale of 1:25 000 and, with some limitations for linear structures, also 1:10 000. Some problems, however, remain. First of all it has to be stressed that with the H-α-λ_{L1} approach, it is only possible to extract different scattering mechanisms from the data set. This means that - see statement at the beginning - some relevant cartographic classes can only be separated if their scattering behaviour is different. In cases where the same scattering mechanisms occur for different object classes (e.g. buildings and forest borders) a separation is only possible with additional information.

Another crucial point is the signal-to-noise ratio (SNR). In areas with a low SNR, i.e. smooth surfaces/water, where the signal is reflected away from the sensor, entropy rises due to the noise contribution and can therefore decrease the classification accuracy.

10 REFERENCES

Cloude, S. R & E. Pottier (1997): An Entropy Based Classification Scheme for Land Applications of Polarimetric SAR. *IEEE Trans. Geosc. & Remote Sensing*, Vol. 35, No. 1.

Hellmann, M. (1997): Classification using Polarimetric and Interferometric SAR Data. *Proc. IGARSS '97*, Singapore.

Hellmann, M., Cloude, S. R. & Papathanassiou, K. P. (1997): Interpretation of SAR-Data using Polarimetric and Interferometric Techniques. *Proc. SPIE '97*, San Diego, USA.

Hellmann, M. Jäger, G., Kraetzschmar, E. & Habermeyer, M.: Classification of Full Polarimetric SAR Data Using Artificial Neural Networks and Fuzzy algorithms. *Proc. IGARSS '99*, Hamburg.

Hellmann, M. & Kraetzschmar, E. (1998a): A New Approach for Interpretation of Full-Polarimetric SAR-Data. *Proc. PIERS Workshop on Advances in Radar Methods*. Baveno, Italy.

Hellmann, M. & Kraetzschmar, E. (1998b): Interpretation of SAR-Data using Polarimetric Techniques. *Proc. of the 2nd International Workshop on Retrieval of Bio- & Geo-physical Parameters from SAR Data for Land Applications*, 21-23 October 1998, ESTEC, Nordwijk, The Netherlands.

Kruse, R., Gebhard, I. & Klawonn, F. (1993): *Fuzzy Systeme*. Teubner Publ., Stuttgart.

Rojas, R.: *Theorie der neuronalen Netze*. Springer Publ., 1993, p. 149-172.

Remote Sensing in the 21st Century: Economic and Environmental Applications, Casanova (ed.)
 ISBN 90 5809 096 5

A model for atmospheric correction of DAIS hyperspectral imager sensor based on experimental optical measurements

V.E.Cachorro, R.Vergaz & A.M.De Frutos
GOA-UVA, Grupo de Optica Atmosférica-Universidad de Valladolid & Departamento de Optica y Física Aplicada, Facultad de Ciencias, Valladolid & Departamento de Física Aplicada I, E.T.S.I. Agrarias de Palencia, Spain

ABSTRACT: Airborne imaging spectrometers, like DAIS (Digital Airborne Imaging Spectrometer) sensor, give a new powerful tool for the monitoring of atmospheric and surface biophysical parameters, in order to assess vegetation status. Although the final objective is to provide operational algorithm to retrieve the earlier parameters, the current developing of these instruments require at the same time the developing and validation of new procedures. Here we present the developing and validation of an atmospheric correction model applied to the DAIS images (DLR) during a flight campaign in the Mediterranean area of Les Gavarres in Gerona (Catalonia, NE Spain). Experimental atmospheric measurements of direct and global irradiances together with the downward radiance of high spectral resolution are ideal to be used together with airborne imager spectrometers. These data allow us to determine the radiative properties of the atmosphere in order to carry out the validation of the model and the atmospheric correction procedure. Problems in the correction of atmospheric water vapor have been clearly pointed out in this experimental flight, and are partially due to the current status of the atmospheric modelization and the inaccuracy of good telemetric ancillary data.

1 INTRODUCTION

Broad-band, high-spatial resolution satellite Remote Sensing (i.e., Lansat-TM, SPOT-HIV, LISIII-IRS) has proven to be an essential tool for land cover change mapping but insufficient to (i) resolve subtle categories like the ones often needed for ecological research, and (ii) measure biophysical magnitudes. Narrow-band-hyper-spectral airborne remote sensors (Green et al., 1995) provides data that can be used for that purposes, but also generates specific processing needs. The atmospheric correction effect is an important processing step to enhance the data quality, help the interpreter, and improve the performance of subsequent processing, i,e., classification algorithms.

Images provided by airborne spectrometers are a powerful tool for monitoring vegetation biophysical parameters, but require atmospheric correction to be compared to reference hand-held spectral measurements, like field reflectance spectroradiometry.

The aim of this work is the atmospheric correction of the DAIS (Digital Airborne Imaging Spectrometer) images acquired during an airborne campaign carried out by DLR in the area of Gerona (NE Spain) on July 9th, 1997. The atmospheric correction model is based on experimental optical measurements carried out at ground level in order to characterize the optical properties of the atmosphere. The high spectral resolution of these data is ideal to be used for imaging spectrometer images.

We have developed in earlier woks a methodology to retrieve water vapor from spectral radiation measurements at ground level (Cachorro et al, 1998a). This methodology is well assessed compared with the water vapor retrieval procedures by satellite or airborne platform sensors.

Water vapor is the main parameter to be corrected in the DAIS images and problems have been detected in this correction when the value retrieved from our ground espectroradiometric experimental measurements has been applied to the correction model. Therefore, an inversion procedure is being developing in order to retrieve the water vapor from the DAIS images.

We think that these problems in the correction can be dued to the problem linked to the telemetry of the observation angles during the recording images but also to the very method we are developing. Therefore, this work tries to be a contribution to the development of atmospheric correction for hyperspectral images in order to retrieve surface reflectance, mainly that due to the interference atmospheric water vapor.

2 DAIS 7915 SENSOR CHARACTERISTICS

The characteristics of DAIS 7915 (Digital Airborne Imaging Spectrometer) system are described in detail in Mueller and Oertel (1997). DAIS is a 79 channels optical imager-spectrometer that collects information from the earth surface on the 0.4-12.3 μm range. Here, we refer to the 32 bands in the 400-1000 nm spectral range with bandwidths varying from 15-30 nm and the 8 bands of 1500 to 1800 nm with 45 nm of bandwidth. Size pixel of 5x5 m at nadir observation. DAIS images were provided by DAIS processing team as calibrate at-sensor radiances in 0.001*mW/(sr cm^2 μm) for band 1-72.

An auxiliary data base contained information on flight parameters were also provided, but unfortunately, essential information for users, such are view zenithal and azimuthal angles, and ground resolution, are not in the database, although the data base contains the necessary lower-level parameters for the users to make their own computations. In our case, a mechanical failure of auxiliary DAIS instruments made the PIT angles were not recorded and we had to estimate them from the records of the airplane navigation instruments. However the computation of the view angles by these telemetric data seems to have errors of importance because of the a posteori problems observed in the water vapor correction procedure.

3 CHARACTERISTICS OF THE AREA OF STUDY

Seven images of 512 samples by 2120 lines were acquired and three were selected over our area of study in Les Gavarres (41 54'N, 2 58'E), between 11h. and 11h. 32' local time at a height of 3200 m. Les Gabarres is a humid Mediterranean forest area (250 km^2) in Catalonia (NE, Spain), between the city of Girona and the Costa Brava tourist area. Les Gabarres is an old, small and low mountain zone (maximum height is 535 m). Vegetation has a 100% cover and is dominated by evergreen oaks, pines, and riparian forest. Shrubs species of typical Mediterranean forest have also an important cover.

We have also considered the agricultural area around this mediterranean forest in order to check and validate our atmospheric correction model. The reason is that the parcel surface albedo can be easily identified providing a first assessment of the surface albedo. Furthermore, these agriculture zones are relatively homogeneous and flat compared with the forest area, allowing a better application of the atmospheric correction model.

4 EXPERIMENTAL ATMOSPHERIC OPTICAL MEASUREMENTS

Ground atmospheric measurements were simultaneously carried out in the same area during the flights. We acquired a total of 6 spectra of direct normal solar irradiance and 3 of global horizontal irradiance between 300-1100 nm with a Licor 1800 spectroradiometer, whose characteristics are described elsewhere (Durán, 1997; Cachorro et al., 1998a). We used these spectra to determine the total optical depth, and aerosol optical depth to derive the aerosol optical parameters. Furthermore atmospheric transmittance can also be derived, which allows to be used directly in the algorithm or to derive ozone and water vapor vertical contents. Three spectra of vertical downward radiance were also measured and used as complementary information of the aerosol properties derived from wavelength dependence of the aerosol optical depth.

5 ATMOSPHERIC CORRECTION MODEL

5.1 Physics of the algorithm

In order to derive surface reflectance from the bidirectional sensor radiance, we follow current methodologies of atmospheric correction (Kaufman 1989; Vermote 1995; Vermote et al., 1997), adapting them to our experimental measurements and application to hyperspectral DAIS atmospheric correction images (Cachorro et al., 1997).

The spectral upward radiance at the top of the atmosphere (the satellite sensor radiance) $R(\lambda)$ is related to the spectral surface reflectance $\rho(\lambda)$ by

$$R(\theta_s, \theta_v, \phi_s\text{-}\phi_v)=R_o(\theta_s, \theta_v)\, t_{abs}(\theta_s, \theta_v)+ [\rho F(\theta s)\, T^{\uparrow}(\theta_v)/\pi(1- \rho S)]t_{abs}(\theta_s, \theta_v) \quad (1.1)$$

Hereafter and for simplicity we omitted the wavelength dependence of these quantities. R_o is the atmospheric path radiance, due to solar radiation that was scattered by the atmosphere to the sensor without being reflected from the earth surface. It contains the molecular and aerosol contributions. t_{abs} is the absorption transmittance of atmospheric gases evaluated with the corresponding atmospheric components and contents for each term. F is the global downward irradiance at ground level assuming a dark surface and $T^{\uparrow}$ is the total upward transmittance. S is the spherical albedo of the atmosphere, which can be evaluated from parameterized expressions from the molecular and aerosol effects, based on the corresponding optical depths. θ_s, θ_v are the solar and viewing zenith angles and (ϕ_s-ϕ_v) the relative azimuth. This radiance is normalized to reflectance units and referred to as normalized radiance or sensor apparent reflectance ρ^* by using the current expression $\rho^*=\pi R/ F_o \mu_o$, where F_o is the extraterrestrial solar irradiance and $\mu_o= \cos \theta_s$.

The above expression is applied to lambertian surfaces and homogeneous terrain. This approach has been considered as a first step in our atmospheric correction process. Other important approach is that the radiative transfer model (and hence the above quantities) takes an unique atmospheric equivalent layer, therefore atmospheric profiles are not considered. For sensors with high spatial resolution (250 m and less), correction for adjacency effects (Vermote et al., 1997) is necessary over heterogeneous areas but here only homogeneous areas were considered. On the other hand surface-BRDF coupling has not been considered (Roujean et al., 1992; Wanner et al., 1997).

5.2 Determination of atmospheric parameters and aerosol model from measured data.

We can take into account that the global measured irradiance at ground level is given by

$$F_{exp} = F/(1- \rho S) \quad (1.2)$$

Introducing it in the above expression, we give a more experimental character to this formulation. In this case the atmospheric downward absorption transmittance is accounted for. The $T^{\uparrow}$ transmittance has been evaluated by means of

$$T^{\uparrow} = t_d + \exp\{-(\tau_r+\tau_a)\} \quad (1.3)$$

where τ_r and τ_a are the molecular and aerosol optical depths. t_d is the diffuse transmittance of the atmosphere, which can be determined theoretically by means of a two flux model, taken the necessary atmospheric parameters determined from the global and direct measured irradiances. t_{abs} is also evaluated from the ozone and water vapor contents, previously retrieved from the direct irradiance spectra taking into account the absorption features of the Chappuis (ozone) and the 940-nm water vapor absorption bands (Cachorro et al., 1996; Cachorro et al., 1998). Due to the short spatial range of our images, the retrieved values were applied to the whole image.
Therefore two approaches have been considered in the atmospheric correction model, one based on the introduction of the input derived atmospheric parameters and the other approach called by us 'experimental approach', where the global ground measured irradiance is introduced directly in the model.

Details like how the aerosol parameters were obtained have been explained elsewhere (Cachorro et al., 1998b).

One of the main problem in atmospheric correction is the evaluation of the path reflectance of the atmosphere. We have taken the solution given in the 5S Code (Tanré et al., 1995) with the above derived aerosol optical depth and aerosol parameters. Obviously, in our case the sensor is placed at 3200 m and not at the top of the atmosphere, however all the boundary layer is included. Therefore we have discounted a 5% from the earlier aerosol optical depth and the correct value for the upward Rayleigh optical depth to introduce in the upwards transmittance (ozone was not included) and path radiance.).

5.3 Simulation of the reflectance at the top of the atmosphere based on experimental data.

In a previous study (Durán 1997; Cachorro et al., 1997) we have carried out a set of simulations of the reflectance at the top of the atmosphere (TOA) based on our site surface reflectance and atmospheric measurements (Valladolid-Palencia, in Castilla y León flat areas) and our atmospheric correction model. In this way we have evaluated the sensitivity of the (TOA) reflectance or satellite sensor bidirectional reflectance to aerosol parameters under typical geometrical conditions, i.e.: range of sensor reflectance variations to be expected in the processing of the DAIS imagery. An example of these simulations can be seen in figures 1 and 2, where we have evaluated the variations of the bidirectional sensor reflectance for representative solar and view angles as function of aerosol optical depth τ_a, asymmetry parameter g and single scattering albedo w_o. These simulations have been performed taking a typical albedo of green vegetation surfaces, like that measured for a sample of green wheat (dashed line).

The first simulated case shows the dependence of the spectral TOA reflectance (from 400 to 1000 nm) on the solar zenith angle, from 0 to 80 degrees (figure 1a). An aerosol optical depth of 0.23 at λ=0.55 μm with varying wavelength according α and β values of the Ångström turbidity parameters and a viewing zenith angle of 30 degrees in the same plane of observation were considered. It was also taken for these simulations the values for g=0.65 and wo=0.88 of typical values founded in our area of study in Castilla y León. As it is well known the atmospheric effect increases in the visible region and decreases in the infrared for increasing solar zenith angles. The infrared region is strongly affected by the absorption of water vapor and oxygen. Increasing values reduce the satellite signal as can be observed with detail in the figure 1. The absorption due to ozone in the visible was also taken into account. Although the extreme values of the solar zenith angle are not real cases, all figures are sufficient relevant for illustration purposes.

Figure 1b shows the variation of the TOA reflectance as a function of the aerosol optical depth, from very low values, 0.01 (practically a Rayleigh atmosphere) to very high haze conditions, as a value of 1. As it is well known the atmosphere enhances the satellite signal in the visible, meanwhile decreases it in the infrared. Besides the effect of geometrical conditions, the aerosol optical depth is the most important atmospheric parameter for the atmospheric effect taking into account the whole spectral range.

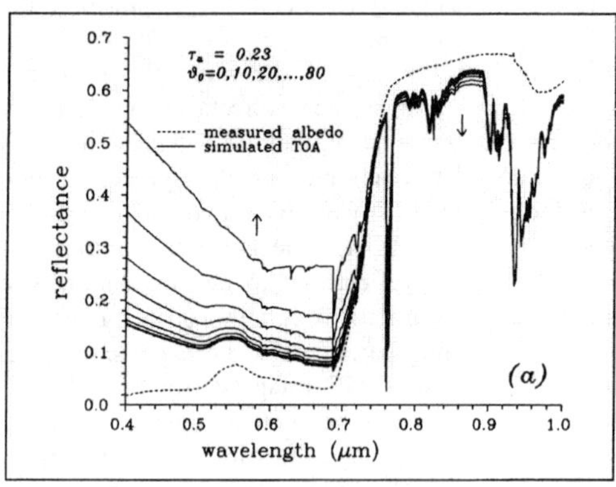

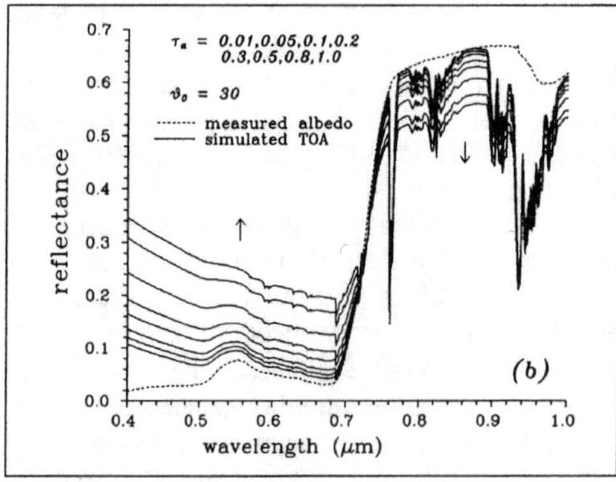

Figure 1. Simulated behaviour of the spectral TOA reflectance as a function of the solar zenith angle a); and for the aerosol optical depth with a solar zenith angle of 30 degrees b). The zenith angle of observation is 30 degrees and the relative azimuth 0 degrees. (See text)

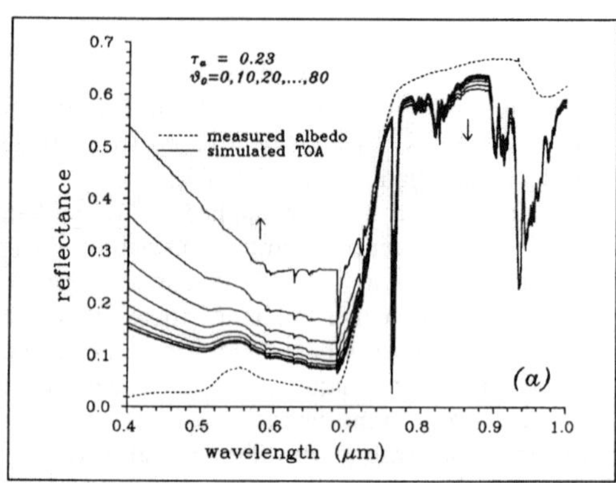

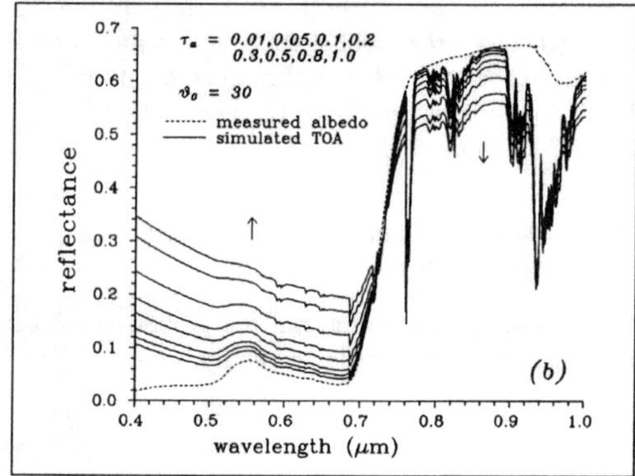

Figure 2. Simulated behaviour of the spectral TOA reflectance as a function of the single scattering albedo w_0 a); and the assymetry parameter *g* b). The zenith angles for solar and observation are 30 degrees with a relative azimuth of 0 degrees. (See text)

Figure 2 shows the spectral TOA reflectance for varying values of the single scattering albedo (figure 2a) and the asymmetry parameter (figure 2b). As can be seen the TOA reflectance increases for increasing single scattering albedo. For the single scattering albedo the atmospheric effect in the visible is very similar to that produced by the aerosol optical depth but less relevant, however its influence in the infrared is very important. Therefore absorbing aerosols (lower w_o values) must be properly detected at the same time of the aerosol optical depth. In figure 2b we can observe a decreasing of the atmospheric effect in the visible when the asymmetry parameter increases, meanwhile the infrared region is practically unaffected.

Although these simulations are only of formal interest, they permit to be familiar with the spectral behavior of TOA reflectances as function of atmospheric parameters and also with the radiative transfer model.

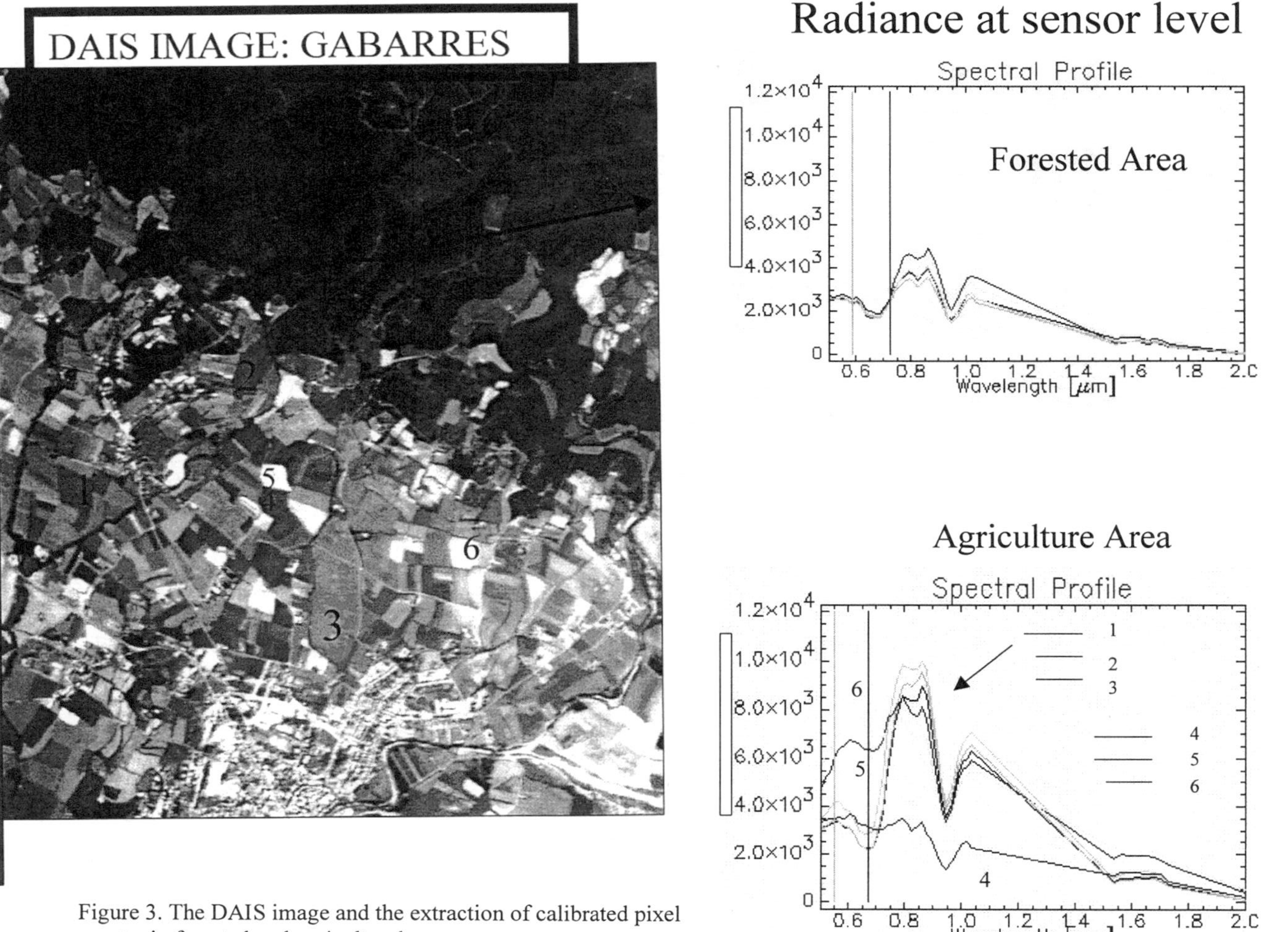

Figure 3. The DAIS image and the extraction of calibrated pixel spectra in forested and agricultural areas

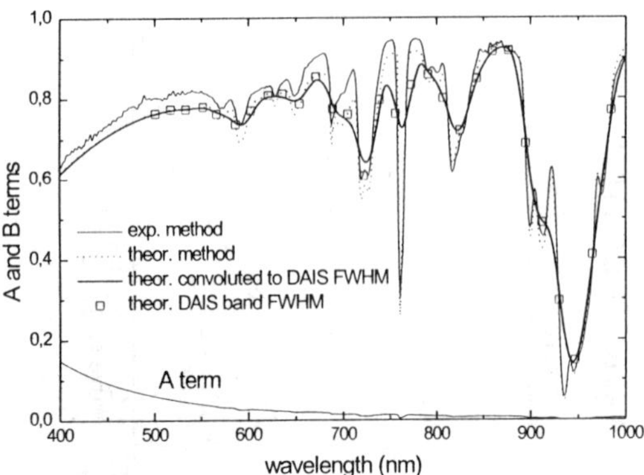

Figure 4. Evaluation of the A and B terms of the algorithm applying various procedures (see text).

6 RESULTS

6.1. Retrieval of sensor measured reflectance samples.

Surface reflectance calibrated values have been extracted form the pixel samples of the DAIS image. This can be seen in figure 3 for an agricultural and forested area respectively. These values correspond to calibrated non-corrected radiances values where the effect of water vapor absorption is clearly observed. Over these specific spectra we have been working in order to assess out atmospheric correction model and at the same time to develop a procedure to retrieve the water vapor content.

6.2. Operational algorithm

The operation algorithm, for a quick image processing, is based on look-up tables that follows the above approaches. As can be seen from expression (1.1) the correction has a linear form

$$\rho^{*}=A+B\rho \tag{1.4}$$

where A and B are constants, which depends on the solar and observation angles. The A and B values have been determined for a given solar zenith angle and nadir observation due to the above mentioned problems at the DAIS wavelengths.

However in figure 4 we show the evaluation of A and B terms needed for the applications of the atmospheric correction process for real condition of image registration, (θs=33 degrees and nadir observation) for a given spectra in the visible and NIR spectral range.

We can observe the low value of the correction A term that is relatively important in the visible region, representing the intrinsic reflectance of the atmosphere. The B term have been evaluated for testing consideration under four different cases. The two method, called by us theoretical and experimental at a step of 1 nm wavelength and also the theoretical one considering only the DAIS bands with their FWHM and considering a convolution with a triangle function of FWHM=32 nm.

As can be seen in figure 4 a strong absorption due to the water vapor is obtained for the atmospheric correction. Their introduction on the image pixel spectra corresponding to different surface reflectance shows clear results of a bad retrieval surface reflectance due to an overestimation of the water vapor absorption.

6.3 Water vapor retrieval from image data.

Due to the above observed problem, we decide to invert the problem and tried to retrieve the water vapor from the image pixel spectra. We are now testing different methods used to retrieve the water vapor amount from the image itself (Gao & Goetz 1990; Carrere & Conel, 1993) and no conclusions about the results can here be shown. We try to make a contribution in this research task.

However we think that part of the observed problems with the DAIS images can be due to the malfunctioning of telemetric instrument and therefore to the associated data.

Next task will be to verify these results taking the whole atmospherically corrected images. The comparison of DAIS spectra with hand-held field spectroradiometer data carried out at specific selected sites (Gamon et al., 1992; Peñuelas & Filella, 1998) allows us to measure the discriminant power of DAIS imagery to detect different plant canopies and assess vegetation conditions within an spectral unmixing approach (Lobo, 1996).

7 ACKNOWLEDGEMENT

We thank Dr. A Lobo and N. Pineda belonging to the "Instituto de Ciencias de la Tierra" (CSIC, Barcelona) for the collaboration in this work. Also thank to the DLR-DAIS processing team. The aerial campaign and the image pre-processing was funded by the Large Scale Facility DAIS program of the European Union. We also thank the "Consejería de Educación y Cultura de la Junta de Castilla y León" by their support to the "Grupo de Optica Atmosférica de la Universidad de Valladolid" by the project VA98/96 & the grant of R. Vergaz.

8 REFERENCES

Cachorro, V.E., P. Durán, R. Vergaz, A.M. De Frutos & S. Hernandez 1997. A study of atmospheric effect on satellite signal based on ground solar radiation measurements of high spectral resolution. In G. Guyot & T. Phulpin (ed), *In hysical Measurements and Signatures in Remote Sensing*: 108-120, Vol. 1. Rotterdam: Balkema.

Cachorro, V.E., Utrillas, M.P. Durán P. Vergaz, R. De Frutos, A.M. & Martinez-Lozano, J.A. 1998a. An study about the-water vapor content determination in the 940 nm band by using using moderated spectral resolution measurements of direct solar irradiance. *Appl. Opt.* 37, 4678-4689.

Cachorro, V.E., A Lobo, N. Pineda, R. Vergaz & A.M, De Frutos, 1998b. Atmospheric correction for hyperspectral Dais imagery of a forested zone. *European Symposium on Remote Sensing, SPIE vol, 3495*. Barcelona, 1998.

Carrere V & J.E Conel, 1993. Recovery of atmospheric water vapor total column abundance from imaging spectrometer data around 940 nm-Sensitivity analysis and applications to airborne visible/infrared imaging spectrometer (AVIRIS) data. *Remote Sensing of Environment* 44, 179-204.

Conese, C M.A. Gilabert, F. Maselli, & L. Bottai, 1993. Topographic normalization of TM scenes through the use of an atmospheric correction method and digital terrain models. Photogr. *Eng. Remote Sensing* 59: 1745-1753.

Durán, P. Medidas espectroradiometricas para la determinación de componentes atmosféricos (ozono, vapor de agua y aerosoles) y modelización del intercambio radiativo en la atmósfera. 1997. *Ph. D. Dissertation*. E.T.S.I. Agrarias de Palencia. University of Valladolid.

Gao & Goetz 1990. Column atmospheric water vapor and vegetation liquid water vapor retrievals from airborne imaging spectrometer data. *J. Geophys. Res.* 95:3549-3564.

Gamon, J.A. ,J. Peñuelas, and C.B. Field 1992. A narrow-waveband spectral index that tracks diurnal changes in photosynthetic efficiency. *Remote Sensing of Environment* 41, 35-44.

Green, O.R., C.M. Sarture, C.J. Chivit, J.A. Faust, P. Hajeck, & H.I. Novak 1995. AVIRIS: A new approach to erath remote sensing. *Optics and Photonics News*, 6: 1-30.

Kaufman, Y.J. 1989. The atmospheric effect on remote sensing and its correction. *In G. Asrar, (ed). Optical Remote Sensing, Technology and Applications*. New York, Wiley.

Lobo, A. 1996. Image segmentation and discriminant analysis for the identification of land cover units in ecology. *IEEE Translational on Geosscience and Remote Sensing,* 35: 1136-1144.

Mueller, A & D. Oertel 1997. *Proc. 3rd Inter. Airborne Remote Sensing Conference and Exhibition,* Vol 1, 684.

O' Neill, N.T., F. Zagolski, M. Bergeron, A. Royer, J.A. Miller & J. Freemandtle, "Atmospheric correction validation of casi images acquired over the Boreas southern study area. *Canadian Journal of Remote Sensing*, 23: 14-162.

Peñuelas, J & I. Filella 1998. Visible and near-infrared reflectance techniques for diagnosing plant physiological status. *Trend in Plant Sciences* 3(**4**), 151-156.

Roujean, J.L., M. Leroy & P.Y. Deschamps 1992. A bidirectional reflectance model of the earth's surface for the correction of remote sensing data. *J. Geophys. Res.* 97: 20455-20468.

Tanré, D., C. Deroo, P. Duhaut, J.J. Morcrette, J. Perbos, P.Y. Deschamps. "5S user Guide". Available at the Laboratoire d' Optique Atmospherique (LOA). University of Lille-1. France, 1986.

Tebo, A. 1997. Imaging spectrometer for the mission to planet Earth. OE Report No. **157**.

Vermote, E.F. 1995. 6S User Guide. *Laboratoire d' Optique Atmospherique (LOA)*. University of Lille-1. France..

Vermote, E.F. N. El Saleous, C.O. Justice, Y.J. Kaufman, J.L. Privette, L. Remer, J.C. Roger, & D. Tanré 1997. Atmospheric correction of visible to middle-infrared EOS-MODIS data over land surfaces: Background, operational algorithm and validation. *J. Geophys. Res.* 102:17131-17141.

Wanner, W., A.H. Strahler, B. Hu, P. Lewis, J.P. Muller, X. Li, C.L. Barker Schaaf, & M.J. Barsley 1997. Global retrieval of bidirectional reflectance and albedo over land from EOS MODIS and MISR data: Theory and algorithm. *(published in J. Geophys. Res. (private communication).*

Remote Sensing in the 21st Century: Economic and Environmental Applications, Casanova (ed.)
© 2000 Balkema, Rotterdam, ISBN 90 5809 096 5

Refined adaptive speckle filtering for SAR images

D.Cherifi & Y.Smara
Image Processing Laboratory (IPL), Electronic Institute, Houari Boumediene University of Sciences and Technology (USTHB), Alger, Algeria

ABSTRACT: The presence of speckle in SAR images reduces the ability to perceive fine details and rules out a visual perception of systematic structures that would make it possible to describe a specific feature of the earth's surface and makes the radiometric and textural aspects less efficient for class discrimination.
Many adaptive filters have been developed for speckle reduction and they are based on a test related to the local coefficient of variation of the observed image, which describes the heterogeneity of the scene. In this paper, the well known non linear adaptive filters such as Frost, Wu, and Nezry filters are presented, and considering the limitations of the analysed filters and the good results achieved using the ratio edge extraction, a modified adaptive speckle filter has been implemented based on both geometrical and statistical properties (coefficient of variation) of the local area. The procedure has depended on several parameters such as window size, the width of the linear features, the thresholds, the values of Cu (coefficient of variation of multiplicative noise) and C_{max}(the maximum value of the distribution of C_u).
The successful use of the ratio edge and line detectors has been employed in adaptive speckle filter, in order to apply the enhanced frost version in every window or subwindow selected.
As was expected, the new filter performs better, i.e., we can achieve very good smoothing in the homogeneous areas preserving, at the same time, structural and textural information, edges, linear features. The filters are tested on a ERS-1 SAR image corresponding to an area of southern Algeria called Laghouat. The good results obtained and the improvements with respect to the classic algorithms are presented and discussed.

1. INTRODUCTION

The speckle, appearing in synthetic aperture Radar(SAR) images as granular noise is due to the interference of waves reflected from many elementary scatters. This speckle in SAR images complicates the problem of image interpretation by reducing the effectiveness of image segmentation and classification. Various ways have been devised to eliminate it.

The primary goal of the filtering ought to be reduction of the speckle noise without sacrificing the information content. The ideal speckle filter should adaptively smooth the speckle noise, retain the edge and the sharpness of the boundaries of the feature, and also preserve the subtle but distinguishable details, such as thin linear features and point targets.

In this paper, we propose a modification of the filtering process presented by Nezry and Desnos [Nez, 91] [Des, 93]. This modified filter is driven from the geometrical operators, used in an interesting edge and line detector proposed by Touzi et al [Tou, 88]. An adaptive window of variable shape and of sufficient size can then be determined, to compute the coefficient of variation and apply an Enhanced generalised Frost filter version.

2. SPECKLE FILTERING TECHNIQUES

The speckle noise, which appears in SAR images is generally modelled as a multiplicative noise:

$$g(i,j) = f(i,j).\, n(i,j) \qquad (1)$$

Where $g(i,j)$ is the intensity of an observed image pixel, $f(i,j)$ is the noise-free image pixel we wish to recover and $n(i,j)$ is the noise, characterised by a distribution with a unit mean (E[n]=1) and a standard deviation σ_n.

Much work has been carried out in the past concerning noise reduction. In an homogeneous area the Box filter (i.e. the moving window averaging filter) or the geometrical Crimmins filter [Cri, 85] are very efficient. However, they have many disadvantages.

The Box filter blurs edges, strong point targets or high frequency texture variations and the geometrical filter preserves edges well but blurs point targets, thin curvilinear structures, and high frequency texture variation, because the same process is always applied whatever the case.

The most well known filters are adaptive filters based on the local statistics such as the Lee filter [Lee, 86], Frost filter [Fro, 82], and Kuan filter [Kuan, 85]. They reduce the speckle noise as a function of degree of heterogeneity measured by the local coefficient of variation. In the presence of point targets and structural features such as lines or edges, if fixed windows are used, the speckle filtering is weak.

Hence refined algorithms have been proposed first by Lee [Lee, 81] for the edge problem only. Edge denoising and preservation are performed by redefining the neighbourhood near the high contrast region.

Later, Wu et Maitre [WU, 90] considered the largest homogeneous window containing the pixel under consideration, using various standard deviation thresholds (quite difficult to determine experimentally) to test the windows homogeneity.

After these studies, Nezry et al [Nez, 91] proposed an adaptive procedure using local statistics but also taking into account geometrical information.

3. METHODOLOGY

3.1. Filtering process

Nezry et al [Nez, 91] proposed a process of filtering controlled both by the coefficient of variation and by the geometrical ratio operators (ratio edge, ratio line) and applied the Gamma MAP filter in selected window. Desnos and al [Des, 93] used the same scheme used in [Nez, 91] but they replaced the Gamma MAP filter by the Enhanced Lee filter used in [Lopes, 90] on fixed window.

The scheme [Nez, 91] is based on the computation of the coefficient of variation. This parameter is a very efficient and robust index of textural information which measures the image homogeneity. For an amplitude image, as a function of the observed coefficient of variation C_a, we can distinguish the following cases:[Lop, 90].

i) $C_a \le C_u$	Homogenous area
ii) $C_u < C_a < C_{max}$	Textural area with possible presence of edge and linear feature.
iii) $C_a \ge C_{max}$	Presence in the area of strong scatters or edges or linear features.

where $C_a = \frac{\sigma_g}{\bar{g}}$ is the observed coefficient of variation.

C_u is usually taken for an amplitude image equal to $C_u = \frac{0.523}{\sqrt{L}}$ where L is the number of looks. The upper threshold C_{max} is more difficult to determine and has been chosen as the maximum value of the coefficient of variation computed in an homogeneous area corresponding to the value beyond which the speckle can no longer be considered as fully developed.

There is obviously some need to check the presence of a feature (edge or line) to be preserved in case ii), and to identify the nature of heterogeneity in case iii). In the latter case the neighbourhood of the pixel to be filtered is to be redefined and also in case ii) if an edge or a line is detected.

In order to have a best spatial adaptivity, the statistical parameter C_a is computed on variable windows determined thanks to the geometrical ratio operators (ratio edge, ratio line), the new adaptive windows are selected, fixed the threshold values.

Edges between extended areas of different $\sigma°$ and curvilinear features swinging within a quasi homogeneous area are modelled by 2 strips or 3 strips of mean intensities P1, P2 respectively with N1 or N2 pixels.

The corresponding detectors are:

The ratio edge $r_2 = \min\left(M_1/M_2, M_2/M_1\right)$

The ratio line $r_3 = \min\left(M_1'/M_2', M_2'/M_1'\right)$

Where M_1, M_2, M_1', M_2' are the averaged pixel values within subwindows of mean intensity P_1, P_2.

The most probable orientation of the possible edge or linear feature is obtained by tacking the minimum value of the ratio r_2 or r_3 computed over usual four principal directions [Tou, 88].

An edge is detected if $r_2 < r_{2min}$ where the threshold r_{2min} is obtained from the conditional pdf defined as [Tou, 88]:

$$p(r_2/C) = \frac{n\Gamma(2NL)}{(\Gamma(NL))^2}\left[\frac{p^{NL}}{\left(r_2^n + p\right)^{2NL}} + \frac{(1/p)^{NL}}{\left(r_2^n + 1/p\right)^{2NL}}\right] r_2^{2NL-1} \quad (2)$$

Where :

$P = \frac{P_1}{P_2}$ is the ratio between the mean powers P_1 and P_2.

n=1 for a power image, n=2 for a amplitude image.

C: contrast ratio of the homogenous areas.

N: number of pixels in each homogeneous area .

L: number of Looks.

The conditional probability of detection (pd) within a boundary between two homogeneous regions of a contrast ratio C is:

$$Pd(r_{2\,min}, C) == \int_0^{r_{2\,min}} P\,(r_2/C)\;dr \qquad (3)$$

The probability of false alarm P1(fa) is defined by:

$$P1(fa)(T) = P(d/C=1) \qquad (4)$$

Considering the four directions, an experimental relation has been established, is given by:

$$P_4(fa) = 1 - (1 - P_1(fa))^3 \qquad (5)$$

Likewise of r2, the distribution probability function of $r_3 \in [0,1]$ is computed and here after the r_{3min} value is derived [Nez, 91]. For an amplitude image we have:

$$f(r_3) = \frac{2\Gamma(N_1L+N_2L)}{\Gamma(N_1L)\Gamma(N_2L)}\left[\left(\frac{P_2}{P_1}\right)^{N_1L}\left(\frac{N_2}{N_1}\right)^{N_2L}\frac{r_3^{(2N_1L-1)}}{\left(r_3^2\frac{P_2}{P_1}+\frac{N_2}{N_1}\right)^{N_1L+N_2L}} + \left(\frac{P_1}{P_2}\right)^{N_2L}\left(\frac{N_1}{N_2}\right)^{N_1L}\frac{r_3^{(2N_2L-1)}}{\left(r_3^2\frac{P_1}{P_2}+\frac{N_1}{N_2}\right)^{N_1L+N_2L}}\right] \qquad (6)$$

Like for the edge detector the threshold r3min is computed numerically by using the contrast ratio value $C=\min\left(p_1/p_2, p_2/p_1\right)$ and the probability of detection $P(d/C)$ is given by:

$$P(d/C) = \int_0^{r_{3\,min}} f(r_3/C)\,dr3 \qquad (7)$$

The probability of false alarm for one direction is given by:

$$P_1(fa) = P(d/C=1) \qquad (8)$$

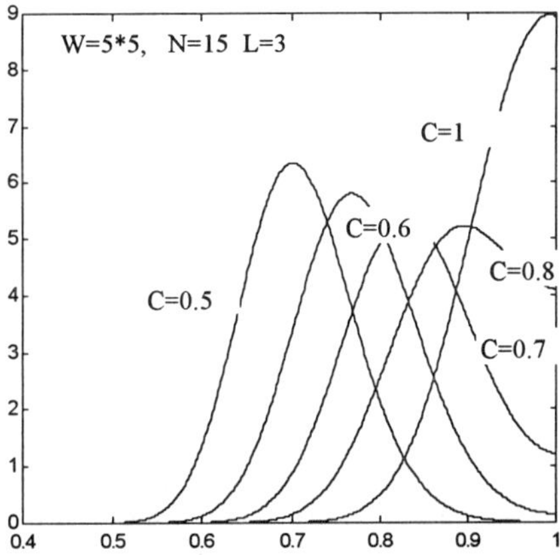

Figure 1 : edge probability distribution for a window 5*5

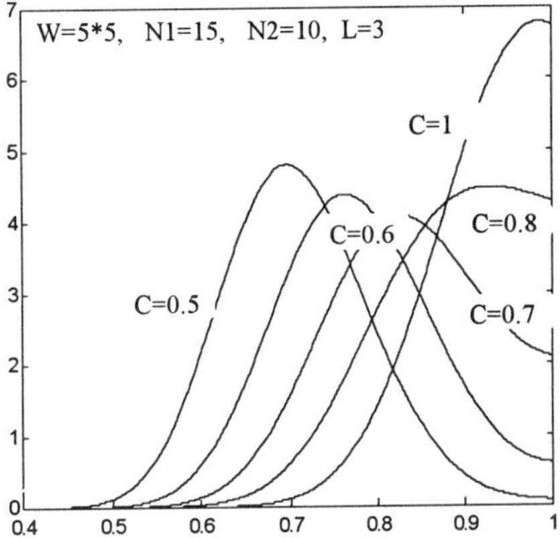

Figure 2 : line probability distribution for a window 5*5

Table 1 : Contrast ratio and threshold values for pd=95%

W=5*5 ; N1=15; N2=10					
C	r_{2min}	$P_4(fa)$	r_{3min}	$P_4(fa)$	
1.0	0.76	0.029	0.728	0.017	p1fa=1%
0.8	0.98	0.99	0.98	0.997	
0.7	0.97	0.98	0.975	0.991	Pd=95%
0.6	0.92	0.81	0.93	0.90	
0.5	0.84	0.27	0.75	0.3	

In figures 1 and 2, some cases of ratio edge and ratio line for N2=10 and N1=15 (Corresponding to 5*5 processing windows) are shown.

With probability of detection pd=95%, the ratio thresholds and the corresponding false alarm probabilities for some contrast values C and 5*5 spatial operator are indicated in table 1.

Threshold values must be chosen looking both at the probability of detection pd and at the Pfa. We can see for example that, if we choose r2min=0.84 and r3min=0.75, we obtain contrast ratio $C \le 0.5$ and pfa4 $\approx 3\%$.

3.2. Enhancement of the Frost Filter

Frost et al [Fro, 82] proposed a spatial domain adaptive wiener filter based on the local statistics and the multiplicative noise model. It averages less in the edge areas to preserve the edge structure.

The speckle noise model adopted for this filter has the form:

$$g(i,j) = [f(i,j)n(i,j)] * h(i,j) \tag{9}$$

where h(i,j) is the system response function, and * denotes the convolution operator. The algorithm has been developed for any systems response function. In practical applications, h(i,j) is assumed constant.

The impulse response m(t) obtained by minimising the mean square error (MMSE) ε given by:

$$\varepsilon = E\left[(f(t) - g(t) * m(t))^2\right] \tag{10}$$

Where t=(i,j) is the spatial coordinate.

$$m(t) = K_1 . e^{-KC_a . d(t)} \tag{11}$$

Where K_1 is a normalising constant, K is the filter parameter, C_a is the local observed local coefficient of variation ($C_a = \frac{\sigma_g}{\bar{g}}$)is computed over uniform moving window (σ_g^2 and $\bar{g}$ are the local variance and average over uniform moving window), $d(t)=d(i,j)=\sqrt{i^2+j^2}$ is the distance between the central pixel and considered one.

The filter is theoretically valid only within homogeneous areas where f(t) is stationary, it can be even applied in practice to the areas where the stationary is not verified [Lop, 90].

The filter must theoretically average well homogeneous areas and preserve well edges and texture at the same time. This result is not conformed practically for different variation coefficient value and for different K values.

The previous impulse response must be replaced by the following one:

$$m(t) = K_1 . e^{-K\frac{C_a - C_u}{C_{max} - C_a} . d(t)} \tag{12}$$

Where C_{max}, C_u are constants satisfying the following one, hence the filter estimated is:

1- $\tilde{f} = \bar{g}$ for $C_a \le C_u$

2- frost filter for $C_u < C_a < C_{max}$

3- $\tilde{f} = g$ for $C_a \ge C_{max}$

3.3. The implemented algorithm

In the implemented algorithm, we have applied the enhanced Frost filter in every adequately window or subwindow selected by the ratio operator (ratio edge, ratio line). The Euclidean distance d(t) calculated in accordance with the orientation and size of selected window.

The implemented statistical and geometrical approach filter uses coefficient of variation and ratio operators, depends from several parameters, such as window size (N), Cu and Cmax, line wide (N_1), threshold values (ratio edge(r_{2min}), ratio line (r_{3min})), K parameter of the Frost filter.

An adaptive window of variable shape and of sufficient size can be determined and filtered by the enhanced Frost filter [Lop, 90] (see figure 3).

The scheme is the same developed by Nezry [Nez, 91] but here we check two times the value of C_a in the window, one inside the general algorithm, the other applying the enhanced Frost filter in every window or subwindow selected. In the Frost filter application, we must take into account the value of euclidean distance d(t) for different size and directions of selected subwindows.

The process allows the preservation of both textural and structural informations and a very high speckle filtering with the adaptive windows.

For a 3 looks amplitude SAR image an area is considered homogeneous if $C_u \cong 0.26-0.29$. So, the choice of C_{max} is more difficult. We have selected $C_{max} \cong 0.34$ [Lopes , 91].

The general scheme of the implemented algorithm is given by figure 3:

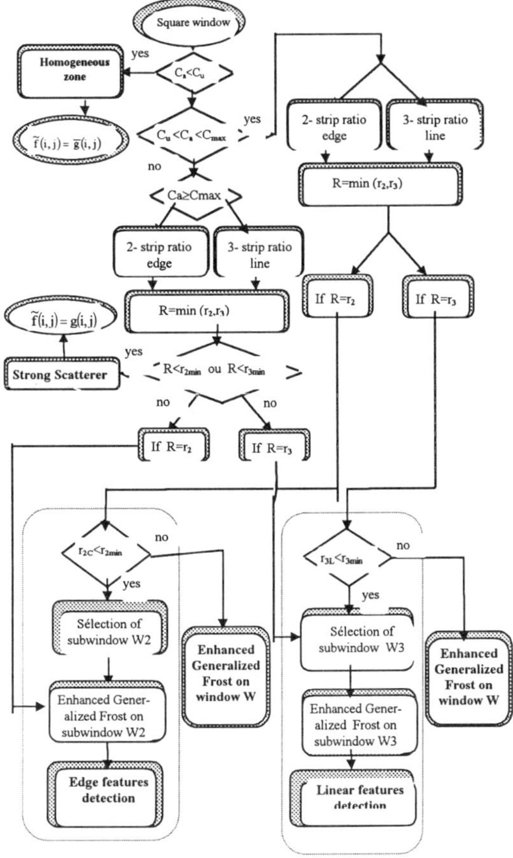

Figure 3: general scheme of the implemented algorithm

4. APPLICATIONS AND RESULTS

The modified statistical and geometrical filter has been applied on an ERS-1 SAR image corresponding to an area of southern Algeria called Laghouat with a 5*5 processing window .We have taken C_{max}=0.34 and C_u= 0.26-0.29, r2min (edge detector)=0.8, r3min (line detector) = 0.7, K = 1-4 (Frost parameter). Comparing its performance with the Frost filter [Fro, 82] and [Lop, 90], we can notice considerable compartments with respect to some important parameter.

To compare qualitatively speckle filtering techniques, a set of criteria will be established to measure the retention of the mean value in homogeneous areas, the speckle reduction capability, edge sharpness and thin feature preservation, and point target retention.

In the homogeneous areas, the ratio of standard deviation to the mean is a good measure of speckle strength for the filtered SAR images, this ratio is also frequently used to measure the amount of speckle reductions. We shall define the ratio as the speckle index.

$$B = \sqrt{\mathrm{var}(\hat{f})}/E[\hat{f}]$$

Where $\hat{f}$ is the speckle value of the filtered SAR image.

Another commonly used criterion is the equivalent number of Looks (ENL). The ENL for the amplitude image is defined as : $ENL = \left(\frac{0.522}{B}\right)^2$

We present in the table 2 and in the figures 4,5, and 6 some results obtained by the implemented algorithm.

Table 2 : Statistics of obtained results

Image	$\frac{\sigma}{\mu}$	ENL = $((\mu/\sigma)0.523)^2$
Original image of Laghouat (3 Looks)	0.257	4.41
Frost 5*5 C_u=0.26, C_{max}=0.34, K=4	0.1737	9.06
Implemented filter 5*5 C_u=0.26, C_{max}=0.34, K=1, r_{2min}=0.8, r_{3min}=0.7, N1=5	0.1617	10.46
Implemented filter 5*5 C_u=0.26, C_{max}=0.34, K=4, r_{2min}=0.8, r_{3min}=0.7, N1=15	0.1598	10.71
Implemented filter 5*5 C_u=0.26, C_{max}=0.34, K=4, r_{2min}=0.8, r_{3min}=0.7, N1=5	0.1708	9.37
Implemented filter 5*5 C_u=0.28, C_{max}=0.43, K=4, r_{2min}=0.8, r_{3min}=0.7, N1=15	0.133	15.46

Figure 4 : Original image of Laghouat region

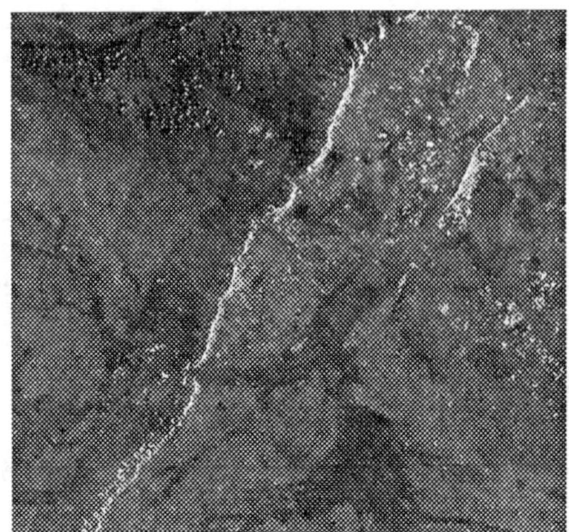

Figure 5-Frost filter C_u=0.26, C_{max}=0.34, K=4

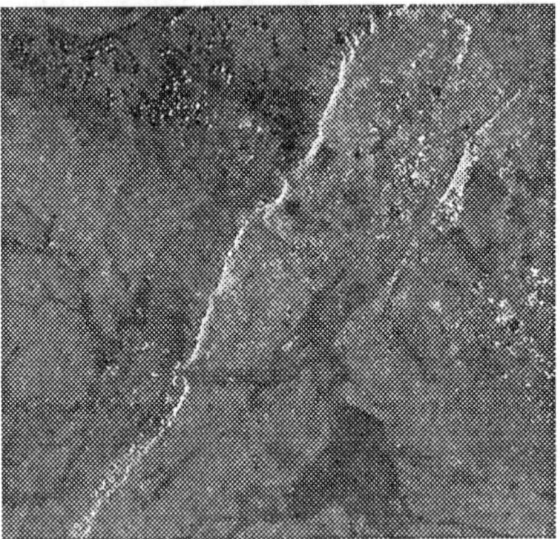

Figure 6 :Statistical & Geometrical modified filter 5*5
C_u=0.26, C_{max}=0.34, K=4
r_{2min}=0.8, r_{3min}=0.7, N1=5

5. DISCUSSION AND CONCLUSION

A refined filter based on ratio detector has been presented in this communication. Assessment of performance of these different non linear filters have been proposed.
Considering the disadvantages of the classical speckle filtering analysed, a different adaptive speckle has been implemented and applied on the ERS-1 images. This filter controlled by some parameters reduces better speckle keeping intact the response of points targets and preserving details.
The good filter performances obtained from the tests suggest the use of the algorithm for speckle filtering.

6. ACKNOWLEDGMENTS

The authors would like to acknowledge the European Space Agency (E.S.A) for providing the SAR ERS-1 data of the area of interest.

7. REFERENCES

1- CRIMMINS, T.R. 1985. *Geometric filter for speckle reduction* . Applied optics. Vol.24, N°10, pp. 1438-1445.

2- CURLANDER, J.C. and McDONOUGH, R.N. 1991. *Synthetic aperture radar: systems and signal processing.* Wiley Series in Remote Sensing, Wiley & Sons, New-York, 647 p.

3- Y.DESNOS, V.MATTEINI,1993. *Review on Structure Detection and Speckle Filtering on ERS-1 Images.* Earsel Advances in Remote Sensing, vol2, N°2, 1993.

4- FROST, V.S., STILES, J.A., SHANMUGAN, K.S. and HOLTZMAN, J.C. 1982. *A model for radar images and its application to adaptative digital filtering of multiplicative noise.* IEEE transactions on pattern analysis and machine intelligence, vol. Pami-4, n° 2, p. 157-165.

5- GOODMAN, J.W., 1976. *Some fundamental properties of speckle.* Journal of Optical Society of America, vol 66, n° 11, p. 1145-1149.

6- KUAN, D.T., SAWCHUK, A.A., STRAND, T.C., and CHAVEL, P., 1985. *Adaptative noise smoothing filter for images with signal dependent noise.* IEEE Transactions on Pattern Analysis and Machine Intelligence, vol. pami 7, n°2, p. 165-177.

7- LEE, J.S. 1981. *Speckle analysis and smoothing of synthetic aperture radar images.* Computer Graphics and Image Processing, vol. 17, p. 24- 31.

8- LEE, J.S. 1986. *Speckle suppression and analysis for synthetic aperture radar images.* Optical Engineering, vol. 25 , N°5, p. 636-643.

9- LOPES, A., TOUZI, R. & NEZRY, E. 1990. *Adaptative speckle filters and scene heterogeneity.* IEEE Transactions on Geoscience and Remote Sensing, vol. 28, n°6, p. 992-1000.

10- LOPES, A., NEZRY, E. & TOUZI, R. 1991. *Structure detection and statistical adaptive speckle filtering in SAR images.* IEEE Transactions on Geoscience and Remote Sensing, vol. 29.

11- METTERNICHT, G., 1993. *A comparative study on the performance of spatial filters for speckle reduction in ERS-1 SAR data.* Proceedings of the 13th EARSeL symposium, Dundee, U.K., p. 275-282.

12- NAGAO, M. and MATSUYAMA, T., 1980. *A structural analysis of complex aerial Photographs.* Plenum press, New-York.

13- NARENDRA, P.M. 1981. *A separable median filter for image noise smoothing.* IEEE transactions on pattern analysis and machine intelligence, vol.pami-3, N°1, January 1981,p. 20-29

14- NERZY, E., LOPES, A. & TOUZI, R.1991. *Dectection of structural and textural fearures for SAR images filtering*. Proceedings of IGARSS'91, Espoo, Vol. 4, pp. 2169-2172

15- SMARA, Y., BELHADJ-AISSA,A., SANSAL,B., 1996. *Speckle filtering of ERS-1 SAR images for multisource classification applied for semi-arid region of southern Algeria.* 16ème Symposium de l'EARSeL, 20-23 Mai 1996, MALTE.

16- TOUZI. R, LOPES.A,BOUSQUET.P, 1988. *A statistical and Geometrical Edge Detector for SAR Images*. IEEE Transaction On Geoscience and Remote Sensing, vol. 26, N°6,Nov. 88.

17- WU, Y. & MAITRE, H. 1990. *A speckle suppression method for SAR images using maximum homogeneous region filter.* IGARSS'1990, PP. 2413-2416.

Remote Sensing in the 21st Century: Economic and Environmental Applications, Casanova (ed.)
© 2000 Balkema, Rotterdam, ISBN 90 5809 096 5

Textural classification using textural signatures

A. Kourgli & A. Belhadj-Aissa
Laboratoire de Traitement d'Images, Institut d'Electronique (USTHB), Algiers, Algeria

ABSTRACT: This paper describes an approach to textural classification of remote sensing images. This study concerns the potential use of high spatial resolution textural features to identify different covers and discriminate between various surface types. These textural features are computed from grey-level cooccurrence matrix statistics at various pixel displacement distances. Textural classification is then performed using textural signatures. The approach is applied to a SPOT multispectral scanner data of the north of Algeria.

1 INTRODUCTION

As a main step in the understanding and description of satellite imagery, texture analysis is one of the most important techniques used in analysis, segmentation and classification of textured images (Kourgli 1998). Texture is simply defined as the relationships between grey levels in neighboring pixels which contribute to the overall appearance of an image . It represents the first level of spatial properties that can be extracted from a digital image.

In texture analysis, the most important task is to extract texture features which completely embody information about the spatial distribution of grey levels in the image. The two major approaches are : the statistical approach and the structural approach.

The first approach is related to statistical parameters which characterize the stochastic properties of the spatial distribution of neighboring grey levels of an image.

In the second approach, texture is described by the substructures, which are called primitives, their organization and their relationships.

A number of techniques for texture discrimination, texture analyzing and image segmentation have been proposed. See (Haralick 1979), (Gool et al. 1984), (Reed & Hans 1993) and (Kourgli & Behadj-Aissa 1997).

Our interest her will be in a purely statistical approach for textural classification which is based upon the grey level cooccurrence matrices. This approach has been successfully used on a number of texture analysis problems and these matrices are known to contain enough texture-context information to match a level of human texture perception. The aim is to develop a texture classification algorithm based on grey-level cooccurrence matrices (GLCM).

All statistical texture analysis algorithms share a common problem : providing a features set which measures a visually perceivable quality of texture patterns. The success of the GLCM representation is tied directly to the fidelity with which it capture the structure of the underlying texture. It has been shown to contain texture information, histogram information, shape information and edge information (Julesz 1972). In tests by Conners and al, comparing discrimination abilities of texture analysis tools the GLCM method performed much better than the grey level run length method, the grey level difference method and the power spectral method (Conners & Harlow 1980).

The GLC matrix is essentially a two dimensional histogram of the number of times that pairs of intensities values occur in a given spatial relationship. That it forms a summary of the subpatterns that could be formed by grey level pairs and the frequency with which they occur. This approach been used successfully on number of real-world texture analysis (Haralick 1986).

2 GREY LEVEL COOCCURRENCE MATRICES FEATURES

Cooccurrence matrices count how often pairs of grey level pixels, that are separated by a certain distance and lie along a certain direction, occur in a digital image.

They are based on the estimation of second-order joint conditional probability density functions $p(i,j \mid d,\theta)$. Each $p(i,j \mid d,\theta)$, ($\theta = 0°, 45°, 90°, 135°, 180°, 225°, 270°, 315°$) is the probability of going from grey level 'i' to grey level 'j', given that the intersample spacing is 'd' and the direction is given by the angle θ.For a subimage with a window size $M \times N$, $p(i,j \mid d,\theta)$is given by :

$$p(i,j \mid d,\theta) = \text{card}\{ (m,n), (k,l) \in M \times N / \; I(m,n) = i \; I(k,l) = j \quad \text{with}(d,\theta) \} \tag{1}$$

These density functions can be written in matrix form

$$\phi(d,\theta) \; (\theta = 0°,45°,90°,135°,180°,225°,270°,315°) \tag{2}$$

An element p_{ij} of the cooccurrence matrix $\phi(d,\theta)$ counts how often each pair of grey levels of pixels (i,j) occurs in a window containing N pairs.

$$p_{ij} = p(i,j \mid d,\theta) \; /N$$

If one choose to ignore the distinction between opposite directions; then symmetric probability matrices can be employed, then the GLCM , $S_{0°}(d)$, $S_{45°}(d)$, $S_{90°}(d)$ and $S_{135°}(d)$ can be used where :

$$\begin{aligned} S_\theta(d) &= 1/2\left(\phi(d,\theta) + \phi(d,\theta+180°)\right), \\ &= 1/2\left(\phi(d,\theta) + \phi^t(d,\theta)\right), \end{aligned} \tag{3}$$

The average of these four matrices is an average matrix $S_a(d)$ which provide an information about the overall repartition of grey levels of the image.

$$Sm(d) = \frac{1}{4}\begin{pmatrix} 1/2\left(\phi(d,0°) + \phi^t(d,0°)\right) + 1/2\left(\phi(d,90°) + \phi^t(d,90°)\right) \\ +1/2\left(\phi(d,45°) + \phi^t(d,45°)\right) + 1/2\left(\phi(d,135°) + \phi^t(d,135°)\right) \end{pmatrix} \tag{4}$$

Several measures are computed from the GLC matrix to describe specific textural characteristics of an image. A set feature is defined by Haralick (Haralick an al. 1973) (angular second moment, contrast, correlation, variance, inverse difference moment, sum average, sum variance, sum entropy, entropy, difference variance, difference entropy, information measures of correlation and maximal correlation coefficient), and by Conner (Trivedi and al. 1984) (Inertia, cluster shade, cluster prominence, local homogeneity, energy and entropy).

Commonly used statistics from the GLCM include mean, variance, entropy, energy, local homogeneity, correlation, inertia, cluster shade and cluster prominence. For a subimage containing L grey levels, they are defined as :

2 .1 *Mean (Average)*

$$\mu x = \sum_{i=0}^{L-1} i \sum_{j=0}^{L-1} pij$$

$$\mu y = \sum_{j=0}^{L-1} j \sum_{i=0}^{L-1} pij \tag{5}$$

2 .2 *Variance*

$$\sigma x^2 = \sum_{i=0}^{L-1} (i-\mu x)^2 \sum_{j=0}^{L-1} pij$$

$$\sigma y^2 = \sum_{j=0}^{L-1} (j-\mu y)^2 \sum_{i=0}^{L-1} pij \tag{6}$$

2 .3 *Energy or angular second moment*

$$E = \sum_{i=0}^{L-1} \sum_{j=0}^{L-1} (pij)^2 \tag{7}$$

2 .4 *Entropy*

$$H = \sum_{i=0}^{L-1} \sum_{j=0}^{L-1} pij \; \log(pij) \tag{8}$$

2 .5 *Inertia (Contrast)*

$$I = \sum_{i=0}^{L-1} \sum_{j=0}^{L-1} (i-j)^2 \; pij \tag{9}$$

2 .6 *Local homogeneity(inverse difference moment)*

$$L = \sum_{i=0}^{L-1} \sum_{j=0}^{L-1} \frac{1}{1+(i-j)^2} \; pij \tag{10}$$

2 .7 *Correlation*

$$C = \sum_{i=0}^{L-1} \sum_{j=0}^{L-1} \frac{(i-\mu x)(j-\mu y) \; pij}{\sigma x \; \sigma y} \tag{11}$$

2 .8 *Cluster Shade*

$$A = \sum_{i=0}^{L-1} \sum_{j=0}^{L-1} (i+j-\mu x-\mu y)^3 \; pij \tag{12}$$

2 .9 *Cluster Prominence*

$$B = \sum_{i=0}^{L-1} \sum_{j=0}^{L-1} (i+j-\mu x-\mu y)^4 \; pij \tag{13}$$

In this study, six of the measures above were used. they are : average, variance, inertia, correlation, cluster shade, cluster prominence. It is known that these measures do not gauge all the important texture-context information contained in a GLCM The principal drawback of many texture methods and the GLCM approach is the requirement of large memory and computation time.

3 NEW ALGORITHM FOR COMPUTING GLCM FEATURES

The practical problems encountered in calculating the GLCM are those of memory required an CPU time. Some users deal with the memory problem by limiting the number of grey levels in their data to reduce the size of GLCM (Marceau and al. 1990) or by proposing an alternative to GLCM (Unser 1986) (Peckinpaugh 1991). But these solutions may lead to a loose of texture information.

We propose to use a new algorithm (Kourgli 1997) which permit to compute some of the measures by the use of an improved method based on the computation of different histograms. So that, the double summation is reduced to a simple summation and instead of storing a matrix, we store one, two or three vectors.

From this algorithm six measures of the nine measures commonly used can be computed. They are : mean (1), variance (2), inertia (3), correlation (4), cluster shade (5) and cluster prominence (6) . Table 1 gives the corresponding histogram equations of these measures, for a symmetric matrix.

Table 1. : Texture Features.

	histogram equations
1	$\mu x = \mu y = \sum_{i=0}^{L-1} i \times occ(i)$
2	$\sigma x^2 = \sigma y^2 = \sum_{i=0}^{L-1} (i - \mu x)^2 \; occ(i)$
3	$I = 2 \sum_{i=0}^{L-1} i^2 \; occ(i) - 2 \sum_{i=0}^{L-1} i \; occ'(i)$
4	$C = 1/(\sigma x \; \sigma x) \left(\sum_{i=0}^{L-1} i \; occ'(i) - \mu x \; \mu x \right)$
5	$A = 2 \sum_{i=0}^{L-1} \left((i - \mu x)^2 (i - 4\mu x) \; occ(i) + 3(i - \mu x)^2 \; occ'(i) \right)$
6	$B = \sum_{i=0}^{L-1} (i-\mu x)^2 \left(2(i-\mu x)^2 - 4 \mu x (i-\mu x) + 6 \mu x^2 \right) occ(i)$ $+4 \sum_{i=0}^{L-1} (i-\mu x)^2 (2(i-\mu x) - 12 \mu x) \; occ'(i) + 6 \sum_{i=0}^{L-1} (i-\mu x)^2 \; occ''(i)$

Where :

$$occ(i) = \underset{(MxN)}{Card} \{j\} \quad \text{according to} \quad R:(d,\theta)$$

$$occ'(i) = \sum_{(MxN)} j \quad \text{according to} \quad R:(d,\theta)$$

$$occ''(i) = \sum_{(MxN)} j^2 \quad \text{according to} \quad R:(d,\theta)$$

The others measures : Energy, entropy and local homogeneity represent special cases and the equation can not be written in terms of histograms (Kourgli 1997).

4 TEXTURAL SIGNATURES AND TEXTURAL CLASSIFICATION

Sometimes, multispectral classification schemes are unable to discriminate between different covers. This phenomenon is du to the similarity of spectral signatures in visible and infrared wavelengths. Welch and al 1990 conclude that automatic classification techniques probably need to incorporate the textural information in the scene.

For classification purpose, we use a supervised clustering which consist in partitioning the image into groups according to textural features. So, each measure of the measures above is computed for various pixel displacement distances to constitute a textural signature.

Figure 1 shows variance signatures as functions of pixel distance *d* for three selected samples taken from a SPOT scene of the north of Algeria.

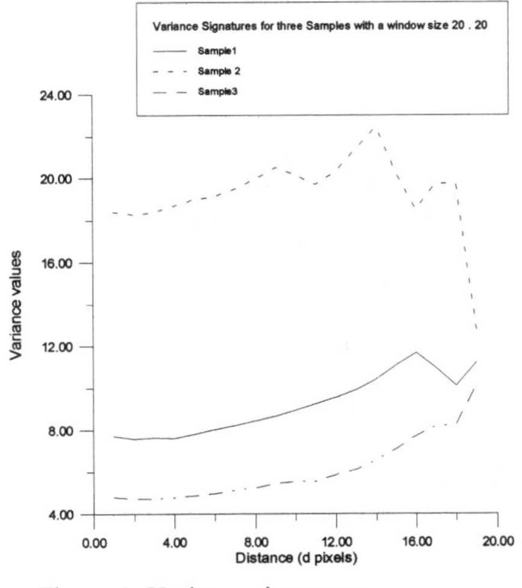

Figure 1. Variance signatures

Figure 2. Original Image

Figure 3. Ten selected samples 20 × 20

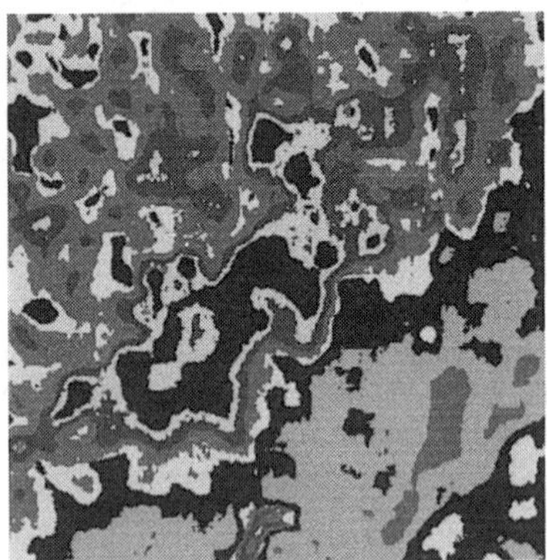

4.1 Image classified by 'mean' signatures

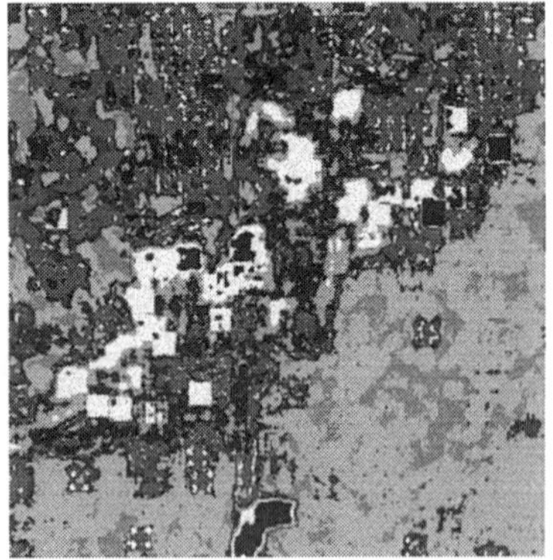

4.2 Image classified by 'variance' signatures

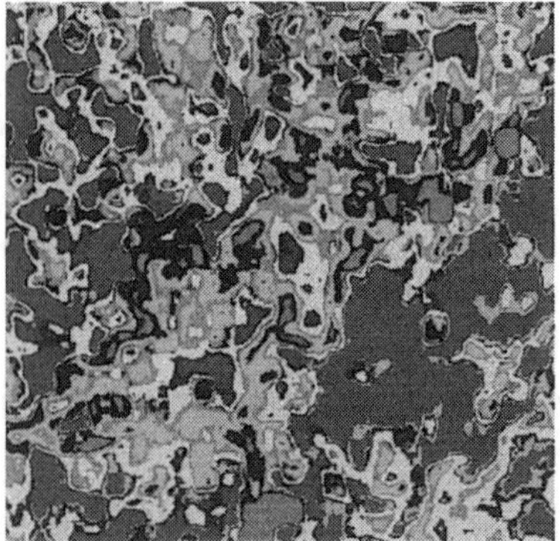

4.3 Image classified by 'correlation' signatures

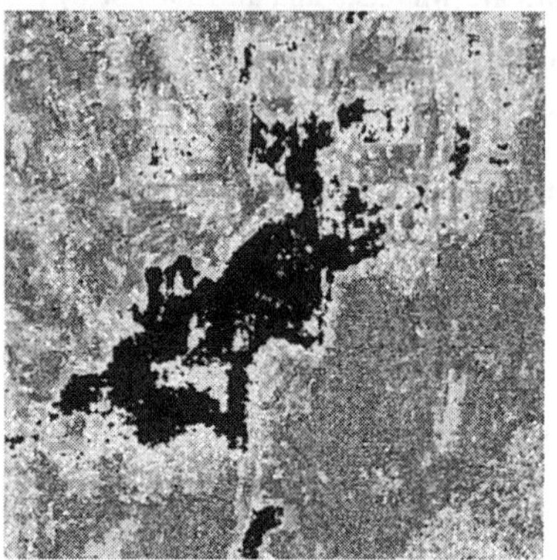

4.4 Image classified by 'inertia' signatures

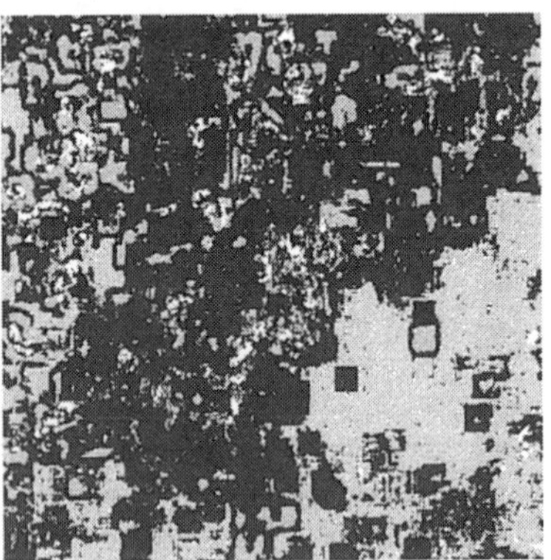

4.5 Image classified by 'cluster shade' signatures

Figure 4. Results of textural classification

The procedure of classification is described as follows :

1. 'C' samples with a window size T×T are selected from the image to constitute 'C' classes

1. For each sample, one of the six textural signatures is computed.

$S_i(d)$ for i=1,2,...,C and d=1,2,...,T-1

2. The global image is scanned with a window size T×T with a step of one pixel in the row and column directions, then, for each pixel :

- The textural signature S(d) of the moving window is computed.

- The textural feature S(d) is compared to the textural feature of each sample.

$$- D_i = \sum_{d=1}^{T-1} (S(d) - S_i(d))^2 \qquad i=1,2,...,C \quad (14)$$

-Then the pixel concerned is affected to the class which minimize D_i.

The method chosen produces then an image partition according to the textural signature chosen..

We applied these measures to a SPOT scene of an area of the north of Algeria. From the full scene an image window 512 × 512 was extracted (Fig.2). The classification is performed using an optimal moving window size 20 × 20 and ten classes (Fig. 3). For each pixel of the original image, the textural signature is computed according to the moving window. This signature is then compared to the ten classes signature and the concerned pixel is affected to the most similar class using the decision rule.

Figure 4 show the results of textural classification of the original image

5 CONCLUSION

This paper has described the approach for classifying textured images. The processing involves the computation of textural features using GLCM.

These results demonstrate the potential usefulness of the textural features. This approach can be easily used to provide simple algorithms for digital image processing. Indeed, the internal spectral variance of classes can be considered as a valuable source of spatial information that can be used as an additional feature in characterizing an identifying land covers.

Open problems that are still under investigation are :

- The unsupervised determination of the optimal window size for processing.

- The interpretation of the images obtained.

- The incorporation of each texture image created ,as an additional band, in spectral feature space for spectral classification purposes to study the possibility of improving classification accuracy.

REFERENCES

Conners, R.W. & C.A.Harlow 1980. A theoretical comparison of texture algorithms. *IEEE trans pattern Anal.Mach.Intell* .PAMI-2:,204-222 .

Gool, L.V. et al. 1985. Texture analysis anno 1983. SURVEY. *Computer Vision Graphics and Image Processing* 29 :336-357.

Haralick, R.M. et al. 1973. Textural features for image classification. *IEEE Trans.Syst.Man.Cybern.*SMC-8:601-621.

Haralik, R.M. 1979. Statistical and structural approaches to texture. *Proc.IEEE.*67:786-804.

Haralick, R.M. 1986. Statistical image texture analysis, *Chapter 11 of Handbook of pattern recognition and image processing.*

Julesz, B. 1972. Visual pattern discrimination . *IRE Trans. Inform.Theory* IT8:84-92.

Kourgli, A. 1997 Analyse de la texture des images satellitaire. *Master Degree U.S.T.H.B. Algiers.*

Kourgli, A. & A.Belhadj-Aissa 1997. Approche structurale de génération d'images de texture. *International Journal of Remote Sensing.* 18. No.17:3611-3627.

Kourgli, A. & A.Belhadj-Aissa 1998. Approche Statistique de Segmentation Texturale d'Images Satellitaires. Int.Conf. 'CNES Statellite-Based : A Tool for the study of the mediterranean', 23-27 November 1998, Tunis.

Marceau, D.J. et al. 1990. Evaluation of the grey-level cooccurence matrix method for land cover classification using SPOT imagery. *IEEE Trans. Geoscience. Rem. Sensing.* 28.N°4:513-518.

Peckinpaugh, S.H. 1991. An improved method for computing grey-level co-occurence matrix based texture measures. *CVGIP : Graphical models and image processing.* 53. N°6:574-580.

Reed, T.R. & J.LM.Hans du buf 1993. A review of recent texture segmentation and feature extraction techniques . *CVGIP:Image Understanding* 57.No.3:359-372.

Trivedi, M.M et al. 1984. Object detection based on grey level cooccurrence. *Computer Graphics an Image Processing* 28:199-219.

Unser, M. 1986. Sum and difference histogram for texture classification. *IEEE Trans.Pattern Anal.Mach.Intell.* PAMI-8. N°1:118-125.

Welch, R.M. et al. 1990. Cloud and surface textural feature in polar regions. *IEEE Trans.Geoscience.Rem.Sensing.* 28. N°4:520-527.

Remote Sensing in the 21st Century: Economic and Environmental Applications, Casanova (ed.)
© 2000 Balkema, Rotterdam, ISBN 90 5809 096 5

SAR imagery in urban area

L. Basly, F. Cauneau, T. Ranchin & L. Wald
Groupe Télédétection et Modélisation, Ecole des Mines de Paris, Sophia Antipolis, France

ABSTRACT: This paper deals with the major factors influencing the detection of urban fabrics in ERS-SAR images. Data, in SLC and PRI format, were acquired over the city of Nantes (West of France). The analysis of the data was performed in conjunction with aerial photographs, maps, meteorological data, as well as a field survey. The ways in which urban fabrics and their environment, data processing techniques, and weather conditions affect the radar signal are examined and discussed. It is found that many effects originate from the geometry of urban features, as well as their orientation relative to the spacecraft orbit. This investigation adds evidence to the fact that specular scattering prevails over the diffuse scattering in an urban area. Practical conclusions concerning the capabilities and the limitations of ERS-SAR imagery are drawn.

1 INTRODUCTION

Almost 70% of Europe's population live in cities, and urbanization is still continuing (EEA 1998). That is why urban planners and decision makers need tools for a comprehensive assessment of the urban environment.

A means for a better understanding of urban areas are remote sensing techniques (Bonn 1996). Satellite imagery can significantly improve the monitoring of cities in a wide range of applications, e.g., the detection of urban changes, mapping roads and streets (Couloignier 1998, Tupin *et al.* 1996), mapping urban demarcation (Weber 1995), mapping urban air pollution (Wald & Baleynaud 1999 ; Basly *et al.* 1999).

Nevertheless the number of studies using SAR data is much lower than that using optical ones (visible and infrared) notably concerning investigations on urban areas (Henderson & Xia 1997, Guerre 1995), though these studies brought to the fore the advantages of using radar images. Indeed, the radar signal is sensitive to the geometry of urban fabrics, and is used for detecting built-up areas as man-made features (Fellah *et al.* 1994, Carlotto 1996, Dousset 1997, Xia & Henderson, 1997, Lortic & Couret 1998). In addition, radar imagery may provide third dimensional information (Dong *et al.* 1997). Hence, SAR imaging techniques are a useful tool to analyse urban land use patterns.
In this perspective, the paper examines the major factors influencing the detection of urban fabrics, that is the orientation, the building materials, the meteorology, the SAR image format, and the quality of the data.

2 SAR DATA AND THE STUDY SITE

2.1 *Data and Pre-processing*

Two sets of images were used. One set of 5 ERS-SAR images in Single Look Complex (SLC) format, and one set of 7 ERS-SAR images in Precision Image (PRI) format. That represents 12 images spanning a period of almost 2 years, from October 1994 to October 1996. All data were taken in descending pass: radial direction is almost East-West.

It is worth noting that SLC data strongly differ from PRI data. On the one hand, SLC data are complex data, in slant range geometry (i.e. the range resolution varies with the incidence angle), and on the other hand, PRI data are system corrected, multilook processed (in order to reduce speckle noise), and in ground range geometry. Furthermore, the pixel size is also different (ESA 1993). After reducing the SLC data to magnitude images, they were remapped to a PRI image taken as a reference.
The speckle was not filtered out in each image. Several tools are available which are respectful of the most pronounced structures (see e-g- Ranchin & Cauneau 1994, Mathieu *et al.* 1998).

However, we have preferred not to alter the

structures any further after the geometrical rectification, and decided to reduce the noise by making a multitemporal average for each kind of dataset. Furthermore, averaging the images decreases the level of speckle preserving the structures, and this simple processing decreases the effects of the variability of the multi-temporal SAR images (Stroobants *et al.* 1996; Basly *et al.* 1997, 1999).

2.2 *The study area*

The city of Nantes is located in western France 54 km from the Atlantic ocean. The geographical co-ordinates are 47.23 N and 1.55 W. The location explains the oceanic climate which is quite mild. The city of crossed by the Loire river, and a natural island, the so-called "Beaulieu Island", is part of the city. Nantes has about 470 000 inhabitants and used to have a very active harbour some centuries ago with many exchanges with America and Africa.

3 INTERPRETATION OF THE ERS-SAR MEAN IMAGES

Screening the two mean images, we notice that the PRI one (Fig. 1) is smoother than the SLC one, due to the multi-look processing of the PRI which reduces the speckle noise. In both images, we observe a high variability of the backscatter due to the complexity of the urban landscape. Indeed, a town contains a wide variety of length scales. Large scales are provided by the ground and buildings themselves, and the small scales by the roughness of the surfaces (Taket *et al.* 1991).

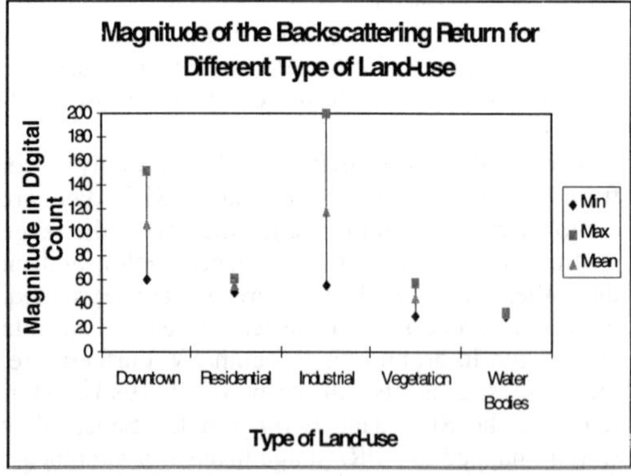

Figure 1. The return is large and highly variable in downtown and the industrial area. The backscattering echo is very low for the vegetation areas and water bodies. Concerning the residential areas, the values of the backscatter return are slightly larger and less variable that that of the vegetation area.

In urban areas, the specular reflection prevails over the diffuse scattering because of the geometrical properties of the objects (Dong *et al.*, 1997). Three mechanisms of specular scattering dominate:

- the simple bounce from tilted roof,
- the double bounce from wall-ground,
- the triple bounce from wall-wall-ground.

Figure I shows the values of the backscatter magnitude according to different types of area.
Both images were analyzed in detail together with with aerial photographs, maps, meteorological data, as well as a field survey, in order to assess the influence of the factors mentioned above.

3.1 *Artificial Areas*

3.1.1 *Downtown and Dense Urban*

The centre of town is comprised mainly of tall buildings with flat and tilted roofs. It is the ancient part of the city, and the density of buildings is large (Fig. 2 area (A)). The backscatter is large on average, but highly variable. The variability is due to the changes in street orientation and in building density.

A strong backscattering-echo comes from two groups of buildings located on either side of a green park (Fig. 2 (B)), enhanced by the low backscattering of the park. Nevertheless, one cannot conclude either on the major orientation of the buildings, or on the number of buildings.

Large anisotropic backscattering may occur when particular conditions are combined (Fig. 2 (C)). If a street is aligned with the flight direction of the satellite (i.e. when the illumination azimuth is orthogonal to the street direction), and if the ratio building height/street width satisfies a condition relative to the incidence angle, favourable multiple reflections occur, resulting in a high backscatter return (Gouinaud 199§j. The street direction may be thus determined.

Metal elements on the roofs generate enhanced diffuse backscattering, but this is an isotropic backscattering and the major building orientation cannot be deduced.

When the density is high, some buildings may be hidden from the radar illumination by the other buildings. This shading effect results into a weak backscatter.

In addition, on flat roofs, the radar signal is spectrally reflected, so there is no return to the satellite.

3.1.2 *Residential and Mean Urban*

Most of the residential houses are single or two storey-buildings with tilted roofs (Fig. 2 (D)). Like downtown there is no major orientation. On average the backscattering is lower than downtown.

High backscattering may arise from a tilted roof facing the radar wave. The roof can be considered as a rectangular smooth surface (i.e, a mirror, by analogy with the optical physics). Such a surface gives a very low radar return except in specular conditions at near vertical incidence (i.e. when the incidence angle is close to 0°) (Simonett & Davis 1983).

In a low density area, if a wall faces the illumination azimuth, a double bounce mechanism can occur resulting in a high return signal. In that case, the residential house can be compared to a dihedral corner reflector. The backscattering-echo of the double bounce mechanism is lower than a single bounce from a tilted roof, because there is a loss of energy for each reflection.

In both cases, the orientation and the bulk size of the buildings greatly affect the return of the backscattered energy. For example, a residential area made up of eleven buildings with twenty storeys, covering a surface of approximately 13 km2, has a weak return (Fig. 2 (E), Fig. 2). The low return is mainly due to the orientation of the

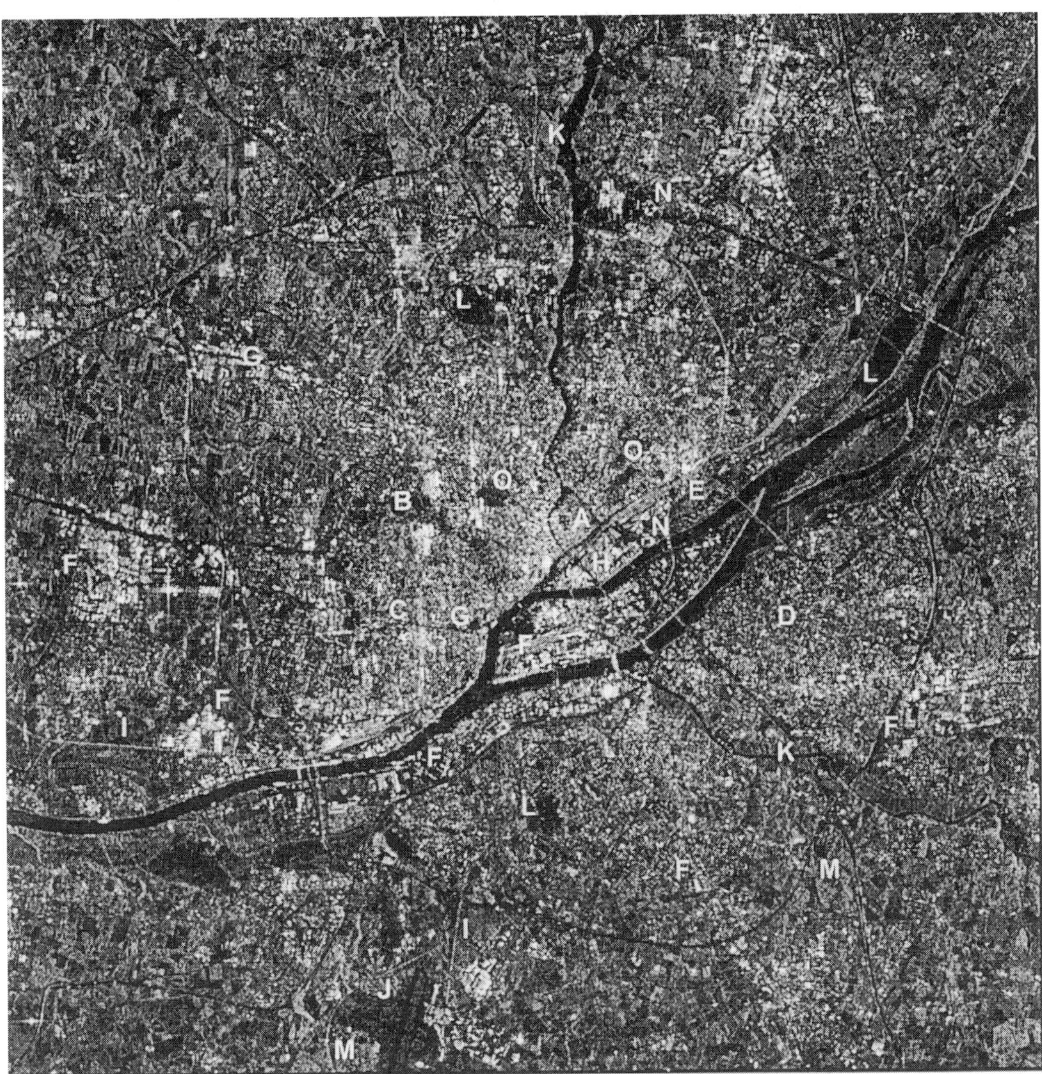

Figure 2. The city of Nantes viewed by ERS-SAR system imaging. This is a mean image derived from 7 PRI images. All the images were acquired in descending pass. The Loire river appears in dark. The airport is located to the South of the river. The island is part of the city.

buildings: firstly the frontages are perpendicular to the ground with a wave incident angle far from 0°, secondly they cover a large surface and do not face the radar beam. Accordingly, neither the single bounce mechanism nor the double bounce one can occur.

3.1.3 *Industrial and Commercial*

In those areas, buildings are often large, tall, with a flat metal roof, and density is low (Fig. 2 (F)). For such buildings, specular reflection occurs on the flat roof, and double or triple bounce mechanisms occur on the walls. Large returns occur if buildings face the illumination azimuth. In this case, they are well detected and their orientation and length (only one dimension) can be measured (Fig. 3). Otherwise, the buildings may not be detected or their dimensions remain unknown.

3.1.4 *Other Urban Features*

3. 1.4. 1 *Bridges*

All the bridges crossing the Loire river are visible. The radar return is large, because bridges are transversally oriented according to the radar wave. In addition, bridges are in contrast with their environment, i.e. the Loire river. We note that two bridges built after the aerial photographs were taken (1993), appear on the ERS-SAR images (Fig. 4).

Nevertheless, most of the bridges crossing the Erdre and the Sèvre river are longitudinally oriented, so they do not appear unless they are made of metal.

3.1.4.2 *Roads, Streets and Railway Tracks*

Roads and streets are smooth surfaces, and the backscatter return is always low. They appear in dark in the images when oriented along the wave vector (Fig. 2 (G)). Otherwise, they can be detected if their width is large enough, the critical width being a function of the height of the lining buildings through the shading effect, the type of building, the density of buildings, and the presence of vegetation, water bodies, car-park ...

Railway tracks appear in bright because large

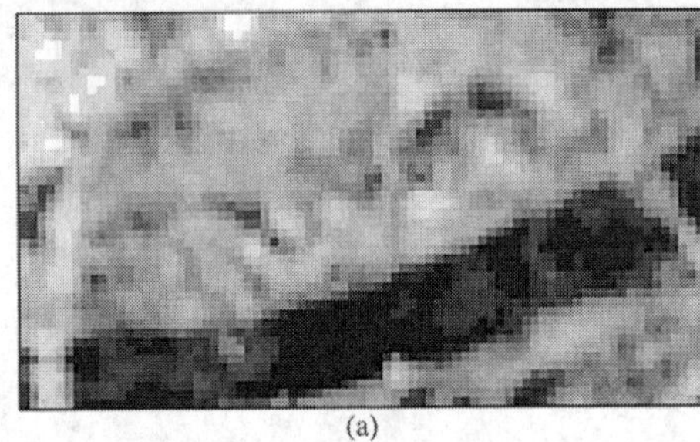

(a)

(b)

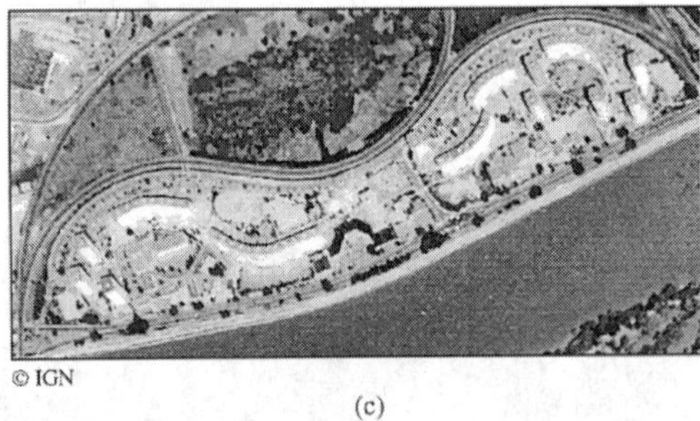

© IGN

(c)

Figure 3. The Malakoff group. Figure 2-a, detail of the SAR image. The Loire river appears in dark. The Vendée bridge is clearly visible on the right. Figure 2-b, photograph taken from the Beaulieu island. We have estimated the height of the higher buildings, and we have found 20 stores, so about 60 meters. Figure 2-c, aerial photograph enables the estimation of the surface covered and also the buildings count. We find approximately 13 km^2 of covered surface, eleven 20-storey buildings, five long buildings and some other buildings.

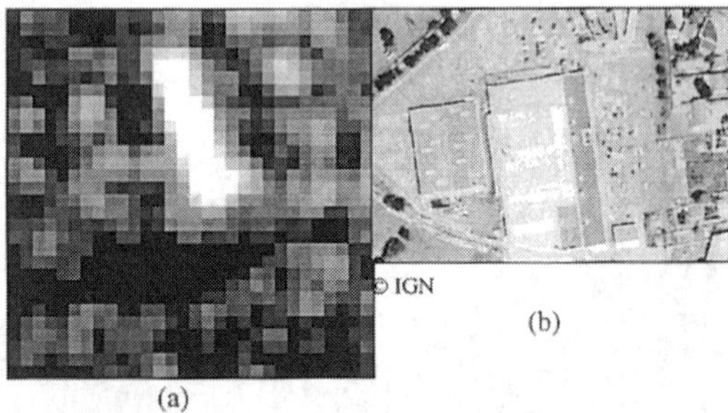

© IGN

(a) (b)

Figure 4. Here is a shopping center in the South of Rezé, a town on the South of the Loire river. Its dimension are approximately 150 meters long and 80 meters wide. Figure 5-a is a detail of the SAR image, the size of the pixel is about 13m. Figure 5-b is an aerial photograph taken in 1993 by IGN, the spatial resolution is about 0,7 meter. Although the shopping center don't exactly face the radar wave, the backscattering is very large. Indeed, the frontage has a considerable surface, and the car-park, in front of it, enable double bounce reflection. The detail of the SAR image features an anisotropy that allows the knowledge of the orientation of the building. The building on the left of the shopping center is not well perceived because of the shadowing effect, so the backscattering is caused by the diffuse backscattering of the roof which is certainly a metal one.

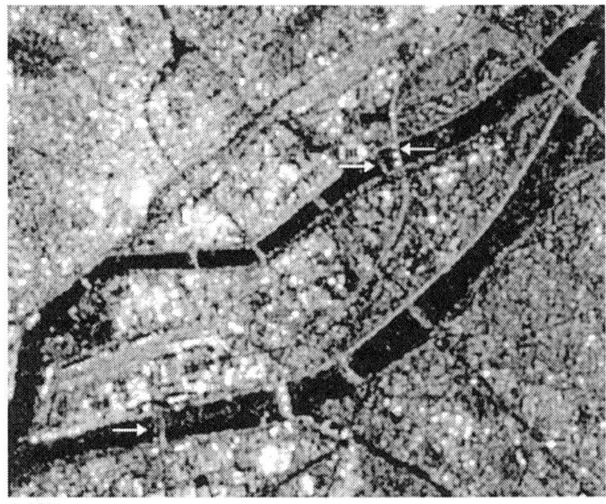

Figure 5. View of the Beaulieu island which is part of the city of Nantes. Detail of the PRI-SAR mean image showing the eleven bridges crossing the Loire river. The brighter one is made of metal. We notice that the pillars of one bridge in the Northern of the island are visible, that bridge is also made of metal (right arrow). The bridges showed with the right arrows are not visible on the aerial photographs (not shown here).

diffuse backscatter arise from electrical wiring and rail made of metal (Fig. 2 (1)). The detection of railway tracks is subjected to their neighbourhood, whatever their orientation. If the backscatter return of the neighbourhood is equal or larger than that of the railways, the latter are undetected. Otherwise, the railways may be shaded by high buildings.

3.1.4.3 *Airport*

The airport appears clearly in the image (Fig. 2 (J)). The runway and taxi-way are visible, and appear in dark because of the specular reflection.

3.2 *Natural areas*

3.2.1 *Inland Water*

Rivers appear in dark on the image because of the specular reflection on a smooth surface (Fig. 2 (K)). However, by analysing several multi-temporal images, we noticed a strong variability in the backscatter of the Loire river, both in time and space. We found that this variability varies according to the wave regime that modified the roughness of the river, and that the wind direction has a greater effect than speed.

3.2.2 *Vegetation*

Parks, sports fields and grass areas, have a very low diffuse backscatter return (Fig. 2 (L)). On the contrary, areas with trees, small shrubs, and tall vegetation may cause a considerable isotropic backscattering-echo (Fig. 2 (M)). Though the intensity of this diffuse backscatter is lower than for a built-up area, vegetation can preclude the detection of small urban fabrics.

We can note that stadium are well detected when surrounded by terraces. The contrast between the terraces and the field helps in determining the major orientation (Fig. 2 (N)).

The cemeteries are also well detected by contrast with the built-up neighbourhood which is brighter (Fig. 2 (0)).

4 CONCLUSIONS AND PERSPECTIVES

Different types of urban areas can be detected in SAR images once properly processed.

In high density built-up areas (downtown) the backscattering-echo is large and rarely anisotropic except in the case of favourable multiple reflections.

In low density areas (residential areas industrial areas), the backscattering-echo may be large but is inferior, on average, to high density areas. Anisotropic backscattering especially occur in industrial areas where there are large buildings.

Although diffuse backscattering is not a dominant mechanism in urban areas, that kind of backscatter can be larger than the triple-bounce mechanism, notably when metal elements are encountered.

Areas with vegetation are generally well distinguished (parks, fields), but the presence of tall vegetation within urban fabrics can lead to misinterpretation.

Streets, highways and rail networks are incompletely detected.

It is worth noting that the orientation of urban fabrics greatly affect the backscatter return. Hence, making use of both an ascending and a descending pass will be useful to determine the orientation of observed objects.

This study underlines the complexity of interpretation of SAR data over urban areas. SAR images cannot be analysed at the pixel level as in optical imagery. Each pixel has to be analysed within its environment by a statistical or textural approach (Carlotto 1996, Basly *et al.* 1997).

As in optical imagery, urban fabrics should be studied by means of multi-source SAR imagery. For example, future studies making use of multi-incidence angles, multi-polarization and multi-pass orbiting will be more powerful in extracting valuable urban information from SAR data. That will be performed with the Advanced Synthetic Aperture Radar of the ENVISAT satellite (ESA 1998).

ACKNOWLEDGMENTS

The authors are indebted to the European Space Agency which has provided the ERS data under the investigation of the A02.Fl13 project. The authors also thank Météo-France for providing the meteorological data.

REFERENCES

BASLY L., CAUNEAU F., RANCHIN T. & WALD L. 1997. ERS SAR imagery for urban climate studies, in : European Space Agency ed., Proc. of the 3rd ERS Symposium ESA SP-414 : Space at the service of our Environment. Florence, Italy, 17-21 March 1997, Noordwijk : ESA-ESTEC, 233236.

BASLY L., BALEYNAUD J.M. & WALD L. 1999. Satellite Data for Air Pollution Mapping, in : Proc. of the 18th EARSeL Symposium on Operational Remote Sensing for Sustainable Development, Enschede, Netherlands, 11-14 May 1998, 133-139.

BASLY L., CAUNEAU F., COUVERCELLE C., RANCHIN T. & WALD L. 1999. SAR Imagery for Urban Air Quality, in : Proc. of the 18th EARSeL Symposium on Operational Remote Sensing for Sustainable Development, Enschede, The Netherlands, 11-14 May 1998, 165-170.

BASLY L., CAUNEAU F., COUVERCELLE C., RANCHIN T. & WALD L. 1998. *Capabilities of ERS Data Applied to Urban Climate Studies*, Ecole des Mines de Paris, Centre d'Energétique (Sophia Antipolis), Final Report A02.Fl13 . project, European Space Agency, TM/98/R/1, 37 p.

BONN F. 1996. *Précis de Télédétection*, Saint-Foy (Québec, Canada): Presses de l'Université de Québec/AUPELF, 633 P.

CARLOTTO J.M. 1996. Detecting Man-Made Features in SAR Imagery, in : Proc. IGARSS, IEEE.

COULOIGNIER I., RANCHIN T. & VALTONEN V.P. & WALD L. 1998. Benefit of the future SPOT-5 and of data fusion to urban roads mapping, *Int. J. of Remote Sensing*, 19 (8): 1519-1532.

DONG Y., FORSTER B. & TICEHURST C. 1997. Radar backscatter analysis for urban environments, *Int. J. of Remote Sensing, (18), 6,1351-1364.*

DOUSSET B. 1997. Interpretation of ERS-SAR Images over Urban Surfaces, in : ESA SP-414, Proc. of the 3rd ERS Symposium : Space at the service of our Environment, Florence, Italy, 17-21 March 1997, Noordwijk : ESA-ESTEC, pp. 53-57.

MATHIEU P., DIRAT L., DUPUIS X. & BARLAUD M. 1998. Speckle Reduction with Edge Preserving, ICASSP'97, Munich Germany, April 21-24 1997. Vol. 4, 2785-2788.

EEA 1998. *Europe's Environment: The Second* Assessment. Copenhagen (Denmark) : European Environment Agency, 293 p.

ESA 1998. SP-1218, *Envisat Mission Opportunities for Science and Applications*, Noordwijk : ESA-ESTEC, 59 p.

SIMONETT D.S. & ROBERT E.D. 1983. *Image Analysis-Active Microwave,* in : Manual of Remote Sensing, 2nd Edition, Vol. 1, Chap. 25, pp. 1125-1181.

STROOBANTS W., LEDEZ C., PENICAND C., VALLET P. & RUDANT J.P. 1996. Une approchc pragmatique dc la l'usion multitemporelle des images ERS-1, in Thierry Ranchin and Lucien Wald eds., Proc. of the conference "Fusion of Earth data: merging point measurements, raster maps and remotely sensed images", Cannes, France, February 6-8 1996, Nice (France) : SEE/URISCA (1996), 131-142.

TACKET N-D-, HOWARTH S.M. & BURGE R.E. I991. A Model for the Imaging of Urban Areas by Synthetic Aperture Radar, *IEEE Transactions on Geoscience and Remote Sensing,* 19 (3), 432-443.

TUPIN F., GOUINAUD C., MAITRE H. & CRETTEZ J.P. 1996. Détection du réseau routier sur des images radar ERS-1, *FRIA AFCET 96 (Rennes)*, 43-52.

WALD L. & BALEYNAUD J.M. 1999. Observing Air Quality over the City of Nantes by Means of Landsat Thermal IR Data, *Int. J. Remote Sensing*, 20 (5): 947-959.

WEBER C. 1995, Images satellitaires et milieur urbain, Paris: Hermès, 185 p.

Remote Sensing in the 21st Century: Economic and Environmental Applications, Casanova (ed.)
© 2000 Balkema, Rotterdam, ISBN 90 5809 096 5

Scale evolution due to hydrodynamic modulation of SAR imaging

I-Fan Shen, Min-Jia Liou & Yun-fei Li
Department of Mechanics and Engineering Science, Fudan University, Shanghai, People's Republic of China

ABSTRACT: Action balance equation describes hydrodynamic modulation of short Bragg waves by non-uniform velocity fields. This modulation shows the results of cross section variation is low-pass filtering of tidal current gradient and of the gradient of underwater bottom topography. Morphological filtering determined by evolutionary partial differential equations, covers linear low-pass filters and a wide class of non-linear filters. The study shows a possible empirical approach to explore the inversion of bottom topography in scale evolution.

1 INTRODUCTION

Owing to the occurrence of new technique in remote sensing, it has become possible to detect, with Synthetic Aperture Radar (SAR), radar waves Bragg-scattered by the free sea surface roughness, ranging from a few centimetres (X band) to a few metres (L band). The ocean features including ocean wave spectrum, underwater topography and the inversion of wind mapping, can be obtained from radar imageries. Since sea roughness, considered as small waves, is strongly modulated by large-scale waves, therefore, the demand for a sound knowledge of wave-wave interaction has increasingly generated research (Shen *et al.* 1994). For simplicity of the wave-wave interaction model, a two-scale model of the radiation balance equation is introduced for describing the weakly nonlinear hydrodynamic modulation of short waves riding on non-uniform velocity backgrounds. In the balance equation a conservation equation of spectrum action is balanced by a source function S. As Hughes (1978) introduced, the source function is given by $S = \beta N(1-N/N_0)$, where N is the action spectrum which is defined by the surface height spectrum, N_0 is the equilibrium action spectrum and β is the spectral relaxation rate, i.e. the wind induced growth rate, for the frequency range $g/2\pi U_{10}$ to 20 Hz. The well-accepted expression form, Eqn. (1) for β was given by Plant (1982).

$$\beta = \frac{(0.04 \pm 0.02)u_*^2 \omega \cos\theta}{C^2} \qquad (1)$$

Here g is gravitational acceleration, U_{10} is wind speed at 10 m above the sea level, u_* is air friction velocity, $\omega(= 2\pi f)$ is radian wave frequency, C is phase speed and θ is the angle between wind and wave directions. However, the expression of the spectral relaxation rate, which depends on the angle between wind and wave direction, does not seem to be very clear. Mitsuyasu & Honda (1982) also explained that there is an almost linear relation between dimensionless growth rate of water waves and dimensionless friction velocity of wind due to the logarithmic plotting coordinates. However, the data is still scattering, even when the experiment was carefully controlled. As the equilibrium action spectrum is introduced, so it is a straightforward way to obtain either a time-dependent solution (Ouchi et al. 1997) or a time-independent solution (Alpers & Hennings 1984) for the perturbed action spectrum of the radiation balance equation. If a time-dependent solution is explored while the angle between wind and wave directions is perpendicular, then the spectral relaxation rate reduces to zero. Then there will be a need to pay much attention to the time-dependent perturbed solution when a sinusoidal variation leads to being mathematically deficient, if the integration interval is from $-\infty$ to t. Besides, the relaxation time is the order of one minute. So the validation of Eqn. (1) applied to action balance equation needs to be further verified.

Based on steady-state action balance equation and Fourier perturbation solution, Alpers and Hennings (1984) proved that SAR imaging is a process of low-pass filtering to tidal current gradient and also to the depth gradient of bottom topography. In image processing filtering means an operator applying to an image; it is an operation in order to smooth out asperity of high frequencies. Traditionally, linear filtering methods are often based on different diversities of Fourier transform, which attenuates high frequencies. Obviously, Fourier treatment is a global operation, whose local result depends on global events such as the size and location of the observation window of the image. Furthermore, the data of most images is nonlinear and non-stationary, so Fourier treatment may be limited to nonlinear and non-stationary data (Huang *et al.* 1998). On the other hand, state-of-art tendency in image processing is a way to define nonlinear iterated filtering methods which are as independent as possible of the observation window. It is well known that most local filtering operations can be translated into differential operators and all local linear low-pass iterated filters are equivalent to a heat equation (Guichard and Morel 1995). Morphological filtering covers linear low-pass filters and a wide class of nonlinear filters. So in this study we would like to explore SAR imaging of bottom topography by means of scale-space analysis of morphological filters.

2 MODULATION THEORY

In this section, basing on the methodology developed by Alpers and Hennings (1984), we would like to extend filtering mechanism of underwater bottom topography in SAR imaging and to investigate the application of morphological filter in the inversion of bottom topography.

2.1 *Hydrodynamic Modulation*

It is assumed that tidal current variation leads only to small deviations of the action density from equilibrium. The theory, which describes the variation of spectral energy density of linear short waves in a slowly varying current field, is the action balance or radiation balance equation, Eqn. (2). The equation is given as follows:

$$\left[\frac{\partial}{\partial t}+\left(C_g+\mathbf{U}\right)\frac{\partial}{\partial \mathbf{x}}-\mathbf{k}\cdot\frac{\partial \mathbf{U}}{\partial \mathbf{x}}\frac{\partial}{\partial \mathbf{k}}\right]N=S \tag{2}$$

Eqn. (2) describes the energy transport and refraction and the interaction with a varying surface current. The source function S describes the rate of change due to the energy input by wind and the loss of energy by wave dissipation such as wave breaking or white capping. In a special case where the source function is null, Eqn. (2) describes that the action density is conserved.

Short Bragg scattering waves are subject not only to tidal current variation but also to the action of wind. The short surface waves would be in local equilibrium, and their spectrum would be also 'equilibrium' or 'saturation' when the tidal current is uniform. Tidal current variation perturbs the action density from equilibrium and causes a modulation of the energy spectrum of the short waves. Therefore, the action density and the surface current are taken as sums of a constant equilibrium term and a time-dependent perturbation term. The source term is also included. When wind action is taken into consideration, the relaxation time, the inverse of the spectral relaxation rate β, is the response time of the short waves modulated by the current variation in the order of tens of short wave periods. It is determined by the combined effect of wind excitation, energy transfer to other waves due to conservative resonant wave-wave interaction, and energy loss due to dissipative processes like local breaking. It may be expected that the relaxation time be of the order of 10-100 Bragg wave periods. If Bragg waves have a wave period of the order of 0.5 second, the relaxation time ranges from 5 seconds to 50 seconds (Alpers and Hennings 1984). So the relaxation time is treated as a free parameter or is determined by Eqn. (1). Alpers and Hennings (1984) also explained that the time derivative term can be ignored in Eqn. (2). Then Eqn. (3) is given as follows,

$$\left[\left(C_g+\mathbf{U}\right)\frac{\partial}{\partial \mathbf{x}}-\mathbf{k}\cdot\frac{\partial \mathbf{U}}{\partial \mathbf{x}}\frac{\partial}{\partial \mathbf{k}}\right]N=S \tag{3}$$

where $S=\beta N\,(1\text{-}N/N_0)$. The unperturbed terms and perturbation terms for $N(\mathbf{x},\mathbf{k})$ and $\mathbf{U}(\mathbf{x})$ are shown in Eqns. (4) and (5) where $\mathbf{k}$ is the wave number of Bragg waves.

$$N(\mathbf{x},\mathbf{k})=N_0(\mathbf{k})+\delta N(\mathbf{x},\mathbf{k}) \tag{4}$$

$$\mathbf{U}(\mathbf{x})=\mathbf{U}_0+\delta\mathbf{U}(\mathbf{x}) \tag{5}$$

Then Eqn. (3) becomes:

$$\left[\left(C_g+\mathbf{U}\right)\frac{\partial}{\partial \mathbf{x}}+\beta\right]\delta N=\mathbf{k}\cdot\frac{\partial \mathbf{U}}{\partial \mathbf{x}}\frac{\partial N_0}{\partial \mathbf{k}} \tag{6}$$

If Fourier expression is introduced to Eqns. (4) and (5), then the perturbations for $\delta N(\mathbf{x},\mathbf{k},t)$ and $\delta\mathbf{U}(\mathbf{x},t)$ are given as Eqns. (7) and (8) where $\mathbf{K}$ is the wave number of bottom topography.

$$\delta N(\mathbf{x},\mathbf{k})=\int_{-\infty}^{\infty}\eta(\mathbf{K},\mathbf{k})\exp(i\mathbf{K}\mathbf{x})d\mathbf{K} \tag{7}$$

$$\delta\mathbf{U}(\mathbf{x})=\int_{-\infty}^{\infty}\mu(\mathbf{K})\exp(i\mathbf{K}\mathbf{x})d\mathbf{K} \tag{8}$$

Equations (7) and (8) substitute into Eqn. (6) and then Eqn. (9) is obtained as follows:

$$\frac{\eta(\mathbf{K},\mathbf{k})}{N_0}= -\frac{\mathbf{K}\cdot\left(C_g+\mathbf{U}_0\right)+i\beta}{\left[\mathbf{K}\cdot\left(C_g+\mathbf{U}_0\right)\right]^2+\beta^2}(\alpha+\gamma)\frac{\mathbf{K}\cdot\mathbf{k}}{\mathbf{k}^2}\left[\mathbf{k}\cdot\mu(\mathbf{K})\right] \tag{9}$$

For simplicity, detailed derivation was shown in Alpers and Hennings (1984). σ and σ_0 are cross section and normalized cross section respectively, and Eqn. (10) gives the difference of both. Equation (11) shows cross section variation is a low-pass filtered replica of the current gradient (Alpers and Hennings 1984).

$$\delta\sigma = \sigma - \sigma_0 \tag{10}$$

$$\frac{\delta\sigma(\mathbf{x})}{\sigma_0} = -(\alpha+\gamma)\frac{\mathbf{K}\cdot(C_g+\mathbf{U}_0)+i\beta}{[\mathbf{K}\cdot(C_g+\mathbf{U}_0)]^2+\beta^2}\cdot\frac{\beta}{i}\frac{\partial U_x(\mathbf{x})}{\partial x} \tag{11}$$

where $\mathbf{U}_0=(U_x,U_y)$, γ is the ratio of group velocity and phase velocity; and α is a constant determined by a power law of short-wave energy spectrum. The x direction is the antenna look axis. Alpers and Hennings (1984) explained that cross-section modulation represents low-pass filtering of the tidal current gradient in the antenna look direction. Besides, cross-section modulation also designates low-pass filtering of d'/d^2, where d' is the gradient of the depth d of bottom-topography in the direction perpendicular to the sandbank crest, if the bottom topography is quasi-two-dimensional and the continuity equation is introduced. Nevertheless, a full three-dimension hydrodynamic model for the bottom-topography-current interaction is more complex. Concerning SAR imaging of hydrodynamic modulation, however, most variables including relaxation time, flow velocity field and etc. need to be determined. Besides, action balance is only applicable to linear-and-equilibrium cases of short waves riding on non-uniform velocity fields. If one single SAR image is applicable and less-effort field investigation is desired, then a suitable adaptive nonlinear low-pass filtering can provide coherent results to Alpers and Hennings' results (1984). Besides, it could be more permissive to obtain the inversion of bottom-topography on the basis of one single SAR image. So morphological filter, which is an iterated filtering operation, represents an evolutionary smoothing process in every different scale and, therefore, it provides another view-angle to SAR imaging. According to the iterated low-pass filtering property in scale, we wish to explore the inversion of bottom topography in the empirical approach of scale evolution. However, field investigation is also needed for confident validation.

2.2 *Morphological Filter*

Here we would like to look at linear and nonlinear morphological filters. The cases shown here are canonical iterated smoothing filters. The explicit finite-difference numerical scheme is chosen with scale step $\Delta\tau$.

1. Linear Filter

Let $u_0(\mathbf{x})$ be the initial image and the heat equation, which is a linear way of iterated low-pass filtering, given as following:

$$\frac{\partial u(\tau,\mathbf{x})}{\partial \tau} = (\Delta u)(\tau,\mathbf{x}) \tag{12}$$

where $u(0,\mathbf{x})=u_0(\mathbf{x})$ and τ is the scale space.

If $G_\tau(\mathbf{x}) = (4\pi\tau)^{-\frac{n}{2}}\exp\left(-\frac{|\mathbf{x}|^2}{4\tau}\right)$ is given, then G_τ itself is solution of the heat equation $\frac{\partial G_\tau}{\partial \tau} = \Delta G_\tau$. There is also a straight-forward way to proof that the solution of Eqn. (12) can be given by $u(\tau,\mathbf{x}) = u_0 * G_\tau$ and $*$ is the convolution. So Eqn. (12) is a typical linear low-pass filter.

2. Nonlinear Filter of Affine Invariant Curvature Motion

Equation (13) gives a canonical example of affine invariant curvature motion.

$$\frac{\partial u(\tau,\mathbf{x})}{\partial \tau} = |Du|\left(\tau div\frac{Du}{|Du|}\right)^{\frac{1}{3}} \tag{13}$$

where $div\left(\frac{Du}{|Du|}\right) = \frac{u_x^2u_{yy} - 2u_xu_yu_{xy} + u_y^2u_{xx}}{|u_x^2+u_y^2|^{-\frac{3}{2}}}$

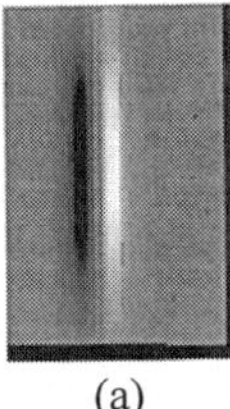

(a)

(b)

Fig. 1(a) and (b): linear filter in scales τ=2 and τ=6.

(a)

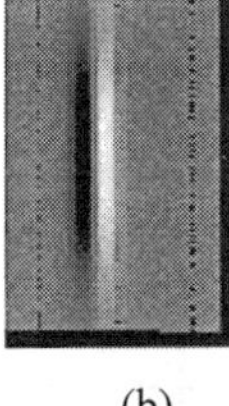

(b)

Fig. 2(a) and (b): nonlinear filter of affine invariant curvature motion in scales τ=2 and τ=6.

3 IMAGE SIMULATION

As to scale evolution simulation of iterated low-pass filtering, Eqns. (12) and (13) are applied, shown in Figs. 1(a) and (b); and 2(a) and (b) in scales $\tau = 2$ and $\tau = 6$, when an artificial simulated bottom topography is created. The slope of bottom topography divided by the depth square is taken.

4 DISCUSSION

In this study, according to low-pass filtering property based on action balance equation, we have shown the application of iterated low-pass linear filtering of heat equation and a fundamental iterated morphological filtering of affine invariant smoothing process in SAR imaging. Velocity bunching modulation is not included, since the magnitude of hydrodynamic modulation is much larger than that of velocity bunching modulation (Alpers and Hennings 1984) and hydrodynamic modulation is mainly concerned. Figures 2 and 3 show the gray-level variation in different scales due to hydrodynamic modulation of bottom topography. The interface along the dark and bright streaks shows the peak of bottom topography. In the near future field investigation combining verification of affine invariant curvature motion can be done.

REFERENCES

Alpers, W. & I. Hennings 1984. A theory of the imaging mechanism of underwater bottom topography by real and synthetic aperture radar. *J. Geophys. Res.* 89:C6:10529-10546.

Guichard, F. & Jean-Michel Morel 1995. *Partial Differential Equations and Image Iterative Filtering.*

Huang, N. E., *et al.* 1998. The empirical mode decomposition and the Hilbert spectrum for nonlinear and non-stationary time series analysis. *Proc. R. Soc. Lond. A.* 454:903-995.

Hughes, B. A. 1978. The effect of internal waves on surface wind waves, 2. Theoretical analysis. *J. Geophys. Res.* 83:455-465.

Holliday, D, G. St-Cyr & N. E. Woods 1987. Comparison of a new ocean imaging model with SARSEX internal wave image data. *Int. J. Remote Sensing.* 8:1423-1430.

Mitsuyasu, H. & T. Honda 1982. Wind-induced growth of water waves. *J. Fluid Mech.* 123:440.

Ouchi, K., N. R. Stapleton & B. C. Barber 1997. Multi-frequency SAR images of ship-generated internal waves. *Int. J. Remote Sensing.* 18:3709-3718.

Plant, W. J. 1982. A relationship between wind stress and wave slope. *J. Geophys. Res.* 87:C3:1961-1967.

Shen, I-F., W. J. Easson, C. A. Greated & W. A. B. Evans 1994. The modulation of short waves riding on solitary waves. *Physics of Fluids.* 6:10:3317-3323.

Remote Sensing in the 21st Century: Economic and Environmental Applications, Casanova (ed.)
 ISBN 90 5809 096 5

A unified approach to change analysis and despeckle of multitemporal ERS-1

L.Alparone
Department of Electronic Engineering, University of Florence, Italy

S.Baronti
'Nello Carrara' Institute on Electromagnetic Waves IROE-CNR, Florence, Italy

M.Falugi & A.Garzelli
Department of Information Engineering, University of Siena, Italy

ABSTRACT: Local-statistics speckle filtering has been extended to multitemporal SAR data by exploiting the space-varying temporal correlation of speckle noise between two images of the same scene taken at different times. A nonlinear transformation aimed at decorrelating the data across time while retaining the multiplicative noise model is defined from the pixel geometric mean and ratio of a couple of spatially overlapped observations. The average temporal correlation coefficient is estimated from the scatter-plots of local standard deviation to local mean calculated on transformed couples of images, through an unsupervised clustering procedure. The images are filtered in the transformed domain and reversely transformed to yield despeckled observations in which seasonal changes are preserved, or even highlighted, and texture analysis is expedited. Tests on SAR images from repeat-pass ERS-1 are presented to corroborate the underlying assumptions.

1 INTRODUCTION

Speckle noise typically occurring in imaging systems operating with coherent radiation, such as Synthetic Aperture Radar (SAR), is originated by the random scattering of objects within elementary resolution cells and appears as a *signal-dependent granular* noise, i.e. spatially correlated or *non-white*, whose effect is to visually degrade the appearance of images and to severely diminish the performances of automatic image segmentation and classification algorithms. Several techniques have been developed for speckle suppression (Lee 1994): adaptive spatial filtering based on local statistics is widespread.

In multi-frequency polarimetric SAR, images in several polarizations and/or frequencies can be combined to produce a less noisy version like, e.g., in the polarimetric whitening filter (Novak 1990), and in the principal components (PC) analysis (Lee & Hoppel 1992). Whenever registered multitemporal images of the same area are available, speckle processing can be designed in order to take advantage of its *temporal* correlation (Alparone et al. 1998b), besides of that of the signal (Villasenor et al. 1993). In fact, a viable approach to multitemporal processing, which is established in the literature for analysis of seasonal changes, is image ratioing (Bruniquel & Lopes 1997, Dekker 1998, Rignot & van Zyl 1993). The pixel ratio between two observations exhibits nonzero mean, thus retaining the properties of the multiplicative noise model.

In this work, a new despeckling algorithm is proposed, that exploits the time correlation of a couple of SAR images, taken with the same orbital parameters and incidence angles, that can be thus overlapped without any spatial warping. A reversible transformation maintaining the properties of the multiplicative noise model, consisting of pixel geometric mean and ratio, is applied to the couple of images to extract their common signal and the change between them. The *time correlation coefficient* (TCC) of speckle is estimated (Alparone et al. 1998a) for deriving the noise variance of the transformed data. Since larger TCCs are associated with higher similarities of the vegetation cover which rules the surface scattering properties (Rignot & van Zyl 1993), a relationship is empirically found out between the *local* TCC of the speckle and the intensity of changes occurred between two passes, which may be modelled by the local variance of the rational change term.

Local-statistics filtering (Kuan et al. 1985) is carried out in the transformed domain. Since the change term may be estimated by filtering the ratio image, an iterative procedure is set-up in which better and better estimates are produced thanks to a refined space-varying TCC of the speckle which is recalculated at each step from the previously estimated change term. Eventually, the data are inversely transformed to yield despeckled observations. Results will be presented and discussed by considering couples of images from repeat-pass ERS-1.

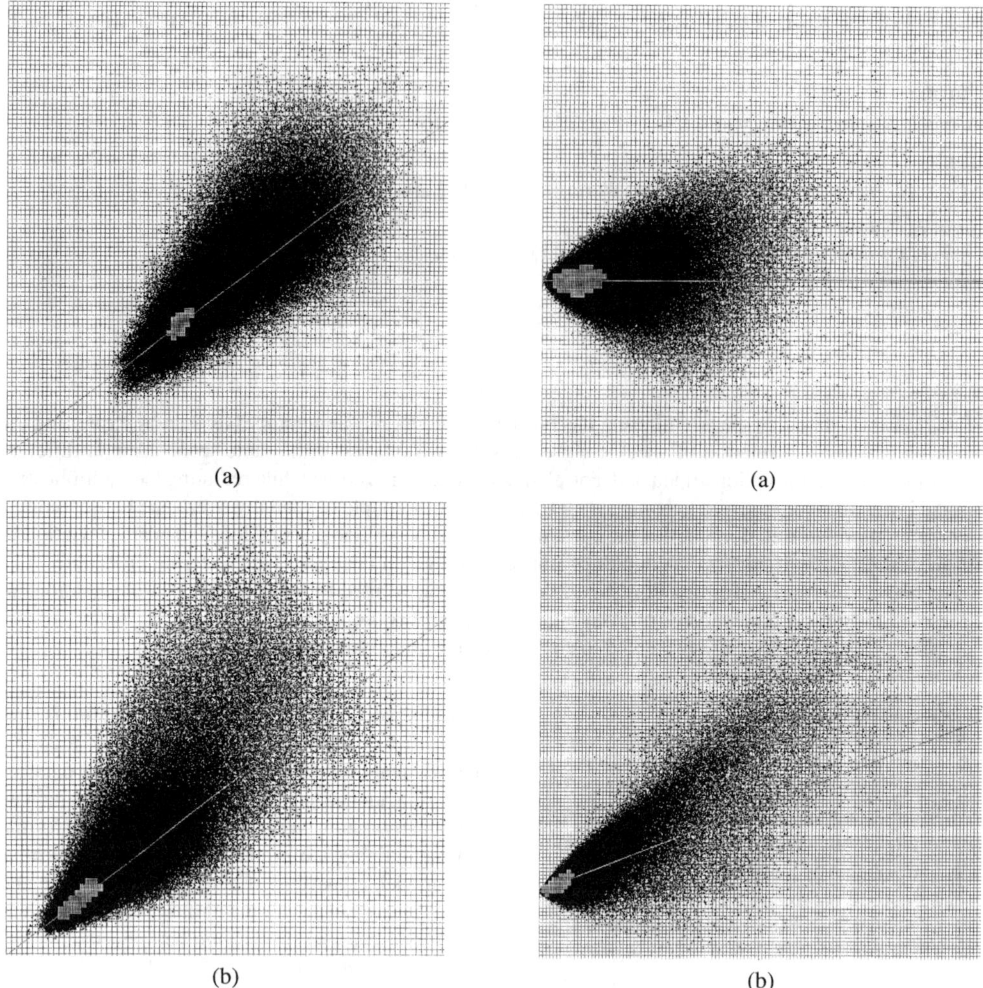

Figure 1. Scatter-plots of local standard deviation vs. local mean of (a) speckled SPOT-P and (b) ERS-1, with block partition and regression straight lines calculated on the highlighted points recognized as belonging to uniform areas.

Figure 2. Scatter-plots of local one-lag covariance vs. local variance of (a) speckled SPOT-P and (b) ERS-1, with superimposed block partition and regression straight lines calculated on the highlighted block clusters.

2 TEMPORAL TRANSFORMATION

A suitable transformation is given by pixel geometric mean and ratio of a couple of overlapped images:

$$\begin{aligned} F_0(i,j) &= \sqrt{f_0(i,j) \cdot f_1(i,j)} \\ F_1(i,j) &= \frac{f_0(i,j)}{f_1(i,j)} \end{aligned} \tag{1}$$

in which $f_0(i,j)$ and $f_1(i,j)$ are the original images, while $F_0(i,j)$ and $F_1(i,j)$ the transformed ones. The inverse transformation is given by

$$\begin{aligned} f_0(i,j) &= F_0(i,j) \cdot \sqrt{F_1(i,j)} \\ f_1(i,j) &= \frac{F_0(i,j)}{\sqrt{F_1(i,j)}}. \end{aligned} \tag{2}$$

The geometric mean strengthens the time-correlated component of the observed signal, thus resulting in a signal to noise ratio (SNR) that is higher than in each individual observation. Instead, the difference removes the time-correlated signal component and highlights seasonal changes as well as the random noise. The multiplicative noise model still holds.

3 MODELLING OF SEASONAL CHANGES

Let us assume that the radar cross-sections in a couple of observations may be expressed as a common signal $s(i,j)$ modulated by the square root of a change term $c(i,j)$, and by its reciprocal, respectively. Both $s(i,j)$ and $c(i,j)$ are spatially correlated space-varying processes, yielding positive values and thus having nonzero mean. Two speckle terms $u_0(i,j)$ and $u_1(i,j)$, having mean one, variance σ_u^2, and being correlated with each other with a CC ρ_{01}, multiplicatively affect the noise-free signals:

$$\begin{aligned} f_0(i,j) &= s(i,j)\cdot\sqrt{c(i,j)}\cdot u_0(i,j) \\ f_1(i,j) &= \frac{s(i,j)}{\sqrt{c(i,j)}}\cdot u_1(i,j) \end{aligned} \tag{3}$$

Eq. (3) defines a multiplicative *temporal change* model, analogous to the multiplicative *texture* model proposed by Ulaby et al. (1986). If (1) is applied to (3), the outcome images may be written as

$$\begin{aligned} F_0(i,j) &= s(i,j)\cdot u_G(i,j) \\ F_1(i,j) &= c(i,j)\cdot u_R(i,j) \end{aligned} \tag{4}$$

in which

$$\begin{aligned} u_G(i,j) &\triangleq \sqrt{u_0(i,j)\cdot u_1(i,j)} \\ u_R(i,j) &\triangleq \frac{u_0(i,j)}{u_1(i,j)} \end{aligned} \tag{5}$$

The variances of the noise terms $u_G(i,j)$ and $u_R(i,j)$ can be well approximated by taking the first-order developments of the root and of the ratio in (5):

$$\begin{aligned} \sigma_G^2 &= \frac{\sigma_u^2}{2}\cdot(1+\rho_{01}) \\ \sigma_R^2 &= 2\cdot\sigma_u^2\cdot(1-\rho_{01}) \end{aligned} \tag{6}$$

Notice that σ_G^2 may be larger than σ_R^2, depending on the ρ_{01}. Taking the ratios of the squares of (6) yields ρ_{01} as

$$\rho_{01} = \frac{4\sigma_G^2-\sigma_R^2}{4\sigma_G^2+\sigma_R^2} \tag{7}$$

Thus, according to (7), the average TCC, ρ_{01}, may be estimated from the variances of the noise terms in $F_0(i,j)$ and $F_1(i,j)$, which are σ_G^2 and σ_R^2, respectively. According to the change model, ρ_{01} is independent of $s(i,j)$, reflecting the average signal underlying between the observations, whereas it will be found out to depend on the change term $c(i,j)$.

4 PARAMETRIC NOISE ESTIMATION

Let us calculate the variance of the multiplicative noise model $g(i,j) = f(i,j)\cdot u(i,j)$:

$$\sigma_g^2(i,j) = \sigma_f^2(i,j)(1+\sigma_u^2)+\mu_f^2(i,j)\cdot\sigma_u^2 \tag{8}$$

in which $\mu_f(i,j)$ denotes the space-varying mean of f and equals that of g: $\mu_f(i,j)=\mu_g(i,j)$. Out of the two terms on right-hand of (8), the former is dominant on textured areas and null on homogeneous areas; conversely, the latter is negligible on high-signal areas. By considering these two effects separately one obtains:

$$\sigma_g(i,j) = \sigma_u\cdot\mu_g(i,j) \tag{9}$$

$$\sigma_g(i,j) = \sigma_f(i,j)\cdot\sqrt{1+\sigma_u^2} \tag{10}$$

Equation (9) yields an estimate of σ_u, namely $\hat{\sigma}_u$, as the regression coefficient on the scatter-plot of σ_g versus μ_g, calculated on homogeneous areas. Equation (10) is of little use since both σ_f and σ_u are unknown.

Let us calculate also the space-varying covariance of lag one along a coordinate direction, say j,

$$\begin{aligned} cov_g(i,j+1) &= cov_f(i,j+1)(1+\rho_x\cdot\sigma_u^2) \\ &+ \rho_x\cdot\sigma_u^2\cdot\mu_f^2(i,j) \end{aligned} \tag{11}$$

in which ρ_x denotes the correlation coefficient of the noise along j. Out of the two terms on right-hand of (11) the former vanish on homogeneous areas, where the variance of the observed signal $g(i,j)$ becomes $\sigma_u^2\cdot\mu_f^2(i,j)$. Thus, ρ_x may be estimated from those points, lying on the covariance-to-variance scatter-plot, corresponding to signal having constant value.

In order to overcome the drawback of manually identifying homogeneous areas, an automatic procedure was developed, based on considerations that such areas tend to produce clusters of points on the scatter-plot, which are aligned along the regression straight line. The scatter-plot plane is partitioned into a number of rectangular blocks; the finer the partition, the more accurate the estimation. Such blocks are sorted and labeled for decreasing number of scatter-points. Thus, the first blocks are those which are the more populated. A succession of regression lines, corresponding to a succession of as many values of parameters, i.e. σ_u^2, ρ_x and ρ_y, is calculated on the points lying inside an increasing number of blocks on the scatter-plots. Thus, the first term of the succession is calculated on the points inside the most dense block, the second inside the two most dense blocks, and so on. The succession attains the true value after a number of terms which depends on the actual percentage of homogeneous points. A stop criterion was devised based on that both the σ_u and the ρs (both ρ_x

and ρ_y) are asymptotically overestimated throughout the succession: after an initial adjustment followed by a flat region, an increasing trend is detected along the plots; the procedure is stopped and the median of the last terms is chosen in order to reject the beginning of the trend. The first terms of the succession, say 1% of blocks, must always be discarded because of statistical fluctuations due to excessively small sample. The size of the local windows on which local statistics are measured is non-critical, at least in the range 5×5 to 9×9, assuming that the noise is white. Otherwise, small windows tend to underestimate the actual values; however, a 7×7 window works better, at least for moderate spatial correlation of speckle.

A speckled version of a SPOT-P optical images with $\rho_x = \rho_y = 0$ and $\sigma_u = 0.261$ was generated and utilized together with a true ERS-1 image.

Table 1. True and measured parameters of the synthetic (SPOT-P) and true (ERS-1) speckled images, 4- and 3-look amplitude, respectively.

	ρ_x	$\hat{\rho}_x$	ρ_y	$\hat{\rho}_y$	σ_u	$\hat{\sigma}_u$
SPOT-P	0	0.006	0	0.001	0.261	0.266
ERS-1	-	0.52	-	0.55	0.302	0.277

Table 1 summarizes the results obtained. The accuracy on the simulated noisy images is quite impressive. The results on ERS-1 data show a slight discrepancy of the measured $\hat{\sigma}_u$ with the theoretical σ_u. Actually, ERS-1 data are resampled to exhibit uniform ground resolution; thus, the true σ_u is lower than the

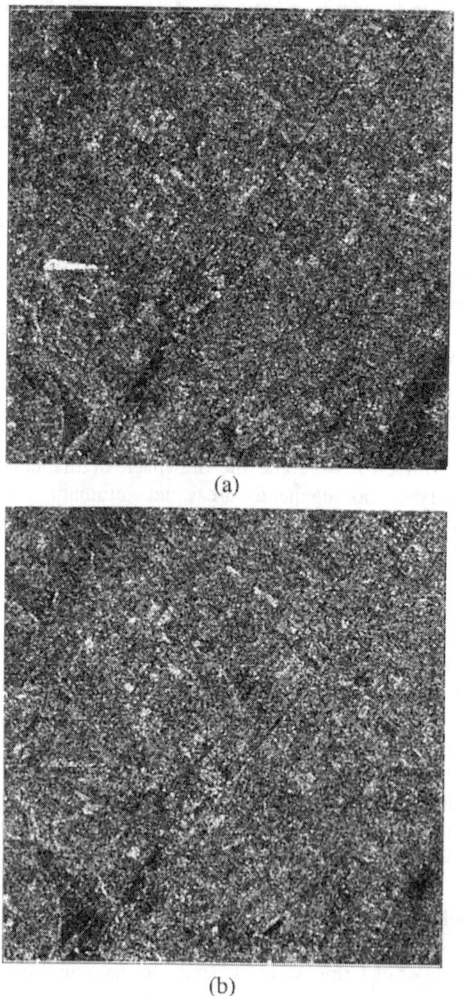

Figure 3. Observations of the test area (512 × 512) acquired by ERS-1 on (a) 23/10/95, (b) 30/7/96.

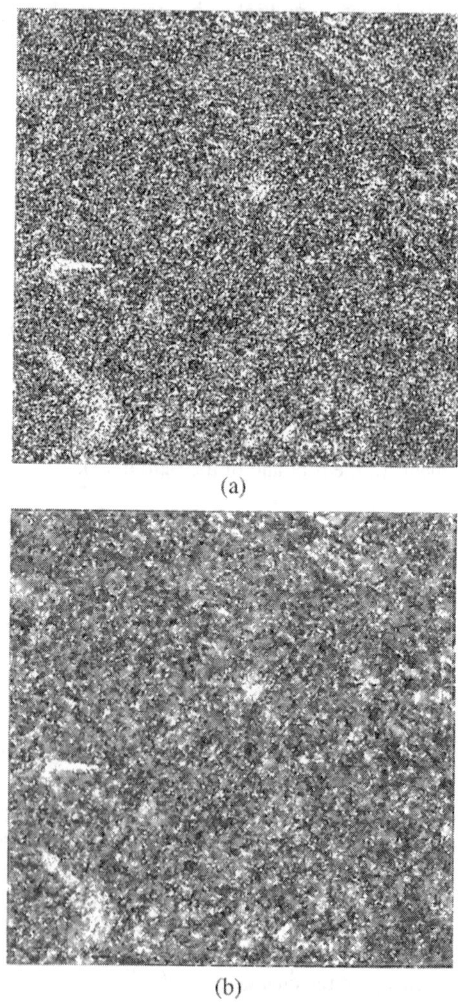

Figure 4. (a) rational component $F_1(i, j)$ calculated from the images of Figure 3; (b) estimated change $\hat{c}(i, j)$ obtained from filtering.

theoretical one. The large spatial CC of speckle depend on that partially overlapped cells are processed to yield a constant 12.5 m ground resolution.

Figures 1-2 show the scatter-plots from which σ_u and ρ_x are retrieved, respectively, for the cases of (a) simulated and (b) true data. Note that the parameters would be overestimated if they were calculated on the whole set of points, due to image textures.

5 MULTITEMPORAL SPECKLE FILTERING

The *local linear minimum mean squared error* (LLMMSE) (Kuan et al. 1985) estimates of s and c, namely $\hat{s}$ and $\hat{c}$, may be approximated as

$$\begin{aligned} \hat{s}(i,j) &= \bar{F}_0(i,j) + [F_0(i,j) - \bar{F}_0(i,j)] \\ &\times \left(1 - \frac{\sigma_G^2 \cdot \bar{F}_0^2(i,j)}{\sigma_0^2(i,j)}\right) \cdot \left(\frac{1}{1+\sigma_G^2}\right) \end{aligned} \quad (12)$$

$$\begin{aligned} \hat{c}(i,j) &= \bar{F}_1(i,j) + [F_1(i,j) - \bar{F}_1(i,j)] \\ &\times \left(1 - \frac{\sigma_R^2 \cdot \bar{F}_1^2(i,j)}{\sigma_1^2(i,j)}\right) \cdot \left(\frac{1}{1+\sigma_R^2}\right) \end{aligned} \quad (13)$$

in which the pairs $\bar{F}_k(i,j)$ and $\sigma_k^2(i,j)$, $k = 0, 1$, denote local average and local variance of F_k, $k = 0, 1$. Such features are evaluated in local neighbourhoods.

LLMMSE filtering in (12)-(13), however, was devised in the assumption of *white* noise. When the speckle is spatially correlated, even though with same variance, the noise power measured on the local window is lower than when the noise is white. Hence, filtering is no longer adjusted to the actual SNR and becomes too strong. An empirical solution consists of defining an *equivalent* σ of the *colored* speckle as

$$\tilde{\sigma}_u = \sigma_u \sqrt{1 - \rho_u} \quad (14)$$

in which ρ_u is the *spatial* CC of the noise affecting the data, which can be stated as

$$\rho_u \triangleq \sqrt{\rho_x \cdot \rho_y}. \quad (15)$$

Thus, the parameters σ_R^2 and σ_G^2 in (12)-(13), which are estimated with the scatter-plot method or are given by (6), once the average TCC ρ_{01} has been otherwise obtained (e.g. measured or modelled) should be corrected by means of their spatial CCs.

Furthermore, large TCCs may be associated with high similarities of the vegetation cover which rules the surface scattering properties (Rignot & van Zyl 1993). We assume that a relationship exists between the *local* TCC of the speckle $\rho(i,j)$ (those given by (7) are *average* values) and the amount of changes occurred across two passes, which may be modelled by $\sigma_c^2(i,j)$, i.e. by the local variance of the change term $c(i,j)$. By denoting with $\bar{\rho}_{01}$ the average TCC given by (7) and with $\bar{\sigma}_c^2$ the average of $\sigma_c^2(i,j)$, the following model may be stated:

$$\rho_{01}(i,j) = \bar{\rho}_{01} \cdot \log_2 \left[1 + \frac{\bar{\sigma}_c^2}{\sigma_c^2(i,j)}\right] \quad (16)$$

Since $c(i,j)$ may be estimated by filtering the ratio image, an iterative procedure was set-up in which better and better estimates of $c(i,j)$ are produced from filtering with (13) thanks to a refined $\rho_{01}(i,j)$ of the speckle which is recalculated at each step from the local variance of the outcome $\hat{c}(i,j)$, by using (16). The σ_G^2 and σ_R^2 in (12) and (13) are given by (6), with the space-varying TCC (16) replacing ρ_{01}. Noise-free versions of $f_0(i,j)$ and $f_1(i,j)$ are obtained from (2).

6 EXPERIMENTAL RESULTS

ERS-1 images of the same area, the archaeological site of Pisticci in Southern Italy, were available. Since the orbital parameters were the same, the registration was a simple shift at least on areas without significant relief. Details from a couple of original images are shown in Figure 3. Due to the close pass dates, only seasonal changes of moderate intensity are recognizable; thus, we expect a TCC of speckle significantly nonzero. In fact, the average TCC was measured as $\bar{\rho}_{01} = 0.114$, by using (7) with σ_G and σ_R estimated through the scatter-plot method.

The originals of Figure 3 were transformed according to (1). The components were processed by Kuan's filter (12)-(13) with a sliding window of size 7×7. The first iteration assumes that the TCC is constant over the whole scene and equal to the value previously estimated. In the next steps, the variance of the noise is calculated for each component from its $\tilde{\sigma}_u$ (14) and the space-varying TCC model (16) ($\sigma_c^2(i,j)$ was calculated on 3×3 windows) by means of (6).

Figures 4(a) and (b) show the rational component before filtering and after filtering with a space-varying noise variance, $\sigma_R^2(i,j)$, adjusted at the 5-th iteration of the procedure. As it appears, the noisiness of the ratio, which is larger than for the single images, is considerably diminished. Thanks to the space-varying model, filtering is more aggressive where the noise is larger. Seasonal changes hardly recognizable are enhanced highlighted in the change image.

The filtered images are shown in Figure 5. Although their quality is noteworthily improved with respect to Figure 3, the images still retain their temporal features, despite of spatio-temporal filtering: seasonal changes are not damped, while speckle is adequately reduced. Linear features that are steady across time are preserved better than by conventional LLMMSE filtering, whose results are portrayed in Figure 6.

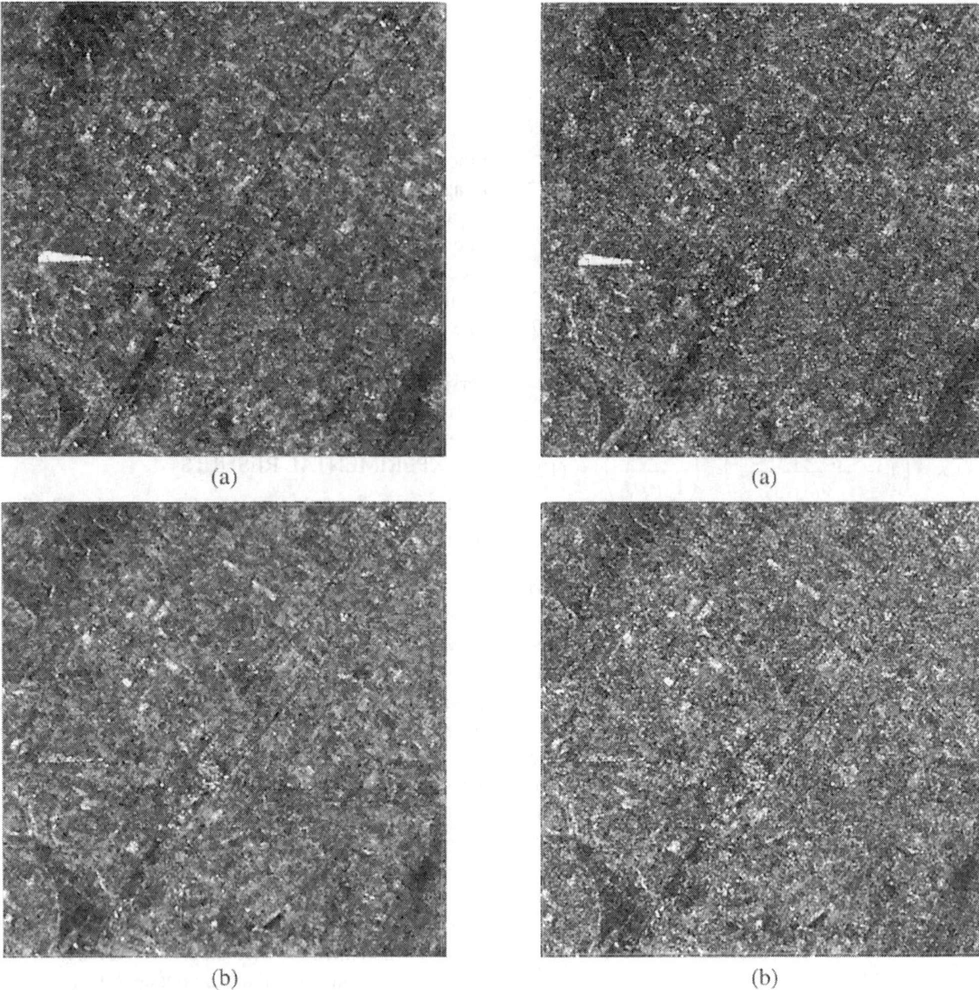

Figure 5. Filtered versions of the images of Figure 3 obtained from reverse transformation of the filtered transformed ones.

Figure 6. Filtered versions of the images of Figure 3 obtained from conventional LLMMSE filtering with $\bar{\sigma}_u = 0.19$.

7 CONCLUSIONS

Local-statistics speckle filtering has been extended to multitemporal ERS-1 observations by exploiting the temporal correlation of the speckle noise, which can expedite the detection and assessment of seasonal variations, being locally related to the amount of changes. A robust procedure was set-up for measuring the average TCC from the variances of the noise affecting the geometric mean and the ratio of a couple of images. Experiments on SAR images from repeat-pass ERS-1 have corroborated the underlying theoretical assumptions and shown the enhanced filtering performances. In addition, filtering is expected to further benefit from a multiresolution approach, like the pyramidal one recently proposed by two of the authors (Aiazzi et al. 1998).

8 ACKNOWLEDGEMENT

Work supported by the Italian Space Agency (ASI).

REFERENCES

Aiazzi, B., Alparone, L. and Baronti, S. 1998. Multiresolution local-statistics speckle filtering based on a ratio Laplacian pyramid. *IEEE Transactions on Geoscience & Remote Sensing* 36(5): 1466–1476.

Alparone, L., Baronti, S. and Garzelli, A. 1998a. Evaluating time correlation of speckle in ERS-1 SAR images. In Tammy Stein (ed.) *Proc. IEEE International Geoscience & Remote Sensing Sympo-*

sium IGARSS'98, Seattle, WA: 30–32.

Alparone, L., Baronti, S. and Carlà, R. 1998b. Texture analysis and despeckle of multitemporal SAR images. In S.B. Serpico (ed.) *"Signal and Image Processing for Remote Sensing IV" Proc. Europto Series, Barcelona, Spain*, SPIE 3500: 135–144.

Bruniquel, J. & Lopes, A. 1997. Multivariate optimal speckle reduction in SAR imagery. *International Journal of Remote Sensing* 18(3): 603-627.

Dekker, R.J. 1998. Speckle filtering in satellite SAR change detection images. *International Journal of Remote Sensing* 19(6): 1133-1146.

Kuan, D.T., Sawchuk, A.A, Strand, T.C. and Chavel, P. 1985. Adaptive noise smoothing filter for images with signal-dependent noise. *IEEE Transactions on Pattern Analysis & Machine Intelligence* 7(2): 165–177.

Lee, J.-S. & Hoppel, K. 1992. Principal component transformation of multifrequency polarimetric SAR imagery. *IEEE Transactions on Geoscience & Remote Sensing* 30(4): 686–696.

Lee, J.-S., Jurkevich, I., Dewaele, P., Wambacq, P. and Oosterlinck, A. 1994. Speckle filtering of synthetic aperture radar images: a review. *Remote Sensing Reviews* 8: 313–340.

Novak L.M. & Burl, M.C. 1990. Optimal speckle reduction in SAR imagery. *IEEE Transactions on Aerospace & Electronic Systems* 26(2): 293–305.

Rignot, E.M. & van Zyl, J.J. 1993. Change detection techniques for ERS-1 SAR data. *IEEE Transactions on Geoscience & Remote Sensing* 31(4): 896–906.

Ulaby, F.T., Kouyate, F., Brisco, B. and Lee Williams, T.H. 1986. Textural information in SAR images. *IEEE Transactions on Geoscience & Remote Sensing* 24(2): 235–245.

Villasenor, J.D., Fatland, D.R. and Hinzman, L.D. 1993. Change detection on Alaska's north slope using repeat-pass ERS-1 SAR images. *IEEE Transactions on Geoscience & Remote Sensing* 31(1): 227–236.

Miscellaneous

Remote Sensing in the 21st Century: Economic and Environmental Applications, Casanova (ed.)
© 2000 Balkema, Rotterdam, ISBN 90 5809 096 5

List of participants

ABAD CERDA Francisco Mr.
Universidad de Castilla La Mancha
Sección de Teledetección y GIS
Instituto de Desarrollo Regional
02071 Albacete (SPAIN)
T: +34 967 599200 F:+34 967 599233
fabad@idr-ab.uclm.es

AIT-BACHIR Saliha Ms.
Université Jean Monnet. CRENAM
6 rue Basse des Rives
42023 Saint-Etienne Cedex 2 (FRANCE)
T: +04 77 42 1922 F:+04 77 42 1924

ALAVI PANAH Seyed K. Dr.
University of Tehran
Iran Desert Research Centre
P.O. Box 14185-354
Tehran (IRAN)
F:+98 21 6704144
salavipa@chamran.ut.ac.ir

AL-RAWI Kamal Dr.
Universidad de Valladolid
Laboratorio de Teledetección, LATUV
Prado de la Magdalena, s/n
47071 Valladolid (SPAIN)
T: +34 983 423132 F:+34 983 423130
kamal@latuv.uva.es

ARIZA LOPEZ Francisco J.
Universidad de Jaén
Dpto. Ingeniería Cartográfica, Geodésica y Fotogrametría
Virgen de la Cabeza, 2
23071 Jaén (SPAIN)
T: +34 953 212469 F:+34 953 212343
fjariza@ujaen.es

AROZARENA VILLAR Antonio Dr.
Instituto Geográfico Nacional
General Ibáñez de Ibero, 3
28003 Madrid (SPAIN)
T: +34 91 5979575 F:+34 91 5979770
aarozarena@mfom.es

ARQUERO HIDALGO Agueda Dr.
Universidad Politécnica de Madrid
Dpto. de Arquitectura y Tecnología de S.I.
Campus de Montegancedo
28660 Boadilla del Monte (Madrid) (SPAIN)
T: +34 91 3367385 F:+34 91 336 66 01
aarquero@fi.upm.es

BALTSAVIAS Emmanuel Dr.
Institute of Geodesy and Photogrammetry
ETH-Hoenggerberg
8093 Zurich (SWITZERLAND)
T: +41 1 6333042 F:+41 1 6331101
manos@geod.ethz.ch

BARON MARTINEZ Julio Diego
Universidad de Almería
Campus Universitario
Cañada de San Urbano, s/n
04120 Almería (SPAIN)
T: +34 950 215418
jbaron@ualm.es

BARONTI Stefano Dr.
IROE-CNR
Via Panciatichi, 64
I-50127 Florence (ITALY)
T: +39 55 4235295 F:+39 55 4235204
baronti@iroe.fi.cnr.it

BASLY Ludovic
Ecole des Mines de Paris
Groupe Télédétection et Modélisation
B.P. 207
06904 Sophia Antipolis Cedex (FRANCE)
T: +33 04 93 95 7453 F:+33 04 9395 7535
ludovic.basly@cenerg.cma.fr

BENES Tomas Dr.
UHUL Lesprojekt
Nabrezni 1326
250 44 Brandys-nad-Labem (CZECH REPUBLIC)
T: +210 202804 F:+420 202 803371
benes@uhul.cz

BINZA-MOUSSIROU M.
Shell International (NAM)
Scheperrsmaat, 2
Postbus 28000
9400 Assen (NETHERLANDS)
F:+31 592 362753
m.binzamoussirou@openmail.xgm-ms.namass.simis.com

BLACH Barbara Ms.
University of Warsaw
Remote Sensing of Environment Laboratory
ul. Krakowskie Przedmieoecie 30
00-927 Warsaw (POLAND)
T: +48 22 8368358 F:+48 22 8261965
telegeo@mercury.ci.uw.edu.pl

BOCHENEK Zbigniew Dr.
Inst. of Geodesy & Cartography
Jasna 2/4
00-950 Warsaw (POLAND)
T: +22 48 8270328 F:+22 48 8270328
zbig@igik.edu.pl

BRADSHAW Roy Dr.
University of Nottingham
School of Geography
Nottingham NG7 2RD (U.K.)
T: +44 115 9515433 F:+44 115 951 5249
roy.bradshaw@.nottingham.ac.uk

BUCHROITHNER Manfred Prof. Dr.
Technical University of Dresden
Institute for Geosciences
D-01062 Dresden (GERMANY)
T: +49 351 463 4809 F:+49 351 463 7028
buc@karst9.geo.tu-dresden.de

BUHK Rainer Mr.
University of Freiburg
Dept. of Remote Sensing & Landscape Information Systems
Tennenbacherstr. 4
D-79085 Freiburg (GERMANY)
T: +49 761 203 3697 F:+49 761 203 3697
buhkrain@felis.uni-freiburg.de

BUSCH Andreas Mr.
Federal Agency for Cartography & Geodesy (BKG)
Richard-Strauss-Allee, 11
D-60598 Frankfurt-am-Main (GERMANY)
T: +49 69 6333 312 F: +49 69 6333 441
busch@ifag.de

CACHORRO REVILLA Victoria Dr.
Universidad de Valladolid
Dpto. de Física Aplicada I
Avda. de Madrid, 57
34071 Palencia (SPAIN)
T: +34 979 729048 F:+34 979 712099
eupap@cpd.uva.es

CALLE MONTES Abel Dr.
Universidad de Valladolid
Laboratorio de Teledetección, LATUV
Prado de la Magdalena, s/n
47071 Valladolid (SPAIN)
T: +34 983 423132 F:+34 983 423130
abel@latuv.uva.es

CAMBOU Francis Prof.
C.E.S.R.
29 avenue du Col. Roche
31029 Toulouse Cédex (FRANCE)
T: +33 05 61 55 66 60 F:+33 05 61 55 66 09
annie.liaut@cesr.fr.

CARMONA-MORENO Cesar Prof.
Joint Research Centre EC
Space Applications Institute/GVM, TP 263
21020 Ispra (Va) (ITALY)
T: +39 0332 789654 F:+39 0332 789073
cesar.carmona-moreno@jrc.it

CASANOVA ROQUE José Luis Prof. Dr.
Universidad de Valladolid
Laboratorio de Teledetección, LATUV
Prado de la Magdalena, s/n
47071 Valladolid (SPAIN)
T: +34 983 423130 F:+34 983 423130
jois@latuv.uva.es

CASELLES MIRALLES Vicente Prof.
Universidad de Valencia
Dpto. Termodinámica
Dr. Moliner, 50
46100 Burjassot (Valencia) (SPAIN)
T: +34 96 3864350 F:+34 96 364 2345
vicente.caselles@uv.es

CASTERAD SERAL Mª Auxiliadora Dr.
DGA-CSIC. Unidad de Suelos y Riegos
Apartado 727
50080 Zaragoza (SPAIN)
T: +34 976 576411 F:+34 976 575501
auxi@syrsig.mizar.csic.es

CATALAN Manuel Prof. Dr.
Universidad de Cádiz/C.A.S.E.M.
Dpto. Física Aplicada
Polígono Río San Pedro s/n
11510 Puerto Real (Cádiz) (SPAIN)
T: +34 956 470866 F:+34 956 470866
manuel.catalan@uca.es

CAVES Ron Dr.
University of Sheffield
Sheffield Centre for Earth Observation Science
Hicks Building
Sheffield S3 7RH (U.K.)
T: +44 114 2223704 F:+44 114 222 3809
r.caves@sheffield.ac.uk

CHARNECA Nuno Mr.
CNIG
Tagus Park-Nucleo Central
Sala 301
2780 Oeiras (PORTUGAL)
T: +351 1 4219800 F:+351 1 421 9856
ncharneca@hotmail.com

CHECA Mª Jesús Ms.
TRAGSATEC
Avda. Ciudad de Barcelona, 120
28007 Madrid (SPAIN)
T: +34 91 3963463 F:+34 91 396 3410

CHERIFI Dalila Ms.
Ecole des Mines de Paris
Groupe Télédétection et Modélisation
B.P. 207
06904 Sophia Antipolis Cedex (FRANCE)
T: +33 04 93957488 F:+33 04 93957535
dalila.cherifi@cenerg.cma.fr

CHUVIECO Emilio Prof.
Universidad de Alcalá de Henares
Dpto. Geografía
Colegios 2
28801 Alcalá de Henares (SPAIN)
T: +34 91 885 4438 F:+34 91 885 4439
ggecs@geogra.alcala.es

CONDE PARDO Patricia Ms.
Universidad de Vigo
Facultad de Ciencias. Dpto. Física Aplicada
Lagaos Marcosende,
Vigo (SPAIN)
T: +34 970 505168 F:+24 968 253086
jesu@uvigo.es

CRUZ MARTINEZ Manuel F.
Universidad de Almería
Campus Universitario
Cañada de San Urbano, s/n
04120 Almería (SPAIN)
T: +34 950 215688
mfcruz@ualm.es

CUEVAS GOZALO José M. Dr.
I.N.I.A.
Sección de Proceso de Datos
Carretera de la Coruña Km 7
28040 Madrid (SPAIN)
T: +34 91 3476817 F:+34 91 3572293
cuevas@inia.es

DE KOK Roeland Mr.
University of Munich
Lehrstuhl fuer Landnutzungsplanung und Naturschutz
Am Hochanger 13
D-85354 Freising (GERMANY)
T: +49 8161 71 5661 F:+49 8161 71 4671
Roeland.DeKok@lrz.uni-muenchen.de

DELGADO DE LA MATA José A. Dr.
Universidad de Valladolid
Dpto. Física Aplicada
Avda. Madrid, 57
39071 Palencia (SPAIN)
T: +34 979 729048 F:+34 979 712099
joseant@fa1.cie.uva.es

DENORE Bernard J. Mr.
Independent Consultant
Pinzón, 2-4
46003 Valencia (SPAIN)
T: +34 963 922469
bernard.denore@ctv.es

DUCHOSSOIS Guy Dr.
European Space Agency
8-10 rue Mario Nikis
75738 Paris Cedex 15 (FRANCE)
Tel: +33 (0)1 53 69 72 85 Fax: +33 (0)1 53 69 76 74
ctapp@hq.esa.fr

EL GHAZZAWI Ihab Mr.
ICL Central Europe
36 B Fortune Green Road
West Hampstead, London (U.K.)
ihab@breathemail.net

ESCUDERO BARBERO Rosario Ms.
TRAGSATEC
Avda. Ciudad de Barcelona, 120
28007 Madrid (SPAIN)
T: +34 91 3963463 F:+34 91 396 3410
reb@tragsatec.es

FALENCKA-JABLONSKA Malgorzata Prof. Dr.
Institute of Forest Research
Laboratory of Ecology & Environmental Protection
Bitwy Warszawskiej 1920 r.3
00-973 Warsaw (POLAND)
T: +48 022 8462016 F:+48 022 8462018

FEHRENBACH Ute Ms.
University of Basel, MCR Laboratory
Spalenring 145
CH-4055 Basel (SWITZERLAND)
T: +41 61 272 6923 F:+41 61 272 6480
fehrenba@ubaclu.unibas.ch

FERNANDEZ MEJUTO Miguel Mr.
Universidad de Castilla La Mancha
Instituto de Desarrollo Regional
Sección de Teledetección y GIS
02071 Albacete (SPAIN)
T: +34 967 599200 F:+34 967 599233
mfmejuto@idr-ab.uclm.es

FERNANDEZ-TURIEL José Luis Dr.
Institute of Earth Sciences "Jaume Almera"
CSIC
Sole I Sabaris s/n
08028 Barcelona (SPAIN)
T: +34 93 4095410 F:+34 93 411 0012
jlfernandez@isa.csic.es

FLOREZ CATON Marta Ms.
Universidad de Valladolid
Laboratorio de Teledetección, LATUV
47071 Valladolid (SPAIN)
T: +34 983 423132 F:+34 983 423130
marta@latuv.uva.es

FOLBRIER Alicja Ms.
University of Warsaw
Remote Sensing of Environment Laboratory
ul. Krakowskie Przedmieoecie 30
00-927 Warsaw (POLAND)
T: +48 22 8368358 F:+48 22 8261965
telegeo@mercury.ci.uw.edu.pl

GARCIA-ESCOBAR Felipe Mr.
Indra Espacio, S.A.
c/Mar Egeo, 4
San Fernando de Henares
288830 Madrid (SPAIN)
T: +34 91 396 3911 F: +34 91 396 3912
fgar@mdr.indra-espacio.es

GODEFROY Madeleine Mrs.
EARSeL Secretariat
2, ave Rapp
75340 Paris Cedex 07 (FRANCE)
T: +33 1 45 567360 F:+33 1 45 567361
earsel@meteo.fr

GOMEZ Heriberto Mr.
University of Nottingham, School of Geography
Nottingham NG7 2RD (U.K.)
T: +44 115 9515433 F:+44 115 951 5249
gomez@geography.nottingham.ac.uk

GONZALEZ YAGÜE Miguel Angel Mr.
TYPSA
Plaza del Liceo, 3
28043 Madrid (SPAIN)
T: +34 91 722 7300 F:+34 91 88 16 86
madrid@typsa.es

GONZALEZ-ALONSO Federico Dr.
I.N.I.A. - Sección de Proceso de Datos
Carretera de la Coruña Km 7
28040 Madrid (SPAIN)
T: +34 91 2070564 F:34 91 442 35 87
alonso@inia.es

GONZALO MARTIN Consuelo Ms.
Universidad Politécnica de Madrid
Dpto. de Arquitectura y Tecnología de S.I.
Campus de Montegancedo
28660 Boadilla del Monte (Madrid) (SPAIN)
T: +34 91 336 7382 F:+34 91 336 7412
chelo@fi.upm.es

GOOSSENS Rudi Prof.
Uníversity of Gent. Dept. of Geography
Krijgslaan 281 (S8-A1)
B-9000 Gent (BELGIUM)
T: +32 9 2644709 F:+32 9 264 4985
rudi.goossens064rug.ac.be

GORROCHATEGUI GONZALEZ Illán
Universidad del Pais Vasco
Arcocha, 8 2° Izda
48200 Durango (Vizcaya) (SPAIN)
illan@clientes.euskaltel.es

GROS AYMERICH Juan Mr.
Ministerio de Defensa
Argentona, s/n
28023 Madrid (SPAIN)
T: +34 91 372 5477 F:+34 91 3725850

GRUSZCZYNSKA Maria Ms.
Inst. of Geodesy & Cartography
Jasna 2/4
00-950 Warsaw (POLAND)
T: +48 22 827 033 28 F:+48 22 827 033 28
zbig@igik.edu.pl

GUDMANDSEN Preben Prof. (em.)
Technical University of Denmark
Building 348
2800 Lyngby (DENMARK)
T: +45 45 253788 F:+45 45 931634
pg@emi.dtu.dk

HAALA Norbert Dr.
University of Stuttgart
Institute of Photogrammetry
Geschwister-Scholl-Str. 24
70174 Stuttgart (GERMANY)
T: +49 711 1214113 F:+49 711 121 3297
norbert.haala@ifp.uni-stuttgart.de

HABOUDANE Driss Mr.
CARTEL
2500 Boulevard de l'Université
Université de Sherbrooke
Sherbrooke
Québec, J1K 2R1 (CANADA)
+1 819 823 9816 F: +1 819 821 7944
92800615@hermes.usherb.ca

HALLIKAINEN Marti Prof.
Helsinky University of Technology
Laboratory for Space Technology
P.O. Box 3000
FIN-02015 HUT (FINLAND)
T: 358 9 4512371 F:358 9 4512898
Hallikainen@avanet.hut.fi

HILD Heiner Dr.
University of Stuttgart
Institute of Photogrammetry
Geschwister-Scholl-Str. 24
70174 Stuttgart (GERMANY)
T: +49 711 1214113 F:+49 711 121 3297
heiner.hild@ifp.uni-stuttgart.de

HOUZELLE Stephane Dr.
ALCATEL SPACE INDUSTRIES
100, Bd. du Midi
06156 Cannes la Bocca Cedex (FRANCE)
T: +33 (0) 4 32927476 F:+33 (0) 4 92927660
stephane.houzelle@space.alcatel.fr

HUDIER Eric Prof.
University of Québec/CIRES, Univ. Colorado
Dept. of Oceanography, 300 des Ursulines
Rimouski, Quebec G5L 3A1 (CANADA)
T: +1 303 2589628 F:+1 303 492 5070
eric_hudier@uqar.uquebec.ca

IBAÑEZ Manuel Dr.
Universitat de Lleida
Dept. Medi Ambient i Ciencies del Sol
Av. Alcalde Rovira Roure, 177
Lleida (SPAIN)
T: +34 973 70 25 38 F:+34 973 23 82 64
m.ibanez@macs.udl.es

IBRAHIM Hussein Dr. Ing.
General Organization of Remote Sensing (GORS)
P. O. Box 12586
DAMASCUS (SYRIAN ARAB REPUBLIC)
F: +963 11 2218700
gors@syriatel.net

ILLERA GUTIERREZ Pilar Dr.
Universidad de Valladolid, Dpto. Física Aplicada
Avda. Madrid, 57
39071 Palencia (SPAIN)
T: +34 979 729048 F:+34 979 712099
illera@cpd.uva.es

JAKOMULSKA Anna Ms.
University of Warsaw
Remote Sensing of Environment Laboratory
ul. Krakowskie Przedmieoecie 30
00-927 Warsaw (POLAND)
T: +48 22 8368358 F:+48 22 8261965
telegeo@mercury.ci.uw.edu.pl

JÖNSSON Lennart Dr.
University of Lund
Dept. of Water Ressources Engineering
P.O. Box 118
S-221 00 Lund (SWEDEN)
F:+46 46 222 4435
lennart.jonsson@tvrl.lth.se

JORGE SANCHEZ Juan Dr.
Universidad Politécnica Cataluña - Dpto. Física Aplicada
Av. Bases de Manresa, 61-73
08240 Manresa (Barcelona) (SPAIN)
T: +34 93 8777275 F:+34 93 8777202
jorge@fa.upc.es

KANTO Kimmo
University of Helsinki. Dept. of Geophysics
P.O. Box 4
FIN-00014 University of Helsinki (FINLAND)
T: +358 9 19122026 F:+358 9 191 23385
kkanto@paju.oulu.fi

KAYA Sinasi Mr.
Istanbul Technical University. Factulty of Civil Engineering
Remote Sensing Division
80626 Maslak/Istanbul (TURKEY)
T: +212 285 6100 F:+212 285 6587
skaya@srv.ins.itu.edu.tr

KERENYI Judit Ms.
Hungarian Meteorological Service
P.O. Box 39
H-1675 Budapest (HUNGARY)
T: +36 1 290 0163 F:+36 1 290 7387
kerenyi@met.hu

KONECNY Gottfried Prof. (em.) Dr.
University of Hannover
Inst. f. Photogrammetry & Engineering Surveys
Nienburgerstr. 1
D-30167 Hannover (GERMANY)
T: +49 511 762 2487 F:+49 511 463 2483
gko@ipi.uni-hannover.de

KORZENIEWSKI Karen Ms.
Remote Sensing Society
c/o Dept. of Geography, University of Nottingham
Nottingham NG7 2RD (U.K.)
T: +44 115 951 5435 F:+44 115 951 5249
rss@nottingham.ac.uk

KRAWCZYK Rodolphe Mr.
ALCATEL SPACE INDUSTRIES - Space Division
100, Bd. du Midi
06156 Cannes la Bocca Cedex (FRANCE)
T: +33 (0)4 9292 7422 F:+33 (0)4 9292 7160
rodolphe.krawczyk@cannes.aerospatiale.fr

KUITTINEN Risto Prof.
Finnish Geodetic Institute
Geodeetinrinne 2
02430 Masala (FINLAND)
T: +358 9 21555307 F:+358 9 295 21555200
risto.kuittinen@fgi.fi

KUNZ Dietmar Mr.
University of Karlsruhe
Inst. for Remote Sensing and Photogrammetry
Englerstr. 7
D-76131 Karlsruhe (GERMANY)
T: +49 721 608 3676 F:+49 721 694568
kunz@ipf.bau-verm.uni-karlsruhe.de

KUNZ Mieczyslaw Mr.
Nicholas Copernicus University
Institute of Geography, Laboratory of Remote Sensing
Gagarina 5
87-100 Torun (POLAND)
T: +48 56 6114698 F:+48 56 6227308
met@cc.uni.torun.pl

LECHI Giovanni Prof.
Politecnico di Milano
Piazza L. Da Vinci, 32
20133 Milano (ITALY)
T: +39 02 23996526 F:+39 02 23996550
giovanmaria.lechi@polimi.it

LINDFORS Antti
University of Helsinki
Dept. of Geophysics
P.O. Box 4
FIN-00014 University of Helsinki (FINLAND)
T: +358 9 191 22 026 F:+358 9 191 23 385
aplinfo@pcu.helsinki.fi

LLORENS-BENITO José F. Dr.
Institute of Earth Sciences "Jaume Almera"
CSIC
Sole I Sabaris s/n
08028 Barcelona (SPAIN)
T: +34 93 4095410 F:+34 93 411 0012
jfllorens@isa.csic.es

LOPEZ Fernando Dr.
Universidad Carlos III de Madrid
Escuela Politécnica Superior. Dpto. Física
Avda. Universidad, 30
28911 Leganés (Madrid) (SPAIN)
T: +34 91 6249468 F:+34 91 6249430
flm@fis.uc3m.es

LOPEZ RAYA Juan T.
Universidad de Almería
Campus Universitario. Cañada de San Urbano, s/n
04120 Almería (SPAIN)
T: +34 950 215011 F:+34 950 215481
jtlopez@ualm.es

LOUAKFAOUI Mostafa
Universidad de Valladolid
Laboratorio de Teledetección, LATUV
Prado de la Magdalena, s/n
47071 Valladolid (SPAIN)
T: +34 983 423132 F:+34 983 423130
mostafa@latuv.uva.es

MANGOLINI Marc Dr.
ALCATEL SPACE INDUSTRIES
100, Bd. du Midi, B.P. 99
06156 Cannes la Bocca Cedex (FRANCE)
T: +33 4 92 92 62 39 F: +33 4 92 92 34 70
Marc.Mangolini@space.alcatel.fr

MARTINEZ BELTRAN Cecilia Ms.
Universidad de Castilla La Mancha
Instituto de Desarrollo Regional
Sección de Teledetección y GIS
02071 Albacete (SPAIN)
T: +34 967 599200 F:+34 967 592333
cbeltran@idr-ab.uclm.es

MARTINEZ BENJAMIN Juan José Prof.
Universidad Politécnica Cataluña
Dpto. Física Aplicada
Jordi Girona Salgado, s/n
Campus Nord - modul B4
08034 Barcelona (SPAIN)
T: +34 934017067 F:+34 93401090
benjamin@etseccpb.upc.es

MARTINEZ CAÑADAS Angel Mr.
Universidad de Castilla La Mancha
Instituto de Desarrollo Regional
Sección de Teledetección y GIS
02071 Albacete (SPAIN)
T: +34 967 599200 ext. 2627 F:+34 967 599233
acaddnadas@idr-ab.uclm.es

MARTINEZ ZQUIERDO M. Estíbaliz
Universidad Politécnica de Madrid
Dpto. de Arquitectura y Tecnología de S.I.
Campus de Montegancedo
28660 Boadilla del Monte (Madrid) (SPAIN)
T: +34 91 3367385 F:+34 91 3366601
emartinez@fi.upm.es

MARTIN-ORDOÑEZ Mª Teresa
DGA-CSIC. Unidad de Suelos y Riegos
P.O. Box 727
50080 ZARAGOZA (SPAIN)
T: +34 976 576311 F:+34 976 575501
terete@syrsig.mizar.csic.es

MEGIER Jacques Dr.
Joint Research Centre. Space Applications Institute
T.P. 263
21020 Ispra (Va) (ITALY)
T: +39 0332 789333 F:+39 0332 789469
jacques.megier@jrc.it

MICHAEL Abraham Mr.
ELTA Electronics Industries Ltd
POB 330 - Dep. 3424
Ashdod 77102 (ISRAEL)
T: +972 08 8572851 F:+972 08 8572844
mike@is.elta.co.il

MILLER John R. Prof.
York University / Petrie Science Building
Dept. of Physics & Astronomy
4700 Keele St.
Toronto, ONT M3J 1P3 (CANADA)
T: +1 416 7362100 F:+1 416 736 5626
jrmiller@yorku.ca

MOLINA SANCHEZ Iñigo Mr.
E.U.I.T - Topográfica
Campus Sur
Carretera de Valencia, km 7
28031 Madrid (SPAIN)
T: +34 91336 77918 F:+34 91 332 2580
ig-mol@nivel-euitto.upm.es

MONDRE Erwin Dr.
Austrian Space Agency
Garnisongasse 7
A-1090 Wien (AUSTRIA)
T: +31 1 4038177 F:+43 1 4058227
a.s.a.@ping.at

MURICKY Eduard Dr.
Land Survey Office
Kostelni 42
170 00 Prague 7 (CZECH REPUBLIC)
T: +420 2 57325009 F:+420 2 57322010

MUSAOGLU Nebiye Ms.
University of Istanbul
Remote Sensing Division
80626 Maslak
Istanbu (TURKEY)
nmusaoglu@srv.ins.itu.edu.tr

NEGRU Cristina Ms.
EUR-OPA Major Hazards Agreement
Council of Europe
F-67075 Strasbourg Cedex (France)
T: +33 3 88 41 29 36 F: +33 3 88 41 27 87
cristina.negru@coe.int

NETZBAND Maik Dr.
University of Dresden
Institute for Ecological and Regional Development
Weberplatz 1
D-01217 Dresden (GERMANY)
T: +49 351 4679 274 F:+49 351 4679 212
maik.netzband@pop3.tu-dresden.de

O'CONNOR Eugene Dr.
CUSG
British Geological Survey
Keyworth
Nottingham NG12 5GG (U.K.)
T: +44 115 936 3227 F:+44 115 936 3474
EAO@wpo.nerc.ac.uk

OLEDZKI Jan R. Prof.
University of Warsaw
Remote Sensing of Environment Laboratory
ul. Krakowskie Przedmieoecie 30
00-927 Warsaw (POLAND)
T: +48 22 8368358 F:+48 22 8261965
telegeo@mercury.ci.uw.edu.pl

ORMEÑO Santiago Prof.
E.U.I.T - Topográfica
Campus Sur
Carretera de Valencia, km 7
28031 Madrid (SPAIN)
T: +34 91 336 6483 F:+34 91 332 2580
santi@nivel-cuitts.upm.es

ORTEGA HORTIGÜELA Félix Mr.
Universidad de Valladolid
Laboratorio de Teledetección, LATUV
Prado de la Magdalena, s/n
47071 Valladolid (SPAIN)
T: +34 983 423132 F:+34 983 423130

PARLOW Eberhard Prof. Dr.
University of Basel
MCR Laboratory. Dept. Geography
Spalenring 145
CH-4055 Basel (SWITZERLAND)
T: +41 61 272 6503 F:+41 61 272 6923
parlow@ubaclu.unibas.ch

PAUL José Mr.
CNIG
Tagus Park-Nucleo Central. Sala 301
2780 Oeiras (PORTUGAL)
T: +351 1 4219800 F:+351 1 421 9856
uva@cnig.pt

PEREZ BURGOS Ana Mª Dr.
Universidad de Valladolid
Laboratorio de Teledetección, LATUV
Prado de la Magdalena, s/n
47071 Valladolid (SPAIN)
T: +34 983 423132 F:+34 983 423130
ana@latuv.uva.es

PIETRANERA Luca Mr.
TELESPAZIO. Business Development
Via Tiburtina 965
00156 Roma (ITALY)
T: +39 06 4079 6213 F:+39 06 4079 6202
luca_pietranera@telespazio.it

PIPPI Ivan Dr.
IROE-CNR
Via Panciatichi, 64
I-50127 Florence (ITALY)
T: +39 55 4235277 F:+39 55 4235204
pippi@iroe.fi.cnr.it

QUINTANILLA RODENAS Antonio Mr.
Universidad de Castilla La Mancha
Instituto de Desarrollo Regional
Sección de Teledetección y GIS
02071 Albacete (SPAIN)
T: +34 967 599200 F:+34 967 599233
arodenas@pol-ab.uclm.es

RAST Mike Dr.
European Space Agency/ESTEC, Postbus 299
NL-2200 Noordwijk (NETHERLANDS)
Tel: +31 71 565 5676
mrast@estec.esa.nl

RECONDO GONZALEZ Carmen Ms.
Universidad de Oviedo
E.U.I.T. Minera y Topográfica
Reinerio Garcia, s/n
33600 Mierres, Asturias (SPAIN)
T: +34 98 5458034 F:+34 98 5458000
crecondo@relay.etsimo.uniovi.es

RIAZA GARCIA Asunción Dr.
Instituto Tecnológico Geominero de España
Ríos Rosas, 23
28003 Madrid (SPAIN)
T: +34 91 3495907 F:+34 91 4426216
ariaza@lander.es

RIGOLLIER Christelle Ms.
Ecole des Mines de Paris
Groupe Télédétection et Modélisation
B.P. 207
06904 Sophia Antipolis Cedex (FRANCE)
T: +33 04 93 95 7453 F:+33 04 9395 7535
christelle.rigollier@cenerg.cma.fr

RUFO Roberta Ms.
EURIMAGE SPA
Via e. D' onofrio 212
00155 Roma (ITALY)
T: +39 06 40694222 F:+39 06 40694232
rufo@eurimage.com

RUIZ VERDU Antonio
Centro de Estudios Hidrográficos
P° Bajo Virgen del Puerto, 3-5
50080 Zaragoza (SPAIN)
T: +34 91 3358012 F:+34 91 3357994
averdu@cedex.es

SACAU CUADRADO Mª Mar Ms.
Universidad de Vigo
Facultad de Ciencias. Dpto. Física Aplicada
Lagaos Marcosende,
Vigo (SPAIN)
T: +34 970505168 F:+24 968 253086
jesu@uvigo.es

SALETA BORDERAS José Luis
Universidad Pública de Navarra
Dpto. Proyectos e Ingeniería Rural
Campus Arrosadia, s/n
31006 Pamplona (SPAIN)
T: +34 948 169021 F:+34 948 169187
saleta@upna.es

SANCHEZ GOMEZ Mª Luisa
Universidad de Valladolid
Prado de la Magdalena, s/n
47071 Valladolid (SPAIN)
T: +34 983 423132 F:+34 983 423013

SANCHEZ TRIGO Mª Carmen Ms.
Universidad de Córdoba
Escuela Téc. Sup. de Ingenieros Agrónomos y de Montes
Avda. Menéndez Pidal s/n. Apdo. Correos 3.048
14080 Córdoba (SPAIN)
T: +34 957 218515 F:+34 957 218563
ir2satrm@uco.es

SANTOS Teresa Dr.
CNIG
Tagus Park-Nucleo Central. Sala 301
2780 Oeiras (PORTUGAL)
T: +351 1 4219824 F:+351 1 421 9856
tags@cnig.pt

SANZ JUSTO Julia Prof.
Universidad de Valladolid
Laboratorio de Teledetección, LATUV
Prado de la Magdalena, s/n
47071 Valladolid (SPAIN)
T: +34 983 423130 F:+34 983 423130
julia@latuv.uva.es

SANZ REDONDO Ana Mª Dr.
Universidad de Ciudad Real. Dpto. de Expresión Gráfica
Paseo de la Universidad, 4
13071 Ciudad Real (SPAIN)
T: +34 926 295396 F:+34 926 295391
asanz@ccp-cr.uclm.es

SCHERER Dieter
University of Basel. MCR Laboratory
Spalenring 145
CH-4055 Basel (SWITZERLAND)
T: +41 61 272 6923 F:+41 61 272 6480
scherer1@ubaclu.unibas.ch

SCHNEIDER Thomas Prof.
University of Munich
Lehrstuhl f. Landnutzungsplanung und Naturschutz
Am Hochanger 13
D-85354 Freising (GERMANY)
T: +49 8161 71 4666 F:+49 8161 71 4671
tomi.schneider@LRZ.uni-muenchen.de

SCHNEIDER Werner Prof. Dr.
University of Agriocultural Sciences
Peter-Jordan-Str. 82
A-1190 Vienna (AUSTRIA)
T: +43 1 476 545 111 F: +43 1 476 545 143
schneiwe@mail.boku.ac.at

SEIDEL Klaus Dr.
Image Science Division
CommunicationsTechnology Laboratory ETH
Gloriastrasse 35
CH-8092 Zurich (SWITZERLAND)
T: +41 1 632 5284 F: +41 1 632 1251
salavipa@chamran.ut.ac.ir

SHAROV Aleksey Dr.
TU Graz
Dept. for Photogrammetry. Institute of Applied Geodesy
Steyrergasse 30
A-8010 Graz (AUSTRIA)
T: +43 316 8761729 F:+43 316 8736337
sharov@pdib47.joanneum.ac.at

SMARA Youcef Dr.
Univ. STHB
Laboratoire de Traitement d'Images
B.P. 32. El Alia
Bab-Ezzouar. Alger 16111 (ALGERIA)
F:+213 2 242802

SOSPEDRA Francisco
University of Valencia
Dept. Thermodynamics
Dr. Moliner, 50
46100 Burjassot (Valencia) (SPAIN)
T: +34 96 3983121 F:+34 96 3642345
francisco.sospedra@uv.es

SOUTHWOOD David Prof.
European Space Agency
8-10 rue Mario Nikis
75738 Paris Cedex 15 (FRANCE)
Fax: +33 1 5369 7674
dsouthwood@ha.esa.fr

SUBERVIOLA SERRANO Pedro Luis
Universidad de Valladolid. Dpto. Física
Rgmto Villarrobledo 15, 1A
34005 Palencia (SPAIN)
T: +34 979 74 91 83
98ci0139@dirac.dec.cie.uva.es

TAPIADOR FUENTES, Fco. Javier
Universidad de Valladolid
Laboratorio de Teledetección, LATUV
Prado de la Magdalena, s/n
47071 Valladolid (SPAIN)
T: +34 983 423132 F:+34 983 423130
tapiador@latuv.uva.es

TELECHEA PRIETO Mónica Ms.
Universidad del Pais Vasco
Paseo Manuel de Lardizabal, no. 1
20009 San Sebastian (SPAIN)
T: +34 943 018070 F: +34 943 219 306
ccbteprm@si-ehu.es

TORRES PALENZUELA Jesús Mr.
Universidad de Vigo
Facultad de Ciencias. Dpto. Física Aplicada
Lagaos Marcosende,
Vigo (SPAIN)
T: +34 970 505168 F:+34 968 253086
jesu@uvigo.es

TROVATO Alessandro Ing.
Università di Catania
Istituto di Macchine, Fac. di Ingegneria
Viale Andrea Doria
I-95125 Catania (ITALY)
T: +39 095 330238 F:+39 095 330258
atrovato@im.ing.unict.it

VASILIEV Leonid Prof.
Institut of Geography323
Staromonetny 29
109017 Moscow (RUSSIA)
T: +70 95 433 3723 F:+70 95 959 0033
vasiliev@igras.geonet.ru

VAUGHAN Patrick Mr.
Universidad de Alcalá de Henares
Dpto. Geografía
Colegios 2
28801 Alcalá de Henares (SPAIN)
T: +34 91 885 44 5257 F:+34 91 885 4439
patrick.vaughan@geogra.alcala.es

VAUGHAN Robin Dr.
Remote Sensing Group. APEME
University of Dundee
Dundee DD1 4HN (U.K.)
T: +44 1382 344 557 F:+44 1382 345415
r.a.vaughan@dundee.ac.uk

VAZQUEZ Antonio Dr.
I.N.I.A.
Sección de Proceso de Datos
Carretera de la Coruña Km 7
28040 Madrid (SPAIN)

VELA MAYORGA Alicia Ms.
Universidad de Castilla La Mancha
Instituto de Desarrollo Regional
Sección de Teledetección y GIS
02071 Albacete (SPAIN)
T: +34 967 599200 F:+34 967 599233
avela@idr-ab.uclm.es

VERDU ARNAL Joan M. Mr.
University of Lleida
Dpt. Environment & Soil Science
Rovira Roure, 177
25198 Lleida (SPAIN)
T: +34 973 702879 F:+34 973 702613
jaj.verdu@macs.udl.es

VIDAL-PANTALEONI Ana Ms.
Universidad Politécnica de Valencia
Dpto. de Comunicaciones
Camino de Vera, s/n
46022 Valencia (SPAIN)
T: +34 963 879710 F:+34 963 877309
avidal@dcom.upv.es

VILLARES DURAN Pilar Dr.
Universidad de Cádiz. C.A.S.E.M.
Dpto. Física Aplicada
Polígono Río San Pedro, s/n
11510 Puerto Real (Cádiz) (SPAIN)
T: +34 956 470866 F:+34 956 470866
pilar.villares@uca.es

WAHLEN Jürgen Mr.
WEU Satellite Centre
Apartado de Correos 511
28850 Torrejón de Ardoz (Madrid) (SPAIN)
T: +34 91 678 6035 F:+34 91 678 6000

WALD Lucien Dr.
Ecole des Mines de Paris
Groupe Télédétection et Modélisation
B.P. 207
06904 Sophia Antipolis Cedex (FRANCE)
T: +33 (0)4 93 95 7449 F:+33 (0)4 9395 7535
wald@cenerg.cma.fr

WINKLER Peter Dr.
FÖMI Remote Sensing Centre
Bonsyak ter 5
H-1149 Budapest (HUNGARY)
T: +361 363 6669 F:+361 252 8282
peter.winkler@rsc.fomi.hu

WOLK-MUSIAL Elzbieta Dr.
University of Warsaw
Remote Sensing of Environment Laboratory
ul. Krakowskie Przedmieoecie 30
00-927 Warsaw (POLAND)
T: +48 22 8368358 F:+48 22 8261965
telegeo@mercury.ci.uw.edu.pl

ZAGAJEWSKI Bogdan Mr.
University of Warsaw
Remote Sensing of Environment Laboratory
ul. Krakowskie Przedmieoecie 30
00-927 Warsaw (POLAND)
T: +48 22 8368358 F:+48 22 8261965
bzag@wgsr.uw.edu.pl

ZARCO-TEJADA Pablo J. Mr.
York University / Petrie Science Building
Centre for Research in Earth & space Science (CRESS)
4700 Keele St.
Toronto, ONT M3J 1P3 (CANADA)
T: +1 416 7362100 Ext. 33510 F:+1 416 736 5626
pzarco@yorku.ca

Author index

Colour plates

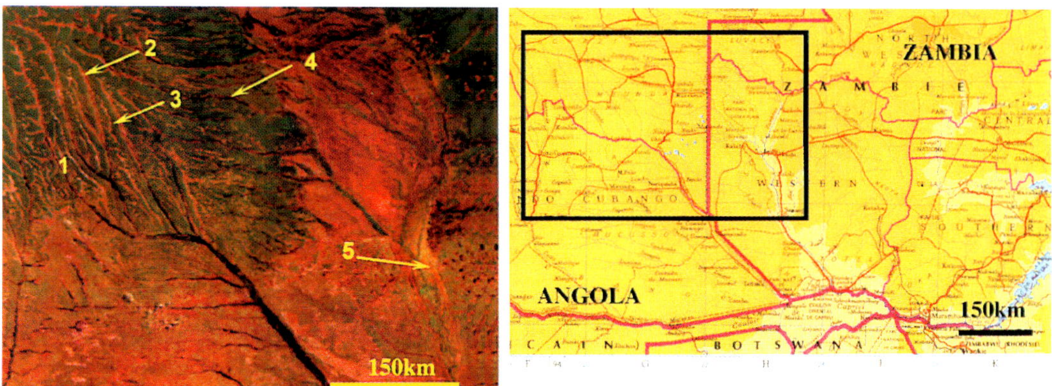

Figure 3. Image from the VEGETATION instrument. This image was taken on 28th July, 1998 over the area along the border between Angola and Zambia. At that time of the year dry season conditions prevail in the area. The color combination is: Red for Short Wave Infrared, Green for Near Infrared and Blue for Red.

The greenish area in the NW area is the so-called 'Miombo' woodland (1).

The yellowish corridor marked 2 is a savanna strip extending along a swampy valley bottom, while the one under 3 was left dry.

The same contrast appears between the alluvial plain along the Luanguinga (4), a tributary of the Zambezi River, that shows dark tones, and the Barotse floodplain (5), along the Zambezi itself. (Carmona-Moreno et al., see p. 18).

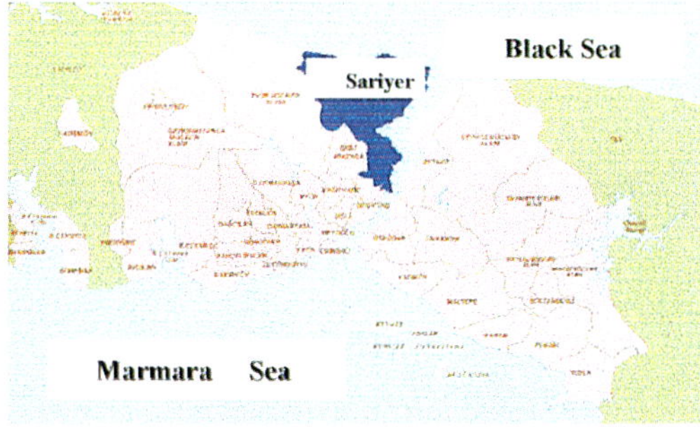

Figure 1. Location of study area. (Musaoglu & Örmeci, see p. 41).

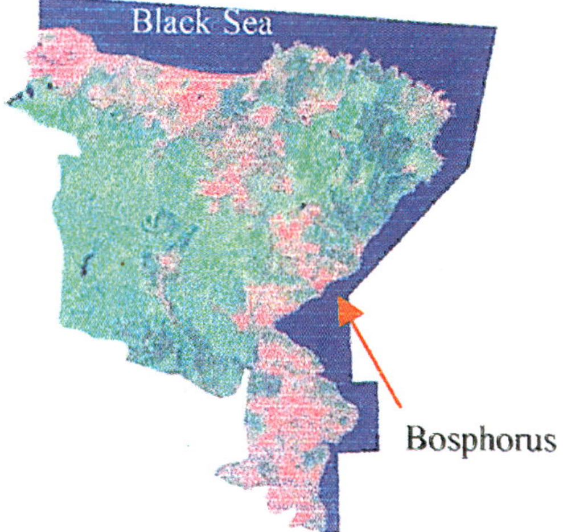

Figure 2. 8 July 1997 dated Landsat TM (Band 7/4/1). (Musaoglu & Örmeci, see p. 42).

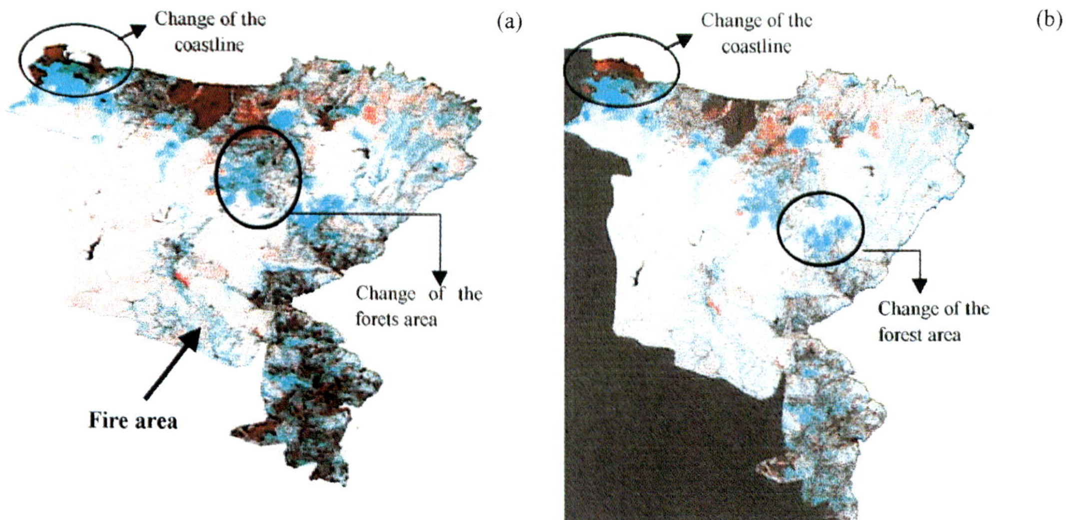

Figure 5. New data groups: a) Vegetation index, b) Normalized vegetation index. (Musaoglu & Örmeci, see p. 43).

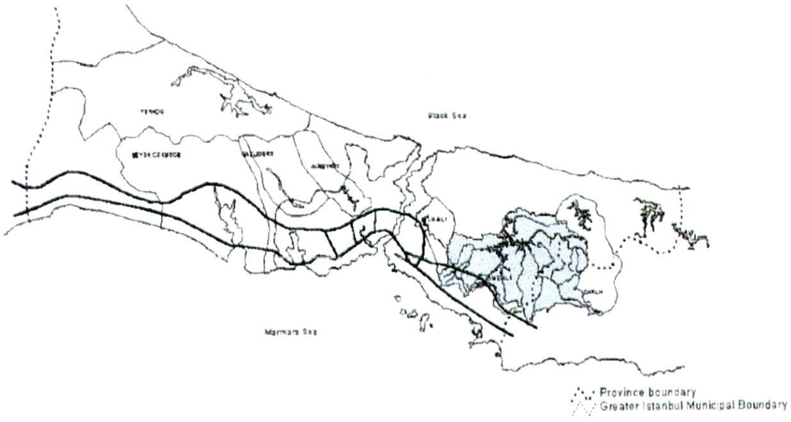

Figure 1. The location of Omerli Water Basin in Istanbul. (Goksel & Turkoglu, see p. 88).

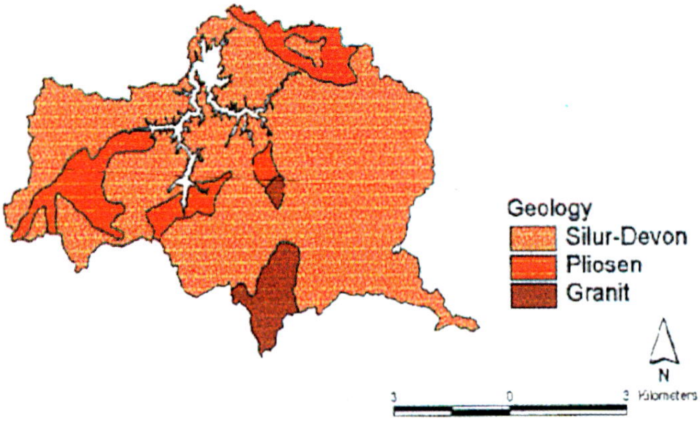

Figure 2. Geological characteristics of the area. (Goksel & Turkoglu, see p. 88).

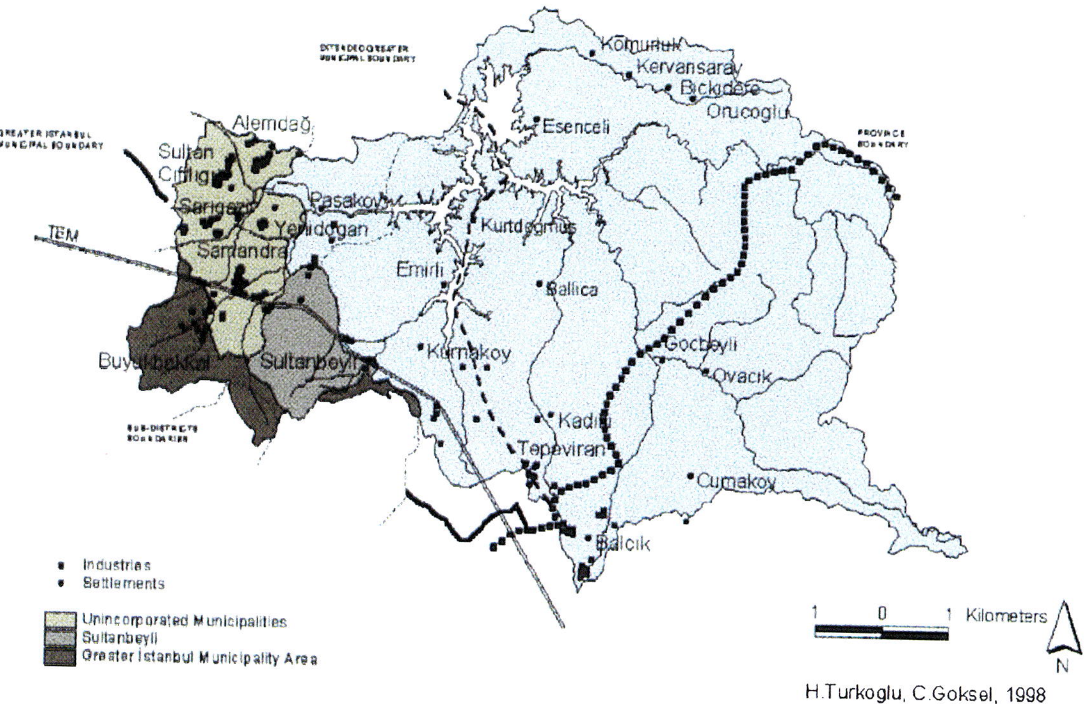

Figure 3. Settlements in Omerli Region. (Goksel & Turkoglu, see p. 90).

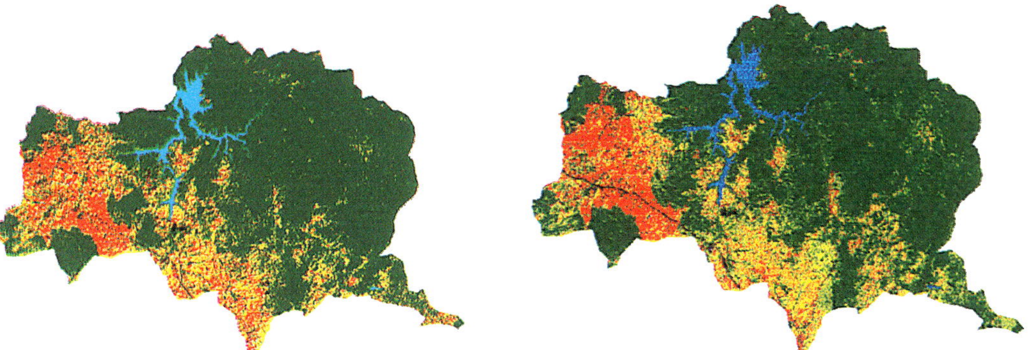

Figure 4. Classified image of 1992. (Goksel & Turkoglu, see p. 90).

Figure 5. Classified image of 1997. (Goksel & Turkoglu, see p. 90).

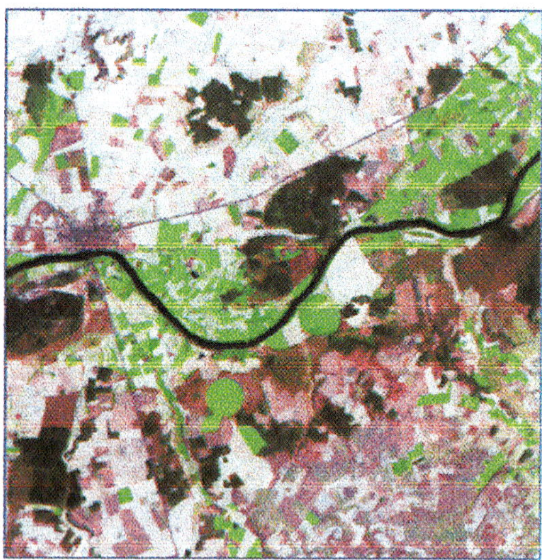

Figure 3. This 5-4-3 image shows agrarian land use in the city of Tordesillas. The effect of land plot concentration is clearly visible in the radial structure top, as is the prevalence of dry land over irrigated land and forest. (Tapiador & Casanova, see p. 96).

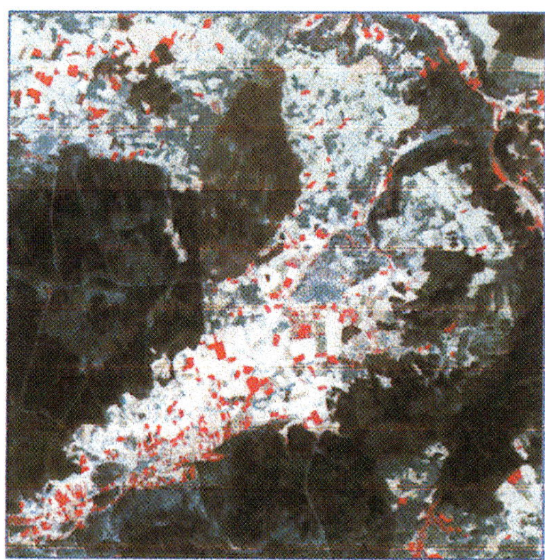

Figure 5. There are only a few spots where forest is the main land cover due to the late re-population in the XVIII century and beyond. These are now leisure and second home areas. (Tapiador & Casanova, see p. 96).

Figure 4. One can see here the small size of the plots in the province as well as other kinds of landscape: Irrigated and dry land alternates with an open field structure, with regular plots in a concentrated village shape. The secondary importance of saltus in this area can be seen. (Tapiador & Casanova, see p. 96).

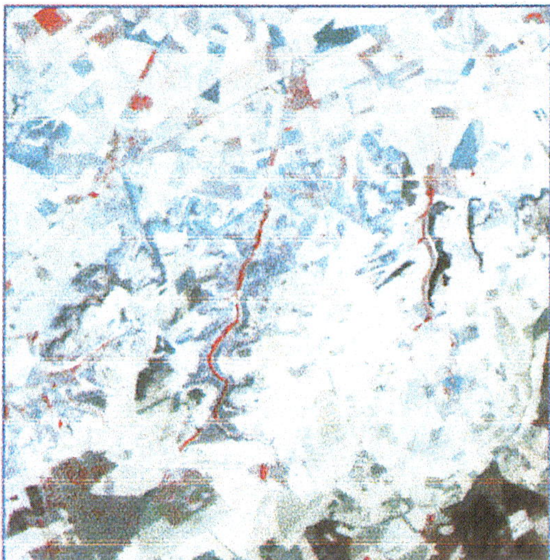

Figure 6. The different substratum between the various levels of plateau does not lead to any special bias in land use, but does in productivity. The structure of property remains similar in both cases. (Tapiador & Casanova, see p. 96).

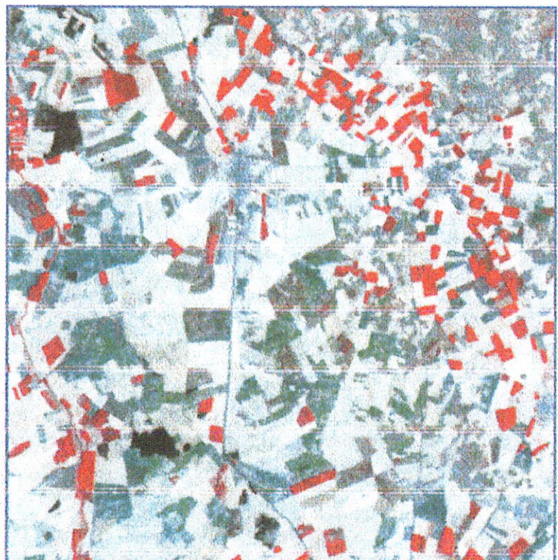

Figure 7. The balance between saltus and ager in the province is tilted in favor of ager – plain land, with a historical exploitation of resources. Thus, the primary description can be considered a set of little plots with cerealist orientation. (Tapiador & Casanova, see p. 97).

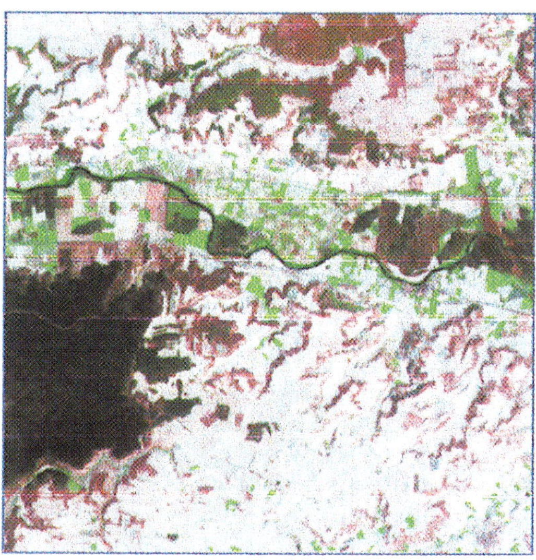

Figure 8. The area of 'Ribera del Duero' wines covers the Duero valley with highly cultivated sectors. The slopes of the bleak plateau remain bare or, in some cases, with reforested Pinus, whereas some woodland is used for second homes, losing its primary function. (Tapiador & Casanova, see p. 97).

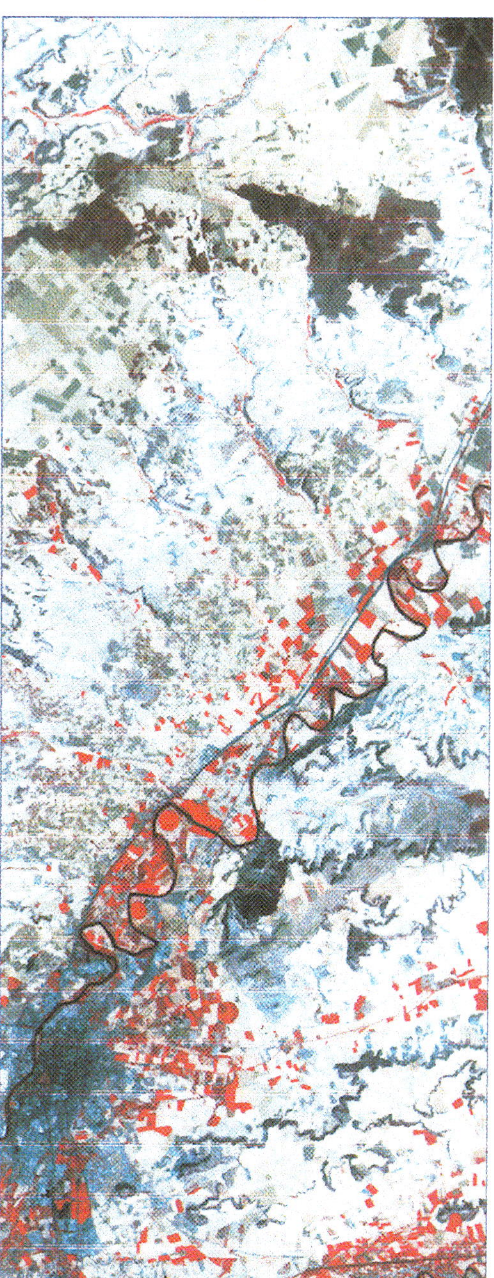

Figure 9. This sector of the river Pisuerga shows the characteristic land use in the bank: Pivot systems for irrigating drought land and other means of transporting water to herbaceous crops such as suger-beet – justifying investment in infrastructure – and wheat when it is amortized, some little greenhouses and market gardens to supply not only the local but regional market. The city of Valladolid city is beneath. (Tapiador & Casanova, see p. 97).

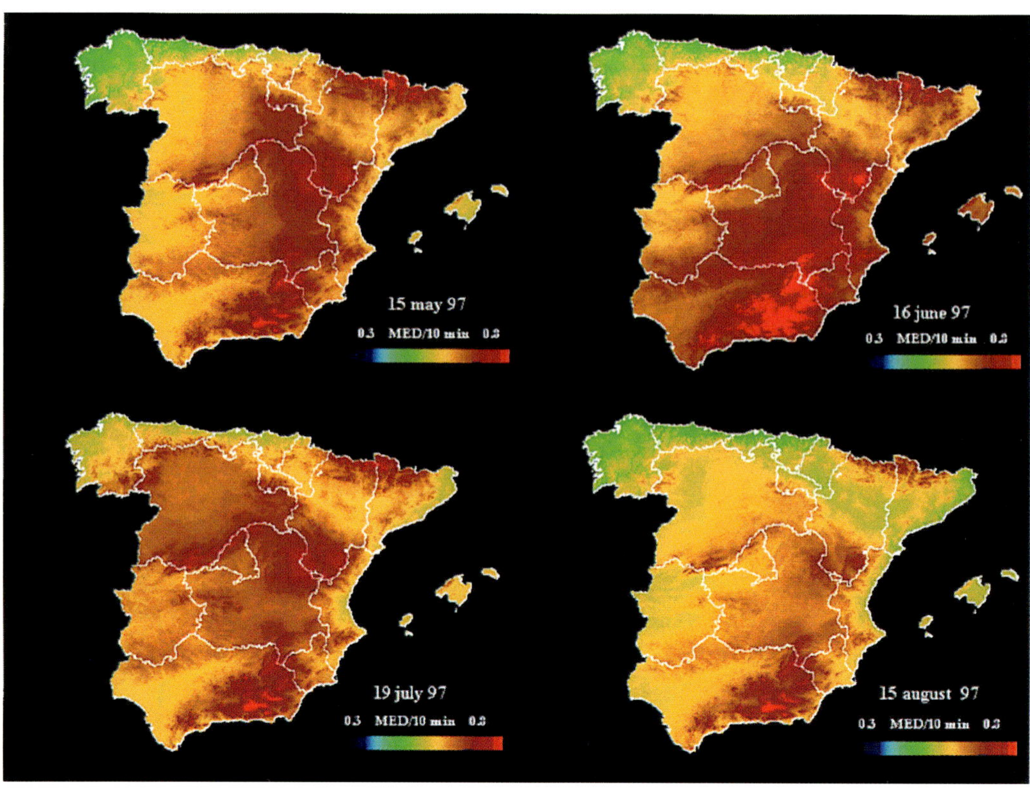

Figure 10. Erythematic Irradiance maps for summer 1997. (Calle et al., see p. 171).

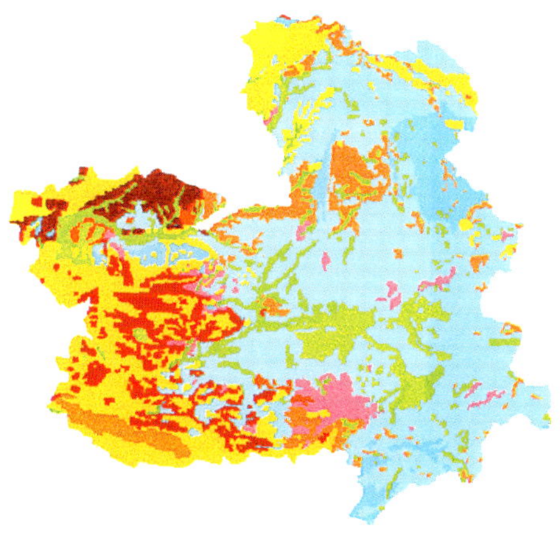

Figure 1. Soil map of Castilla La Mancha. (Ormeño & Molina, see p. 204).

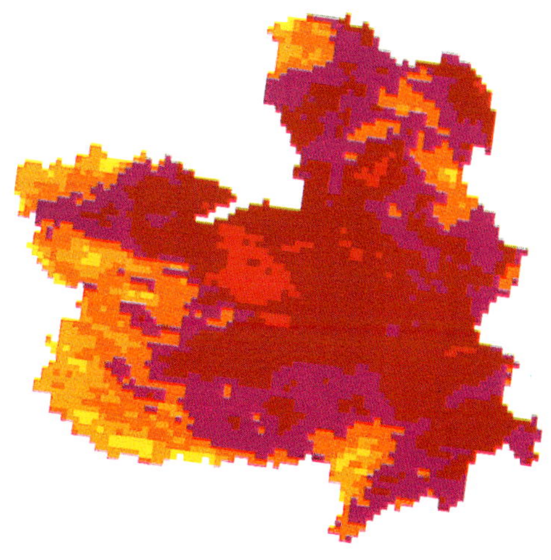

Figure 2. Apparent thermal inertia. (Ormeño & Molina, see p. 204).

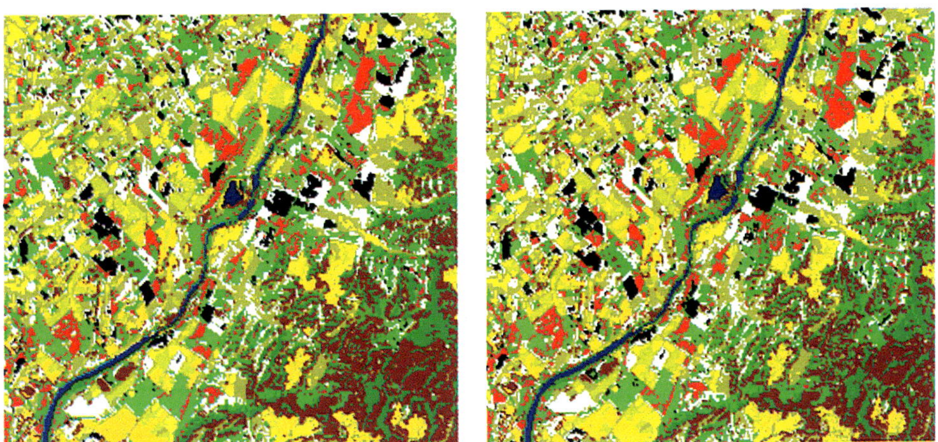

Figure 3. The left image represents the reference data. The right one is the classified image using Supervised ART-II, with vigilance parameter ρ=0.98, dynamic learning rate β=0.50, and training set of 9000 exemplars. The classification accuracy is 85.82%. Classes are assigned colours as follow: 1-white, 2-brown, 3- bright yellow, 4-red, 5-black, 6-dark blue, 7-dark green, 8-light green, 9-green, 10-yellow, 11-dark yellow, 12-yellowish green, and 13-blue. (Al-Rawi et al., see p. 233).

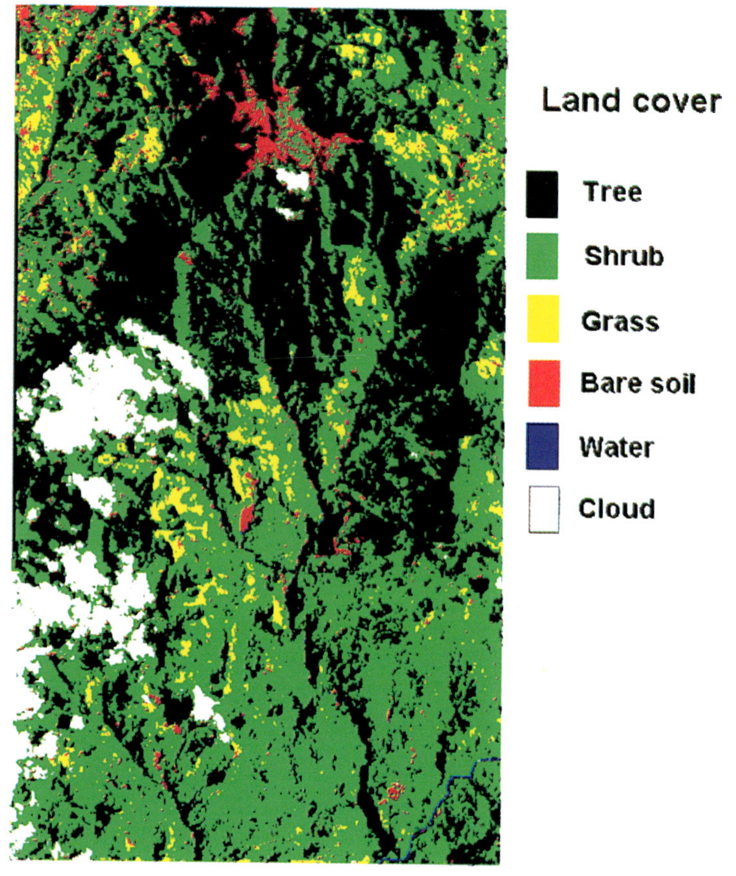

Figure 5. The Jabonosa basin. Land cover classification using Landsat TM data. (Gómez et al., see p. 400).

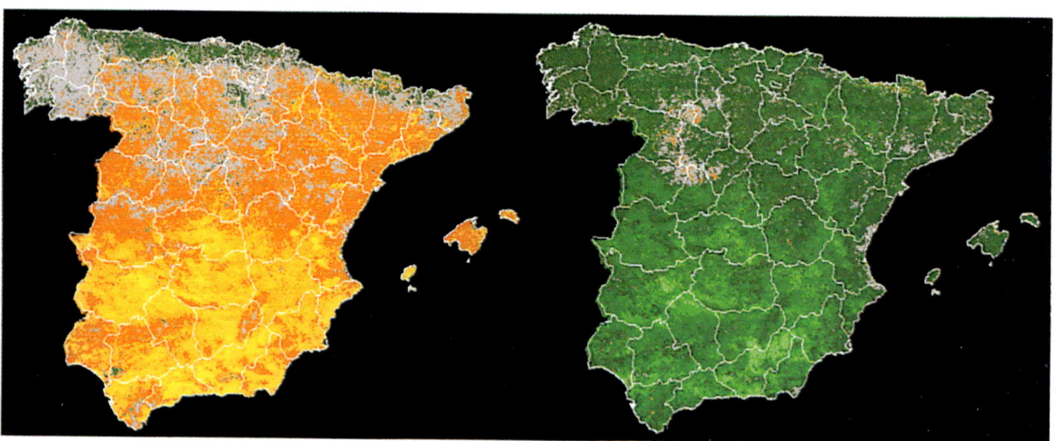

Figure 1 and 2: Mean value of the MNDVI anomalies during de period from February to July. Figure 1 is 1995 and figure 2 is 1998 year. (González-Alonso et al., see p. 404).

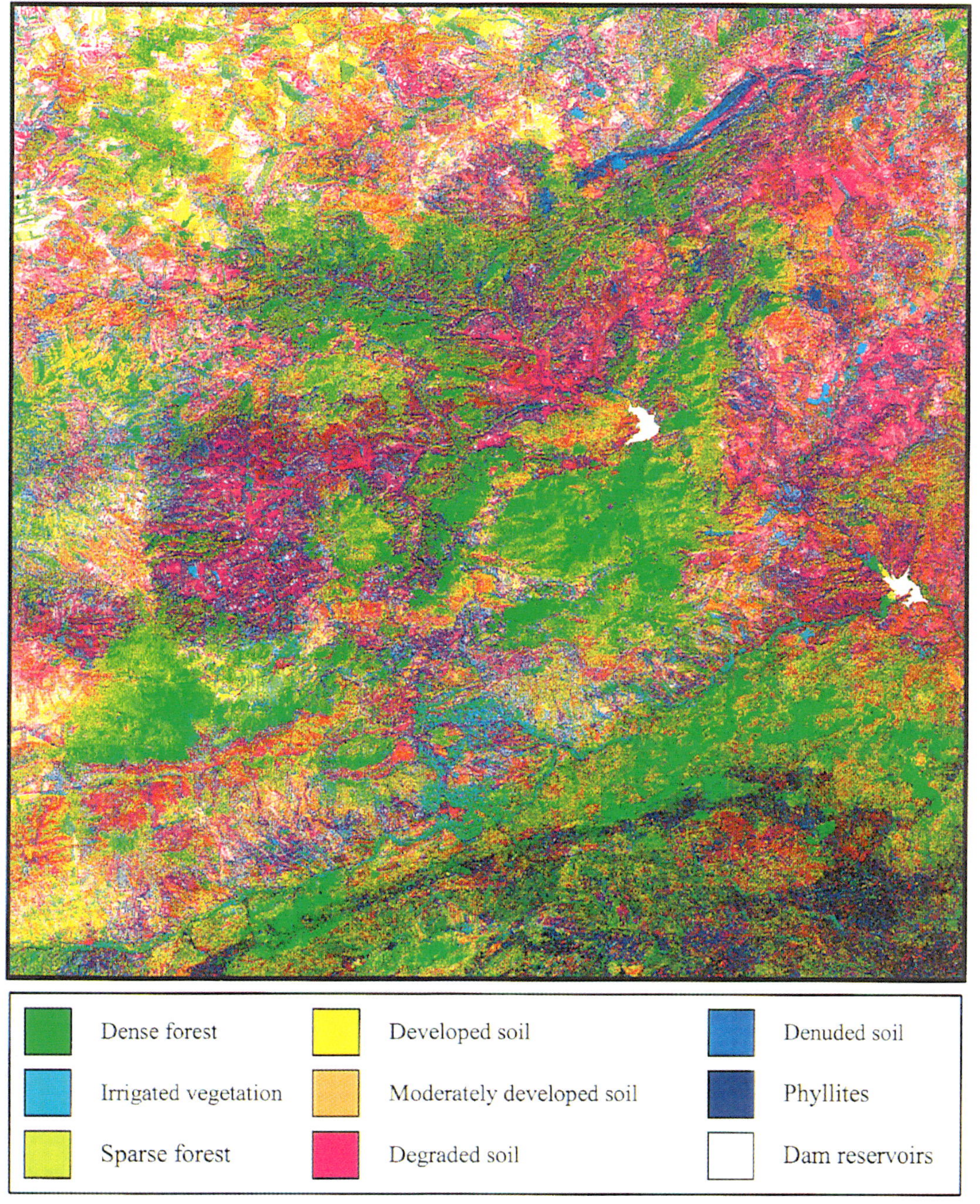

Figure 5. Colour composite obtained by encoding shape index *IF* in red, coloration index *IC* in green, and intensity *I* in blue, for TM bands 7, 4 and 1. Image of 45 km by 45 km. (Haboudane et al., see p. 412).

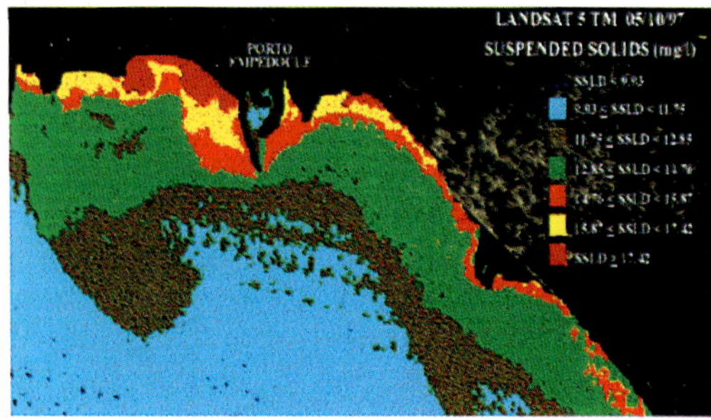

Figure 1. Image generated for marine currents, as measured by suspended solids, in front of Porto Empedocle. (Geraci et al., see p. 417).

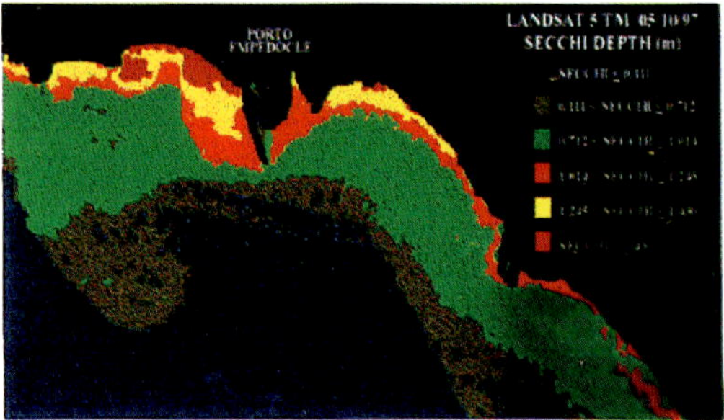

Figure 2. Image generated for water transparency, as measured by Secchi disk, in front of Porto Empedocle. (Geraci et al., see p. 417).

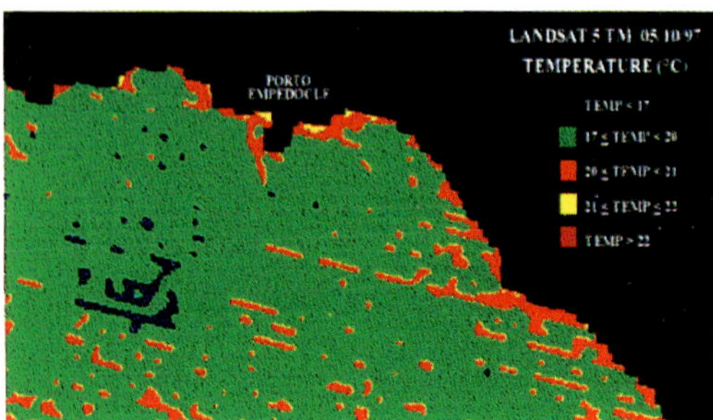

Figure 3. Image generated for surface water temperatures, in front of Porto Empedocle. (Geraci et al., see p. 417).

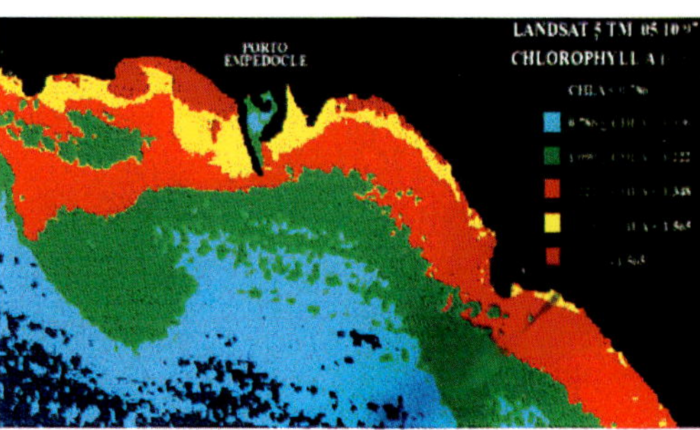

Figure 4. Image generated for chlorophyll a content of surface waters in front of Porto Empedocle. (Geraci et al., see p. 417).

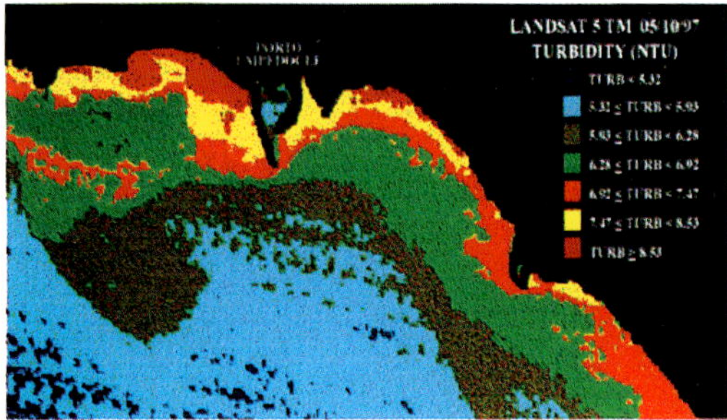

Figure 5. Image generated for water transparency, as measure by turbidity, in front of Porto Empedocle. (Geraci et al., see p. 417).

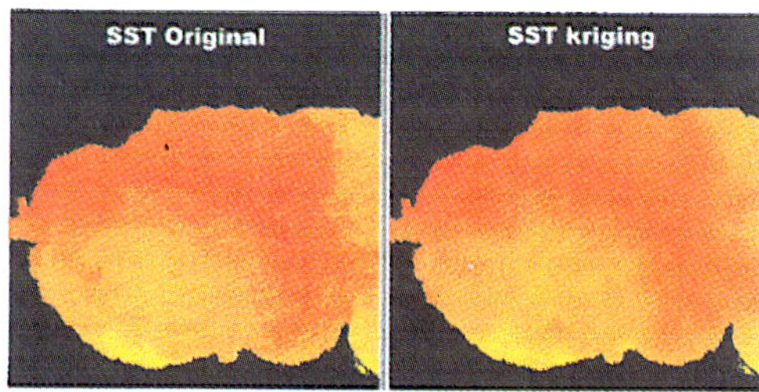

Figure (2). SST original and SST Kriging. (Romo et al., see p. 435).

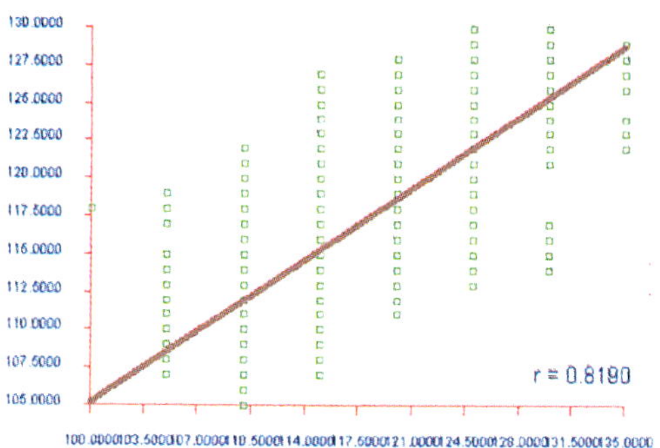

Figure (3). Lineal regression between the real image and that estimated by Kriging. A correlation value of r= 0.8190 is obtained. (Romo et al., see p 435).

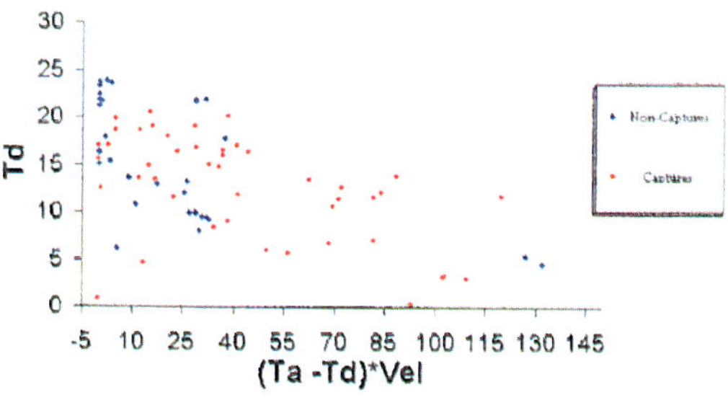

Figure (6). Graph for the prediction of tuna in the Mediterranean. Data from the 1996 and 1997 fishing campaigns in our pilot boat. (Romo et al., see p. 436).

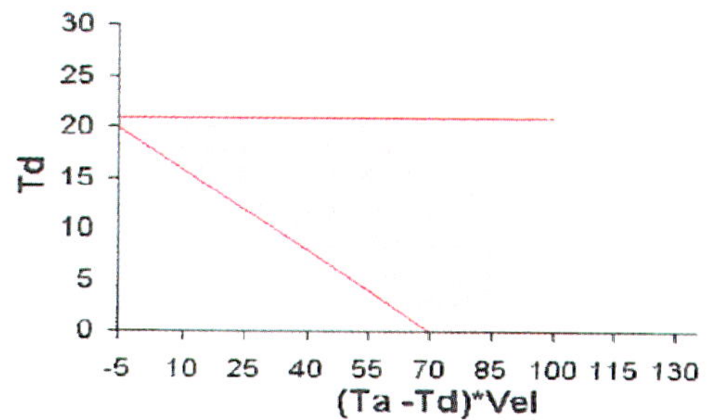

Figure (7). Representation of the most significant areas defining meteorological conditions favouring the presence of tuna. (Romo et al., see p. 436).

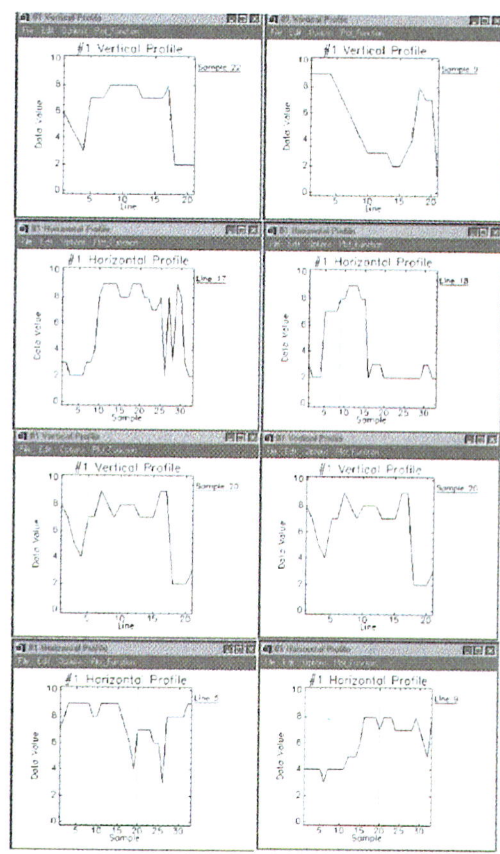

Figure (5). Variability of relative humidity in pixels where thermal fronts can be observed in the SST of the same day and at the same time. (Romo et al., see p. 436).

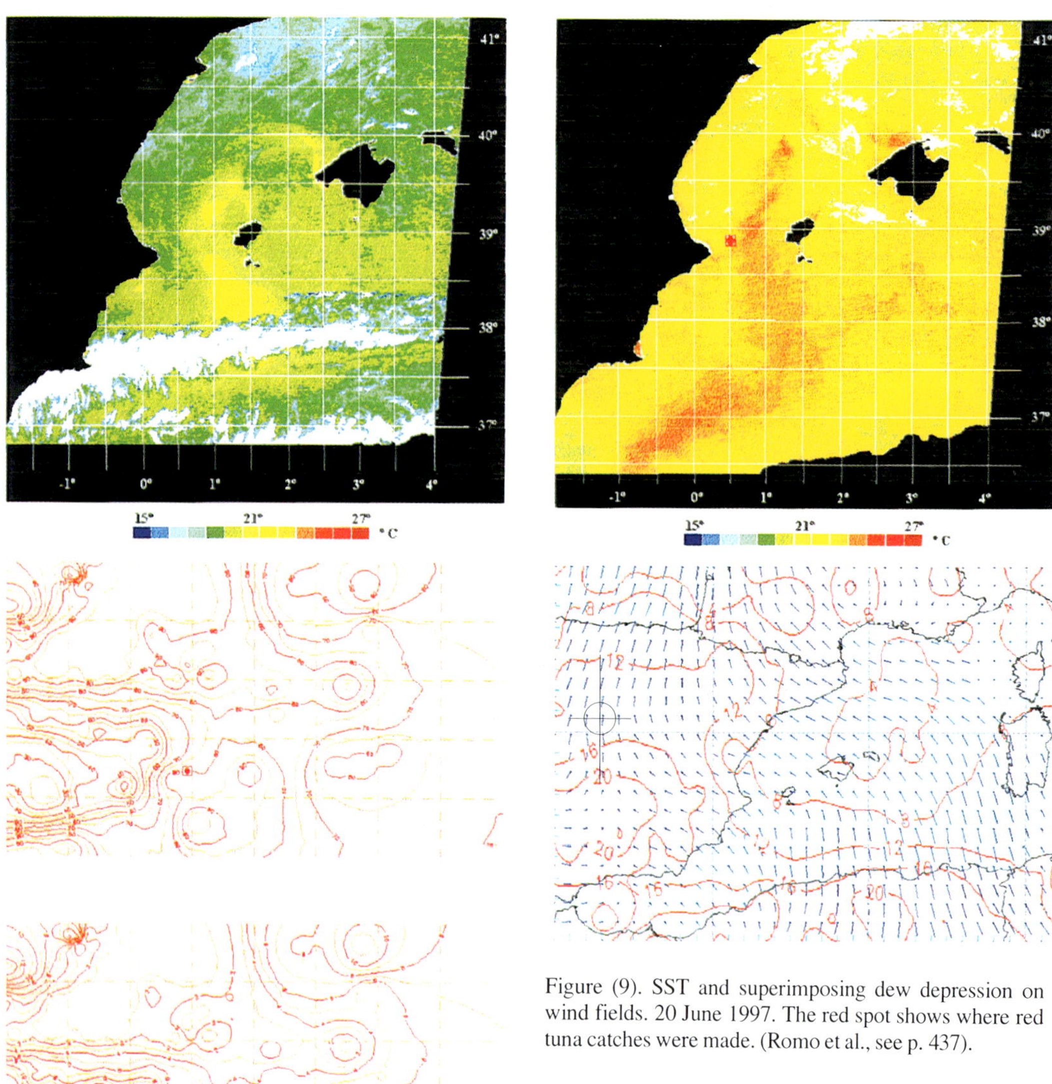

Figure (8). Situation corresponding to 22 May 1997. Upper. On this day red tuna catches were made at the points marked with a grey cross. Middle. Dew depression. Left overlay. Western Mediterranean SST. Lower. Relative humidity at surface level. (Romo et al., see p. 437).

Figure (9). SST and superimposing dew depression on wind fields. 20 June 1997. The red spot shows where red tuna catches were made. (Romo et al., see p. 437).

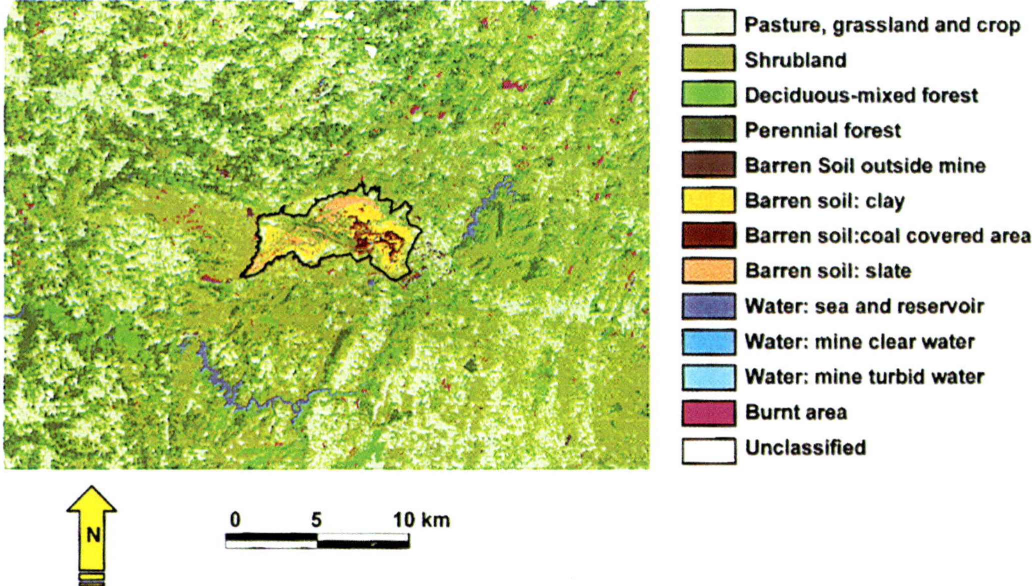

Figure 3. Stepwise linear supervised classification of As Pontes region. Landsat TM imagery 05-04-1997. Bands 1-2-3-4-5-7. (Llorens et al., see p. 495).

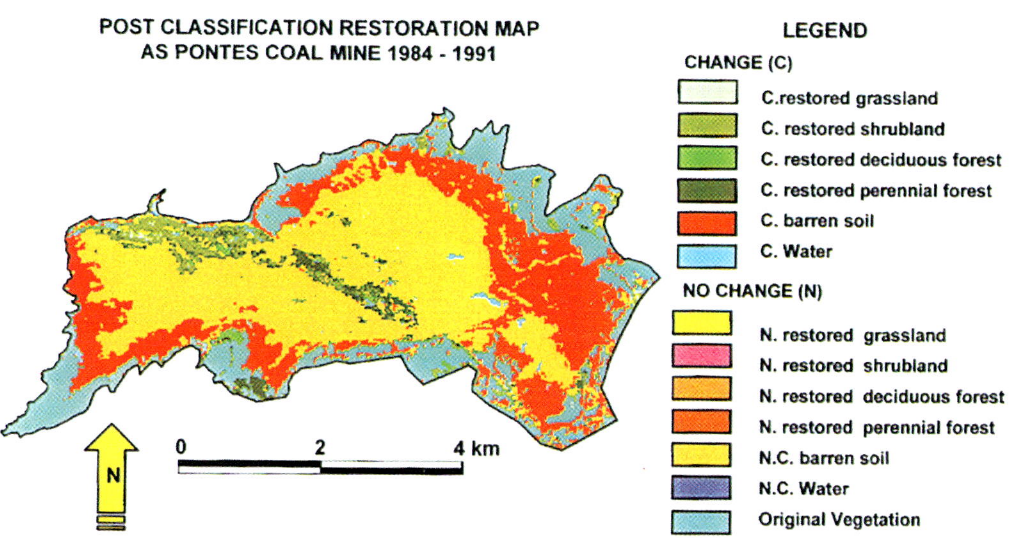

Figure 4. Restoration Map of As Pontes region for 1984-1991 obtained with Post Classification automatic logical comparison algorithm. Stepwise Linear Classification. Landsat TM imagery 08/09/1984 and 11/08/1991. Bands 1-2-3-4-5-7. (Llorens et al., see p. 495).

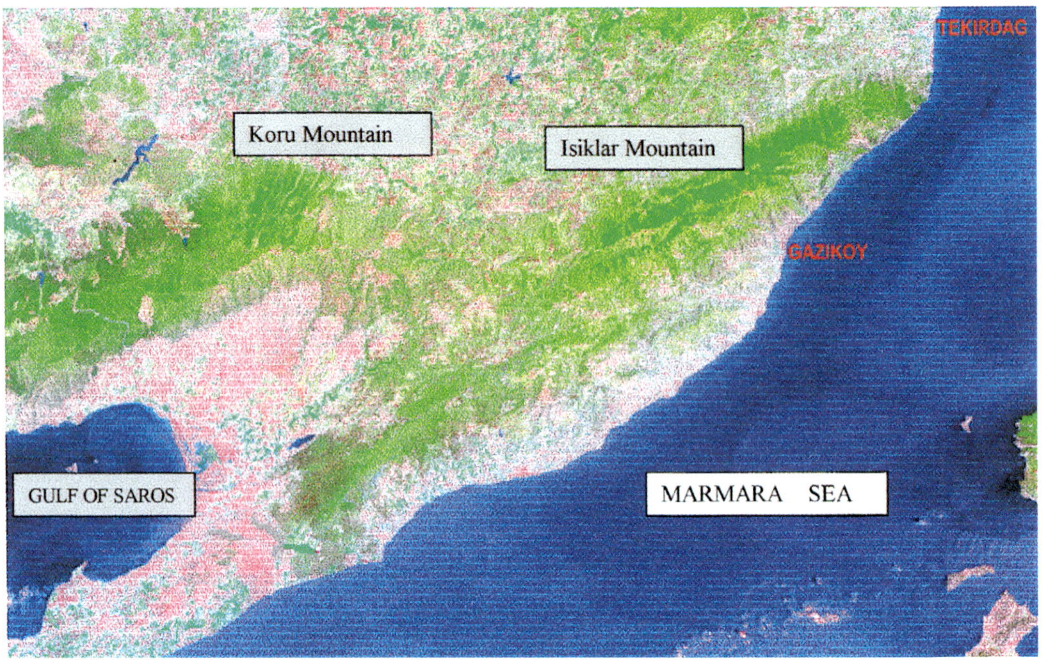

Figure 1. Study area (Landsat 5 TM image, band combination 7, 4, 1). (Kaya & Muftuoglu, see p. 510).

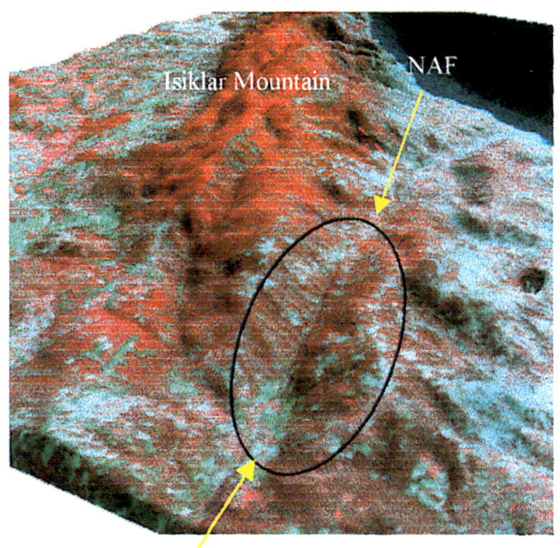

Figure 2. Rift valleys. (Kaya & Muftuoglu, see p. 511).

Figure 4. Merge image (Kaya & Muftuoglu, see p. 511).

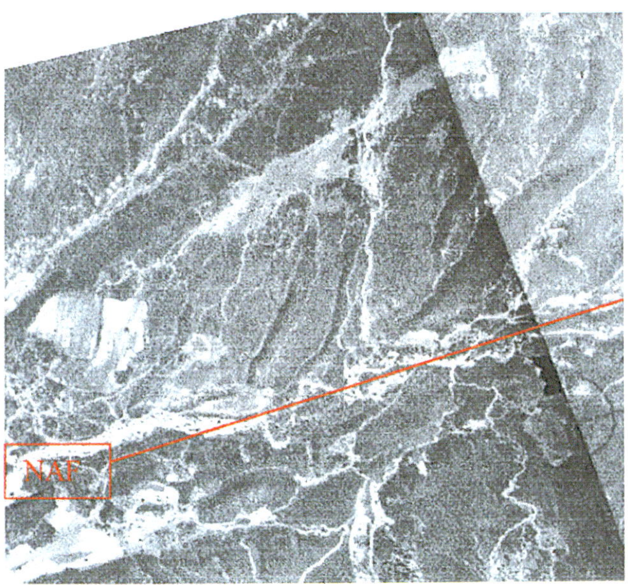

Figure 3. Aerial photographs (Scale 1/10 000). (Kaya & Muftuoglu, see p. 511).

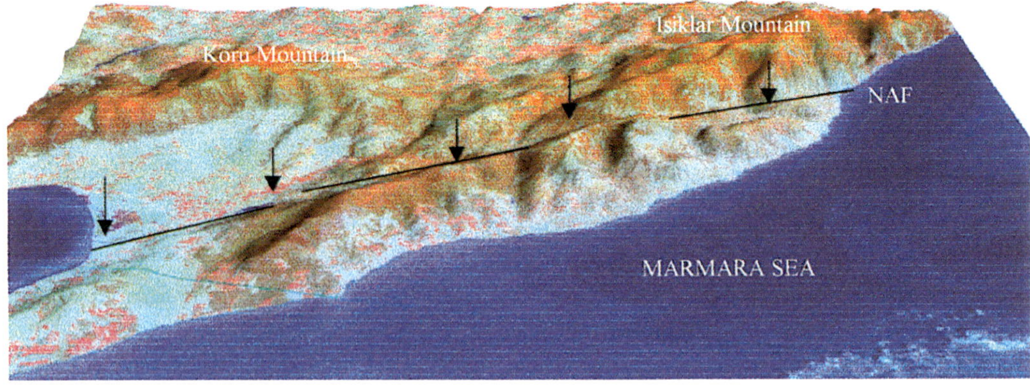

Figure 5. General view of the study area (Digital Elevation Model). (Kaya & Muftuoglu, see p. 512).

Figure 6. Digital Elevation Model overlaid with merge image. (Kaya & Muftuoglu, see p. 512).